VLSI
DESIGN
SECOND EDITION

DEBAPRASAD DAS

Associate Professor and Head
Department of Electronics and Telecommunication Engineering
Assam University
Silchar

OXFORD
UNIVERSITY PRESS

OXFORD
UNIVERSITY PRESS

Oxford University Press is a department of the University of Oxford.
It furthers the University's objective of excellence in research, scholarship,
and education by publishing worldwide. Oxford is a registered trade mark of
Oxford University Press in the UK and in certain other countries.

Published in India by
Oxford University Press
YMCA Library Building, 1 Jai Singh Road, New Delhi 110001, India

ISBN-13: 978-0-19-809486-9
ISBN-10: 0-19-809486-8

Typeset in Times New Roman
by Ideal Publishing Solutions, Delhi
Printed and bound in India by Repro India Ltd

To my parents
(Late) Sh. Birendranath Das
and
Smt. Pratima Das
my wife
Joyita
and my daughters
Adrija and Adrisha

Features of

Key topics: Highlights the topics and concepts discussed in each chapter

Colour plates: Provides coloured illustrations to depict topics, such as stick diagrams, 3D integration, nMOS and pMOS under different operating conditions, layout of CMOS inverter, multilevel utilization to allow better visualization of devices and processes

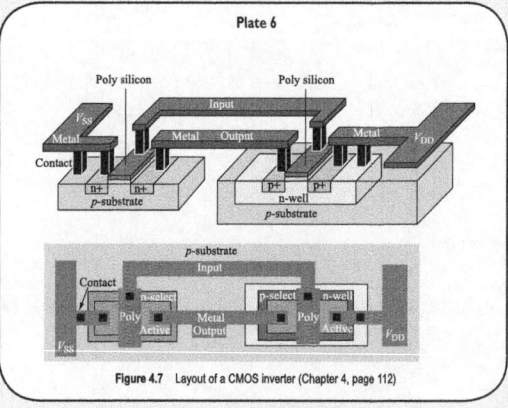

Figure 4.7 Layout of a CMOS inverter (Chapter 4, page 112)

Example 2.2 Calculate the drain current of an n-channel MOSFET with the following parameters.

$$L = 2\ \mu m,\ W = 10\ \mu m,\ \mu_n = 0.06\ m^2/Vs,\ C_{ox} = 15 \times 10^{-4}\ F/m^2,\ V_{T0} = 0.4\ V$$

Solution The drain current of an n-channel MOSFET is given by

$$I_D = \frac{\mu_n C_{ox} W}{L} \times \left[(V_{GS} - V_t)V_{DS} - \frac{V_{DS}^2}{2} \right] \text{ for the linear region}$$

and

$$I_{D,sat} = \frac{\mu_n C_{ox} W}{2L} \times (V_{GS} - V_t)^2 \text{ for the saturation region, } V_{DS} \geq (V_{GS} - V_t)$$

Substituting the given parameters in the above equation, we can calculate

$$\beta_n = \mu_n C_{ox}(W/L) = 0.06 \times 15 \times 10^{-4} \times \frac{10 \times 10^{-6}}{2 \times 10^{-6}} = 4.5 \times 10^{-4}\ A/V^2$$

Examples: Includes examples to demonstrate the applicability of concepts discussed

the Book

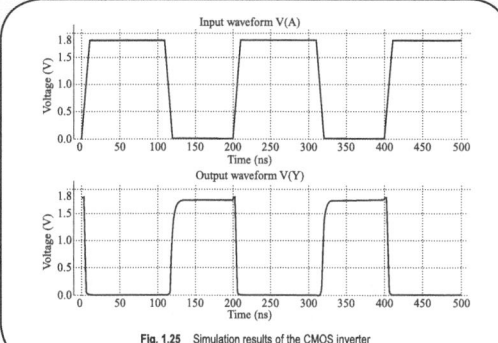

Input waveform V(A)

Output waveform V(Y)

Fig. 1.25 Simulation results of the CMOS inverter

Illustrations: Important topics have been well supported with suitable illustrations

## CHAPTER 9

Logic Synthesis

KEY TOPICS

- Introduction to synthesis
- Synthesis at various levels
- Logic synthesis
- Design styles
- Algorithms
- Boolean space
- Binary decision diagram

- Logic synthesis advantages
- Disadvantages of logic synthesis
- Sequential logic optimization
- Building blocks for logic design

Coverage: Provides comprehensive coverage of the basic concepts, methodologies, and algorithms of VLSI design

EXERCISES

Fill in the Blanks

1. Subthreshold operation of MOSFET is very much useful in _____
   (a) biomedical applications   (b) memory
   (c) charge coupled devices   (d) none of these

2. Th
   (a)
   (c)

True or False

1. Body effect is not a second order effect.
2. In full scaling, magnitude of electric field is constant.
3. In full scaling, the power density remains constant.
4. In constant electric field scaling, the oxide capacitance is scaled do 1/s.
5. In c

Multiple Choice Questions

1. Si is preferred over Ge because
   (a) Si is cheaper   (b) Si band gap is large
   (c) Si technology is matured (d) All of the above

2. Polysilicon is used for gate in MOSFET because
   (a) it is semi-metal   (b) it has lattice matching wi
   (c) it is easy to fabricate (d) none of these

Objective questions: Has a rich set of end-chapter exercises with close to 240 review questions, more than 270 MCQs, and 180 unsolved problems.

SUMMARY

- A CMOS inverter has minimum static power dissipation, highest swing in the logic levels (from 0 to V_{DD}), and maximum noise margin.
- In a symmetric CMOS inverter, the pMOS occupies more area in a chip than the nMOS transistor.
- The propagation delay of the gate is determined by the size of the transistors and the load capacitance.
- Dynamic power dissipation occurs only during the transition periods and is independent of the size of the transistors.
- The CMOS inverter can exhibit inverting VTC characteristics even when the power supply voltage is scaled down below the threshold voltage of the transistors. Under this condition, the transistors are operated in the subthreshold region.

Summary: A list of key topics at the end of each chapter to revise all the important concepts explained

Preface to the Second Edition

Very large scale integration (VLSI) refers to the level of integration in manufacturing an integrated circuit, a small semiconductor chip on which a pattern of electronic components and their interconnections are fabricated. Since the invention of integrated circuit in 1959, its manufacturing process has evolved over the decades, from small scale integration (SSI) in the early 1960s to VLSI in the late 1970s. VLSI started with integrating tens of thousands of transistors on a single chip in the 1980s to hundreds of thousands, then millions, and now billions of components. The advances in VLSI technology have revolutionized the world of electronics and contributed to the development of present day computers and digital systems that are faster, smaller, and more power efficient than their predecessors.

About the Book

VLSI design is a technique or methodology to design the layout or footprint of an integrated circuit. It is a multidisciplinary field which involves understanding of semiconductor device physics, knowledge of circuit design, concept of analog and digital logic design, algorithms, and familiarity of electronic design automation (EDA) tools circuit. This textbook has been developed to provide a comprehensive coverage of the basic concepts and methodologies of VLSI design. This edition has been revised and updated so as to cover the syllabi of technical universities in India as well as to incorporate the faculty feedback.

Key Features

- All the steps of VLSI design have been incorporated keeping in mind the industry perspective
- Terminologies used exclusively in the real design process have been followed throughout the book
- Connection between VLSI design and CAD tools is also drawn in this book
- Different state-of-the-art electronic design automation (EDA) tools have been explained thoroughly
- Numerous review questions and unsolved problems provided as end-chapter exercises to help readers assess their understanding of the concepts covered

New to the Second Edition

- Chapters on MOS inverters, semiconductor memories, sub-system design, and low power logic circuits
- Sections on resonant tunnelling diodes, single electron transistors, spin transistors, ballistic electron devices, organic field effect transistors, carbon nanotubes, molecular transistors, small signal analysis of single-stage amplifiers, working principle of bistable circuits, and design of CMOS D flip-flops

- Revised and expanded coverage of topics such as PLD, Elmore delay model, Domino and NORA CMOS logic, and gate and device sizing

Extended Chapter Material

Chapter 1 The chapter has been expanded to include new sections on solid-state nano-electronic devices such as resonant tunnelling diodes, single electron transistors, spin transistors, and ballistic electron devices, as well as molecular nano-electronic devices such as organic field effect transistors, carbon nanotubes, and molecular transistors.

Chapter 2 Three new solved examples have been included to further explain the working of *n*-channel MOSFET.

Chapter 3 This is a new chapter which discusses MOS inverters in detail.

Chapter 4 The section on layout design of CMOS inverter has been elaborated, and a new figure has been included which depicts cross-sectional view of a CMOS inverter. New examples have been included on SPICE modelling of two-input CMOS NAND and NOR gates and calculating the Elmore delay at a given node in a tree network.

Chapter 5 A new section on small signal analysis of different configurations of single-stage amplifiers has been added to the chapter. Five new examples have also been included.

Chapter 6 The section on sequential logic circuits has been expanded to include the working principle of bistable circuits and design of CMOS D flip-flops. Sections on Domino and NORA CMOS logic have been strengthened, and a new example has been added which shows how to design a logic using domino CMOS logic.

Chapter 7 This edition dedicates an independent chapter to semiconductor memories which was covered in Chapter 5 of the first edition.

Chapter 10 The sections on timing-driven logic synthesis and gate and device sizing have been rewritten in order to further strengthen the explanation.

Chapter 14 The section on stuck-at-fault has been elaborated and a new table has been included which shows fault-free and faulty outputs for all possible input combinations.

Chapter 16 A brief overview of CMOS technology has been included in this chapter.

Chapters 17 and 18 Two new chapters on subsystem design and low-power logic circuits have been added in this edition.

As a result of reorganization, the chapters on VHDL and Verilog have been moved to the end of the book as Appendices A and B.

Content and Coverage

The book comprises 18 chapters and four appendices.

Chapter 1 begins with describing the evolution of VLSI technology over the last few decades. It explains the VLSI design methodology, the styles of VLSI design, and necessity of VLSI CAD tools. Towards the end of the chapter, the IC chip industry and recent advancements of VLSI and its future projections are also described.

Chapter 2 deals with basics of MOS transistors. In this book, we have presented VLSI circuits implemented using only MOS transistors. Hence, a detailed discussion is included for MOS transistors. The chapter covers the structure and operation of MOS transistors, types of MOS transistors, $I–V$ characteristics, and several important MOSFET parameters. It also describes the MOSFET scaling and its effects, different MOSFET capacitances, and MOSFET models.

Chapter 3 introduces different types of inverters such as resistive load inverter, enhancement-type nMOS load inverter, depletion-type nMOS load inverter, and CMOS inverter. It then discusses in detail the static and transient characteristics of CMOS inverters.

Chapter 4 explains the process of developing standard cell library and its characterization process. It explains the components of standard cell library, the design of components, and their characterization. This chapter gives an idea about main aspects of VLSI circuits in terms of power, speed, and area.

Chapter 5 deals with analog CMOS circuit design. It includes the design of small analog blocks such as current source/sink, current mirrors, switched capacitor circuits, and MOS voltage and current reference circuits. Further, it explains the design of CMOS amplifier, differential amplifier, OPAMP, and complex analog circuits such as DAC, ADC, SD modulator, and PLL. It also covers the analog design using the reconfigurable analog array FPAA.

Chapter 6 introduces different types of digital circuit design using MOS and CMOS transistors. It covers design of static CMOS, pseudo-nMOS, transmission gate, dynamic CMOS, domino CMOS, NORA, Zipper, and TSPC CMOS circuits. It also introduces the design of CPL, CVSL, differential CMOS, adiabatic logic, and dual-threshold CMOS logic circuits.

Chapter 7 discusses the design and operation of different semiconductor memories which are used for storing digital information.

Chapter 8 describes the concept of BiCMOS circuits, its technology, and design of different BiCMOS logic circuits.

Chapter 9 deals with logic synthesis, its main concepts, theorems, and algorithms behind logic synthesis with examples. Several design examples are also provided using universal building blocks multiplexers and decoders.

Chapter 10 is fully dedicated to explain one of the main aspects of VLSI design—timing. It discusses several important terms and definition about timing, and the important parameters that critically affects timing. It also describes the timing analysis in combinational and sequential circuits, and timing models.

Chapter 11 explains the main steps of physical design such as partitioning, floor planning, placement, and routing. Each of these steps is described along with their main objectives, different techniques, and important algorithms.

Chapter 12 introduces the process of verifying the physical design and analysis of reliability effects in VLSI circuits. It includes verification steps such as DRC, LVS, ERC, and extraction. The important reliability issues such as electromigration, TDDB, hot carrier, NBTI, ESD, latch-up, IR drop, and soft errors in VLSI circuits are also discussed.

Chapter 13 discusses IC packaging. IC package characterization methodology and package models are introduced in this chapter.

Chapter 14 deals with VLSI testing. It introduces important testing techniques in VLSI circuits, fault models, and different fault simulation techniques. It also covers the different DFT approaches such as scan-test, BST, BIST, followed by ATPG and IDDQ test. Towards the end it introduces DFM, design economics, and yield.

Chapter 15 covers field programmable gate array-based VLSI design. It introduces the concepts of programmable hardware with ROM, PAL, PLA, and PLDs. The architecture of FPGA and its design strategy are discussed at the end.

Chapter 16 describes the technological aspects of VLSI design. The integrated circuit manufacturing process technology is discussed in this chapter. It starts with crystal growth, followed by photolithography, oxidation, diffusion, ion implantation, etching, epitaxial growth, metallization, and finally IC packaging.

Chapter 17 deals with design of small subsystems such as adders, multipliers, drivers, and divider circuits.

Chapter 18 explains why low-power circuits have become important in recent days. The concepts of low power circuits are discussed at a basic level in order to understand the low power issues and the basic techniques to design low power circuits.

Appendices A and B provide the fundamentals for understanding the digital VLSI design using hardware description languages such as VHDL and Verilog, respectively.

Appendix C on SPICE explains the design and simulation of a circuit. The general syntax of writing SPICE netlist and several examples are also included.

Appendix D provides chapter-wise answers to all the objective questions.

Acknowledgements

I would like to thank my institute, Department of Electronics and Telecommunication Engineering, Assam University, Silchar, for their support. I am also thankful to my Ph D guide Prof. H. Rahaman, Indian Institute of Engineering Science and Technology (IIEST) (formerly Bengal Engineering and Science University), Shibpur, for his constant inspiration and guidance. My sincere thanks to Prof. Bhargab B. Bhattacharya, ISI, Kolkata, for his suggestions and inspirations.

I thank all my students especially those who have given their constructive feedback since beginning. Finally, I thank the editorial team of Oxford University Press, India, for providing me all the support during the development of this edition.

I am grateful to my wife, Joyita, and my daughters, Adrija and Adrisha, for their continuous support and encouragement. Without their support it would not have been possible for me to write the second edition of the book.

It has been our utmost endeavour to publish an error-free book and we sincerely apologize to the readers for any unintentional mistakes that may be present in the book. All suggestions for further improvement of the book are welcome.

Debaprasad Das

Preface to the First Edition

All electronic devices, from home appliances to computers, have one thing in common—the integrated circuit or IC. An IC is defined as a circuit containing millions of transistors having interconnections on a piece of semiconductor, which has an area of only a few square millimetres. Each transistor, in turn, is of the size of about a few nanometres. Let us take an example to appreciate the dimensions. Let the transistor size be zoomed to centimetre scale, i.e., the size of the order of a ludo coin. With the same zoom factor of 10^7, the chip area will now be a few hundred square kilometres. Now, the goal is to place a few million ludo coins in this area and connect them such that we get the desired function. This challenging methodology, known as VLSI, is the key enabler for the state-of-the-art integrated circuits present in our common day-to-day appliances, such as toys, personal computers, mobile phones, washing machines, microwave ovens, automobiles, etc. A typical VLSI circuit contains millions of transistors and their interconnections designed on a very small area, say, on a finger tip.

A usual challenge in designing a VLSI circuit involves solving matrices of a very large order. The only feasible solution for handling such a large matrix is through the usage of computers and building efficient algorithms. This is why most chips or ICs nowadays are designed with the help of electronic design automation or EDA softwares and the process is known as computer-aided design or CAD. VLSI design, therefore, deals with circuits, CAD tools, algorithms, and methodologies to design, verify, and test very large scale integrated circuits.

About the Book

VLSI Design will serve as a useful textbook for undergraduate engineering students. The book provides a thorough understanding of the basic concepts, methodologies, and algorithms of VLSI design. Beginning with an introduction to VLSI systems and basic concepts of MOS transistors, the book gives an exhaustive coverage of standard cell library design and characterization, CMOS analog and digital designs, and BiCMOS logic circuits. Hardware description languages such as VHDL and Verilog are covered before taking up an exhaustive step-wise discussion on the various stages involved in designing a VLSI chip (which includes logic synthesis, timing analysis, floorplanning, placement and routing, verification, and testing). The book also contains separate chapters on FPGA architecture and VLSI process technology. All the tools that are used to design VLSI circuits starting from behavioural netlist to the final layout of the chips to be manufactured have been discussed in the book.

Acknowledgements

I would like to thank my institute, Meghnad Saha Institute of Technology (MSIT) for enabling me with all the support that was needed for my work. I am thankful to my colleagues for all the help. I feel blessed getting the support of my teacher, guide, and philosopher, Prof. D. Mukhopadhyay, Jadavpur University, Kolkata. He inspired me to look at the subject from a different perspective. I am very much thankful to my PhD guide Prof. H. Rahaman, Bengal Engineering and Science University, Shibpur, for his constant inspiration and guidance in every aspect of my research work.

Finally, I am grateful to my wife, Joyita, and my daughter, Adrija, for being a constant source of inspiration. Without their support it would not have been possible for me to write this book. I would also like to thank my family members, parents-in-law, and other well wishers for their moral support.

I thank Oxford University Press, India for providing me all the support during the development of this book. My sincere thanks also go to all the peer reviewers of this book.

I sincerely apologize to the readers for any unintentional mistakes that may be present in this book. All suggestions for further improvement of the book are welcome.

Debaprasad Das

Brief Contents

Detailed Contents

List of Colour Plates

Plate 1

- Silicon-on-insulator (SOI) process *(Chapter 1, Fig. 1.34, p. 33)*
- Example of 3D integration: (a) several active layers implementing logic, memory, and I/O, (b) die stack *(Chapter 1, Fig. 1.35, p. 34)*
- (a) Normal capacitive structure, (b) MOS capacitor *(Chapter 2, Fig. 2.1, p. 45)*

Plate 2

- MOS capacitor under different bias conditions *(Chapter 2, Fig. 2.2, p. 45)*
- Cross-sectional view of MOSFET: (a) enhancement-type, (b) depletion-type *(Chapter 2, Fig. 2.4, p. 47)*
- Cross-sectional view of enhancement-type MOSFET: (a) *n*-channel, (b) *p*-channel *(Chapter 2, Fig. 2.5, p. 48)*

Plate 3

- nMOS under different operating conditions: (a) accumulation, (b) depletion, (c) inversion *(Chapter 2, Fig. 2.7, p. 49)*
- pMOS under different operating conditions: (a) accumulation, (b) depletion, (c) inversion *(Chapter 2, Fig. 2.8, p. 50)*
- Gradual channel approximation (GCA) model *(Chapter 2, Fig. 2.10, p. 54)*

Plate 4

- nMOS transistor under saturation *(Chapter 2, Fig. 2.11, p. 56)*
- Cross-sectional view of MOSFET (across the channel) *(Chapter 2, Fig. 2.20, p. 69)*
- MOSFET junction capacitances *(Chapter 2, Fig. 2.22, p. 72)*

Plate 5

- Three-dimensional structural view of a CMOS inverter *(Chapter 4, Fig. 4.5, p. 110)*
- Different mask layers *(Chapter 4, Fig. 4.6, p. 111)*

Plate 6

- Layout of a CMOS inverter *(Chapter 4, Fig. 4.7, p. 112)*
- (a) Lines used to draw stick diagram, (b) stick diagram of CMOS inverter, (c) stick diagram of two-input NAND gate *(Chapter 4, Fig. 4.8, p. 112)*

Plate 7

- 3-input NOR gate: (a) CMOS logic, (b) colour palate, (c) stick diagram *(Chapter 4, Fig. 4.9, p. 113)*
- (a) Colour palate used, (b) stick diagram implementing $F = \overline{(A + B + C)D}$ *(Chapter 4, Fig. 4.10, p. 113)*
- Photolithography process: creating patterns on photoresist, pattern transfer from photoresist to SiO_2 layer by etching *(Chapter 16, Fig. 16.11, p. 508)*

Plate 8

- Multilevel metallization: (a) 3-level metal connection to n-active without stacked vias, (b) 3-level metal connection to n-active with stacked vias, (c) 6-level metal interconnect *(Chapter 16, Fig. 16.24, p. 522)*
- Schematic of (a) wire bonding, (b) tape automated bonding, (c) flip-chip bonding *(Chapter 16, Fig. 16.27, p. 525)*

Plate I

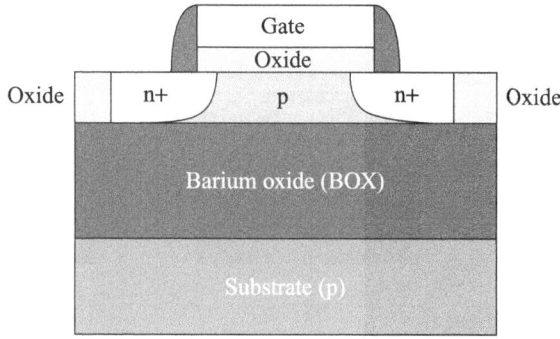

Figure 1.34 Silicon-on-insulator (SOI) process (Chapter 1, page 33)

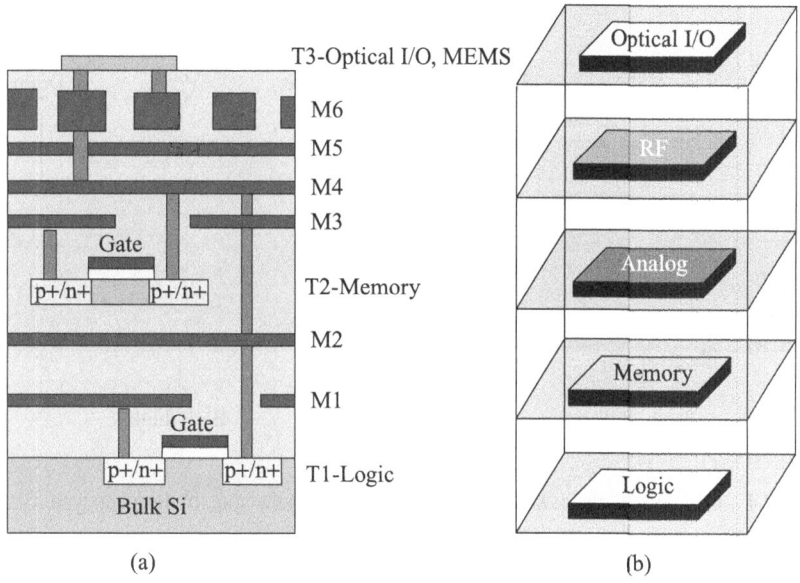

(a) (b)

Figure 1.35 Example of 3D integration: (a) several active layers implementing logic, memory, and I/O, (b) die stack (Chapter 1, page 34)

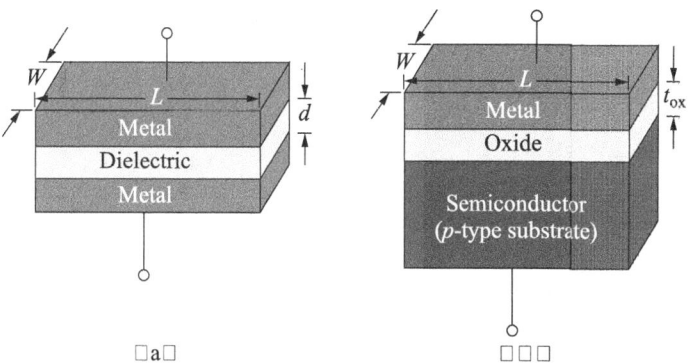

(a) (b)

Figure 2.1 (a) Normal capacitive structure, (b) MOS capacitor (Chapter 2, page 45)

Plate 2

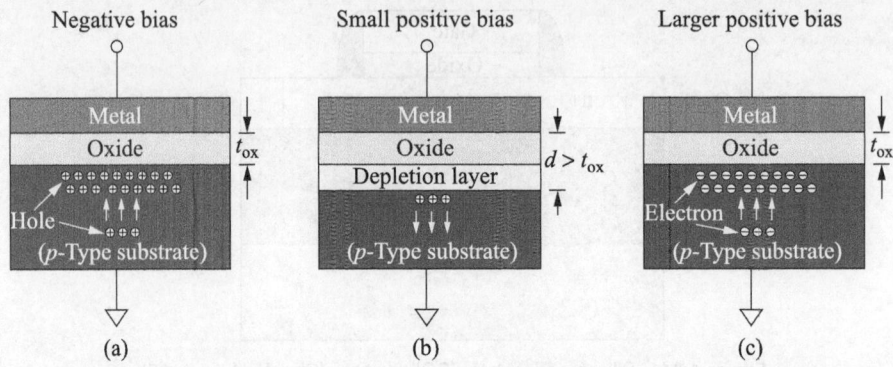

Figure 2.2   MOS capacitor under different bias conditions (Chapter 2, page 45)

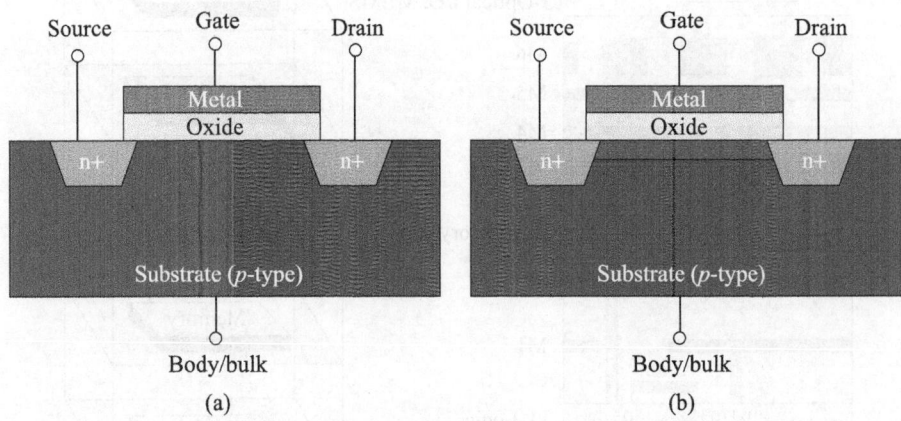

Figure 2.4   Cross-sectional view of MOSFET: (a) enhancement type, (b) depletion type (Chapter 2, page 47)

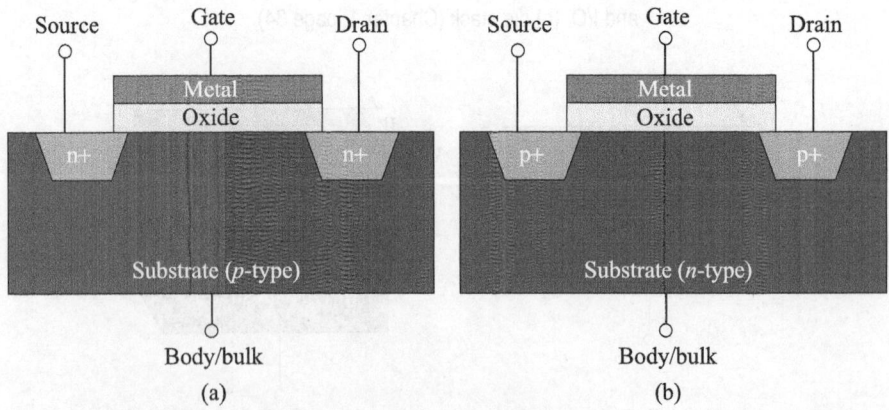

Figure 2.5   Cross-sectional view of enhancement-type MOSFET: (a) *n*-channel, (b) *p*-channel (Chapter 2, page 48)

Plate 3

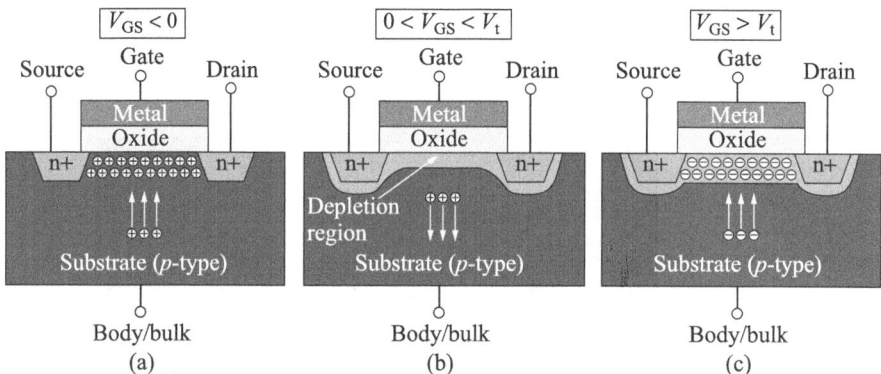

Figure 2.7 nMOS under different operating conditions: (a) accumulation, (b) depletion, (c) inversion (Chapter 2, page 49)

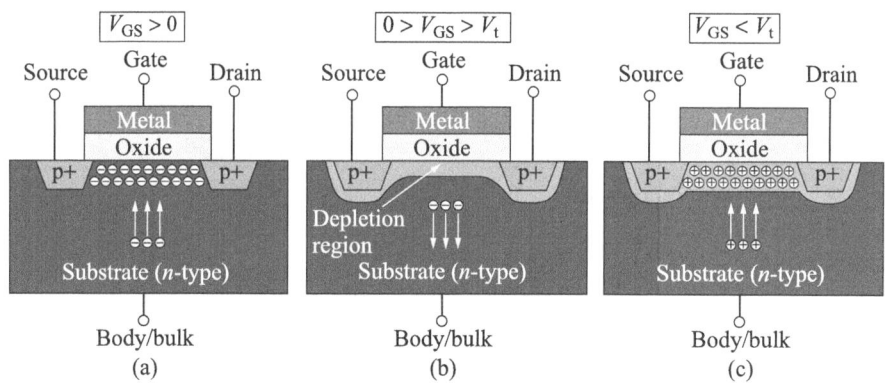

Figure 2.8 pMOS under different operating conditions: (a) accumulation, (b) depletion, (c) inversion (Chapter 2, page 50)

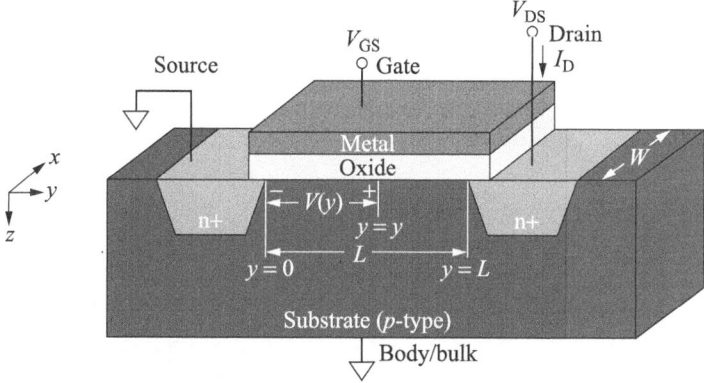

Figure 2.10 Gradual channel approximation (GCA) model (Chapter 2, page 54)

Plate 4

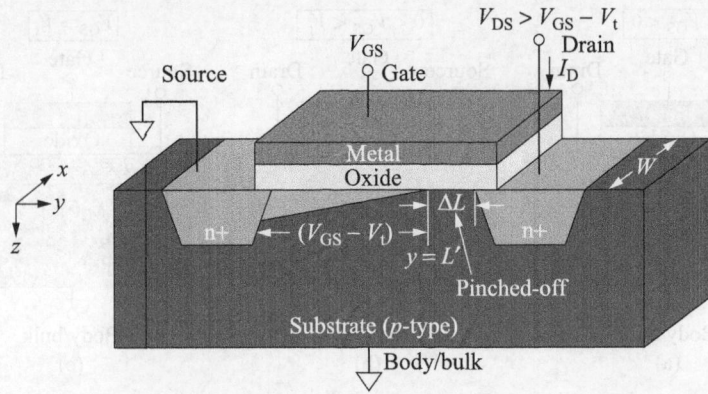

Figure 2.11 nMOS transistor under saturation (Chapter 2, page 56)

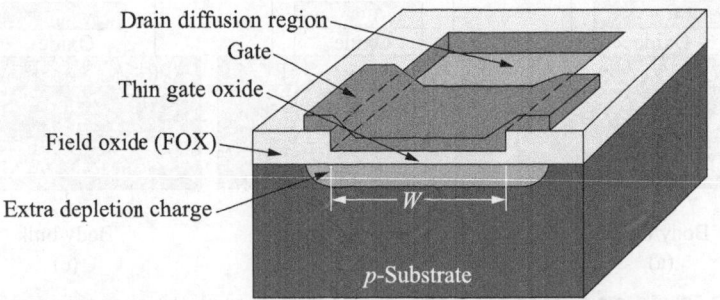

Figure 2.20 Cross-sectional view of MOSFET (across the channel) (Chapter 2, page 69)

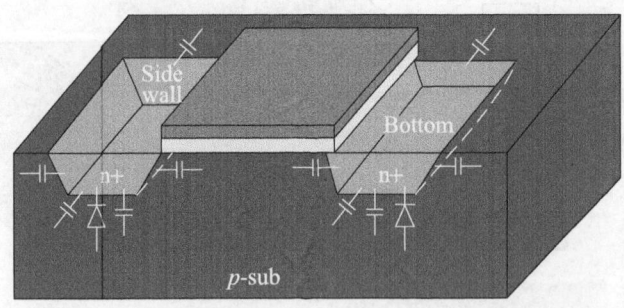

Figure 2.22 MOSFET junction capacitances (Chapter 2, page 72)

Plate 5

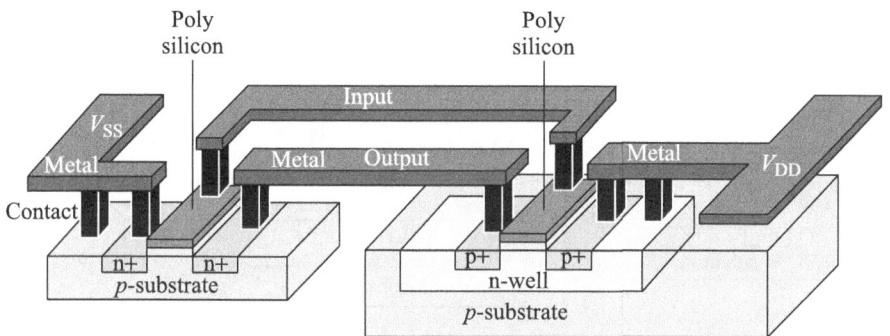

Figure 4.5 Three-dimensional structural view of a CMOS inverter (Chapter 4, page 110)

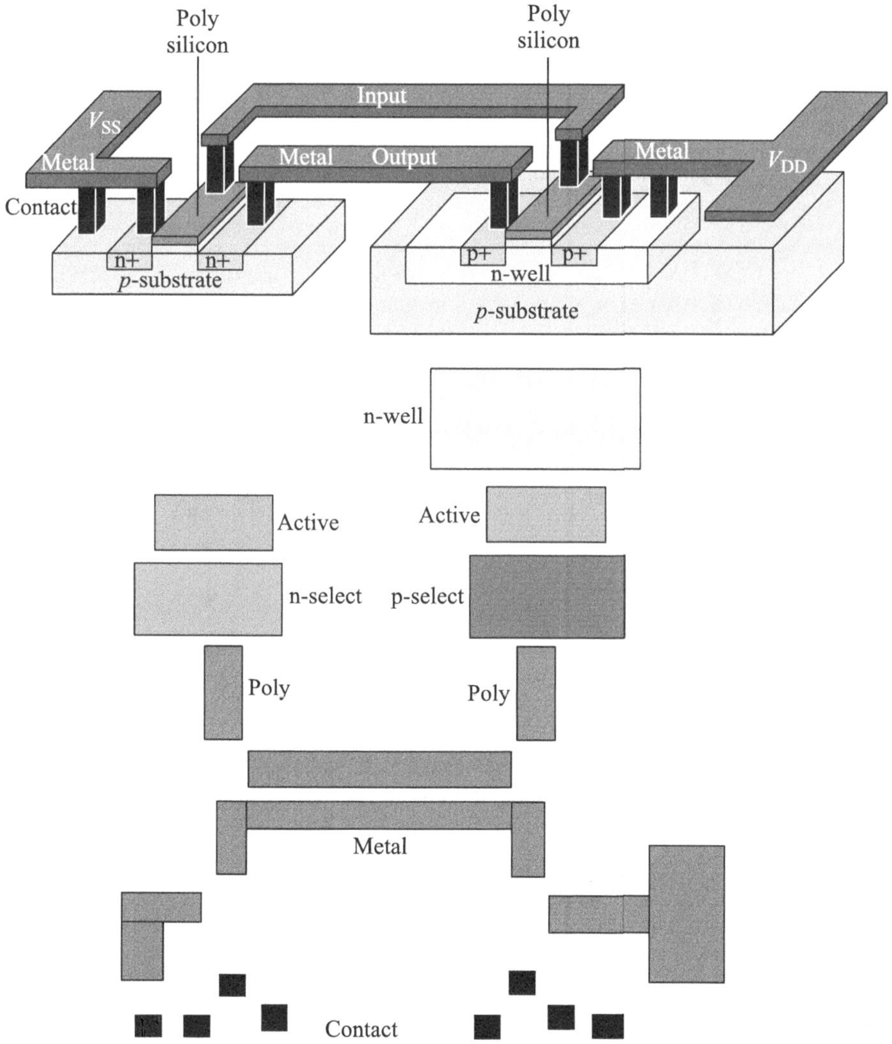

Figure 4.6 Different mask layers (Chapter 4, page 111)

Plate 6

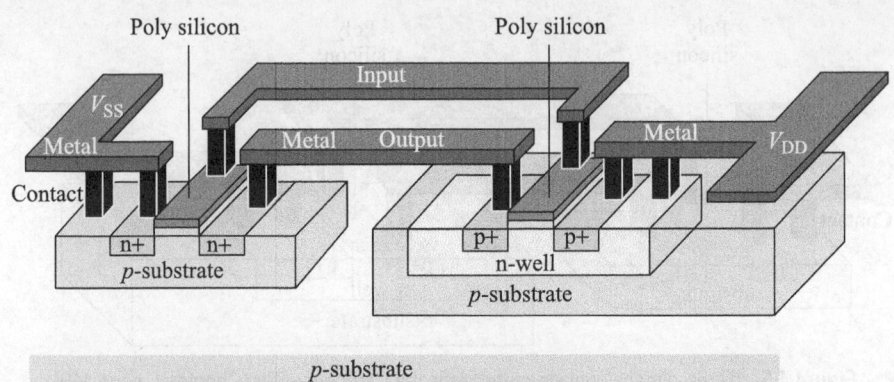

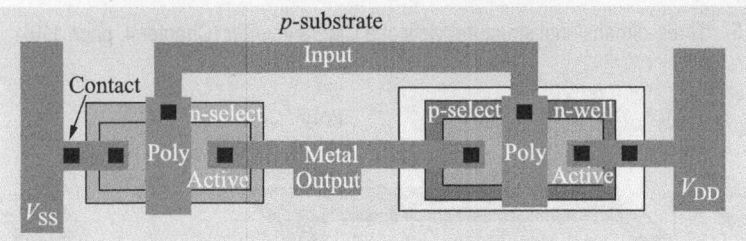

Figure 4.7   Layout of a CMOS inverter (Chapter 4, page 112)

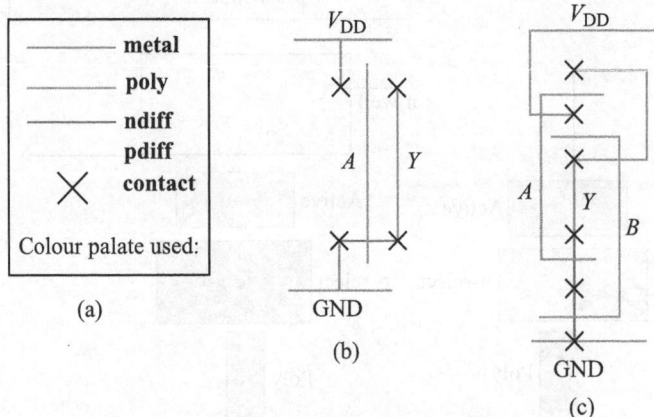

Figure 4.8   (a) Lines used to draw stick diagram, (b) stick diagram of CMOS inverter, (c) stick diagram of two-input NAND gate (Chapter 4, page 112)

Plate 7

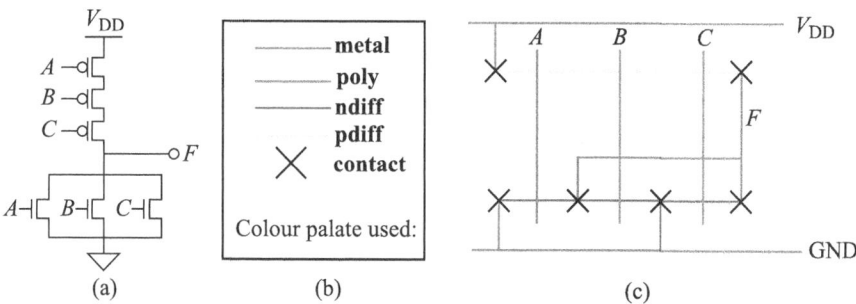

Figure 4.9 3-input NOR gate: (a) CMOS logic, (b) colour palate, (c) stick diagram
(Chapter 4, page 113)

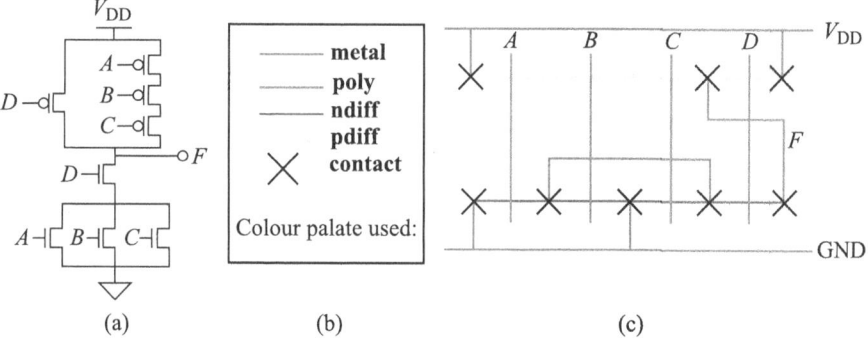

Figure 4.10 (a) Colour palate used, (b) stick diagram implementing $F = \overline{(A + B + C)D}$
(Chapter 4, page 113)

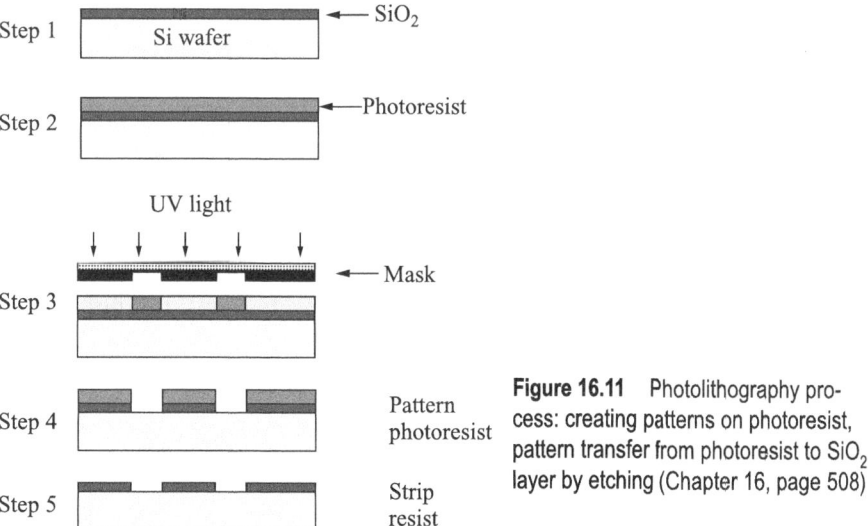

Figure 16.11 Photolithography process: creating patterns on photoresist, pattern transfer from photoresist to SiO_2 layer by etching (Chapter 16, page 508)

Plate 8

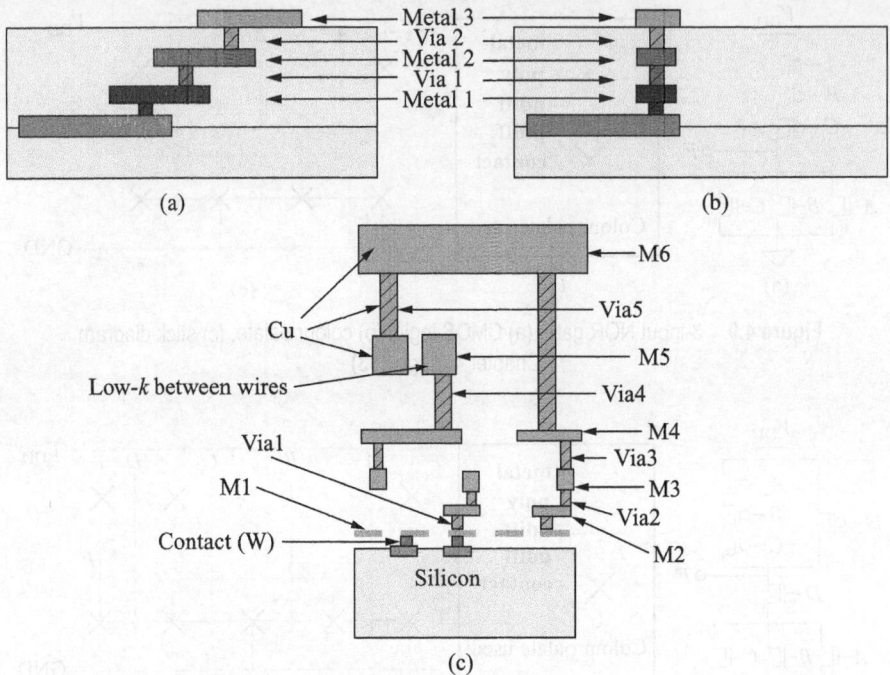

Figure 16.24 Multilevel metallization: (a) 3-level metal connection to n-active without stacked vias, (b) 3-level metal connection to n-active with stacked vias, (c) 6-level metal interconnect (Chapter 16, page 522)

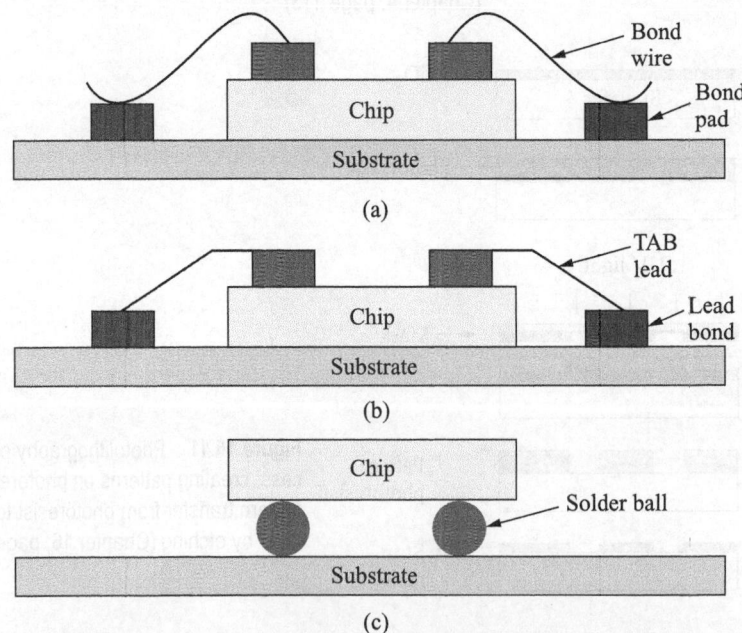

Figure 16.27 Schematic of (a) wire bonding, (b) tape automated bonding, (c) flip-chip bonding (Chapter 16, page 525)

Introduction to VLSI Systems

1.1 Historical Perspective

In the beginning of the twenty-first century, we find ourselves surrounded by different machines and appliances that are impossible to build without applying the principles of electronics onto them. Moreover, the communication boom that we see now would not have been possible without the advancement in the electronics industry and without using integrated circuits (ICs). Integrated circuits can simply be described as a large circuit manufactured on a very small semiconductor chip. Starting with a few transistors on a single chip in the early 1970s, it has increased to a billion transistors within a span of 40 years. The ever-increasing demand of humans has pushed the technology to integrate more and more components into a single chip. And it is expected that the trend will continue.

The journey started when the first transistor was invented in 1947 by Bardeen, Brattain, and Schockley at the Bell Telephone Laboratory. It was followed by the introduction of the bipolar junction transistor (BJT) in 1949 by Schockley. Then it took seven years to build a digital logic gate. In 1956, Harris first introduced the bipolar digital logic gate using discrete components. The biggest revolution happened when Jack Kilby at Texas Instruments first made the monolithic integrated circuit in 1958. This was a significant breakthrough in the semiconductor technology by Kilby, for which he was awarded Nobel Prize in 2000. The first commercial IC was introduced by Fairchild Corporation in 1960 followed by TTL IC in 1962. The ECL logic family has come up in 1974. Another breakthrough in the IC technology is the introduction of the first microprocessor 4004 by Intel in 1972. Since then, there has been a steady progress in the IC industry resulting in high density chips such as Pentium processors.

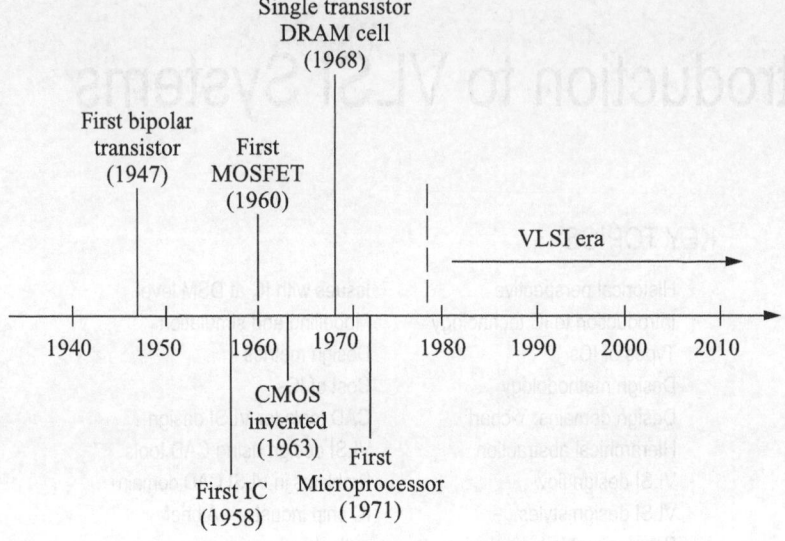

Fig. 1.1 Evolution of VLSI (major milestones)

The evolution of very large scale integration (VLSI) in the last half century is illustrated in Fig. 1.1, indicating major milestones.

Depending on the level of integration of components into a single chip, the IC technology is classified into several categories, among which VLSI is an advanced integration level. The acronym VLSI is used in the context of integrated circuit (IC) design and manufacturing. Table 1.1 lists the different IC technologies that have evolved over the last 50 years.

Table 1.1 Evolution of integration level in integrated circuits

Year	Era	Level of integration
1958	Single transistor	–
1960	Monolithic IC	1
1962	Multi-function	2–4
1964	Complex function	5–20
1967	Medium scale integration (MSI)	20–200
1972	Large scale integration (LSI)	200–2000
1978	Very large scale integration (VLSI)	2000–20,000
1989	Ultra large scale integration (ULSI)	Above 20,000

The main target of integrated circuit design and fabrication is to achieve more functionality at higher speed using less power, less area, and low cost. In the early 1960s, Intel cofounder Gordon Moore had predicted that *the number of devices on a single chip will double in every eighteen months*. Over the last 50 years it has been found that the semiconductor industry has really followed the prediction of Moore, and hence it has become a law which is famously known as the Moore's law. Figure 1.2 depicts how the microprocessor technology has evolved over the last 40 years.

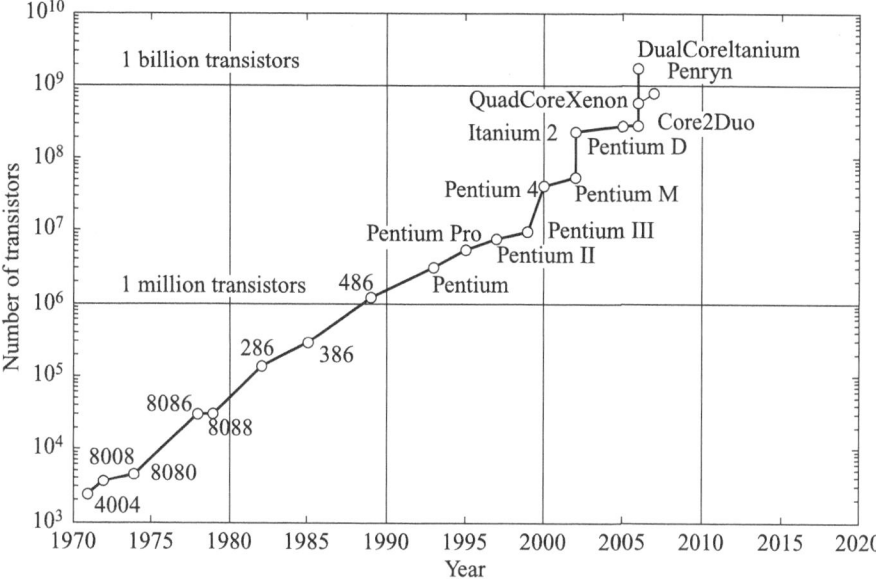

Fig. 1.2   Level of integration for Intel microprocessors (*Source:* Intel Corporation)

1.2 Introduction to IC Technology

VLSI technology has been the enabler for advancement of the present electronics and communication age. Silicon has been a natural choice of material for IC manufacturing because of its abundance in nature, and more importantly its native oxide which is most suitable for IC fabrication. Almost 90% of the electronic circuits fabricated worldwide are made of silicon using CMOS technology. CMOS technology is the most popular technology because of its low power and less area requirement.

Figure 1.3 describes the IC manufacturing process in a pictorial format. First, the single crystal silicon ingot is grown, which is sliced to obtain wafers. The wafers are the substrate of the IC chip. Now on the wafer, the devices and their interconnections are patterned using the lithography technique. A processed wafer contains several identical ICs called a *die*. The processed wafer is then tested to identify

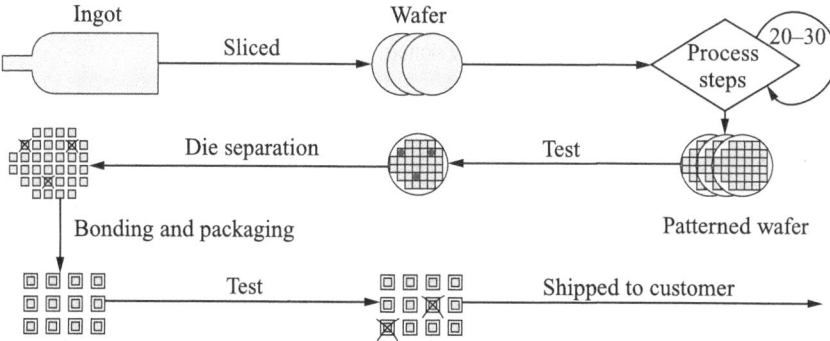

Fig. 1.3   Schematic view of IC technology

faulty circuits. The dies are separated from the wafer and each die is bonded and packaged to form IC chips. These chips are tested to sort out the faulty ones from the manufactured chips, and then the good chips are shipped to the customer site.

The IC is patterned on the wafer using the photolithography technique which is a combination of several process steps, namely, oxidation, coating of photoresist, exposure to UV light, etching, diffusion, ion implantation, film deposition, and metallization. These process steps are discussed in detail in Chapter 16.

The major advantages of the IC over discrete component-based circuits are as follows:

- Smaller physical size
- Higher speed
- Lower power consumption
- Reduced cost

1.3 Types of Integrated Circuits

Depending on the types of substrates, devices, signal processing, and applications, ICs are classified as shown in Table 1.2.

The material used for substrate of the IC can be either of silicon, GaAs, or Si–Ge. Table 1.2 shows the classification of ICs on the basis of types of substrates.

Table 1.2   Classifications of integrated circuits

Based on types of substrates	Based on types of devices	Based on the type of signals such as analog, digital, mixed, and radio frequency (RF)	Based on the applications
Silicon	Bipolar	Digital IC: These ICs are operated with only digital signals.	Standard parts (TTL, or TTL-equivalent ICs at SSI, MSI, and LSI levels). These are general purpose ICs, e.g., microprocessor, memory, etc.
GaAs	CMOS	Analog IC: These ICs are operated with only analog signals.	Application-specific standard products (ASSP). Examples are: a controller chip for PC, a chip for modem, etc.
Si–Ge	BiCMOS	Mixed-signal IC: These ICs are operated with both analog and digital signals.	Application-specific integrated circuits (ASIC). Examples are: a chip for mobile phone, for satellite, a chip for toy, etc.
	MESFET	RF IC: These ICs are operated with radio frequency signals.	
	HBT		
	HEMT		

Based on the devices used for implementing the IC, they are classified as follows:

- *Bipolar*—In these ICs, bipolar junction transistors (BJTs) are used to implement the IC.
- *CMOS*—In the CMOS ICs, combination of nMOS and pMOS transistors are used to implement the IC.

- *BiCMOS*—In these ICs, combination of BJT and CMOS transistors are used.
- *MESFET*—Metal semiconductor field effect transistors are used in these ICs.
- *HBT*—These ICs use hetero-junction bipolar transistors.
- *HEMT*—High electron mobility transistors are used in these ICs.

1.4 Design Methodology

VLSI design is a sequential process of generating the physical layout of an IC, starting from the specification of that circuit. It can be fully or semi-automated using numerous softwares called *electronic design automation* (EDA) or *computer-aided design* (CAD) tools. The designers first get an idea of a new system or device for a particular application. This new idea is translated in the form of an integrated circuit chip using the VLSI design flow.

The concept or idea is first documented in a formal language and then translated into register transfer level (RTL) using hardware description languages (HDL) such as Verilog and VHDL. The RTL netlist is then complied and tested to check if the functionality expected is correctly described. Usually, an iterative process is used to describe the circuit behaviour in the RTL netlist. Once the behavioural netlist is finalized, the constraints are imposed on the design. Some major constraints are less area, low power, and high speed. Taking the behavioural netlist and design constraints, different synthesis styles have come up with optimum hardware, which meet all the constraints and generate correct functional output.

The complexity of the VLSI circuits is usually large. The design complexity is handled using several approaches. First, the design is viewed at different abstraction levels and then is decomposed hierarchically. Automated tools are used for synthesis, simulation, design, testing, and verification. If possible, the design is reused.

Typically, an electronic system has two parts—hardware and software. These can be designed concurrently in a manner so as to save the design cycle time. It also helps the hardware and software design teams to work in sync to sort out any incompatibility issues that might arise in the later stages of the design cycle.

There are two design styles used in VLSI design. One is the top–down approach and another is the bottom–up approach. In the top–down approach, the system is built starting from the top up to the bottom. While in the bottom–up approach, the basic building blocks are built first, and they are combined or assembled to build the entire system. Both approaches have their merits and demerits.

1.5 Design Domains—Y-Chart

The IC design can be described in the following three domains:
- Behavioural
- Structural
- Physical

In the behavioural domain, a circuit is described fully by its behaviour without describing its physical implementation or structure. In the structural domain, a circuit is described by its components and their interconnections. The physical domain deals with actual geometry of the circuit and describes the shape, size, and locations of its components.

The three design domains are represented pictorially by a chart called the *Y-chart* (Fig. 1.4) introduced by Gajski–Kuhn.

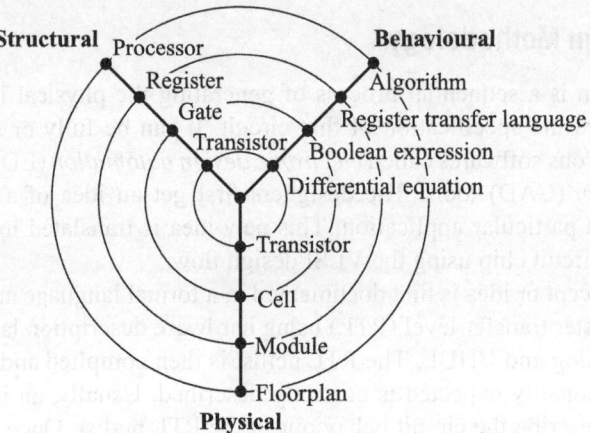

Fig. 1.4 Gajski–Kuhn Y-chart

1.6 Hierarchical Abstraction

Hierarchical decomposition or 'divide and conquer' is a useful methodology that partitions the entire system into its components. The components are again partitioned into modules and this process continues until the basic building blocks are reached. This methodology is illustrated in Fig. 1.5.

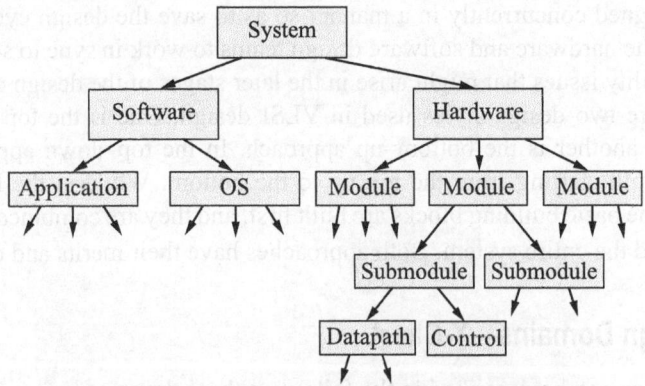

Fig. 1.5 Hierarchical decomposition of an electronic system

For example, let us consider a CPU as a system at the top level. It can be partitioned into two main components: datapath and control logic. Again, the datapath can be divided into modules such as ALU, registers, and shift registers. The ALU can be further divided into basic arithmetic units such as full adders, and so on. This is illustrated in Fig. 1.6.

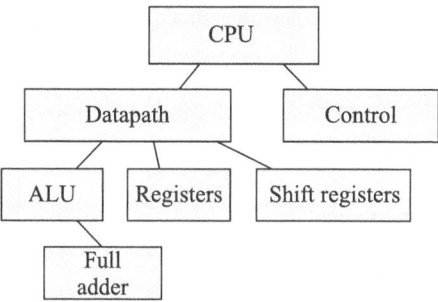

Fig. 1.6 Hierarchical decomposition of CPU

1.6.1 Regularity, Modularity, and Locality

The large complexity of the VLSI design is handled by the hierarchical decomposition of a system into several functional blocks. While hierarchically decomposing, a large system regularity must be followed. Regularity indicates that the decomposition process must not produce a large number of blocks, and the blocks need to be similar as much as possible. An array structure normally has a good regularity. The array multiplier is a perfect example of a regular structure. Regularity avoids a number of different blocks to be designed and verified and can be maintained at all levels of abstraction.

Modularity is another important aspect of hierarchical decomposition. It means that the functional blocks must have well-defined interfaces and functionality. Modular design allows different modules to be designed concurrently and also enables design reuse.

Hierarchical decomposition must also consider the locality of the functional blocks. The decomposition should be such that the blocks, exchanging signals frequently, must be close to each other in order to reduce the interconnect length. Again, the internal wiring must be local and should not affect other modules.

1.7 Design Flow

The VLSI design flow is a sequence of steps followed to translate the idea of a system into a chip. The flow is based on the standard design automation tools. The basic steps are shown in Fig. 1.7. It starts with system specifications such as area, speed, and power. Then the functional design is done followed by functional verification to check if the design is correct. In this phase, the design is described at the behavioural level. Next, the design is implemented at the logic level and verified for its correctness. This is followed by the transistor level or circuit design verification. Up to this step, the flow is known as *logical design*. The next phase is the *physical design* which actually deals with the geometry of the chip. Once the physical layout is generated, it must be verified to check if the layout really implements the actual design. The last and final step is the fabrication and testing of the chip. Figure 1.7 illustrates the VLSI design flow at the top level.

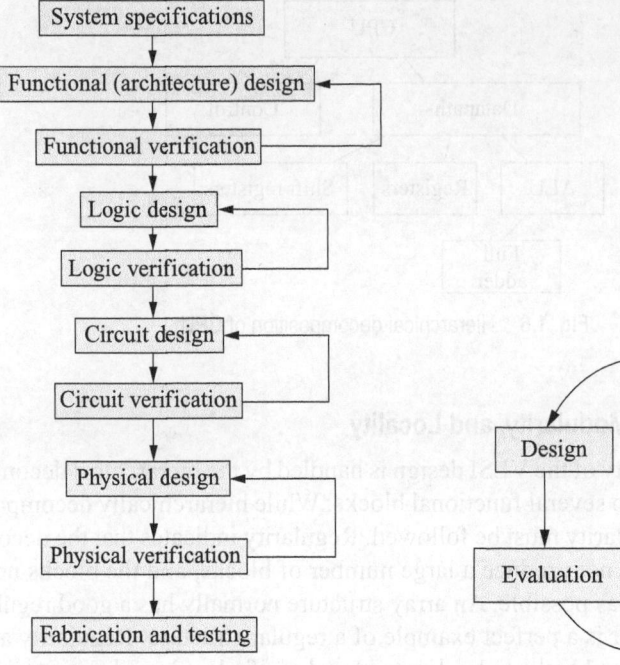

Fig. 1.7   VLSI design flow

Fig. 1.8   Design iteration loop

Though a VLSI design flow looks entirely sequential, there are feedback loops at each stage of the design flow. After each stage, the design is evaluated to see if the functionality and specifications are met. Otherwise, the designers go back and redesign, and the process is repeated in an iterative manner till the specifications are satisfied. The typical design loop is shown in Fig. 1.8.

Generally, top–down is the preferred approach in VLSI design. The bottom–up approach is infeasible for large systems.

1.8 Design Styles

Depending on the application, cost of production, performance, and the volume of production, there are different VLSI design styles that are followed to implement a chip. Each of the styles has its own advantages and disadvantages and is chosen on the basis of the target application. The commonly used design styles are as follows:

- Field programmable gate array (FPGA) design
- Gate array design
- Standard cell-based design
- Full-custom design
- Semi-custom design
- Programmable logic device (PLD)

The following subsections discuss each of the design styles in brief.

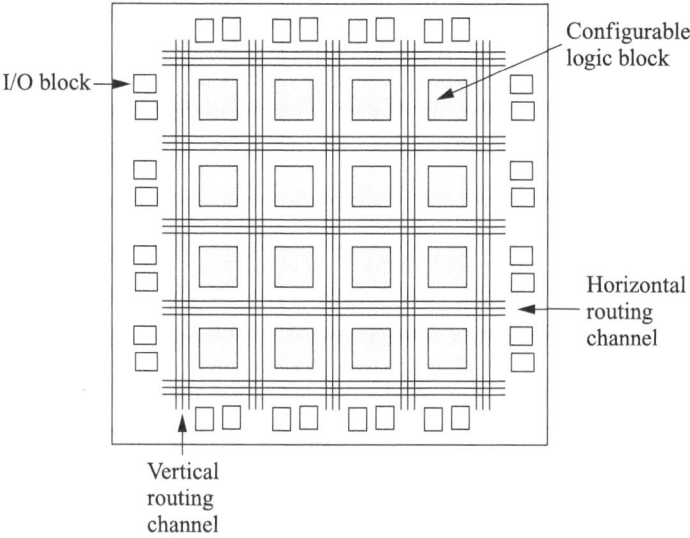

Fig. 1.9 Typical FPGA architecture

1.8.1 Field Programmable Gate Array (FPGA) Design

Field programmable gate array (FPGA) is a fully fabricated IC chip in which the interconnections can be programmed to implement different functions. An FPGA chip has thousands of logic gates which are to be connected to implement any logic function. A typical FPGA architecture is shown in Fig. 1.9. It has the following three main components:

- I/O buffers
- Array of configurable logic blocks (CLBs)
- Programmable interconnects

In the FPGA-based design, first a behavioural netlist is written to describe the functionality of the design. This is done using the hardware description languages such as Verilog or VHDL. Then the netlist is synthesized to come up with the gate level design. The next step is to map the logic blocks into available logic cells. This process is called the *technology-mapping*. This is followed by placement and routing, which configures the CLBs and defines interconnections. The next step is to generate the bit-stream and download the bit-stream into an FPGA chip with the help of a software interface. Then the FPGA chip can function as desired as long as the power is ON, or it is reprogrammed.

The benefits of using the FPGA-based design are that the design cycle time is very short and very much suitable for low volume prototype development with low cost. A concept can be quickly implemented in the hardware and checked if it is worth implementing in a large volume. But FPGA-based design is not optimized for performance. The typical design time is few hours to a couple of days for an FPGA-based design.

1.8.2 Gate Array Design

In a gate array (GA) structure, the transistors are fabricated on the silicon wafer. But the interconnections are not fabricated. The metal mask layers are customized to define the interconnections between the transistors for a targeted functionality. It can also be used for the prototype development in short time, ranked after the FPGA. GA-based design time typically varies from a few days to a few weeks. Depending on the array structure, the GA are of the following three types:

- Channelled
- Channel-less
- Structured

In the channelled gate array architecture, there are rows of transistors called arrays and channels are provided between the rows of transistors for their interconnections. In the channel-less gate array there are no channels between the rows. As there are no channels in the channel-less architecture, the interconnections are made by drawing metal lines through the unused transistors. In case of the structured GA architecture, either channelled or channel-less structure can be used, but the only difference is that it includes custom blocks.

Figure 1.10 explains the basic architectures for channelled, channel-less, and structured GA architecture.

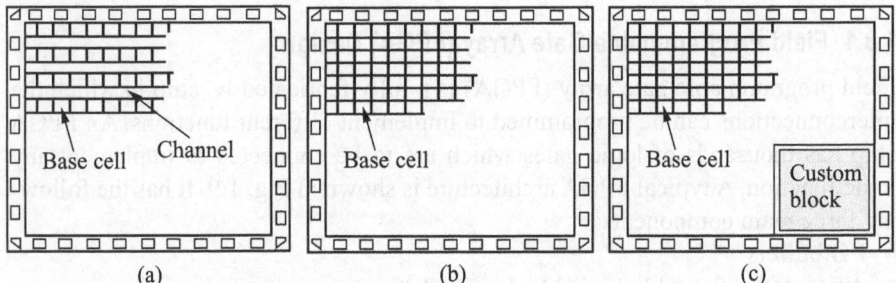

Fig. 1.10   Gate array (GA) architecture: (a) channelled; (b) channel-less; (c) structured

1.8.3 Standard Cell-based Design

The standard cell-based integrated circuit refers to a class of integrated circuits which uses the pre-designed, pre-tested, and pre-characterized standard cells. The standard cells include basic logic gates (AND, OR, NAND, NOR, XOR, XNOR, NOT, etc.), some mega cells (such as multiplexer, full-adder, decoder, etc.), sequential elements (such as D flip-flop, scan-FF, flip-flop with direct set/reset/clear inputs, registers, etc.), input–output buffers (I/O cells), and some special cells. All these standard cells are designed, tested, and characterized and put in a database which is known as a *standard cell library*.

In the standard cell-based architecture, the standard cells are placed in rows to build the integrated circuit chip. However, this design style also includes the already designed mega modules or fixed blocks. Figure 1.11 describes the architecture of a standard cell-based design.

1.8.4 Full-custom Design

In the full-custom design, the designers do not use the pre-designed standard cell library. Instead, they design the entire chip from the scratch. As each and every

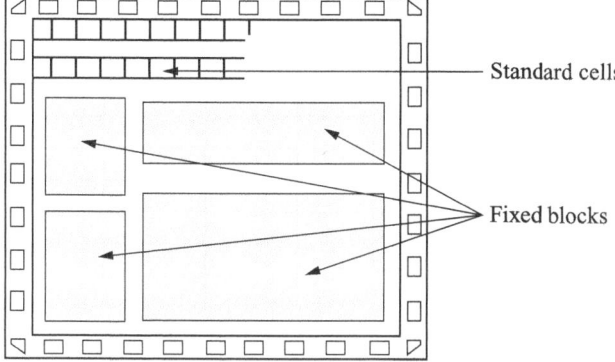

Fig. 1.11 Standard cell-based architecture

part is designed in this approach, the chips are highly optimized for area, power, and delay. Hence, a full-custom design is always superior to any other design style. However, full-custom design cycle time is higher compared to other design styles. Full-custom design style is used for high performance and high volume products.

1.8.5 Semi-custom Design

In this style of design, almost all the basic building blocks are used from the standard cell library. Only few cells are designed from the beginning, which are not available in the standard cell library or to be optimized for a specific target. This approach is faster compared to the full-custom style but slower than the standard cell-based design. Performance-wise also, it is superior to the standard cell-based design but inferior to the full-custom design.

1.8.6 Programmable Logic Device (PLD)

Programmable logic devices (PLDs) are standard products, which can be programmed to obtain the desired functionality required for a specific application. The programming can be done either by the end user or by the manufacturer. The PLDs which are programmed by the manufacturer are known as mask-programmable logic devices (MPLDs). The PLDs, which are programmed by the end user are called field-programmable logic devices (FPLDs). The architecture of PLDs is very regular and fixed. It cannot be changed by the end user. The PLDs have wide range of applications and have low risk and cost in manufacturing in large volume. Hence, the PLDs are cheaper. As the PLDs are premanufactured, tested, and placed in inventory in advance, the design cycle time is very short. The PLDs are classified into three categories based on the architecture and programmability.

- Read only memory (ROM)
- Programmable array logic (PAL)
- Programmable logic array (PLA)

Read Only Memory

Read only memory (ROM) is a memory chip which can be programmed once to store binary data. The simplest architecture of 8 × 4 ROM is shown in Fig. 1.12.

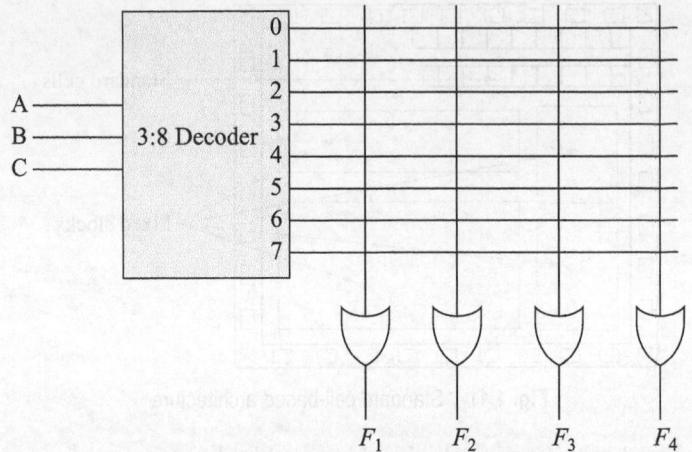

Fig. 1.12 An 8 × 4 ROM architecture

It has a decoder with k (=3) inputs and 2^k (=8) output lines. The output lines of the decoder are connected to the inputs of OR gates with programmable interconnections. All the interconnections (in this case 32) are programmable. The interconnections have fuses which can be opened or blown by applying high-voltage pulses into the fuses.

In general, $2^k \times n$ ROM has $k{:}2^k$ decoder and n OR gates. ROM can be programmed by suitably blowing the fuses to implement any logic functions. There are different types of ROM available such as:

- Programmable read only memory (PROM)
- Erasable PROM (EPROM)
- Electrically erasable PROM (EEPROM or E^{2}PROM)

Programmable Array Logic

Programmable array logic (PAL) is another programmable architecture which can be programmed to implement the desired function. The PAL architecture is shown in Fig. 1.13. It has a programmable AND array or plane followed by a fixed OR array or plane.

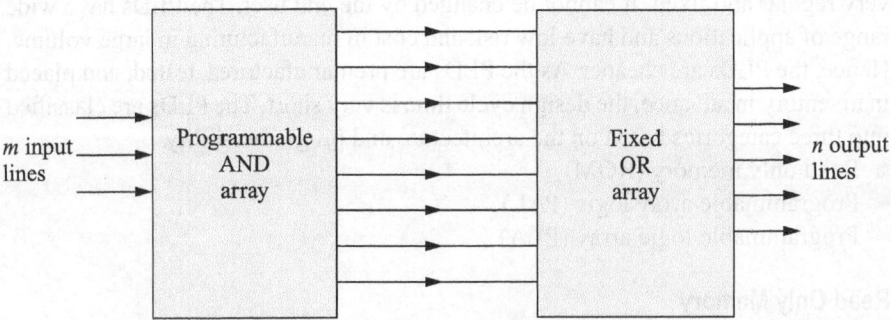

Fig. 1.13 Block diagram of PAL architecture

PAL also includes flip-flops so that it can be used to design state machines. The AND array is programmed to generate the product terms or literals of the Boolean functions to be implemented and then ORed by the OR plane to generate the functions.

Programmable Logic Array

In contrast to the PAL, programmable logic array (PLA) has both programmable AND array and OR array. The AND array is programmed to generate the product terms and then ORed by programming the OR array. The structure of PLA is shown in Fig. 1.14.

A comparison between different design styles is shown in Tables 1.3 and 1.4.

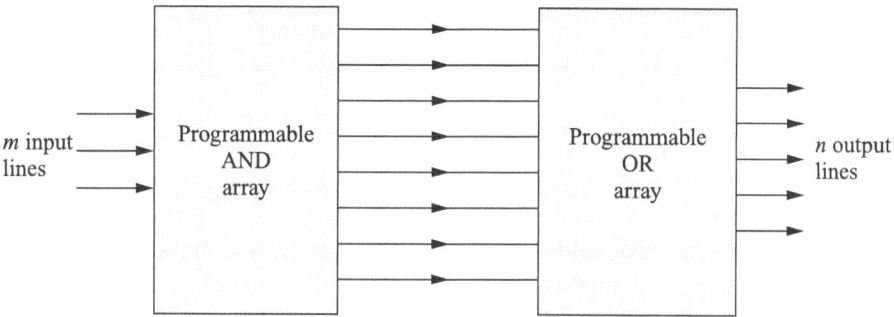

Fig. 1.14 Block diagram of PLA architecture

Table 1.3 Comparison of design styles

	Full-custom	Standard cell-based	Gate array	FPGA
Area	Small	Small to moderate	Moderate	Large
Performance	High	High to moderate	Moderate	Low
Fabrication layers	All	All	Metal layers	None
Design time	High	High to moderate	Moderate	Low

Table 1.4 Comparison of different design styles based on logic cells and interconnections

	Full-custom	Standard Cell-based	Gate array	FPGA	PLD
Cell size	Variable	Fixed height	Fixed	Fixed	Fixed
Cell type	Variable	Variable	Fixed	Programmable	Programmable
Cell placement	Variable	In row	Fixed	Fixed	Fixed
Interconnection	Variable	Variable	Variable	Programmable	Programmable

1.9 Computer-aided Design

The complexity of VLSI design is growing as per Moore's law (see Fig. 1.2). As the technology is advancing, more and more transistors are put in a smaller area to achieve high speed at less power consumption. It is not the problem of handling only the number of the transistors, but the dimensions of the transistors are

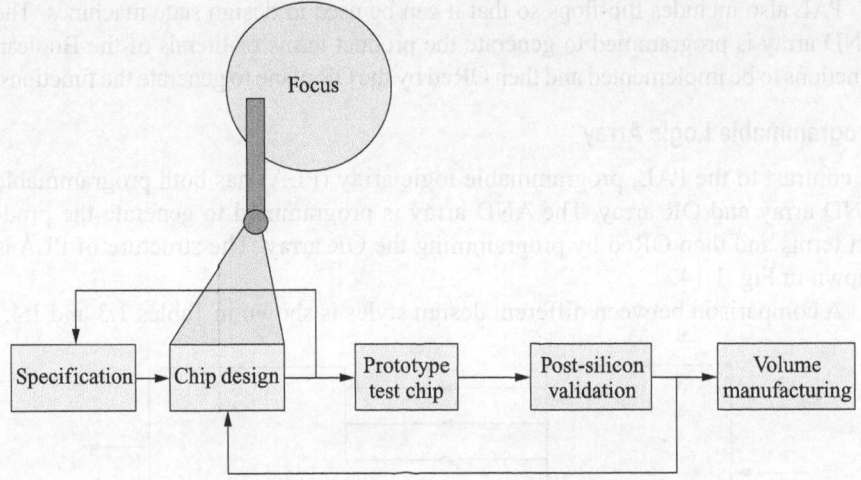

Fig. 1.15 Integrated circuit development process

shrunk from micron to nanometer range. This small dimension has created a new challenge of low-dimensional issues. So VLSI design deals with the two opposite things: one is large and another is small. Large is the number of the transistors and small is their dimensions. Hence, manual design has become impractical and computer-aided design (CAD) tools have become essential in VLSI design.

The typical IC development process is illustrated in Fig. 1.15 where the first prototype or test chip is developed and evaluated to check if the designed chip has met the desired functionality and specifications. If not, the design needs to be corrected or modified until the desired functionality and specifications are met. But more the number of iterations, more is the development cost and chance of losing the market. Hence, most of the focus is on the design cycle.

The target is to create the device right at the first time. Hence, the CAD tools must be highly efficient in creating design, checking all its functionality, and verifying the layout. The tool must be capable of handling multi-million gate designs with reasonable computer resources (i.e., memory and execution time).

Any problem in the design found early in the design cycle always reduces re-iterations. Hence, tools must have checking capability at each stage in the design flow. More importantly, the tools must not have bugs; otherwise, it may miss a critical design issue or flag issues incorrectly.

Mathematically, most CAD problems are either combinatorial decisions or optimization problems. *Decision problems* have a binary (true or false) solution: 'Are the two functions equivalent?' The *optimization problems* are targeted to finding a minimum cost solution: 'Find a logic implementation of a function which will have minimum delay'. Most of these problems are *intractable* (NP-hard or NP-complete). The algorithms have exponential complexity or higher. Sometimes *heuristics* (approximation algorithms) are used to get inexact but practical solutions, using reasonable computer time and memory.

Table 1.5 describes how MOSFET parameters and supply voltage scale with advancement of VLSI technology.

Table 1.5   MOSFET technology projection (*Source*: SIA01)

Year	2001	2003	2005	2007	2010	2013	2016
Drawn channel length (nm)	90	65	45	35	25	18	13
Physical channel length (nm)	65	45	32	25	18	13	9
Gate oxide thickness (nm)	2.3	2.0	1.9	1.4	1.2	1.0	0.9
V_{DD} (V)	1.2	1.0	0.9	0.7	0.6	0.5	0.4

1.10 CMOS Integrated Circuit

Almost 90% of the integrated circuits fabricated today use the CMOS technology. CMOS has outperformed the BJT due to their following superior performances:

- Low power dissipation
- Low area due to less device requirement
- Easy scaling down of MOS device dimensions
- Low fabrication cost

The semiconductor world market is counted in billions of dollars per year. Almost 80% of the market is dominated by CMOS; interestingly, silicon (Si) devices account for 97% of all microelectronics.

The industry is projected to be 25 times the present size after 2020. Silicon is used for mostly all IC fabrication, because of the following reasons:

- Abundance of Si in nature
- High quality native oxide of silicon (SiO_2)
- Appropriate mechanical strength—up to 12-inch wafer

The desirable properties of materials used for CMOS integrated circuits are as follows:

- Si is used as a substrate
- Any other material should be compatible with Si
- Incorporation of new material should not increase production cost unexpectedly

The materials for gate and the insulator must have the following properties:

- Be compatible with Si technology
- Provide very good interface with Si
- Be reliable and reproducible

1.11 Issues with Integrated Circuits at the DSM Level

With the advancement of the CMOS IC technology, the dimensions have reduced from the micron to submicron to nanometer regime. This aggressive device scaling has manifested new issues for the IC designers.

To understand the physical dimensions of the devices, let us compare the device dimensions with that of a typical human hair. A human hair is typically of the order of 75 micron in diameter. Cross-section of a single human hair can contain thousands of transistors! At 65 nm technology node, around 40,000 transistors could fit on the area equivalent to the cross-section of a human hair.

Table 1.6 DSM issues for device and interconnect

DSM devices	DSM interconnects
Velocity saturation	Interconnect RC delays
Short-channel effects	IR drop and *Ldi/dt* effect
Subthreshold current	Capacitive coupling
DIBL, GIDL	Inductive coupling
Hot-carrier effects	Electromigration
Thin gate oxide tunnelling current	Antenna effects

With such small dimensions of the devices, as well as interconnects, a number of issues, typically known as deep-sub-micron (DSM) issues, crop up as summarized in Table 1.6.

The DSM issues of the CMOS devices are discussed in detail in Chapters 2 and 12 under the reliability section.

Another important issue is increase in power dissipation of the integrated circuits. With a large number of devices packed into a small area, the power dissipation in a chip is almost 1 kW which is equivalent of 10 filament bulbs of 100 W! The active and standby power dissipation of the chip has already been almost equal.

The performance of the ICs is measured in terms of how many million operations can be performed per unit time. This indicates that the performance is dependent upon how fast circuits can respond to an input. In other words, the propagation delay through the circuits or devices must be reduced from microseconds to nanoseconds for migrating a design from an operating frequency of MHz to GHz.

Figure 1.16 illustrates how the propagation delay through the circuits depends on process (P), voltage (V), and temperature (T) conditions, which are very often called *PVT* or *PTV* conditions.

Typically, when the ICs are fabricated using a particular technology node, the devices are intended to have the target dimensions. For example, if a circuit is fabricated using 180 nm technology node, the gate length of the transistors must be 180 nm. But due to process variations during manufacturing, the fabricated transistors would not have the same gate length. Rather, there will be a Gaussian (typical) distribution of the dimensions with the mean as the target gate length.

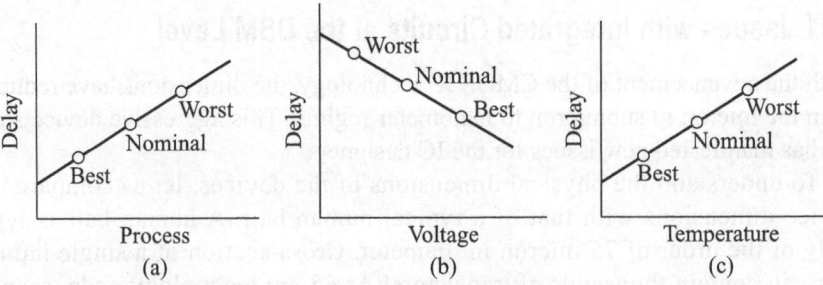

Fig. 1.16 (a) Delay vs process technology; (b) delay vs power supply voltage; (c) delay vs operating temperature

Hence, the transistors with gate length lesser than the nominal value will be faster, and the transistors with gate length greater than the nominal value will be slower.

Similar to the delay variation due to process variation, the delay also varies with the power supply voltage and operating temperature. As the voltage is increased, delay through the devices reduces; and if the temperature is increased, delay through the devices increases.

More importantly, optimizing all three parameters, i.e., area, speed, and power cannot be achieved simultaneously. Reducing the area must be traded off with increase in delay, or speed can be achieved with the sacrifice of chip area and power. A typical delay vs chip area plot is shown in Fig. 1.17.

Another issue which has become critical is the interconnect delay. For older technologies, the main design effort was to reduce the device delay only. However, the advanced technology nodes have forced the designers to think over the interconnect delay that has started dominating the gate delay, as the feature size is reduced from micron to submicron technologies. Figure 1.18 shows how the interconnect delay has become dominating over the gate delay in the submicron regime.

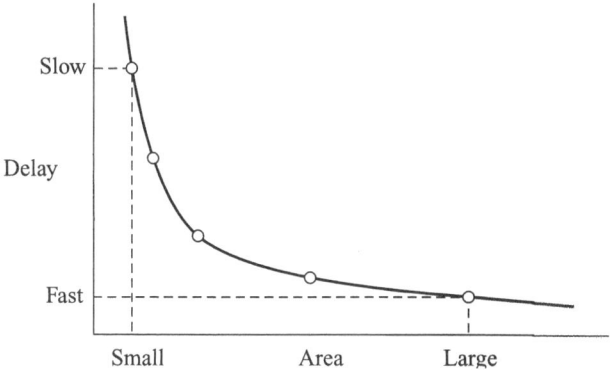

Fig. 1.17 Delay vs chip area

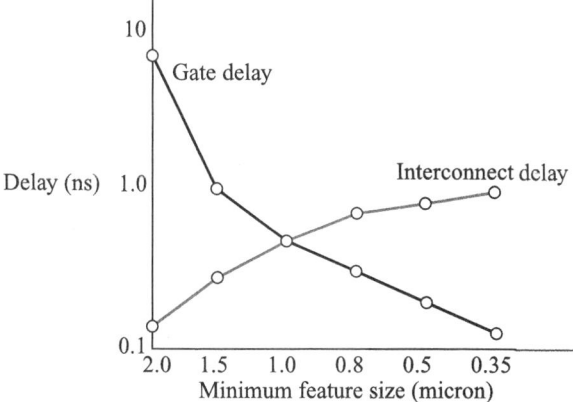

Fig. 1.18 Delay vs minimum feature size

1.12 Modelling and Simulation

It is always followed that when a circuit is to be developed, it is first designed and simulated. The job of a circuit designer is to achieve better speed, area, and power with some innovative configuration of the transistors. But they must also ensure that the innovative idea really works in silicon. If the idea does not work, it is required to find out the reasons. All these need to be done early in the design stage so that any vague idea does not get propagated down the design cycle. In order to design and simulate the circuit early in the flow, the circuit must be modelled first. The model is an approximation of the real circuit. The model is not the actual device but it captures the important properties of the circuit which is relevant. For example, a timing model of a circuit would have all timing parameters which represent the timing behaviour of the circuit, leaving out all other details of the circuit. More accurate the model, more accurate is the simulation result. Hence, modelling is very important early in the design stage.

There are basically two aspects of the modelling and simulation process: first, we need to generate the correct model of the circuit; second, we need to stimulate that circuit in a manner so that it exercises the functionality of the circuit. But the question that now arises is: what needs to be modelled? Typically, in a circuit, the transistors, wires, and the circuit conditions are to be modelled.

- *Transistors*   Both nMOS and pMOS transistors must be modelled.
- *Wires*   The wires are not ideal connectors. They have
  - Resistance effects, IR drop effects
  - Capacitance effects
  - Coupling, inductance effects

  Hence, the wires must be modelled to capture all these effects.
- *Circuit conditions*   The circuit operating conditions such as temperature, power supply, substrate voltage, chip ground vs board ground, etc., must be modelled.

1.13 Design Metrics

The design quality is measured in terms of the following design metrics:
- *Functionality*   A design size and complexity can be measured by its functionality. For example, what are the different operations it can perform; or how many number of inputs and outputs the system has.
- *Cost (non-recurring and recurring)*   The cost of designing and manufacturing of a design directly gives a measure about the design quality. For example, more metal layers in a design means more processing steps, and hence more cost.
- *Reliability (noise margin/immunity)*   A good design must have large noise margins.
- *Performance (speed, power, energy)*   A good design must have high speed, low power dissipation, and energy consumption.
- *Speed (delay, operating frequency)*   The processing speed of the design must be high so that high frequency inputs can be applied to its input.
- *Time-to-market*   The design must be designed, verified, and finally implemented in the form of the integrated circuit as quick as possible to become available first in the market to beat the competitors.

1.14 Cost of Integrated Circuits

In general, the cost of ICs has two parts: the fixed cost and the variable cost. All the non-recurring expenses are part of the fixed cost. All the recurring expenses are part of the variable cost.

■ *NRE (non-recurrent engineering) costs or fixed costs* The fixed cost is independent on the number of products sold. The fixed costs include.
   − Design cost (in terms of effort in time and manpower)
   − Research and development (R&D) cost
   − Infrastructure building cost
   − Costs involved in marketing and sales
   − Cost of manufacturing equipment
   − Cost involved in training the manpower
   − Cost of VLSI CAD tools

With increase of production volume, the fixed cost per product sold is reduced.

■ *Recurrent costs or variable costs* The variable cost depends on the components manufactured, cost of assembly, and testing cost. It is directly proportional to the number of components manufactured and the chip area.

Hence, the total product cost can be expressed as

Total product cost = Fixed cost + Variable cost × Number of products sold

Or, we can write the cost per IC as

Cost per IC = Variable cost per IC + Fixed cost/Number of ICs

1.15 CAD Tools for VLSI Design

Computer-aided design (CAD) tools are the programs that fully or partly automate the VLSI design steps. With the integration density of devices in a chip increased from thousands to millions or even billions today, CAD tools are inevitable in the VLSI design process. Just to understand the necessity of CAD tools, let us consider a simple example of solving a set of linear equations. To find out the voltage or current in a circuit with three loops, we need to derive three independent linear equations and solve them. However, with the set containing more and more independent equations, the problem becomes harder and impossible for solving a set with more than 10 equations within a reasonable time. Now think of ICs containing millions of circuit components. The only possible approach is to write a computer program to solve it. This not only reduces the time for getting the solution, but also reduces the chances of human errors and can be used as many times as required.

Specific VLSI design styles require specific CAD tools. Also different CAD tools must be there for solving problems in different domains.

Figure 1.19 illustrates different VLSI design styles and different aspects of VLSI design. For example, a full-custom design style would require a different set of CAD tools as compared to the standard cell-based design. Similarly, the FPGA-based design style requires another different set of CAD tools. Again, the different problem domains require different set of CAD tools. Typically, the VLSI design

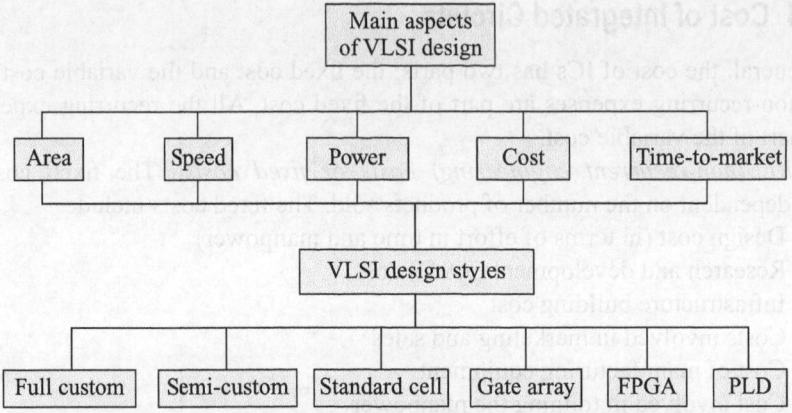

Fig. 1.19 Different VLSI design styles and different aspects of VLSI design

domain can be broadly divided into two: the implementation and the verification. In the implementation domain, the tools are required to design the circuit, synthesize the circuit, and draw the layout. In the verification domain, the tools are required to verify the functionality of the circuit, checking the timing specifications, doing the power analysis, verifying the layout, and checking the design rules, etc.

In summary, based on the design flows, CAD tools are classified as:

- Implementation tools
  - Logic and physical synthesis
  - Design for test
  - Full custom layout
  - Floorplanning
  - Place and route
- Verification tools
  - Simulation
  - Timing analysis
  - Formal verification
  - Power analysis
  - Signal integrity
  - DRC and LVS

Table 1.7 illustrates how the CAD tools have evolved in the last 60 years.

Table 1.7 Evolution of VLSI CAD tools

Year	Design tools
1950–65	Manual design
1965–75	Layout editors, automatic routers for printed circuit board (PCB), efficient partitioning algorithm
1975–85	Automatic placement tools, well-defined phases of design of circuits, significant theoretical development in all phases
1985–95	Performance driven placement and routing tools, parallel algorithms for physical design, logic synthesis, high level synthesis, testing and design for testability (DFT)
1995–Present	Parasitic extraction, yield estimation and enhancement, CAD tools for FPGA and SOC, design for manufacturability (DFM), analog and mixed-signal (AMS) CAD tools

1.16 VLSI Design using Computer-aided Design Tools

VLSI design is basically a design and verification process at different levels of abstraction. A flow diagram of VLSI design at different levels of abstraction is shown in Fig. 1.20. It starts with the architecture of the overall system to be designed. First, the functional specifications are documented in a formal hardware description language, which is verified. Next, the logic is designed for the system to get the desired functionality and the logic is verified. Once the logic is verified, the circuit is designed using circuit components such as transistors, and verified to check for functionality. Then the physical layout of the circuit is designed and verified for correctness.

1.16.1 Functional Specification and Verification

In the very first step, the functionality is specified in a very formal hardware description language (HDL). For example, let us consider that we have to build a VLSI circuit which will multiply binary numbers. Then the multiplier needs to be specified in details. For example, whether the multiplier will multiply signed numbers or unsigned numbers. Depending on the type of operands, the architecture of the multiplier can be decided. Again the size of the operands must be specified,

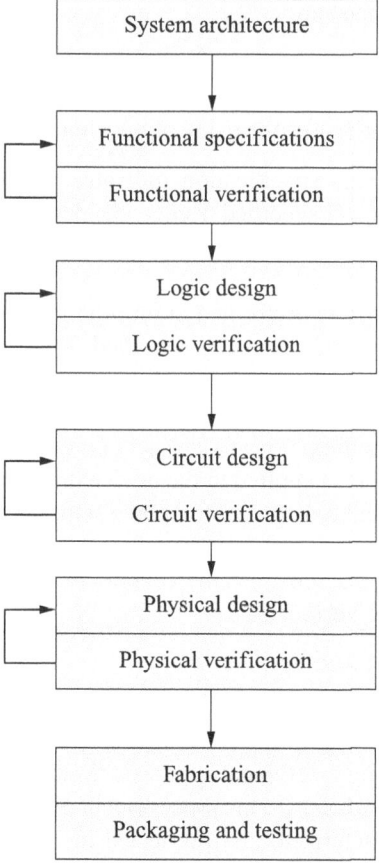

Fig. 1.20   VLSI design flow

that is the number of bits; 4-bit, 8-bit, or 16-bit multiplier. Then the constraints can be imposed on the design, such as area, power, and timing constraints. The complete set of specifications can be written as:

Algorithm　The algorithm to be used for implementing the circuit must be specified. For example, a signed binary multiplier can be implemented using different algorithms, such as Booth's algorithm, Baugh–Wooley's algorithm, etc. Hence, specifications must say which algorithm should be used.

Number of Inputs and Outputs and their Sizes　These specifications state the number of I/O pins in the circuit.

Number of Bits to be Used for Internal Operations　This specification indicates the bit-size of the internal operands. Very often the internal results overflow and in such cases they are truncated. This causes inaccuracy in the final results. Specifying internal operand bit-size ensures that truncation error is not significant.

Number of Clock Signals to be Used　This specification is absolutely necessary as the number of clock inputs determines the clock routing in the design.

Maximum Clock Frequency to be Applied　This specification indicates how fast the circuit has to respond to the input signal changes. This ultimately determines the speed of the system.

Chip Area　Specifying chip area indicates that the circuit must not exceed the specified chip area. This has significant impact on the device performance. Typically, reduction of chip area increases the delay of the circuit.

Power Dissipation　This specification indicates that the chip can dissipate a maximum amount of power specified. Hence, the designer must take care of power dissipating factors, such as switching activity, power supply voltage, etc., while designing the circuit.

Functional simulation is performed before the logic design. This ensures that the functionality is fully and correctly specified. A set of bit patterns called test vectors are used to test the functionality of the circuit. If the simulated output bit patterns match with the expected bit patterns, then the functional description is said to be correctly specified.

The set of test vectors is applied to the device under test (DUT) and the output is evaluated. The evaluated output is compared with the expected output to check for its functional correctness.

Figures 1.21 and 1.22 illustrate the functional description, test vectors, and output of the functional verification.

1.16.2 Logical Design and Verification

The next step after the functional verification is the logic design step. In this step, the design is synthesized from the HDL specifications using a set of design constraints and the cell library. A typical synthesis flow is shown in Fig. 1.23.

It basically maps the design to the process technology and the logic cells already pre-designed, pre-characterized, and pre-tested in the cell library. The synthesis process always tries to meet the design constraints such as area, speed, and power.

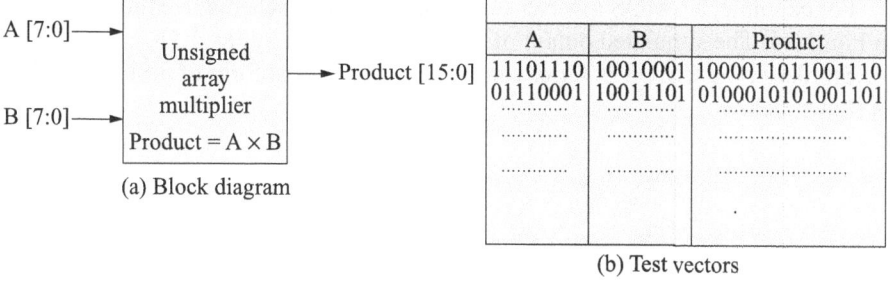

A [7:0] ⟶ Unsigned array multiplier · Product [15:0]
B [7:0] ⟶ Product = A × B

(a) Block diagram

A	B	Product
11101110	10010001	1000011011001110
01110001	10011101	0100010101001101
....	....	....
....	....	....
....	....	....

(b) Test vectors

```
entity Mult8 is
  port (A, B : in STD_LOGIC_VECTOR(7 downto 0);
    Product : out STD_LOGIC_VECTOR(15 downto 0);
end Mult8;
architecture behv of Mult8 is
begin
  Product <= A * B;
end behv;
```

(c) HDL specification

Fig. 1.21   (a) Block diagram; (b) test vectors; (c) functional specification

⊞ /mult4/a	12	0	1		12			3	
⊞ /mult4/b	8	0		1	8	3			7
⊞ /mult4/product	96	0		1	12	96	36	9	21

Fig. 1.22   Typical output of a function simulation

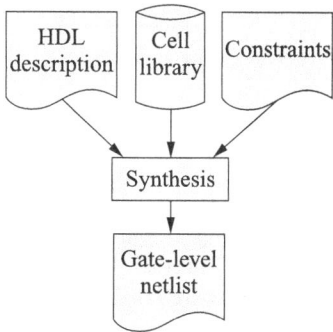

HDL description   Cell library   Constraints

↓   ↓   ↓

Synthesis

↓

Gate-level netlist

Fig. 1.23   Design synthesis step

1.16.3 Circuit Design and Verification

In the circuit design phase, the design is implemented at the transistor level. The transistor level design can be either a schematic design or a SPICE netlist to describe the circuit. SPICE is circuit simulation software which is widely used for circuit design and simulation. After describing the circuit components and their connectivity, the input stimulus are applied to the circuit and then simulated to check the output voltage, current, or waveform.

For example, a CMOS inverter can be designed in a schematic editor as shown in Fig. 1.24. The simulated output of the same is shown in Fig. 1.25.

Similarly, the CMOS inverter can be described in the form of a SPICE netlist as shown below:

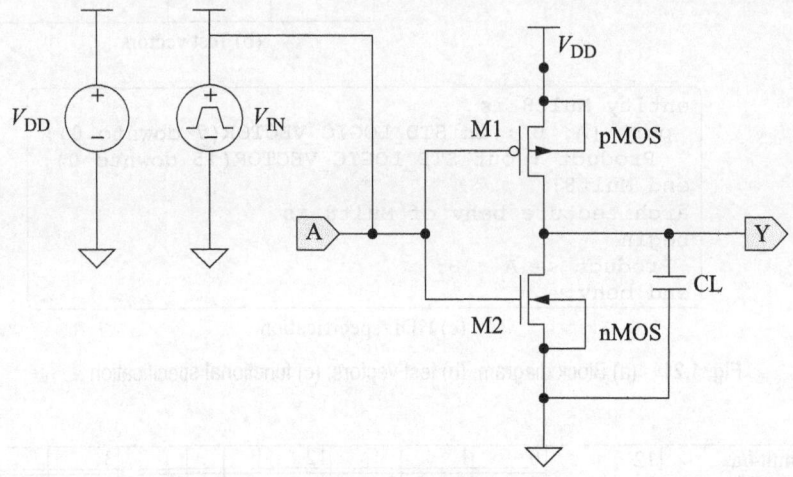

Fig. 1.24 Schematic of CMOS inverter

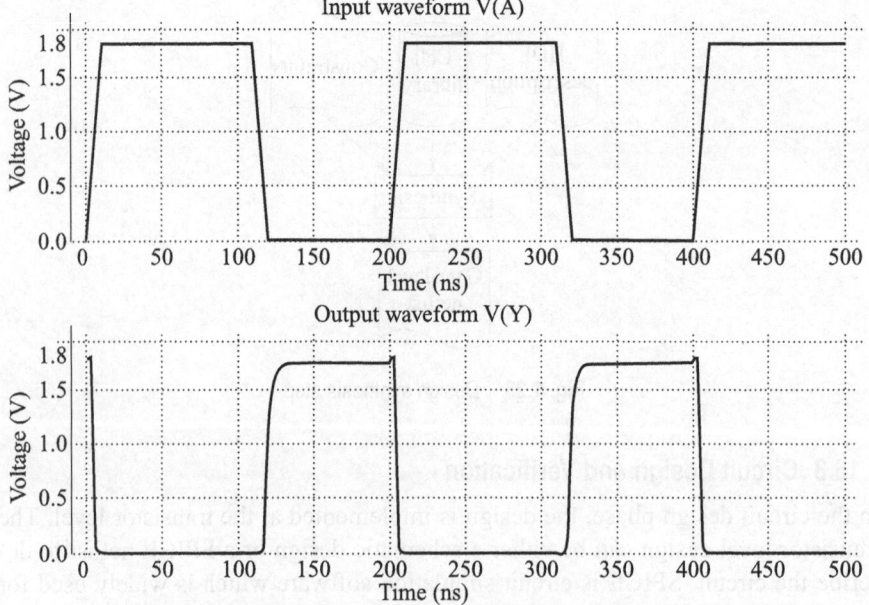

Fig. 1.25 Simulation results of the CMOS inverter

```
****** SPICE netlist of CMOS inverter *******
.INCLUDE "E:\CMOS-CELLS\SPICE MODELS\MY_MODEL.TXT"
M1 Y A VDD VDD PMOS W=5U L=1.8U AS=10u PS=5u AD=10u PD=5u
M2 Y A 0   0   NMOS W=2U L=1.8U AS=10u PS=5u AD=10u PD=5u
CL Y 0 0.1PF

VDD VDD 0 1.8V
VIN A 0 PULSE 0 1.8V 0N 1N 1N 5N 10N

.TRAN 0.01N 20N
.PRINT TRAN V(A) V(Y)
.PLOT TRAN V(A) V(Y)

.END
********** end of SPICE netlist **************
```

1.16.4 Physical Design and Verification

The VLSI design flow can be broadly classified into two phases: logical and physical. In the logical design phase, the design is first created in the behavioural level based on the functionality of the design. Then the behavioural design, often called RTL netlist, is synthesized to generate a gate-level design/netlist. The design constraints, such as power, speed, and area are considered in the process of logic synthesis. The netlist is simulated for correctness in both the behavioural and gate level.

The next phase is the physical design. The physical design phase converts the netlist into a geometric representation. The outcome is called a *layout*. A typical layout of IC is shown in Fig. 1.26.

The physical design phase mainly involves the following five steps:
- Circuit or logic partitioning
- Floorplanning
- Placement
- Routing
- Compaction and verification

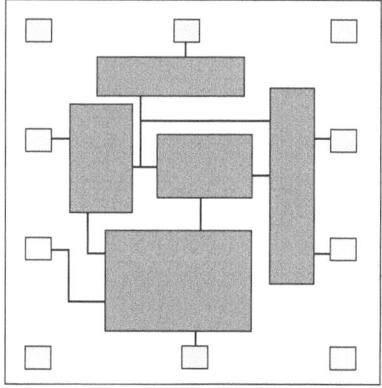

Fig. 1.26 A typical IC layout

Circuit or Logic Partitioning

In the partitioning process, the large circuit is partitioned into a number of sub-circuits (called blocks). Factors such as number of blocks, block sizes, interconnection between blocks, critical delay, etc., are considered during the circuit partitioning. A small example of circuit partitioning is shown in Fig. 1.27.

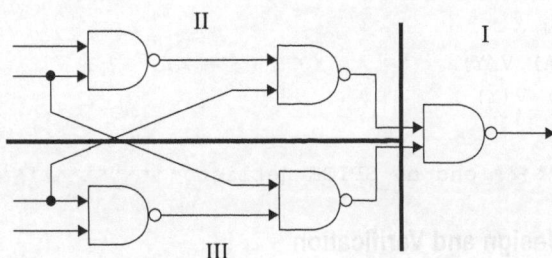

Fig. 1.27 A circuit is partitioned into three sub-circuits

Floorplanning

This step sets up a plan for a good layout. It tentatively places the modules (modules can be blocks, functional units, etc.) at an early stage when details such as shape, area, I/O pin positions of the modules, etc., are not yet fixed. A typical floorplan of an integrated circuit is shown in Fig. 1.28.

Placement

In this step, the exact placement of the modules (modules can be gates, standard cells) are done. The details of the module design are known in this phase. The main goal of placement is to minimize the total area, delay, congestion, interconnect metrics, etc. An example of placement is shown in Fig. 1.29.

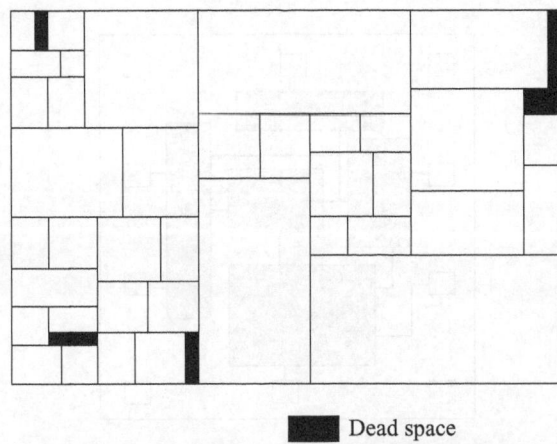

■ Dead space

Fig. 1.28 A typical floorplan of integrated circuit

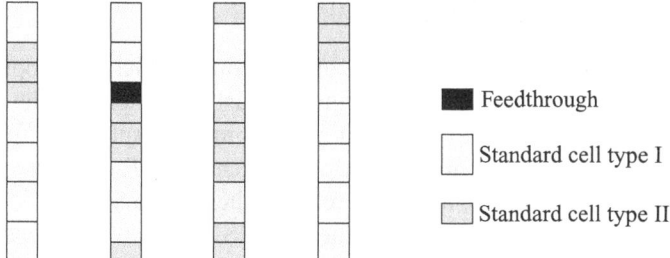

Fig. 1.29 A typical placement of an integrated circuit

Routing

This step completes the interconnections among modules. Factors such as critical path, clock skew, crosstalk, congestion, repeater placement, wire spacing, etc., are considered during the routing. The routing problem is divided into two problems, *global routing* and *detailed routing*. In the global routing steps, the plan for the interconnection is prepared and in the detailed routing step, the actual interconnections are made. An illustration of routing in IC is shown in Fig. 1.30.

Compaction and Verification

Compaction is the process of compressing the layout from all directions to minimize the chip area. In the verification process, the layout is checked for its correctness. The verification step includes *design rule checking* (DRC), *circuit extraction* (generates a circuit from the layout to compare with the original netlist), and *performance verification* (extracts geometric information to compute resistance, capacitance, and delay).

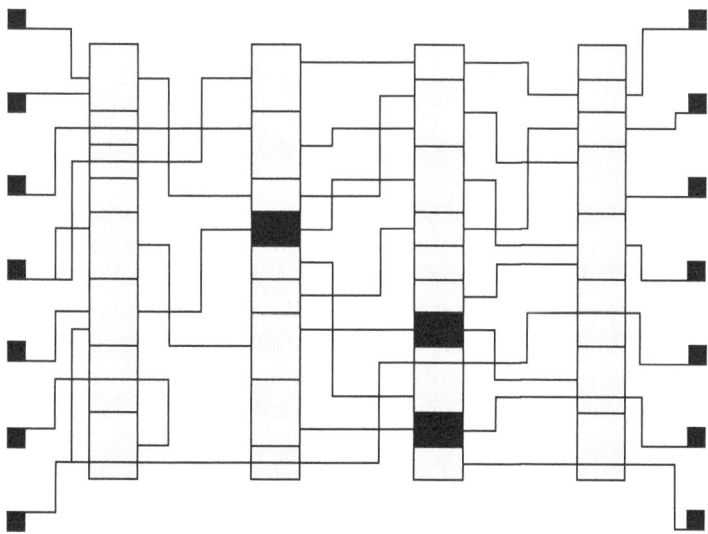

Fig. 1.30 An example of routing in an integrated circuit

1.17 Problems in VLSI CAD Domain

Most of the problems in VLSI design are optimization problems. An optimization problem is the one in which we want to find, not just *a* solution, but the *best* solution. How can computers help in circuit partitioning, floorplanning, placement, etc.? These are *optimization problems*. Most of them are *NP-complete* problems, i.e., there is not yet any efficient method to solve them optimally.

Commonly used algorithmic approaches in VLSI CAD are as follows:

- Greedy algorithm
- Stochastic algorithm
- Graph methods
- Mathematical programming
- Dynamic programming
- Computational geometry

Greedy Algorithm A *greedy algorithm* is the one that works well for optimization problems. At each stage, it finds the locally optimal solution in the hope to find the global optimum. For example, applying the greedy algorithm to the travelling salesman problem yields the following decision: 'At each stage, visit the unvisited city nearest to the current city'. The greedy algorithm works in phases. At each phase:

- Take the best you can get right now, without taking future consequences into consideration.
- Hope that by choosing a *local* optimum at each step, it will end up at a *global* optimum.

In every step, try to make the choice that looks like the best at that moment. This approach may lead to an optimal solution at the end for some problems. An example is the *left edge algorithm* for the *channel routing* problem.

Stochastic Algorithm There are many search techniques and most of them are specialized to solve specific types of problems. An important class of search techniques is stochastic algorithm where certain steps are based on random choice. The stochastic search algorithms are effective and efficient in finding near-optimal solutions to complex problems, but with no guarantee of finding true global optima. There are mainly two types of stochastic search: (a) simulated annealing and (b) genetic algorithms.

Simulated annealing is a popular stochastic algorithm designed in analogy with the physical process of cooling a molten material where condensing a material forms a crystalline solid form. In analogy, searching for an optimal solution is similar to finding a configuration of the cooled system with minimum free energy. Because of its ability of rejecting local optima, simulated annealing is a powerful algorithm for numerical and combinatorial optimization problems.

- In every step, try randomly a neighbouring solution.
- Accept or reject the new solution depending on its quality and the temperature.
- This approach does not lead to the optimal solution, but it is very useful in solving NP-complete problems. An example is the gate array placement by simulated annealing.

Genetic algorithms are simplified models of the search processes of natural evolution. They use the techniques inspired by Darwin's theory of evolution, such as inheritance, mutation, selection, and crossover. The algorithms start with a set of solutions (represented by chromosomes) called *population*. The solutions from one population are taken and used to form a new population. This is motivated by a hope, that the new population will be better than the old one. Solutions which are selected to form new solutions (offspring) are selected according to their fitness— the more suitable they are, the more are the chances that they will reproduce. This is repeated until some condition (e.g. number of populations or improvement of the best solution) is satisfied. One example is the application of genetic algorithm in test pattern generation.

Graph Methods Graph algorithms are used very often in VLSI CAD. Minimum spanning tree, network flow, graph colouring, etc., are useful tools. An example is the *rectilinear Steiner tree method* for the *routing* problem. A *graph* is an abstract representation of a set of objects where some pairs of objects are connected by links. The interconnected objects are represented by mathematical abstractions called *vertices*, and the links that connect some pairs of vertices are called *edges*. Typically, a graph is depicted in diagrammatic form as a set of dots for the vertices, joined by lines or curves for the edges. Mathematically, a graph is represented as

$$G = (V, E)$$

where V is the set of vertices and E is the set of edges. Figures 1.31(a) and (b) illustrate a graph without and with edge weights, respectively.

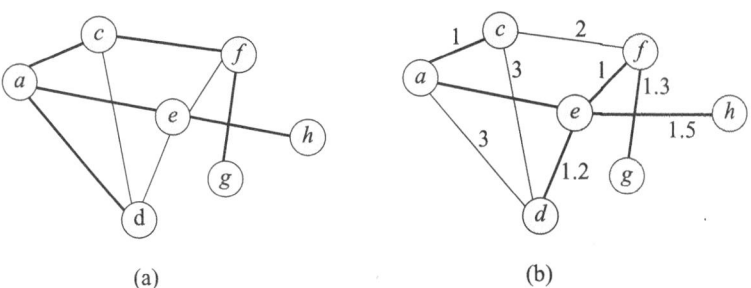

(a)   (b)

Fig. 1.31 (a) A graph without edge weights; (b) a graph with edge weights

Tree A tree is a graph with no loops. The following two definitions of a tree are equivalent:

Definition 1: A tree is an acyclic graph of n vertices that has $(n - 1)$ edges.

Definition 2: A tree is a connected graph such that $\forall u, v \in V$, there is a unique path connecting u to v. A tree is shown in Fig. 1.32(b).

Spanning Tree A *spanning tree* of a graph is just a subgraph that contains all the vertices and is a tree as shown in 1.33(a). Given a graph $G(V, E)$, a spanning tree T is a tree that connects all the vertices in V. A graph may have many spanning trees.

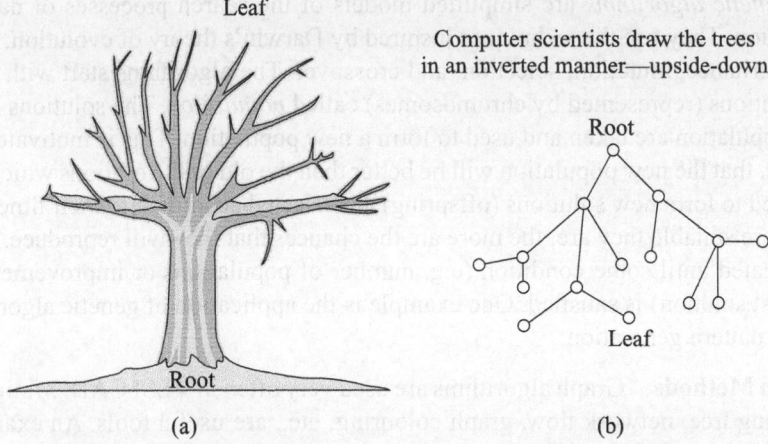

Fig. 1.32 (a) A natural tree; (b) a tree that represents a design hierarchy

Mathematically, it can be expressed as: A spanning tree of a graph $G(V, E)$ is a tree $T(V, E')$, where $E' \subseteq E$, that connects all the vertices in G [Fig. 1.33(a)].

$$E' = \{(a, c),\ (c, f),\ (f, g),\ (a, e),\ (e, h),\ (a, d)\}$$

Minimum Spanning Tree A *minimum spanning tree* (MST) or *minimum weight spanning tree* is a spanning tree with a weight less than or equal to the weight of every other spanning tree. The weight of a tree is just the sum of the weights of its edges. If we label each edge by the distance between the two terminals, we can find a minimum spanning tree efficiently.

Steiner Tree For a multi-terminal net, we can easily construct a spanning tree to connect the terminals together. But the wire length will be unnecessarily large. In such cases, it is better to use a Steiner tree. A Steiner tree is a tree connecting all the terminals and some additional nodes (called Steiner nodes) as shown in Fig. 1.33(b). A rectilinear Steiner tree is a Steiner tree in which all the edges run horizontally or vertically as shown in Fig. 1.33(c).

The minimum Steiner tree problem is NP-complete. It may need to route millions of nets simultaneously without creating over-congested regions. The obstacles may exist in the routing region.

Mathematical Programming We can formulate a problem as an objective function to be minimized or maximized with some constraints to be satisfied. An example is formulating the floor-planning problem as a *linear mathematical program*.

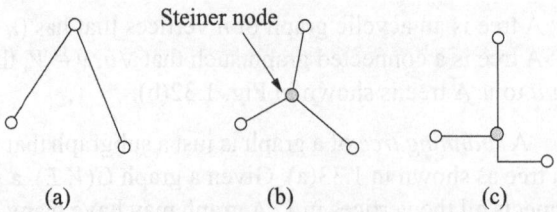

Fig. 1.33 (a) Spanning tree; (b) Steiner tree; (c) rectilinear Steiner tree

1.17.1 What is Important in Solving CAD VLSI Problems?

The very first step is to know the problems and then formulate these problems and their constraints. The second step is to solve the problem using a suitable algorithm. However, the world keeps changing very fast. For example, the technology, objectives, constraints, required solution quality, computational power, etc., will keep on changing before coming to a state where everything is frozen. It is more important to learn the techniques to formulate and solve problems.

In summary, we can write,

- Challenges in design of VLSI chips and system-on-chips (SoCs) are multifaceted. Therefore, it is required to develop solutions to the multi-objective optimization problems.
- Design automation is indispensable.
  Application of efficient computational techniques must be employed to develop automated tools.
  Focus so far has been on digital systems, however analog and mixed signal systems are far more complex. Tools must be capable of handling mixed signal systems.
- Quest for new materials and fabrication technologies.
- VLSI design and efficient CAD tools for
  Multimedia applications (such as digital signal processing, image processing, video processing)
  Telecommunication systems
  Fuzzy logic systems
- Desiderata for newcomers to the area of VLSI-CAD
  Basic VLSI design (essential)
  Algorithms and data structures (essential)
  Linear and nonlinear optimization techniques
  Graph theory and combinatorics

1.18 IC Chip Industry—A Brief Outlook

There are a number of companies involved in the process of IC design and manufacturing. Electronic design automation (EDA) vendors are the companies who develop EDA tools for IC designers. There are a number of companies who have fabrication facility for IC manufacturing. There are some other fabless companies which do not have any fabrication facility. They are just involved in designing the IC. The IP (intellectual property) vendors are those companies who develop IPs and license the users to use them. The example of different semiconductor companies and their involvements in IC design and manufacturing process is listed in the following text.

EDA Vendors   They provide various tools for various steps of VLSI design, but do not own hardware products, their products are software, and the methods to make them work together.

Example: Synopsys, Cadence, Mentor, Magma

Chip Manufacturers They have their own process, product, and fabrication plants, may or may not have their own tools.

Example: Intel, AMD, TI, IBM

Fabless Semiconductor Companies They have their own hardware products (RTL), but no fabrication plants. They may or may not have their own/third party EDA tools and methodology. If they have their own flow/tools they are called to have 'Customer Owned Tool model'.

Example: Broadcom, Marvel

ASIC Vendors They have the chip design methodology process and fabrication plants, but no hardware product; caters to fabless semiconductor companies.

Example: TI, IBM

Fabrication Plants They deliver process and manufacturing ability.

Example: TSMC, IBM, TI, UMC

IP Providers They provide custom blocks doing specific functions, e.g., memory, clock generators (PLL).

Example: ARM

1.19 Recent Developments and Future Projections

The trend of miniaturization has come to such an extent that a normal device and its interconnect structure simply do not work. For example, shrinking of gate oxide thickness has reduced it to almost a few atomic layers of SiO_2. At this low dimension, the quantum mechanical tunnelling starts dominating and the ultra-thin gate oxide cannot work as an insulator anymore. So, technologists are trying for alternative material for gate oxides having *high-K* or high dielectric constant. This solves the purpose of reducing the threshold voltage of the MOS device. At the same time, it prevents quantum mechanical tunnelling due to increased thickness.

On the other hand, the field oxide which is used to isolate the metal wires creates high parasitic coupling capacitances. The coupling capacitances have important effects on timing and functionality. It can introduce crosstalk noise which may violate the functionality of a circuit. It can also lead to delay degradation. Hence, to reduce the parasitic capacitance, materials with *low-K* or a low dielectric constant are used for insulating metal layers.

Traditionally, aluminium has been the metal used for interconnection wires. But as an alternate metal, copper which has a better conductivity than Al is found more suitable for interconnects. Cu interconnect will have lesser parasitic resistances and hence lesser RC wire delay. But there is a problem with Cu interconnects. Cu can easily diffuse into Si and hence, it degrades the device characteristics. To prevent the Cu diffusion into Si, IBM researchers have developed a process called the *Dual Damascene process* which uses a special metallization step. In this metallization process, Cu is filled into the trenches etched into the insulator and then a chemical mechanical polishing (CMP) step is used to remove the extra material from the top.

1.19.1 Silicon-on-Insulator

The silicon-on-insulator (SOI) technology has been developed and found to an alternative to conventional CMOS technology. The main advantage of SOI technology is the reduction of the parasitics associated with the CMOS devices. This provides better transistor characteristics as compared to a conventional CMOS device. The SOI process is shown in Fig. 1.34.

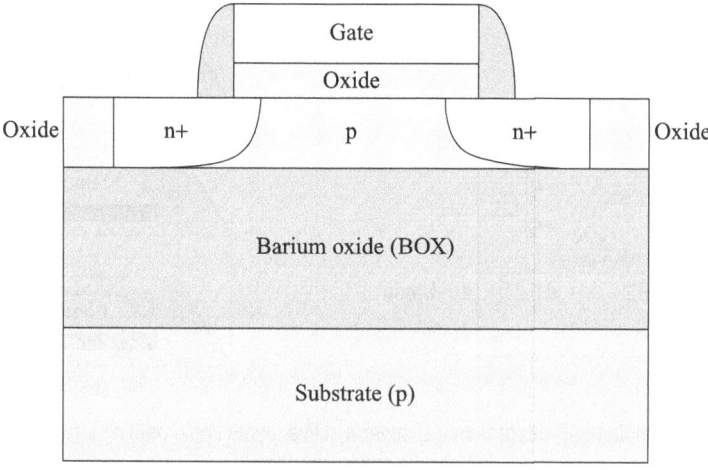

Fig. 1.34   Silicon-on-insulator process (see Plate 1)

SOI transistors are constructed over a thin layer of Si deposited on a thick layer of insulator (SiO_2).

1.19.2 Three-dimensional Chips

The concept of the three-dimensional chips has been introduced with the help of Fig. 1.35.

In the 3D integration, extra active layers (such as T1, T2, T3) added in between the metal layers can implement logic at one level, memory at another level, and I/O at another level. This reduces the interconnect length between active layers and hence the delay. Another approach is to bond fully processed wafers on which the circuits are fabricated. Only the interconnections between the dies have to be made and packaged. This type of integration is not fully 3D, but 2.5D, and is called *systems on a package*.

1.19.3 Nanoelectronic Devices

Over the last few decades, the MOS devices are scaled down aggressively to achieve higher level of integration. The device dimensions have been reduced from micrometer to sub-micrometer to nanometer scale. But this down-scaling cannot sustain further when the dimension reaches the molecular dimensions. At this scale, the conventional CMOS device structure cannot be achieved and hence the conventional IC design and technology have to be completely changed. Scientists have proposed many new devices at the nanometer dimensions which

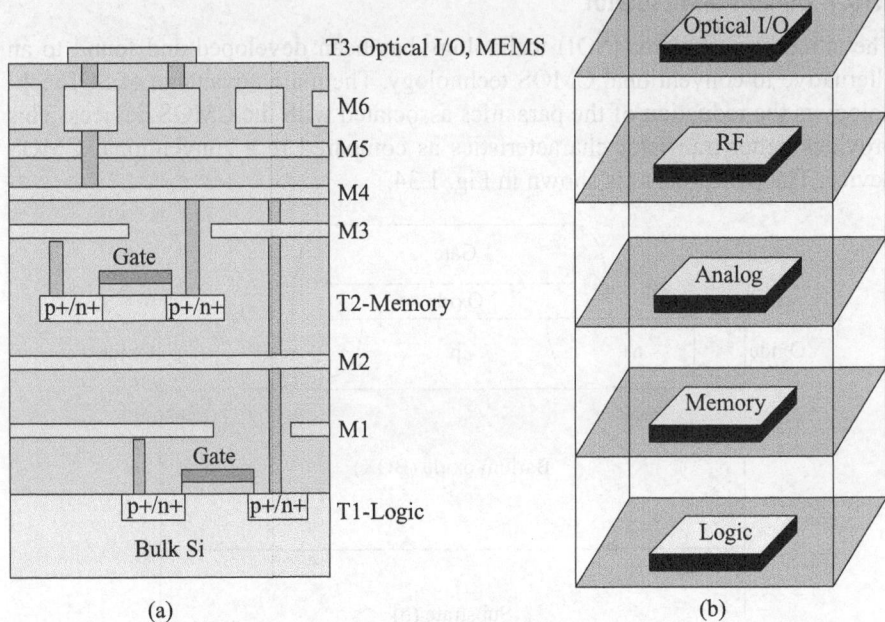

Fig. 1.35 Example of 3D integration: (a) several active layers implementing logic, memory, and I/O; (b) die stack (see Plate 1)

are often called *nanoelectronic devices*. The different nanoelectronic devices can be classified into either solid-state or molecular.

- Solid-state devices
  - Resonant tunnelling diodes (RTDs)
  - Single electron technology (SET)
  - Spin transistors
  - Ballistic electron devices
- Molecular devices
  - Organic transistors
  - Carbon nanotubes (CNTs)
  - Monomolecular transistors

As the control of these nanodevices is obtained by utilizing a single (or few) electron(s), the device size and power consumption are expected to be just a fraction of those in current CMOS devices.

Resonant Tunnelling Diode Resonant tunnelling diode (RTD) is an advanced form of the tunnelling diode. The tunnelling diode is constructed, using two degenerately doped (10^{19} cm^{-3}) *p*- and *n*-type semiconductors as shown in Fig. 1.36(a).

There are a number of electrons in the conduction band in the *n*-type material and there is a large number of empty states in the valence band in the *p*-type material. The electrons from the *n*-type material tunnel through the thin potential barrier and reach the *p*-type material giving rise to a large current. The current–voltage characteristics exhibit negative differential resistance (NDR).

The RTD is basically a double barrier quantum well (DBQW) structure made off five regions as shown in Fig. 1.36(b). Two contact regions I and V are heavily

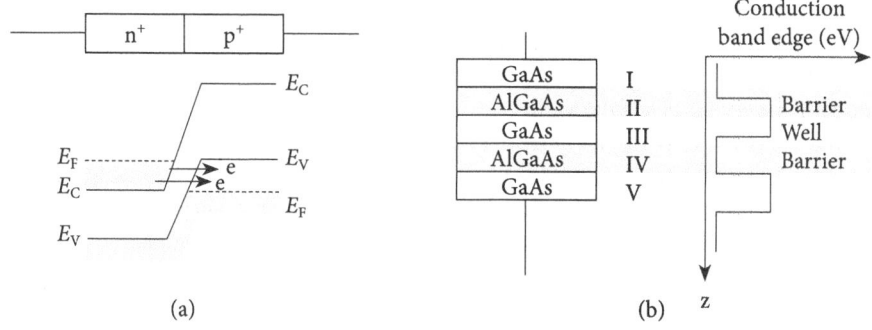

Fig. 1.36 (a) Tunnel diode and (b) double barrier quantum well (DBQW) structure

doped narrow bandgap semiconductors (e.g., GaAs). Two barrier layers II and IV are made of wide bandgap semiconductors such as AlGaAs. Region III is the quantum well using narrow bandgap material (GaAs). The device is grown using metal organic chemical vapor deposition (MOCVD) technique. The typical width of the quantum well is 5 nm and the barrier thickness is 1.5–5 nm.

When a forward voltage bias is applied, an electric field is created, that causes electrons to move from the source to the drain by tunnelling through the quantized states within the quantum well. As the applied voltage is increased, more and more electrons in the source have the same energy as the quantized state and they are able to tunnel through the well, which results in an increase in the current. When the applied voltage is increased to the point (V_P) where the energy level of the electrons in the source coincides with the energy level of the quantized state of the well, the current reaches a maximum (I_P), as shown in Fig. 1.37(b).

As the applied voltage continues to increase, the energy of the electrons misaligns with the quantized state. Hence, the current continues to decrease after reaching the

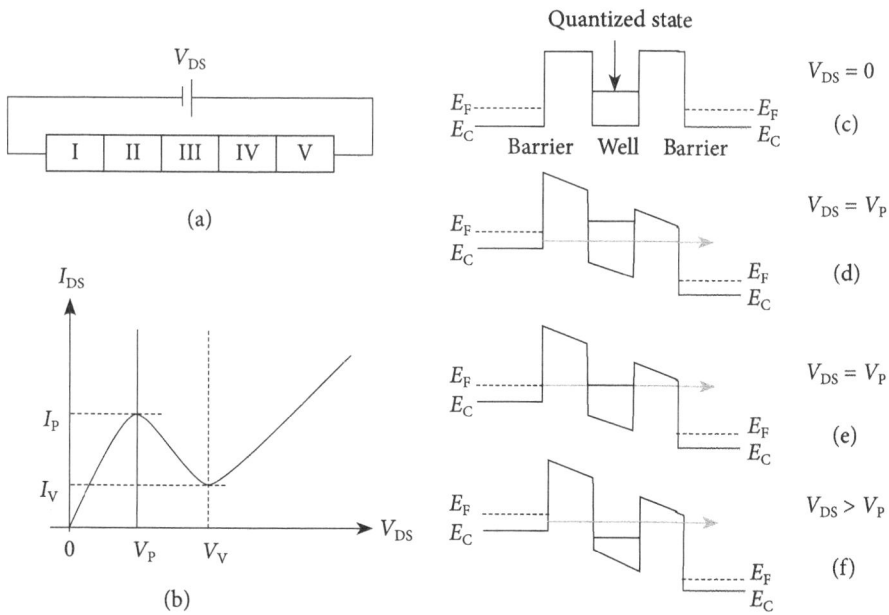

Fig. 1.37 (a) Biased RTD, (b) *I–V* characteristics of RTD and (c)–(f) energy-band diagram of RTD

peak. At $V = V_V$, the current reaches minimum called valley current (I_V). After the valley voltage, as the applied voltage increases, current begins to rise again because of substantial thermionic emission where the electrons can tunnel through the non-resonant energy levels of the well.

The bandwidth of the RTD is in THz range. The higher tunnel current density and shorter transit time make RTD to operate at extremely high frequencies. RTDs can work at room temperature. Small device dimension and very low power consumption make the RTD a good choice at the nanometer dimensions.

Single Electron Transistor   A single electron transistor (SET) is based on the quantum mechanical tunnelling. In classical physics, no electron can pass through an energy barrier if its energy is less than the height of the barrier. However, in quantum mechanics the electron can cross a barrier even if its energy is not more than the barrier height provided the barrier thickness is small. This phenomenon is known as quantum mechanical tunnelling.

A SET is formed by two tunnel junctions connected in series as illustrated in Fig. 1.38. The tunnel junction consists of two metal electrodes separated by a thin (~1 nm) insulator. The electrons from one electrode tunnel through the insulator to go to the other electrode. As the tunnelling is a discrete process, the electric charge that crosses the tunnel junction is always multiple of electronic charge. The two tunnel junctions create a Coulomb island where electrons can reach by tunnelling through either of the tunnel junctions. The gate terminal is isolated from the island by a thick insulator through which no tunnelling can occur.

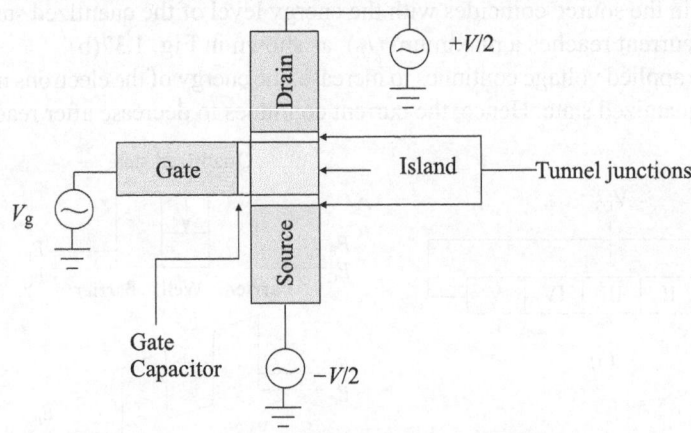

Fig. 1.38   Schematic of a single electron transistor

When no bias is applied to drain and source terminal, the electrons do not have sufficient energy to tunnel through the barrier. The Coulomb energy is given by $E_c = e^2/2C$, where C is the total capacitance of the drain and source junctions and the gate capacitor. When the applied bias between the source and the drain is greater than e/C then electron can tunnel through the tunnel junctions. This causes a current flow through the device, which is independent of the gate bias. The voltage e/C is known as the Coulomb gap voltage. When the applied bias between the source and the drain is less than the Coulomb gap voltage the transistor action starts.

Spin Transistors   Magnetoelectronic devices show great promising application in the nanoscale dimensions. The spin of the electron and its charge are utilized in these devices. Alternate layers of ferromagnetic and nonmagnetic materials are stacked together. The resistance of the multilayer stack depends on the relative alignment of the magnetizations of the ferromagnetic layers due to the spin-dependent scattering at the interface or the bulk of the ferromagnetic layer.

A spin transistor utilizes two ferromagnetic layers—one as a spin injector and the other as a spin analyzer. The output characteristics are controlled by the relative magnetization configuration of the ferromagnets and the bias conditions. The magnetization configuration in spin transistors can be used as nonvolatile binary data. Spin transistors are potentially applicable to integrated circuits for ultrahigh-density nonvolatile memory whose memory cell is made of a single spin transistor and for the nonvolatile reconfigurable logic is based on the functional spin transistor gates.

Various spin transistors have been proposed by several research groups. The spin-valve transistor was proposed by Monsma et al. and the spin field-effect-transistor (FET), was proposed by Datta and Dass.

Ballistic Electron Devices   In every semiconductors the electrons undergo several collisions with the lattice defects, impurities, and phonons while travel through the material. The process is also known as scattering. The distance travelled by an electron between two successive collisions is termed as free path or scattering length. It is one of the most important parameters that determine the quality of the semiconductor in terms of electron transport. The scattering process is purely random in nature. The average of all the free paths is known as the mean-free path. It varies from material to material. The electron mean-free path in silicon is typically a few nanometers and in GaAs it is about 100–200 nm. The electrical resistance of a semiconductor device is determined by the scattering length. The conventional semiconductor devices are much larger in dimensions as compared to the electron scattering length. As a result, electrons undergo several collisions while travelling through the device.

With the advancement of semiconductor processing technology, it has been possible to fabricate devices with dimensions smaller than the electron mean-free path. In such devices, the electrons travel through the device without any scattering except the scattering at the boundaries. This phenomenon is known as ballistic transport. The ballistic devices are those devices which utilize the ballistic transport phenomena. The ballistic devices show great speed and quick response owing to the ballistic transport of electrons. The phonon scattering which strongly depends on the temperature is absent in ballistic devices. Thus, the ballistic devices are temperature independent which is a great advantage.

Organic Field Effect Transistor (OFET)   Organic materials are being explored to make field effect transistor (FET) which can be used to make disposable electronic circuits. The self-assembly technique and the new soft-lithography are the techniques that may be used to create the nanoscale structures using organic semiconductors. The emissive properties of these materials also allow the optical elements or devices to be integrated electronic circuits. Though the OFET may not be able to completely replace the existing CMOS-based integrated circuit technology but it definitely shows exciting opportunity for new and emerging applications possibly made by this technology.

Figure 1.39 illustrates the conventional MOSFET and organic FET structure. The conventional metal oxide–semiconductor (MOS) FET consists of a semiconducting substrate (*p*- or *n*-type), two oppositely doped electrode regions (*n*⁺ or *p*⁺) called source and drain, and metal oxide double layer. When a sufficient voltage (+ve or −ve) is applied on the metal electrode (also called gate), a conducting layer is induced between the source and the drain through the substrate. Then current flows from drain-to-source or source-to-drain when bias is applied between drain and source.

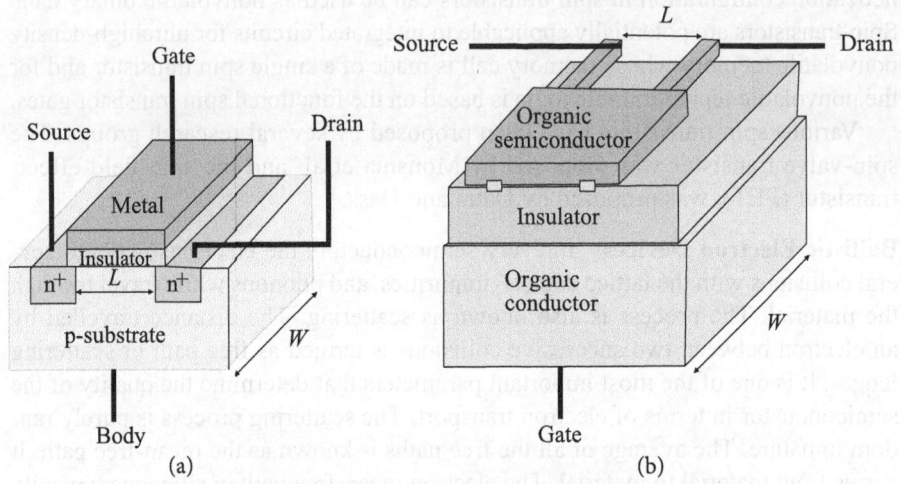

Fig. 1.39 (a) Conventional MOSFET and (b) organic FET

The basic operation of organic FET is also similar to that of the conventional MOSFET. The organic FET consists of an organic conducting substrate at the bottom, an insulating layer in the middle, and an organic semiconducting layer on the top. The source and drain electrodes are directly connected to the semiconducting layer on the top.

When no gate bias is applied, there is no charge density connecting the source and the drain. Thus, the resistance between the source and the drain is very high. When a gate bias greater than a threshold value is applied, a high charge density layer is formed next to the insulator interface. Thus, the resistance between the source and the drain is reduced significantly. Hence, current flows through the very thin region known as the channel.

OFETs may be used in flexible displays made of light emitting diode or liquid crystals where the OFET can be used to switch the pixels ON/OFF. OFETs may find applications in smart-cards or smart tags that require a relatively low density of transistors and flexibility in circuit design. The organic nature of OFET may also be used to combine the devices with detection capabilities of chemical or biological moieties thereby making an impact in the pharmaceutical applications.

Carbon Nanotubes A carbon nanotube (CNT) is one-dimensional structure made of carbon atoms. The hexagonal array of carbon atoms forms a two-dimensional sheet called graphene. CNT can be thought of made by rolled graphene sheet in the form of a cylinder. The excellent electrical, mechanical, and thermal properties of CNT make it an important material for nanoscale regime. A CNT exhibits both metallic and semiconducting properties based on the geometry. The extremely high current density (10^{10} A/cm^2) and long mean-free path (~1 µm) of CNT can lead to extremely high speed nanoelectronic devices and nanointerconnects. Metallic CNTs are being explored for making interconnects whereas semiconducting CNTs are being explored for making field-effect-transistor (CNTFET).

Molecular Transistor   The molecular electronics targets electronic devices made of specific molecules. The small dimensions of molecules are a great advantage. The variety of molecules can be explored to synthesize devices with great variety in their optical, mechanical, and electrical properties. The molecules can itself as logic gate. The dynamics of the non-stationary molecular state created by the electrical, electrochemical, or optical inputs can be combined to form the logic circuit. The addressed molecular states and their dynamics can be used to implement multi-value logic or finite state machines. The monomolecular device consists of the molecule and external wires.

Unlike the conventional top-down approach of semiconductor technology, molecular electronics requires bottom-up approach where the molecules are first synthesized and assembled to form circuits of increasing complexity.

SUMMARY

- A VLSI design contains millions of components integrated over a small area of silicon.
- MOSFETs are mostly used for fabricating ICs worldwide. Its structure is very well suited to IC fabrication technology which is also referred to as planar technology.
- Silicon is the mostly used material for manufacturing ICs.
- FPGA-based VLSI design is suitable for prototype development as the design cycle time is very short.
- Full-custom design is used for high performance and high volume products. The chips are highly optimized in design but require long design cycle time.
- Computer-aided design (CAD) tools are must for state-of-the-art VLSI design.
- With advancement of technology, the dimensions are shrunk to almost 0.7× per generation and the integration density continues to increase.
- VLSI design is a multidisciplinary domain where all branches of science and technology play key roles.
- The ideas of system-on-chip (SOC) and three-dimensional (3D) IC are the future VLSI projections.

SELECT REFERENCES

Alpert, C., D. Mehta, and S.S. Sapatnekar 2007, *The Handbook of Algorithms for Physical Design Automation*, CRC Press.

Anil Kumar, P. S., and J. C. Lodder 2002, The spin-valve transistor, *J. Phys. D: Appl. Phys.* 33 2911–2920.

Aimin M. Song 2004, Room-Temperature Ballistic Nanodevices, *Encyclopedia of Nanoscience and Nanotechnology*, Vol. 9, pp. 371–389.

Anisur, R., J. Guo, S. Datta, and M. S. Lundstrom 2003, *Theory of Ballistic Nanotransistors*, IEEE Transactions on Electron Devices, Vol. 50, No. 9, 1853–1864.

Ciletti, M.D. 2005, *Advanced Digital Design with the Verilog HDL*, Pearson Education, New Delhi.

Datta, S. and B. Dass, *Appl. Phys. Lett* 56 (1990) 665.

Gerez, S.H. 2007, *Algorithms for VLSI Design Automation*, Wiley, New York.

Hakim, R., E. D. Mentovich, and S. Richter 2013, 'Towards Post-CMOS Molecular Logic Devices'. In *Architeture and Design of Molecule Logic Gates and Atom Circuits*, Springer Berlin Heidelberg, pp. 13–24.

Kang, S.M. and Y. Leblebici 2003, *CMOS Digital Integrated Circuits: Analysis and Design*, 3rd edn, McGraw-Hill, New Delhi.

Lengauer, T. 1990, *Combinatorial Algorithms for Integrated Circuit Layout*, Wiley, New York.

Mano, M.M., and M.D. Ciletti 2008, *Digital Design*, 4th edn, Pearson Prentice-Hall.

Monsma, D.J., J.C. Lodder, Th. J.A. Popma and B. Dieny, *Phys. Rev. Lett.* 74 (1995) 5260.

Nir, T., and Y. Roichman, *Organic Field Effect Transistors*, available at webee.technion.ac.il (http://webee.technion.ac.il/orgelect/PapersAndPatents_Files/Transistors_tutorial_Nit_Tessler.pdf).

Ohtsuki, T. 1986, *Layout Design and Verification*, Elsevier Science, Amsterdam.

Preas, B.T., and M. Lorenzetti 1988, *Physical Design Automation of VLSI Systems*, Benjamin-Cummings, San Francisco.

Rabaey, J.M., A. Chandrakasan, and B. Nikolic 2008, *Digital Integrated Circuits: A Design Perspective*, 2nd ed., Pearson Education.

Sarrafzadeh, M., and C.K. Wong 1996, *An Introduction of VLSI Physical Design*, McGraw-Hill International.

Sherwani, N.A. 1993, *Algorithms for VLSI Physical Design Automation*, Kluwer Academic Publishers.

Smith, M.J., and Sebastian 2002, *Application Specific Integrated Circuits*, Pearson Education.

Sugahara, S., and N. Junsaku 2010, *Spin-Transistor Electronics: An Overview and Outlook*, pp. 2124–2154.

Sun, J.P., G.I. Haddad, P. Mazumder, J.N. Schulman 1998, *Resonant Tunnelling Diodes: Models and Properties,* Proceedings of the IEEE, Vol. 86, No. 4, pp. 641–661.

Tessler, N. and Y. Roichmai, webee.technion.ac.il

Weste, Neil H.E., D. Harris, and A. Banerjee 2009, *CMOS VLSI Design: A Circuits and Systems Perspective*, 3rd ed., Pearson Education.

EXERCISES

Fill in the Blanks

1. Delay _____ with the increase in the power supply voltage.
   (a) increases   (b) decreases   (c) remains constant   (d) none of these
2. Delay _____ with the increase in the operating temperature.
   (a) increases   (b) decreases   (c) remains constant   (d) none of these
3. VLSI design is _____
   (a) a sequential process with feedback loops
   (b) a parallel process with no feedback loops
   (c) both a sequential and parallel process with feedback loops
   (d) sequential process with no feedback loops

4. CAD tools _____ the VLSI design.
   (a) automate
   (b) reduce design cycle time
   (c) reduce the chance of errors
   (d) all of these
5. Standard cell-based design takes _____ time as FPGA-based design.
   (a) more   (b) less   (c) equal   (d) same

Multiple Choice Questions

1. The advantage of IC over discrete component-based circuits is
   (a) low power   (b) small size   (c) low cost   (d) all of these
2. With the advancement of technology
   (a) channel length is reduced
   (b) gate oxide thickness is reduced
   (c) power supply voltage is reduced
   (d) all of these
3. FPGA-based design is more suitable for
   (a) prototype development
   (b) large scale product development
   (c) low power application
   (d) high speed application
4. The interconnect delay is
   (a) always less than the gate delay
   (b) always more than the gate delay
   (c) always equal to the gate delay
   (d) can be more or less than the gate delay, depending on the technology used
5. According to Moore's law, the number of components doubles in every
   (a) 10 months   (b) 20 months   (c) 22 months   (d) 18 months

True or False

1. Delay can be decreased by increasing the chip area.
2. MOSFET was fabricated earlier than bipolar junction transistor.
3. CMOS technology is used for majority of the integrated circuits.
4. In the Y-chart, the details of design information increases when moved from centre to the periphery.
5. FPGA-based design has less area than full-custom design.

Short-answer Type Questions

1. What are the different types of integrated circuits?
2. Describe the IC manufacturing process in brief.
3. Draw the Y-chart and explain the VLSI design process.
4. What do you mean by the hierarchical abstraction? Explain the concepts of regularity, modularity, and locality.
5. Draw the flow diagram of typical VLSI design flow and explain.
6. What are different VLSI design styles? Explain each of them in brief.
7. Explain the FPGA-based VLSI system design.
8. Explain the gate array-based VLSI system design.
9. Explain the standard cell-based VLSI system design.
10. Explain the semi-custom and full-custom styles of VLSI system design.
11. What are the programmable logic devices? Explain them in brief.
12. Discuss the ROM-based system design.
13. Discuss the PAL-based system design.
14. Discuss the PLA-based system design.
15. Compare different VLSI design styles.
16. Discuss the CMOS integrated circuits and its issues with IC at DSM Level.
17. Explain how the cost of IC is calculated.
18. What are the different CAD tools used in VLSI design?

19. Discuss the functional specification and functional verification.
20. Discuss the logic design and verification.
21. Discuss the circuit design and verification.
22. Discuss the physical design and verification.
23. Discuss the VLSI CAD problems with mathematical formulae.
24. Discuss the IC chip industry with examples.
25. What is SOI technology? What are its advantages?
26. What is a 3D IC? Explain with diagrams.
27. Discuss the recent advanced technologies and future projections.
28. Discuss the nanoelectronic devices.

Long-answer Type Questions

1. Describe the IC manufacturing process in brief. Draw the Y-chart and explain the VLSI design process. What do you mean by hierarchical abstraction? What are concepts of regularity, modularity, and locality?
2. Draw the flow diagram of typical VLSI design flow and explain. What are the different VLSI design styles? Explain each of them in brief. Explain the FPGA-based VLSI system design.
3. Explain the gate array-based VLSI system design. Explain the standard cell-based VLSI system design. Explain the semi-custom and full-custom styles of VLSI system design. What are the programmable logic devices? Explain them in brief.
4. Discuss the ROM-, PAL-, and PLA-based system design. Compare their merits and demerits.
5. Discuss the functional specification and functional verification, the logic design and verification, the circuit design and verification, and the physical design and verification.
6. Write short notes on
   (a) 3D integrated circuit   (b) SOI technology   (c) Nanoelectronic devices

MOS Transistors

2.1   Introduction to Semiconductor Devices

Semiconductor devices have become important over the last few decades. They have electrical conductivity which is less than the conductors and greater than the insulators. The most important fact about semiconductors is that their conductivity can be changed to a large extent by various parameters such as electrical, optical, and thermal excitations. For example, a semiconductor is like an insulator at a certain temperature and like a metal at some other temperature, i.e., the resistance varies from giga ohm to a few milli ohm. If a device can be made using the semiconductor material, it can conduct fully under certain input condition; at some other input condition, it may not conduct at all. It behaves like a switch and is, therefore, very suitable for a digital circuit design. The large change in conductivity is achieved by a small change in the input condition. So it can amplify the input condition and transform it to the output. Hence, semiconductor devices, if properly designed, can be used for digital logic design as well as for analog designs such as amplifiers.

Transistors are semiconductor devices having three terminals, where current between the two terminals can be controlled by voltage or current at the third terminal. Scientists have come up with a large number of transistor structures (e.g., BJT, JFET, MOSFET, UJT), but the metal oxide semiconductor (MOS) field effect transistor (FET) or simply MOSFET is a semiconductor device which is most suitable for digital and analog ICs. Almost 90% of the ICs are fabricated worldwide using MOSFET. Therefore, we have focused on the concentration of the MOSFET device in this book. This chapter introduces the MOSFET structure, its operation, and characteristics. It also introduces MOSFET device parameters, MOSFET scaling, and modelling of a MOSFET.

2.2 Charge Carriers in Semiconductors

Semiconductors are classified into two types: (a) intrinsic semiconductors where no impurity is introduced; and (b) extrinsic semiconductors where there is some impurity introduced. There are two types of mobile charge carriers in semiconductors: negative charge carrier called electron (e) and positive charge carrier called hole (h). Depending on the majority of electrons or holes in the extrinsic semiconductors, they are classified into two types: n-type and p-type. If a trivalent material is introduced into the tetravalent semiconductor material, the latter becomes p-type semiconductor. Similarly, if a pentavalent material is introduced into the tetravalent semiconductor material, the latter becomes n-type semiconductor. The trivalent and pentavalent materials used to dope the semiconductors are called *dopants*. A semiconductor when introduced with dopants is called *extrinsic* semiconductor. Similarly, a semiconductor without any dopant is called *intrinsic* semiconductor.

If n and p are electrons and hole concentrations in a semiconductor, from the law of mass action under equilibrium, we have

$$np = \text{constant} \tag{2.1}$$

For an intrinsic semiconductor,

$$n = p = n_i \tag{2.2}$$

Hence, we can write

$$np = n_i^2 \tag{2.3}$$

2.3 Parallel Plate Capacitor

A parallel plate capacitor as shown in Fig. 2.1(a) has two metallic plates separated by a dielectric material. If a charge Q is placed on one plate, the equal and opposite charge $-Q$ is induced on the other plate of the capacitor. If ε is the dielectric constant of the material, d is the separation between the plates, and A is the area of the plates, then the capacitance can be written as

$$C = \varepsilon \frac{A}{d} \tag{2.4}$$

The capacitance is expressed in farad (F), while other dimensions are in SI units. The dielectric constant can be expressed as

$$\varepsilon = \varepsilon_r \times \varepsilon_0 \tag{2.5}$$

where ε_r is the relative permittivity of the dielectric and $\varepsilon_0 = 8.85 \times 10^{-12}\,\text{F/m}$.

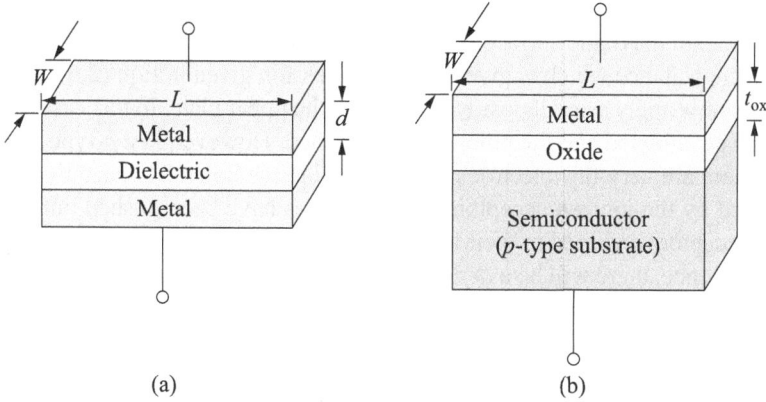

Fig. 2.1 (a) Normal capacitive structure; (b) structure of the MOS capacitor (see Plate 1)

2.4 MOS Capacitor

In a MOS capacitor, the lower plate is replaced by a semiconductor material which is often called substrate. The structure of a MOS capacitor is shown in Fig. 2.1(b). For example, let us consider that the substrate material is *p*-type. The *p*-type substrate has majority carrier holes distributed over the entire bulk material. If the upper plate is negatively charged, holes will be attracted towards the upper plate and will accumulate near the oxide–semiconductor interface. This situation is the same as that of a parallel plate capacitor. This metal oxide semiconductor structure is known as *MOS capacitor*. The expression for MOS capacitance can be written as

$$C_{\mathrm{MOS}} = \varepsilon_{\mathrm{ox}} \frac{A}{t_{\mathrm{ox}}} = \varepsilon_{\mathrm{ox}} \frac{W \times L}{t_{\mathrm{ox}}} \qquad (2.6)$$

where area of the MOS capacitor is $A = (W \times L)$ and t_{ox} is the oxide thickness, and $\varepsilon_{\mathrm{ox}}$ is the dielectric constant of the oxide material.

Let us now understand the behaviour of MOS capacitor under different bias conditions as shown in Fig. 2.2. First, we apply a negative voltage to the upper plate. The

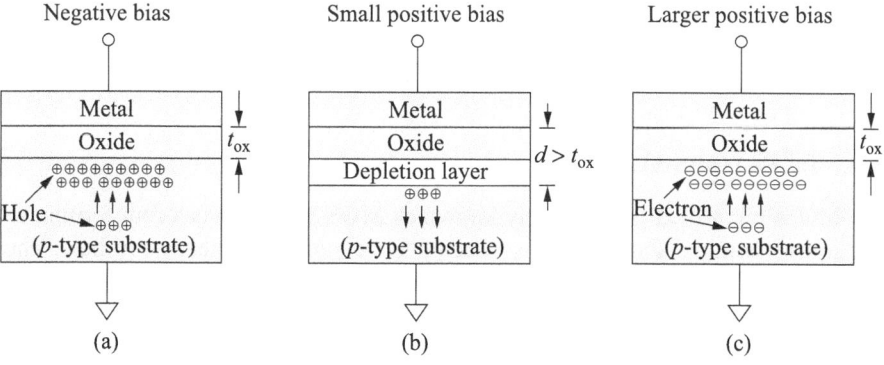

Fig. 2.2 MOS capacitor under different bias conditions (see Plate 2)

holes in the substrate will be attracted towards oxide surface and they will accumulate underneath the oxide surface. This phenomenon is called *accumulation*. Hence, we get a capacitance which is given by the expression given in Eqn (2.6).

Let us now apply a small positive voltage to the upper plate. In this case the electrons will be attracted and the holes will be repelled. However, in a *p*-type semiconductor, there are very few electrons as compared to the holes. The negative charges are created by the ionized acceptors after the holes have been pushed out of them. But the acceptors are fixed in their locations and cannot be driven to the edge of the insulator. Hence, there will be a *depletion* region created underneath the oxide layer. Therefore, the distance of the induced charge from the upper plate increases. As a result the capacitance is lower as compared to the parallel plate capacitor.

As more and more positive voltage is applied to the upper plate, the thickness of the depletion layer continues to increase. Thus, the capacitance continues to decrease. This does not, however, continue indefinitely. We know from the law of mass action that as hole density reduces, the electron density increases. At some point, the hole density is reduced and electron density is increased to such an extent that electrons now become the majority carriers near the silicon dioxide interface. This phenomenon is known as *surface inversion*. Beyond this point, a large positive voltage on the upper plate induces more electrons in the semiconductor. The induced electrons are mobile, and will be attracted to the silicon insulator interface. The induced electrons accumulate underneath the oxide layer, which is known as *inversion layer*. Therefore, the MOS capacitance again increases to the parallel plate capacitance value. The MOS capacitance variation with the applied bias is depicted in Fig. 2.3.

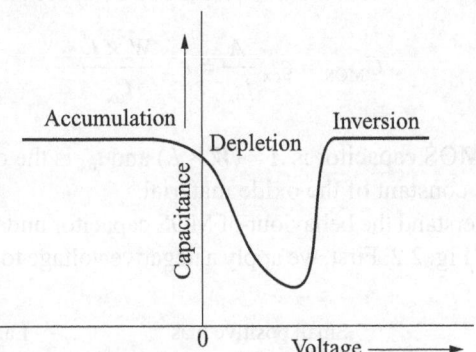

Fig. 2.3   MOS capacitance–voltage characteristics

2.5   MOS Transistor

The MOSFET, which is commonly known as MOS transistor has four terminals. The source terminal serves as the source of carriers of either electron or hole. The drain terminal collects the carriers flown from the source terminal. The carriers flow from the source to the drain terminal through a conducting path called *channel*. The flow of carriers in the channel is controlled by applying voltage at a third terminal called gate of the MOSFET. The channel can be created either physically or electrically. Depending on how the channel is created, MOSFETs are classified into two types: (a) enhancement type and (b) depletion type.

Enhancement-type MOSFET   If the MOSFET is normally OFF, and is turned ON by applying voltage at the gate terminal, then the MOSFET is known as *enhancement-type MOSFET*.

Depletion-type MOSFET   If the MOSFET is normally ON, and is turned OFF by applying voltage at the gate terminal, then the MOSFET is known as *depletion-type MOSFET*.

In the enhancement-type MOSFET, the channel is created electrically by applying voltage at the gate terminal; in the depletion-type MOSFET, the channel is created physically by introducing a physical layer between the source and drain at the time of manufacturing. Figure 2.4 shows the cross-sectional view of the enhancement-type and depletion-type MOSFETs.

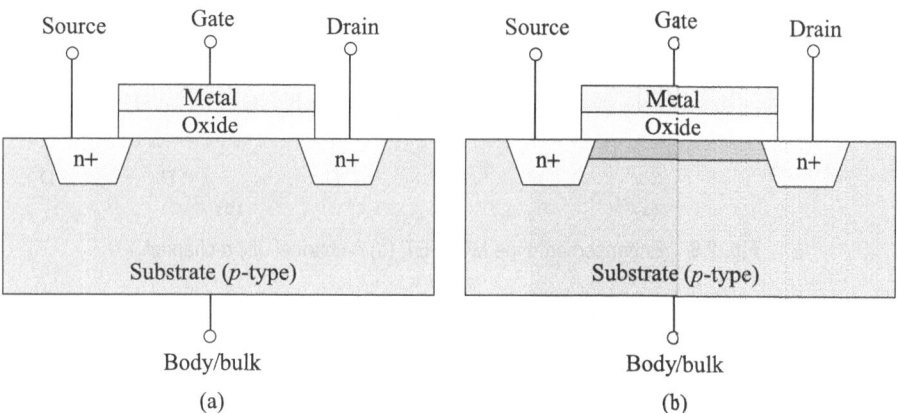

Fig. 2.4   Cross-sectional view of MOSFET: (a) enhancement type; (b) depletion type (see Plate 2)

2.5.1   Structure of Enhancement-type MOS Transistor

The structure of the enhancement MOSFET is shown in Fig. 2.5. A semiconductor, either p-type or n-type, is taken as a substrate. On the substrate, two diffusion regions are created. If the substrate is p-type, the diffusion regions are n^+-type (n^+ indicates heavily doped n-type). If the substrate is n-type, the diffusion regions are p^+-type (p^+ indicates heavily doped p-type). The diffusion regions are known as source and drain. Between the source and drain, an oxide layer is formed on top of the substrate. The oxide layer is known as gate oxide and it plays a very important role in determining the MOSFET characteristics. On top of the oxide layer a metal or polysilicon is deposited. Then four metal contacts are taken from the source, gate, drain, and the bulk to form the electrodes. Essentially, a MOSFET is a four-terminal device with the terminals drain, gate, source, and bulk.

There are different symbols used to represent the MOSFETs, and these are illustrated in Fig. 2.6.

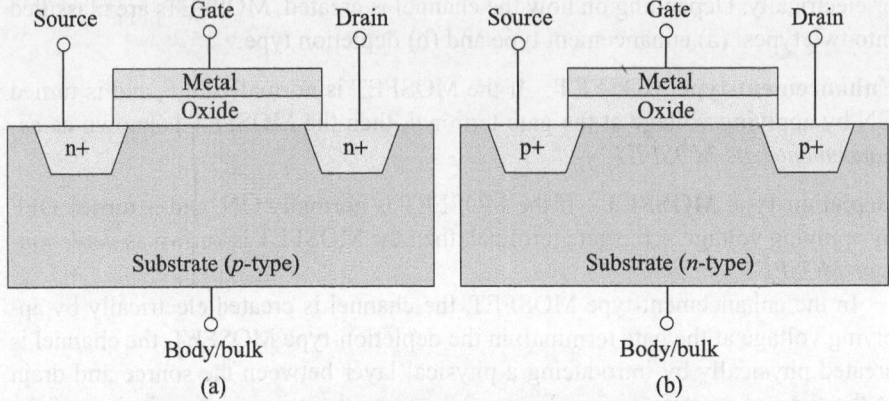

Fig. 2.5   Cross-sectional view of enhancement-type MOSFET: (a) *n*-channel; (b) *p*-channel (see Plate 2)

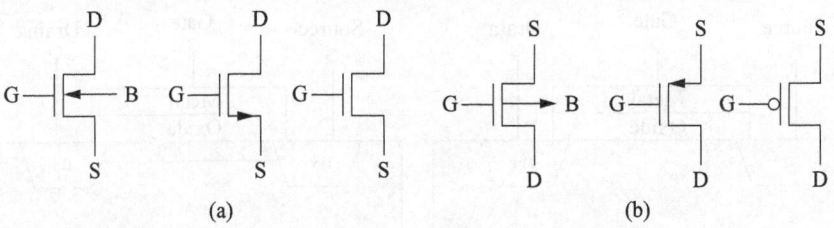

(a)   (b)

Fig. 2.6   Enhancement-type MOSFET: (a) *n*-channel; (b) *p*-channel

The four terminals of MOSFET are abbreviated as D for drain, G for gate, S for source, and B for bulk or body terminal.

2.5.2   Operation of Enhancement-type nMOS Transistor

If a voltage is applied at the gate terminal, a conducting channel is formed underneath the oxide layer between the source and drain. A MOSFET is called *n*-channel or simply nMOS when the channel is formed with electrons, whereas, it is called *p*-channel or simply pMOS when the channel is formed with holes. Let us first explain the operation of an *n*-channel enhancement-type MOSFET. The structure of the *n*-channel enhancement-type MOSFET is shown in Fig. 2.5(a). Depending on the voltage applied at the gate terminal, the operation of MOSFET is divided into three conditions:

■ Accumulation
■ Depletion
■ Inversion

Accumulation (V_{GS} is negative)

If a negative voltage is applied at the gate terminal, the majority carriers (holes) from the *p*-type substrate are attracted towards the gate terminal and are accumulated underneath the gate oxide layer. This condition is known as *accumulation*.

Depletion (V_{GS} is small positive)

Now, if a small positive voltage is applied, the holes will be repelled into the substrate. The repelled holes will move into the substrate and create negatively charged fixed acceptor ions underneath the gate oxide layer. These fixed negatively charged ions form the *depletion* layer.

Inversion (V_{GS} is large positive)

If the positive gate voltage is large enough, the majority carrier holes are repelled into the substrate and the small numbers of minority carrier electrons are attracted towards the gate oxide surface. The attracted electrons accumulate underneath the gate oxide layer and form a conducting path between the source and drain. This conducting path is called channel and the condition is known as *strong inversion*.

Figure 2.7 illustrates the different operating conditions of an *n*-channel enhancement-type MOSFET.

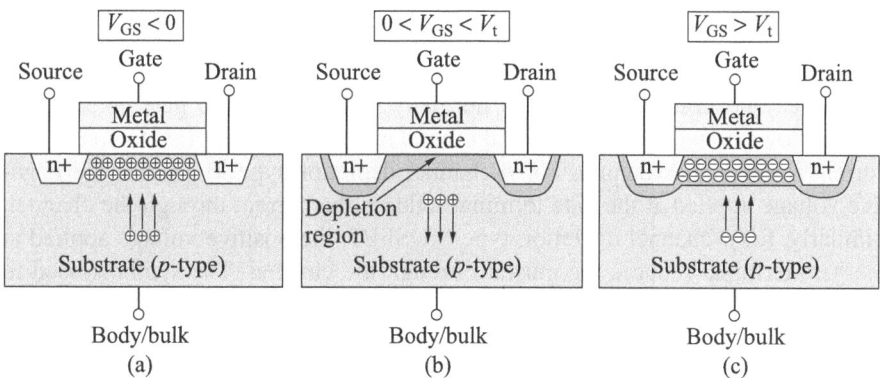

Fig. 2.7 nMOS under different operating conditions: (a) accumulation; (b) depletion; (c) inversion (see Plate 3)

2.5.3 Operation of Enhancement-type pMOS Transistor

The structure of a pMOS transistor is shown in Fig. 2.5(b). The structure and operating principle of pMOS are fully complementary to that of an nMOS transistor. When a positive voltage is applied at the gate, the majority carrier electrons are attracted towards the gate oxide surface and get accumulated underneath the gate oxide layer. These accumulated electrons form the accumulation layer underneath the gate oxide layer. If a small negative voltage is applied at the gate, the electrons are repelled into the substrate and a depletion region below the gate oxide layer is created. If the negative gate voltage is increased further, the majority carrier electrons are repelled deep into the substrate and the minority carrier holes are attracted towards the gate oxide surface and accumulated under the gate oxide layer. These accumulated holes below the gate oxide layer form the inversion layer or a conducting channel between the source and drain.

Figure 2.8 illustrates the different operating conditions of the *p*-channel enhancement-type MOSFET.

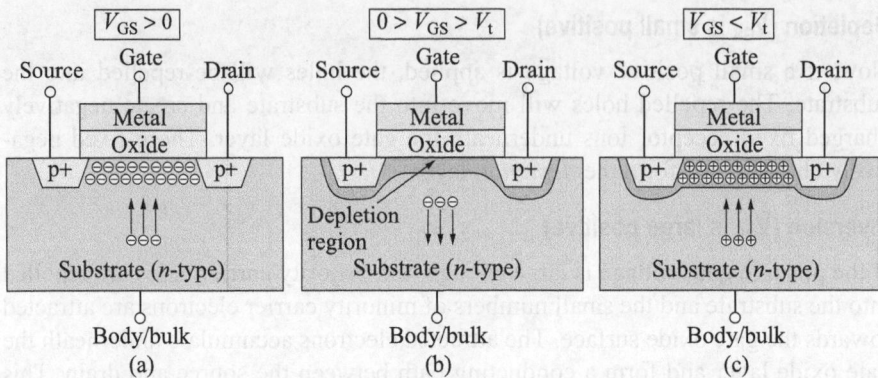

Fig. 2.8 pMOS under different operating conditions: (a) accumulation; (b) depletion; (c) inversion (see Plate 3)

2.5.4 Depletion-type MOSFET

Depletion-type MOSFET is another type in which the channel is created physically using the ion implantation process. As the channel is already present between the source and drain, the device is normally ON, i.e., at zero gate voltage, the current flows between the source and drain. Gate voltage is applied to control the current through the channel. For n-channel depletion-type MOSFET, the negative voltage applied at the gate terminal reduces the current through the channel. Similarly, for p-channel depletion-type MOSFET, the positive voltage applied at the gate terminal reduces the current through the channel. The symbols used to represent the depletion-type MOSFET are shown in Fig. 2.9.

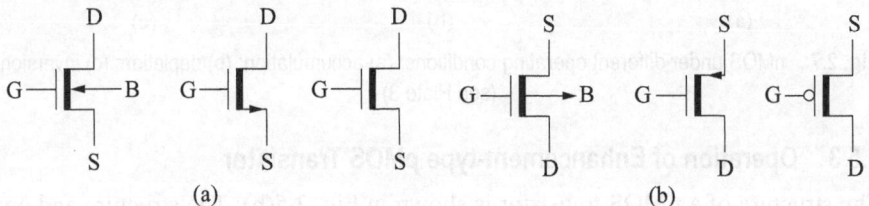

Fig. 2.9 Depletion-type MOSFET: (a) n-channel; (b) p-channel

2.6 MOSFET Threshold Voltage

Threshold voltage (V_t) of a MOSFET is defined as the gate voltage with respect to the source terminal at which the inversion layer is formed underneath the gate oxide layer. When a large gate voltage is applied, the minority carriers are attracted towards the gate oxide surface and they are accumulated below the gate oxide. Creation of this inversion layer is called *inversion*. Depending on the carrier concentration of the inversion layer, inversion is classified into two types: strong inversion and weak inversion. In strong inversion, the p-type substrate is completely inverted into an n-type material for nMOS transistors whereas, for the pMOS transistors, the n-type substrate is completely inverted into a p-type material. In case of weak inversion, the inversion happens partially.

To derive the expression for the threshold voltage, we identify the components of the gate voltage required to create the inversion layer under strong inversion as shown below:

- The difference between the work function of the gate and channel
- The voltage required to create strong inversion
- The voltage required to offset the depletion layer charge
- The voltage required to offset the fixed oxide charge

Combining the above four components, the threshold voltage of the MOSFET can be written as

$$V_t = \Phi_{ms} - 2\varphi_F - \frac{Q_{b0}}{C_{ox}} - \frac{Q_{ox}}{C_{ox}} \tag{2.7}$$

where

$\Phi_{ms} = \varphi_s - \varphi_m$ = work function difference
φ_F = equilibrium electrostatic potential (Fermi potential)
Q_{b0} = depletion layer charge density per unit area
Q_{ox} = fixed oxide charge density per unit area
$C_{ox} = \varepsilon_{ox}/t_{ox}$ = gate oxide capacitance per unit area

The depletion layer charge density can be derived by solving the Poisson equation as follows:

$$\nabla \times \vec{E} = \frac{\rho}{\varepsilon_{Si}} \tag{2.8}$$

where charge density, $\rho = q(-N_A)$ and ε_{Si} = permittivity of the silicon substrate. Equation (2.8) can be written in one dimension as

$$\frac{dE}{dx} = \frac{\rho}{\varepsilon_{Si}} \quad \text{or} \quad dE = -\frac{qN_A}{\varepsilon_{Si}} dx \tag{2.9}$$

Integrating both sides of Eqn (2.9), we get

$$E(x) = -\frac{qN_A}{\varepsilon_{Si}} x \tag{2.10}$$

If the potential be φ, Eqn (2.10) can be rewritten as

$$-\frac{d\varphi}{dx} = E(x) = -\frac{qN_A}{\varepsilon_{Si}} x$$

or

$$d\varphi = \frac{qN_A}{\varepsilon_{Si}} x \, dx \tag{2.11}$$

Now integrating both sides of Eqn (2.11), we obtain

$$\int_{\varphi_F}^{\varphi_s} d\varphi = \int_0^{x_d} \frac{qN_A}{\varepsilon_{Si}} x \, dx$$

or
$$\varphi_s - \varphi_F = \frac{qN_A x_d^2}{2\varepsilon_{Si}} \tag{2.12}$$

Hence, the thickness of the depletion region is given by

$$x_d = \sqrt{\frac{2\varepsilon_{Si} |\varphi_s - \varphi_F|}{qN_A}} \tag{2.13}$$

The depletion layer charge per unit area can be written as

$$Q = q(-N_A)x_d = -\sqrt{2qN_A \varepsilon_{Si} |\varphi_s - \varphi_F|} \tag{2.14}$$

Under strong inversion condition, the surface potential φ_s must be equal to $-\varphi_F$. Hence, the depletion layer charge density per unit area under strong inversion condition can be written as

$$Q_{b0} = -\sqrt{2qN_A \varepsilon_{Si} |-2\varphi_F|} \tag{2.15}$$

If the substrate (bulk) potential is not same as that of the source, the depletion layer charge density becomes a function of the source-to-bulk voltage, as given by

$$Q_b = -\sqrt{2qN_A \varepsilon_{Si} |-2\varphi_F + V_{SB}|} \tag{2.16}$$

Hence, the threshold voltage under *substrate bias* can be written as

$$V_t = \Phi_{ms} - 2\varphi_F - \frac{Q_b}{C_{ox}} - \frac{Q_{ox}}{C_{ox}} = \Phi_{ms} - 2\varphi_F - \frac{Q_{b0}}{C_{ox}} - \frac{Q_{ox}}{C_{ox}} - \frac{Q_b - Q_{b0}}{C_{ox}}$$

$$V_t = V_{t0} - \frac{Q_b - Q_{b0}}{C_{ox}} = V_{t0} + \frac{\sqrt{2qN_A \varepsilon_{Si}}}{C_{ox}} \left(\sqrt{|-2\varphi_F + V_{SB}|} - \sqrt{|-2\varphi_F|} \right)$$

$$V_t = V_{t0} + \gamma \left(\sqrt{|-2\varphi_F + V_{SB}|} - \sqrt{|-2\varphi_F|} \right) \tag{2.17}$$

where
$$\gamma = \frac{\sqrt{2qN_A \varepsilon_{Si}}}{C_{ox}} \tag{2.18}$$

γ = substrate-bias coefficient or body-effect coefficient
V_{t0} = threshold voltage without substrate bias

Example 2.1 Calculate the threshold voltage for a polysilicon gate nMOS transistor with the following parameters:

$N_A = 2 \times 10^{16}$ cm^{-3}, $\quad\quad t_{ox} = 300$ Å

$N_D = 2 \times 10^{19}$ cm^{-3}, $\quad\quad N_{ox} = 10^{10}$ cm^{-2}

Solution   Let us calculate the Fermi potential for p-substrate as

$$\varphi_F(\text{substrate}) = \frac{kT}{q}\ln\left(\frac{n_i}{N_A}\right) = 0.026 \times \ln\left(\frac{1.45 \times 10^{10}}{2 \times 10^{16}}\right) = -0.367\,\text{V}$$

Similarly, we can calculate the Fermi potential for n-type polysilicon gate as

$$\varphi_F(\text{gate}) = \frac{kT}{q}\ln\left(\frac{N_D}{n_i}\right) = 0.026 \times \ln\left(\frac{2 \times 10^{19}}{1.45 \times 10^{10}}\right) = 0.547\,\text{V}$$

Then the work function difference is given by

$$\varphi_{ms} = \varphi(\text{substrate}) - \varphi(\text{gate}) = -0.367 - 0.547 = -0.914\,\text{V}$$

Let us now calculate the oxide capacitance per unit area

$$C_{ox} = \frac{\varepsilon_{ox}}{t_{ox}} = \frac{3.97 \times 8.85 \times 10^{-14}}{300 \times 10^{-8}} = 1.17 \times 10^{-7}\,\text{F/cm}^2$$

The depletion layer charge is calculated as

$$Q_{b0} = -\sqrt{2qN_A\varepsilon_{Si}\left|-2\varphi_F\right|}$$

$$= -\sqrt{2 \times 1.6 \times 10^{-19} \times 2 \times 10^{16} \times 11.7 \times 8.85 \times 10^{-14}\left|-2 \times -0.367\right|}$$

$$= -6.97 \times 10^{-8}\,\text{C/cm}^2$$

The fixed oxide charge is calculated as

$$Q_{ox} = qN_{ox} = 1.6 \times 10^{-19} \times 10^{10} = 1.6 \times 10^{-9}\,\text{C/cm}^2$$

Substituting the values in Eqn (2.7), we can calculate the threshold voltage as

$$V_{t0} = \varPhi_{ms} - 2\varphi_F - \frac{Q_{b0}}{C_{ox}} - \frac{Q_{ox}}{C_{ox}}$$

$$= -0.914 - 2 \times (-0.367) - \left(\frac{-6.97 \times 10^{-8}}{1.17 \times 10^{-7}}\right) - \left(\frac{1.6 \times 10^{-9}}{1.17 \times 10^{-7}}\right)$$

$$V_{t0} = -0.914 + 0.734 + 0.594 - 0.0138 = 0.4002\,\text{V}$$

2.7   MOSFET Current Equation

Figure 2.10 shows an nMOS transistor with terminal voltages. The source and bulk terminals are connected to the ground. The voltage V_{DS} is applied at the drain terminal and V_{GS} is applied at the gate terminal. Let us assume that $V_{GS} > V_t$ so that the channel is formed underneath the gate oxide layer. Because of voltage difference between the drain and source, the current I_D flows from drain to source. Let us derive an expression for the drain current as a function of V_{GS} and V_{DS}. We assume that the channel length L is extended along y-axis and the channel width W is extended along x-axis.

The simplest model for the current in MOS transistor is the charge sheet model. In this model, the inversion layer charge per unit area is given by

$$Q_n = C_{ox}(V_{GC} - V_t) \tag{2.19}$$

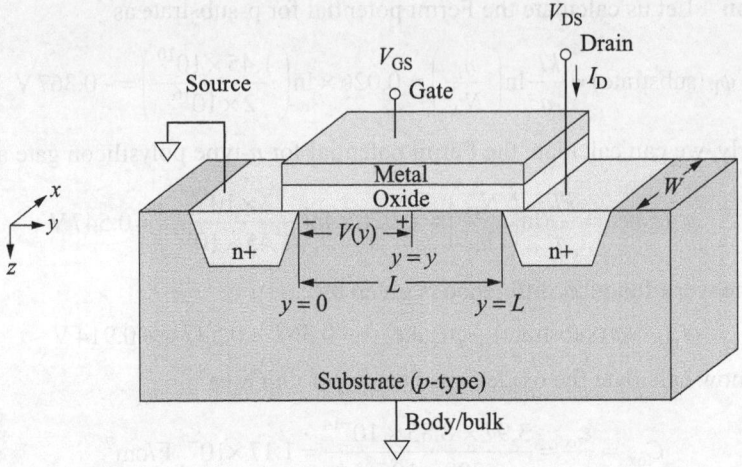

Fig. 2.10 Gradual channel approximation model (see Plate 3)

where V_{GC} is the gate-to-channel voltage drop and V_t is the threshold voltage of the nMOS transistor. The gate-to-channel potential at any distance y from the source end can be written as

$$V_{GC}(y) = V_{GS} - V(y) \tag{2.20}$$

where $V(y)$ is the voltage drop along the channel due to drain current I_D.

Hence, the inversion layer charge at any distance y is written as

$$Q_n(y) = C_{ox}(V_{GS} - V(y) - V_t) \tag{2.21}$$

The drain current is given by

$$I_D = Q_n W v_n \tag{2.22}$$

where W = channel width and v_n = electron drift velocity in the channel.

The electron drift velocity is a function of electric field in the channel and is given by

$$v_n = \mu_n E(y) = \mu_n \frac{dV(y)}{dy} \tag{2.23}$$

Combining Eqs (2.22) and (2.23), we can write

$$I_D = Q_n W \mu_n \frac{dV(y)}{dy} \tag{2.24}$$

Substituting Eqn (2.21) in Eqn (2.24), we get

$$I_D = C_{ox}(V_{GS} - V(y) - V_t) \times W \mu_n \frac{dV(y)}{dy} \tag{2.25}$$

Integrating both sides of Eqn (2.25), we get

$$\int_0^L I_D \, dy = \int_0^{V_{DS}} C_{ox}(V_{GS} - V(y) - V_t) \times W \mu_n \times dV(y)$$

$$I_D \int_0^L dy = \mu_n C_{ox} W \times \int_0^{V_{DS}} (V_{GS} - V(y) - V_t) \times dV(y)$$

$$I_D \times L = \mu_n C_{ox} W \times \left[(V_{GS} - V_t) V_{DS} - \frac{V_{DS}^2}{2} \right]$$

$$I_D = \frac{\mu_n C_{ox} W}{L} \times \left[(V_{GS} - V_t) V_{DS} - \frac{V_{DS}^2}{2} \right] \tag{2.26}$$

Equation (2.26) describes the MOSFET current expression. The term $\mu_n C_{ox} = k'_n$ is called the *transconductance parameter* which entirely depends on the process technology. Using the transconductance parameter, Eqn (2.26) can be rewritten as

$$I_D = \frac{k'_n W}{L} \times \left[(V_{GS} - V_t) V_{DS} - \frac{V_{DS}^2}{2} \right]$$

$$I_D = \beta_n \times \left[(V_{GS} - V_t) V_{DS} - \frac{V_{DS}^2}{2} \right] \tag{2.27}$$

where $\beta_n \left(= k'_n \dfrac{W}{L} \right)$ is known as the *gain factor* and W/L ratio is known as the *aspect ratio* of the MOSFET.

Case 1 When V_{DS} is small, i.e., $V_{DS} \ll (V_{GS} - V_t)$ the MOSFET drain expression reduces to

$$I_D = \frac{\mu_n C_{ox} W}{L} \times (V_{GS} - V_t) V_{DS} \tag{2.28}$$

The channel resistance can be written as

$$R_{ch} = \frac{V_{DS}}{I_D} = \frac{L}{\mu_n C_{ox} W (V_{GS} - V_t)} \tag{2.29}$$

The MOSFET simply acts as a voltage-dependent linear resistor. The channel resistance is increased as the channel length is increased, and decreased as the channel width is increased. Under this condition, the nMOS transistor is said to be operating in linear region.

Case 2 When V_{DS} is increased, the drain current must follow Eqn (2.26) and the channel resistance dynamically varies with V_{DS} as

$$r_{ch} = \frac{\partial V_{DS}}{\partial I_D} = \frac{L}{\mu_n C_{ox} W (V_{GS} - V_t - V_{DS})} \tag{2.30}$$

As V_{DS} increases, the dynamic channel resistance increases; hence, I_D does not increase linearly with V_{DS}.

Case 3 When V_{DS} is large, i.e., $V_{DS} \geq (V_{GS} - V_t)$; but when $V_{DS} = (V_{GS} - V_t)$, the gate-to-channel voltage drop V_{GC} at the drain end is given by

$$V_{GC} = V_{GS} - V(y = L) = V_{GS} - V_{DS} = V_{GS} - (V_{GS} - V_t) = V$$

Hence, the inversion layer just begins to disappear at the drain end. Under this condition the channel is pinched off. The drain current does not increase; rather it remains constant. This nMOS transistor operates in saturation region and the expression for drain current as in Eqn (2.20) does not hold. Substituting $V_{DS} = (V_{GS} - V_t)$ in Eqn (2.26), we get

$$I_D = \frac{\mu_n C_{ox} W}{L} \times \left[(V_{GS} - V_t)(V_{GS} - V_t) - \frac{(V_{GS} - V_t)^2}{2} \right]$$

or

$$I_{D,sat} = \frac{\mu_n C_{ox} W}{2L} \times (V_{GS} - V_t)^2 \tag{2.31}$$

$$I_{D,sat} = \frac{k_n W}{2L} \times (V_{GS} - V_t)^2 \tag{2.32}$$

$$I_{D,sat} = \frac{\beta W}{2L} \times (V_{GS} - V_t)^2 \tag{2.33}$$

At saturation, the channel is pinched off at the drain end and the channel length is reduced by a small amount as illustrated in Fig. 2.11. Hence, the effective channel length is expressed as

$$L' = L - \Delta L \tag{2.34}$$

where ΔL is the length of the pinched-off region. The drain current under saturation considering reduction in channel length is given by (see Section 2.10.1 for derivation)

$$I_{D,sat} = \frac{\beta_n}{2}(V_{GS} - V_t)^2(1 + \lambda V_{DS})$$

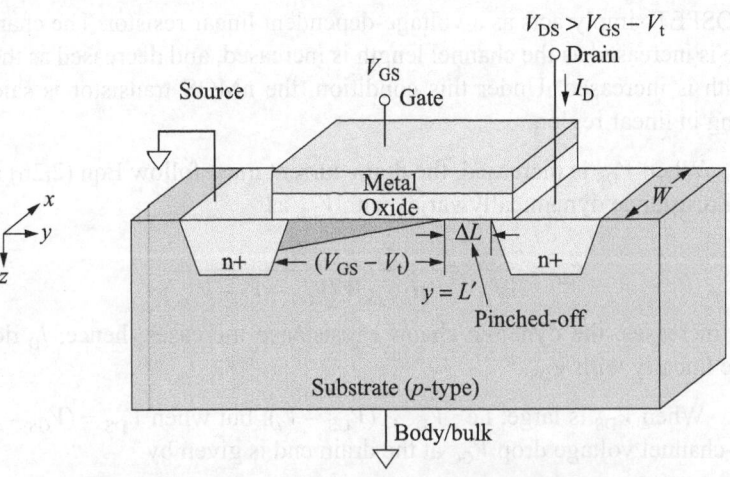

Fig. 2.11 nMOS transistor under saturation (see Plate 4)

2.7.1 Transconductance (g_m), Output Conductance (g_{DS}), and Figure of Merit of the MOS Transistor

The transconductance (g_m) of a MOS transistor is given by

$$g_m = \left.\frac{\partial I_D}{\partial V_{GS}}\right|_{V_{DS} = \text{constant}} = \frac{\mu_n C_{ox} W}{2L} \times 2(V_{GS} - V_t) = \frac{\mu_n C_{ox} W (V_{GS} - V_t)}{L} \tag{2.35}$$

The output conductance (g_{DS}) of a MOS transistor is given by

$$g_{DS} = \left.\frac{\partial I_D}{\partial V_{DS}}\right|_{V_{GS} = \text{constant}} = \frac{\mu_n C_{ox} W}{2L}(V_{GS} - V_t)^2 \times \lambda = \frac{\lambda I_D}{(1 + \lambda V_{DS})} \cong \lambda I_D \tag{2.36}$$

We can express the transconductance as

$$g_m = \frac{\mu_n C_{ox} W (V_{GS} - V_t)}{L} = \frac{\mu_n C_g (V_{GS} - V_t)}{L^2} \tag{2.37}$$

where C_g is the gate capacitance given by

$$C_g = \varepsilon_{ox} \frac{W \times L}{t_{ox}} = C_{ox} \times WL \tag{2.38}$$

Again, in saturation,

$$V_{GS} - V_t = V_{DS} \tag{2.39}$$

Hence, Eqn (2.37) can be written as

$$g_m = \frac{\mu_n C_g (V_{GS} - V_t)}{L^2} = \frac{\mu_n C_g V_{DS}}{L^2} \tag{2.40}$$

Or the ratio of transconductance and the gate capacitance can be written as

$$\frac{g_m}{C_g} = \frac{\mu_n V_{DS}}{L^2} = \frac{\mu_n}{L} \times \frac{V_{DS}}{L} = \frac{\mu_n E_{DS}}{L} = \frac{v_{dn}}{L} = \frac{1}{\tau_n} \tag{2.41}$$

As shown in Eqn (2.41), the ratio of transconductance and the gate capacitance is proportional to the transit time of the electron (τ_n) in traversing the channel length L and has a dimension of (sec^{-1}). Hence, this parameter gives an indication of the frequency response of the MOS transistor, and is known as figure-of-merit (FOM) of the MOS transistor. Thus, FOM of an MOS transistor is expressed as

$$\text{FOM} = \frac{g_m}{C_g} \tag{2.42}$$

There is another definition of FOM of an MOS transistor. The product of the gate charge (Q_G) and the channel resistance at ON condition of the MOSFET is also known as FOM, and is given by

$$\text{FOM} = Q_G \times R_{ch} \tag{2.43}$$

The gate charge is expressed as the total amount of charge consumed by the gate capacitances.

2.8 MOSFET *V–I* Characteristics

Figures 2.12 and 2.13 depict the nMOS transistor *V–I* characteristics. For small V_{DS}, the drain current linearly increases with V_{DS}. As V_{DS} is increased, the drain current saturates and becomes independent on V_{DS}.

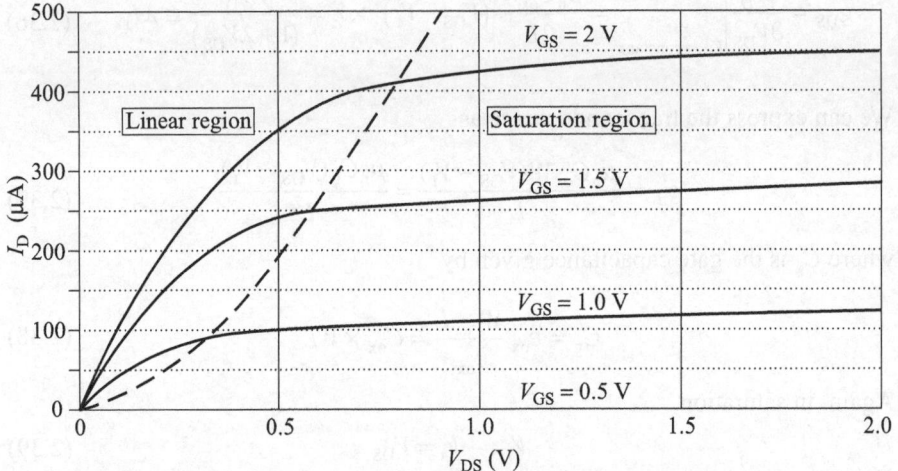

Fig. 2.12 I_D vs V_{DS} characteristics of the nMOS transistor

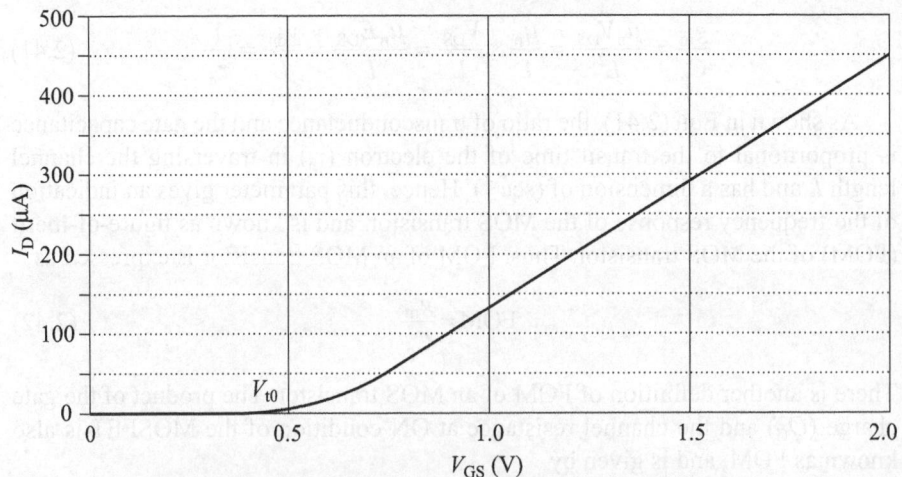

Fig. 2.13 I_D vs V_{GS} characteristics of the nMOS transistor

Example 2.2   Calculate the drain current of an n-channel MOSFET with the following parameters.

$$L = 2 \text{ } \mu\text{m}, W = 10 \text{ } \mu\text{m}, \mu_n = 0.06 \text{ m}^2/\text{V s}, C_{ox} = 15 \times 10^{-4} \text{ F/m}^2, V_{T0} = 0.4 \text{ V}$$

Solution   The drain current of an n-channel MOSFET is given by

$$I_D = \frac{\mu_n C_{ox} W}{L} \times \left[(V_{GS} - V_t)V_{DS} - \frac{V_{DS}^2}{2} \right] \text{ for the linear region}$$

and

$$I_{D,sat} = \frac{\mu_n C_{ox} W}{2L} \times (V_{GS} - V_t)^2 \text{ for the saturation region, } V_{DS} \geq (V_{GS} - V_t)$$

Substituting the given parameters in the above equation, we can calculate

$$\beta_n = \mu_n C_{ox}(W/L) = 0.06 \times 15 \times 10^{-4} \times \frac{10 \times 10^{-6}}{2 \times 10^{-6}} = 4.5 \times 10^{-4} \text{ A/V}^2$$

Therefore, the expression for the drain current becomes

$$I_D = 4.5 \times 10^{-4} \times \left[(V_{GS} - 0.4)V_{DS} - \frac{V_{DS}^2}{2} \right] \text{ for the linear region}$$

and

$$I_{D,sat} = 0.5 \times 4.5 \times 10^{-4} \times (V_{GS} - 0.4)^2 \text{ for the saturation region}$$
$$V_{DS} \geq (V_{GS} - 0.4)$$

Now, to calculate the drain current we assume the values of drain-to-source voltage and gate-to-source voltage as follows.

V_{GS} (V)	0.45	0.5	0.6	0.7	0.8	0.9			
V_{DS} (V)	0.0	0.1	0.2	0.3	0.4	0.5	0.6	0.7	0.8	0.9

Substituting the values of V_{GS} and V_{DS} in the above equation, we get the following values of I_D in µA (Table 2.1).

Table 2.1

V_{DS} (V)→ V_{GS} (V)↓	0.0	0.1	0.2	0.3	0.4	0.5	0.6	0.7	0.8	0.9
0.45	0	0.5625	0.5625	0.5625	0.5625	0.5625	0.5625	0.5625	0.5625	0.5625
0.5	0	2.2500	2.2500	2.2500	2.2500	2.2500	2.2500	2.2500	2.2500	2.2500
0.6	0	6.7500	9.0000	9.0000	9.0000	9.0000	9.0000	9.0000	9.0000	9.0000
0.7	0	11.2500	18.0000	20.2500	20.2500	20.2500	20.2500	20.2500	20.2500	20.2500
0.8	0	15.7500	27.0000	33.7500	36.0000	36.0000	36.0000	36.0000	36.0000	36.0000
0.9	0	20.2500	36.0000	47.2500	54.0000	56.2500	56.2500	56.2500	56.2500	56.2500

The variation of I_{DS} for different values of V_{GS} and V_{DS} is shown in Fig. 2.14.

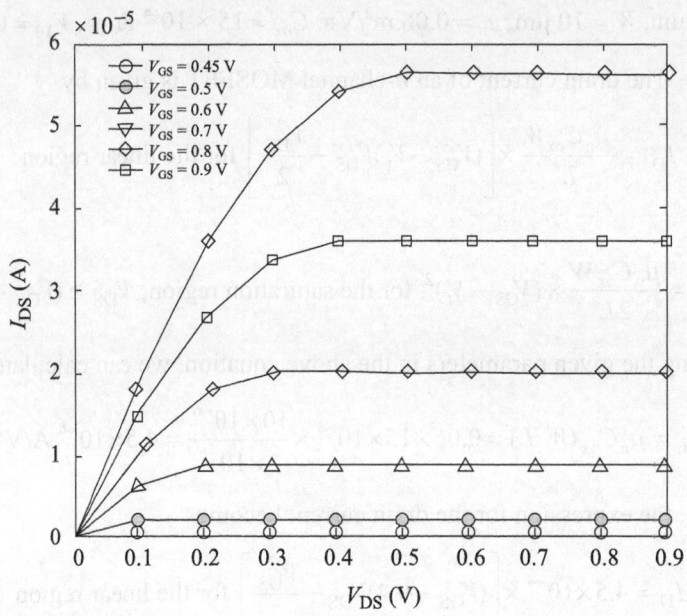

Fig. 2.14

Example 2.3　An *n*-channel MOSFET has a threshold voltage of 0.4 V. In the saturation region, at one point $I_{DS} = 2.25$ µA when $V_{GS} = 0.5$ V. What will be the value of I_{DS} if V_{GS} is increased to 0.7 V while it is operated in the saturation region?

Solution　In the saturation region, the drain current is given by

$$I_{D,sat} = \frac{\mu_n C_{ox} W}{2L} \times (V_{GS} - V_t)^2 = \frac{\beta_n}{2} \times (V_{GS} - V_t)^2$$

From the given parameters, we can write

$$\beta_n = \frac{2 \times I_{D,sat}}{(V_{GS} - V_t)^2} = \frac{2 \times 2.25 \times 10^{-6}}{(0.5 - 0.4)^2} = 4.5 \times 10^{-4} \text{ A/V}^2$$

When $V_{GS} = 0.7$ V, the drain current becomes
$$I_{D,sat} = 0.5 \times 4.5 \times 10^{-4} \times (0.7 - 0.4)^2 = 20.25 \times 10^{-6} \text{ A}$$

Example 2.4　An *n*-channel MOSFET has the following *I–V* characteristics (Fig. 2.15).

(a)　What are the values of β_n and V_{t0}?
(b)　If $V_{GS} = 0.65$ V, what is the value of I_{DS} in the saturation region?

(c) What is the minimum value of V_{DS} for operating in the saturation region if $V_{GS} = 0.75$ V?

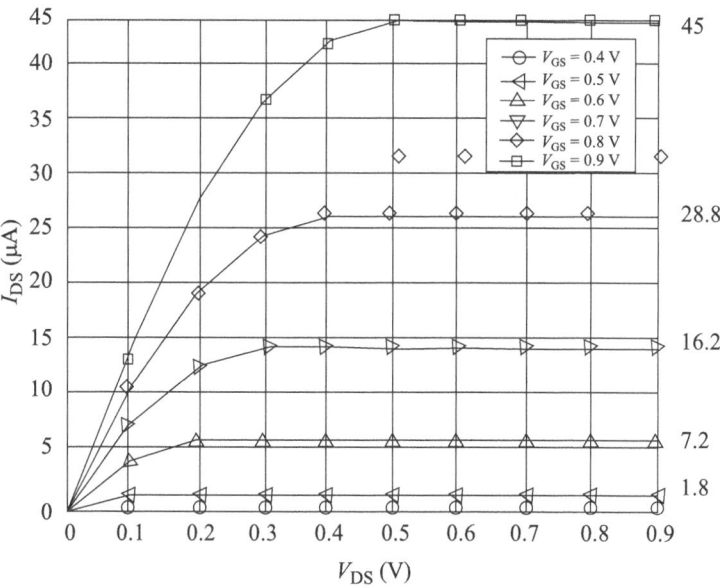

Fig. 2.15

Solution

(a) The values of I_{DS} lie on the V_{DS} axis when $V_{GS} = 0.4$ V. Therefore, the threshold voltage of the MOSFET is 0.4 V.

In the saturation region the drain current is given by

$$I_{D,sat} = \frac{\mu_n C_{ox} W}{2L} \times (V_{GS} - V_t)^2 = \frac{\beta_n}{2} \times (V_{GS} - V_t)^2$$

From the given *I–V* characteristics we find that $V_{GS} = 0.8$ V and $I_{D,sat} = 28.8$ μA. Hence, we can write

$$\beta_n = \frac{2 \times I_{D,sat}}{(V_{GS} - V_t)^2} = \frac{2 \times 28.8 \times 10^{-6}}{(0.8 - 0.4)^2} = 3.6 \times 10^{-4} \text{ A/V}^2$$

Considering another point at $V_{GS} = 0.7$ V and $I_{D,sat} = 16.2$ μA, we obtain

$$\beta_n = \frac{2 \times I_{D,sat}}{(V_{GS} - V_t)^2} = \frac{2 \times 16.2 \times 10^{-6}}{(0.7 - 0.4)^2} = 3.6 \times 10^{-4} \text{ A/V}^2$$

(b) When $V_{GS} = 0.65$ V, the drain current in the saturation region is given by

$$I_{D,sat} = 0.5 \times 3.6 \times 10^{-4} \times (0.65 - 0.4)^2 = 11.25 \times 10^{-6} \text{ A}$$

(c) The condition for saturation region is $V_{DS} \geq (V_{GS} - V_t)$

For $V_{GS} = 0.75$ V, the minimum value of $V_{DS} = V_{GS} - V_{t0} = 0.75 - 0.4 = 0.35$ V.

Example 2.5 An *n*-channel MOSFET has the following *I–V* characteristics (Fig. 2.16). Plot the I_{DS} vs V_{DS} characteristics in the saturation region.

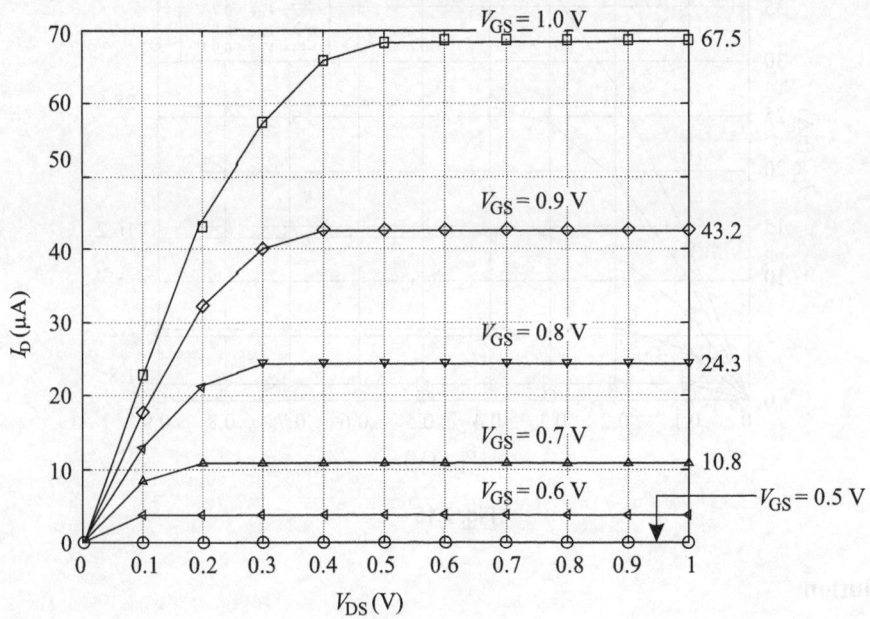

Fig. 2.16

Solution From the *I–V* characteristics we find the saturated drain current values for different values of V_{GS} as given below. As the $I_{DS} = 0$ µA for $V_{GS} = 0.5$ V, therefore, the threshold voltage of the MOSFET is 0.5 V.

V_{GS} (V)	0.5	0.6	0.7	0.8	0.9	1.0
$I_{D,sat}$ (µA)	0.0	2.7	10.8	24.3	43.2	67.5

The value of the parameter β_n can be obtained as

$$\beta_n = \frac{2 \times I_{D,sat}}{(V_{GS} - V_t)^2} = \frac{2 \times 10.8 \times 10^{-6}}{(0.7 - 0.5)^2} = 5.4 \times 10^{-4} \text{ A/V}^2$$

Then the drain current in the saturation region can be written as

$$I_{D,sat} = 0.5 \times 5.4 \times 10^{-4} \times (V_{GS} - 0.5)^2 = 2.7 \times 10^{-4} \times (V_{GS} - 0.5)^2$$

Using the above equation, the values of I_{DS} as a function of $V_{GS} \geq V_{t0}$ are obtained as shown in Fig. 2.17.

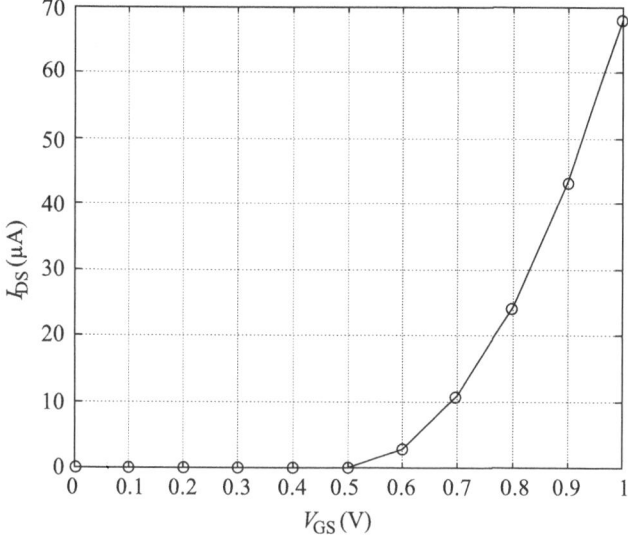

Fig. 2.17

2.9 MOSFET Scaling

The ever-increasing demand of the features of the electronic appliances has forced to put more and more transistors in a small IC chip. The number of devices doubles almost in eighteen months, as per the famous Moore's law. The integration of more devices into a small area has essentially pushed the VLSI process technology to be advanced from micron to submicron, submicron to deep submicron (DSM), and to nanometre regime. As the technology advances, the area occupied by the transistors is just halved per technology node. According to the ITRS roadmap, the shrink factor is almost 0.7 per technology node. When a square is halved, its dimensions are shrunk by $1/\sqrt{2} = 0.707$.

Hence, the reduction of MOSFET dimensions is known as *MOSFET scaling*. MOSFET dimensions are scaled by a scale factor (*S*), which is approximately 1.4 per technology node. Table 2.2 shows different technology nodes and scaling with the year. There are two main scaling techniques defined as follows:

- Constant field scaling or full scaling
- Constant voltage scaling

Table 2.2 Different VLSI technology nodes

Year	Technology node	Shrink factor	Scale factor
1997	350 nm	$250 \div 350 = 0.7142$	1.40
1999	250 nm	$180 \div 250 = 0.72$	1.39
2001	180 nm	$130 \div 180 = 0.7222$	1.38
2003	130 nm	$90 \div 130 = 0.6923$	1.44
2006	90 nm	$65 \div 90 = 0.7222$	1.38
2008	65 nm	$45 \div 65 = 0.6923$	1.44
2010	45 nm	$32 \div 45 = 0.7111$	1.41
2012	32 nm	$22 \div 32 = 0.6875$	1.45

Both the scaling techniques have significant effects on the MOSFET device characteristics. Depending on the target application, suitable scaling techniques are used. Typically, combinations of both scaling are used to migrate from one technology to a newer technology node.

2.9.1 Constant Field Scaling (Full Scaling)

In the constant field scaling, the dimensions of the MOSFET and the terminal voltages are scaled with the same scale factor so that the electric field remains constant. Table 2.3 summarizes the scaling parameters.

Scaling of the above-mentioned parameters has an effect on the MOSFET characteristics as illustrated in Table 2.4.

It is evident that full scaling significantly effects the power dissipation. It reduces the power dissipation by a factor S^2 at the cost of reduction in drain current by a factor of S.

Table 2.3 Full scaling parameters

Parameter	Before scaling	After scaling
Channel length	L	$L' = L/S$
Channel width	W	$W' = W/S$
Gate oxide thickness	t_{ox}	$t'_{ox} = t_{ox}/S$
Junction depth	x_j	$x'_j = x_j/S$
Power supply voltage	V_{DD}	$V'_{DD} = V_{DD}/S$
Threshold voltage	V_t	$V'_t = V_t/S$
Doping concentration	N_A	$N'_A = S \times N_A$
	N_D	$N'_D = S \times N_D$

Table 2.4 Effects of full scaling

Quantity	Before scaling	After scaling
Gate oxide capacitance	C_{ox}	$C'_{ox} = SC_{ox}$
Drain current	I_D	$I'_D = I_D/S$
Power dissipation	P	$P' = P/S^2$
Power density	$P/Area$	$P'/Area' = P/Area$

2.9.2 Constant Voltage Scaling

In the constant voltage scaling, the dimensions of the MOSFET are scaled but the terminal voltages are kept constant. This type of scaling increases the electric field. Table 2.5 summarizes the scaling parameters.

This type of scaling is used where the signal levels of the IC cannot be scaled to match the signal requirements by other chips in the printed circuit board where it is used. Otherwise, it will require level shifting buffers at each I/O pads of the chip and hence, complicate the design. Scaling of the above-mentioned parameters has effect on the MOSFET characteristics as illustrated in Table 2.6.

Table 2.5 Constant voltage scaling parameters

Parameter	Before scaling	After scaling
Channel length	L	$L' = L/S$
Channel width	W	$W' = W/S$
Gate oxide thickness	t_{ox}	$t'_{ox} = t_{ox}/S$
Junction depth	x_j	$x'_j = x_j/S$
Power supply voltage	V_{DD}	$V'_{DD} = V_{DD}$
Threshold voltage	V_t	$V'_t = V_t$
Doping concentration	N_A	$N'_A = S^2 \times N_A$
	N_D	$N'_D = S^2 \times N_D$

Table 2.6 Effects of constant voltage scaling

Quantity	Before scaling	After scaling
Gate oxide capacitance	C_{ox}	$C'_{ox} = S \times C_{ox}$
Drain current	I_D	$I'_D = I_D \times S$
Power dissipation	P	$P' = P \times S$
Power density	$P/Area$	$P'/Area' = S^3 \times P/Area$

In the constant voltage scaling, the drain is increased by a factor of S which increases the device speed. However, it has significant effect on the power dissipation. It increases the power dissipation by a factor S and power density by S^3. This large power density could exaggerate the reliability issues.

2.9.3 Advantages and Disadvantages of MOSFET Scaling

Scaling of MOSFET has many advantages and disadvantages. Following are the advantages of scaling:

- More transistors can be integrated per chip; means more capability
- Improvement in speed
  - Due to decrease in channel length L, and hence due to decrease in transit times
- Increase in current
  - Hence improved parasitic capacitance charging time
- Improved 'throughput' of the chip

Following are the disadvantages:

- Short channel effects
- Complex process technology
- Parasitic effects dominate over transistor effects

The MOSFET scaling is ultimately limited to the following reasons:

- Lithography
- Quantum effects
- Oxide tunnelling

2.10   Small Geometry Effects

Due to scaling, the MOSFET dimensions are aggressively shrunk. This has imposed the small geometry effects which introduce potential problems in device characteristics and modelling of these effects. These effects are classified into two types:
- Short channel effects
- Narrow channel effects

2.10.1   Short Channel Effects

A MOSFET with a channel length comparable to the source/drain junction depth is known as short channel MOSFET. These short channel devices suffer from several issues that change the characteristic of the device. These issues are known as *short channel effects* or DSM (deep submicron) issues. Following subsections explain the short channel effects.

Channel Length Modulation

When the MOSFET operates under saturation region, the effective channel length of the MOSFET reduces by an amount ΔL, and the effective channel length is given by Eqn (2.25). The reduction in channel length ΔL is a function of the drain-to-source voltage V_{DS}. For long channel MOSFET, this reduction in channel length is insignificant; and hence it can be ignored. But for short channel MOSFET, this reduction is significant compared to the physical channel length of the device. A MOSFET is called short channel MOSFET if its channel length is comparable to the source/drain junction depth. Using the effective channel length L', the MOSFET current expression under saturation condition can be written as follows:

$$I_{D,sat} = \frac{\mu_n C_{ox} W}{2L'} \times (V_{GS} - V_t)^2$$

$$I_{D,sat} = \frac{\mu_n C_{ox} W}{2(L - \Delta L)} \times (V_{GS} - V_t)^2$$

$$I_{D,sat} = \frac{\mu_n C_{ox} W}{2L} \times (V_{GS} - V_t)^2 \times \frac{L}{L - \Delta L}$$

$$I_{D,sat} = \frac{\mu_n C_{ox} W}{2L} \times (V_{GS} - V_t)^2 \times \left(1 - \frac{\Delta L}{L}\right)^{-1}$$

As $\Delta L \ll L$, we can write

$$I_{D,sat} = \frac{\mu_n C_{ox} W}{2L} \times (V_{GS} - V_t)^2 \times \left(1 + \frac{\Delta L}{L}\right)$$

Since ΔL is a function of V_{DS}, we can write

$$I_{D,sat} = \frac{\mu_n C_{ox} W}{2L} \times (V_{GS} - V_t)^2 \times (1 + \lambda V_{DS}) \qquad (2.44)$$

where λ is an empirical model parameter and is known as channel length modulation coefficient. Due to the channel length modulation, the saturated drain current slowly increases with V_{DS}, rather than being constant.

Threshold Voltage Lowering

In short channel MOSFET, the threshold voltage is reduced due to the increase in drain voltage. This phenomenon is called drain-induced barrier-lowering (DIBL) (Troutman, 1979). While deriving the expression for the threshold voltage of the MOSFET, we have assumed that the entire channel depletion region charge is created by the applied gate voltage. But this is not true when the drain voltage is applied. Near the drain end the depletion region charge is created by the applied drain voltage. While the amount of gate voltage to be applied to create this depletion charge is reduced, the threshold voltage is reduced with the increase in drain-to-source voltage. This phenomenon is present in long channel MOSFET as well, but can be ignored as the amount of depletion charge created by the drain voltage is insignificant compared to the entire depletion region charge. However, this is significant in short channel MOSFET as a significant portion of the depletion charge is created by the drain voltage.

Punch-Through

For large drain voltage, the depletion region can be extended from the drain to the source of a MOSFET. This causes short circuit between the source and drain and thus, normal operation of the MOS device ceases to exist. Due to the short circuit, a sharp increase in current leads to permanent damage to the device. This phenomenon is known as *punch-through*.

Subthreshold Conduction

When the MOSFET is operated in cut-off region, $V_{GS} < V_t$, a small drain current flows through the device. When $V_{GS} = V_t$, the substrate underneath the oxide layer gets completely inverted and a full conduction is possible between the drain and source. But, when $V_{GS} < V_t$, the minority carriers start accumulating near the oxide–semiconductor interface and a channel is partially created between the source and drain. This phenomenon is called *weak inversion*. Under the weak inversion condition, a small current flows through the channel which is known as subthreshold conduction. The subthreshold current can be expressed as

$$I_D = I_0 e^{q(V_{GS} - V_t)/nkT} = I_0 e^{(V_{GS} - V_t)/nV_T} \tag{2.45}$$

where $V_T = \dfrac{kT}{q} = 0.026$ at $T = 300$ K and n is an empirical constant ($n \approx 1.5$).

Subthreshold current is not desirable in digital circuits as this leads to leakage when the MOS device is supposed to be OFF ($V_{GS} < V_t$). Thus, a device quality with respect to subthreshold conduction is measured by measuring the rate of decrease in subthreshold current with V_{GS} below V_t. This is quantified by measuring the slope of the $\log(I_D)$ vs V_{GS} curve. A typical $\log(I_D)$ vs V_{GS} curve is shown in Fig. 2.18.

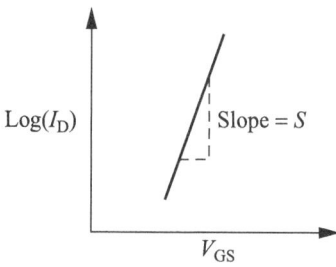

Fig. 2.18 Log (I_D) vs V_{GS} curve

The slope of the log(I_D) vs V_{GS} curve is known as subthreshold slope and is defined as

$$S = \frac{dV_{GS}}{d(\log I_D)} = \frac{dV_{GS}}{d(\ln I_D)} \times \ln 10 \tag{2.46}$$

From Eqn (2.45), we can write

$$\ln I_D = \ln I_0 + \frac{V_{GS}}{nV_T} - \frac{V_t}{nV_T}$$

Differentiating both sides, we can write

$$d(\ln I_D) = \frac{dV_{GS}}{nV_T}$$

$$\frac{d(\ln I_D)}{dV_{GS}} = \frac{1}{nV_T} \tag{2.47}$$

Therefore, subthreshold slope can be written as

$$S = nV_T \times \ln 10 \tag{2.48}$$

Subthreshold slope is expressed in mV/decade. For $n = 1$ (ideal transistor), the subthreshold slope is 60 mV/decade at room temperature. This indicates that 60 mV reduction in V_{GS} leads to decrease in subthreshold current by a factor of 10. Practically n is greater than 1 ($n \approx 1.5$) and hence, subthreshold current drops by a factor of 10 with 90 mV decrease in V_{GS}.

Velocity Saturation

The drift velocity of the carriers increases as the electric field increases. However, this does not happen continually. At some higher electric field, the carrier velocity saturates as illustrated in Fig. 2.19. This phenomenon is known as *velocity saturation*. This velocity of the carriers saturate because at high electric field, the carriers undergo collisions and lose the extra energy gained due to the increased electric field. The electric field beyond which the carriers start to saturate is called critical electric field (E_c). The critical electric field typically varies from 5 V/μm to 7.5 V/μm. If the power supply voltage is 1 V, MOSFET with channel length 0.2 μm or lesser experiences the velocity saturation effect.

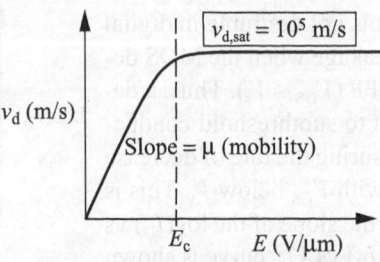

Fig. 2.19 Drift velocity vs electric field

In the short channel MOSFET, Eqn (2.22) overestimates the drain current under velocity saturation. Using the saturated drift velocity in Eqn (2.22), we get

$$I_D = Q_n \times W v_{D,Sat} \qquad (2.49)$$

Hence, the drain current under velocity saturation is given by

$$I_D = C_{ox} V_{D,Sat} \times W \times v_{D,Sat} \qquad (2.50)$$

2.10.2 Narrow Channel Effects

The MOS devices are called narrow channel devices when the channel width is comparable to the maximum depletion region thickness. Similar to the short channel effects, the narrow channel MOSFET also exhibits the following changes in the device characteristics.

Increase in Threshold Voltage

Figure 2.20 shows a narrow channel MOSFET in which there are two thick field oxides (FOX) at both ends of the thin gate oxide layer. Gate electrode partially overlaps the FOX. When gate voltage is applied, relatively shallow depletion regions are formed underneath the FOX. Hence, additional gate voltage is required to support this extra depletion region charge. Thus, the threshold voltage of the MOSFET increases for narrow channel devices.

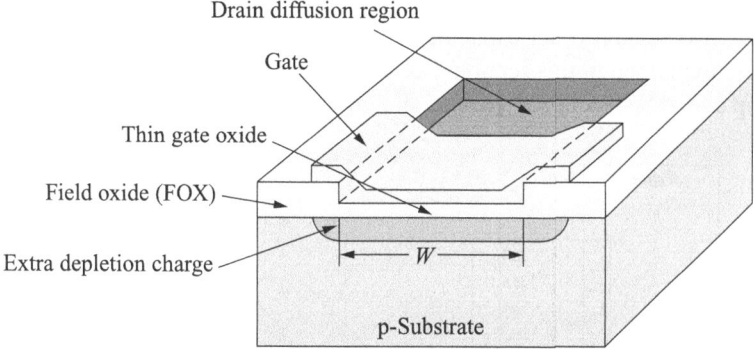

Fig. 2.20 Cross-sectional view of the MOSFET (across the channel) (see Plate 4)

2.10.3 Gate-induced Drain Leakage

When gate voltage is zero, the MOSFET is OFF and hence, the drain current must be zero. But even if $V_{GS} = 0$, for large drain voltage there is a leakage current in the device due to band-to-band tunnelling of carriers. This leakage current is known as *gate-induced drain leakage* (GIDL).

2.10.4 Other Small Geometry Effects

There are several other issues in reducing the MOSFET dimensions. These are discussed in the following text.

Hot Electron Effect

As the technology advances, the MOSFET dimensions are scaled at a faster rate compared to the supply voltage. As a result, the electric field in the channel region increases. At increased horizontal and vertical electric field, electrons and holes gaining high kinetic energy (*hot carriers*) are injected into the gate oxide, and cause permanent changes in the oxide interface charge distribution, degrading the *I–V* characteristics of the MOSFET.

Time-dependent Dielectric Breakdown

The gate oxide thickness of the MOS devices is shrunk aggressively in DSM designs. So the vertical electric field ($E_{ox} = V_{DD}/t_{ox}$) puts excessive electrical stress on the ultra thin gate oxides. When the thin oxide is operated at high electric field over a period of time, depending on how long the device is operated, there is degradation in the gate oxide insulator. Eventually, the gate oxide may breakdown and the MOS device may get completely damaged. This problem is known as *time-dependent dielectric breakdown* (TDDB).

2.11 MOSFET Capacitances

The MOSFET structure is basically a capacitive structure in which the thin gate oxide layer (dielectric) is separated by metal or polysilicon gate electrode at the top and semiconductor substrate at the bottom. The gate oxide capacitance per unit area is expressed as $C_{ox} = \varepsilon_{ox}/t_{ox}$. Not only is there the gate oxide capacitance, but there are also several other capacitances associated with the MOSFET structure. These are explained in this section. Let us first consider a MOSFET structure both from top and cross-sectional view as shown in Fig. 2.21.

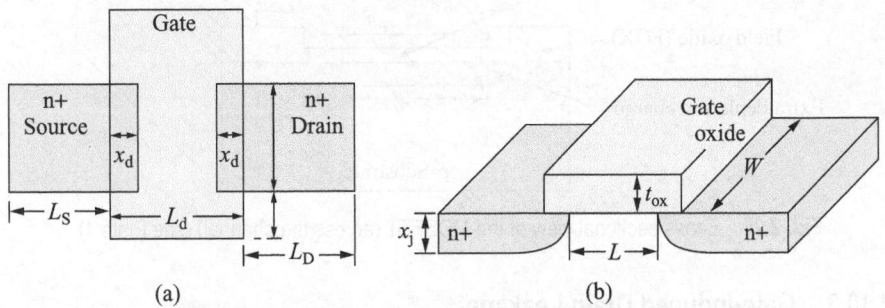

Fig. 2.21 MOSFET: (a) top view; (b) cross-sectional view

The capacitances associated with the MOSFET are not the intended circuit components, but rather, they exist due to the physical structure of the MOSFET. These capacitances are called parasitic device capacitances, and based on their origins, they are broadly classified into two types:

- Oxide-related capacitances
- Junction capacitances

2.11.1   Oxide-related Capacitance

The oxide-related capacitances are purely due to the gate oxide structure and hence, are called gate capacitances. There are three components of the gate capacitance with respect to the other three terminals of the MOSFET:

- Gate-to-source capacitance (C_{GS})
- Gate-to-drain capacitance (C_{gd})
- Gate-to-bulk capacitance (C_{gb})

There are also gate-source and gate-drain overlap capacitances. As shown in Fig. 2.17, the gate terminal has some overlap at the source and drain end. This overlap is required for the device to be fabricated properly, considering the fabrication system mask alignment tolerance. But because of the overlap, there are two capacitances: gate-source overlap capacitance (C_{GSO}) and gate-drain overlap capacitance (C_{GDO}). The value of the overlap capacitance is given by

$$C_{GSO} = C_{GDO} = C_{ox} \times W \times x_d \tag{2.51}$$

where

$$C_{ox} = \frac{\varepsilon_{ox}}{t_{ox}} \tag{2.52}$$

C_{ox} = gate oxide capacitance per unit area, ε_{ox} = dielectric constant of the gate oxide material, t_{ox} = gate oxide thickness, W = channel width, and x_d = gate-source/drain overlap length.

The overlap capacitances are independent of the terminal voltages and are fixed for transistor dimensions.

When the MOS transistor operates under *cut-off region*, there is no channel region. Hence, the gate-to-source and gate-to-drain capacitances are zero, i.e., $C_{GS} = C_{gd} = 0$. Hence, the gate capacitance is entirely determined by the gate-to-bulk capacitance as given by

$$C_{gb} = C_{ox} \times W \times L \tag{2.53}$$

In the *linear region* of operation, the channel is formed, and it shields the bulk from the gate. Hence, the gate-to-bulk capacitance is zero, i.e., $C_{gb} = 0$. The gate-to-channel capacitance is shared equally by the gate-to-source and gate-to-drain capacitances. Therefore, we can write

$$C_{gs} = C_{gd} \cong \frac{1}{2} W \times L \times C_{ox} \tag{2.54}$$

When the MOS transistor operates under *saturation region*, the channel is pinched off at the drain end. Hence, the gate-to-drain capacitance is zero ($C_{gd} = 0$). Also the gate-to-bulk capacitance is zero ($C_{gb} = 0$) under this condition. The gate capacitance is entirely determined by the gate-to-source capacitance, which is given by

$$C_{gs} \cong \frac{2}{3} W \times L \times C_{ox} \tag{2.55}$$

The gate capacitance and its components can be summarized as shown in Table 2.7.

Table 2.7   Gate capacitances of the MOS transistor

Region of operation	Cutoff	Linear	Saturation
C_{gb}	$C_{ox} \times W \times L$	0	0
C_{gs}	0	$\frac{1}{2} W \times L \times C_{ox}$	$\frac{2}{3} W \times L \times C_{ox}$
C_{gd}	0	$\frac{1}{2} W \times L \times C_{ox}$	0
C_G (overlap)	$C_{ox} \times W \times xd$	$C_{ox} \times W \times x_d$	$C_{ox} \times W \times x_d$
C_{GS} (total)	$C_{ox} \times W \times xd$	$\frac{1}{2} C_{ox} \times W \times L + C_{ox} \times W \times x_d$	$\frac{2}{3} C_{ox} \times W \times L + C_{ox} \times W \times x_d$
C_{GS} (total)	$C_{ox} \times W \times L + 2 \times C_{ox} \times W \times x_d$	$C_{ox} \times W \times L + 2 \times C_{ox} \times W \times x_d$	$\frac{2}{3} C_{ox} \times W \times L + 2 \times C_{ox} \times W \times x_d$

2.11.2   Junction Capacitances

The junction capacitances are also called *diffusion capacitances*. In the MOS transistor, there are two PN junctions between the source and bulk, and the drain and bulk. These PN junctions have depletion layer capacitances. The depletion layer capacitance again has two components: one is due to the bottom wall and other is due to the side wall, as illustrated in Fig. 2.22.

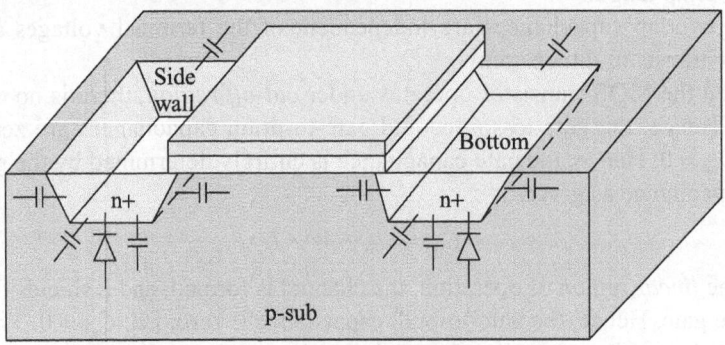

Fig. 2.22   MOSFET junction capacitances (see Plate 4)

The vertical junctions form capacitances with bottom plate of the source and the drain regions which is given by

$$C_{jb} = C_j \times W \times L_S = C_j \times W \times L_D \qquad (2.56)$$

where C_j is the vertical junction capacitance per unit area.

The lateral junctions also form capacitances with side walls of the source and drain regions which is given by

$$C_{jsw} = C_{js}x_j \times (W + 2 \times L_S) = C_{js}x_j \times (W + 2 \times L_D) \qquad (2.57)$$

where C_{js} is the side-wall junction capacitance per unit length. Hence, the total junction capacitance is the sum of the vertical junction capacitance and the side-wall junction capacitance. It can be written as follows:

$$C_{\text{S-junc}} = C_{\text{S-diff}} = C_{jb} \times \text{AREA} + C_{jsw} \times \text{Perimeter}$$
$$= C_{jb} \times W \times L_S + C_{jsw} \times (2 \times L_S + W)$$

(2.58)

Equation (2.58) gives expression for the junction capacitance at the source end. Similar expression can be written for the junction capacitance at the drain end just by replacing L_S with L_D. Note that the side wall at the gate side is not considered for side-wall junction capacitance as this side is not a PN junction, but rather a conducting channel.

2.11.3 Representation of MOSFET Capacitances

The four terminal MOS transistor capacitance model is shown in Fig. 2.23.

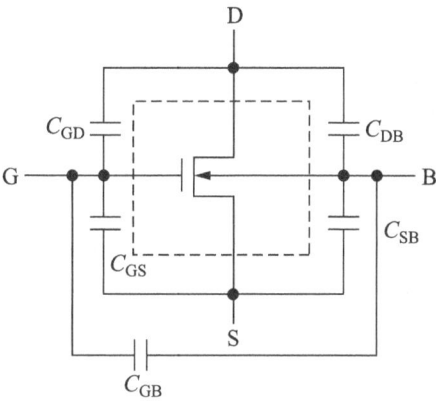

Fig. 2.23 MOS transistor capacitances

The gate and the junction capacitances are combined to form lumped capacitances between the four terminals of the MOS transistors. These capacitances are as follows:

$$C_{GS} = C_{gs} + C_{GSO}$$
$$\acute{C}_{GD} = C_{gd} + C_{GDO}$$
$$C_{GB} = C_{gb}$$
$$C_{SB} = C_{\text{S-diff}}$$
$$C_{DB} = C_{\text{D-diff}}$$

(2.59)

2.11.4 MOS C–V Characteristics

The C–V characteristic of an ideal MOS capacitor is shown in Fig. 2.24. The total capacitance is due to the gate oxide capacitance C_{ox} when the MOS device is OFF ($V_{GS} < V_t$). When the gate voltage is increased to the threshold voltage, the depletion layer capacitance C_d arises in series with the gate oxide capacitance. Hence, the total capacitance decreases to a value $C_{ox}C_d/(C_{ox} + C_d)$. With the increase in gate voltage, the depletion region thickness continues to increase and the depletion layer capacitance continues to increase. Hence, the total capacitance continues to

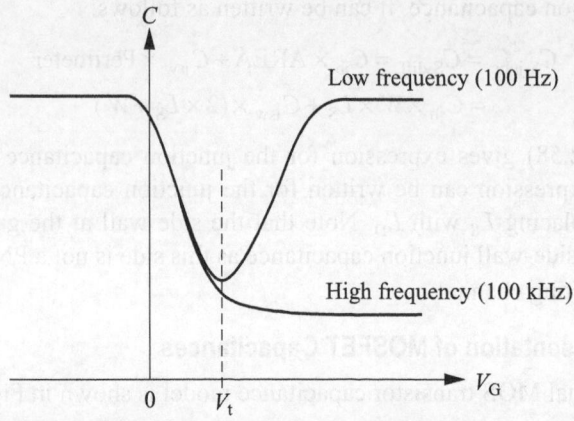

Fig. 2.24 MOS C–V characteristics

decrease. Once the inversion is reached, the depletion layer thickness no longer grows and the capacitance becomes constant. But at low frequencies, the capacitance again increases as gate voltage is increased beyond the threshold voltage. This is because the AC measurement can sample small variations in inversion layer rather than in depletion region. At high frequencies (1 kHz and above), however, the inversion layer charge cannot respond to these frequencies. This is due to the fact that the inversion layer charge establishment is accomplished by the generation–recombination process. Therefore, it is slow in response.

2.12 MOSFET Modelling

Simulation program with integrated circuit emphasis (SPICE) is the most widely used circuit simulation software in the industry as well as in academia. It simulates the MOSFET characteristics and circuits designed using MOS transistors. MOSFET modelling means that extracting several model parameters of the MOS device when fit in the model equations, the characteristics obtained must resemble those obtained by the experiments.

SPICE has built-in MOSFET models. These are as follows:

- Level-1, where the MOS device is described by square law current–voltage characteristics.
- Level-2, where the MOS device is modelled in detail.
- Level-3 is the semi-empirical model.

Both the level-2 and level-3 incorporate the short channel effects and charge-controlled capacitances. Following sections explain the MOSFET models in details.

2.12.1 Level-1 MOSFET Model

Level-1 model is proposed by Shichman and Hodges which is based on the gradual channel approximation (GCA). The model equations are given as follows:

For linear region:

$$I_D = \frac{k'W}{2L_{eff}} \times \left[2(V_{GS} - V_t)V_{DS} - V_{DS}^2 \right] \times (1 + \lambda V_{DS}) \text{ for } V_{GS} \geq V_t \text{ and } V_{DS} < V_{GS} - V_t$$

For saturation region:

$$I_D = \frac{k'W}{2L_{eff}} \times (V_{GS} - V_t)^2 \times (1 + \lambda V_{DS}) \text{ for } V_{GS} \geq V_t \text{ and } V_{DS} \geq V_{GS} - V_t$$

Threshold voltage

$$V_t = V_{t0} + \gamma \left(\sqrt{|-2\varphi_F| + V_{SB}} - \sqrt{|-2\varphi_F|} \right)$$

The effective channel length

$$L_{eff} = L - 2L_D$$

where L_D is the lateral gate overlap over source and drain region.
In Level-1, the model parameters are listed in Table 2.8.

Table 2.8 Level-1 MOSFET model parameters

Symbol	Value	Model parameter	Value
k'	27.6 µA/V^2	KP	27.6µ
V_{t0}	1.0 V	VTO	1
γ	0.53 V1/2	GAMMA	0.53
$2\varphi_F$	−0.58	PHI	0.58
λ	0	LAMBDA	0
μ_n	800 cm^2/V s	UO	800
t_{ox}	100 nm	TOX	100E−9
N_A	10^{15} cm^{-3}	NSUB	1E15
L_D	0.8 µm	LD	0.8E−6

Level-1 model is very much useful for simple simulation problems with a small set of model parameters.

2.12.2 Level-2 MOSFET Model

Level-2 model is more accurate than the level-1 model. It is based on the device geometry and uses detailed device physics to define the model equations. The model equations are given by

Drain current:

$$I_D = \frac{k'}{(1 - \lambda V_{DS})} \times \frac{W}{L_{eff}} \times \left[\left(V_{GS} - V_{FB} - |2\phi_F| - \frac{V_{DS}}{2} \right) V_{DS} \right.$$
$$\left. - \frac{2}{3} \times \gamma \times \left\{ (V_{DS} - V_{BS} + |2\phi_F|)^{3/2} - (-V_{BS} + |2\phi_F|)^{3/2} \right\} \right]$$

Here, V_{FB} is the flat-band voltage.
Saturation voltage:

$$V_{D,sat} = V_{GS} - V_{FB} - |2\phi_F| + \gamma^2 \times \left(1 - \sqrt{1 + \frac{2}{\gamma^2} \times (V_{GS} - V_{FB})} \right)$$

Saturation current:

$$I_D = I_{D,sat} \times \frac{1}{(1 - \lambda V_{DS})}$$

Threshold voltage:

$$V_{t0} = \Phi_{ms} - \frac{qN_{SS}}{C_{ox}} + |2\phi_F| + \gamma \times \sqrt{|2\phi_F|}$$

The transconductance parameter with considering mobility variation with electric field:

$$k'_{(new)} = k' \left(\frac{\varepsilon_{Si}}{\varepsilon_{ox}} \times \frac{t_{ox}\xi_c}{(V_{GS} - V_t - \xi_t \times V_{DS})} \right)^{\xi_c}$$

where
ξ_c = gate-to-channel critical electric field
ξ_t = represents the contribution of drain voltage to the gate-to-channel field (0 < ξ_t <0.5)
ξ_e = a exponential fitting parameter
Effective channel length:

$$L'_{eff} = L_{eff} - \Delta L$$

where

$$\Delta L = \sqrt{\frac{2\varepsilon_{Si}}{qN_A}} \times \left[\frac{V_{DS} - V_{D,sat}}{4} + \sqrt{1 + \left(\frac{V_{DS} - V_{D,sat}}{4} \right)^2} \right]$$

Channel length modulation coefficient:

$$\lambda = \frac{\Delta L}{L_{eff} \times V_{DS}}$$

Subthreshold current:

$$I_D = I_0 e^{q(V_{GS} - V_t)/nkT}$$

2.12.3 Level-3 MOSFET Model

The level-3 model is the semi-empirical model which is based both on analytical and empirical expressions. The drain current in linear region is modelled as follows:

$$I_D = \mu_s \times C_{ox} \times \frac{W}{L_{eff}} \times \left(V_{GS} - V_t - \frac{1 + F_B}{2} \times V_{DS} \right) \times V_{DS}$$

where

$$F_B = \frac{\gamma F_s}{4\sqrt{|2\phi_F| + V_{SB}}} + F_n$$

The effective mobility is

$$\mu_{\text{eff}} = \frac{\mu_{\text{s}}}{1 + \mu_{\text{s}} \times \dfrac{V_{\text{DS}}}{v_{\text{max}} L_{\text{eff}}}}$$

where surface mobility is given by

$$\mu_{\text{s}} = \frac{\mu}{1 + \theta\left(V_{\text{GS}} - V_{\text{t}}\right)}$$

Capacitance models

$$C_{\text{SB}} = \frac{C_{\text{j}} \times \text{AS}}{\left(1 - \dfrac{V_{\text{BS}}}{\phi_0}\right)^{M_{\text{j}}}} + \frac{C_{\text{jsw}} \times \text{PS}}{\left(1 - \dfrac{V_{\text{BS}}}{\phi_0}\right)^{M_{\text{jsw}}}}$$

$$C_{\text{DB}} = \frac{C_{\text{j}} \times \text{AD}}{\left(1 - \dfrac{V_{\text{BD}}}{\phi_0}\right)^{M_{\text{j}}}} + \frac{C_{\text{jsw}} \times \text{PD}}{\left(1 - \dfrac{V_{\text{BD}}}{\phi_0}\right)^{M_{\text{jsw}}}}$$

$$C_{\text{jsw}} = \sqrt{10} \times C_{\text{j}} \times x_{\text{j}}$$

2.12.4   Berkley Short Channel IGFET Model

Berkley short channel IGFET model (BSIM) is based on a smaller number of parameters extracted from experimental data. It is accurate and efficient in modelling MOSFET behaviour and is used in almost all microelectronic industries. BSIM3 model which is denoted as level-49 is being used worldwide. A more advanced model, BSIM4 has also been developed and some companies have already started using the BSIM4 model. The SPICE model parameters for MOS transistors are shown in Tables 2.9–2.11.

Table 2.9   SPICE model parameters

Parameter category	Parameter	Description
Control	LEVEL	Level selection
	MOBMOD	Mobility model selection
	CAPMOD	Capacitance model selection
DC	VTHO	Threshold voltage
	K1	Transconductance
	U0	Mobility
	VSAT	Saturation voltage
	RSH	Drain-source diffusion sheet resistance
AC	CGS(D)O	Gate-source/drain overlap capacitance
	CJ	Bulk p-n zero-bias bottom capacitance
	MJ	Bulk p-n bottom grading
	CJSW	Bulk p-n zero-bias perimeter capacitance
	MJSW	Bulk p-n sidewall grading

(contd)

Table 2.9   (*contd*)

Parameter category	Parameter	Description
Process	TOX	Oxide thickness
	XJ	Metallurgical junction depth
	GAMMA	Bulk threshold parameter
	NCH	Channel doping concentration
	NSUB	Substrate doping density
Temperature	TNOM	Nominal temperature
Binning	LMIN	These parameters are used for binning which
	LMAX	is the process of adjusting model parameters
	WMIN	for different values of drawn channel length
	WMAX	and width.

Table 2.10   SPICE transistor parameters

Parameter	Symbol	SPICE name	Unit	Default value
Drawn length	L	L	m	–
Effective width	W	W	m	–
Source area	AREA	AS	m^2	0
Drain area	AREA	AD	m^2	0
Source perimeter	PERIM	PS	m	0
Drain perimeter	PERIM	PD	m	0
Squares of source diffusion		NRS	—	1
Squares of drain diffusion		NRD	—	1

Table 2.11   Other transistor parameters

L	Channel length
W	Channel width
LD	Lateral diffusion length
WD	Lateral diffusion width
VTO	Zero-bias threshold voltage
KP	Transconductance
GAMMA	Bulk threshold parameter
PHI	Surface potential
LAMBDA	Channel-length modulation
RD	Drain ohmic resistance
RS	Source ohmic resistance
RG	Gate ohmic resistance
RB	Bulk ohmic resistance
RDS	Drain-source shunt resistance
RSH	Drain-source diffusion sheet resistance
IS	Bulk p-n saturation current
JS	Bulk p-n saturation/current area
PB	Bulk p-n potential
CBD	Bulk-drain zero-bias p-n capacitance
CBS	Bulk-source zero-bias p-n capacitance
CJ	Bulk p-n zero-bias bottom capacitance

(*contd*)

Table 2.11 *(contd)*

CJSW	Bulk p-n zero-bias perimeter capacitance
MJ	Bulk p-n bottom grading
MJSW	Bulk p-n sidewall grading
FC	Bulk p-n forward-bias
CGSO	Gate-source overlap capacitance
CGDO	Gate-drain overlap capacitance
CGBO	Gate-bulk overlap capacitance
NSUB	Substrate doping density
NSS	Surface-state density
NFS	Fast surface-state density
XJ	Metallurgical junction depth
TPG	Gate material type
TOX	Oxide thickness
UCRIT	Mobility degradation critical
UEXP	Mobility degradation exponent
UTRA	(Not Used) mobility degradation
VMAX	Maximum drift velocity
NEFF	Channel charge coefficient
XQC	Fraction of channel charge
DELTA	Width effect on threshold
THETA	Mobility modulation
ETA	Static feedback
KAPPA	Saturation field factor
KF	Flicker noise coefficient
AF	Flicker noise exponent

SUMMARY

- Conductivity of semiconductors can be modulated by a large extent by application of external excitation such as optical, thermal, or electrical.
- MOS devices are mostly used for IC design and fabrication because the device structure is well suited for VLSI technology, large packing density, most importantly low power requirement.
- An nMOS is turned ON if its gate is connected to high voltage. On the other hand, a pMOS is turned ON if its gate is connected to low voltage.
- MOS device dimensions are reduced by a factor of almost 0.7 per technology node.

SELECT REFERENCES

Allen, P.E. and D.R. Holberg 2010, *CMOS Analog Circuit Design*, Oxford University Press, New Delhi.

Kang, S.M. and Y. Leblebici 2003, *CMOS Digital Integrated Circuits: Analysis and Design*, 3rd ed., Tata McGraw-Hill, New Delhi.

Martin, K. 2004, *Digital Integrated Circuit Design*, Oxford University Press, New Delhi.

Streetman, B.G. 1995, *Solid State Electronics Devices*, 3rd ed., Prentice-Hall of India, New Delhi.

Troutman, R.R. 1979, 'VLSI Limitations from Drain-induced Barrier Lowering', *IEEE Journal of Solid-State Circuits*, vol. SC-14, no. 2, April.

EXERCISES

Fill in the Blanks

1. Subthreshold operation of MOSFET is very much useful in _____.
   (a) biomedical applications   (b) memory
   (c) charge coupled devices   (d) none of these
2. The main advantage of short channel devices is _____.
   (a) its power consumption is low   (b) it has good output characteristics
   (c) it has high speed   (d) it is easy to fabricate
3. The phenomenon in MOSFET like early effect in BJT is _____.
   (a) body effect   (b) hot carrier effect
   (c) channel length modulation   (d) subthreshold conduction
4. Subthreshold operation occurs in _____.
   (a) strong inversion region   (b) weak inversion
   (c) saturation region   (d) cut-off region
5. The ON-resistance of a MOSFET _____.
   (a) linearly increases with V_{GS}
   (b) linearly decreases with V_{GS}
   (c) exponentially increases with V_{GS}
   (d) non-linearly decreases with V_{GS}

Multiple Choice Questions

1. Si is preferred over Ge because
   (a) Si is cheaper   (b) Si band gap is large
   (c) Si technology is matured   (d) All of the above
2. Polysilicon is used for gate in MOSFET because
   (a) it is semi-metal   (b) it has lattice matching with Si
   (c) it is easy to fabricate   (d) none of these
3. The threshold voltage of an enhancement nMOS transistor is
   (a) greater than 0 V   (b) less than 0 V
   (c) equal to 0 V   (d) none of these
4. Main advantage of depletion load nMOS inverter circuit over enhancement-type nMOS load is
   (a) fabrication process is easier
   (b) sharp VTC transitions and better noise margins
   (c) less power dissipation
   (d) none of these
5. Which one is not second order effect?
   (a) body effect   (b) channel length modulation
   (c) subthreshold conduction   (d) hot carrier effect
6. In constant voltage scaling, the doping density
   (a) remains unchanged   (b) increases by a factor s
   (c) increases by a factor s^2 (d) increases by a factor s^3
7. In full scaling, the power dissipation
   (a) decreases by a factor s (b) decreases by a factor s^2
   (c) remain unchanged   (d) increases by a factor s
8. In constant voltage scaling, the power dissipation
   (a) increases by a factor s (b) remain unchanges
   (c) decreases by a factor s^2 (d) decreases by a factor s^3

True or False

1. Body effect is not a second order effect.
2. In full scaling, magnitude of electric field is constant.
3. In full scaling, the power density remains constant.
4. In constant electric field scaling, the oxide capacitance is scaled down by a factor of $1/s$.
5. In constant voltage scaling, the doping density is increased by a factor s^2.

Short-answer Type Questions

1. Discuss the MOS system with C–V characteristic.
2. Draw the structure of a MOSFET and explain the operating principle.
3. Differentiate the enhancement-type and depletion-type MOSFET.
4. What do you mean by threshold voltage in a MOSFET? Derive an expression for the same.
5. Derive the expression for current in a MOSFET.
6. What is channel resistance? Discuss how it varies with the gate-to-source voltage.
7. What is the dynamic resistance of a MOSFET? Derive an expression for the same.
8. Draw the MOSFET I–V characteristic and explain the different regions.
9. What do you mean by MOSFET scaling? What are the different types of scaling techniques?
10. Show how the MOSFET parameters are scaled in full-scaling.
11. Show how the MOSFET parameters are scaled in constant voltage scaling.
12. What are the short channel effects? Discuss them in brief.
13. What do you mean by narrow channel effects? Explain.

Long-answer Type Questions

1. Consider an MOS system with the following parameters:
$$\varphi_{ms} = -0.8 \text{ V}$$
$$t_{ox} = 200 \text{ Å}$$
$$N_A = 1.5 \times 10^{16} \text{ cm}^{-3}$$
$$Q_{ox} = 3.2 \times 10^{-9} \text{ C/cm}^2$$
   Determine the threshold voltage under zero bias at $T = 300$ K. Assume $\varepsilon_{ox} = 3.97\varepsilon_0$ and $\varepsilon_{Si} = 11.7\varepsilon_0$.
2. Calculate the threshold voltage for a polysilicon gate nMOS transistor with the following parameters:
   $N_A = 1 \times 10^{16} \text{ cm}^{-3}$ $t_{ox} = 200 \text{ Å}$
   $N_D = 1.5 \times 10^{19} \text{ cm}^{-3}$ $N_{ox} = 8 \times 10^9 \text{ cm}^{-2}$
3. Calculate the drain current of an nMOS transistor with following parameters:
$$\beta_n = 60 \text{ μA/V}^2$$
$$V_{GS} = 1.0 \text{ V}$$
$$V_{DS} = 1.5 \text{ V}$$
$$V_{t0} = 0.6 \text{ V}$$
4. Calculate the channel resistance of an nMOS transistor with following parameters:
$$\beta_n = 50 \text{ μA/V}^2$$
$$V_{GS} = 1.0 \text{ V}$$
$$t_0 = 0.6 \text{ V}$$

5. An nMOS with the following parameters is scaled by a factor 1.4. Calculate how the current will be scaled. Assume full scaling.

$$\beta_n = 60 \, \mu A/V^2$$
$$V_{GS} = 1.0 \, V$$
$$V_{DS} = 1.5 \, V$$
$$V_{t0} = 0.6 \, V$$

6. An nMOS with the following parameters is scaled by a factor 1.4. Calculate how the current will be scaled. Assume constant voltage scaling.

$$\beta_n = 50 \, \mu A/V^2$$
$$V_{GS} = 1.0 \, V$$
$$V_{DS} = 1.5 \, V$$
$$V_{t0} = 0.6 \, V$$

7. Derive an expression for saturated drain current considering channel length modulation.
8. What do you mean by drain-induced barrier lowering (DIBL)?
9. What are the short channel effects? Discuss them in detail.
10. Show that the subthreshold slope is 60 mV/decade for an ideal transistor at room temperature.
11. Discuss the narrow channel effects in brief.
12. What are the different MOSFET capacitances? Discuss each of them with their origins.
13. Draw and explain the MOS $C-V$ characteristic.
14. Discuss the MOSFET modelling with level-1 model parameters.
15. Discuss the MOSFET modelling with level-2 model parameters.
16. Discuss the BSIM model.
17. To design the width of a MOSFET such that a specified current is induced for a given applied bias. Consider an ideal n-channel MOSFET with parameters $L = 1.25$ μm, $\mu_n = 650 \, cm^2/V \, s$, $C_{ox} = 6.9 \times 10^{-8} \, F/cm^2$ and $V_t = 0.65 \, V$. Design the channel with W such that $I_D(sat) = 4 \, mA$ for $V_{GS} = 5 \, V$.
18. If Si is doped with 10^{23} atoms/m³, what are the electron and hole concentrations in the doped Si at room temperature? Assume that intrinsic carrier concentration, $n_i = 1.5 \times 10^{16} \, m^{-3}$.
19. Find the drain current for an nMOS having a substrate concentration of $N_A = 1.5 \times 10^{23} \, m^{-3}$ with $\mu_n C_{ox} = 180 \, \mu A/V^2$, $W = 5 \, \mu$m, $L = 0.25 \, \mu$m, $V_{GS} = 1.5 \, V$, $V_{tn} = 0.5$ V, $V_{DS} = 1.0 \, V$.
20. Calculate the figure-of-merit (FOM) of an nMOS with following parameters. $W = 2 \, \mu$m, $L = 0.2 \, \mu$m, $t_{ox} = 200 \, \varepsilon_{SiO_2} = 3.9$, $V_{GS} = 1.0 \, V$, $V_{tn} = 0.6 \, V$, $\mu_n = 660 \, cm^2/V \, s$.
21. A chip fabricated with 180 nm process technology has a power density of 1 μW. Calculate the same if the chip is fabricated with 90 nm technology node. Assume constant voltage scaling.
22. What is the sub-threshold slope of an ideal transistor at 100°C?
23. Calculate the saturated drain current of an nMOS considering the velocity saturation effect. Given $W = 1.2 \, \mu$m, $t_{ox} = 100$Å, $\varepsilon_{SiO_2} = 3.9$, $V_{DS,sat} = 1.0 \, V$, $V_{D,Sat} = 10^7 \, cm/s$.

MOS Inverters

3.1 Introduction

An inverter is a basic logic circuit with single input and output. It is also known as the NOT gate. The output of the inverter is the complement of its input. The symbol of an inverter and its truth table are shown in Figs 3.1(a) and (b).

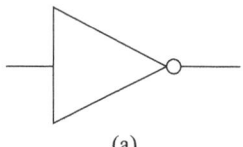

Input	Output
0	1
1	0

(a)   (b)

Fig. 3.1   Inverter: (a) symbol and (b) truth table

3.2 Voltage Transfer Characteristics

The voltage transfer characteristics (VTC) of an inverter represent its output voltage as a function of the input voltage. Figure 3.2 shows the VTC of an ideal inverter.

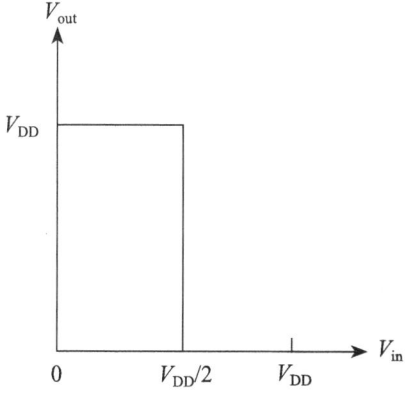

Fig. 3.2   Voltage transfer characteristics of an ideal inverter

In an ideal inverter, the output voltage (V_{out}) is maintained at the maximum value (V_{DD}) when the input voltage (V_{in}) varies from 0 to $V_{DD}/2$, and becomes 0 for input voltages from $V_{DD}/2$ to V_{DD}.

3.2.1 Critical Voltages

We have seen that in an ideal inverter circuit, the input voltage for logic '0' ranges from 0 to $V_{DD}/2$; and the input voltage for logic '1' ranges from $V_{DD}/2$ to V_{DD}. The same is true for the output voltage as well. However, in practical inverter circuits the input voltage for logic '0' ranges from 0 to V_{IL}, and the input voltage for logic '1' ranges from V_{IH} to V_{DD}. Similarly, the output voltage for logic '0' ranges from 0 to V_{OL}, and that for logic '1' ranges from V_{OH} to V_{DD}; where the variables are defined as follows:

V_{IL} = maximum input voltage that can be treated as logic '0'
V_{IH} = minimum input voltage that can be treated as logic '1'
V_{OL} = maximum output voltage that can be treated as logic '0'
V_{OH} = minimum output voltage that can be treated as logic '1'

> **Note**: The first letter in the suffix above indicates input (I) or output (O) voltage and the second letter indicates logic low (L) or high (H). For example, the suffix OL indicates output voltage corresponding to logic low.

The four critical voltage points are illustrated in Fig. 3.3.

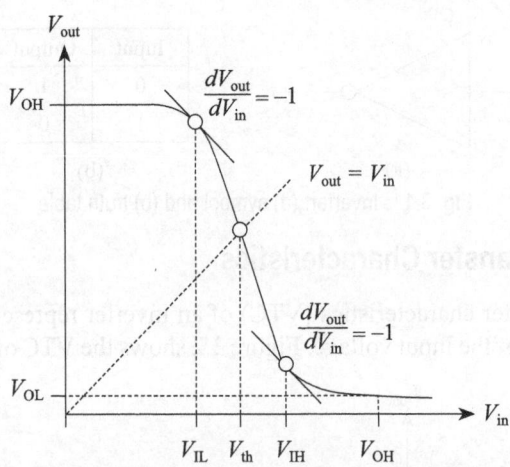

Fig. 3.3 Typical voltage transfer characteristics of a practical inverter

The values of V_{IL} and V_{IH} are determined from the VTC curve, as the input voltages for which the value of $\dfrac{dV_{out}}{dV_{in}} = -1$.

The threshold voltage of the inverter is represented by V_{th} and is determined by the input voltage for which $V_{in} = V_{out}$.

3.2.2 Noise Margins

Practical inverter circuits experience external disturbances leading to extra voltage on the signal lines. Such unwanted additional voltage is termed as *noise*. If the noise in a digital circuit exceeds certain margins, known as *noise margins*, the desired logic levels are changed.

We derive expressions to define noise margins for logic '0' and logic '1' as follows:

In practical inverter circuits, the maximum input low level is V_{IL} and the maximum output low level is V_{OL}. Therefore, if $V_{IL} > V_{OL}$, then there is a margin in the input voltage for logic '0', which can be allowed without causing any change in the output voltage. Hence, the noise margin for logic '0' (NM_L) is defined as follows:

$$NM_L = V_{IL} - V_{OL} \tag{3.1}$$

Similarly, the noise margin for logic '1' (NM_H) is defined as follows:

$$NM_H = V_{OH} - V_{IH} \tag{3.2}$$

Example 3.1   Calculate the noise margin of a digital logic circuit having the following information:

$$V_{IL} = 0.6 \text{ V}, \ V_{IH} = 1.5 \text{ V}, \ V_{OL} = 0.2 \text{ V, and } V_{OH} = 1.8 \text{ V}$$

The power supply voltage is 2.0 V.

Solution   The noise margin for logic '0' is calculated as follows:

$$NM_L = V_{IL} - V_{OL} = 0.6 - 0.2 = 0.4 \text{ V}$$

Similarly, the noise margin for logic '1' is calculated as follows:

$$NM_H = V_{OH} - V_{IH} = 1.8 - 1.5 = 0.3 \text{ V}$$

The results indicate that an input pulse having voltage levels 0.6 V and 1.5 V can produce an output pulse having voltage levels 0.2 V and 1.8 V. Therefore, at the next stage input for logic '0', a noise of 0.4 V can be allowed in the input voltage without affecting the results. Similarly, for logic '1', a noise of 0.3 V can be allowed in the input voltage without affecting the results. _____

3.3 Resistive Load Inverter

A MOS inverter can be designed in several ways though the basic structure remains the same. Figure 3.4 shows the basic structure of a MOS inverter. It consists of an nMOS transistor connected between the output node and the ground, and is controlled by the input voltage applied at its gate terminal. A load is connected between the power supply (V_{DD}) and the output terminal. The nMOS transistor acts as a switch between the output and ground terminals. When the input voltage is low, the nMOS transistor is OFF, and the output is connected to the power supply through the load. The load can either be a resistor or a MOS transistor. On the other hand, when the input is high, the nMOS transistor becomes ON, and the output is connected to ground terminal.

We now analyse the output voltage levels for logic '0' and logic '1':

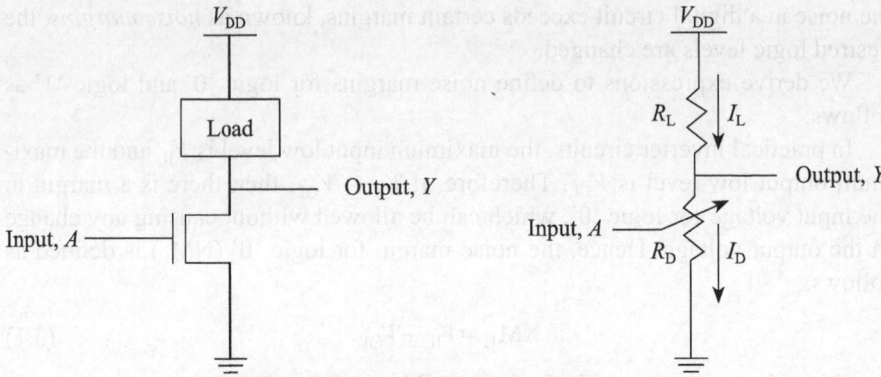

Fig. 3.4 Basic structure of a MOS inverter

Let us consider the load as a resistor of value R_L; and let the resistance of the driver nMOS transistor be R_D. The output voltage is calculated using the following expression:

$$V_{\text{out}} = \frac{R_D}{R_D + R_L} \times V_{DD} \tag{3.3}$$

Since the output is logic high for a logic low input, therefore, ideally, either R_L is zero or R_D is infinite, so that $V_{\text{out}} = V_{DD}$.

On the other hand, for logic high input, the output is logic low. Therefore, either R_L is infinite or R_D is zero, so that $V_{\text{out}} = 0$.

3.3.1 Calculation of V_{OH} and V_{OL}

The output voltage is expressed as follows:

$$V_{\text{out}} = V_{DD} - I_L R_L \tag{3.4}$$

Now, if $V_{\text{in}} = 0$, the nMOS transistor is OFF; therefore, $I_D = I_L = 0$.

Therefore $$V_{OH} = V_{DD} \tag{3.5}$$

When, $V_{\text{in}} =$ high, the nMOS transistor is ON. Therefore, the current through the nMOS is expressed as follows:

$$I_L = I_D \Rightarrow \frac{V_{DD} - V_{OL}}{R_L} = \frac{\beta_n}{2}[2(V_{DD} - V_{tn})V_{OL} - V_{OL}^2] \tag{3.6}$$

Or, $$V_{OL}^2 - V_{OL} 2(V_{DD} - V_{tn} + 1/\beta_n R_L) + 2V_{DD}/\beta_n R_L = 0 \tag{3.7}$$

Or, $V_{OL} = (V_{DD} - V_{tn} + 1/\beta_n R_L) \pm \sqrt{(V_{DD} - V_{tn} + 1/\beta_n R_L)^2 - 2V_{DD}/\beta_n R_L}$ (3.8)

For maximum V_{OL}, we consider only the positive sign in Eqn (3.8). Therefore, the equation is rewritten as follows:

$$V_{OL} = (V_{DD} - V_{tn} + 1/\beta_n R_L) + \sqrt{(V_{DD} - V_{tn} + 1/\beta_n R_L)^2 - 2V_{DD}/\beta_n R_L} \tag{3.9}$$

3.3.2 Calculation of V_{IL} and V_{IH}

By definition, V_{IL} and V_{IH} are input voltages such that $\dfrac{dV_{out}}{dV_{in}} = -1$.

At V_{IL}, the nMOS transistor is in its saturation region as $V_{DS} > V_{GS} - V_{tn}$. Therefore, we have the following equation:

$$\frac{V_{DD} - V_{out}}{R_L} = \frac{\beta_n}{2}(V_{in} - V_{tn})^2 \tag{3.10}$$

Or,

$$-\frac{1}{R_L}\frac{dV_{out}}{dV_{in}} = \beta_n(V_{in} - V_{tn}) \tag{3.11}$$

Or,

$$\frac{1}{R_L} = \beta_n(V_{IL} - V_{tn}) \tag{3.12}$$

Or,

$$V_{IL} = V_{tn} + \frac{1}{\beta_n R_L} \tag{3.13}$$

At V_{IH}, the nMOS transistor is in the linear region. Therefore, we have the following equation:

$$\frac{V_{DD} - V_{out}}{R_L} = \beta_n\left[(V_{in} - V_{tn})V_{out} - \frac{1}{2}V_{out}^2\right] \tag{3.14}$$

Or,

$$-\frac{1}{R_L}\frac{dV_{out}}{dV_{in}} = \beta_n\left[(V_{in} - V_{tn})\frac{dV_{out}}{dV_{in}} + V_{out} - V_{out}\frac{dV_{out}}{dV_{in}}\right] \tag{3.15}$$

i.e.,

$$\frac{1}{R_L} = \beta_n[-(V_{IH} - V_{tn}) + V_{out} + V_{out}] \tag{3.16}$$

Hence,

$$V_{IH} = 2V_{out} + V_{tn} - \frac{1}{\beta_n R_L} \tag{3.17}$$

Now, at $V_{in} = V_{IH}$, V_{out} can be determined as follows:

$$\frac{V_{DD} - V_{out}}{R_L} = \beta_n\left[\left(2V_{out} - \frac{1}{\beta_n R_L}\right)V_{out} - \frac{1}{2}V_{out}^2\right] \tag{3.18}$$

Or,

$$\frac{V_{DD}}{\beta_n R_L} - \frac{V_{out}}{\beta_n R_L} = 2V_{out}^2 - \frac{V_{out}}{\beta_n R_L} - \frac{1}{2}V_{out}^2 \tag{3.19}$$

Or,

$$\frac{3}{2}V_{out}^2 = \frac{V_{DD}}{\beta_n R_L} \tag{3.20}$$

Or,
$$V_{\text{out}} = \sqrt{\frac{2}{3}\frac{V_{\text{DD}}}{\beta_{\text{n}}R_{\text{L}}}}$$
(3.21)

Therefore, we arrive at the following expression:

$$V_{\text{IH}} = \sqrt{\frac{8}{3}\frac{V_{\text{DD}}}{\beta_{\text{n}}R_{\text{L}}}} + V_{\text{tn}} - \frac{1}{\beta_{\text{n}}R_{\text{L}}}$$
(3.22)

Example 3.2 Write a SPICE netlist to design a resistive inverter and obtain the transfer characteristics for different values of load resistance.

Solution

```
* Resistive MOS Inverter
.include "D:\backup\CMOS-CELLS\SPICE-MODELS\model.txt"
.param RL=10k
MN Out In 0 0 NMOS L=0.18u W=0.45u
RL VDD Out 'RL'
Vin In 0 DC 0
VDD VDD 0 1.8
.dc lin source Vin 0 1.8 0.05
.step param RL lin 5 10k 110k
.plot V(Out)
.end
```

The output of the above SPICE netlist is shown in Fig. 3.5. We see that as the load resistance increases, the transition region becomes steep.

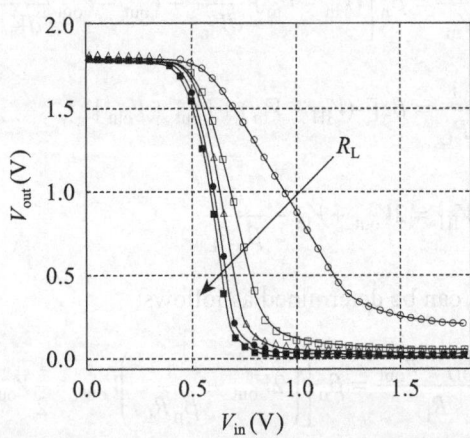

Fig. 3.5 Voltage transfer characteristics of a resistive load inverter

---

Example 3.3 Calculate the critical voltages and noise margins of a resistive load inverter, using the following information: $V_{\text{DD}} = 5.0$ V, $R_{\text{L}} = 100$ kΩ, $\beta_{\text{n}} = 50$ μA/ V^2, $V_{\text{tn}} = 0.5$ V.

Solution   For a resistive load inverter, $V_{OH} = V_{DD} = 5.0$ V. Therefore, we calculate the values of the critical voltages as follows:

$$V_{OL} = V_{DD} - V_{tn} + 1/\beta_n R_L + \sqrt{(V_{DD} - V_{tn} + 1/\beta_n R_L)^2 - 2V_{DD}/\beta_n R_L}$$

$$= \left(5.0 - 0.5 + \frac{1}{50 \times 10^{-6} \times 100 \times 10^3}\right)$$

$$+ \sqrt{(5.0 - 0.5 + \frac{1}{50 \times 10^{-6} \times 100 \times 10^3})^2 - \frac{2 \times 5.0}{50 \times 10^{-6} \times 100 \times 10^3}}$$

$$= (5.0 - 0.5 + 0.2) - \sqrt{(5.0 - 0.5 + 0.2)^2 - 2.0}$$

$$= 4.7 - 4.48 = 0.22 \text{ V}$$

$$V_{IL} = V_{tn} + \frac{1}{\beta_n R_L}$$

$$= 0.5 + \frac{1}{50 \times 10^{-6} \times 100 \times 10^{-3}}$$

$$= 0.5 + 0.2 = 0.7 \text{ V}$$

$$V_{IH} = \sqrt{\frac{8}{3} \frac{V_{DD}}{\beta_n R_L}} + V_{tn} - \frac{1}{\beta_n R_L}$$

$$= \sqrt{\frac{8}{3} \times \frac{5.0}{50 \times 10^{-6} \times 100 \times 10^3}} + 0.5 - \frac{1}{50 \times 10^{-6} \times 100 \times 10^3}$$

$$= \sqrt{\frac{8}{3}} + 0.5 - 0.2$$

$$= 1.63 + 0.5 - 0.2 = 1.33 \text{ V}$$

Similarly, the noise margins are calculated as follows:

$$NM_L = V_{IL} - V_{OL} = 0.7 - 0.22 = 0.48 \text{ V}$$

$$NM_H = V_{OH} - V_{IH} = 5.0 - 1.33 = 3.67 \text{ V}$$

3.4 Enhancement-type nMOS Load Inverter

The disadvantage of a resistive inverter is that it requires a large area on silicon in order to implement the resistive load. Therefore, as an alternative to resistive load, it is possible to design a MOS inverter with an enhancement-type nMOS transistor acting as a load, as shown in Fig. 3.6.

The load nMOS transistor ML can operate both in the saturation region and in the linear region, as explained in following: In Fig. 3.6(a), ML operates in the saturation region as its gate and drain are shorted together and tied to the power supply, V_{DD}. On the other hand, in Fig. 3.6(b), ML operates in the linear region depending on the bias voltage V_{DD1}.

When the input is at logic low, the driver nMOS transistor MD is OFF, and the output is at logic high through the load transistor. On the other hand, when the input is at logic high, the driver nMOS transistor becomes ON, and the output is connected to ground through MD.

The circuit configuration as shown in Fig. 3.6(b) is not generally preferred as it requires dual power supply voltages. Nevertheless, both the circuit configurations have a drawback as they lead to significant static power dissipation.

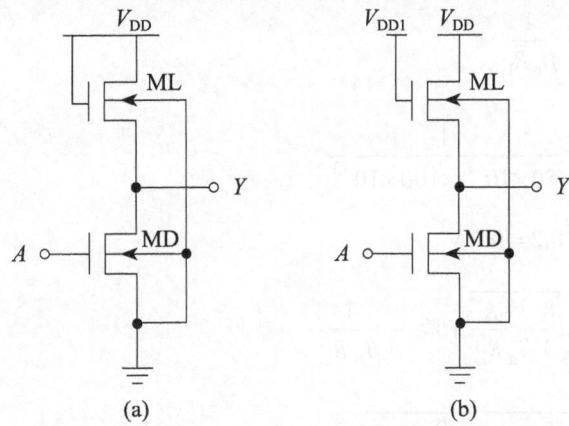

(a)   (b)

Fig. 3.6   MOS inverter with enhancement-type nMOS load

3.5 Depletion-type nMOS Load Inverter

It is also possible to design an MOS inverter with a depletion-type nMOS transistor acting as load, as shown in Fig. 3.7. The depletion-type nMOS transistor has a built-in channel region. Therefore, with zero gate voltage, the nMOS is ON, whereas for large gate voltage, the channel region gets depleted and the nMOS transistor becomes OFF. In Fig. 3.7, we see that ML is always ON as its gate is shorted to the source terminal.

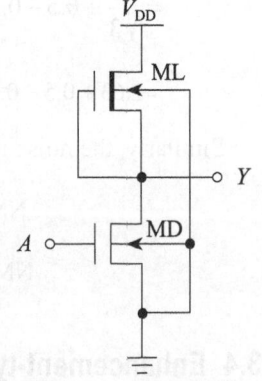

In this type of circuit, when the input is at logic low, the driver nMOS transistor MD is OFF, and the output is at logic high through the load transistor ML. On the other hand, when the input is at logic high, the driver nMOS transistor becomes ON, and the output is connected to the ground through MD.

Fig. 3.7   MOS inverter with depletion-type nMOS load

3.6 CMOS Inverter

The CMOS, or complementary MOS, inverter uses a pMOS transistor in the pull-up network, and nMOS transistor in the pull-down network, as illustrated in Fig. 3.8.

When the input is at logic low, the pMOS transistor is ON, and the nMOS transistor is OFF. Therefore, the output is at logic high. On the other hand, when the input is at logic high, the pMOS transistor is OFF, and the nMOS transistor is ON, and hence the output is at logic low.

The advantage of a CMOS inverter is that as the nMOS and pMOS transistors functionally complement each other, they are never ON together; and hence, there is no direct path from V_{DD} to the ground for static current to flow. Therefore, the static power dissipation is zero, if we neglect the leakage current. Another advantage of the CMOS inverter is that the logic swing is from 0 to V_{DD}, which is not possible in both resistive and nMOS load inverter circuits.

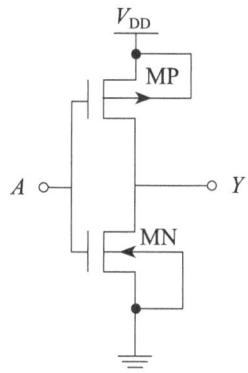

Fig. 3.8 CMOS inverter

3.6.1 Voltage Transfer Characteristics

The VTC of a CMOS inverter are shown in Fig. 3.9. We see that the VTC curve is divided into five different parts, depending on the operating region of the nMOS and pMOS transistors.

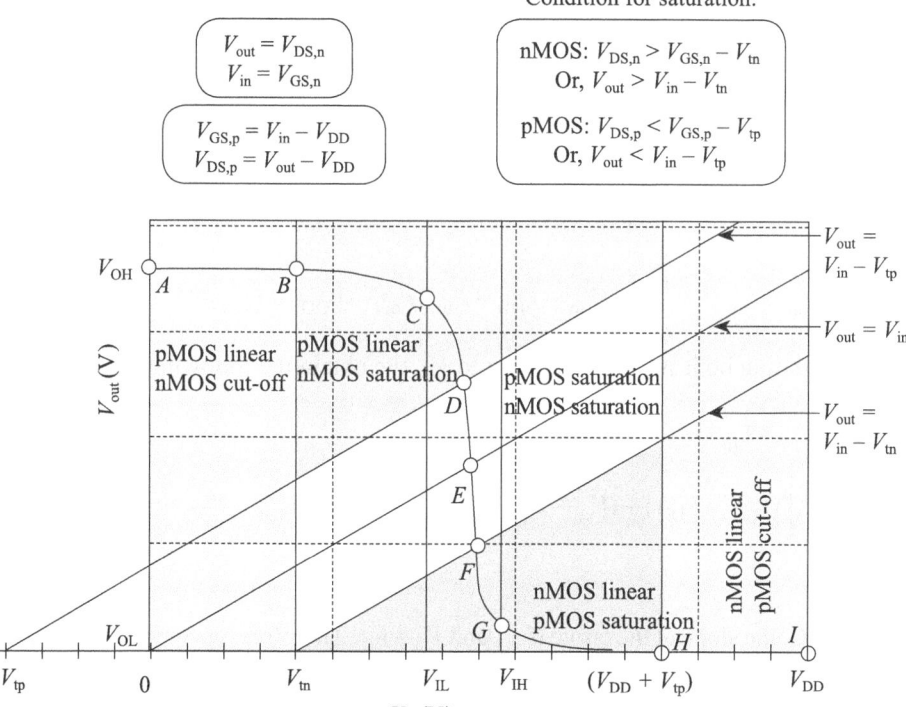

Fig. 3.9 Voltage transfer characteristics of a CMOS inverter

In the region between points A and B, the nMOS is in the cut-off region and the pMOS is in the linear region. In the region between points B and C, the nMOS is in the saturation region and the pMOS is in the linear region.

Further, in Fig. 3.9, for the region above the line defined by $V_{out} = V_{in} - V_{tp}$, the pMOS operates in the linear region. Below this line, the pMOS operates in the saturation region as it satisfies the condition $V_{DS} \leq V_{GS} - V_{tp}$ or $V_{out} \leq V_{in} - V_{tp}$. At point C, the slope of the curve is -1; this point defines the critical point V_{IL}, which is the maximum value of input voltage that can be treated as logic low.

Similarly, in the region between points D and F, both the nMOS and pMOS are in the saturation region; and at point C, the input and output voltages are equal. This point defines the threshold voltage of the inverter (V_{th}).

Likewise, in the region between points F and H, the nMOS is in the linear region and the pMOS is in the saturation region; and at point G, the slope of the curve is -1. This point defines the critical point V_{IH}, which is the minimum value of input voltage that can be treated as logic high.

Moving on in Fig. 3.9, we see that in the region between points H and I, the nMOS is in the linear region and the pMOS is in the cut-off region. In the region above the line defined by $V_{out} = V_{in} - V_{tn}$, the nMOS is in the saturation region as it satisfies the condition $V_{DS} \geq V_{GS} - V_{tn}$ or, $V_{out} \geq V_{in} - V_{tn}$; and below this line, the nMOS operates in the linear region.

We observe that the VTC curve of a CMOS inverter is very sharp compared to the resistive load and nMOS load inverter circuits.

3.6.2 Calculation of V_{IL}

In the region between points B and D, the nMOS operates in saturation and pMOS operates in the linear region. Therefore, we have the following equation:

$$\frac{\beta_n}{2}(V_{GS,n} - V_{tn})^2 = \frac{\beta_p}{2}\left[2(V_{GS,p} - V_{tp})V_{DS,p} - V_{DS,p}^2\right] \qquad (3.23)$$

Or,

$$\frac{\beta_n}{2}(V_{in} - V_{tn})^2 = \frac{\beta_p}{2}\left[2(V_{in} - V_{DD} - V_{tp})(V_{out} - V_{DD}) - (V_{out} - V_{DD})^2\right] \quad (3.24)$$

Differentiating both sides with respect to V_{in}, we obtain the following equation:

$$\beta_n(V_{in} - V_{tn}) =$$
$$\beta_p\left[(V_{out} - V_{DD}) + (V_{in} - V_{DD} - V_{tp})\left(\frac{dV_{out}}{dV_{in}}\right) - (V_{out} - V_{DD})\left(\frac{dV_{out}}{dV_{in}}\right)\right]$$

$$(3.25)$$

At point C, the slope of the curve is -1 and $V_{in} = V_{IL}$. Therefore, we write Eqn (3.25) as follows:

$$\beta_n(V_{IL} - V_{tn}) = \beta_p[(V_{out} - V_{DD}) - (V_{IL} - V_{DD} - V_{tp}) + (V_{out} - V_{DD})] \quad (3.26)$$

Or, $$\beta_n (V_{IL} - V_{tn}) = \beta_p [2V_{out} - V_{DD} - V_{IL} + V_{tp}] \tag{3.27}$$

i.e., $$\left(\frac{\beta_n}{\beta_p} + 1 \right) V_{IL} = 2V_{out} - V_{DD} + V_{tp} + \frac{\beta_n}{\beta_p} V_{tn} \tag{3.28}$$

Or, $$V_{IL} = \frac{2V_{out} - V_{DD} + V_{tp} + \frac{\beta_n}{\beta_p} V_{tn}}{\left(\frac{\beta_n}{\beta_p} + 1 \right)} \tag{3.29}$$

3.6.3 Calculation of V_{IH}

In the region between points F and H, we see that the nMOS operates in saturation and the pMOS operates in the linear region. Therefore, we have the following equation:

$$\frac{\beta_n}{2} [2(V_{GS,n} - V_{tn})V_{DS,n} - V_{DS,n}^2] = \frac{\beta_p}{2} (V_{GS,p} - V_{tp})^2 \tag{3.30}$$

Or, $$\frac{\beta_n}{2} [2(V_{in} - V_{tn})V_{out} - V_{out}^2] = \frac{\beta_p}{2} (V_{in} - V_{DD} - V_{tp})^2 \tag{3.31}$$

Differentiating both sides with respect to V_{in}, we obtain the following equation:

$$\beta_n \left[V_{out} + (V_{in} - V_{tn}) \left(\frac{dV_{out}}{dV_{in}} \right) - V_{out} \left(\frac{dV_{out}}{dV_{in}} \right) \right] = \beta_p (V_{in} - V_{DD} - V_{tp}) \tag{3.32}$$

At point G, the slope of the curve is -1 and $V_{in} = V_{IH}$. Therefore, we rewrite Eqn (3.32) as follows:

$$\beta_n (2V_{out} - V_{IH} + V_{tn}) = \beta_p (V_{IH} - V_{DD} - V_{tp}) \tag{3.33}$$

Or, $$\left(\frac{\beta_n}{\beta_p} + 1 \right) V_{IH} = V_{DD} + V_{tp} + \frac{\beta_n}{\beta_p} (2V_{out} + V_{tn}) \tag{3.34}$$

Or, $$V_{IH} = \frac{V_{DD} + V_{tp} + \frac{\beta_n}{\beta_p} (2V_{out} + V_{tn})}{\frac{\beta_n}{\beta_p} + 1} \tag{3.35}$$

3.6.4 Calculation of V_{OL}

The output voltage of a CMOS inverter is given by the following equation:

$$V_{out} = V_{DS,n} \tag{3.36}$$

When the input is at logic '1' (i.e., $V_{in} > V_{DD} + V_{tp}$), the pMOS transistor is turned OFF, and the nMOS transistor operates in the linear region. Since there is no static current in the circuit, there is no voltage drop across the nMOS transistor. Therefore, we have the following expression:

$$I_{D,n} = \frac{\beta_n}{2}[2(V_{GS,n} - V_{tn})V_{DS,n} - V_{DS,n}^2] = 0 \tag{3.37}$$

Or,

$$2(V_{GS,n} - V_{tn})V_{DS,n} - V_{DS,n}^2 = 0 \tag{3.38}$$

By solving, we get

$$(2V_{GS,n} - 2V_{tn} - V_{DS,n})V_{DS,n} = 0 \tag{3.39}$$

Or,

$$V_{DS,n} = 0 \tag{3.40}$$

Thus, the output voltage is derived as follows:

$$V_{out} = V_{OL} = 0 \tag{3.41}$$

3.6.5 Calculation of V_{OH}

When the input is at logic '0' (i.e., $V_{in} < V_{tn}$), the nMOS transistor is turned OFF, and the pMOS transistor operates in the linear region. As there is no static current in the circuit, there is no voltage drop across the pMOS transistor. Hence, the output voltage is given by the following equation:

$$V_{out} = V_{OH} = V_{DD} \tag{3.42}$$

3.6.6 Calculation of V_{th}

The threshold voltage of a CMOS inverter (V_{th}) is defined when the input and output voltages are equal:

i.e.

$$V_{th} = V_{in} = V_{out} \tag{3.43}$$

In the VTC curve of a CMOS inverter (Fig. 3.9), we see that V_{th} is defined at point E, where both the nMOS and pMOS transistors operate in the saturation region. Therefore, we write the following equation:

$$\frac{\beta_n}{2}(V_{GS,n} - V_{tn})^2 = \frac{\beta_p}{2}(V_{GS,p} - V_{tp})^2 \tag{3.44}$$

Or,

$$\sqrt{\frac{\beta_n}{\beta_p}}(V_{in} - V_{tn}) = \pm(V_{in} - V_{DD} - V_{tp}) \tag{3.45}$$

i.e.,

$$V_{th} = \frac{V_{tn}\sqrt{\dfrac{\beta_n}{\beta_p}} + V_{DD} + V_{tp}}{\sqrt{\dfrac{\beta_n}{\beta_p}} + 1} \tag{3.46}$$

Example 3.4 Calculate the critical voltages and noise margins for a CMOS inverter, given the following values:

$$V_{DD} = 5.0 \text{ V}, V_{tn} = 0.4 \text{ V}, V_{tp} = -0.4 \text{ V}, \beta_n = 50 \text{ μA/V}^2 \text{ and } \beta_p = 20 \text{ μA/V}^2$$

Solution

$$V_{IL} = \frac{2V_{out} - V_{DD} + V_{tp} + \dfrac{\beta_n}{\beta_p} V_{tn}}{\left(\dfrac{\beta_n}{\beta_p} + 1\right)}$$

$$= \frac{2V_{out} - 5.0 + (-0.4) + \dfrac{50}{20} \times 0.4}{\dfrac{50}{20} + 1}$$

$$= \frac{2V_{out} - 5.0 - 0.4 + 1.0}{2.5 + 1} = \frac{2V_{out} - 4.4}{3.5} = 0.57V_{out} - 1.26$$

Using Eqn (3.24), we can write the following equation:

$$\frac{50}{2}(0.57V_{out} - 1.26 - 0.4)^2 =$$

$$\frac{20}{2}[2(0.57V_{out} - 1.26 - 5.0 - (-0.4))(V_{out} - 5.0) - (V_{out} - 5.0)^2]$$

Or,

$$25(0.57V_{out} - 1.66)^2 = 10[2(0.57V_{out} - 5.86)(V_{out} - 5.0) - (V_{out}^2 - 10V_{out} - 25)]$$

Or,

$$25(0.3249V_{out}^2 - 1.8924V_{out} + 2.7556) =$$
$$10[2(0.57V_{out}^2 - 2.85V_{out} - 5.86V_{out} + 29.3) - V_{out}^2 + 10V_{out} + 25]$$

Or,

$$8.1225V_{out}^2 - 47.31V_{out} + 68.89 - 11.4V_{out}^2 + 57V_{out}$$
$$+ 117.2V_{out} - 586 + 10V_{out}^2 - 10V_{out} - 25 = 0$$

Or,

$$6.7225V_{out}^2 + 116.89V_{out} - 542.11 = 0$$

Or,

$$V_{out}^2 + 17.388V_{out} - 80.64 = 0$$

Or,

$$V_{out} = \frac{-17.388 \pm \sqrt{302.34 + 322.56}}{2} = \frac{-17.388 \pm 24.998}{2}$$
$$= 3.805 \text{ V and } -28.193 \text{ V}$$

The acceptable value of V_{out} is 3.805 V, since negative V_{out} is invalid.

Hence, V_{IL} can be calculated as follows:

$$V_{IL} = 0.57V_{out} - 1.26 = 0.57 \times 3.805 - 1.26 = 0.90885 \text{ V}$$

$$V_{IH} = \cfrac{V_{DD} + V_{tp} + \cfrac{\beta_n}{\beta_p}(2V_{out} + V_{tn})}{\cfrac{\beta_n}{\beta_p} + 1}$$

$$= \cfrac{5.0 + (-0.4) + \cfrac{50}{20}(2V_{out} + 0.4)}{\cfrac{50}{20} + 1}$$

$$= \cfrac{5.0 - 0.4 + 5V_{out} + 1.0}{2.5 + 1}$$

$$= \cfrac{5V_{out} + 5.6}{3.5} = 1.43V_{out} + 1.6$$

Using Eqn (3.31), we write the following equation:

$$\frac{\beta_n}{2}[2(V_{in} - V_{tn})V_{out} - V_{out}^2] = \frac{\beta_p}{2}(V_{in} - V_{DD} - V_{tp})^2$$

Or, $$\frac{50}{2}[2(1.43V_{out} - 0.4)V_{out} - V_{out}^2] = \frac{20}{2}(1.43V_{out} + 1.6 - 5.0 - (-0.4))^2$$

Or, $$35.75V_{out}^2 - 10V_{out} - 25V_{out}^2 - 10(2.0449V_{out}^2 - 8.58V_{out} + 9) = 0$$

Or, $$9.699V_{out}^2 - 75.8V_{out} + 90 = 0$$

Or, $$V_{out}^2 - 7.815V_{out} + 9.279 = 0$$

Or, $$V_{out} = \frac{7.815 \pm \sqrt{61.074 - 37.116}}{2} = \frac{7.815 \pm 4.895}{2} = 6.355 \text{ V and } 1.46 \text{ V}$$

The acceptable value of V_{out} is 1.46 V, as the other value (>power supply) is invalid. Therefore, V_{IH} can be calculated as follows:

$$V_{IH} = 1.43V_{out} + 1.6 = 1.43 \times 1.46 + 1.6 = 3.6878 \text{ V}$$

Therefore, the maximum value for logic '0' and minimum value of logic '1' of the output voltage are as follows:

$$V_{OL} = 0 \text{ and } V_{OH} = 5.0 \text{ V}$$

Using these values, the noise margins are calculated as follows:

$$NM_L = V_{IL} - V_{OL} = 0.90885 - 0 = 0.90885 \text{ V}$$

$$NM_H = V_{OH} - V_{IH} = 5 - 3.6878 = 1.3122 \text{ V}$$

3.7 Design of Symmetric Inverter

The threshold voltage for an ideal inverter is given by the following expression:

$$V_{th} = \frac{V_{DD}}{2} \tag{3.47}$$

Using Eqn (3.46), we write the following expression:

$$V_{th} = \frac{V_{tn}\sqrt{\dfrac{\beta_n}{\beta_p}} + V_{DD} - V_{tp}}{\sqrt{\dfrac{\beta_n}{\beta_p}} + 1} = \frac{V_{DD}}{2} \tag{3.48}$$

Or,
$$\left(\sqrt{\frac{\beta_n}{\beta_p}} + 1\right)\frac{V_{DD}}{2} = V_{tn}\sqrt{\frac{\beta_n}{\beta_p}} + V_{DD} - V_{tp} \tag{3.49}$$

Or,
$$\sqrt{\frac{\beta_n}{\beta_p}}\left(\frac{V_{DD}}{2} - V_{tn}\right) = \frac{V_{DD}}{2} - V_{tp} \tag{3.50}$$

Or,
$$\sqrt{\frac{\beta_n}{\beta_p}} = \frac{\dfrac{V_{DD}}{2} - V_{tp}}{\dfrac{V_{DD}}{2} - V_{tn}} \tag{3.51}$$

For a symmetric inverter, when the nMOS and pMOS transistors are fully complementary to each other, we have the following condition:

$$V_{tn} = V_{tp} \tag{3.52}$$

As a result, the above equation can be rewritten as follows:

$$\sqrt{\frac{\beta_n}{\beta_p}} = 1 \tag{3.53}$$

Or,
$$\frac{\beta_n}{\beta_p} = \frac{\mu_n C_{ox}\left(\dfrac{W}{L}\right)_n}{\mu_p C_{ox}\left(\dfrac{W}{L}\right)_p} = 1 \tag{3.54}$$

Or,

$$\frac{\left(\dfrac{W}{L}\right)_{\mathrm{p}}}{\left(\dfrac{W}{L}\right)_{\mathrm{n}}} = \frac{\mu_{\mathrm{n}}}{\mu_{\mathrm{p}}} \neq 1 \tag{3.55}$$

Therefore, for a symmetric inverter, the channel width-to-channel length ratio (*W/L* ratio) for the pMOS and nMOS transistors are not identical. Rather, it depends on the mobility ratio of the electrons and holes. Considering the mobility ratio to be γ, the condition in Eqn (3.55) can be rewritten as follows:

$$\left(\frac{W}{L}\right)_{\mathrm{p}} = \gamma\left(\frac{W}{L}\right)_{\mathrm{n}} \tag{3.56}$$

That is, the channel width-to-channel length ratio (*W/L* ratio) of the pMOS transistor is γ times that of the nMOS transistor. Generally, for a given technology, the pMOS and nMOS transistors have the same channel length. Therefore, for a symmetric inverter, the width of the pMOS transistor is γ times that of the nMOS transistor.

This has an important implication to digital integrated circuit designers—the pMOS transistors occupy a greater area in the chip compared to the nMOS transistors.

3.8 Transient Analysis of CMOS Inverter

So far, we have discussed the static characteristics of a CMOS inverter. Let us now explain the transient characteristics. In order to obtain the transient characteristics, we apply a pulse at the input of the inverter and connect a capacitive load at its output, as illustrated in Fig. 3.10.

The input of the inverter is a pulse (V_{in}) with a finite rise and fall time as illustrated in Fig. 3.11.

We see that when the input pulse makes a 0→1 transition, the nMOS transistor turns ON and the pMOS transistor turns OFF. This causes the dynamic current to flow through the nMOS and pMOS transistors, as illustrated in Fig. 3.11. The output node switches from logic '1' to logic '0' during this transition time. After the transition period, the nMOS transistor remains ON and the pMOS transistor remains OFF. Therefore, the output remains constant at logic '0', and there is no current through either the nMOS transistor or the pMOS transistor.

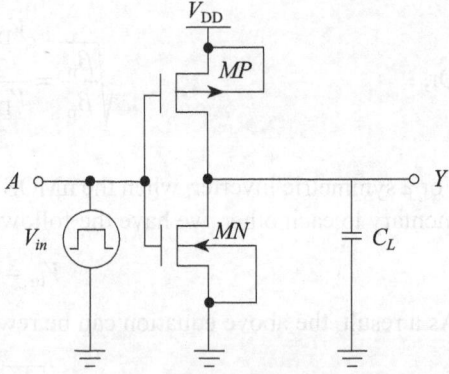

Fig. 3.10 Schematic of a CMOS inverter with a pulse input and capacitive load

After a certain period, the input pulse makes a 1→0 transition. Then the nMOS transistor turns OFF and the pMOS transistor turns ON. This causes the dynamic

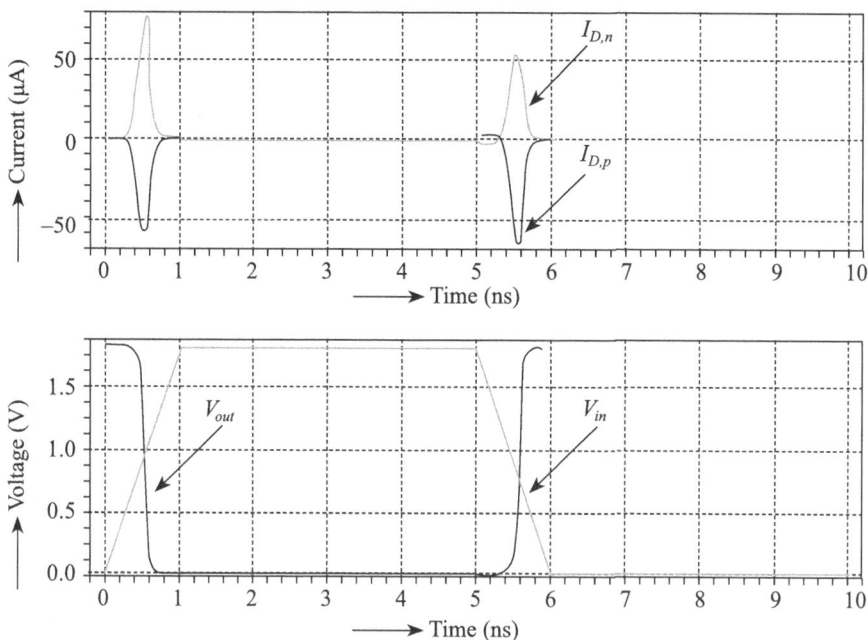

Fig. 3.11 The input/output voltage and nMOS/pMOS drain current waveforms of a CMOS inverter

current to flow again through the nMOS and pMOS transistors as illustrated in Fig. 3.11. The output node switches from logic '0' to logic '1' during this transition time. After the transition period, the nMOS transistor remains OFF and the pMOS transistor remains ON. Therefore, the output remains constant at logic '1' and there is no current either through the nMOS transistor or the pMOS transistor.

The effective load experienced by the inverter when it is used in a digital logic circuit is represented by the load capacitor (C_L). The load capacitor gets discharged through the nMOS transistor during the input 0→1 transition, and gets charged through the pMOS transistor during the input 1→0 transition. The effective load includes the wire capacitance and the input capacitance of the subsequent gate.

When the frequency of the input pulse increases, the rise/fall times also become very sharp. In this condition, the transient characteristics of a CMOS inverter either overshoot or undershoot during the transition period, as illustrated in the output (V_{out}) waveform shown in Fig. 3.12. This is due to the Miller capacitance of the circuit, which comprises the gate-to-drain ($C_{G,D}$) capacitances of the nMOS and pMOS transistors.

3.8.1 Delay Calculation

As explained in the previous section, whenever there is a transition in the input pulse, the output of the CMOS inverter also makes an opposite transition—the transition of the output involves the charging or discharging of the load capacitor. Since a capacitor cannot get charged or discharged instantaneously—it takes some

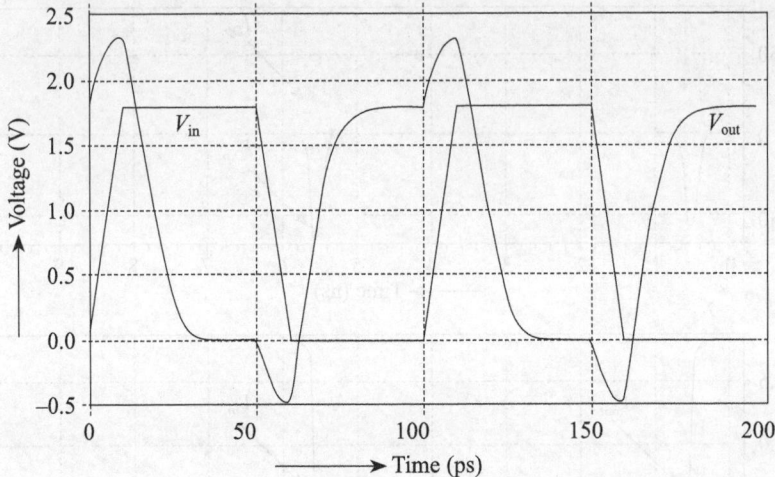

Fig. 3.12 Input and output waveforms of a CMOS inverter for very sharp rise/fall time

time to charge or discharge, the charging and discharging time is determined by the time constant of the circuit.

In case of digital logic circuits, time delay is defined as the time difference between the 50% point of the input waveform and the 50% point of the output waveform, as shown in Fig. 3.13.

There are two time delays between the input and output waveforms: one is due to the delay between the input and output waveforms during the high-to-low transition of the output waveform, termed as TPHL, and the other is due to the delay between input and output waveforms during the low-to-high transition of output waveform, termed as TPLH.

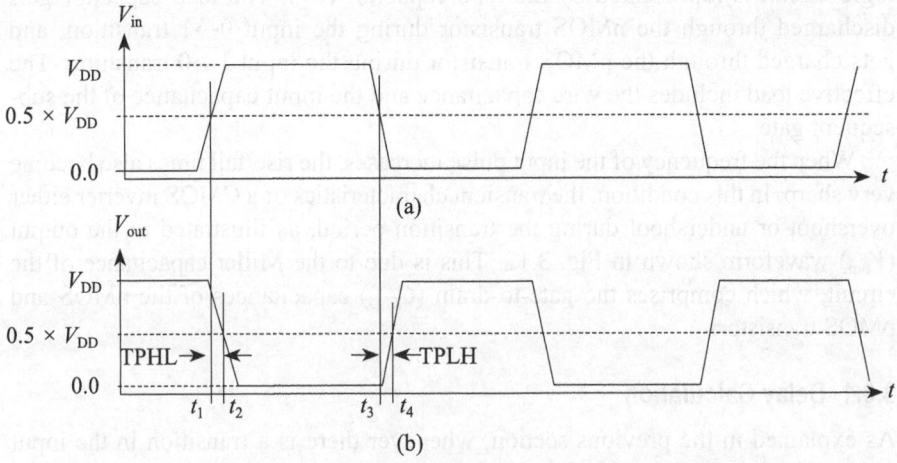

Fig. 3.13 Input and output waveforms of a CMOS inverter: (a) input waveform, (b) output waveform

We see from Fig. 3.13 that TPHL is defined as follows:

$$\text{TPHL} = t_2 - t_1 = (\text{time at which } V_{\text{in}} = 0.5 \times V_{\text{DD}})$$
$$- (\text{time at which } V_{\text{out}} = 0.5 \times V_{\text{DD}}) \tag{3.57}$$

Similarly, TPLH is defined as follows:

$$\text{TPHL} = t_4 - t_3 = (\text{time at which } V_{\text{in}} = 0.5 \times V_{\text{DD}})$$
$$- (\text{time at which } V_{\text{out}} = 0.5 \times V_{\text{DD}}) \tag{3.58}$$

The average of TPHL and TPLH is known as the propagation delay (t_{p}) of the gate and is given by the following expression:

$$t_{\text{p}} = \frac{\text{TPHL} + \text{TPLH}}{2} \tag{3.59}$$

The propagation delay of the gate is determined by the load capacitance and the dimensions of the nMOS and pMOS transistors. A detailed derivation of the propagation delay is provided in Section 10.4.

3.8.2 Calculation of Power Dissipation

In a CMOS inverter, the power dissipation occurs when the load capacitor is charged and discharged. When the load capacitor is charged, it takes some energy from the power supply; part of this energy is stored in the load capacitor and the remainder is dissipated across the pMOS transistor. Next, during the discharge of the load capacitor, the stored energy is discharged through the nMOS transistor.

The average energy dissipated over one switching cycle is calculated as follows:

$$P_{\text{av}} = \frac{1}{T} \int_0^T v(t) i(t)\, dt \tag{3.60}$$

Or,
$$= \frac{1}{T} \left[\int_0^{T/2} V_{\text{out}} \left(-C_L \frac{dV_{\text{out}}}{dt} \right) dt + \int_{T/2}^T (V_{\text{DD}} - V_{\text{out}}) \left(C_L \frac{dV_{\text{out}}}{dt} \right) dt \right] \tag{3.61}$$

Or,
$$P_{\text{av}} = \frac{1}{T} \left[\left(-C_L \frac{V_{\text{out}}^2}{2} \right) \Big|_{V_{\text{DD}}}^0 + \left(V_{\text{DD}} V_{\text{out}} C_L - \frac{1}{2} C_L V_{\text{out}}^2 \right) \Big|_0^{V_{\text{DD}}} \right] \tag{3.62}$$

Or,
$$P_{\text{av}} = \frac{1}{T} C_L V_{\text{DD}}^2 = C_L V_{\text{DD}}^2 f \tag{3.63}$$

Hence, we see that in one switching cycle, the circuit consumes $C_L V_{\text{DD}}^2$ amount of energy, where C_L is the output capacitance and V_{DD} is the power supply voltage.

If the circuit switching frequency is f, then the dynamic power dissipation is determined as follows:

$$P_{av} = C_L V_{DD}^2 f \qquad (3.64)$$

From Eqn (3.64), it is clear that the average switching power (P_{av}) is entirely dependent on the power supply voltage, load capacitance, and the switching frequency; and is independent of the dimensions of the transistors.

Since in digital logic circuits, the switching of the logic gates may not occur at every clock cycle, therefore, in order to take this into consideration, an activity factor is introduced in the expression of the power dissipation. Thus, the power dissipation is now expressed as follows:

$$P_{av} = \alpha C_L V_{DD}^2 f \qquad (3.65)$$

where α ($0 < \alpha < 1$) is the activity factor.

3.9 Power Supply Voltage Scaling in CMOS Inverter

We have seen from Eqn (3.65) that the power dissipation of a CMOS circuit is very sensitive to the power supply voltage as it has square law dependency on it. Therefore, designers always try to reduce the power supply voltage in order to minimize the power dissipation. However, reducing power supply voltage has a profound effect on the performance of the inverter—the propagation delay of the CMOS logic circuits increases with the reduction in power supply voltage.

Theoretically, the minimum value of the power supply voltage should be such that it can turn on the MOS transistors. Therefore, the minimum power supply voltage can be the maximum of the threshold voltages of the nMOS and pMOS transistors, as expressed in the following expression:

$$V_{DD,min} = \max(V_{tn}, |V_{tp}|) \qquad (3.66)$$

This is the minimum value of the power supply voltage required to operate the CMOS inverter; and the CMOS inverter can operate with correct functionality if the power supply voltage is scaled down till the above condition is satisfied.

However, if the power supply voltage is scaled down further, the inverter still demonstrates the VTC characteristics, as illustrated in Fig. 3.14. Even at $V_{DD} = 0.2$ V (= 200 mV), which is less than the threshold voltage of the transistors, it still works as an inverter. The reason for this is that the transistors work in the sub-threshold region, where they can operate even though the operating voltage is below the threshold voltage. However, the VTC curve severely deteriorates at $V_{DD} = 100$ mV. Therefore, the power supply voltage cannot be scaled down further.

At room temperature (300 K), the thermal voltage is approximately 25 mV. Therefore, the thermal noise becomes significant compared to the power supply voltage. Hence, the minimum value of the power supply voltage must be at least two to four times higher than the thermal voltage at room temperature.

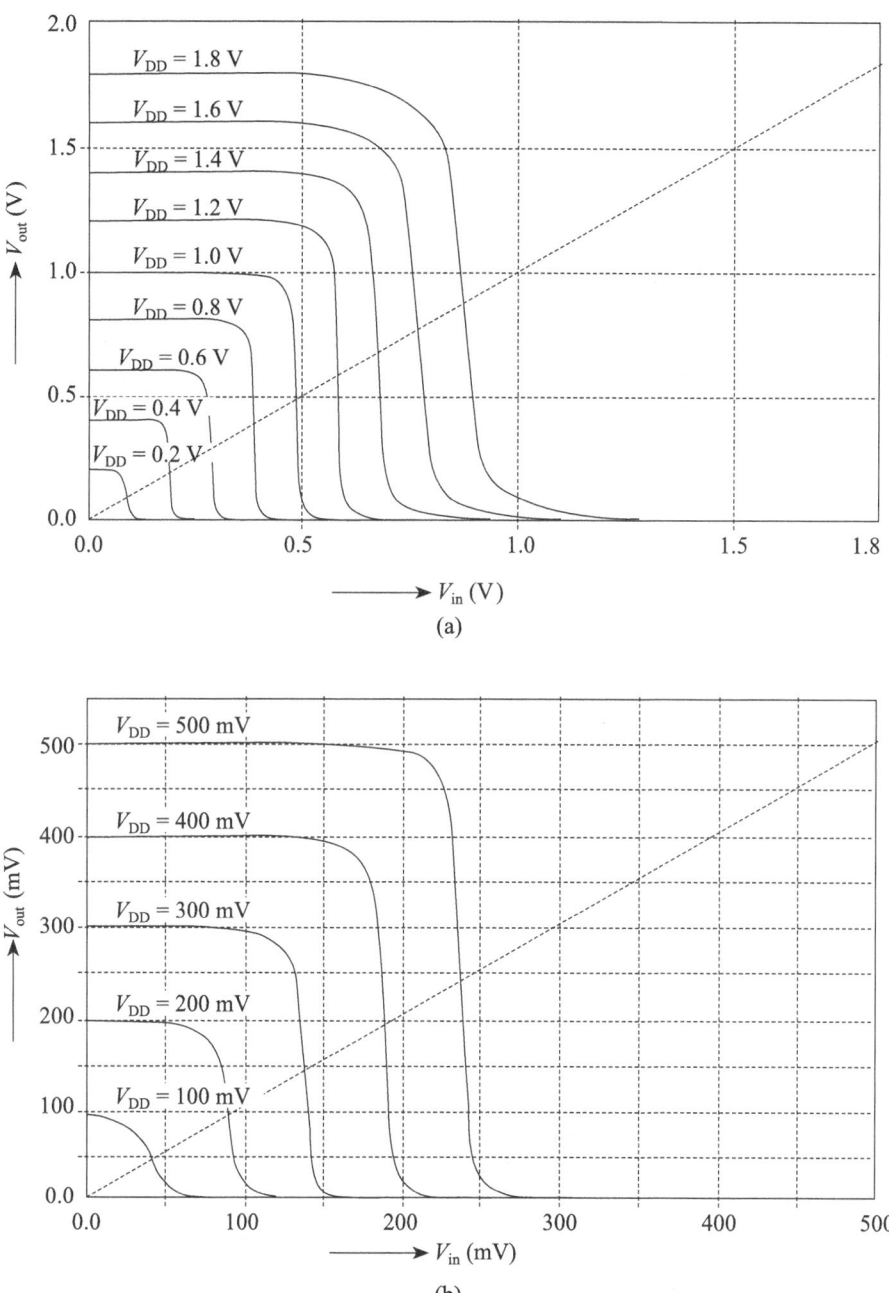

Fig. 3.14   VTC curve for inverter with power supply voltage scaling: (a) V_{DD} = 0.2 V to 1.8 V
(b) V_{DD} = 100 mV to 500 mV

SUMMARY

- A CMOS inverter has minimum static power dissipation, highest swing in the logic levels (from 0 to V_{DD}), and maximum noise margin.
- In a symmetric CMOS inverter, the pMOS occupies more area in a chip than the nMOS transistor.
- The propagation delay of the gate is determined by the size of the transistors and the load capacitance.
- Dynamic power dissipation occurs only during the transition periods and is independent of the size of the transistors.
- The CMOS inverter can exhibit inverting VTC characteristics even when the power supply voltage is scaled down below the threshold voltage of the transistors. Under this condition, the transistors are operated in the sub-threshold region.

SELECT REFERENCES

Sung-Mo, K. and Y. Leblebici, *CMOS Digital Integrated Circuits—Analysis and Design,* 3rd ed., Tata McGraw-Hill Edition, 2003.

Martin, K., *Digital Integrated Circuit Design,* Oxford University Press, 2004.

Rabaey, Jan M., A. Chandrakasan, and B. Nikolic, *Digital Integrated Circuits—A Design Perspective,* 2nd ed., Pearson Education, 2008.

Weste, Neil H. E., D. Harris, and A. Banerjee, *CMOS VLSI Design—A Circuits and Systems Perspective,* 3rd ed., Pearson Education, 2009.

EXERCISES

Fill in the Blanks

1. The CMOS inverter has _____ power dissipation.
2. The nMOS inverter occupies _____ area than the resistive load inverter.
3. The noise margin must be _____ in order to tolerate higher noise.
4. Dynamic power dissipation occurs during _____ activity of the gates.
5. An nMOS transistor is _____ than a pMOS transistor.

Multiple Choice Questions

1. A MOS inverter can be implemented using
   (a) One nMOS transistor and a resistor
   (b) Two pMOS transistors
   (c) One pMOS transistors and a resistor
   (d) Two resistors
2. The threshold voltage (V_{th}) of a CMOS inverter is defined
   (a) When $V_{in} \neq V_{out}$ (b) When $V_{in} = V_{out}$
   (c) When $V_{in} = 0.5V_{out}$ (d) When $V_{in} = 2V_{out}$
3. In order to achieve the same performance as that of the nMOS transistor, a pMOS transistor requires
   (a) More area   (b) Less area
   (c) Same area   (d) None of these
4. Dynamic power dissipation of a CMOS inverter depends on the
   (a) Power supply voltage   (b) Channel width of the nMOS
   (c) Channel width of the pMOS   (d) All of these

5. The VTC curve of a CMOS inverter determines
   (a) Noise margin of the gate   (b) Threshold voltage
   (c) V_{OL}, V_{OH}, V_{IL}, V_{IH} (d) All of these

True or False

1. The dynamic power dissipation of a CMOS inverter does not depend on the load capacitance.
2. The VTC curve of a resistive load inverter becomes steep with increase in resistive load.
3. It is not possible to operate a CMOS inverter with a power supply voltage that is less than the threshold voltage of the nMOS and pMOS transistors.
4. The pMOS transistor acts as pull-up in a CMOS inverter.
5. The propagation delay of the CMOS inverter is independent of the transistor sizes.

Short-answer Type Questions

1. Derive the expression for V_{IL} in a resistive load inverter.
2. What is dynamic power dissipation in CMOS circuits? Does it depend on the power supply voltage? If so, explain how.
3. Show that for a symmetric inverter, the W/L ratio of pMOS and nMOS transistors is equal to the ratio of the mobility of electrons and holes.
4. Explain why a pMOS transistor is used in the pull-up network and the nMOS transistor is used in the pull-down network in CMOS circuits.
5. Define noise margins. Explain how a greater noise margin helps in reducing the effect of external noise in digital circuits.
6. What is Miller effect in transient characteristics of a CMOS inverter? Explain.

Long-answer Type Questions

1. What are the critical voltages of a CMOS inverter? Derive the expressions for the critical voltages of a CMOS inverter.
2. Explain the working principle of a resistive load inverter circuit. Derive the expressions for noise margins of a resistive load inverter.
3. Discuss how the scaling of power supply voltage affects the voltage transfer characteristics of a CMOS inverter. What is the minimum power supply voltage that can be used to operate a CMOS inverter?
4. Derive the expression for dynamic power dissipation in a CMOS inverter. Explain how the switching activity affects the dynamic power dissipation.
5. Write a SPICE netlist for an enhancement-type load nMOS inverter. Calculate the critical voltages of a resistive load inverter. Also calculate the noise margins, if $V_{DD} = 3.0$ V, $R_L = 50$ kΩ, $\beta_n = 60$ μA/V², and $V_{tn} = 0.4$ V.
6. Write a SPICE netlist for a CMOS inverter to obtain its voltage transfer characteristics. How will you perform the transient analysis of a CMOS inverter and calculate propagation delay?
7. Calculate the critical voltages and noise margin for a CMOS inverter, if $V_{DD} = 3.3$ V, $V_{tn} = 0.3$ V, $V_{tp} = -0.3$ V, $\beta_n = 60$ μA/V², and $\beta_p = 20$ μA/V². Also calculate the power dissipation for a load of 0.1 pF and frequency of 100 MHz.

Standard Cell Library Design and Characterization

4.1 Introduction

Integrated circuits (ICs) with few hundreds of components can be designed and the layout can be drawn by skilled custom hardware designers. But as the number of components per chip grow exponentially over time, manual design style is no longer feasible. Electronic design automation (EDA) is needed. The very basic EDA tools are synthesis and place/route tools. Synthesis tools have developed from simple scripting of small logic blocks to the powerful synthesis tools of today. They are capable of synthesizing abstract, high-level coded designs into gate level netlist of predefined logic cells and their interconnections.

The higher level hardware description languages (HDLs) such as Verilog and VHDL are used for describing a gate level netlist. Synthesis is performed by breaking the problem into smaller problems until a level of Boolean functions is reached. The synthesis tool then finds a match between the Boolean expression and the Boolean functions supplied by a cell library consisting of a variety of cells implementing Boolean functions in logic hardware. The synthesis tool uses timing, power, and area specifications from the cell library and optimizes the design according to the constraints specified by the designer. Thus, very large designs are implemented and optimized with a certain level of optimization in a shorter design cycle time.

The automated design flow requires a standard cell library of pre-designed logic circuits implementing a selected range of logic functions, characterized for timing, area, and power. In addition, a model for interconnect wires between internal nodes is required for a timing verification with interconnect. A standard cell library design involves defining the set of logic functions in the cell library, and accurately simulating and modelling of electrical characteristics of logic gates, and exporting all the characterized information in a standard format.

4.2 Standard Cell Library

The standard cell library is a collection of standard logic gates that are used to design a full-chip. The standard cells are much like the biological cells in our body, which are the basic building blocks. The cells provide the basic logic functions, such as, NOT, AND, OR, NAND, NOR, XOR, flip-flop, buffer, etc. Figure 4.1 illustrates the contents of a typical standard cell library. First, a basic set of standard cells is designed for a given process technology node and these cells are characterized to evaluate their properties, such as, area, speed, and power dissipation. These basic cells are packed to form a prototype cell library, and then are used to design a prototype full-chip and requirement criteria are measured. If the requirements are met, the process technology qualifies for production. The prototype cell library is so called because it is a very quickly designed cell library that just contains few basic cells, which are functionally verified, and only timing and power characterization is done. But no other verifications, such as, design rule checking, reliability checking, etc., are not performed. Once the cell library qualifies the customer requirements in terms of area, speed, and power, the cell library is designed to contain all the required cells, characterized and verified.

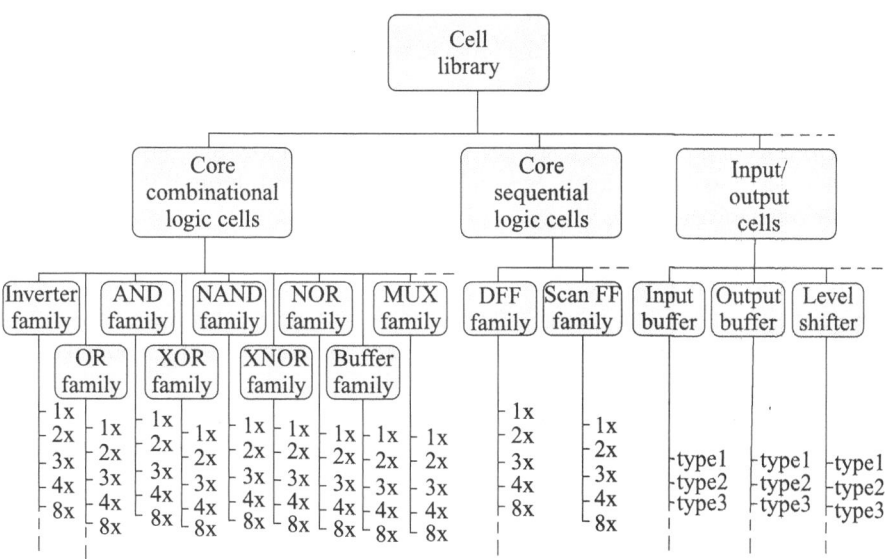

Fig. 4.1 Contents of a typical standard cell library

Typically, a standard cell library contains the following two types of cells:
1. *Standard core cells* These are the basic cells that provide the logical operations, such as, AND, OR, NOT, and so on. These are called core cells as they form the core of the full-chip.

2. *Input/Output (I/O) cells* The I/O cells do not have any functional requirements. They act as the interface between the internal core logic and the outside of the chip. They are sometimes called input/output buffers.

Figure 4.2 shows the block diagram of a typical full-chip design.

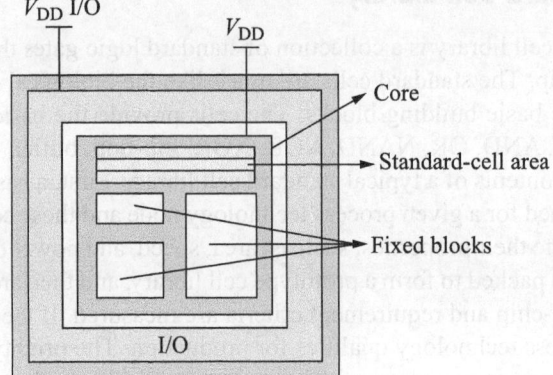

Fig. 4.2   Block diagram of a full-chip

4.2.1   Core Cells

The cells that are used to design the logic of circuit are known as *core cells*. These cells are designed to have the following specific characteristics:
1. They are of fixed height.
2. They have variable width.
3. They are designed to end on the left and right side.
4. Standard cells are placed in rows.
5. Interconnects are routed in channels placed between standard cell rows.
6. Cells are connected to V_{DD} and the ground through horizontal buses.
7. They accommodate Metal2 (second metal layer) feed-through that run vertically through the cells.

4.2.2   Input/Output Cells

The cells that are used to interface the core logic and the external world are called I/O cells. They are placed between the bond pads and core cells at the boundary of the die. They have designed to
- act as interface or buffer between core logic and external signals
- act as voltage level shifter if the core logic and external signals work at different voltage levels
- contain electrostatic discharge (ESD) protection circuit

4.3   Schematic Design

The cells are designed first using a schematic editor. The design steps are as follows:
Step 1:   The circuit components are picked up from symbol browser and placed onto the drawing area.
Step 2:   The components are then connected by drawing wires.
Step 3:   The input and output pins are attached to the input and output nodes.

Step 4: The voltage or current source is connected to the input pins.

Step 5: The capacitor is connected at the output pins to model the load seen by the pin.

Step 6: The SPICE models are inserted or specified.

Step 7: The circuit is then simulated.

Figure 4.3 illustrates the schematic design of the CMOS inverter.

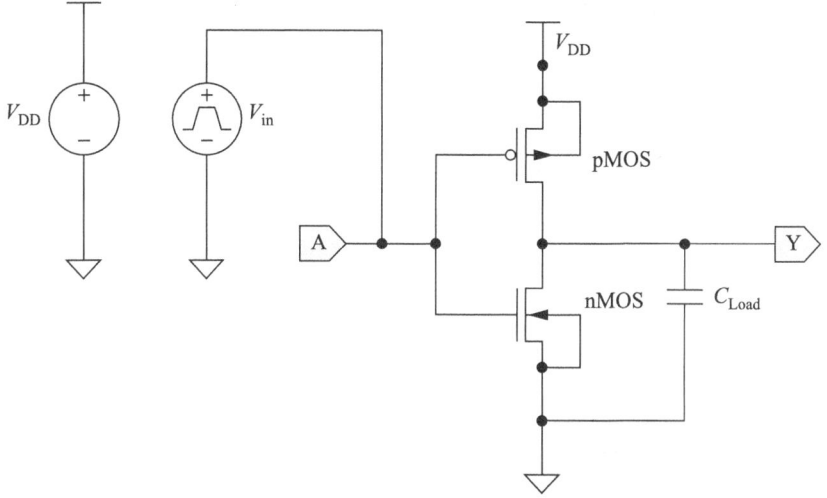

Fig. 4.3   Schematic design of the CMOS inverter

4.4   Layout Design

A layout of an integrated circuit is the footprint of the entire circuit consisting of millions of polygons. Each of the polygons represents some physical component of the devices or circuits. Basically, the layout represents different mask layers required to manufacture the integrated circuit. Different mask layers are drawn using different colours.

To understand the basics of layout drawing, let us consider the cross-sectional view of a CMOS inverter as shown in Fig. 4.4.

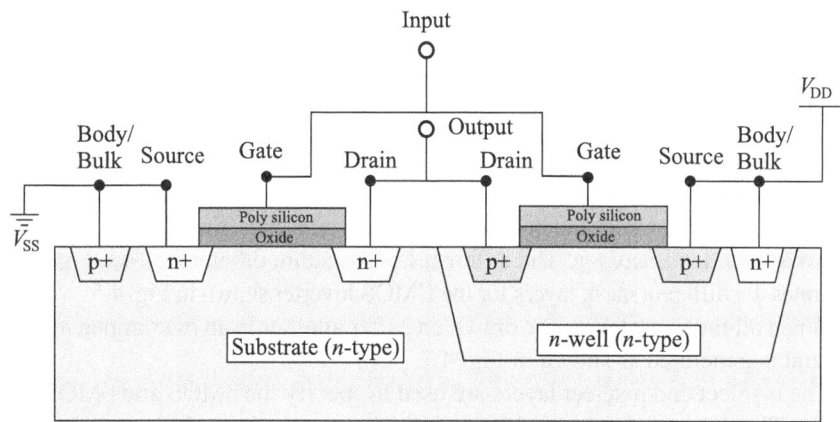

Fig. 4.4   Cross-sectional view of CMOS inverter

The circuit shown in Fig. 4.4 consists of a pMOS and an nMOS transistors connected as illustrated in the schematic shown in Fig. 4.3. There are a number of different device parts as listed below.

- Substrate (the starting material)
- N-well (the region where pMOS is fabricated)
- n^+ diffusion layers (drain and source of nMOS and body/substrate contact region of pMOS)
- p^+ diffusion layers (drain and source of pMOS and body/substrate contact region of nMOS)
- Oxide layer (gate oxide of both pMOS and nMOS)
- Poly silicon layer (instead of metal gate poly silicon is used in both pMOS and nMOS)
- Metal layers (to form the electrical connections)

The three-dimensional view of the CMOS inverter is shown in Fig. 4.5.

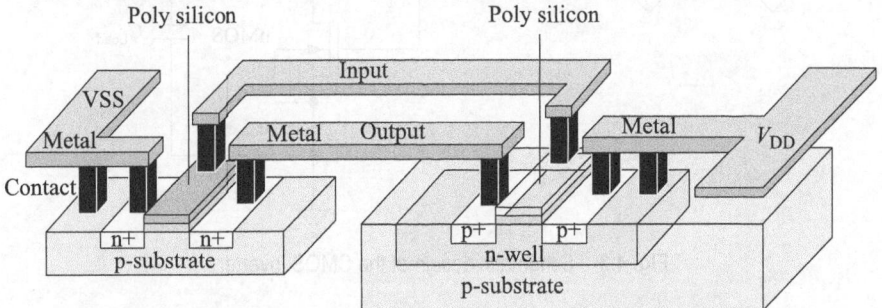

Fig. 4.5 Three-dimensional structural view of a CMOS inverter (see Plate 5)

The fabrication process of a CMOS inverter is described as follows:

Step 1: Create an n-well on the p-substrate to fabricate a pMOS transistor.

Step 2: Create n^+ diffusion regions on the p-substrate to fabricate the nMOS source and drain regions.

Step 3: Create p^+ diffusion regions on the n-well region to fabricate the pMOS source and drain regions.

Step 4: Create poly-Si layers on the oxide layers to fabricate the gate of nMOS and pMOS transistors.

Step 5: Create contact regions required to fabricate the source/drain and substrate contact.

Step 6: Create metal layers to fabricate the interconnect layer.

Each of the six steps requires a photolithography process steps to fabricate the layer. The layers are grown from the bottom, layer by layer. Hence, the VLSI technology is known as *planar technology*. The different layers require different masks. Figure 4.6 illustrates the different mask layers for the CMOS inverter shown in Fig. 4.5.

When all the mask layers are drawn one after another in an overlapped manner, a layout is generated as shown in Fig. 4.7.

The n-select and p-select layers are used to specify the nMOS and pMOS transistors. The active layer is used to specify the source/drain diffusion layers for

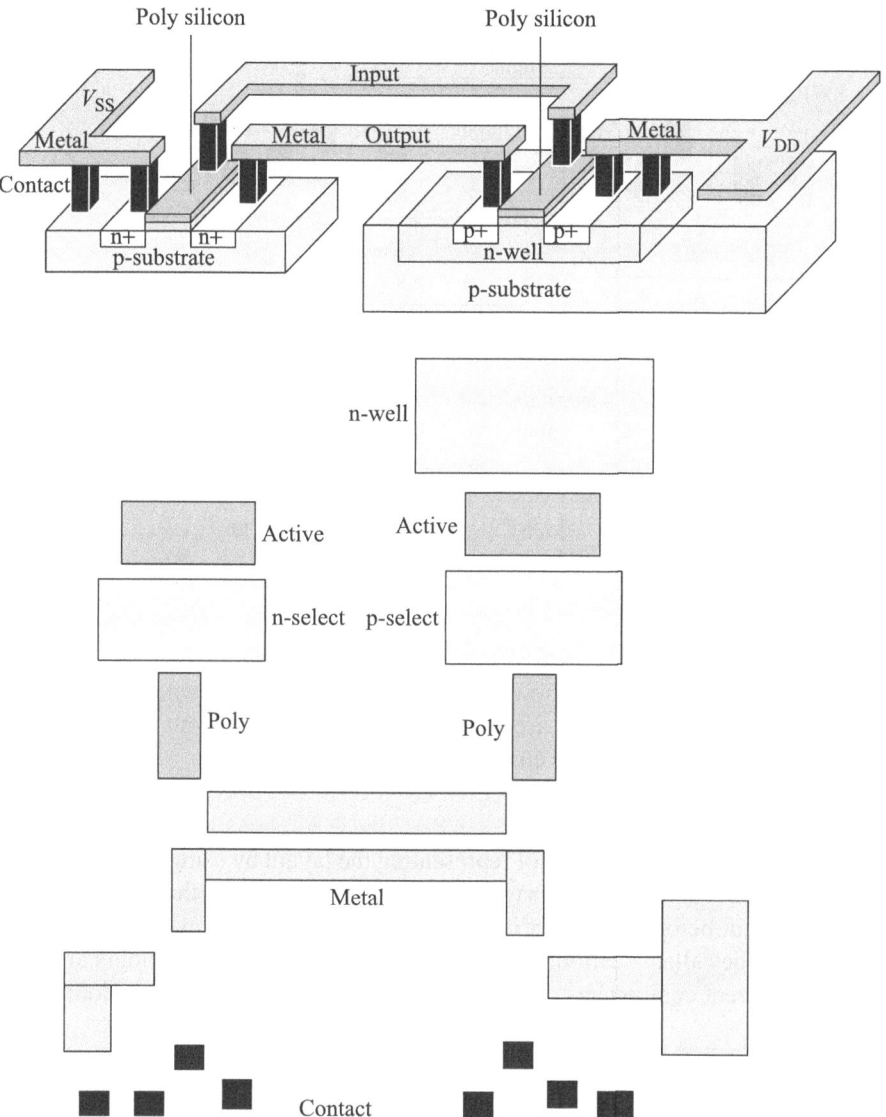

Fig. 4.6 Different mask layers (see Plate 5)

both nMOS and pMOS. When poly overlaps active, it forms a MOS transistor. An n-well layer is drawn on the p-substrate layer to create a pMOS. Contact layers are drawn to specify the contact between poly and metal, and poly and active. A contact between the poly and metal layer is called *poly-contact* and a contact between the active and metal layer is called *active contact*. The metal layers are drawn to specify the interconnect layers.

    ndiff – combination of n-select and active
    pdiff – combination of p-select and active
    nMOS – combination of ndiff and poly
    pMOS – combination of pdiff and poly

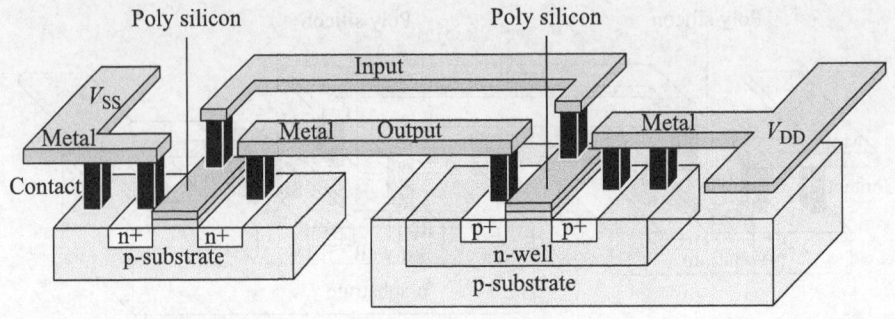

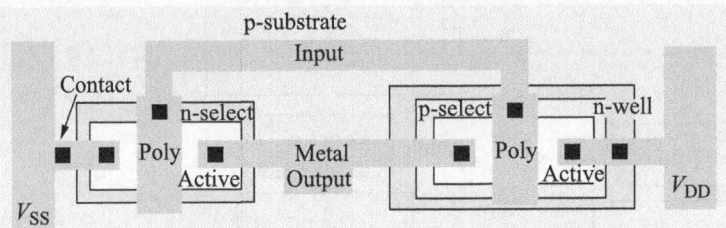

Fig. 4.7 Layout of a CMOS inverter (see Plate 6)

All the drawings that are shown in Figs 4.4–4.7 are not drawn to any scale or using any design rules, as they are just illustrations. But the layout must be drawn using proper scale and with given design rules.

4.4.1 Stick Diagram

A *stick diagram* is a simple way of representing the layout by using thick lines with their interconnections. A stick diagram is useful in estimating the area and planning the layout before the layout is generated within a shorter cycle time. A stick diagram can be called a cartoon of a layout. Lines using different colours are used to draw different components of the layout. Figure 4.8 illustrates the colours used

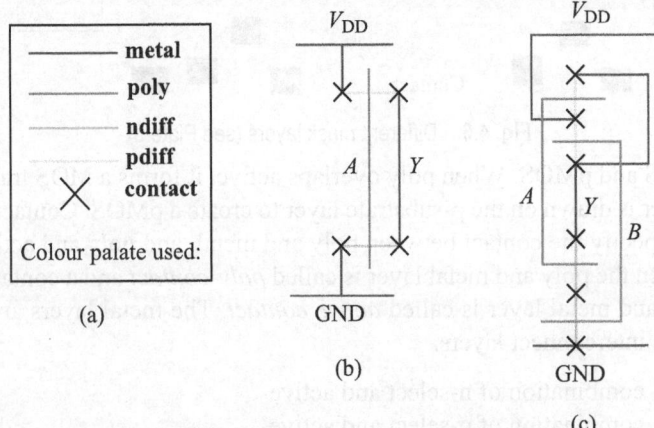

Fig. 4.8 (a) Lines used to draw stick diagram; (b) stick diagram of CMOS inverter; (c) stick diagram of two-input NAND gate (see Plate 6)

for drawing a stick diagram, and the stick diagrams for the CMOS inverter and two-input NAND gate. The active contact is represented by a cross (×) drawn in black.

Example 4.1 Draw a stick diagram for a CMOS circuit implementing three-input NOR function.

Solution The three-input NOR function is given by $F = \overline{A+B+C}$. Its stick diagram can be implemented as shown in Fig. 4.9.

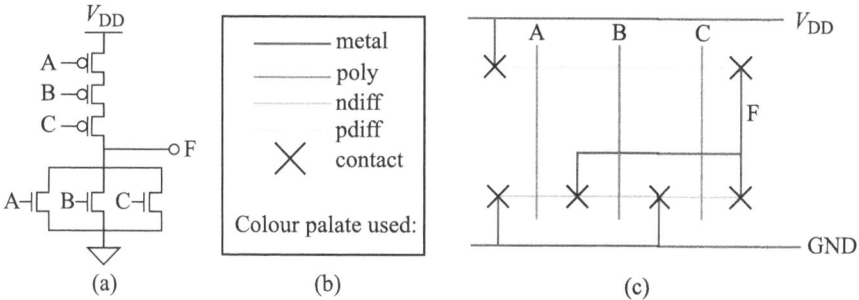

(a)   (b)   (c)

Fig. 4.9 Three-input NOR gate: (a) CMOS logic; (b) colour palate; (c) stick diagram
(see Plate 7)

Example 4.2 Draw a stick diagram for a CMOS circuit implementing function $F = \overline{(A+B+C)D}$.

Solution The stick diagram implementing the given function is shown in Fig. 4.10.

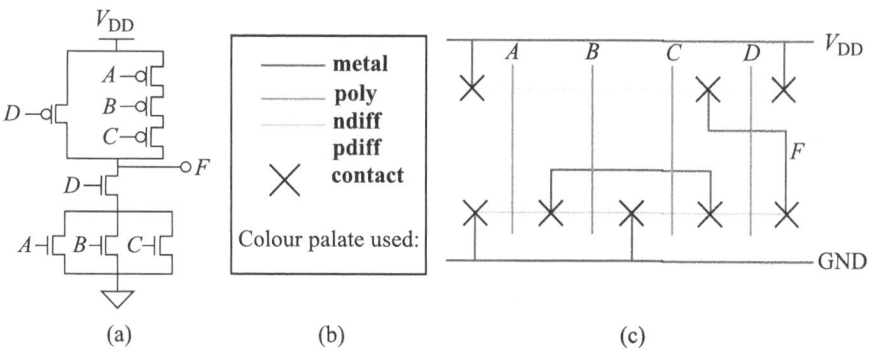

(a)   (b)   (c)

Fig. 4.10 (a) Colour palate used; (b) stick diagram implementing $F = \overline{(A+B+C)D}$
(see Plate 7)

Example 4.3 Draw an alternate stick diagram of CMOS inverter.

Solution The alternate design of stick diagram of CMOS inverter is shown in Fig. 4.11(a).

Example 4.4 Draw the stick diagram of a two-input CMOS NOR gate.

Solution The stick diagram of a two-input CMOS NOR gate is shown in Fig. 4.11(b).

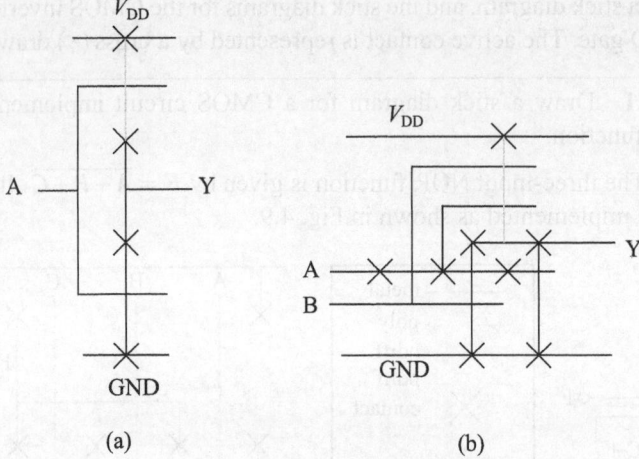

Fig. 4.11   (a) Stick diagram of CMOS inverter and (b) stick
diagram of two-input NOR gate.

4.4.2   Design Rules

The layout must be drawn using a proper scale and a specific set of rules. The set of rules are known as design rules. Typical design rules are: *spacing* between objects, minimum *width*, minimum *overlap*, etc. For example, when a metal line is drawn its width must not be smaller than the minimum width specified. Similarly, when two metal lines are drawn, their spacing should not be lesser than the minimum spacing specified.

- **Micron rule** When the design rules are expressed in absolute values in micron unit.
- **λ rule** Design rules are expressed in terms of a unit λ.

Design rules come from the IC fab due to the limitations of the processing instruments. For example, if we want to drill a hole on a surface, the diameter of the hole is limited to the diameter of the drill bit. Again, while drilling two holes side by side, the gap must exceed certain values; otherwise, the gap material will collapse. Similarly, the photolithography process has inherent limitations on the geometry that is patterned on the silicon surface. These limitations are framed as a rule set which are known as design rules. Design rules are also targeted to achieve a good reasonable yield. As the technology is matured, the design rules become robust. Table 4.1 compares the λ-rule and the micron rule.

The λ rules are first introduced by Mead and Conway in 1980 (Mead and Conway, 1980). Using the concept of scalable λ rules (MOS Implementation Service MOSIS, Information Sciences Institute, University of Southern California, which collects designs from academic, government, and commercial customers, forms one mask to share overhead cost). It has developed a set of λ rules that cater to a variety of manufacturing processes.

4.4.3   MOSIS Design Rules

Table 4.2 shows the design rules specified by MOSIS.

Table 4.1   Comparison between λ-rule and micron rule

λ-rule	Micron rule
Scalable design rules	Absolute design rules
Based on a unit λ where λ is the size of a minimum feature	Based on absolute distances (e.g., 0.5 μm)
Generic for all process technology nodes	Tuned to a specific process technology node
Specifying λ particularizes the scalable rules	Complex rules, especially for deep submicron technology
For each new process technology, λ is reduced, keeping the rules same	For each new process technology, rules must be generated a fresh
Parasitics are generally not specified in λ units	All geometries are expressed using these rules
Layouts are portable—technology migration is easy	Layout are not portable—technology migration is difficult
Key disadvantage is everything does not scale with the same factor	Scaling issue does not arise
These rules are generally conservative as the dimensions are always rounded up to nearest integer that is multiples of λ	These rules are most accurate as all the dimensions are specified in absolute value

Table 4.2   MOSIS design rules

Layer	Dimension	Value
Metal	Minimum width	4λ
Diffusion	Minimum spacing	4λ
Contact	Area	2λ × 2λ
	Surrounded by (on layers above and below)	1λ
	Spacing	3λ
Poly	Minimum width	2λ
	Overlap to diffusion (for transistor)	2λ
	Spacing to diffusion (no transistor)	1λ
	Minimum spacing	3λ
N-well	Surrounds pMOS by	6λ
	Avoids nMOS by	6λ

4.4.4   Micron Rules

Table 4.3 shows the typical micron rules.

4.4.5   Design Rule Checking

The IC layout is the collection of patterns drawn on different layers. These patterns are nothing but polygons. When the IC is manufactured, these patterns are transferred on the silicon wafer. Hence, the correctness of these patterns ultimately decides the quality of the manufactured chips and the yield of manufactured chips. The layouts are generated either manually or by using automated layout generation tools. Once the layout is generated, it must be checked against the layout design rules. For VLSI circuits, the number of rules is large in number, and the number of polygons to be checked is of the order of millions. Hence, there is a

Table 4.3   Micron rules

Layer	Rule	Dimension	Micron
n-well	1.1	Width	3.0
	1.2	Spacing	9.0
Active	2.1	Width	3.0
	2.2	Active to active	3.0
	2.3	n^+ active to n-well	7.0
	2.4	p^+ substrate contact to n-well	4.0
	2.5	n-well to n^+-well tie down	0.0
	2.6	n-well overlap of p^+ active	3.0
Poly 1	3.1	Width	2.0
	3.2	Space	3.0
	3.3	Gate overlap of active	2.0
	3.4	Active overlap of gate	3.0
	3.5	Field poly1 to active	1.0
Poly 2	4.1	Width	3.0
	4.2	Space	3.0
	4.3	Poly1 overlap of poly2	2.0
	4.4	Space to active or well edge	2.0
	4.5	Space to poly1 contact	3.0
Contact	5.1	Contact size	2.0×2.0
	5.2	Spacing	2.0
	5.3	Poly overlap	2.0
	5.4	Active overlap	2.0
	5.5	Poly contact to active edge	3.0
	5.6	Active contact to gate	3.0
Metal 1	6.1	Width	3.0
	6.2	Spacing	3.0
	6.3	Overlap to contact	1.0
	6.4	Overlap of via	2.0
Via	7.1	Space to contact	2.0
	7.2	Size (except for pads)	2.0×2.0
	7.3	Spacing	3.0
Metal 2	8.1	Width	3.0
	8.2	Space	3.0
	8.3	Metal 2 overlap of via	2.0
Pad	9.1	Maximum pad opening	90×90
	9.2	Pad size	100×100
	9.3	Separation	75
P-base	10.1	p-base active to n-well	5.0
	10.2	Collector n^+ active to p-base active	4.0
	10.3	p-base active overlap of n^+/p^+-active	4.0
	10.4	p^+-active to n^+-active	7.0

need of an automated design rule checker (DRC). A DRC is a software program which reads the layout information and the rule set and checks each of the polygons with respect to the design rules. If there is any violation to the rule, it flags violation error and highlights the violated polygon in the layout drawing window. Then the violated polygon is corrected to abide by the design rule and the DRC is run again to verify the layout.

4.5   Layout versus Schematic

Once the layout is generated, it is checked against the design rules. Design rule check ensures that the layout is drawn as per the manufacturing rules and the layout is manufacturable. A layout is called DRC clean if there are no DRC errors in the layout. But a DRC clean layout does not ensure that the layout is drawn according to the schematic or circuit diagram. For example, in the case of the CMOS inverter, if the layout is drawn by connecting V_{DD} to the gate of the pMOS and nMOS transistors and the input pin is connected to the source of the pMOS, the DRC will not show any violation. But this is a serious problem from the circuit design point of view. Hence, layout must be checked against the schematic or circuit diagram. This check is known as layout versus schematic (LVS) check.

4.6   Extraction from Layout

Extraction is a process of extracting the circuit information from the drawn layout. There are two types of extraction performed: (a) device extraction and (b) parasitic extraction. These are as discussed in the following text.

4.6.1   Device Extraction

This is a process of extracting circuit or netlist from the layout. This is required for many purposes. For example, to check for LVS, the circuit must be extracted from the drawn layout and checked against the schematic. The other important applications are power analysis, timing analysis, and signal integrity analysis. From the drawn layout, the devices can be extracted from the geometry and layer information. For example, if there is overlap between poly- and n-diffusion layer, it forms an nMOS transistor. Similarly, if a poly overlaps a p-diffusion layer within an n-well layer, it forms a pMOS transistor. Then, from the metal and contact layer, it can find out the connectivity information.

4.6.2   Parasitic Extraction

Interconnects are metal lines that are used to connect output of a driver to an input of the load. In VLSI circuits, the interconnect lines exhibit parasitic resistance, capacitance, inductance, and conductance. But it has been investigated that the parasitic resistance and capacitance are the most dominating parasitic effects of interconnect. In order to estimate the effect of these parasitic resistance and capacitance, they need to be found out from the geometry information and material properties. Estimating the resistance and capacitance values from the geometry information and material properties is known as *parasitic extraction*.

Parasitic Capacitance Estimation

Parasitic capacitance is estimated on the basis of geometrical information. In general, when two metal plates are separated by a dielectric, it forms a capacitor. But in VLSI interconnects the geometry is not always like a parallel plate capacitor. So we need to consider different capacitive structures as explained in the following sections.

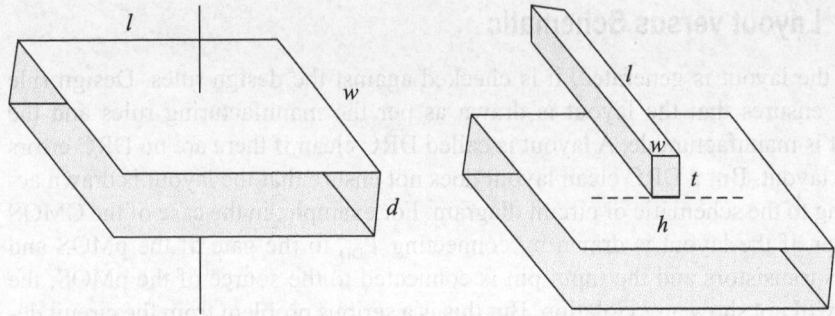

Fig. 4.12 Parallel plate capacitor   **Fig. 4.13** A conductor over a ground plane

Parallel Plate Capacitor   The capacitance of a parallel plate capacitive structure is shown in Fig. 4.12,

$$C = \varepsilon \frac{l \times w}{d}$$

Conductor above a Ground Plane   Consider a conductor over a ground plane as shown in Fig. 4.13. Capacitance for a conductor over a ground plane is given by

$$C = \varepsilon \times \left[1.13 \frac{w}{h} + 1.44 \left(\frac{w}{h} \right)^{0.11} + 1.46 \left(\frac{t}{h} \right)^{0.42} \right]$$

Two Conductors above a Ground Plane   Consider two conductors over a ground plane as shown in Fig. 4.14. Capacitance of the conductors with respect to ground is given by

$$C = \varepsilon \times \left[1.10 \frac{w}{h} + 0.79 \left(\frac{w}{h} \right)^{0.1} + 0.46 \left(\frac{t}{h} \right)^{0.17} \left(1 - 0.87 h^{\left(\frac{-s}{h} \right)} \right) \right]$$

Coupling capacitance between the conductors is given by

$$C_c = \varepsilon \left[\frac{t}{s} + 1.2 \left(\frac{s}{h} \right)^{0.1} \left(\frac{s}{h} + 1.15 \right)^{-2.22} + 0.253 \ln \left(1 + 7.17 \frac{w}{s} \right) \left(\frac{s}{h} + 0.54 \right)^{-0.64} \right]$$

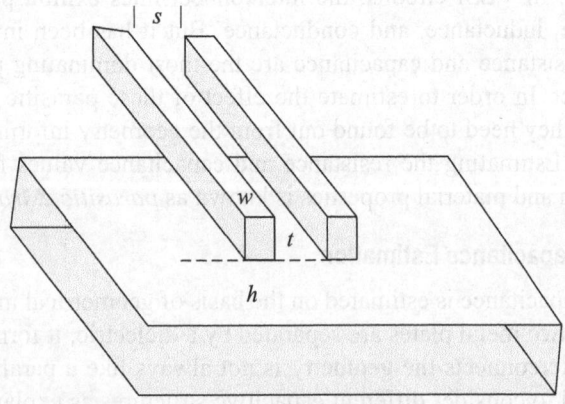

Fig. 4.14 Two conductors over a ground plane

Parasitic Resistance Estimation

The parasitic resistance of a wire is given by

$$R = \rho \frac{l}{wt} = R_s \frac{l}{w}$$

where $R_s = \frac{\rho}{t}$ is called the sheet resistance and is measured in Ω/square.

The sheet resistance is used to estimate the interconnect resistance in a simple way. An interconnect is divided into number of squares and counting the number of squares multiplied by the sheet resistance gives the interconnect resistance. This is illustrated in Fig. 4.15.

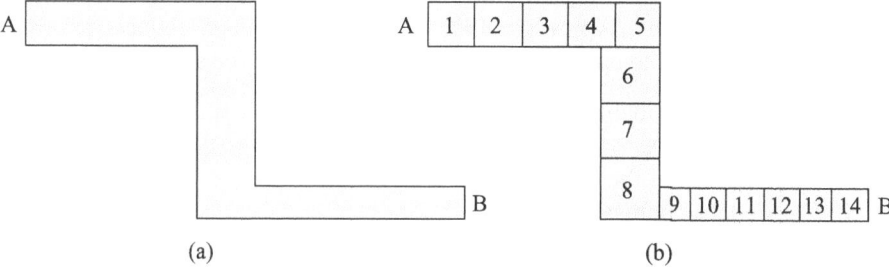

(a)   (b)

Fig. 4.15   (a) A interconnect from A to B; (b) interconnect AB is divided into number of squares

The resistance of the interconnect of length AB is calculated using sheet resistance as:

$$R = 14 \times R_s$$

4.7   Antenna Effect

In the process of VLSI interconnect fabrication, the partially processed metal lines collect static charge in a plasma environment. The amount of collected charge is proportional to the length of the metal lines. Now, if the metal line is connected to the gate of a MOSFET without any path to the diffusion layer, the collected charge may discharge through the gate oxide causing damage to the gate oxide. This effect is known as *antenna effect*.

4.7.1   Antenna Rule

There is a rule which decides what is the maximum length of interconnect that can be connected to a gate of a MOSFET in one metal layer. The antenna rule is given by

$$A_{\text{interconnect}} \leq A \times R_{\text{antenna}}$$

where $A_{\text{interconnect}} = w \times l = $ interconnect surface area

$A = W \times L = $ gate area of MOS device

$R_{\text{antenna}} = $ antenna ratio (typically 100)

If there is an antenna rule violation, there are two ways to solve it:
1. Break the metal line into smaller parts and make some portions of the broken line in the higher metal layer. This is called jumper insertion.
2. Insert an antenna protection diode which serves a low resistance discharge path and protects the gate oxide.

Antenna fixing mechanisms are discussed in detail in Chapter 12.

4.8 Standard Cell Library Data

A standard cell library contains two main components:
1. *Timing information* This is typically in the Synopsys Liberty format, and provides functional definitions, timing, power, and noise information for each cell.
2. *Layout information* Typically in LEF format, which contains reduced information about the cell layouts, required by the automated 'Place and Route' tools.

Other than these two main components, a cell library also contains the following information:
- Complete layout of the cells
- SPICE model or netlist of the cells
- Verilog or VHDL model of the cells
- Reliability models of the cells

4.9 Library Characterization

Library characterization is a process of simulating the designed standard cells and extracting all information about the cells. It also represents these extracted information in a standard format or model so that the chip design flow can access all details about any standard cell.

4.9.1 Design Margin

Normally, the standard cell library is characterized at different design corners. The design corners are a function of three parameters as shown below:
- Supply voltage (V)
- Temperature (T)
- Process technology (P)

The above parameters are together termed as *PTV*. We must design the circuits that will reliably operate over all the extreme *PTV* corners.

4.9.2 Supply Voltage

When a system is designed, it is targeted to operate at specified supply voltage. This *supply voltage* is known as nominal voltage (nom V_{DD}). But the nominal V_{DD} varies due to several reasons as shown below:
- Power supply variations
- IR drops in the power rails
- *Ldi/dt* noise

These problems are practically impossible to eliminate. So the specification of supply voltage includes nominal voltage with a ±10% variation around the nominal voltage. Hence, designers need to take care of the effect of supply variations while designing the circuit. Speed has direct dependency on the power supply voltage. Any variations in the power supply leads to variations in the propagation delay.

4.9.3 Temperature

The increase in junction temperature of the MOS device causes degradation of drain current. The junction temperature is the sum of the ambient temperature and the rise in temperature due to power dissipation in the device. So the device operating temperature varies widely depending on the working environment.

Table 4.4 shows the typical temperature ranges for different applications. A device that is targeted for a particular application needs to ensure it works in the specified temperature range.

Table 4.4 Temperature ranges

Standard	Minimum (°C)	Maximum (°C)
Commercial	0	70
Industrial	−40	85
Military	−55	125

4.9.4 Process Variation

During the manufacturing of devices and interconnects there is always a variation in the lateral dimension, doping concentration, and thickness. The variations are different between wafers, between a die on the same wafer, and across an individual die. These variations are known as *inter-die* and *intra-die* variations. These variations manifest ultimately change in the device parameters, such as:

- Channel length
- Oxide thickness
- Threshold voltage

Similarly, interconnects are also undergoing following changes:

- Line width
- Spacing
- Metal and dielectric thickness
- Contact resistance

The process variations cause the device characteristics to be different within a die or within a wafer. Hence, this effect needs to be taken care of while designing the VLSI circuits so that reliable performance is achieved across all devices fabricated in a batch. Transistors are classified according to the following process variations:

- *Strong (Fast)* A transistor which has least delay.
- *Nominal (Typical)* A transistor which has nominal delay.
- *Weak (Slow)* A transistor which has highest delay.

4.9.5 Design Corners

The design margins are grouped together to decide the ultimate behaviour of the device. The group is categorically divided into three types and each type is known as a design corner. The most commonly used design corners are shown in Fig. 4.16. The four extreme corners are joined together to form a box within which the circuit is characterized and performance is ensured. The double letter indicates:

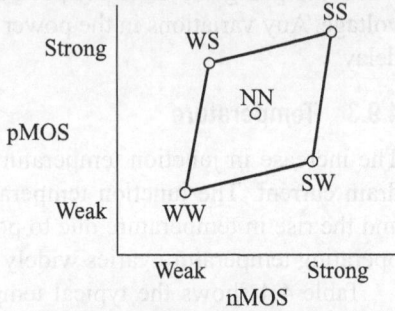

Fig. 4.16 Design corners

WW : weak weak corner
SW : strong weak corner
WS : weak strong corner
SS : strong strong corner

Table 4.5 illustrates the best, nominal, and worst timing analysis corners. V_{DD} is assumed to be 1.8 V with a ±10% variation, i.e., 1.98 V and 1.62 V. The operating temperature range is –40°C (low temperature), 27°C (room temperature), and 105°C (high temperature).

Table 4.5 Design corners for timing analysis

Transistor	V_{DD}	Temperature	For timing analysis	Nomenclature
Strong	High	Low	Best	S_1.98_–40
Weak	Low	High	Worst	W_1.62_105
Nominal	Nominal	Room	Nominal	N_1.8_27

A combination of a strong transistor, highest voltage, and lowest temperature will produce least delay through a circuit, whereas a combination of a weak transistor, lowest voltage, and highest temperature will produce highest delay through a circuit. A combination of a nominal transistor, nominal voltage, and room temperature will produce nominal delay through a circuit. If a circuit works at the three corners, it can be ensured that the circuit is immune to the process, voltage, and temperature variations. However, in order to be 100% sure that the circuit is immune to these variations, the circuit is characterized at all possible corners ($3 \times 3 \times 3 = 27$) and ensured the functionality. The characterization at the 27 corners is a very time-consuming task for a full cell library, because it involves multiple simulations using circuit simulator such as SPICE. So the characterization process is automated using a cell characterization flow that internally invokes SPICE, simulates the circuit, measure the timing parameters, and store in a special format (e.g., .lib).

4.10 Cell Characterization

It is always seen that a logic synthesis tool results in better design if the standard logic cells are characterized with greater accuracy and details. During the synthesis process, the synthesizer should have enough choice of picking up a cell from the cell library. For example, consider the three different cases as shown in Fig. 4.17.

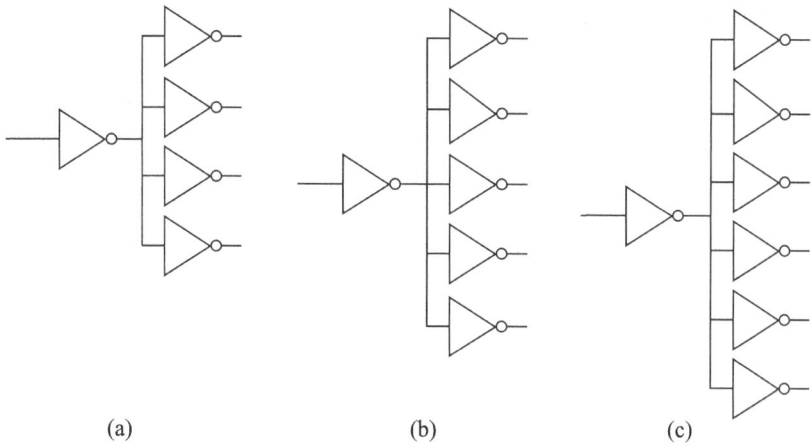

Fig. 4.17   Inverter having different fan-out capability

The output load to the driver inverter is different in the three cases. The inverter in Fig. 4.17(a) is driving a least load whereas the inverter in Fig. 4.17(c) is driving the highest load. Now, while picking up inverter cells from the library, the synthesizer will search for the optimum one for each of these cases. Hence, the cell library must be characterized at different load for timing so that the synthesizer can choose the appropriate inverter.

The cell library should have different versions of the same cell. For example, there should be different inverters with different drive or fan-out capability. The set of same functionality cells is often called logic families. More the number of cells in a family, better is the synthesis results from power, delay, and area perspective. For example, having one inverter with highest drive capability will oversize the design for the circuit shown in Fig. 4.17(a). Similarly, having one inverter with least drive capability may not suffice the drive requirement for the circuit as shown in Fig. 4.17(c).

The delay of a design or cell depends on the following parameters:

- Input slew (rise/fall time) of the input waveform
- Output load or fan-out (input capacitance of load cells and wire capacitance)

There is a trade-off between the number of characterization points and accuracy of using timing analysis. Accuracy will be better if characterization is done at finer points. But characterizing at finer points would increase the library development cycle time and hence impact the time-to-market.

The most popular synthesis tools are listed in Table 4.6.

Table 4.6   Synthesis tools

Sl. No.	Vendor	Tool
1.	Synopsys	Design compiler
2.	Cadence	Ambit PKS
3.	Magma	Blast Create/Blast Fusion

4.11 Circuit Simulation

Transistor level circuits are simulated at different design corners to characterize the cells. The most popular simulator is SPICE which simulates circuits at very small time steps, and the $I-V$ characteristics are obtained. SPICE was initially developed at the University of California at Berkeley in 1975. It is almost used in all academic institutes and R&D organizations because of its accuracy. The simulator has internally three main parts. The first part is called the *pre-processor* which reads the input netlist, builds the data structure using the circuit parameters and connectivity information, and checks for its correctness. The second or core part is the *simulation engine* which simulates the circuit and calculates voltage and current at each circuit node at very small time steps. The third part is called the *post-processor* which outputs the specified parameters such as voltage, current, or power at the specified circuit nodes.

Example 4.3 Design and simulate a CMOS inverter circuit.

(a) Obtain the transient response.
(b) Obtain the transfer characteristics.

Solution

(a) Transient response
The SPICE netlist for the CMOS inverter as shown in Fig. 4.18 can be written as:

```
.INCLUDE "E:\CMOS-CELLS\SPICE MODELS\model.txt"
M1 Y A VDD VDD PMOS  W=5U L=0.18u
M2 Y A 0 0 NMOS W=2U L=0.18u
CL Y 0 0.1PF

VDD VDD 0 1.8V
VIN A 0 PULSE 0 1.8V 0N 1N 1N 5N 10N

.TRAN 0.01N 20N
.PRINT TRAN V(A) V(Y)
.PLOT TRAN V(A) V(Y)

.END
```

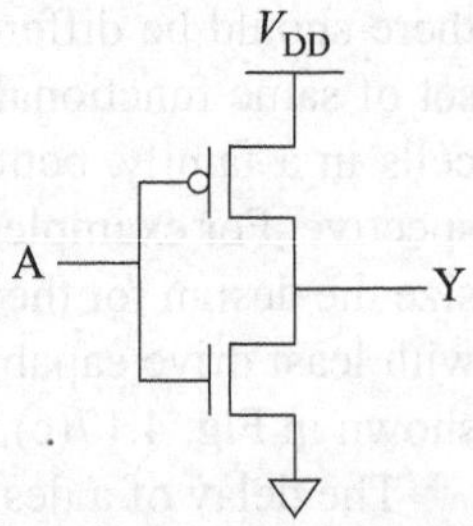

Fig. 4.18 CMOS inverter

The simulated input and output waveforms are shown in Fig. 4.19.
(b) Transfer characteristics

The SPICE netlist for obtaining the transfer curve is

```
.include "E:\CMOS-CELLS\SPICE MODELS\model.txt"
C1 y Gnd 0.1pF
M2 y a Gnd Gnd NMOS L=0.18u W=2u
M3 y a Vdd Vdd PMOS L=0.18u W=5u

vdd Vdd Gnd 1.8
vin a 0 0.0

.dc vin 0 1.8 0.01
.plot dc v(y)
.end
```

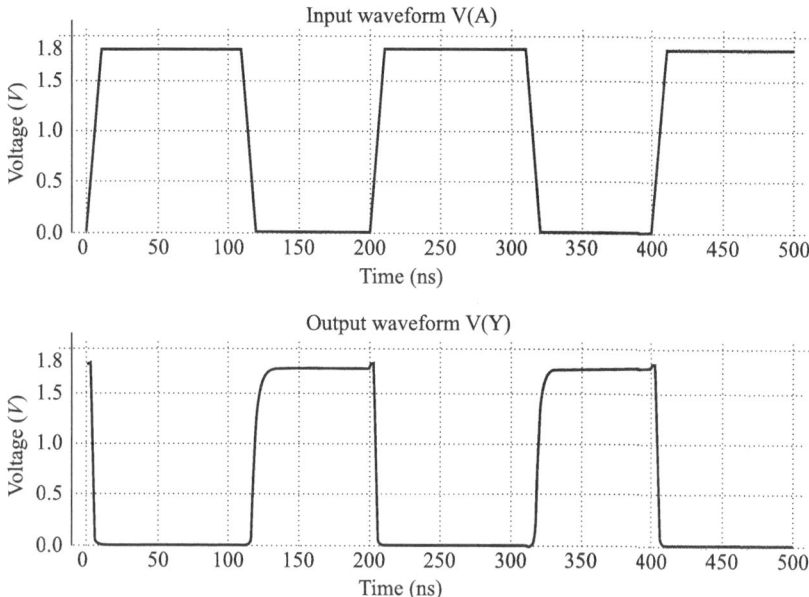

Fig. 4.19 Simulation results of CMOS inverter

The transfer curve is shown in Fig. 4.20.

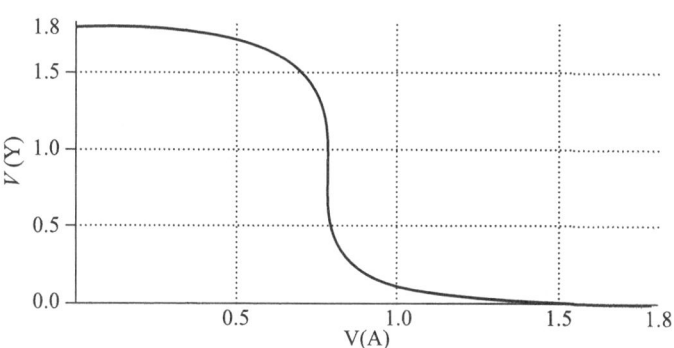

Fig. 4.20 Transfer curve of the CMOS inverter

Example 4.4 Write a SPICE netlist to design a two-input CMOS NAND gate and obtain its output response.

Solution

```
*2-input CMOS NAND gate
MP1 VDD A OUT VDD PMOS W=1u L=0.18u
MP2 VDD B OUT VDD PMOS W=1u L=0.18u
MN1 OUT A N1 VSS NMOS W=1u L=0.18u
MN2 N1 B VSS VSS NMOS W=1u L=0.18u
VDD VDD 0 1.8
VSS VSS 0 0
VA A 0 PULSE 0 1.8 0n 1n 1n 3n 10n
```

```
VB B 0 PULSE 0 1.8 0n 1n 1n 8n 20n
CY OUT 0 0.1pF
.TRAN 1p 40n
.PLOT TRAN V(A) V(B) V(OUT)
.END
```

Example 4.5   Write a SPICE netlist to design a two-input CMOS NOR gate and obtain its output response.

Solution

```
* 2-input CMOS NOR gate
MP1 VDD A N1 VDD PMOS W=1u L=0.18u
MP2 N1 B OUT VDD PMOS W=1u L=0.18u
MN1 OUT A VSSVSS NMOS W=1u L=0.18u
MN2 OUT B VSS VSS NMOS W=1u L=0.18u
VDD VDD 0 1.8
VSS VSS 0 0
VA A 0 PULSE 0 1.8 0n 1n 1n 3n 10n
VB B 0 PULSE 0 1.8 0n 1n 1n 8n 20n
CY OUT 0 0.1pF
.TRAN 1p 40n
.PLOT TRAN V(A) V(B) V(OUT)
.END
```

4.12   Measuring the Propagation Delay

Propagation delay of a gate is defined as the time delay between the input and the output waveforms of a gate. Consider the input and output waveforms of an inverter as shown in Fig. 4.21.

There are two time delays between the input and output waveforms. One is due to the delay between the input and output waveforms for high-to-low transition of the output waveform. This delay is termed as TPHL. The other is due to delay

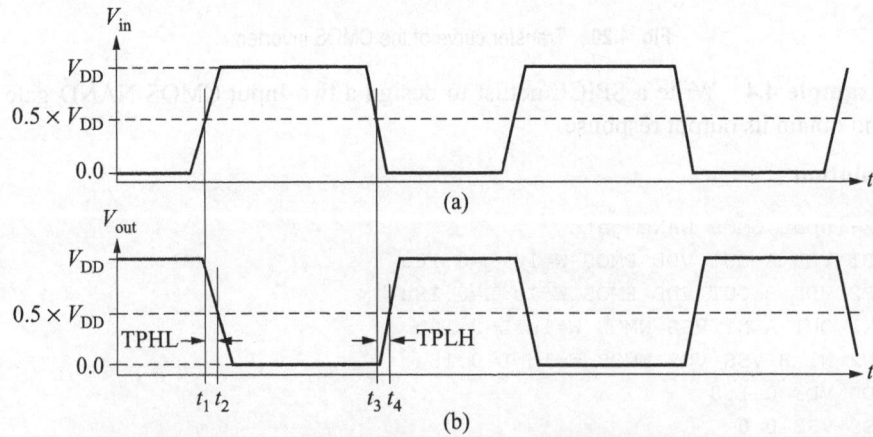

Fig. 4.21   Input and output waveforms of CMOS inverter

between the input and output waveforms for low-to-high transition of the output waveform. This delay is termed as TPLH. As shown in Fig. 4.21, TPHL is defined as

$$\text{TPHL} = t_2 - t_1 = (\text{time at which } V_{\text{out}} \text{ is } 0.5 \times V_{\text{DD}}) - (\text{time at which } V_{\text{in}} \text{ is } 0.5 \times V_{\text{DD}})$$

Similarly, TPLH is defined as

$$\text{TPHL} = t_4 - t_3 = (\text{time at which } V_{\text{out}} \text{ is } 0.5 \times V_{\text{DD}}) - (\text{time at which } V_{\text{in}} \text{ is } 0.5 \times V_{\text{DD}})$$

The average of TPHL and TPLH is the propagation delay (t_p) of a gate and given by

$$t_\text{p} = \frac{\text{TPHL} + \text{TPLH}}{2}$$

The TPHL and TPLH values can be measured by doing SPICE simulation of a gate. The `.measure` statement is used to measure the delay values.

```
.measure tran TPHL trig v(a) val=0.9 cross=1 targ v(y)
+ val=0.9 cross=1
.measure tran TPLH trig v(a) val=0.9 cross=2 targ v(y)
+ val=0.9 cross=2
```

4.13 Measuring Rise and Fall Time

The output waveform takes finite time to make low-to-high and high-to-low transitions. Consider Fig. 4.22. For the high-to-low transition of the output waveform there is a finite fall time which is known as fall time (t_fall) of the output waveform. Similarly, for the low-to-high transition of the output waveform, there is a finite rise time which is known as rise time (t_rise) of the output waveform.

The rise and fall times are defined as

$$t_\text{fall} = t_2 - t_1 = (\text{time at which } V_{\text{out}} \text{ is } 0.1 \times V_{\text{DD}}) - (\text{time at which } V_{\text{out}} \text{ is } 0.9 \times V_{\text{DD}})$$
$$t_\text{rise} = t_4 - t_3 = (\text{time at which } V_{\text{out}} \text{ is } 0.9 \times V_{\text{DD}}) - (\text{time at which } V_{\text{out}} \text{ is } 0.1 \times V_{\text{DD}})$$

The rise and fall times can be measured by doing SPICE simulation of a gate. The `.measure` statement is used to measure these values as shown below:

```
.measure tran Trise trig v(y) val=1.62 cross=1 targ v(y)
+ val=0.18 cross=1
.measure tran Tfall trig v(y) val=0.18 cross=2 targ v(y)
+ val=1.62 cross=2
```

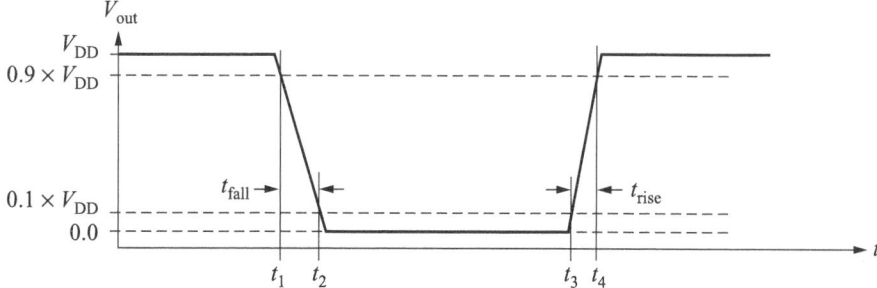

Fig. 4.22 Output waveform of a CMOS inverter showing rise and fall times

4.14 Ring Oscillator

A technology is often characterized by estimating the speed of the design that would be fabricated using this technology. This is done by a ring oscillator circuit. A ring oscillator circuit is a chain of odd numbers (typically 3 or 5) of the minimum size inverters. The last inverter's output is connected to the input of the first inverter as shown in Fig. 4.23 to form a feedback loop.

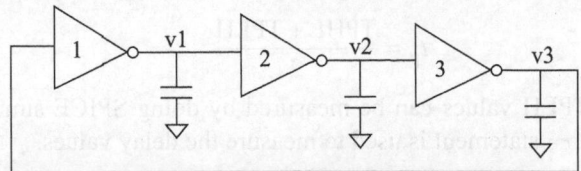

Fig. 4.23   Three-stage ring oscillator circuit

The input is inverted at the third stage output as odd numbers of stages are used. Since the last stage output is fed back to the first stage, the output of each stage keeps on changing after each cycle and results in oscillations. As each inverter has intrinsic propagation delay and the output appears after finite time period, the circuit does not have a stable operating point. The inverter threshold voltage is the only DC operating point where the input and output voltage are equal. Any noise voltage on any of the nodes forces the output to be oscillating. That is why the circuit is known as ring oscillator.

SPICE netlist:

```
.include "E:\CMOS-CELLS\SPICE MODELS\model.txt"
.global vdd gnd
.ic v(v3,0)=0.5

.subckt inv1 a y
M2 y a Gnd Gnd NMOS L=2u W=2u
M3 y a Vdd Vdd PMOS L=2u W=5u
.end
x1 v3 v1 inv1
x2 v1 v2 inv1
x3 v2 v3 inv1

vdd Vdd Gnd 1.8

.tran 0.1n 20n
.plot tran v(v1) v(v2) v(v3)
.end
```

The simulated waveform is shown in Fig. 4.24.
The oscillation period T of the three-stage ring oscillator circuit is given by

$$T = 2\tau_p + 2\tau_p + 2\tau_p = 6\tau_p$$

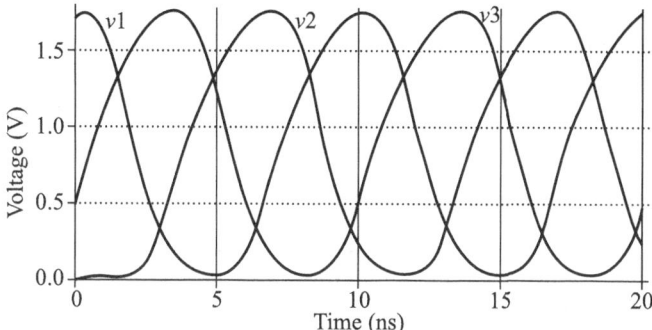

Fig. 4.24 Three-stage ring oscillator's output waveform

where τ_p is the average propagation delay through the inverter. In general, the oscillation frequency of the n-stage ring oscillator circuit is given by

$$f = \frac{1}{T} = \frac{1}{2 \times n \times \tau_p}$$

The ring oscillator is utilized to characterize the maximum operating speed of new fabrication process technology.

4.15 Load/Slew Characterization Table

The standard cells of a library are characterized for timing for different values on input slew and output capacitance. A typical load/slew characterization table is shown in Table 4.7.

Table 4.7 Load-slew characterization table

Load/Slew	0.1 pF	0.2 pF	0.3 pF	0.4 pF
0.5 ns	1.1 ns	1.2 ns	1.3 ns	1.4 ns
1.0 ns	2.1 ns	2.2 ns	2.3 ns	2.4 ns
1.5 ns	3.1 ns	3.2 ns	3.3 ns	3.4 ns

The dimension of the matrix depends on the characterization points. The entries in the table are the propagation delay values. This timing characterization data is used at the chip level timing analysis. At the chip level, any gate's fan-out load can be estimated. Depending on the input slew, the propagation delay is just read from the load-slew characterization table. The output slew is read from a similar table which is used for the next stage timing delay analysis.

4.16 Power Characterization

Power dissipation is one of the three primary design parameters of the VLSI circuit design. The other two parameters are area and speed. We have discussed the speed or timing in the previous sections. Let us now discuss the *power characterization*, which is a process of analyzing the power dissipation of standard cells for various in-put and output conditions. For example, for different values of output capacitances, the power dissipation is characterized by running circuit simulator such as SPICE. There are different sources of power dissipation in static CMOS circuits. Following sections discuss the different power dissipation mechanisms in static CMOS circuits.

4.16.1 Dynamic Power Consumption

In a CMOS logic circuit, power is dissipated during the transition of the output node capacitance. During the low–high transition, the output capacitor is charged through the pMOS transistor and during the high–low transition, the output capac-itor is discharged through the nMOS transistor. Let us consider a CMOS inverter circuit and its input/output waveforms as shown in Fig. 4.25.

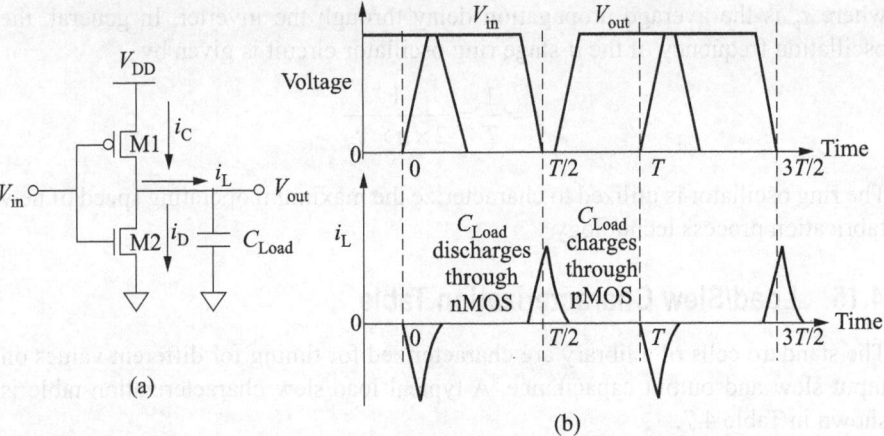

Fig. 4.25 (a) CMOS inverter; (b) typical input (V_{in}) and output (V_{out}) waveform; and load current (i_L) waveform

We can calculate the average power dissipated over one period as

$$P_{av} = \frac{1}{T} \int_0^T v(t)i(t)\, dt$$

$$= \frac{1}{T} \left[\int_0^{T/2} V_{out} \left(-C_{Load} \frac{dV_{out}}{dt} \right) dt + \int_{T/2}^T (V_{DD} - V_{out}) \left(C_{Load} \frac{dV_{out}}{dt} \right) dt \right]$$

or

$$P_{av} = \frac{1}{T} \left[\left(-C_{Load} \frac{V_{out}^2}{2} \right) \Big|_0^{T/2} + \left(V_{DD} \times V_{out} \times C_{Load} - \frac{1}{2} C_{Load} V_{out}^2 \right) \Big|_{T/2}^T \right]$$

or

$$P_{av} = \frac{1}{T} C_{Load} V_{DD}^2 = C_{Load} V_{DD}^2 f$$

In one switching cycle, the circuit consumes $C_{Load}V_{DD}^2$ amount of energy, where C_{Load} is the output capacitance and V_{DD} is the power supply voltage. If the circuit switching frequency is f, then the dynamic power dissipation is given by

$$P_{av} = C_{Load}V_{DD}^2 f$$

From the expression of the average switching power, it is clear that the switching power is entirely dependent on the power supply voltage, the load capacitance, and the switching frequency. It is independent of the dimensions of transistors.

Example 4.6   Calculate the dynamic power dissipation of a CMOS inverter operating at 200 MHz. The power supply voltage is 2.0 V and the load capacitance is 10 fF. If the delay through the inverter is 10 ps then calculate the power delay product (PDP).

Solution   The dynamic power dissipation of a CMOS inverter is given by

$$P_{avg} = C_L V_{DD}^2 f$$
$$= 10 \times 10^{-15} \times 2.0^2 \times 200 \times 10^6 \text{ W}$$
$$= 8.0 \times 10^{-6} \text{ W}$$

The power-delay-product is given by

$$\text{PDP} = P_{avg} \times \tau_p$$
$$= 8.0 \times 10^{-6} \times 10 \times 10^{-12}$$
$$= 8.0 \times 10^{-17} \text{ J}$$

4.16.2   Power Dissipation Due to Short-circuit Current

During the transition of the input signal due to the finite rise or fall times, there is always a short-circuit path between V_{DD} and the ground. So there is a short-circuit power dissipation and it is given by

$$P_{sc} = t_{sc}V_{DD}I_{peak}f$$

where t_{sc} is the time for which both nMOS and pMOS transistors are ON simultaneously, and I_{peak} is the maximum saturation current that flows through the transistors.

4.16.3   Static or Leakage Power Dissipation

The static power dissipation is always present due to the leakage current of the transistors even if the circuit is not switching. This is also known as leakage power dissipation and is given by

$$P_{stat} = I_{leakage}V_{DD}$$

where $I_{leakage}$ is the leakage current that flows between V_{DD} and the ground.

The total power dissipation of a CMOS circuit is the sum of these three power dissipations and can be expressed as

$$P_{total} = P_{dyn} + P_{sc} + P_{stat}$$

4.16.4   Power Delay Product (PDP)

The power delay product (PDP) is the product of average power dissipation and the propagation delay of a CMOS circuit and is given by

$$PDP = P_{avg}\tau_p$$

PDP is a measure of energy dissipation.

4.16.5   Energy Delay Product (EDP)

The energy delay product of a CMOS circuit is the product of PDP and the propagation delay time. It is given by

$$EDP = PDP \times \tau_p$$

4.16.6   Power and Energy Measurement

By measuring the average current drawn from the power supply (I_{DD}), we can measure the power dissipation. The following statements in SPICE are very useful in measuring the power dissipation:

```
.MEASURE TRAN I(VDD) AVG
```

OR

```
.MEASURE CHARGE INTEGRAL I(VDD) FROM=0NS TO=100NS
.MEASURE ENERGY PARAM='CHARGE*SUPPLY'
```

OR

```
.MEASURE PWR AVG P(VDD) FROM=0NS TO=100NS
```

4.17   Reliability and Noise Characterization

In the library development cycle, the cells are characterized for reliability analysis to be done at the chip level. The characterized data at the cell level can be done with greater accuracy by doing SPICE simulation. This data is used to avoid any SPICE simulation and speed up the analysis at the chip level.

4.17.1   Crosstalk Noise Analysis

For example, let us consider crosstalk noise analysis at the chip level. In the VLSI circuits there will be cases where several metal lines run parallel to each other. These will produce large parasitic coupling capacitances between the metal lines. As any of the signal line makes a switching or logic transition, it will affect the neighbouring lines. The line that is affected is called the *victim*, and the line that affects is called the *aggressor*. Now, if there is crosstalk noise, it is always important. It depends on where the noise is occurring. If it occurs at the input of a combinational logic, it is not a problem. But, if it occurs at the input of the sequential element, it can cause incorrect logic value to be stored in the flip-flop, and hence can violate the functionality. A noise appearing at the input of a combinational logic may propagate through the cell if the noise exceeds the *noise threshold* of the cell. Even if the noise has occurred at the input of combinational cell, it may propagate through the cell and reach to the input of a sequential logic. Hence, at the chip level, the noise analysis requires the information of the noise threshold value of each of the combinational logic cells. So, it can prune out noises based on the noise threshold value (Arvind et al. 2001).

Figure 4.26 illustrates the crosstalk noise in VLSI designs. Three metal lines or interconnects run parallel to each other over a long distance, and they suffer from crosstalk noise. The parallel lines have coupling capacitances between themselves. If the top and bottom lines switch from logic 0 to logic 1, they will induce a glitch in middle wire as shown in Fig. 4.26. The affected wire is called victim net while the others are known as aggressor net.

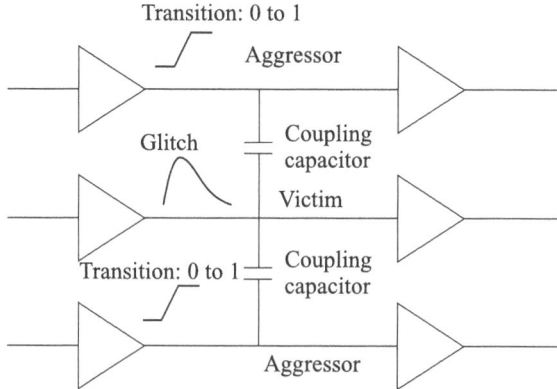

Fig. 4.26 Crosstalk noise due to coupling capacitances

At the cell level, the glitch threshold of each cell is calculated by doing SPICE simulation. Glitch or noise threshold is a measure of the glitch area which can propagate through the logic. At the chip level, after the detailed routing is done, the parasitic RCs are extracted and depending on their relative switching time the glitch is calculated. The calculated glitch value is compared with the glitch threshold of the cell and error is reported if it is greater than the glitch threshold.

4.17.2 Electromigration Analysis

Electromigration (EM) is a phenomenon which occurs under high current density through the wires. When large current flows through a small cross-sectional wire, the current density through the wire is large. Under this condition, the high-speed electrons flowing through the wire collide with the host atoms of the wire and displace them to create either a void or a hillock. This can cause an open or short circuit in the interconnect wire.

Two types of EM analysis are performed for power and signal lines: (a) power EM for power lines and (b) signal EM for signal lines. EM is a statistical phenomenon and it happens over a long operating time. Hence, its effect is measured in terms of failure-in-time (FIT) rate. If the failure rate is below the allowed values, the design is safe from EM. For the EM analysis at the chip level, the following characterization is performed at the cell level:

- FIT rates due to EM are calculated at the cell level.

4.17.3 Gate Oxide Integrity (GOI) Analysis

As the technology is advanced from submicron to deep submicron (DSM), and DSM to nanometre regime, the gate oxide thickness of the MOS device is scaled down. The power supply voltage (V_{DD}) is also scaled down but at a lesser rate as compared to the oxide thickness (t_{ox}). Hence, the electric field ($E_{ox} = V_{DD}/t_{ox}$) across the thin gate oxide is increased as the technology progresses. When the thin gate oxide is operated at high electric field over a longer period of time, the quality of the oxide degrades, and after some operating lifetime of the gate oxide may break down. This phenomenon is called time-dependent dielectric break down (TDDB).

However, the gate oxide quality is ensured at the operating voltage for the entire device operating lifetime at the technology level. But if any over-voltage appears across the gate oxide over the nominal operating voltage, the gate oxide undergoes severe degradation and chance of failure due to gate oxide breakdown increases. The over-voltage can appear due to several reasons. Some of the causes of over-voltage at the gate are as follows:

- Overshoot and undershoot due to crosstalk noise
- Overshoot due to Miller effect
- Overshoot and undershoot due to *Ldi/dt* effect

To measure the effect of the over-voltage, the FIT rate is calculated and checked if it is under the permissible value.

For the GOI analysis at the chip level, FIT rates are calculated at the cell level for different over-voltage conditions.

4.17.4 Channel Hot Carrier Analysis

Channel hot carrier (CHC) is a similar effect due to over-voltage at the drain node of the MOS device. Under high voltage at the drain node, the lateral electric in the channel region becomes very high. The charge carriers, especially electrons when following under such a large electric field acquire high kinetic energy and collide with atom near the drain end to cause impact ionization. Some of the high energetic electrons can surmount the potential barrier and can get into the oxide and hence, increase the threshold voltage of the MOS device. This increase in threshold voltage in turn causes the device to become slow.

Like GOI, the CHC effect is also measured in terms of FIT rate and checked against the allowed limit.

For the CHC analysis at the chip level, FIT rates are calculated at the cell level for different over-voltage conditions.

4.18 Interconnect Delay Modelling

Interconnects are metal wires that are used to transmit signal from one node of the circuit to other nodes. In VLSI circuits, the interconnect exhibits parasitic resistance, capacitance, and inductance. As the technology advances, the dimension of the transistors and interconnects are scaled down. But with the scaling, the effects of parasitic resistance, capacitance, and inductances become more and more important. According to the recent technology trends, the parasitic resistance and capacitance have become dominant in comparison to the inductance effects. Hence, we will focus mainly on resistance and capacitance of interconnects and analyse their effect on the delay.

4.18.1   *RC* Delay Model

The simple model for the interconnect is the lumped *RC* model as shown in Fig. 4.27(a). The propagation delay of the *RC* model as shown in Fig. 4.27(a) is given by

$$\tau_{\text{wire}} = 0.69\tau$$

where $\tau = RC$ = time constant of the *RC* circuit.

The accuracy of the lumped *RC* model is improved by using *T* network where the resistance is divided equally as *R/2* and a *T* network is modelled as shown in Fig. 4.27(b).

The accuracy of the *RC* model is improved further by representing a distributed *RC* network. The lumped resistance and capacitance are divided by *n*. The distributed *RC* network is formed by a series connection of identical *n* number of *L* network of resistance *R/n* and capacitance *C/n*. This is illustrated in Fig. 4.27(c).

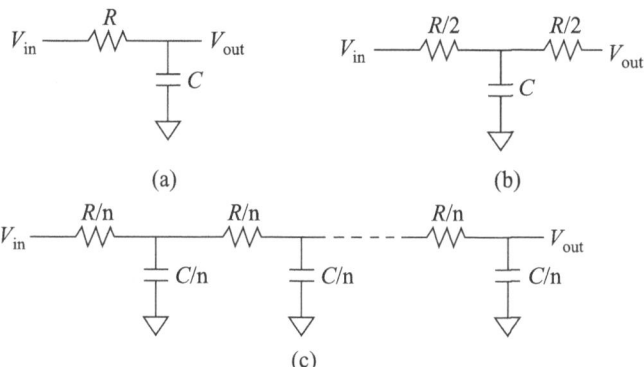

Fig. 4.27   *RC* model: (a) lumped L-network; (b) lumped T-network; (c) distributed L-network

4.18.2   Elmore Delay Model

Elmore delay model is useful for estimating delays of *RC* tree networks. The *RC* interconnect model described in the previous subsection is not useful for tree network. For the tree network, derivation of an expression of delay is very difficult. Otherwise, a circuit simulation is required to calculate the delay which is often time consuming. Elmore delay model is an approximation but still fairly simple and accurate method of estimating delay in the tree network. Let us consider an *RC* tree network as shown in Fig. 4.28.

The Elmore delay at *i*th node is given by

$$\tau_i = \sum_{j=1}^{N} C_j \sum_{\text{for all } k \in P_{ij}} R_k$$

where N = number of nodes, P_{ij} = path between input and *i*th node which is common to the path between input and *j*th node.

For example, let us calculate Elmore delay between input and node 7 which is given by

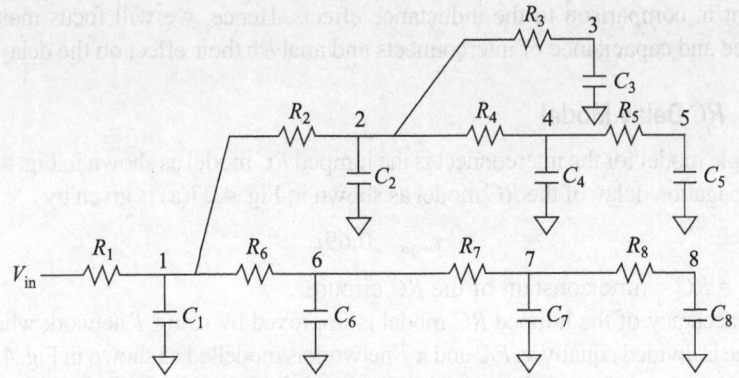

Fig. 4.28 An *RC* tree network

$$\tau_7 = R_1C_1 + R_1C_2 + R_1C_3 + R_1C_4 + R_1C_5 + (R_1 + R_6)C_6$$
$$+ (R_1 + R_6 + R_7)C_7 + (R_1 + R_6 + R_7)C_8$$

Another example, Elmore delay between input and node 5, is given by

$$\tau_5 = R_1C_1 + (R_1 + R_2)C_2 + (R_1 + R_2)C_3 + (R_1 + R_2 + R_4)C_4$$
$$+ (R_1 + R_2 + R_4 + R_5)C_5 + R_1C_6 + R_1C_7 + R_1C_8$$

Example 4.7 Calculate the Elmore delay at node 3 for the network shown in Fig. 4.29.

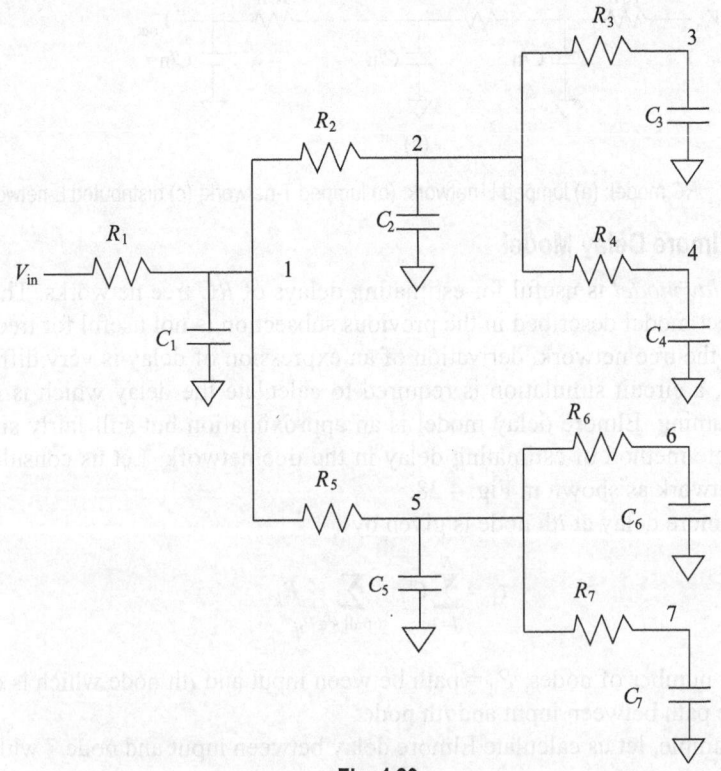

Fig. 4.29

Solution   The Elmore delay between input and node 3 is given by

$$\tau_3 = R_1 C_1 + (R_1 + R_2)C_2 + (R_1 + R_2 + R_3)C_3 + (R_1 + R_2)C_4 + R_1 C_5 + R_1 C_6 + R_1 C_7$$

4.18.3   Transmission Line Model

If the length of the interconnect is such that the transit time through the interconnect is comparable to the rise/fall time of the pulse waveform, then the interconnect should be modelled as a transmission line. A transmission line is modelled as a distributed RLCG network, as shown in Fig. 4.30. RLCG indicates resistance, inductance, capacitance, and conductance, all of which are specified per unit length of the interconnect line.

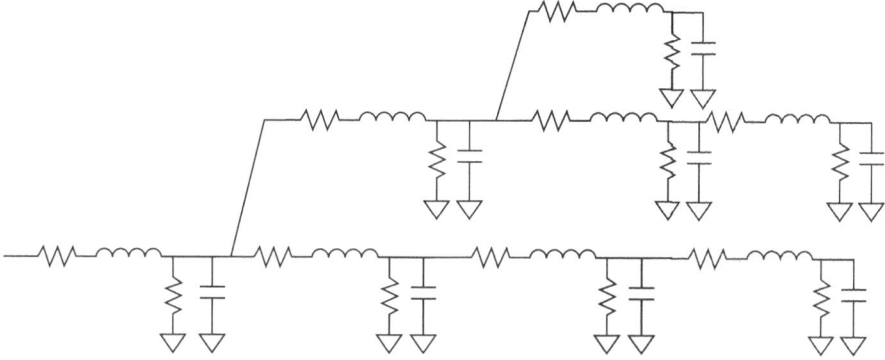

Fig. 4.30   Transmission line model of the interconnect

SUMMARY

- Standard cells are basic logic gates which are pre-designed, pre-characterized, and pretested.
- Standard cell characterization is the process of simulating the cells and extracting the cell attributes, e.g., cell area, pin-to-pin delay, and power dissipation.
- Layout is the physical drawing of the circuit to be implemented in silicon with specified geometry information.
- Design rules are a set of rules that come from the process technology. It specifies the minimum width, spacing, overlap, etc., between different mask layers used to draw the layout.
- Layout versus schematic (LVS) is a process of checking the netlist extracted from the layout and the schematic netlist for one-to-one mapping.
- Extraction is the process of generating circuit parameters from the drawn layout. By extraction, both device and interconnect parameters are generated using which an equivalent circuit can be generated from the layout.
- PTV condition specifies the process, temperature, and voltage conditions to be used for the characterization of a design and library.
- Timing characterization is a process of generating propagation delay, and rise/fall time parameters through a cell for different operating conditions.
- Power characterization is a process of generating power dissipation data for different operating conditions.

- Noise characterization is a process of generating noise threshold for standard cells.
- Reliability characterization process generates the reliability parameters such as failure rate of the cells for different reliability issues such as electromigration, channel hot carrier, TDDB, etc.
- Interconnect modelling is a process of generating equivalent circuit representation for interconnect wires.

SELECT REFERENCES

Arvind, N.V., P.R. Suresh, V. Sivakumar, C. Pal, and D. Das 2001, *Integrated Crosstalk and Oxide Integrity Analysis in DSM Designs*, VLSI Design, Fourteenth International Conference, pp. 518–23.

Aur, S., D.E. Hocevar, and P. Yang 1987, *Circuit Hot Electron Effect Simulation*, Electron Devices Meeting, International, Vol. 33, pp. 498–501.

Baker, R. Jacob, H.W. Li, and D.E. Boyce 2004, *CMOS Circuit Design, Layout, and Simulation*, Prentice-Hall of India, New Delhi.

Berkley SPICE, http://www-device.eecs.berkeley.edu/~bsim3/latenews.html, last accessed on 30 December 2010.

Kang, S.M. and Y. Leblebici 2003, *CMOS Digital Integrated Circuits: Analysis and Design*, 3rd ed., Tata McGraw-Hill, New Delhi.

Martin, K. 2004, *Digital Integrated Circuit Design*, Oxford University Press.

Mead, C. and L. Conway 1980, *Introduction to VLSI Systems*, Addison-Wesley.

MOSIS, http://www.mosis.org, last accessed on 30 December 2010.

Rabaey, J.M., A. Chandrakasan, and B. Nikolic 2008, *Digital Integrated Circuits: A Design Perspective*, 2nd ed., Pearson Education.

Rashid, M.H. 1994, *SPICE for Circuits and Electronics Using pSpice*, 2nd ed., Prentice-Hall, New Jersey.

Sicard, E. and S.D. Bendhia 2005, *Basics of CMOS Cell Design*, Tata McGraw-Hill, New Delhi.

Smith, M.J.S. 2002, *Application Specific Integrated Circuits*, Pearson Education.

Tennakoon, H. and C. Sechen 2008, 'Nonconvex Gate Delay Modeling and Delay Optimization,' *Computer-Aided Design of Integrated Circuits and Systems*, IEEE Transactions, Vol. 27, No. 9, September, pp. 1583–94.

Weste, N.H.E., D. Harris, and A. Banerjee 2009, *CMOS VLSI Design: A Circuits and Systems Perspective*, 3rd ed., Pearson Education.

Windschiegl, A., T. Mahnke, M. Eiermann, W. Stechele, and P. Zuber 2002, "A Wire Load Model Considering Metal Layer Properties," *Electronics, Circuits and Systems*, 9th International Conference, pp. 765–8.

EXERCISES

Fill in the Blanks

1. The smallest feature size (lambda or λ) used to measure an IC is _____ .
   (a) half the length of the smallest transistor
   (b) two-thirds the length of the smallest transistor
   (c) one-fourth the length of the smallest transistor
   (d) none of the above
2. For a 0.5 µm process technology _____ .
   (a) λ = 0.25 µm   (b) λ = 0.5 µm   (c) λ = 1 µm   (d) λ = 0.125 µm

3. Technology that is used to manufacture IC in much greater volume is
   _____ .
   (a) bipolar technology
   (b) MOS technology
   (c) CMOS technology
   (d) BiCMOS technology
4. The MOS transistor was initially difficult to fabricate because of _____
   (a) problems with the oxide interface
   (b) problems with the gate fabrication
   (c) problems with the diffusion technology
   (d) problems with the metallization process
5. Logic gate that is used to measure the gate equivalent/count in an IC is
   _____ .
   (a) NOT gate   (b) two-input NAND gate
   (c) two-input NOR gate   (d) two-input XOR gate

Multiple Choice Questions

1. ASIC cell library contains
   (a) physical layout of the cells   (b) timing model of the cells
   (c) routing model of the cells   (d) all of these
2. ASIC cell library is a collection of
   (a) standard cells with different logic functions
   (b) standard cells with different fan-out capabilities
   (c) standards cells with different fan-in capabilities
   (d) all of the above
3. The earliest IC was manufactured using
   (a) bipolar technology   (b) nMOS technology
   (c) CMOS technology   (d) MOS technology
4. CMOS uses
   (a) positive logic —V_{DD} is logic '1' and V_{SS} is logic '0'
   (b) Negative logic —V_{SS} is logic '1' and >V_{DD} is logic '0'
   (c) None of the above
   (d) both (a) and (b)
5. An ASIC cell library typically contains
   (a) combinational logic cells   (b) sequential logic cells
   (c) datapath logic cells   (d) I/O cells
   (e) all of the above
6. In the VLSI logic design process we can
   (a) minimize both area and delay   (b) minimize area at the cost of delay
   (c) maximize speed by decreasing area (d) minimize delay by reducing area
7. Lowest propagation delay through a gate is due to
   (a) strong transistor, low temperature, high voltage
   (b) weak transistor, high temperature, high voltage
   (c) strong transistor, high temperature, high voltage
   (d) weak transistor, low temperature, low voltage
8. A feed-through cell is an empty cell (with no logic) that is used for
   (a) vertical interconnections   (b) horizontal interconnections
   (c) both (a) and (b)   (d) feeding power to a cell
9. A hard macro refers to a
   (a) flexible block   (b) fixed block
   (c) flexible block with fixed aspect ratio (d) flexible block with fixed pin locations

True or False

1. MOS IC was cheaper than a bipolar IC.
2. SPICE stands for simulation program with integrated circuit emphasis
3. The unit of sheet resistance is Ω/square.
4. A 1x inverter indicates a minimum size inverter.
5. PVT corner in the context of VLSI design indicates process, voltage, and temperature.

Short-answer Type Questions

1. What are the components of a standard cell library? Discuss each of them separately.
2. Discuss the necessity of the input/output cells.
3. What do you mean by design corners?
4. Discuss the layout design rules.
5. What do you mean by DRC, LVS, and extraction?
6. What is a ring oscillator circuit?
7. Draw a stick diagram of CMOS NAND gate.
8. Draw a stick diagram of CMOS XOR gate.
9. What do you mean by reliability characterization?
10. What do you mean by crosstalk noise characterization?

Long-answer Type Questions

1. What is a standard cell library? Explain the design principle of a standard cell library. Does a full-custom designer use a standard cell library? Justify your answer.
2. What is a CMOS process? What is DRC in the context of layout design? Discuss the design rules in details.
3. What is antenna effect in VLSI? What is the fixation mechanism for this? Discuss with necessary diagrams.
4. Consider a 5 mm long, 0.32 μm wide metal 2 wire in a 180 nm process. The sheet resistance is 0.05 Ω/square and the capacitance is 0.2 fF/μm. Construct a 3-segment π-model for the wire.
5. What is crosstalk delay in VLSI? Discuss the origin of crosstalk delay and how does it affect the performance of an IC.
6. What is crosstalk noise in VLSI? Discuss the origin of crosstalk noise and how does it affect the performance of an IC.
7. What do you mean by design margin and design corners in VLSI? Discuss each of them in brief.
8. What are interconnect delay models? Discuss them in detail.
9. Write a SPICE deck to design and simulate a two-input NAND gate.
10. Write a SPICE deck to calculate the propagation delay and dynamic power dissipation for a two-input XOR gate.
11. Calculate the power dissipation of a CMOS circuit operating under 1.2 V. Assume that the switching frequency is 100 MHz and load capacitance is 1 pF.

Analog CMOS Design

5.1 Introduction

The electrical signals are classified into two types: analog and digital. The signals that are continuous in both time and amplitude are *analog signals*, whereas the signals discretized in the time domain and quantized in amplitude are called *digital signals*. The high noise immunity properties of digital systems have outperformed the analog systems for quite sometime. Perhaps, almost everything today is digital, and it has been accelerated by the advent of the integrated circuit (IC) technology. Even though all signals generated naturally are analog, they are converted to digital signals using the analog-to-digital conversion technique.

As most of the designers have started doing digital design, there is a strong demand of analog circuit designers today. Why has analog design become so important? Analog designs are fundamentally necessary in many applications where digital designs simply cannot be used. Some of the applications are as follows:

- *Analog-to-digital conversion* All naturally occurring signals are analog. For processing the signals in the digital domain, they need to be converted to the digital domain by the analog-to-digital converter (ADC). Hence, the design of ADC must be in the analog domain.

- *Digital-to-analog conversion* After processing the signals in the digital domain, they again need to be converted back to the analog domain for reception. Hence, digital-to-analog converter (DAC) must be used. So, DAC design is again an important aspect of analog design.

- *Signal amplification*   The signals occurring from natural sources are of very small amplitude—ranging from a few microvolt to a few millivolt. Also, during the signal transmission, the signals get attenuated, and they must be amplified to recover the signal level.
- *Oscillator*   It generates the alternating signals with the desired frequency of oscillation, and it is an important analog design.
- *Phase-locked loop (PLL)*   It is used to track the incoming frequency of a signal. It is also used as a clock generator. PLL is another very important analog design.

In this chapter, we shall discuss the analog design using CMOS transistors only. As discussed previously, almost 90% of the ICs that are fabricated worldwide use CMOS technology, our main focus in this book has been on CMOS design.

5.2   MOSFET Small Signal Model

In this chapter, we shall discuss the analog circuits designed using MOS transistors. Unlike digital design, in analog design, the transistors are operated in three regions of operation. We shall first discuss the large signal and small signal models of MOSFET. The large signal model is used to find out the DC biasing condition, whereas the small signal model is used for designing analog circuits where signal variation is small, and *I–V* characteristic can be approximated by a straight line.

The large signal model of a MOSFET is represented by the *I–V* equations in the linear and saturation regions as shown in Eqns (5.1) and (5.2) (refer to Eqns (2.26) and (2.44) in Chapter 2).

Linear region: $$I_D = \frac{\mu_n C_{ox} W}{L} \times \left[(V_{GS} - V_t)V_{DS} - \frac{V_{DS}^2}{2} \right] \qquad (5.1)$$

Saturation region: $$I_D = \frac{\mu_n C_{ox} W}{2L} \times (V_{GS} - V_t)^2 (1 + \lambda V_{DS}) \qquad (5.2)$$

These equations are used to analyse MOS circuits when the signal amplitude is large enough causing nonlinear swing across the bias point. However, for small signal amplitudes, simple circuit models cannot be used to analyse the MOS circuits. Mostly, MOSFETs are biased in the saturation region when used in analog circuits; we will derive the small signal model of MOSFET operating under the saturation region. As in saturation region I_D is a function of V_{GS}, it can be modelled as a voltage-dependent current source equal to $g_m V_{GS}$, where g_m is the transconductance gain given by

$$g_m = \left. \frac{\partial I_D}{\partial V_{GS}} \right|_{V_{DS} = \text{constant}} = \frac{\mu_n C_{ox} W}{2L} \times 2(V_{GS} - V_t) \times (1 + \lambda V_{DS})$$

$$= \frac{\mu_n C_{ox} W}{L} \times (1 + \lambda V_{DS}) \sqrt{\frac{2L I_D}{\mu_n C_{ox} W(1 + \lambda V_{DS})}}$$

$$= \sqrt{\frac{2\mu_n C_{ox} W}{L} (1 + \lambda V_{DS}) I_D}$$

$$\cong \sqrt{2\mu_n C_{ox}(W/L) I_D} \qquad (5.3)$$

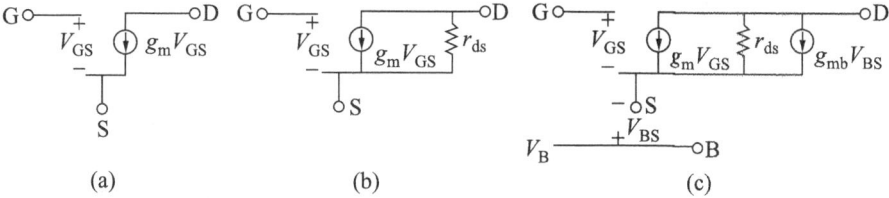

Fig. 5.1 MOSFET small signal model: (a) basic; (b) channel length modulation modelled by a resistor; (c) body/substrate bias modelled by a current source

The basic small signal model of MOSFET is shown in Fig. 5.1(a). However, for short-channel MOSFETs, I_D has a dependency on V_{DS} due to the channel length modulation. To model this dependency, a resistor is introduced in parallel to the current source as shown in Fig. 5.1(b). The resistance is given by

$$r_{ds} = \frac{\partial V_{DS}}{\partial I_D}\bigg|_{V_{GS} = \text{constant}}$$

$$= \frac{1}{\partial I_D / \partial V_{DS}}$$

$$= \frac{1}{\dfrac{\mu_n C_{ox} W}{2L}(V_{GS} - V_t)^2 \times \lambda}$$

$$= \frac{1 + \lambda V_{DS}}{\lambda \times I_D}$$

$$\cong \frac{1}{\lambda I_D} \tag{5.4}$$

Another factor on which I_D depends is the substrate or body bias. This is modelled by a voltage-dependent current source equal to $g_{mb} V_{BS}$ in parallel to r_{ds}, as shown in Fig. 5.1(c). The transconductance g_{mb} is given by

$$g_{mb} = \frac{\partial I_D}{\partial V_{BS}} = \frac{\partial I_D}{\partial V_t} \times \frac{\partial V_t}{\partial V_{BS}}$$

$$= -\frac{\mu_n C_{ox} W}{L}(V_{GS} - V_t) \times (1 + \lambda V_{DS}) \times \frac{\partial V_t}{\partial V_{BS}}$$

$$= \frac{\mu_n C_{ox} W}{L}(V_{GS} - V_t) \times (1 + \lambda V_{DS}) \times \frac{\partial V_t}{\partial V_{SB}} \tag{5.5}$$

The threshold voltage depends on the substrate bias as V_{SB} is given by

$$V_t = V_{t0} + \gamma \left(\sqrt{|-2\varphi_F + V_{SB}|} - \sqrt{|-2\varphi_F|} \right) \tag{5.6}$$

Differentiating both sides of Eqn (5.6) with reference to V_{SB}, we get

$$\frac{\partial V_t}{\partial V_{SB}} = \frac{\gamma}{2\sqrt{|-2\varphi_F + V_{SB}|}}$$

Therefore, Eqn (5.5) can be written as

$$g_{mb} = g_m \frac{\gamma}{2\sqrt{|-2\varphi_F + V_{SB}|}} = \eta g_m \tag{5.7}$$

We have seen in Chapter 2 that MOS devices have associated capacitances between its four terminals. Thus, the complete small signal model of MOSFET, including the capacitances is shown in Fig. 5.2.

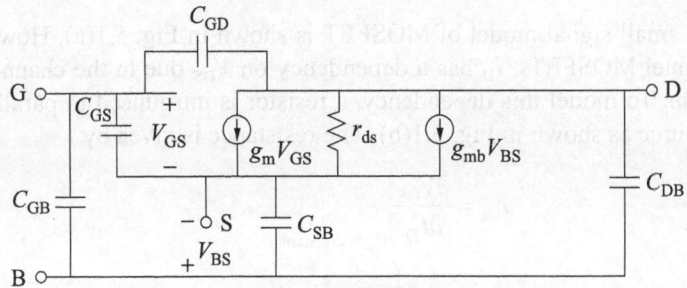

Fig. 5.2   Complete small signal model of MOSFET

5.3   MOSFET as a Switch

MOS devices can be used as simple switches where the switch can be made ON or OFF by applying voltage to the gate terminal. MOS switches are used in several analog designs, such as, switched capacitor circuits, multiplexers, and modulators. The MOS switch is a voltage-controlled resistor, as shown in Figs 5.3(a) and (b). When gate voltage (V_{GS}) is above the threshold voltage (V_t) of the MOSFET, the source and drain terminals are connected, and they are isolated if $V_{GS} < V_t$.

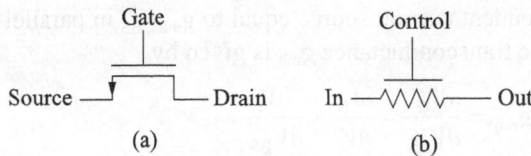

Fig. 5.3   (a) MOS switch; (b) voltage-controlled resistor

The controlled switch or resistor has two resistances, R_{ON} and R_{OFF}, where R_{ON} is the ON resistance and R_{OFF} is the OFF resistance of the MOSFET. Ideally, in a switch, R_{ON} should be zero and R_{OFF} should be infinite. The MOS switch has very small ON resistance and very high OFF resistance, and thus can be used as a switch. Let us now find out the expression for ON resistance of the MOS switch. When the switch is ON, the voltage across the switch must be small, i.e., V_{DS} is small. For small V_{DS}, the MOS device operates in the linear region; hence, we must consider the linear region of the MOSFET. The I–V equation in the linear region is given by

$$I_D = \frac{\mu_n C_{ox} W}{L} \times \left[(V_{GS} - V_t)V_{DS} - \frac{V_{DS}^2}{2} \right] \tag{5.8}$$

The ON resistance can be derived as

$$R_{ON} = \frac{1}{\partial I_D / \partial V_{DS}} = \frac{L}{\mu_n C_{ox} W (V_{GS} - V_t - V_{DS})} \tag{5.9}$$

Example 5.1 Calculate the ON resistance of the MOSFET having the following parameters:

$$\mu_n C_{ox} = 40 \ \mu A/V^2 \qquad W/L = 10$$
$$V_t = 0.7 \ V \qquad\qquad V_{GS} = 1.8 \ V$$
$$V_{DS} = 1 \ V$$

Solution

$$R_{ON} = \frac{L}{\mu_n C_{ox} W (V_{GS} - V_t - V_{DS})}$$

$$= \frac{1}{40 \times 10^{-6} \times 10 \times (1.8 - 0.7 - 1.0)} = 25 \ k\Omega$$

The variation of R_{ON} vs V_{GS} is shown in Fig. 5.4.

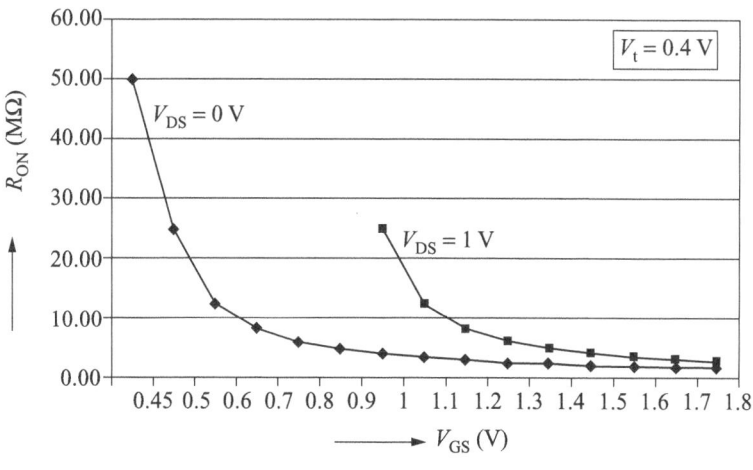

Fig. 5.4 R_{ON} vs V_{GS}

5.4 MOS Diode/Resistor

The MOS device can act as resistor when the drain and gate terminals are connected to each other. This gate–drain connected MOS device is called *MOS diode*, as shown in Figs 5.5(a) and (b). In VLSI circuits, the resistors are replaced by MOS diodes as fabricating a resistor would require a large area in the chip.

When the gate and drain are connected, we can write, $V_{GS} = V_{DS}$, and the transistor operates in the saturation region. The small signal model of an nMOS diode

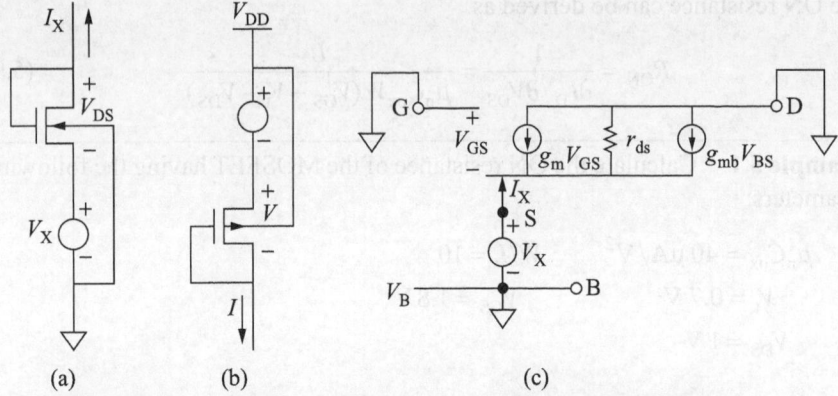

Fig. 5.5　(a) nMOS diode; (b) pMOS diode; (c) small signal model of nMOS diode

is shown in Fig. 5.5(c). We can write, $V_{GS} = -V_X$, $V_{BS} = -V_X$, and $V_{DS} = -V_X$. Applying Kirchhoff's current law at node S, we can write

$$g_m V_{GS} + \frac{V_{DS}}{r_{ds}} + g_{mb} V_{BS} + I_X = 0$$

or

$$-V_X\left(g_m + \frac{1}{r_{ds}} + g_{mb}\right) = -I_X$$

or

$$r_{out} = \frac{V_X}{I_X} = \frac{1}{g_m + g_{ds} + g_{mb}} \cong \frac{1}{g_m} \tag{5.10}$$

Equation (5.10) gives the expression for the small signal resistance of the MOS diode.

5.5　Small Signal Analysis of Single Stage Amplifiers

There are three different configurations of a single stage amplifier. These are as follows:

- common-source stage
- common-drain stage
- common-gate stage

5.5.1　Common-source Stage

A common-source single stage amplifier is shown in Fig. 5.6(a). As the input voltage increases beyond V_{th} of the nMOS transistor (MN) the transistor is turned ON. The drain current increases as long as $V_{DS} > (V_{GS} - V_{th})$ is satisfied,

$$V_{out} = V_{DD} - I_D R_D \tag{5.11}$$

The small-signal equivalent model of the common-source amplifier is shown in Fig. 5.6(b). The expression for the output voltage is given by

$$V_{out} = (r_{ds} \| R_D) i_{out} = (r_{ds} \| R_D)(-g_m V_{GS}) \tag{5.12}$$

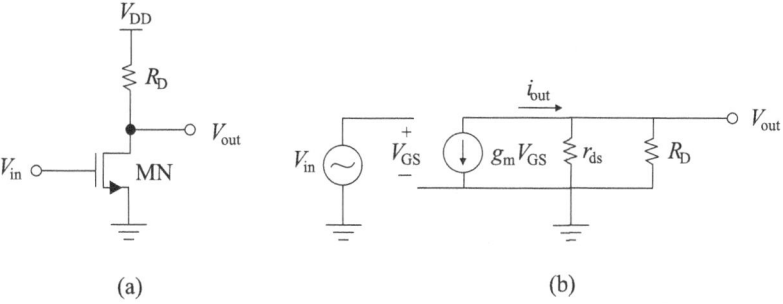

Fig. 5.6 (a) Common-source single stage amplifier and (b) small-signal equivalent model

As $V_{GS} = V_{in}$, Eq. (5.12) becomes

$$V_{out} = (r_{ds} \| R_D)(-g_m V_{in}) \tag{5.13}$$

Therefore, the small-signal voltage gain can be expressed as

$$A_v = \frac{\partial V_{out}}{\partial V_{in}} = -g_m (r_{ds} \| R_D) = -g_m \frac{r_{ds} R_D}{r_{ds} + R_D} \tag{5.14}$$

5.5.2 Common-Drain Stage

The common-drain configuration is also known as source follower. A common-drain single stage amplifier is shown in Fig. 5.7(a) and its equivalent circuit model is shown in Fig. 5.7(b)

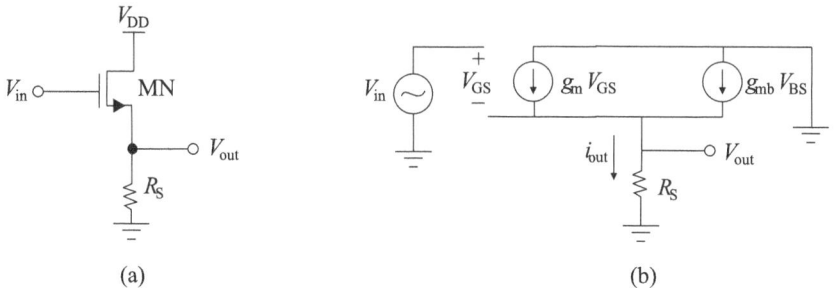

Fig. 5.7 (a) Common-drain single stage amplifier and (b) small-signal equivalent model

The output voltage is given by

$$V_{out} = i_{out} R_S \tag{5.15}$$

where

$$i_{out} = g_m V_{GS} + g_{mb} V_{BS} \tag{5.16}$$

Again,

$$V_{GS} = V_{in} - V_{out} \tag{5.17}$$

and

$$V_{BS} = -V_{out} \tag{5.18}$$

Therefore, Eq. (5.15) becomes

$$V_{out} = i_{out} R_S = [g_m(V_{in} - V_{out}) + g_{mb}(-V_{out})] R_S \tag{5.19}$$

Therefore, the small-signal voltage gain can be expressed as

$$A_v = \frac{\partial V_{\text{out}}}{\partial V_{\text{in}}} = \left[g_m - g_m \frac{\partial V_{\text{out}}}{\partial V_{\text{in}}} - g_{mb} \frac{\partial V_{\text{out}}}{\partial V_{\text{in}}} \right] R_S \tag{5.20}$$

or,

$$A_v = \frac{g_m R_S}{1 + (g_m + g_{mb}) R_S} \tag{5.21}$$

5.5.3 Common-gate Stage

The common-gate (CG) single stage amplifier and its small-signal equivalent circuit are shown in Fig. 5.8.

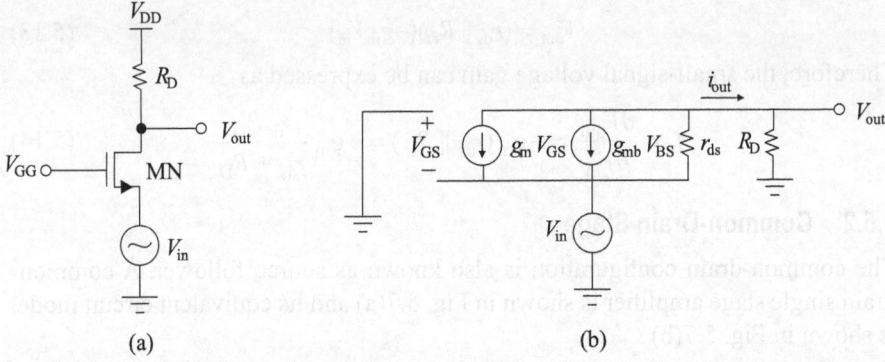

(a)   (b)

Fig. 5.8 (a) Common-gate single stage amplifier and (b) small-signal equivalent model

The output voltage is given by

$$V_{\text{out}} = V_{\text{in}} - \left(g_m V_{GS} + g_{mb} V_{BS} + \frac{V_{\text{out}}}{R_D} \right) r_{ds} \tag{5.22}$$

Again,

$$V_{GS} = -V_{\text{in}} \tag{5.23}$$

and

$$V_{BS} = -V_{\text{in}} \tag{5.24}$$

Therefore,

$$V_{\text{out}} = V_{\text{in}} - (-g_m V_{\text{in}} - g_{mb} V_{\text{in}}) r_{ds} - V_{\text{out}} \frac{r_{ds}}{R_D} \tag{5.25}$$

or,

$$A_v = 1 + (g_m + g_{mb}) r_{ds} - A_v \frac{r_{ds}}{R_D} \tag{5.26}$$

or,

$$A_v \left(1 + \frac{r_{ds}}{R_D} \right) = 1 + (g_m + g_{mb}) r_{ds} \tag{5.27}$$

or,

$$A_v = \frac{1+(g_m + g_{mb})r_{ds}}{1+\dfrac{r_{ds}}{R_D}} = \frac{1+(g_m + g_{mb})r_{ds}}{R_D + r_{ds}} R_D \qquad (5.28)$$

Example 5.2 Find the voltage gain of the common-source single stage amplifier with a diode-connected nMOS load as shown in Fig. 5.9.

Solution The small-signal voltage gain of a CS stage amplifier is given by

$$A_v = -g_{m1}(r_{ds1} \| R_D)$$

For the diode-connected nMOS load (M2), the equivalent load resistance is given by

$$r_{out2} = \frac{1}{g_{m2} + g_{ds2} + g_{mb2}}$$

Therefore, the voltage gain is given by

$$A_v = -g_{m1}\left(r_{ds1} \| \frac{1}{(g_{m2} + g_{ds2} + g_{mb2})}\right)$$

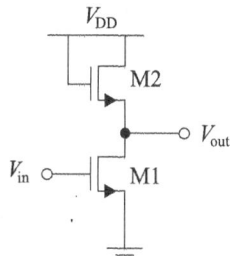

Fig. 5.9 CS stage amplifier with a diode-connected nMOS load

5.6 MOS Current Source and Sink

A current source is a source which delivers constant current irrespective of the terminal voltage. An ideal current source has infinite resistance, whereas a practical current source has very high resistance.

A current sink also has constant current irrespective of its terminal voltage.

MOSFET can be used as current source or sink when it operates in the saturation region. An nMOS can act as a current sink where the negative node is connected to V_{SS}. Alternately, a pMOS can act as a current source where the positive node is connected to V_{DD}.

An nMOS current sink and its $I-V$ characteristics are shown in Fig. 5.10. For nMOS to be operated in saturation region, $V_{DS} \geq V_{GS} - V_{t0}$ or, $V_{OUT} \geq V_{GG} - V_{t0}$.

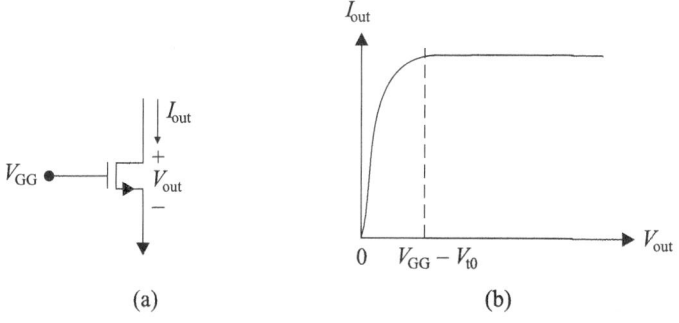

(a)   (b)

Fig. 5.10 (a) nMOS current sink and (b) $I-V$ characteristics of nMOS current sink

Hence, as long as the following condition is satisfied, the nMOS will behave like a current sink.

$$V_{OUT} \geq V_{GG} - V_{t0} \tag{5.29}$$

When the output voltage drops below $(V_{GG} - V_{t0})$, the nMOS transistor goes into linear region and the drain current continues to decrease as the output voltage decreases.

Similarly, a pMOS can act as a current source when it is operated in saturation region. A pMOS current source and its *I–V* characteristics are shown in Fig. 5.11.

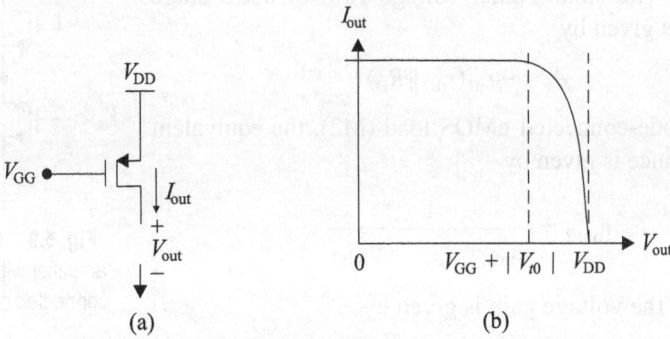

(a)   (b)

Fig. 5.11   (a) A pMOS current source and (b) *I–V* characteristics of pMOS current source.

The pMOS acts as current source as long as the following condition is satisfied.

$$V_{OUT} \leq V_{GG} + |V_{t0}|$$

When the output voltage increases beyond $(V_{GG} + |V_{t0}|)$, the pMOS transistor goes into linear region and the drain current continues to decrease as the output voltage increases further. The current drops to zero when the output reaches V_{DD} as the potential difference between the source and drain becomes zero.

Technique to Increase Output Resistance

The output resistance can be increased using a resistor connected between the source and the ground as shown in Fig. 5.12(a). To calculate the output resistance of the current sink, let us consider the equivalent circuit as shown in Fig. 5.12(b).

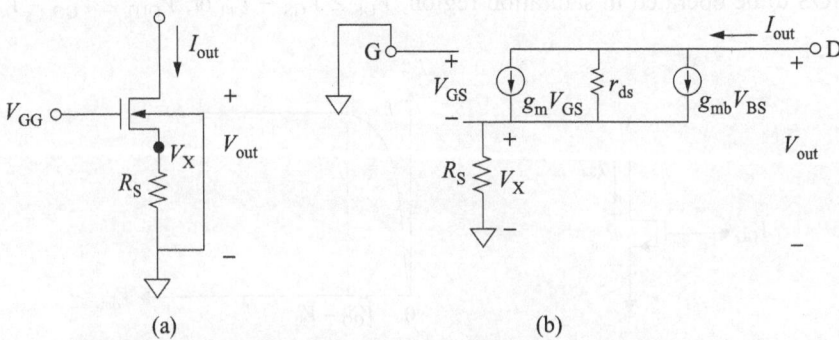

(a)   (b)

Fig. 5.12   (a) Current sink with source degeneration; (b) equivalent circuit of degenerated current sink

The KCL equation at source node can be written as

$$-\frac{V_X}{R_S} + g_m V_{GS} + \frac{(V_{out} - V_X)}{r_{ds}} + g_{mb} V_{BS} = 0 \tag{5.30}$$

But, $V_{GS} = -V_X$, $V_X = V_{SB} = -V_{BS}$, and $V_X = R_S I_{out}$. Therefore, we can write

$$-\frac{R_S I_{out}}{R_S} - g_m R_S I_{out} + \frac{(V_{out} - R_S I_{out})}{r_{ds}} - g_{mb} R_S I_{out} = 0 \tag{5.31}$$

Hence, output resistance can be written as

$$R_{out} = \frac{V_{out}}{I_{out}} = r_{ds}(1 + g_m R_S + \frac{R_S}{r_{ds}} + g_{mb} R_S) \tag{5.32}$$

$$\begin{aligned}
R_{out} &= R_S + r_{ds}(1 + g_m R_S + g_{mb} R_S) \\
&= R_S + r_{ds}[1 + R_S(g_m + g_{mb})] \\
&= r_{ds} + R_S[1 + r_{ds}(g_m + g_{mb})]
\end{aligned} \tag{5.33}$$

If $(g_m + g_{mb})r_{ds} \gg 1$, we can write

$$R_{out} = r_{ds} + R_S r_{ds}(g_m + g_{mb}) = r_{ds}[1 + R_S(g_m + g_{mb})] \tag{5.34}$$

Thus, the output resistance is increased by a factor of $[1 + R_s(g_m + g_{mb})]$.

5.6.1 Cascode Current Sink

The cascode configuration of the current sink increases the output resistance, in which the resistor of the circuit shown in Fig. 5.12(a) is replaced by a common gate MOS transistor. Figure 5.13(a) shows a cascode current sink. Let us now find out the output resistance from the small signal equivalent circuit, as shown in Fig. 5.13(b). We can write the KCL equation at the node S1 = D2 as follows:

$$g_{m1} V_{GS1} + \frac{V_{out} - V_X}{r_{ds1}} + g_{mb1} V_{BS1} - \frac{V_X}{r_{ds2}} = 0 \tag{5.35}$$

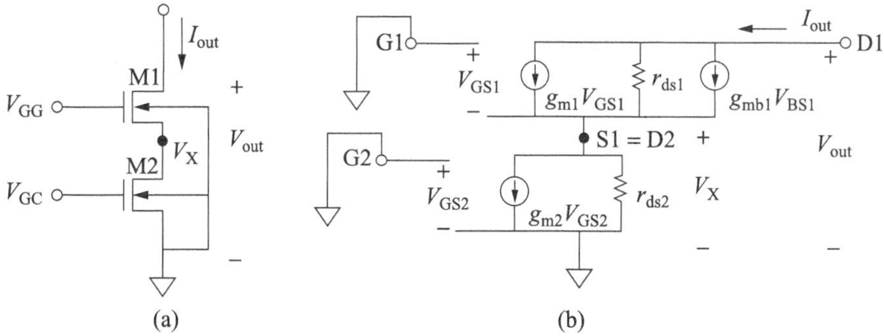

Fig. 5.13 (a) Cascode current sink; (b) equivalent circuit of cascode current sink

The source and bulk are connected together for transistor M2. Hence, $V_{BS2} = 0$ and as gate of M2 is common, V_{GS2} is also zero. From the circuit configuration, we can write $V_X = r_{ds2} I_{out}$, $V_{GS1} = -V_X$, and $V_{BS1} = -V_X$. Hence, we can write Eqn (5.35) as

$$-g_{m1}r_{ds2}I_{out} + \frac{V_{out}}{r_{ds1}} + \frac{r_{ds2}}{r_{ds1}}I_{out} - g_{mb1}r_{ds2}I_{out} - I_{out} = 0 \qquad (5.36)$$

$$\frac{V_{out}}{r_{ds1}} = I_{out}[g_{m1}r_{ds2} + \frac{r_{ds2}}{r_{ds1}} + g_{mb1}r_{ds2} + 1] \qquad (5.37)$$

Hence, output resistance can be written as

$$R_{out} = \frac{V_{out}}{I_{out}} = r_{ds1}[g_{m1}r_{ds2} + \frac{r_{ds2}}{r_{ds1}} + g_{mb1}r_{ds2} + 1] \qquad (5.38)$$
$$= r_{ds1} + r_{ds2} + r_{ds1}(g_{m1} + g_{mb1})r_{ds2}$$

For $g_m > g_{mb1}$ and $g_{m1}r_{ds1} \gg 1$, Eqn (5.38) reduces to

$$R_{out} \cong (g_{m1}r_{ds1})r_{ds2} \qquad (5.39)$$

Hence, the output resistance is increased by a factor of $(g_{m1}r_{ds1})$.

5.7 Current Mirror

The current mirror circuit copies current from a reference. It finds application in analog circuits such as differential amplifiers, digital-to-analog (D/A) converters, etc. A current mirror has very high internal resistance and can act as active current source. It is used to provide constant bias current to the amplifying transistors of an amplifier. It can be used as active load of an amplifier to provide very large voltage gain.

A current mirror circuit is shown in Fig. 5.14. In this circuit, M1 is always operating in the saturation region as its drain and gate terminals are shorted. As M1 and M2 have a common gate, their gate-source voltages are identical, and thus the current through M1 and M2 must be equal if their dimensions are identical. The currents in two transistors are mirrored, hence the name current mirror. If I_{ref} and I_{out} are currents through M1 and M2, respectively, we can write

$$\frac{I_{ref}}{I_{out}} = \frac{\dfrac{\mu_n C_{ox} W_1}{2L_1}(V_{GS} - V_{t1})^2(1 + \lambda V_{DS1})}{\dfrac{\mu_n C_{ox} W_2}{2L_2}(V_{GS} - V_{t2})^2(1 + \lambda V_{DS2})} \qquad (5.40)$$

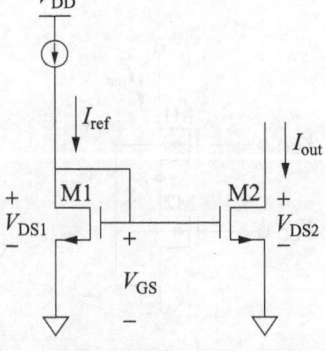

Fig. 5.14 MOS current mirror circuit

M1 and M2 are fabricated in the same integrated circuit under identical process conditions, so their threshold voltage must be same, i.e., $V_{t1} = V_{t2}$. Thus, we can write

$$\frac{I_{ref}}{I_{out}} = \frac{(W_1 L_2)}{(W_2 L_1)}\left(\frac{1 + \lambda V_{DS1}}{1 + \lambda V_{DS2}}\right) \qquad (5.41)$$

If $V_{DS1} = V_{DS2}$, Eqn (5.41) reduces to

$$\frac{I_{ref}}{I_{out}} = \frac{W_1 L_2}{W_2 L_1} \qquad (5.42)$$

or

$$\frac{I_{out}}{I_{ref}} = \frac{W_2 / L_2}{W_1 / L_1}$$

Hence, ratio of the output and reference current is a function of the aspect ratios of the MOS devices which can be controlled by the designer.

Example 5.3　The aspect ratios of the transistors in the circuit shown in Fig. 5.15 are as follows. Find currents I_1, I_2, I_3, and I_4.

Transistor	M0	M1	M2	M3	M4
Aspect ratio *W/L*	10/1	10/1	10/1	5/1	15/1

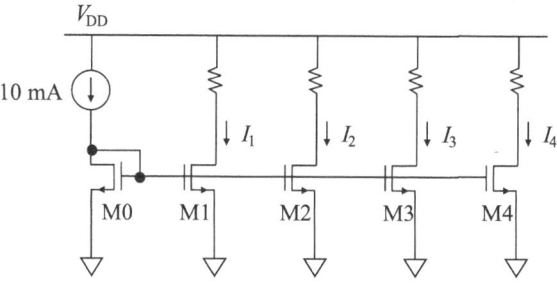

Fig. 5.15　Circuit for Example 5.3.

Solution　The reference current in the given circuit is $I_{ref} = 10$ mA. The transistor M0 operates in saturation region as its gate and drain are connected. For the transistor M0, we can write

$$I_D = I_{ref} = \frac{\mu_n C_{ox}}{2}\left(\frac{W}{L}\right)_0 (V_{GS,0} - V_{t0})^2$$

or,

$$V_{GS,0} = \sqrt{\frac{2 I_{ref}}{\mu_n C_{ox}\left(\dfrac{W}{L}\right)_0}} + V_{t0}$$

As the gate of the transistors M1, M2, M3, and M4 are connected to the gate of the transistor M0, we can write

$$V_{GS,0} = V_{GS,1} = V_{GS,2} = V_{GS,3} = V_{GS,4}$$

Assuming that the transistors have identical threshold voltage and other process parameters remaining the same, we can write

$$I_1 = \frac{\mu_n C_{ox}}{2}\left(\frac{W}{L}\right)_1 (V_{GS,1} - V_{t1})^2 = \frac{\mu_n C_{ox}}{2}\left(\frac{W}{L}\right)_1 \left(\sqrt{\frac{2I_{ref}}{\mu_n C_{ox}\left(\frac{W}{L}\right)_0}} + V_{t0} - V_{t1}\right)^2$$

$$= \frac{(W/L)_1}{(W/L)_0} I_{ref} = \frac{10/1}{10/1} \times 10 = 10\,\text{mA}$$

$$I_2 = \frac{\mu_n C_{ox}}{2}\left(\frac{W}{L}\right)_2 (V_{GS,2} - V_{t2})^2 = \frac{\mu_n C_{ox}}{2}\left(\frac{W}{L}\right)_2 \left(\sqrt{\frac{2I_{ref}}{\mu_n C_{ox}\left(\frac{W}{L}\right)_0}} + V_{t0} - V_{t2}\right)^2$$

$$= \frac{(W/L)_2}{(W/L)_0} I_{ref} = \frac{10/1}{10/1} \times 10 = 10\,\text{mA}$$

$$I_3 = \frac{\mu_n C_{ox}}{2}\left(\frac{W}{L}\right)_3 (V_{GS,3} - V_{t3})^2 = \frac{\mu_n C_{ox}}{2}\left(\frac{W}{L}\right)_3 \left(\sqrt{\frac{2I_{ref}}{\mu_n C_{ox}\left(\frac{W}{L}\right)_0}} + V_{t0} - V_{t3}\right)^2$$

$$= \frac{(W/L)_3}{(W/L)_0} I_{ref} = \frac{5/1}{10/1} \times 10 = 5\,\text{mA}$$

$$I_4 = \frac{\mu_n C_{ox}}{2}\left(\frac{W}{L}\right)_4 (V_{GS,4} - V_{t4})^2 = \frac{\mu_n C_{ox}}{2}\left(\frac{W}{L}\right)_4 \left(\sqrt{\frac{2I_{ref}}{\mu_n C_{ox}\left(\frac{W}{L}\right)_0}} + V_{t0} - V_{t4}\right)^2$$

$$= \frac{(W/L)_4}{(W/L)_0} I_{ref} = \frac{15/1}{10/1} \times 10 = 15\,\text{mA}$$

Example 5.4   Calculate the transconductance and output resistance of the CS stage amplifier shown in Fig. 5.16.

Solution   The small-signal equivalent circuit is shown in Fig. 5.16(b). We can write the input voltage as

$$V_{in} = V_{GS} + R_s i_{out}$$

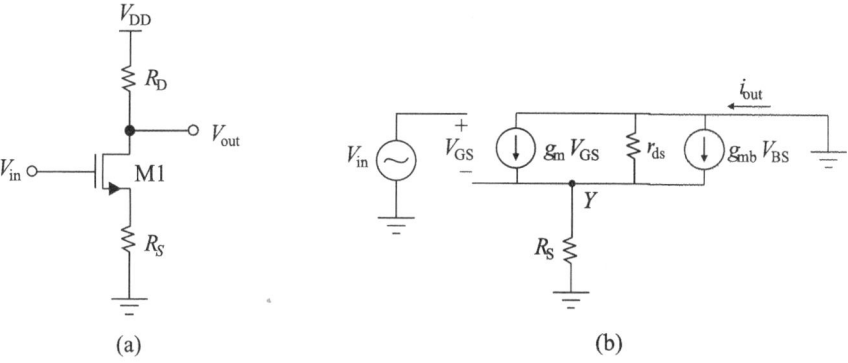

Fig. 5.16   Common-source stage amplifier

Applying KCL at node Y, we can write

$$i_{out} = g_m V_{GS} + g_{mb} V_{BS} - \frac{i_{out} R_S}{r_{ds}}$$

Again,

$$V_{BS} = -i_{out} R_S$$

and

$$V_{GS} = V_{in} - i_{out} R_S$$

Therefore, the expression of i_{out} can be written as

$$i_{out} = g_m (V_{in} - i_{out} R_S) + g_{mb} (-i_{out} R_S) - \frac{i_{out} R_S}{r_{ds}}$$

or,

$$i_{out} = g_m V_{in} - i_{out} \left(g_m R_S + g_{mb} R_S + \frac{R_S}{r_{ds}} \right)$$

Hence, the transconductance can be written as

$$G_m = \frac{i_{out}}{V_{in}} = \frac{g_m}{1 + (g_m R_S + g_{mb} R_S + R_S / r_{ds})} = \frac{g_m r_{ds}}{[1 + (g_m + g_{mb}) R_S] r_{ds} + R_S}$$

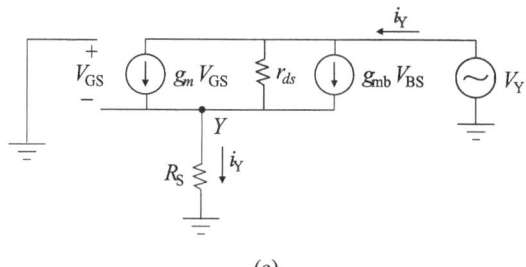

(c)

Fig. 5.16   (c) Small-signal circuit for calculating output resistance

The current flowing through r_{ds} can be written as

$$i_{rds} = i_Y - g_m V_{GS} - g_{mb} V_{BS}$$

Again,

$$V_{GS} = -i_Y R_S$$

and

$$V_{BS} = -i_Y R_S$$

We can write expression of i_{rds} as

$$i_{rds} = i_Y + (g_m + g_{mb}) R_S i_Y$$

We can also write

$$V_Y = i_Y R_S + [i_Y + (g_m + g_{mb}) R_S i_Y] r_{ds}$$

Thus, the output resistance is given by

$$r_{out} = \frac{V_Y}{i_Y} = R_S + [1 + (g_m + g_{mb}) R_S] r_{ds} = R_S[1 + (g_m + g_{mb}) r_{ds}] + r_{ds}$$

Typically, $(g_m + g_{mb}) r_{ds} \gg 1$. Thus, expression of r_{out} becomes

$$r_{out} = R_S(g_m + g_{mb}) r_{ds} + r_{ds} = r_{ds}[1 + (g_m + g_{mb}) R_S]$$

Therefore, the output resistance is increased by a factor $[1 + (g_m + g_{mb}) R_S]$.

5.7.1 Deviation from Ideal Situation

There are three factors which prevent the current mirror to behave ideally:
- Channel length modulation
- Offset in the threshold voltage of the two MOS transistors
- Imperfect geometry of the transistors

1. If drain-to-source voltages of the two transistors are not equal, all other parameters being the same, the current ratio is written as

$$\frac{I_{out}}{I_{ref}} = \frac{1 + \lambda V_{DS2}}{1 + \lambda V_{DS1}} \tag{5.43}$$

If $V_{DS1} \neq V_{DS2}$, the output current is not same as reference current. If λ is small (i.e., high output resistance), ratio approaches unity, and the output current approaches the reference current. Hence, for a good current mirror circuit, the drain-to-source voltage must be identical, and the output resistance must be as high as possible.

2. If the two transistors have offset in their threshold voltages, all other parameters being the same, the ratio of the output and reference currents is written as

$$\frac{I_{out}}{I_{ref}} = \frac{(V_{GS} - V_{t2})^2}{(V_{GS} - V_{t1})^2} \tag{5.44}$$

If $V_{t1} \neq V_{t2}$, the output again deviates from the reference current. But the deviation reduces for large V_{GS}, as for large V_{GS} the threshold offset $(=\Delta V_t = V_{t1} - V_{t2})$ becomes insignificant and hence, output currents approximates to the reference current.

3. If the aspect ratios of the two transistors are identical and all other parameters remain the same, then the output current must be equal to the reference current, and we can write

$$I_{out} = I_{ref} \tag{5.45}$$

In practice, when the transistors are fabricated, there are deviations in the physical dimensions from the drawn dimensions in the layout due to the process variations. This will cause the output current to deviate from the reference current. In order to avoid this deviation, the dimension of the two transistors can be made large so that any variation in geometry due to the process variation becomes insignificant compared to the absolute dimensions.

5.7.2 Cascode Current Mirror

We have seen that due to channel length modulation, the output current deviates from the reference current. This is avoided by increasing the output resistance of the MOS transistors. In the cascode current mirror, the output resistance is increased by using the cascode structure as shown in Fig. 5.17.

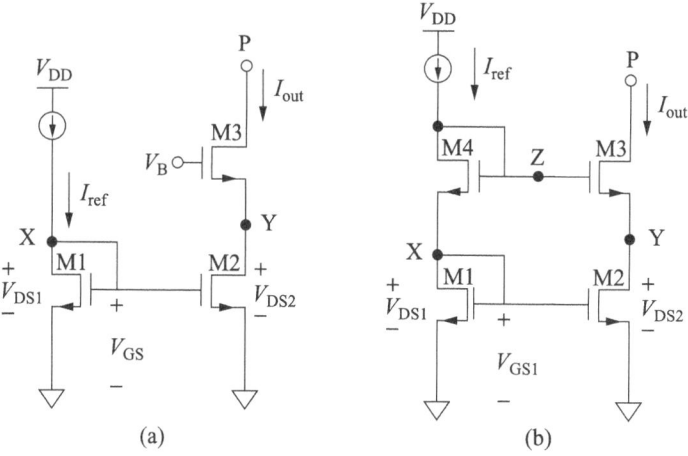

Fig. 5.17 (a) Cascode current source; (b) cascode current mirror

In the current mirror circuit, the drain-to-source voltages may not always be equal. Hence, the channel length modulation causes imperfection in the mirroring action. In order to make the drain-to-source voltages equal, a third transistor is connected in cascode to M2, as shown in Fig. 5.17(a). The voltage V_B is chosen such that $V_X = V_Y$. This ensures that $I_{out} = I_{ref}$. Let us find out the voltage V_B. From Fig. 5.17(a), we can write

$$V_B = V_{GS3} + V_Y \tag{5.46}$$

As the requirement is $V_X = V_Y$, we can write

$$V_B = V_{GS3} + V_X \tag{5.47}$$

This implies that the voltage V_B can be obtained by adding V_X to the gate-to-source voltage of the transistor M3. This can be achieved by connecting a MOS diode M4 in series with the transistor M1 as shown in Fig. 5.17(b). We can write the voltage node Z as

$$V_Z = V_{GS4} + V_X \qquad (5.48)$$

As node Z is connected to the gate of transistor M3, we can write

$$V_{GS4} + V_X = V_{GS3} + V_Y \qquad (5.49)$$

If V_{GS3} can be made equal to V_{GS4}, Eqn (5.49) yields $V_X = V_Y$. This can be achieved if the following condition is met:

$$\frac{(W/L)_3}{(W/L)_4} = \frac{(W/L)_2}{(W/L)_1} \qquad (5.50)$$

This condition can be easily achieved by suitably designing the aspect ratios of the MOS transistors.

5.8 Resistor Realization Using Switched Capacitor

To realize resistors in ICs, a large area is required which is not acceptable. Hence, there is an alternate way of realizing resistors using the switched capacitor technique. Let us introduce the switched capacitor concept first with the help of the circuit shown in Fig. 5.18.

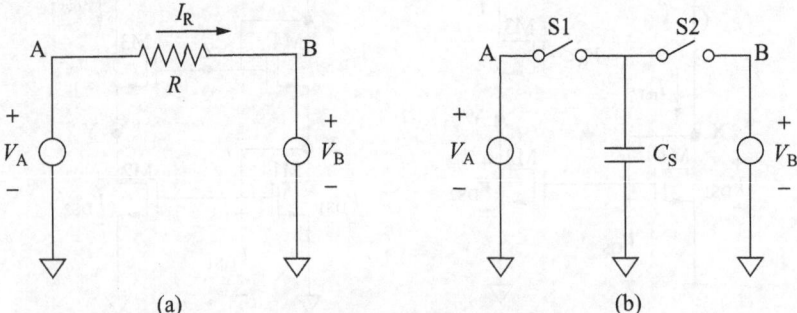

Fig. 5.18   (a) A resistor with two voltage sources; (b) a capacitor with two voltages connected with two switches

The current through the resistor shown in Fig. 5.18(a) can be written as

$$I_R = \frac{V_A - V_B}{R} \qquad (5.51)$$

The resistor in this circuit transfers a certain amount of charge per second from node A to node B. Let us now analyse if the same can be done using a capacitor. For this purpose, we consider the circuit as shown in Fig. 5.18(b). The capacitor C_S is alternately connected to node A and node B at a rate f_{CLK} per second. At any instant of time the stored charge in the capacitor can be written as

$$Q(t) = C_S v(t) \qquad (5.52)$$

Hence, the average current flowing through the capacitor can be written as

$$I_{avg} = \frac{dQ(t)}{dt} = C_S \frac{dv(t)}{dt} = C_S \times f_{CLK}(V_A - V_B) \qquad (5.53)$$

Comparing Eqns (5.51) and (5.53), we note that the resistor R can be replaced by a capacitor which is given by

$$R = \frac{1}{C_S f_{CLK}} \tag{5.54}$$

$$R = \frac{T_{CLK}}{C_S} \tag{5.55}$$

where $T_{CLK} = 1/f_{CLK}$ is the clock period.

Hence, a resistor can be emulated by a capacitor alternately switched between the nodes. This forms the basic concept of a switch capacitor circuit.

Advantages of Switched Capacitor Circuit

There are several advantages of the switch capacitor circuits. Some of them are as follows:

- Very high value resistor can be achieved using a small area
- Temperature and process independent
- Transfer functions with negative resistance can be obtained
- Better resistor value in terms of
  - Good tolerance
  - Good matching
  - Good temperature coefficient
  - Better voltage linearity
  - Wide range

5.9 MOS Voltage and Current References

The reference voltage or current sources are such sources which are more precise and stable. They must produce stable voltage or current independent of the process parameter, power supply, and temperature variations. The stability of the reference source with reference to power supply voltage V_{DD} is defined by

$$S_{V_{DD}}^{V_{REF}} = \frac{(\partial V_{REF}/V_{REF})}{(\partial V_{DD}/V_{DD})} = \frac{V_{DD}}{V_{REF}}\left(\frac{\partial V_{REF}}{\partial V_{DD}}\right) \tag{5.56}$$

If 10% change in V_{DD} causes 10% change in V_{REF}, the sensitivity S will be 1.

The reference sources are required in current mirror circuits, OPAMPs, comparators, analog-to-digital converters, etc. The simplest voltage reference circuit can be a voltage divider implemented using either passive or active circuit elements.

Figure 5.19 shows three different implementations of the voltage divider circuit. In the resistor-based circuit, the reference voltage can be written as

$$V_{REF} = \frac{R_2}{R_1 + R_2} \times V_{DD} \tag{5.57}$$

Equation (5.57) shows that the reference voltage is directly proportional to the power supply voltage, and any change in power supply will cause the reference voltage to change, which is not acceptable for reference voltage source. Hence, this simple voltage divider does not work as a reference voltage source.

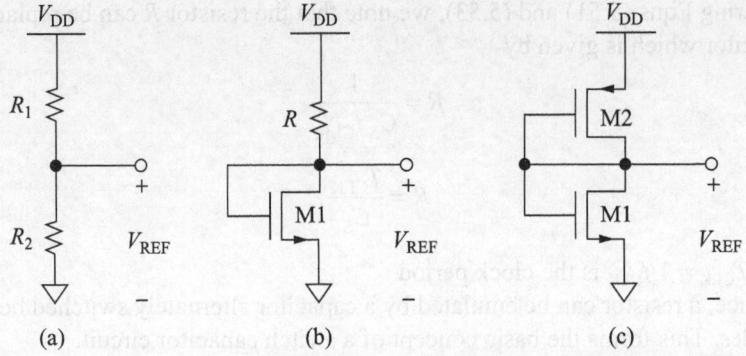

Fig. 5.19   Voltage divider using (a) two resistors; (b) a resistor and an nMOS transistor; (c) a pMOS and an nMOS transistors

Figure 5.19(b) shows an alternate implementation of voltage reference circuit replacing the bottom resistance by an MOS diode. For the resistor and MOSFET-based voltage divider, we can write

$$I_{\mathrm{D}} = \frac{V_{\mathrm{DD}} - V_{\mathrm{REF}}}{R} = \frac{\beta_{\mathrm{n}}}{2}(V_{\mathrm{REF}} - V_{\mathrm{tn}})^2 \tag{5.58}$$

$$V_{\mathrm{REF}} = V_{\mathrm{tn}} + \sqrt{\frac{2I_{\mathrm{D}}}{\beta_{\mathrm{n}}}} = V_{\mathrm{tn}} + \sqrt{\frac{2(V_{\mathrm{DD}} - V_{\mathrm{REF}})}{R\beta_{\mathrm{n}}}} \tag{5.59}$$

or
$$V_{\mathrm{REF}} - V_{\mathrm{tn}} = \sqrt{\frac{2(V_{\mathrm{DD}} - V_{\mathrm{REF}})}{R\beta_{\mathrm{n}}}} \tag{5.60}$$

Taking squares of both sides of Eqn (5.60), we get

$$V_{\mathrm{REF}}^2 - 2V_{\mathrm{REF}}V_{\mathrm{tn}} + V_{\mathrm{tn}}^2 = \frac{2V_{\mathrm{DD}}}{R\beta_{\mathrm{n}}} - \frac{2V_{\mathrm{REF}}}{R\beta_{\mathrm{n}}}$$

Rearranging, we can write

$$V_{\mathrm{REF}}^2 + V_{\mathrm{REF}}\left(\frac{2}{R\beta_{\mathrm{n}}} - 2V_{\mathrm{tn}}\right) + \left(V_{\mathrm{tn}}^2 - \frac{2V_{\mathrm{DD}}}{R\beta_{\mathrm{n}}}\right) = 0 \tag{5.61}$$

Hence, we can find the reference voltage by writing the root of the quadratic equation of V_{REF} as

$$V_{\mathrm{REF}} = -\left(\frac{1}{R\beta_{\mathrm{n}}} - V_{\mathrm{tn}}\right) + \sqrt{\left(\frac{1}{R\beta_{\mathrm{n}}} - V_{\mathrm{tn}}\right)^2 - \left(V_{\mathrm{tn}}^2 - \frac{2V_{\mathrm{DD}}}{R\beta_{\mathrm{n}}}\right)}$$

or
$$V_{\mathrm{REF}} = V_{\mathrm{tn}} - \frac{1}{R\beta_{\mathrm{n}}} + \sqrt{\frac{2(V_{\mathrm{DD}} - V_{\mathrm{tn}})}{R\beta_{\mathrm{n}}} + \frac{1}{R^2\beta_{\mathrm{n}}^2}} \tag{5.62}$$

Differentiating both sides of Eqn (5.59) with reference to V_{DD}, we get

$$\frac{\partial V_{REF}}{\partial V_{DD}} = \frac{1}{2} \times \left(\frac{2(V_{DD} - V_{REF})}{R\beta_n}\right)^{-\frac{1}{2}} \times \frac{2}{R\beta_n} \times \left(1 - \frac{\partial V_{REF}}{\partial V_{DD}}\right)$$

or

$$\frac{\partial V_{REF}}{\partial V_{DD}} = \frac{1}{(V_{REF} - V_{tn})} \times \frac{1}{R\beta_n} \times \left(1 - \frac{\partial V_{REF}}{\partial V_{DD}}\right)$$

or

$$\frac{\partial V_{REF}}{\partial V_{DD}} = \frac{1}{1 + R\beta_n(V_{REF} - V_{tn})} \qquad (5.63)$$

Therefore, sensitivity S can be written as

$$S_{V_{DD}}^{V_{REF}} = \frac{V_{DD}}{V_{REF}}\left(\frac{\partial V_{REF}}{\partial V_{DD}}\right)$$

$$= \left(\frac{1}{1 + R\beta_n(V_{REF} - V_{tn})}\right)\frac{V_{DD}}{V_{REF}} \qquad (5.64)$$

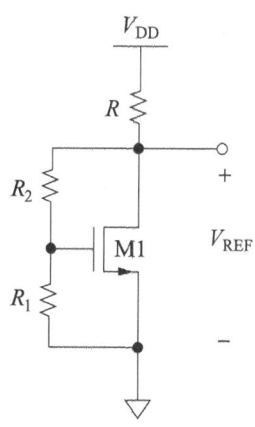

For $V_{DD} = 5$ V, $W/L = 2$, $K' = 110 = \mu A/V^2$ and $R = 100$ k Ω, $V_{REF} = 1.281$ V, and the sensitivity is 0.283.

The value of the reference voltage can be increased in the circuit configuration as shown in Fig. 5.20. We can write

$$V_{REF} = V_{R1} + V_{R2} = V_{GS} + \frac{V_{GS}}{R_1} \times R_2 = V_{GS}\left(1 + \frac{R_2}{R_1}\right) \qquad (5.65)$$

Fig. 5.20   Circuit configuration to increase the reference voltage

5.10   CMOS Amplifier

A single-stage CMOS amplifier is shown in Fig. 5.21.

The current I_D is a function of gate voltage V_{GS} and drain voltage V_{DS}. Hence, we can write the change in I_D as

$$dI_D = \frac{\partial I_D}{\partial V_{GS}}dV_{GS} + \frac{\partial I_D}{\partial V_{DS}}dV_{DS} \qquad (5.66)$$

As the drain current is driven by a current source, it is constant and hence $dI_D = 0$. Then Eqn (5.66) can be written as

$$\frac{\partial I_D}{\partial V_{GS}}dV_{GS} + \frac{\partial I_{DS}}{\partial V_{DS}}dV_{DS} = 0 \qquad (5.67)$$

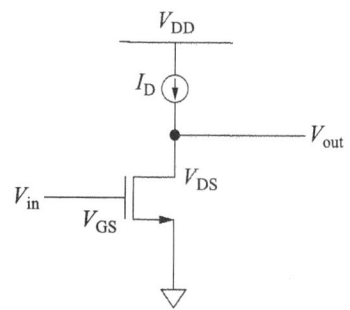

Fig. 5.21   Single-stage CMOS amplifier

With the application of the input AC signal, the change in gate-to-source voltage is the input voltage ($v_{in} = dV_{GS}$), and the change in drain-to-source voltage is the output voltage ($v_{out} = dV_{DS}$). Hence, we can write Eqn (5.67) as

$$g_m v_{in} + \frac{1}{r_{ds}} v_{out} = 0 \tag{5.68}$$

where

$$\text{Transconductance } g_m = \frac{\partial I_D}{\partial V_{GS}}\bigg|_{V_{DS}=\text{constant}} \tag{5.69}$$

$$\text{Output resistance } r_{ds} = \frac{\partial V_{DS}}{\partial I_D}\bigg|_{V_{GS}=\text{constant}} \tag{5.70}$$

Hence, from Eqn (5.68) the voltage gain can be written as

$$A_v = \frac{v_{out}}{v_{in}} = -g_m r_{ds} \tag{5.71}$$

Let us now derive the expressions for transconductance and output resistance to find the voltage gain. From Eqn (2.33), under saturation, we can write

$$(V_{GS} - V_t) = \sqrt{\frac{2I_D}{\beta_n}} \tag{5.72}$$

But under saturation, $V_{D,sat} >= V_{GS} - V_t$. So the minimum saturation voltage can be written as

$$V_{ON} = V_{D,sat} = V_{GS} - V_t = \sqrt{\frac{2I_D}{\beta_n}} \tag{5.73}$$

The drain current under saturation is given by

$$I_D = \frac{\beta_n}{2}(V_{GS} - V_t)^2 = \frac{\beta_n}{2} V_{ON}^2 \tag{5.74}$$

Hence, the transconductance can be written as

$$g_m = \frac{\partial I_D}{\partial V_{GS}} = \beta_n(V_{GS} - V_t) = \beta_n V_{ON} = k_n' \frac{W}{L} V_{ON} \tag{5.75}$$

Substituting V_{ON}, we get

$$g_m = \beta_n \sqrt{\frac{2I_D}{\beta_n}} = \sqrt{2\beta_n I_D} = \sqrt{2k_n' \frac{W}{L} I_D} \tag{5.76}$$

Again

$$\beta_n = \frac{2I_D}{V_{ON}^2} \tag{5.77}$$

Therefore, transconductance can be written as

$$g_m = \beta_n V_{ON} = \frac{2I_D}{V_{ON}} \tag{5.78}$$

The design parameters are listed in Table 5.1.

Table 5.1 Design parameters for the CMOS amplifier

Transconductance (g_m)	Design parameter
$k'_n \dfrac{W}{L} V_{ON}$	W, L, V_{ON}
$\sqrt{2k'_n \dfrac{W}{L} I_D}$	W, L, I_D
$\dfrac{2I_D}{V_{ON}}$	I_D, V_{ON}

To increase g_m should we increase V_{ON} or decrease it? Is g_m linearly dependent on the transistor size? Does it depend on its square root? Or is it independent of transistor size?

In fact, which formula should be applied depends on how the transistor is biased and sized. If size and V_{ON} are known, the first formula applies. If the drain current and size are known, the second one does. If gate voltage and drain current are given, and the transistor is accordingly sized, the third formula should be used.

From Eqn (5.4), we get the expression for the output resistance as

$$r_{ds} = \frac{1}{\lambda I_D} \tag{5.79}$$

where λ is the channel length modulation factor. The channel length modulation increases as the channel length decreases, and becomes significant for the short channel devices. Hence, we can write channel length modulation as

$$\lambda \propto \frac{1}{L} \tag{5.80}$$

Or, $\lambda = \dfrac{\lambda'}{L}$ where λ' is a proportionality constant which depends on the process technology. Equation (5.79) can be written as

$$r_{ds} = \frac{L}{\lambda' I_D} \tag{5.81}$$

Substituting the expression for I_D, we get

$$r_{ds} = \frac{2L}{\lambda' \beta_n V_{ON}^2} = \frac{2L^2}{\lambda' \mu_n C_{ox} W V_{ON}^2} \tag{5.82}$$

Hence, the voltage can be written as

$$A_v = -g_m \times r_{ds} = -\beta_n V_{ON} \times \frac{2L}{\lambda' \beta_n V_{ON}^2} = -\frac{2L}{\lambda' V_{ON}} \qquad (5.83)$$

We see that when biased at constant V_{ON}, the voltage gain depends only on L and is independent of W. In terms of drain current and geometry, voltage gain can be written as

$$A_v = -\frac{2L}{\lambda' \sqrt{2I_D/\beta_n}} = -\frac{2L}{\lambda'} \sqrt{\frac{\mu_n C_{ox}}{2I_D} \times \frac{W}{L}} = -\frac{1}{\lambda'} \sqrt{\frac{2\mu_n C_{ox} WL}{I_D}} \qquad (5.84)$$

Hence, if the transistor is biased at constant current, the DC gain is determined by the square root of the gate area ($= W \times L$).

Frequency Response of the CMOS Amplifier

Let us now find out the frequency response of the CMOS amplifier. For this purpose, we consider the circuit shown in Fig. 5.22. Applying KCL at the drain node, we can write the following equation:

$$sC_{gd}(v_{in} - v_{out}) - g_m v_{in} - \frac{v_{out}}{r_{ds}} - sC_{out} v_{out} = 0 \qquad (5.85)$$

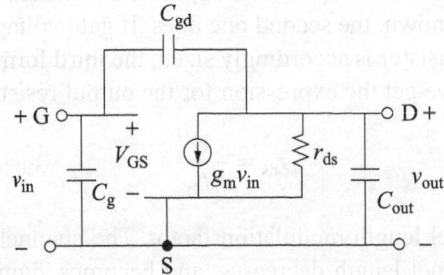

Fig. 5.22 Small signal equivalent model of the CMOS amplifier

or

$$v_{out}\left(-sC_{gd} - \frac{1}{r_{ds}} - sC_{out}\right) = v_{in}(g_m - sC_{gd}) \qquad (5.86)$$

Then we can write the voltage gain as

$$A_v = \frac{v_{out}}{v_{in}} = \frac{g_m - sC_{gd}}{-sC_{gd} - 1/r_{ds} - sC_{out}} = -g_m r_{ds} \frac{1 - sC_{gd}/g_m}{1 + sr_{ds}(C_{gd} + C_{out})} \qquad (5.87)$$

Let us define the zero and pole as

$$z = \frac{g_m}{C_{gd}} \quad \text{and} \quad p = -\frac{1}{r_{ds}(C_{gd} + C_{out})} \qquad (5.88)$$

Thus, in terms of zero and pole, the voltage gain can be written as

$$A_v = -g_m r_{ds} \frac{1 - s/z}{1 - s/p} \tag{5.89}$$

Normally, $\omega C_{gd}/g_m \ll 1$, hence, voltage gain can be written as

$$A_v = -\frac{g_m r_{ds}}{1 + sr_{ds}(C_{gd} + C_{out})} = -\frac{A_o}{1 + sr_{ds}C_{total}} \tag{5.90}$$

where $C_{total} = C_{gd} = C_{out}$.

Equation (5.90) describes the frequency response of the amplifier. This bandwidth (BW) of the amplifier can be written as

$$BW = \frac{1}{r_{ds}C_{total}} = \frac{1}{r_{ds}(C_{gd} + C_{out})} \tag{5.91}$$

Hence, product of gain and bandwidth (GBW) can be written as

$$GBW = A_v \times BW = -g_m r_{ds} \times \frac{1}{r_{ds}(C_{gd} + C_{out})} = -\frac{g_m}{C_{gd} + C_{out}} = -\frac{g_m}{C_{total}} \tag{5.92}$$

The gain–bandwidth product is independent of the output resistance. The variation of the gain–bandwidth product with the frequency is shown in Fig. 5.23.

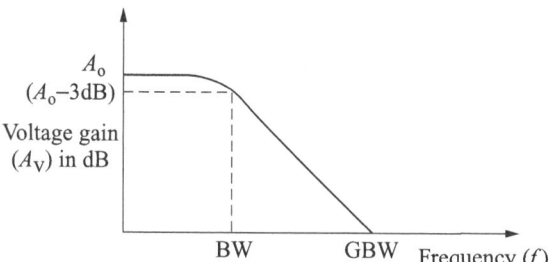

Fig. 5.23   Voltage gain vs frequency

5.11   MOS Differential Amplifier

The differential amplifier is an important analog building block which is often used in various analog circuits. It is used as the input stage in operational amplifier (OPAMP). It has two inputs and one output. One input is called non-inverting and the other is called inverting. The differential amplifier output is a function of the differential mode input and common mode input. If v_1 and v_2 are the two input signals, then output voltage can be written as

$$V_{out} = A_D(v_1 - v_2) \pm A_C \left(\frac{v_1 + v_2}{2} \right) \tag{5.93}$$

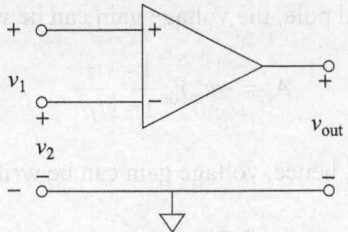

Fig. 5.24 Symbol of a differential amplifier

where A_D is the differential mode voltage gain, and A_C is the common mode voltage gain. The symbolic representation of the differential amplifier is shown in Fig. 5.24.
The differential mode input and common mode inputs are given by

$$v_{id} = v_1 - v_2 \tag{5.94}$$

$$v_{ic} = \frac{v_1 + v_2}{2} \tag{5.95}$$

respectively.

Circuits which amplify the *difference* of two-input voltages (each of which has equal and opposite signal excursions) have many advantages over single-ended amplifiers. The advantages of using differential inputs are:

- Noise picked up by both inputs gets cancelled in the output.
- Input and feedback paths can be isolated.
- If both inputs have the same DC bias, the output is insensitive to changes in the bias.

The main goal of the differential amplifier is to amplify the differential mode input signal. Hence, the differential amplifier is characterized by a parameter called common mode rejection ratio (CMRR), which is defined by

$$\text{CMRR} = 20 \log \frac{A_D}{A_C} \text{ in dB} \tag{5.96}$$

A good differential amplifier must have a large differential mode voltage gain and small common mode voltage gain, and must have high CMRR.

A differential amplifier using MOS transistors is shown in Fig. 5.25. The total current flowing through the transistors is kept constant using a current source. The current equation can be written as

$$I_{D1} + I_{D2} = I_{SS} \tag{5.97}$$

Let us now explain how this circuit will behave like a differential amplifier. Let us first consider $v_1 = v_2$, then current through the transistor M1

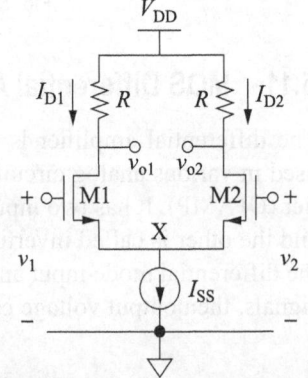

Fig. 5.25 MOS differential amplifier

will be larger than that of M2. This will cause voltage drop across the resistor connected to M1 to be more than the resistor connected to M2. Thus, the voltage v_{o1} will be less than the voltage v_{o2}, and we can write

$$(v_{o2} - v_{o1}) > 0 \text{ or positive, for } v_{id} > 0$$

Similarly, let us now consider that $v_1 = v_2$, then current through M1 will be less than M2. Thus, the voltage v_{o1} will be more than the voltage v_{o2}, and we can write

$$(v_{o2} - v_{o1}) < 0 \text{ or negative, for } v_{id} < 0$$

Hence, the circuit amplifies the difference between the two input signals.

Quantitative Analysis

From Fig. 5.23, we can write the output voltages as

$$\begin{aligned} v_{o1} &= V_{DD} - I_{D1}R \\ v_{o2} &= V_{DD} - I_{D2}R \end{aligned} \tag{5.98}$$

If the voltage at node X is V_X, we can write

$$\begin{aligned} v_1 &= V_X + V_{GS1} \\ v_2 &= V_X + V_{GS2} \end{aligned} \tag{5.99}$$

The difference between the input signals can be written as

$$v_1 - v_2 = V_{GS1} - V_{GS2} \tag{5.100}$$

But from the current expression of the MOS transistors, we can write

$$\left(V_{GS} - V_t\right)^2 = \frac{2I_D}{\beta_n} \tag{5.101}$$

$$V_{GS} - V_t = \sqrt{\frac{2I_D}{\beta_n}} \tag{5.102}$$

Substituting Eqn (5.102) in Eqn (5.100), we get

$$v_1 - v_2 = \sqrt{\frac{2I_{D1}}{\beta_n}} - \sqrt{\frac{2I_{D2}}{\beta_n}} \tag{5.103}$$

Squaring both sides of Eqn (5.103), we get

$$\left(v_1 - v_2\right)^2 = \frac{2I_{D1}}{\beta_n} + \frac{2I_{D2}}{\beta_n} - 2\sqrt{\frac{2I_{D1}}{\beta_n} \times \frac{2I_{D2}}{\beta_n}} \tag{5.104}$$

$$\left(v_1 - v_2\right)^2 = \frac{2I_{SS}}{\beta_n} - 2 \times \frac{2}{\beta_n}\sqrt{I_{D1}I_{D2}} \tag{5.105}$$

$$\frac{\beta_n}{2}\left(v_1 - v_2\right)^2 - I_{SS} = -2\sqrt{I_{D1}I_{D2}} \tag{5.106}$$

Squaring both sides of Eqn (5.106), we get

$$\left(\frac{\beta_n}{2}\right)^2 (v_1 - v_2)^4 + I_{SS}^2 - 2\frac{\beta_n}{2}(v_1 - v_2)^2 I_{SS} = 4I_{D1}I_{D2} \tag{5.107}$$

$$\left(\frac{\beta_n}{2}\right)^2 (v_1 - v_2)^4 + I_{SS}^2 - 2\frac{\beta_n}{2}(v_1 - v_2)^2 I_{SS} = (I_{D1} + I_{D2})^2 - (I_{D1} - I_{D2})^2 \tag{5.108}$$

$$\left(\frac{\beta_n}{2}\right)^2 (v_1 - v_2)^4 + I_{SS}^2 - 2\frac{\beta_n}{2}(v_1 - v_2)^2 I_{SS} = I_{SS}^2 - (I_{D1} - I_{D2})^2 \tag{5.109}$$

$$\left(\frac{\beta_n}{2}\right)^2 (v_1 - v_2)^4 - \beta_n(v_1 - v_2)^2 I_{SS} = -(I_{D1} - I_{D2})^2 \tag{5.110}$$

$$(I_{D1} - I_{D2})^2 = -\left(\frac{\beta_n}{2}\right)^2 (v_1 - v_2)^4 + \beta_n(v_1 - v_2)^2 I_{SS} \tag{5.111}$$

$$I_{D1} - I_{D2} = \frac{\beta_n}{2}(v_1 - v_2)\sqrt{\frac{4I_{SS}}{\beta_n} - (v_1 - v_2)^2} \tag{5.112}$$

$$\Delta I_D = \frac{\beta_n}{2}\Delta V_{in}\sqrt{\frac{4I_{SS}}{\beta_n} - \Delta V_{in}^2} \tag{5.113}$$

where

$$\Delta I_D = I_{D1} - I_{D2}$$
$$\Delta V_{in} = v_1 - v_2 \tag{5.114}$$

$$\frac{\partial \Delta I_D}{\partial \Delta V_{in}} = \frac{\beta_n}{2}\sqrt{\frac{4I_{SS}}{\beta_n} - \Delta V_{in}^2} - \frac{1}{2} \times \frac{\beta_n \Delta V_{in}^2}{\sqrt{\frac{4I_{SS}}{\beta_n} - \Delta V_{in}^2}} \tag{5.115}$$

$$G_m = \frac{\partial \Delta I_D}{\partial \Delta V_{in}} = \frac{\beta_n}{2}\left(\frac{\frac{4I_{SS}}{\beta_n} - 2\Delta V_{in}^2}{\sqrt{\frac{4I_{SS}}{\beta_n} - \Delta V_{in}^2}}\right) \tag{5.116}$$

For $\Delta V_{in} = 0$, $G_m = \sqrt{I_{SS}\beta_n}$. The difference in the output voltages can be written using Eqn (5.98) as

$$\Delta V_{out} = v_{o1} - v_{o2} = R(I_{D1} - I_{D2}) = RG_m\Delta V_{in} \tag{5.117}$$

The voltage gain can be written as

$$A_D = \frac{\Delta V_{out}}{\Delta V_{in}} = RG_m = R\sqrt{I_{SS}\beta_n} \tag{5.118}$$

5.12   Cascode Amplifier

The cascade of a common source (CS) and a common gate (CG) is termed as *cascode* topology. The term is derived from the phrase 'cascaded triodes'. Figure 5.26 shows a cascode amplifier.

The transistor M1 is the input device and the transistor M2 is the cascode device. The transistor M1 operates in the saturation region and the input signal is applied to its gate. The transistor is biased by a voltage source V_B. This topology offers many advantages over normal MOS amplifiers.

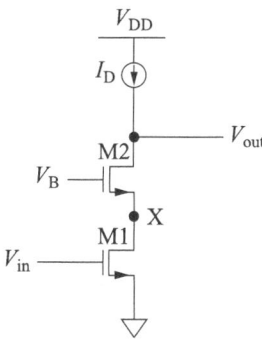

Fig. 5.26   Cascode amplifier

For M1 to be operated in the saturation region, the voltage at node X must be

$$V_X \geq V_{in} - V_{t1} \tag{5.119}$$

The bias voltage V_B can be written as

$$V_B = V_{GS2} + V_X \tag{5.120}$$

From Eqns (5.119) and (5.120), we can write

$$V_B - V_{GS2} \geq V_{in} - V_{t1}$$

$$V_B \geq V_{in} + V_{GS2} - V_{t1} \tag{5.121}$$

Now, if the transistor M2 also operates in the saturation region, we can write

$$V_{out} \geq V_B - V_{t2} \tag{5.122}$$

Using Eqn (5.121), Eqn (5.122) becomes

$$V_{out} \geq V_{in} + V_{GS2} - V_{t1} - V_{t2} \tag{5.123}$$

Let us now find out the voltage gain of the cascode amplifier. For this purpose, we consider the small signal equivalent circuit as shown in Fig. 5.27. From Eqn (5.16), we can straightway write the expression for the output resistance as

$$R_{out} = r_{ds2} + r_{ds1}[1 + r_{ds2}(g_{m2} + g_{mb2})] \tag{5.124}$$

If $g_m r_{ds} \gg 1$, the output resistance becomes

$$R_{out} = r_{ds1} r_{ds2}(g_{m2} + g_{mb2}) \tag{5.125}$$

The cascode transistor increases the output resistance of M1 by a factor of $r_{ds2}(g_{m2} + g_{mb2})$.

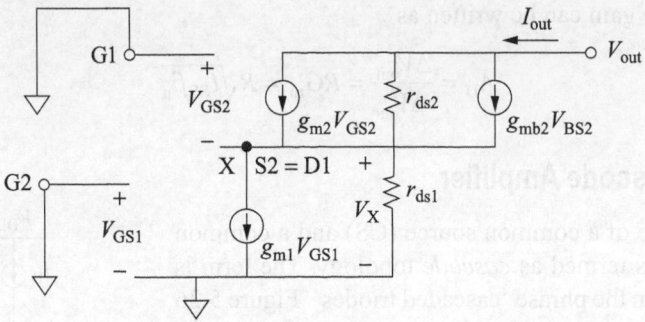

Fig. 5.27 Small signal equivalent circuit of cascode amplifier

The voltage gain is determined by the product of transconductance and output resistance according to Eqn (5.54). The transconductance of M1 is g_m, which is determined by the bias current and the aspect ratio of the transistor with reference to Eqn (5.3). The voltage gain is increased by increasing the output resistance R_{out}. The cascode configuration is very useful in increasing the output resistance, and hence, increasing the voltage gain. The voltage gain of a cascode amplifier can be written as

$$A_v = g_{m1} \times r_{ds1} r_{ds2} (g_{m2} + g_{mb2}) \tag{5.126}$$

5.13 Current Amplifier

A current amplifier is a class of amplifier in which the input impedance of the amplifier is small, and the output impedance is large. It takes current as an input signal and delivers the output signal as current. A current amplifier must be driven by a source with a large impedance and is loaded with a small impedance.

The current mirror circuit can be used as a current amplifier when the input signal is added to the reference current as shown in Fig. 5.28.

The ratio of output current and reference can be written as

$$I_{out} = \left(\frac{W_2/L_2}{W_1/L_1} \right) I_{ref} \tag{5.127}$$

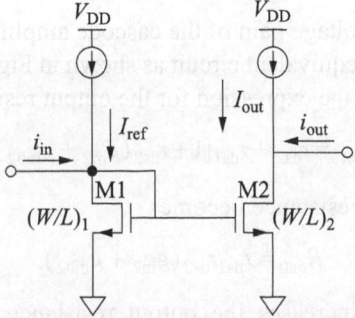

Fig. 5.28 Current amplifier

For a differential change in reference current due to the input signal, the change in output current can be written as

$$i_{out} = \Delta I_{out} = \left(\frac{W_2/L_2}{W_1/L_1} \right) \Delta I_{ref} = \frac{W_2/L_2}{W_1/L_1} i_{in} \qquad (5.128)$$

Hence, current gain can be written as

$$A_i = \frac{i_{out}}{i_{in}} = \frac{(W/L)_2}{(W/L)_1} \qquad (5.129)$$

If $(W/L)_2 = (W/L)_1$, current gain can be achieved. By a proper choice of aspect ratios of the transistors, the current amplifier can be designed for a desired gain.

5.14   Output Amplifier

The output amplifier is a special class of amplifier which is capable of driving a small output resistor (50–1000 Ω) and/or a large capacitive load (5–1000 pF). To drive a small resistor, the output resistance of the output amplifier must be equal or smaller than the load resistance. In order to drive a large capacitive load, the amplifier must have a large output current sink or source capability.

There are several approaches to achieve these requirements. Some of them are as follows:

- Class A amplifier
- Source follower
- Push–pull amplifier
- Negative feedback

In this section, we shall discuss the class A amplifier. Figure 5.29(a) shows a class A amplifier. It is basically an inverter with a current source load. The amplifier drives a resistance R_L in parallel with a capacitance C_L.

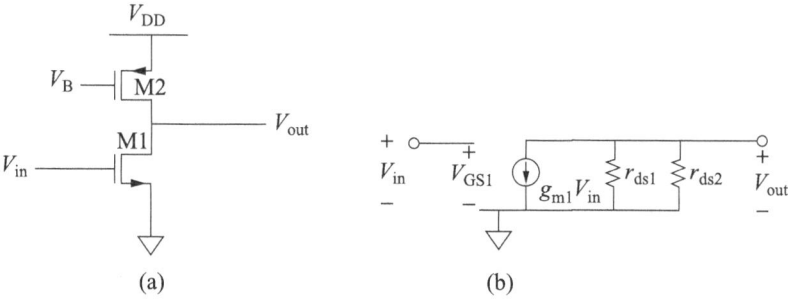

(a)   (b)

Fig. 5.29   (a) Class A amplifier; (b) small signal model of class A amplifier

The small signal equivalent circuit is shown in Fig. 5.29(b). The output resistance of the amplifier is given by

$$R_{out} = r_{ds1} \| r_{ds2} = \frac{r_{ds1} r_{ds2}}{r_{ds1} + r_{ds2}} \qquad (5.130)$$

The gain of the amplifier is given by

$$A_v = -g_{m1} \times R_{out} = -\frac{g_{m1}r_{ds1}r_{ds2}}{r_{ds1}+r_{ds2}} \tag{5.131}$$

5.15   Source Follower

The source follower is a common drain amplifier having a large current gain and low output resistance. It is used as an output amplifier. A source follower circuit is shown in Fig. 5.30(a).

In the source follower circuit, the input signal is applied to the gate of the transistor and the output is taken from the source of the transistor. As the source terminal follows the gate terminal, it is called the source follower. The small signal equivalent circuit is shown in Fig. 5.30(b). Applying KCL at the source node, we can write

$$g_{m1}(v_{in} - v_{out}) - g_{mb1}v_{out} = \frac{v_{out}}{R} \tag{5.132}$$

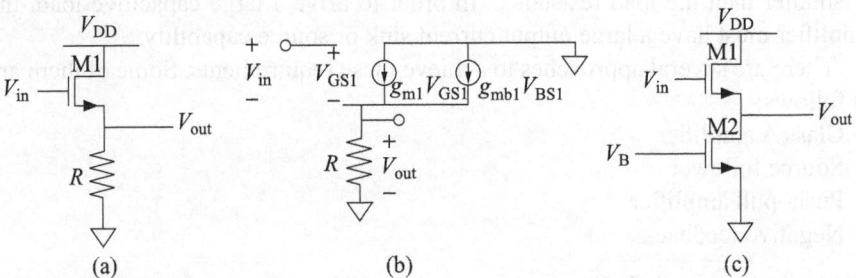

Fig. 5.30   (a) Source follower; (b) equivalent circuit; (c) source follower with a current source

The voltage gain can be written as

$$A_v = \frac{v_{out}}{v_{in}} = \frac{g_{m1}R}{1+(g_{m1}+g_{mb1})R} \tag{5.133}$$

The resistor is replaced by a current sink as shown in Fig. 5.30(c) to avoid the amplifier from going into the non-linear region due to a large change in input voltage. The current sink is implemented by an nMOS transistor operating in the saturation region.

5.16   Voltage Level Shifter

A voltage level shifter is a circuit that shifts the signal logic level. The voltage level shifter is used in the input/output buffer circuit for interfacing the external environment. The level shifting is required if the core logic in an IC operates at a lower voltage level but the chip has to interface other components in the system that operate at a higher voltage level. For example, the core logic of a chip might operate at 1.8 V, whereas the other chips operate at 3.3 V. Hence, the chip operating at 1.8 V must have voltage level shifting arrangement in the input and output buffers for 3.3 V to 1.8 V, and 1.8 V to 3.3 V, respectively.

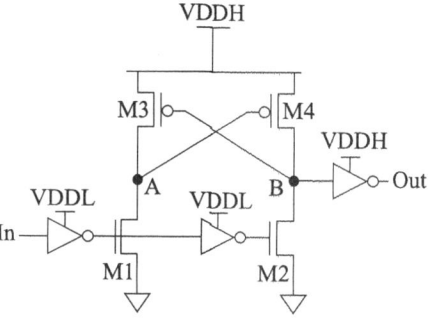

A voltage level shifter circuit which shifts the low voltage level to the high voltage level is shown in Fig. 5.31. This circuit is designed using two types of MOS transistors. One, having high threshold voltage (high V_t), and other having low threshold voltage (low V_t).

Fig. 5.31 Voltage level shifter

There are low power supply voltages, VDDH and VDDL, where VDDH is higher than VDDL. The high V_t transistors are connected to the VDDH power supply, and the low V_t transistors are connected to the VDDL power supply. The voltage level shifter produces the same signal at the output (like a buffer) but only the voltage levels are different.

The transient characteristic of the voltage level shifter is shown in Fig. 5.32. As seen from the voltage waveforms, the input and output logic levels are different.

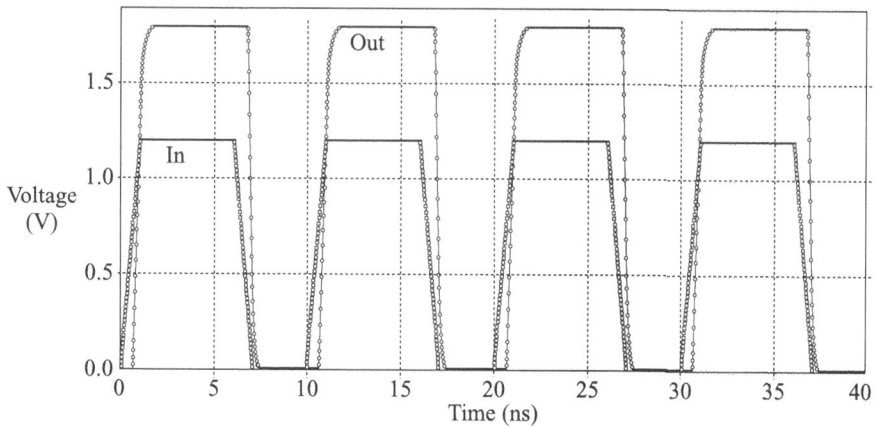

Fig. 5.32 Voltage level shifter: input–output waveform

5.17 CMOS Operational Amplifier

The operational amplifier is one of the most important building blocks in the analog circuit design. The open loop gain in the operational amplifier is very high, and when it is connected in the negative feedback, the closed loop gain becomes almost independent of the open loop gain. There are two operational amplifier structures: (a) buffered, where the output resistance is low, and it is used as voltage operational amplifier; and (b) unbuffered, where the output resistance is high, and it is used as operational transconductance amplifier (OTA). The operational amplifiers are widely known as OPAMP. There are two or more stages in OPAMPs to have a sufficiently large open loop gain. But two-stage OPAMP is the most popular.

In this section, we will introduce the design technique of CMOS OPAMP and its compensation technique for maintaining stability when it is used in negative feedback.

5.17.1 Design of CMOS OPAMP

Let us first explain the design parameters that are used to design a CMOS OPAMP circuit.

- *Gain*   The open-loop gain of the OPAMP must be very high, so that when it is used in negative feedback, the closed-loop gain must be independent of open-loop gain.
- *Small signal bandwidth (BW)*   The open-loop gain decreases as the frequency of the operation increases. Hence, the design must consider the BW of the OPAMP.
- *Large signal bandwidth*   The OPAMP is used in large signal transients. Hence, it must respond to the transient signals that change very fast in time.
- *Output swing*   The OPAMP must have a large output voltage swing.
- *Linearity*   The OPAMP characteristics must be linear.
- *Noise and offset*   The noise and offset must be insignificant.
- *Power supply rejection*   The power supply rejection must be high.

The basic building blocks of CMOS OPAMP are as follows:
- Differential transconductance stage
- High gain second stage
- Output amplifier
- Biasing circuit
- Compensation circuit

The simple one-stage OPAMP is shown in Fig. 5.33.

This is a differential amplifier with a single-ended and differential output. This circuit is also called an operational transconductance amplifier (OTA) because the output is a current. The output resistance is given by

$$R_{out} = r_{ds2} \parallel r_{ds4}$$

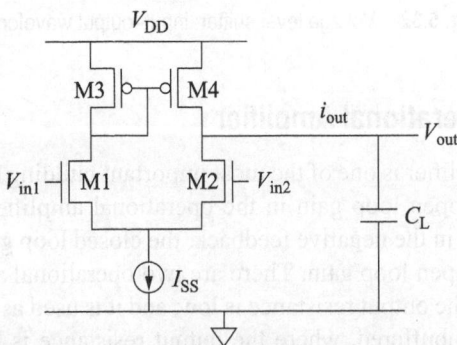

Fig. 5.33 Operational transconductance amplifier

The voltage gain is given by

$$A_v = g_m R_{out} = g_m (r_{ds2} \| r_{ds4})$$ (5.134)

5.17.2 Design of Two-stage OPAMP

A simple two-stage OPAMP can be constructed as shown in Fig. 5.34 by following the differential amplifier by a common source stage with a constant current load. The current source for the differential amplifier is implemented by an n-channel MOS transistor in saturation. The two-stage design permits us to optimize the output stage for driving the load and the input stage for providing a good differential gain and CMRR. A differential amplifier with nMOS transistors, and an output stage with a pMOS driver are shown in Fig. 5.34. However, a pMOS differential amplifier with the nMOS common source stage is better for low noise operation.

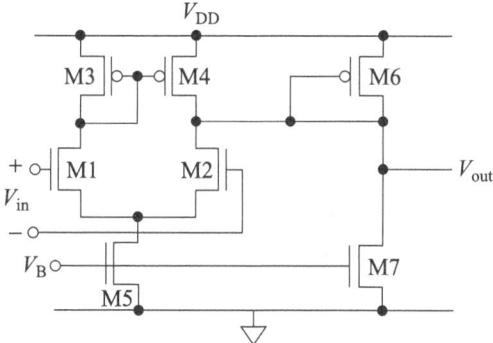

Fig. 5.34 Two-stage OPAMP

5.17.3 Compensation of OPAMP

The high-gain OPAMP is generally used in negative feedback configuration to obtain a very accurate transfer function which can be controlled by the feedback elements only. Figure 5.35 shows a negative feedback configuration in general. The high-gain amplifier has an open-loop gain $A(s)$ and the feedback element has a function $\beta(s)$. The loop gain of the feedback system can be written as

$$L(s) = A(s)\beta(s)$$ (5.135)

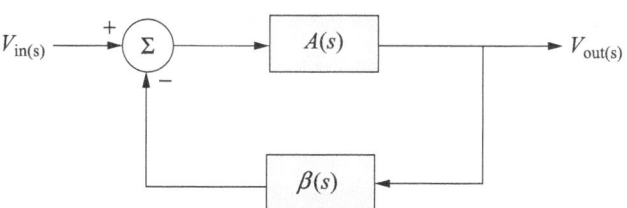

Fig. 5.35 A negative feedback system

The gain of the negative feedback system with the feedback is written as

$$A_f(s) = \frac{A(s)}{1 + A(s)\beta(s)} \qquad (5.136)$$

The magnitude and the phase of the loop gain are important because that decides whether the feedback system is positive or negative. If the following conditions are met, system gain becomes infinite and it becomes an oscillator:

$$|A(j\omega)\beta(j\omega)| = 1 \quad \text{and} \quad \angle A(j\omega)\beta(j\omega) = 0° \qquad (5.137)$$

The above conditions are together called the *Barkhausen criterion*. In order to avoid sustained oscillation, the above conditions must be avoided. The following equation describes how we can avoid the Barkhausen criterion

$$|A(j\omega_{0°})\beta(j\omega_{0°})| < 1 \qquad (5.138)$$

where $\omega_0^°$ is defined as the frequency for which the phase of loop gain is zero, i.e.,

$$\angle A(j\omega_{0°})\beta(j\omega_{0°}) = 0° \qquad (5.139)$$

An alternate condition for avoiding sustained oscillation is

$$\angle A(j\omega_{0\,dB})\beta(j\omega_{0\,dB}) > 0° \qquad (5.140)$$

where $\omega_{0\,dB}$ is defined as the frequency for which the loop gain is unity, i.e.,

$$|A(j\omega_{0\,dB})\beta(j\omega_{0\,dB})| = 1 \qquad (5.141)$$

If the conditions stated by Eqns (5.121) and (5.123) are satisfied, the sustained oscillations cannot occur, and the system becomes stable.

Let us now consider the small signal equivalent circuit of the uncompensated OPAMP. The small signal equivalent circuit of a two-stage uncompensated OPAMP is shown in Fig. 5.36.

Each stage of the OPAMP is considered as a gain stage with a single-pole frequency response. The poles of the differential and output states are given by $p_1 = -1/(R_1 C_1)$ and $p_2 = -1/(R_2 C_2)$, respectively. Notice that the phase of the output of each stage will undergo a phase change of 90 = around its pole frequency. Most

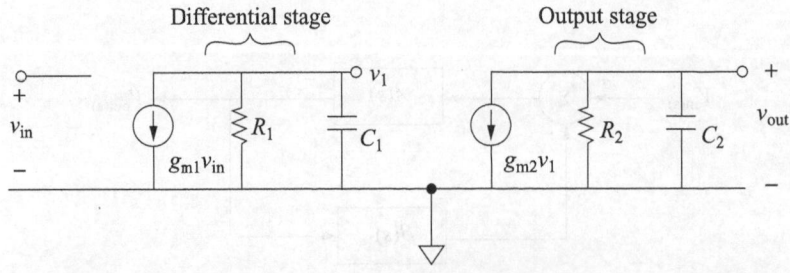

Fig. 5.36   Small signal equivalent circuit of a two-stage uncompensated OPAMP

OPAMPs are used with negative feedback. If the OPAMP stages themselves contribute a phase difference of 180°, the negative feedback becomes positive feedback. If the loop gain at this frequency is greater than 1, the circuit will become unstable. Both stages of the OPAMP have a single-pole frequency response. The poles for both the stages can be quite close together. As a result, they can contribute a total of 180° phase shift over a relatively narrow frequency range. Hence, the OPAMP must be compensated.

Pole Splitting   For compensation, it is required to separate the poles of the OPAMP so that the gain drops below 1 by the time the phase shift through the OPAMP becomes 180°, even if it means that we have to reduce the bandwidth of the OPAMP. This is often achieved by a technique called *pole splitting*. The lower frequency pole is brought to a low enough frequency, so that the gain diminishes to below 1 by the time the second pole is reached. One way of doing this is to use a Miller capacitor.

Miller Compensation   In this technique, a capacitor (C_C) is connected between the input differential stage and the output stage. The small signal equivalent circuit for Miller-compensated two-stage OPAMP is shown in Fig. 5.37. This compensation capacitor increases the effective capacitance parallel to the resistor R_1 by an amount $g_{m2} R_2 C_2$.

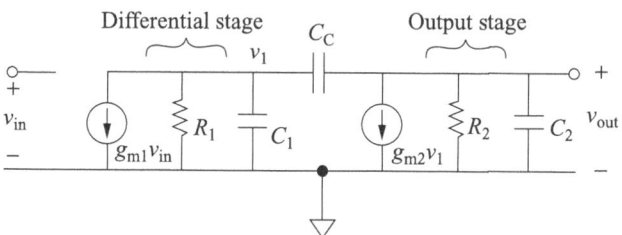

Fig. 5.37   Small signal equivalent circuit of a two-stage compensated OPAMP

Hence, the pole p_1 shifts towards the origin to a location p_1', and is given by

$$p_1' = -\frac{1}{g_{m2} R_1 R_2 C_C} \tag{5.142}$$

The pole p_2 shifts away from the origin to a location p_2', and is given by

$$p_2' = -\frac{g_{m2} C_C}{C_1 C_2 + C_2 C_C + C_1 C_C} \tag{5.143}$$

The Miller-compensated two-stage CMOS OPAMP is shown in Fig. 5.38.

Slew Rate

Miller compensation also sets the slew rate of the OPAMP. For a large signal input, the output current of the OTA is equal to the tail current. The effective load

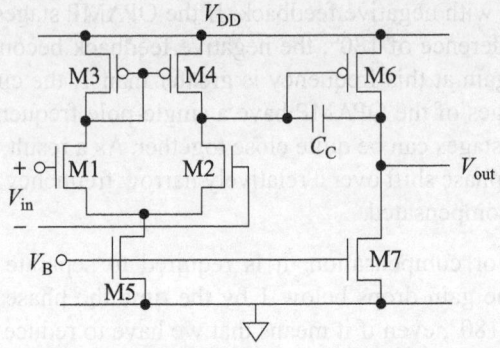

Fig. 5.38 Miller-compensated two-stage CMOS OPAMP

capacitance for this stage is A_2C_2 where A_2 is the gain of the output stage. The current through the transistor M5 can be written as

$$I_5 = A_2 \times C_C \frac{dV}{dt} \tag{5.144}$$

Output of the OTA slews at a rate $I_5/(A_2C_2)$. So the OPAMP slews at a rate which is A_2 times this value. Hence, the slew rate of the OPAMP is

$$\text{SR} = \frac{I_5}{C_C} \tag{5.145}$$

Design Equations

The currents through the transistors M3 and M4 are equal as they form a current mirror.

$$I_3 = I_4 \tag{5.146}$$

Again, the transistors M1 and M3 are in series, hence, they carry the same current.

$$I_1 = I_3 \tag{5.147}$$

Again, we can write

$$I_1 = I_2 = I_5/2 \tag{5.148}$$

The transistor M3 is always saturated as its drain and gate are connected together. The transistors M3 and M4 have the same V_S, V_G, and I_D. As $(W/L)_3 = (W/L)_4$, they must have the same V_D as well; hence, M4 is also saturated.

Again, M3 and M6 has the same V_S and V_G. If $I_3/I_6 = (W/L)_3/(W/L)_4$, M6 has the same V_D as M3 and M6 will be saturated.

The gain–bandwidth product determines g_m of M1 and M2 and can be written as

$$\text{GBW} = \frac{g_{m2}}{C_C} \tag{5.149}$$

Knowing g_m of M1 and M2, W/L ratios can be determined by the expression

$$g_{m2} = \sqrt{2k'_n (W/L)_2 I_2} \qquad (5.150)$$

As currents through M2, M4, M6, and M7 are known, their r_{ds} values can be calculated using the expression

$$r_{ds} = 1/\lambda I_D \qquad (5.151)$$

The gain of the OPAMP can be written as

$$A = g_{m1}(r_{ds2} \| r_{ds4}) \times g_{m6}(r_{ds6} \| r_{ds7}) \qquad (5.152)$$

As g_m of M2 is known and all r_{ds} values are known, g_m of M6 can be determined. Since I_6 is known, $(W/L)_6$ can be calculated.

Example 5.5   OPAMP design

Specifications

$k'_n = 120\ \mu A/V^2$	$V_{tn} = 0.4\ V$	Gain, $A = 80$ dB	SR = 20 V/μs
$k'_p = 60\ \mu A/V^2$	$V_{tp} = -0.4\ V$	GBW = 50 MHz	$\lambda = 0.05\ V^{-1}(L = 1\ \mu m)$

Solution

Let us first choose the value of compensation capacitor, $C_C = 2$ pF.
We shall bias the second stage at five times the tail current of the differential stage. Using the slew rate, let us calculate I_5 as

$$I_5 = SR \times C_C = 20 \times 10^6 \times 2 \times 10^{-12} = 40\ \mu A$$

Therefore,

$$I_1 = I_2 = I_3 = I_4 = I_5/2 = 20\ \mu A$$

The current in the output stage can be calculated as

$$I_6 = I_7 = 5 \times 40 = 200\ \mu A$$

Using GBW, we can calculate g_{m1} and g_{m2} as

$$g_{m1} = g_{m2} = GBW \times C_C = 2\pi \times 50 \times 10^6 \times 2 \times 10^{-12} = 628\ \mu S$$

Hence, $(W/L)_2$ can be calculated as

$$628 \times 10^{-6} = \sqrt{2 \times 120 \times 10^{-6} \times (W/L)_2 \times 20 \times 10^{-6}}$$

or

$$(W/L)_2 \approx 82 = (W/L)_1$$

Let us now calculate output resistance of M2 and M4 as

$$r_{ds2} = r_{ds4} = 1/\lambda I_2 = \frac{1}{0.05 \times 20 \times 10^{-6}} = 1.0 \times 10^6 \; \Omega$$

Therefore,

$$r_{ds2} \| r_{ds4} = 0.5 \times 10^6 \; \Omega$$

Similarly, we can calculate output resistance of M6 and M7 as

$$r_{ds6} = r_{ds7} = 1/\lambda I_6 = \frac{1}{0.05 \times 200 \times 10^{-6}} = 1.0 \times 10^5 \; \Omega$$

Therefore,

$$r_{ds6} \| r_{ds7} = 0.5 \times 10^5 \; \Omega$$

Let us now calculate g_{m6} as

$$10,000 = 628 \; \mu S \times 0.5 \times 10^6 \times g_{m6} \times 0.5 \times 10^5$$

$$g_{m6} = 637 \; \mu S$$

Then we can calculate the aspect ratio of M6 as

$$637 \times 10^{-6} = \sqrt{2 \times 60 \times 10^{-6} \times (W/L)_6 \times 200 \times 10^{-6}}$$

or

$$(W/L)_6 \approx 17$$

But the geometry of M3 and M4 has to be in the current ratio with M6 as

$$I_3/I_6 = (W/L)_3/(W/L)_6$$

$$\frac{20 \; \mu A}{200 \; \mu A} = \frac{(W/L)_3}{17}$$

Therefore, aspect ratio of M3 and M4 should be

$$(W/L)_3 = (W/L)_4 = 1.7$$

Finally, let us assume that an nMOS bias transistor of $W/L = 4$ is available with a current of 10 μA. This will give the W/L of M5 and M7 as 16 and 80, respectively. Hence, in summary, aspect ratios of the MOS transistors are given by

$$\left(\frac{W}{L}\right)_1 = \frac{82 \; \mu m}{1 \; \mu m}, \quad \left(\frac{W}{L}\right)_2 = \frac{82 \; \mu m}{1 \; \mu m}, \quad \left(\frac{W}{L}\right)_3 = \frac{1.7 \; \mu m}{1 \; \mu m}, \quad \left(\frac{W}{L}\right)_4 = \frac{1.7 \; \mu m}{1 \; \mu m},$$

$$\left(\frac{W}{L}\right)_5 = \frac{16 \; \mu m}{1 \; \mu m}, \quad \left(\frac{W}{L}\right)_6 = \frac{17 \; \mu m}{1 \; \mu m}, \quad \left(\frac{W}{L}\right)_7 = \frac{80 \; \mu m}{1 \; \mu m}$$

We have assumed that the channel length of all the MOS transistors is 1 μm. This completes the design for the two-stage CMOS OPAMP.

5.17.4 Cascode OPAMP

The two-stage OPAMP discussed in the previous section is mostly used in the CMOS amplifier. But due to the following reasons it is not suitable for unbuffered applications.

- Insufficient gain
- Limited stable bandwidth
- Poor power supply rejection ratio (PSRR)

The cascode OPAMP discussed in this section has improvement over the above three parameters. The cascoding is applied to both the input differential stage and the output stage. We shall explain the cascoding in each stage separately.

First-stage Cascode OPAMP

The two-stage OPAMP has insufficient gain. Hence, the gain must be increased by some means. As we have seen that cascode configuration increases the output resistance, it can be used to increase the gain of the amplifier. Remember that gain is a product of the transconductance and the output resistance, i.e., $|A_v| = g_m R_{out}$. The cascoded first stage OPAMP is shown in Fig. 5.39. The gain of the cascode OPAMP can be written as

$$A_v = g_{m1}(g_{m2}r_{ds2}^2 \| g_{m4}r_{ds4}^2) \tag{5.153}$$

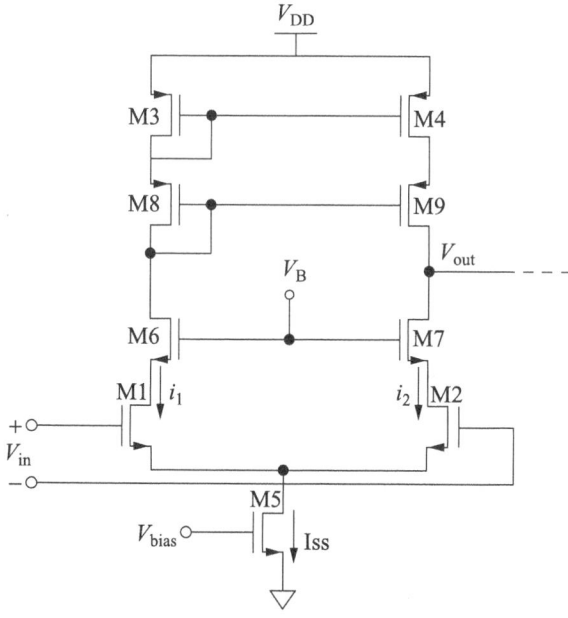

Fig. 5.39 Cascoded first stage of two-stage OPAMP

Cascode OPAMP has higher gain compared to the two-stage OPAMP. But the output swing is limited, and cascode OPAMP cannot be used for the application of unity gain buffer. The output-swing is decreased by the voltage-drop across the common source output transistor.

The circuit shown in Fig. 5.39 can be used as a single-stage OPAMP where very high gain is not required. This circuit has one dominant pole for the output stage. A self-compensation is achieved when the output capacitance is connected at the output. The increase in the output-swing for level-translation stage is used as shown in Fig. 5.40. The pMOS transistors M10 and M11 are used for level translation. M10 operates in the saturation region and acts as a current source. M10 also biases the source follower M11. In this circuit, compensation is achieved by the Miller compensation technique. This circuit also has a better power supply rejection ratio.

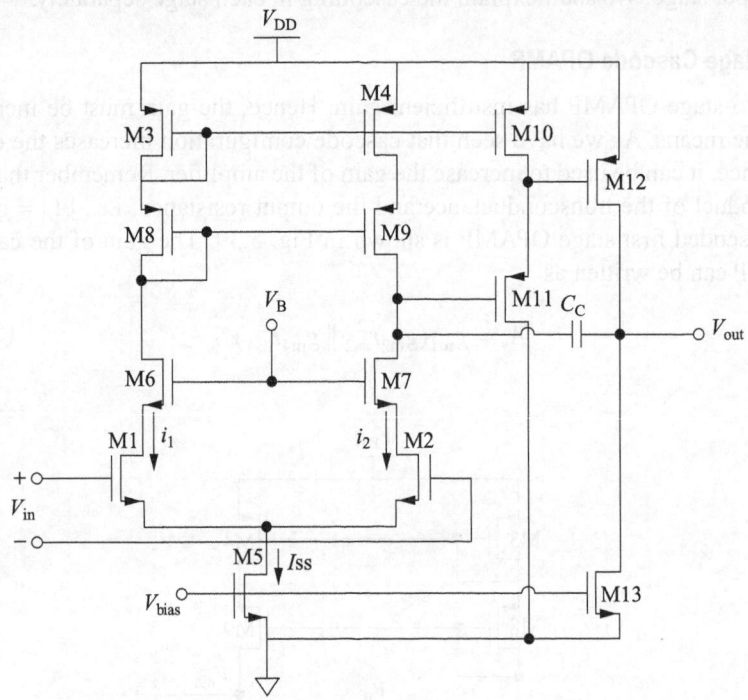

Fig. 5.40 Two-stage OPAMP with cascoded first-stage

Second-stage Cascode OPAMP

In order to increase the gain of the two-stage OPAMP, the second stage can be cascoded. This will eliminate the level-translators and also has better compensation. The cascoded second stage is shown in Fig. 5.41. The resistor R_Z is used in series with the compensation capacitor that controls the right half-plane (RHP) zero.

The gain of the cascoded second stage OPAMP is given by

$$A_v = g_{m1}g_{m6}\left(r_{ds2} \parallel r_{ds4}\right) \times \left(g_{m8}r_{ds8}r_{ds6} \parallel g_{m9}r_{ds9}r_{ds7}\right)$$

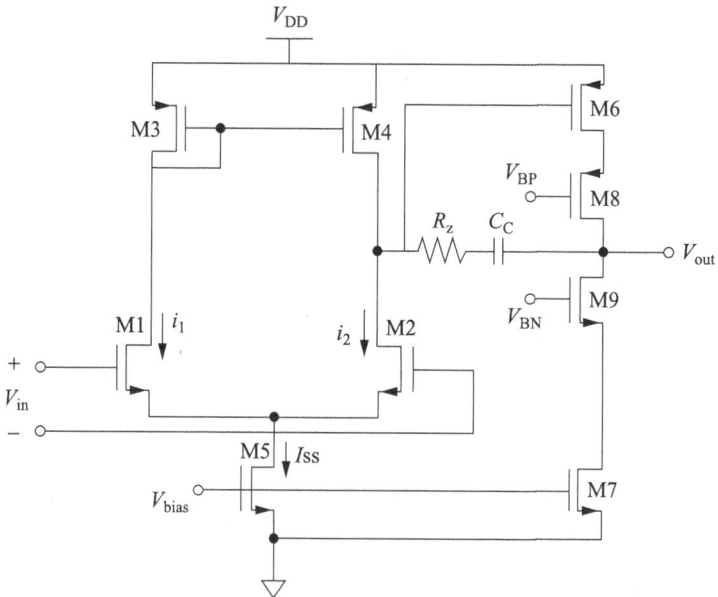

Fig. 5.41 Cascoded second-stage OPAMP

5.18 Comparator

A comparator is used to compare an analog signal with a reference voltage, and produces either a high or low signal. It is used in analog-to-digital converters where the analog input is compared with a reference voltage to produce digital output.

The symbolic representation of a comparator is shown in Fig. 5.42(a).

If the analog input is greater than the reference voltage, the output of the comparator goes high. Conversely, if the analog input is less than the reference voltage, the output of the comparator goes low. The transfer characteristics of the comparator are shown in Fig. 5.42(b), where the dotted line represents the ideal behaviour. The upper and lower limit of the output voltage is defined as V_{OH} and V_{OL}, respectively. As from the characteristics, we can observe that if the

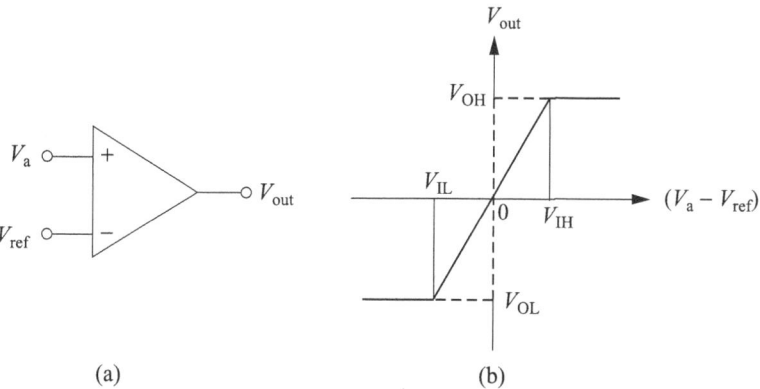

Fig. 5.42 (a) Symbol of a comparator; (b) transfer characteristics of a comparator

difference between input signal and reference voltage (ΔV_{in}) lies between V_{IL} and V_{IH}, the output voltage linearly increases with input signal. Hence, the gain of the comparator is defined as

$$A_v = \frac{V_{OH} - V_{OL}}{V_{IH} - V_{IL}} \tag{5.154}$$

The gain A_v approaches infinity as ($V_{IH} - V_{IL}$) approaches zero. Conversely, we can say that if the gain of the comparator is very high, $V_{IH} = V_{IL}$, the comparator behaves ideally. For a small difference between the input and reference voltage, the output is highly amplified. But the maximum output voltage is limited to the power supply voltages. Hence, we get saturated output for ($V_{IL} < \Delta V_{in} < V_{IH}$).

Dynamic Behaviour of Comparator

If a step signal is applied at the analog input of the comparator, the output of the comparator also shows a step response, but there is finite delay time between the output and input signal. This delay is called the propagation delay of the comparator. The dynamic characteristic of a comparator is shown in Fig. 5.43.

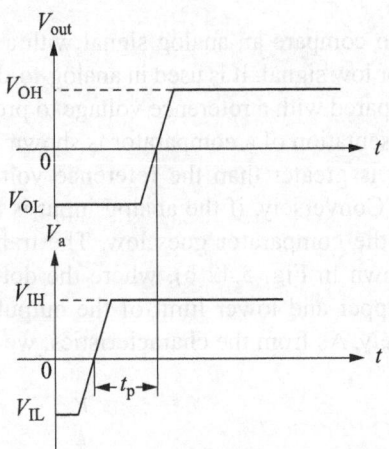

Fig. 5.43 Dynamic characteristics of a comparator

The propagation delay time is determined by the slew-rate of the comparator, and is given by

$$t_p = \frac{\Delta V_{out}}{SR} = \frac{V_{OH} - V_{OL}}{SR} \tag{5.155}$$

A comparator essentially requires a differential input and large gain. So the two-stage CMOS OPAMP can be used as a comparator, as shown in Fig. 5.44. As the comparator is used in open-loop configuration, the compensation is not required.

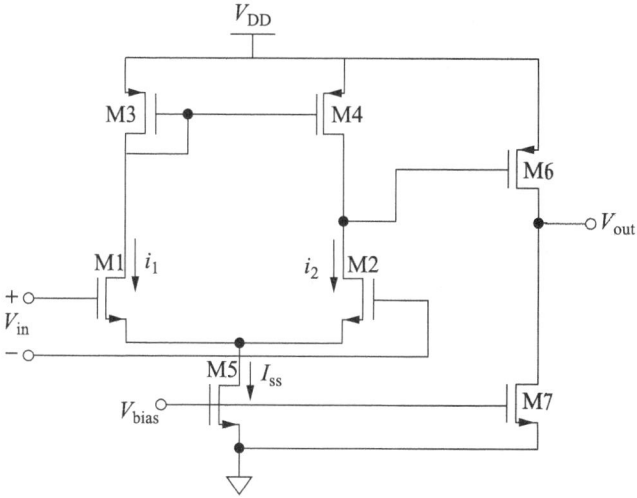

Fig. 5.44   CMOS two-stage comparator

5.19   Switched Capacitor Filter

In Section 5.8, we have introduced switched capacitor circuit to realize a resistor. In this section, we will use the switched capacitor circuit to design filter circuits. A filter is a frequency selective network which passes a band of frequencies from the input to the output, and stops a band of frequencies. The band of frequencies that a filter passes is called the pass band, and the band of frequencies that it stops is called stop band.

Filters are mainly classified into four types depending on the band of frequencies they pass and stop. The classification is as follows:

- Low-pass filter (LPF)
- High-pass filter (HPF)
- Band-pass filter (BPF)
- Band-stop filter (BSF)

The characteristics of the different types of filters are shown in Fig. 5.45.

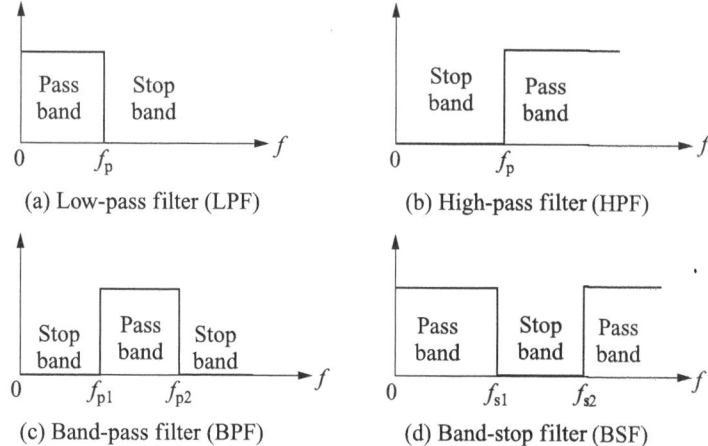

Fig. 5.45   Characteristics of filters: (a) low-pass; (b) high-pass; (c) band-pass; (d) band-stop

5.19.1 Integrator or Low-Pass Filter

Let us consider the first continuous time integrator as shown in Fig. 5.46(a). The output voltage is written as

$$V_{out} = -\frac{1}{RC} \int V_{in} dt \tag{5.156}$$

The resistor R can be replaced by a switched capacitor resistor, as shown in Fig. 5.46(b), when the circuit becomes an integrator in the discrete-time domain. The two switches S1 and S2 are operated by two non-overlapping clock waveforms.

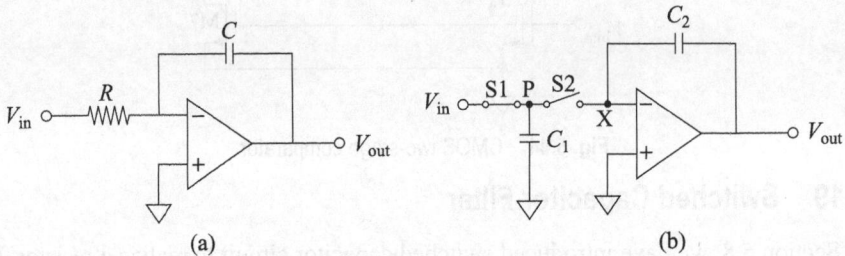

Fig. 5.46 Continuous time integrator

Operation

In every clock pulse, (i) when S1 is ON, the capacitor C_1 absorbs a charge C_1V_{in}, (ii) when S2 is ON, this charge is deposited on C_2. Therefore, if V_{in} is constant, the output changes by $(V_{in}C_1/C_2)$ at every clock cycle. Hence, the output signal is of a staircase shape as shown in Fig. 5.47.

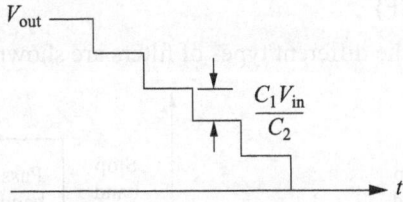

Fig. 5.47 Output response to a constant input voltage

The staircase can be approximated by a ramp if the step height is small; hence, the output voltage is the integration of the input voltage, and the circuit becomes an integrator. The output at the kth clock cycle can be written as

$$V_{out}(kT) = V_{out}[(k-1)T] - V_{in}[(k-1)T] \times C_1/C_2 \tag{5.157}$$

Taking the z-transform on both sides, we get

$$V_{out}(z) = z^{-1}V_{out}(z) - z^{-1}V_{in}(z) \times \frac{C_1}{C_2} \tag{5.158}$$

Hence, the transfer function can be written as

$$H(z) = \frac{V_{out}(z)}{V_{in}(z)} = -\frac{C_1}{C_2} \times \frac{z^{-1}}{1 - z^{-1}} = -\frac{C_1}{C_2} \times \frac{1}{z - 1} \qquad (5.159)$$

This is the expression for transfer function of a low-pass filter in the discrete-time domain. Substituting $z = e^{sT}$ in Eqn (5.142), we get the transfer function s-domain as

$$H(s) = \frac{V_{out}(s)}{V_{in}(s)} = -\frac{C_1}{C_2} \times \frac{1}{e^{sT} - 1} \approx -\frac{C_1}{C_2} \times \frac{1}{sT} \qquad (5.160)$$

$[\because e^{sT} = 1 + sT + (sT)^2/2! + \cdots \approx 1 + sT \text{ for } sT \ll 1.]$
Taking inverse-Laplace transform of Eqn (5.143), we obtain

$$V_{out}(t) = -\frac{C_1}{TC_2} \int V_{in}(t)dt = -\frac{1}{R_{eq}C_2} \int V_{in}(t)dt \qquad (5.161)$$

where $R_{eq} = T/C_1$ is the equivalent resistance of the switched-capacitor circuit. Hence, the circuit acts as an integrator.

5.20 Digital-to-Analog Converter

The analog-to-digital (A/D) and digital-to-analog (D/A) converters are an integral part of any digital signal processing system. A typical block converters of the digital signal processing system is shown in Fig. 5.48.

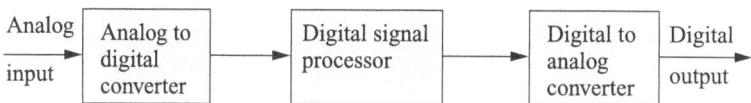

Fig. 5.48 A typical digital signal processing system

Generally, the analog-to-digital converter (ADC) is designed using the digital-to-analog converter (DAC). Hence, we shall discuss the DACs first before discussing ADCs. In a digital system, data is represented by a number of bits called *word*. An n-bit word is represented by n number bits, where the 0th bit is the least significant bit (LSB). As the digital word changes by 1-bit, the corresponding analog value jumps by a step, and the step is given by

$$\Delta V = V_{REF}/2^n \qquad (5.162)$$

where V_{REF} is the analog output for full-scale range (FSR) when all bits of the digital word are 1. A typical characteristic of DAC is shown in Fig. 5.49.

There are two types of DACs: serial and parallel. Serial DAC converts one bit at a time, whereas a parallel DAC converts all the bits at a time. Hence, parallel

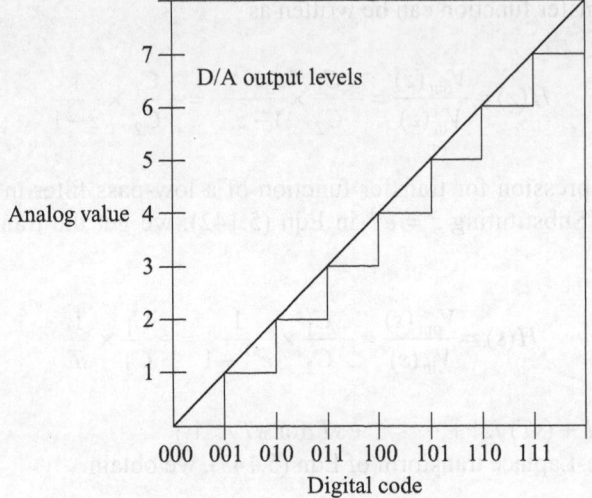

Fig. 5.49 Characteristics of DAC

DACs are *n* times faster than serial DACs for an *n*-bit digital input. Again, depending on scaling method, DACs are classified into the following three types:

- Current scaling
- Voltage scaling
- Charge scaling

5.20.1 Current Scaling DAC

There are two types of current scaling DAC: (a) binary-weighted resistor-type DAC and (b) *R-2R* ladder DAC.

Binary-weighted Resistor-type DAC

The binary-weighted resistor-type DAC is shown in Fig. 5.50. There are *n* numbers of switches operated by the digital inputs. The switches connect a reference voltage source V_{REF} to the inverting input of OPAMP through resistors. The

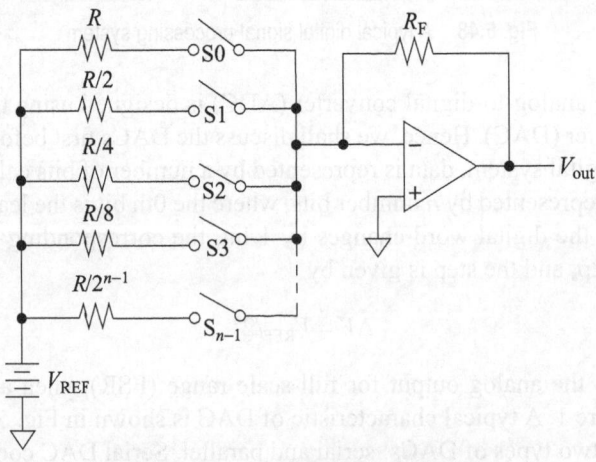

Fig. 5.50 Binary-weighted resistor-type DAC

resistances are halved for the next significant bit. The highest resistance is R in series with the switch operated by the LSB of digital input. Then, applying KCL at the inverting input node of the OPAMP, we get

$$d_0 \frac{V_{REF}}{R} + d_1 \frac{V_{REF}}{R/2} + d_2 \frac{V_{REF}}{R/4} + \cdots + d_{n-1} \frac{V_{REF}}{R/2^{n-1}} + \frac{V_{out}}{R_F} = 0 \qquad (5.163)$$

$$V_{out} = -\frac{R_F}{R}(2^0 d_0 + 2^1 d_1 + 2^2 d_2 + \cdots + 2^{n-1} d_{n-1})V_{REF} \qquad (5.164)$$

One of its disadvantages is that it requires widespread resistances.

R-2R Ladder-type DAC

An R-$2R$ ladder-type DAC is shown in Fig. 5.51. In this DAC, resistors are used having only two values, R and $2R$. In this DAC, the output voltage can be written as

$$V_{out} = -\frac{V_{REF}}{2^n}(2^0 d_0 + 2^1 d_1 + 2^2 d_2 + \cdots + 2^{n-1} d_{n-1}) \qquad (5.165)$$

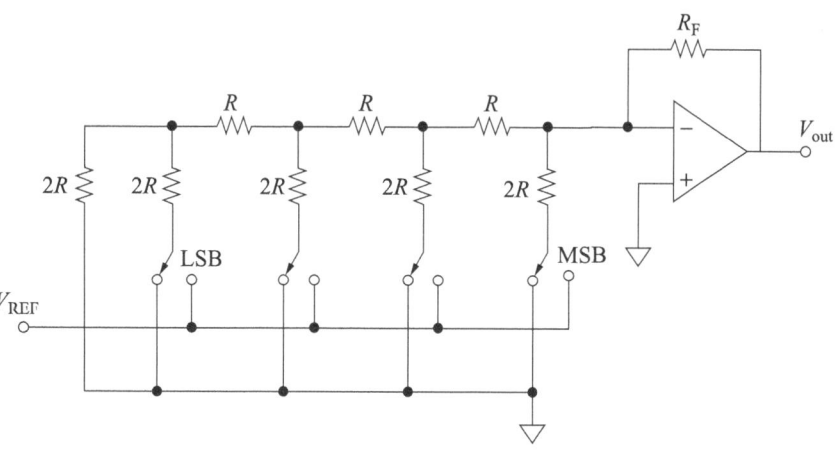

Fig. 5.51   *R-2R* ladder-type DAC

5.20.2   Charge Scaling DAC

A charge scaling DAC is shown in Fig. 5.52. In this DAC, the binary-weighted capacitors are used. The circuit operates in two phases, in the first phase, the switch S_C is connected to the ground, and the other switches S1 through S4 are also connected to the ground. All the capacitors are discharged in this phase. In the next phase, the switch S_C is disconnected from the ground, and other switches are connected to V_{REF} if their corresponding bits are 1, or to the ground if their corresponding bits are 0. The output voltage can be written as

$$V_{out} = \frac{V_{REF}}{2^n}(2^0 d_0 + 2^1 d_1 + 2^2 d_2 + \cdots + 2^{n-1} d_{n-1}) \qquad (5.166)$$

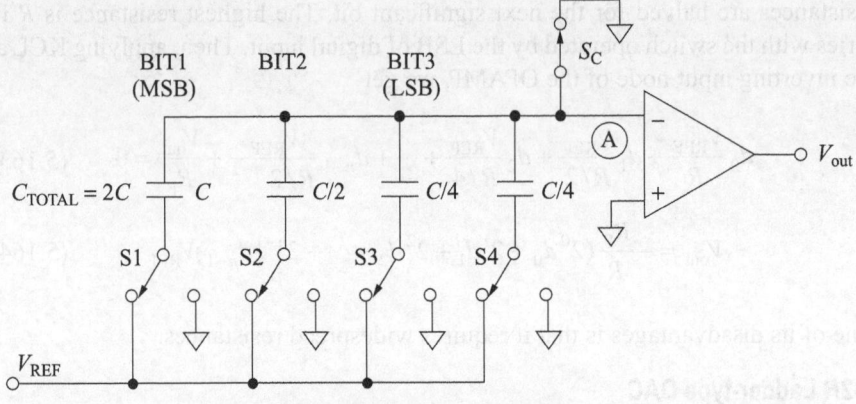

Fig. 5.52 Charge scaling DAC

5.21 Analog-to-Digital Converter

The analog-to-digital conversion is achieved in the following four steps:
- Anti-aliasing filtering of the analog signal is used to remove the high-frequency harmonics that has aliasing effect
- Sampling and holding of the analog signal
- Quantization of the sampled values
- Encoding of quantized levels

At first, the analog signal is sampled at a rate greater or equal to the Nyquist sampling rate. The Nyquist sampling rate is given by

$$f_N \geq 2f_m \tag{5.167}$$

where f_m is the maximum frequency content in the analog signal.

Next, the sampled value must be held for some time for it to be converted to the digital signal before the next sample is to be taken. A typical sample and hold circuit is shown in Fig. 5.53.

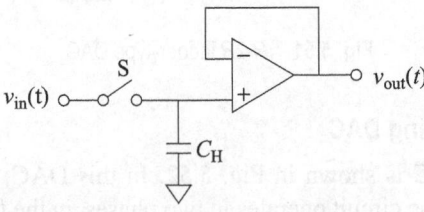

Fig. 5.53 A typical sample and hold circuit

The next step after the sample and hold is the quantization. The complete range of the analog signal amplitude level is divided into a number of quantized levels depending on the number of bits to be used for the digital representation of the analog input. For example, if 3 bits are used, then the signal range is divided into $2^3 - 1 = 7$ quantized levels. But the sampled value of the analog input signal can have any value, and may not equal to the quantized level. Hence, the sampled value is approximated to the closest quantized level. This process is known as *quantization*. Figure 5.54 illustrates the quantization process.

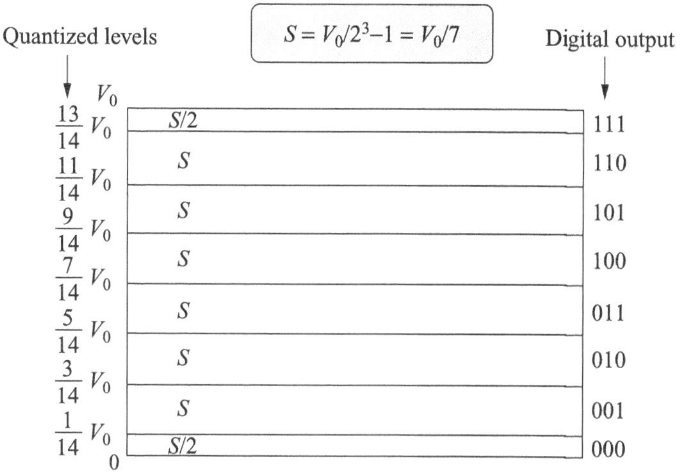

Quantized levels $S = V_0/2^3 - 1 = V_0/7$ Digital output

V_0		
$\frac{13}{14}V_0$	S/2	111
$\frac{11}{14}V_0$	S	110
$\frac{9}{14}V_0$	S	101
$\frac{7}{14}V_0$	S	100
$\frac{5}{14}V_0$	S	011
$\frac{3}{14}V_0$	S	010
$\frac{1}{14}V_0$	S	001
0	S/2	000

Fig. 5.54   Quantization process with quantized levels and digital output

There are different ADC architectures, such as Flash ADC, successive approximation ADC, counting-type ADC, dual-slope ADC, etc. Among all the ADC architectures, Flash ADC is the fastest and we shall discuss the Flash ADC in the following section.

5.21.1   Flash ADC

The Flash ADC is also known as parallel-comparator ADC. The schematic of Flash ADC is shown in Fig. 5.55. It shows a 3-bit parallel ADC. There are $7 = (2^3 - 1)$ comparators in the 3-bit ADC. The comparator outputs are connected to the inputs of an 8:3 encoder. The analog input is applied to the analog input of the comparators and the scaled reference voltages are applied to the reference input of the comparators.

Flash converters are extremely fast. The speed is determined only by the time taken to charge the common inputs to the input voltage and the reaction time of the comparators and the decoding logic. However, the amount of circuitry needed rises exponentially with resolution. This makes the flash architecture impractical above 8–10 bits of resolution.

5.21.2   Sigma–Delta Modulator

The ADCs that we have discussed so far follow the Nyquist sampling rate while sampling the input analog signal. There are other types of ADCs that uses a much higher sampling rate than the Nyquist sampling rate. These ADCs are called the oversampling ADCs. The main purpose of the oversampling ADCs are to obtain a much higher resolution. The block diagram of the oversampling ADC is shown in Fig. 5.56.

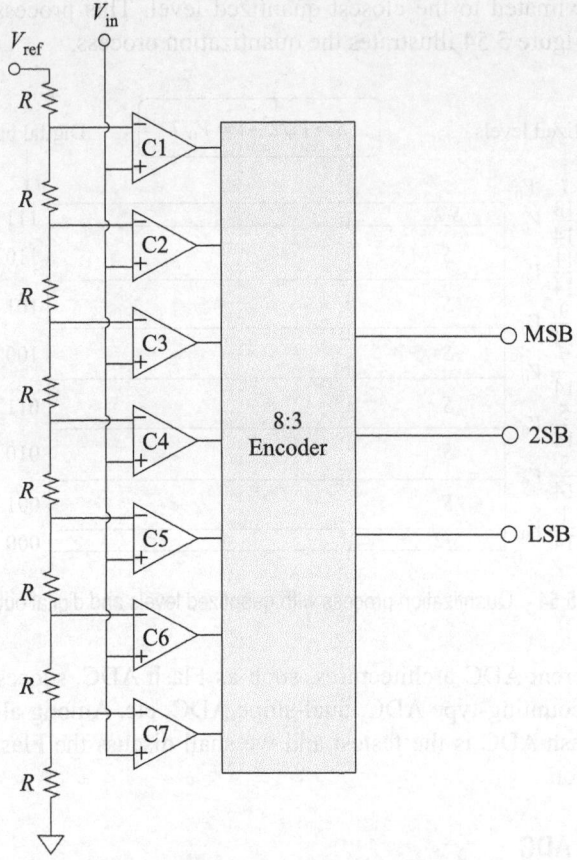

Fig. 5.55 Flash-type ADC

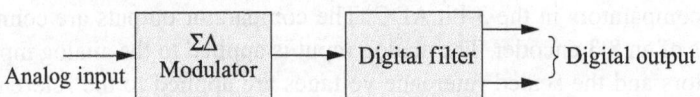

Fig. 5.56 Block diagram of an oversampling ADC

The oversampling converter does not require a sample and holds circuit as it is designed using switched-capacitor circuits which operate in a sampled-data system. Also, the anti-aliasing filter is not needed as it samples at a much higher rate than the Nyquist sampling rate. The quantization is done by the modulator and the encoding is done by a digital filter.

The sigma–delta modulation was first introduced in 1962. But it became popular only after the advancement of VLSI technology to design large digital signal processing ICs. The sigma–delta modulation technique is based on the delta modulation. Let us first describe the delta modulation process before discussing sigma–delta modulation. The delta modulation process is based on the quantization of signal changes from sample to sample, rather than on the absolute value of the signal.

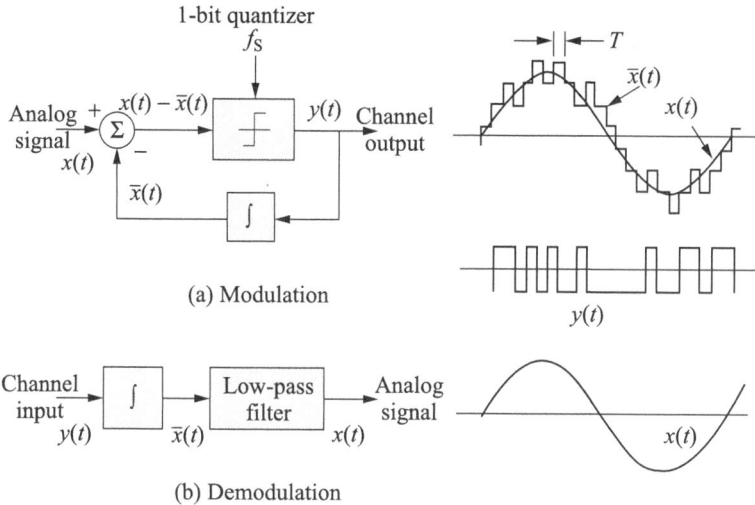

Fig. 5.57 Delta modulation and demodulation

The delta modulation and demodulation schemes are shown in Fig. 5.57. In the modulator, the integrator output signal $\bar{x}(t)$ is subtracted from the input signal $x(t)$ to generate the error signal which generates the quantized output. If the integrated output is more than the signal amplitude, the quantized level is decreased, or else it is increased. In the demodulator, the quantized output is integrated and passed through a low-pass filter to smoothen it. But delta modulation suffers from a serious problem which is known as slope-overloading problem. If the input signal changes rapidly, the modulator output cannot track it.

In the delta modulation–demodulation process, there are two integrators. As the integration is a linear operation, it can be moved from the demodulation to the modulation stage, and two integrators can be combined into a single one as shown in Figs 5.58(a) and (b). This arrangement is known as sigma–delta modulation. The

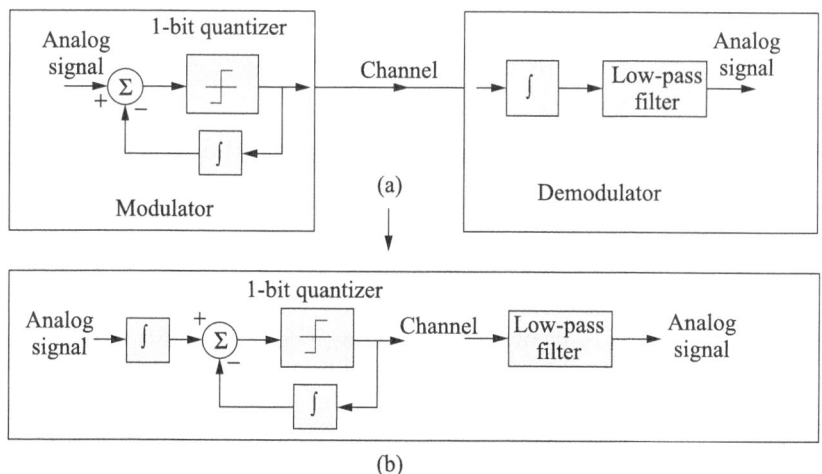

Fig. 5.58 Sigma–delta modulator

name sigma–delta modulator comes from the fact that the integrator (sigma) is put in front of the delta modulator.

The block diagram of a sigma–delta modulator is shown in Fig. 5.59. The 1-bit quantizer is modelled as an error source $Q(s)$.

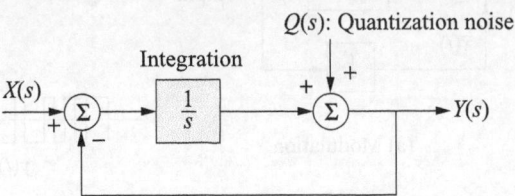

Fig. 5.59   Block diagram of a sigma–delta modulator in the *s*-domain

We can write the output signal in the *s*-domain as

$$Y(s) = Q(s) + \frac{1}{s} \times [X(s) - Y(s)] \qquad (5.168)$$

$$Y(s) = Q(s) \times \frac{s}{s+1} + X(s) \times \frac{1}{s+1} \qquad (5.169)$$

$$\frac{Y(s)}{X(s)} = \frac{Q(s)}{X(s)} \times \frac{s}{s+1} + \frac{1}{s+1} \qquad (5.170)$$

$$\frac{Y(s)}{Q(s)} = \frac{s}{s+1} + \frac{X(s)}{Q(s)} \times \frac{1}{s+1} \qquad (5.171)$$

From Eqns (5.153) and (5.154), we find that the transfer function with respect to the input signal is a low-pass filter when the quantization noise is zero, i.e.,

$$H_X(s) = \frac{Y(s)}{X(s)}\bigg|_{Q(s)=0} = \frac{1}{s+1} \qquad (5.172)$$

Again, the transfer function with respect to the quantization error signal is a high-pass filter when the input signal is zero, i.e.,

$$H_Q(s) = \frac{Y(s)}{Q(s)}\bigg|_{X(s)=0} = \frac{s}{s+1} \qquad (5.173)$$

Hence, this acts as a low-pass filter for the input signal and a high-pass filter for quantization noise. The plot of two transfer functions is shown in Fig. 5.60. It can be seen that the signal has a high value at the lower frequencies where the quantization noise is less; at higher frequencies beyond the bandwidth of the signal, the noise increases. This means that the noise power is pushed to the higher frequencies out-of-bandwidth of the signal. This high-pass characteristic is known as *noise-shaping* of the sigma–delta modulator.

It is seen that after the input signal passes through the modulator, it is fed into the digital filter. The function of the digital filter is to provide a sharp cutoff at the bandwidth of interest which essentially removes out-of-band quantization noise

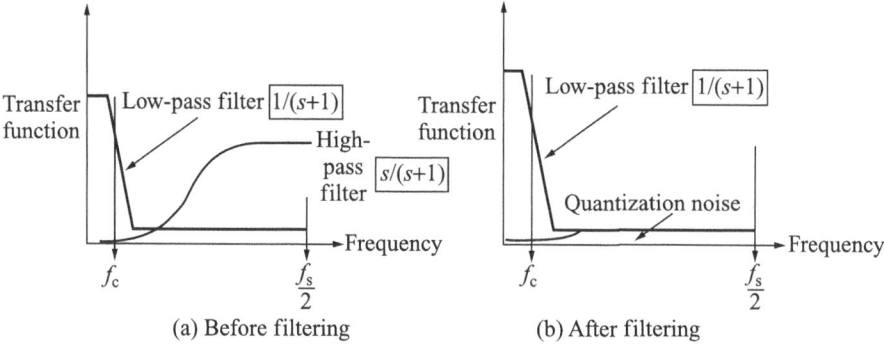

Fig. 5.60 Transfer function of the sigma–delta modulator

and signals. Figure 5.56 shows that the digital filter eliminates the quantization noise that the modulator pushed out to the higher frequencies.

Example 5.6 Calculate the output voltage of an R-2R ladder-type DAC for the digital input 11001. Use the reference voltage as 10 V.

Solution The output voltage can be expressed as

$$V_{out} = -\frac{V_{REF}}{2^5}(2^0.1 + 2^1.0 + 2^2.0 + 2^3.1 + 2^4.1) = -\frac{10}{32}(1 + 8 + 16) = -\frac{250}{32} = -7.8125 \text{ V}$$

Example 5.7 A 3-bit flash-type ADC is to convert analog input ranging from −5 V to +5 V. What will be the reference voltages for each comparator?

Solution An N-bit flash-type ADC requires $(2^N - 1)$ number of comparators. Thus, a 3-bit system requires $(2^3 - 1) = 7$ comparators. The range of input voltage −5 V to +5 V is divided into 7 steps. Therefore, the step size is

$$S = \frac{5 - (-5)}{2^3 - 1} = \frac{10}{7} = 1.4286 \text{ V}$$

The range of analog input, equivalent digital output, and the reference voltage of the comparators are shown in Table 5.2.

Table 5.2

Range of analog input (V)	Equivalent digital output	Reference voltage of the comparators (V)
+25/7 to +35/7	111	+25/7
+15/7 to +25/7	110	+15/7
+5/7 to +15/7	101	+5/7
−5/7 to +5/7	100	−5/7
−15/7 to −5/7	011	−15/7
−25/7 to −15/7	010	−25/7
−35/7 to −25/7	001	−35/7
−45/7 to −35/7	000	

5.22 Phase-locked Loop

Phase-locked loop is commonly known as PLL. PLL circuits are used for frequency control. They can be configured as frequency multipliers, demodulators, tracking generators, or clock recovery circuits. Different applications require different characteristics of PLL, but the basic principle of operation remains the same. The block diagram of a PLL is shown in Fig. 5.61.

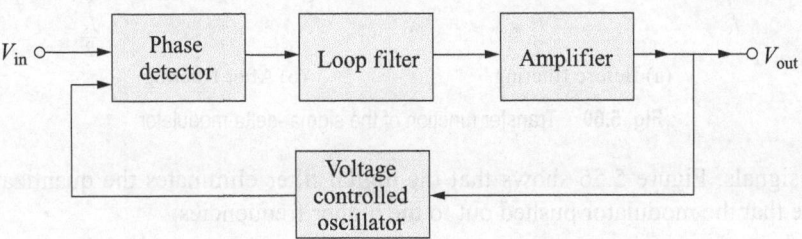

Fig. 5.61 A schematic diagram of the phase-locked loop

There are two types of PLL: analog PLL and digital PLL. Analog PLL is used to track the frequency of the incoming signal. Digital PLL is used as clock recovery circuit. The clock signal is generated to recover the incoming data.

A PLL is a feedback control system, which consists of a phase detector, a loop filter, and an amplifier in the forward path. There is a voltage-controlled oscillator (VCO) in the feedback path.

A phase detector produces an output which is proportional to the phase difference between the input signals. If the two input signals, $v_1(t)$ and $v_2(t)$, of the phase detector have a phase difference $\Delta\varphi$, then we can write the output signal as

$$v_{\text{out}}(t) \propto \Delta\varphi \tag{5.174}$$

$$v_{\text{out}}(t) = K_{\text{PD}}\Delta\varphi \tag{5.175}$$

where K_{PD} is the proportionality constant expressed in V/radian.

The VCO is an oscillator whose output signal frequency varies in proportion to the input signal amplitude. The frequency of oscillation of the VCO output can be written as

$$\omega_{\text{osc}}(t) = \omega_0 + K_{\text{VCO}}V_{\text{out}}(t) \tag{5.176}$$

where V_{out} is the input signal to the VCO and ω_0 is the free running frequency. The free running frequency is defined as the frequency of oscillation for zero VCO input voltage. Or the instantaneous phase of the VCO output signal can be written as

$$\varphi_{\text{osc}}(t) = \int[\omega_0 + K_{\text{VCO}}V_{\text{out}}(t)]dt \tag{5.177}$$

5.22.1 Principle of Operation of PLL

The input signal is applied to one input of a phase detector. The other input is connected to the output of the VCO. The output of the phase detector is a voltage that is proportional to the phase difference between the two inputs. This signal is applied to the loop filter that determines the dynamic characteristics of the PLL. The filtered signal is amplified and applied to the VCO which controls the VCO output. The output of the VCO is at a frequency that is equal to the input signal of the frequency.

Normally, the loop filter is designed to obtain the desired characteristics required by the application of the PLL. The bandwidth of the loop filter must be greater if the PLL is to acquire and track a signal, rather than for a fixed input frequency. The frequency range which the PLL can accept and lock is called the capture range. Once the PLL is locked, the range of frequencies that the PLL can track is called the tracking range or lock range. Generally, the tracking range is larger than the capture range.

Let us now derive the transfer function of a PLL. From the block diagram of PLL as shown in Fig. 5.57, we can write

$$V_{out}(s) = K_{PD}[\varphi_i(s) - \varphi_{osc}(s)] \times F(s) \times A \tag{5.178}$$

where $F(s)$ is the transfer function of the loop filter and A is the voltage gain of the amplifier.

From Eqn (5.160) by taking Laplace transform on both sides, we can write

$$\varphi_{osc}(s) = K_{VCO} \frac{V_{out}(s)}{s} \tag{5.179}$$

Substituting Eqn (5.162) in Eqn (5.161), we can write

$$V_{out}(s) = K_{PD}\left[\varphi_i(s) - K_{VCO}\frac{V_{out}(s)}{s}\right] \times F(s) \times A \tag{5.180}$$

$$V_{out}(s)[s + AK_{PD}K_{VCO}F(s)] = sK_{PD}\varphi_i(s) \times F(s) \times A \tag{5.181}$$

Hence, the transfer function of the PLL can be written as

$$\frac{V_{out}(s)}{\varphi_i(s)} = \frac{sAK_{PD}F(s)}{s + AK_{PD}K_{VCO}F(s)}$$

For first-order loop-filter, the transfer function is $F(s) = 1$. Hence, we can write transfer function of PLL as

$$\frac{V_{out}(s)}{\varphi_i(s)} = \frac{sAK_{PD}}{s + AK_{PD}K_{VCO}} \tag{5.182}$$

For first-order low-pass filter (LPF), the transfer function is $F(s) = \dfrac{1}{1 + s/\omega_{LPF}}$, and the PLL transfer function becomes

$$\frac{V_{out}(s)}{\varphi_i(s)} = \frac{sAK_{PD}\left(\dfrac{1}{1+s/\omega_{LPF}}\right)}{s+AK_{PD}K_{VCO}\left(\dfrac{1}{1+s/\omega_{LPF}}\right)} \tag{5.183}$$

$$\frac{V_{out}(s)}{\varphi_i(s)} = \frac{sAK_{PD}}{s\left(1+s/\omega_{LPF}\right)+AK_{PD}K_{VCO}} \tag{5.184}$$

$$\frac{V_{out}(s)}{\varphi_i(s)} = \frac{sAK_{PD}}{s^2/\omega_{LPF}+s+AK_{PD}K_{VCO}} \tag{5.185}$$

Lock Range   It is the range of frequency about ω_0 for which the PLL keeps track of the incoming input frequency.

Capture Range   It is the range of input frequency for which an initially unlocked loop will lock on an input signal.

The capture range is smaller or equal to the lock range. If $F(s) = 1$, then capture range = lock range. If $F(s) = \dfrac{1}{1+s/\omega_{LPF}}$, then capture range < lock range.

5.22.2 Applications of PLL

Some of the applications of PLL are as follows:
- FM modulation and demodulation
- Frequency synchronization
- Data synchronization and conditioning
- Frequency multiplication and division
- Voltage to frequency conversion
- AM detection

5.23   Field Programmable Analog Array

A field programmable analog array (FPAA) is an IC, which can be configured to implement various analog functions using a set of configurable analog blocks (CABs), and a programmable interconnection network, and is programmed using on-chip memories. It is an analog counterpart of the digital field programmable gate array (FPGA).

Traditionally, the designs of analog circuits follow the full-custom design style, where the designs start from transistor level and each transistor is optimized for best performance. It is also very different from the digital design in which the transistors operate in either cut-off or saturation region. Hence, voltage levels are either ground or V_{DD}. But in case of analog design, the transistor must be biased properly so that it acts in any three (cut-off, linear, or saturation) operating regions, as required in the circuit implementation. So, traditionally, analog design has been a time-consuming process.

With the introduction of FPAA, the analog design has become much simpler and also faster. FPAA has elevated the design and implementation process of analog design to higher levels of abstraction. FPAA provides the analog equivalent of logic gates, which can be used to describe analog functions, such as gain stages and filters without going into details of OPAMPs, resistors, capacitors, transconductance stages, and current mirrors. With the higher level of abstraction, the design process has become simpler, so that any novice designer can perform a complex analog design in a reasonable time.

Reconfigurable analog hardware has progressed much slowly as compared to reconfigurable digital hardware. Though analog ICs with tunable by adjustable biases were available, truly reconfigurable analog circuitry in the form of FPAAs emerged in the late 1980s, and commercial FPAA came in the market in 1996.

Benefits of FPAA are as follows:

- Faster prototyping
- Faster time-to-market
- Shorter design cycles
- Design integration
- Improved component matching

5.23.1 FPAA Architecture

Similar to the configurable logic blocks (CLBs) in FPGA, FPAA contains the configurable analog blocks (CABs), but lesser in number as compared to CLBs in FPGA. The main components of FPAA are as follows:

- Operational amplifiers (OPAMPs)
- Programmable capacitor arrays
- Programmable resistor arrays
- Configurable switches

In addition, modern FPAAs contain analog-to-digital converters (ADCs) and digital-to-analog converters (DACs) that can be used to interface analog systems with digital logic implemented on FPGAs and microcontrollers.

CABs are of two types:

- Continuous time: containing programmable resistor arrays
- Discrete time: either switched capacitor-based design or pulse-based design

A schematic diagram of the FPAA architecture is shown in Fig. 5.62. There is an array of CABs and interconnection network containing a switch matrix. The configuration bit stream is stored in a shift register. Part of the bit stream is used to program the interconnections and others are used to program the functionality of CABs. A CAB can be programmed to function like an amplifier, integrator, voltage-controlled oscillator (VCO), adder, multiplier, etc. The CABs in an FPAA can be identical or different.

5.23.2 An Example Design Using FPAA

Figure 5.63(a) illustrates an implementation of phase-locked loop (PLL) in an FPAA. PLL consists of three main blocks: phase detector, low-pass filter (LPF),

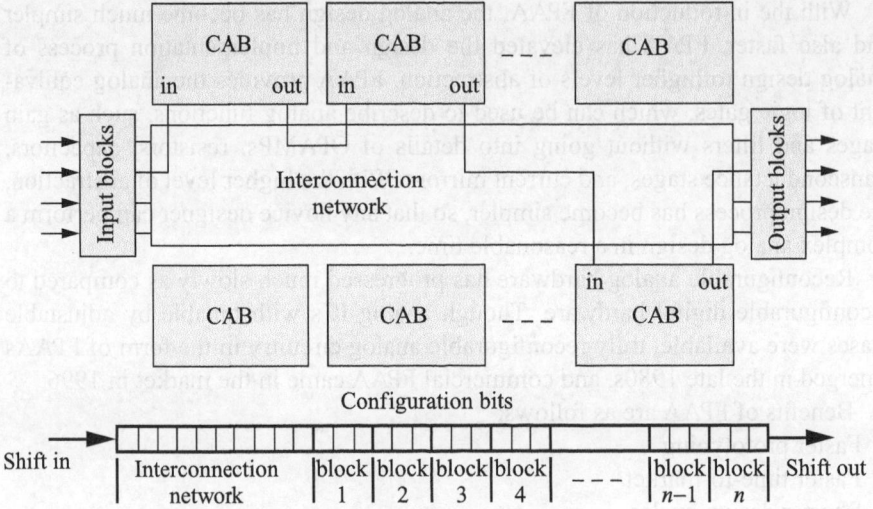

Fig. 5.62 General architecture of FPAA

and VCO. Figure 5.63(b) shows the synthesized blocks for implementing a PLL. The phase detector is implemented by a multiplier and the LPF is implemented by an integrator.

The VCO is implemented by a multiplier, integrator, and amplifier. Then each of the synthesized blocks are mapped to the available CABs in the FPAA as shown in Fig. 5.63(c).

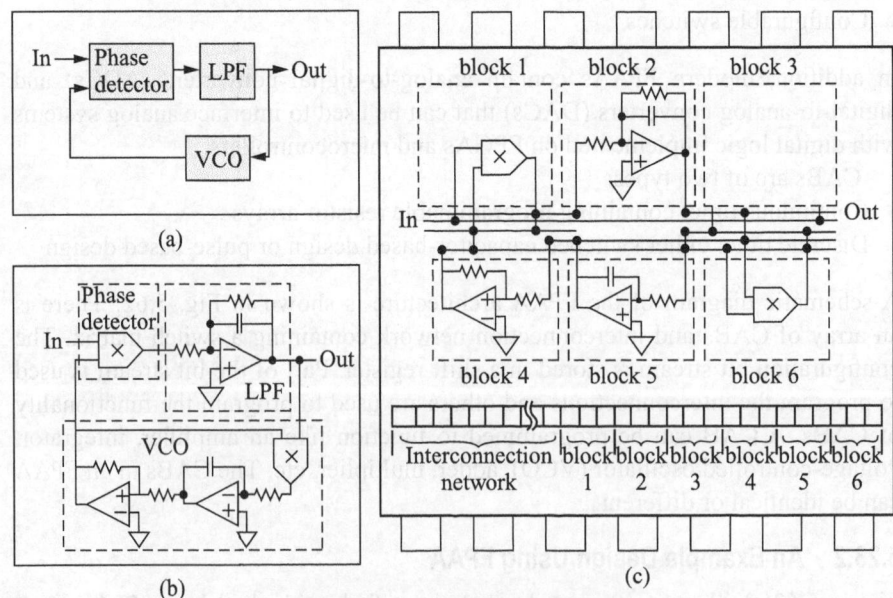

Fig. 5.63 A PLL designed in FPAA

Summary

- The simplest small signal model of a MOSFET is described by a voltage-dependent current source ($g_m V_{GS}$) where g_m is the transconductance and V_{GS} is the gate-to-source voltage.
- The threshold voltage of a MOS device has a dependency on substrate-to-source voltage V_{SB}. Hence, the source and substrate (body) of MOS devices used in a circuit must be shorted together, unless the substrate is biased at different potential than the source intentionally.
- The MOSFET can be considered a voltage-dependent resistor, which is very high (almost infinity) when $V_{GS} = V_t$ and becomes less when $V_{GS} = V_t$.
- The MOSFET can act as a diode called MOS diode when the gate and drain terminals are connected together.
- A pMOS operating in the saturation region can act as current source, whereas an nMOS operating in saturation region can act as current sink.
- A cascode configuration of nMOS current sink has a higher output resistance.
- A switched capacitor can realize a resistor when operated by two non-overlapping clock waveforms.
- A differential amplifier amplifies the difference between the two input signals.
- Compensation of the OPAMP circuit is required in order to increase the stability.
- A two-stage CMOS OPAMP without compensation can act as a comparator.
- Flash-type ADC is the fastest among all other types of ADCs.
- Sigma–delta modulator is an oversampling analog-to-digital converter (ADC).
- The frequency range which the PLL can accept and lock is called the capture range. Once the PLL is locked, the range of frequencies that the PLL can track is called the tracking range.
- FPAA is analogous to FPGA where analog circuits can be reconfigured using a set of configurable analog blocks (CAB), and a programmable interconnection network is available in FPAA.

SELECT REFERENCES

Allen, P.E. and D.R. Holberg 2010, *CMOS Analog Circuit Design*, Oxford University Press, New Delhi.

Baker, R.J., H.W. Li, and D.E. Boyce 2004, *CMOS Circuit Design, Layout, and Simulation*, Prentice-Hall of India, New Delhi.

Hall, T., D. Anderson, and P. Hasler 2002, 'Field-Programmable Analog Arrays: A Floating-gate Approach', *12th International Conference on Field Programmable Logic and Applications*, Montpellier, France, September.

Lee, E.K.F. and P.G. Gulak 1991, 'A CMOS Field-programmable Analog Array', *IEEE Journal of Solid-State Circuits*, vol. 26, no. 12, December, pp. 1860–7.

Pankiewicz, B., M. Wojcikowski, S. Szczepanski, and Y. Sun 2002, 'A Field Programmable Analog Array for CMOS Continuous-time OTA-Cfilter Applications', *IEEE Journal of Solid-State Circuits*, vol. 37, no. 2, February, pp. 125–36.

Ray, B., P.P. Chaudhuri, and P.K. Nandi 2000, *Design of OTA-based Field Programmable Analog Array*, Proceedings of 13th International Conference on VLSI Design, January, pp. 494–8.

Razavi, B. 2002, *Design of Analog CMOS Integrated Circuits*, Tata McGraw-Hill, New Delhi.

EXERCISES

Fill in the Blanks

1. The ON-resistance of a MOSFET _____ .
   - (a) linearly increases with V_{gs}
   - (b) linearly decreases with V_{gs}
   - (c) exponentially increases with V_{gs}
   - (d) non-linearly decreases with V_{gs}
2. Transconductance of a differential amplifier _____ .
   - (a) increases with W/L ratio
   - (b) decreases with W/L ratio
   - (c) does not depend upon W/L ratio
   - (d) none of these
3. CMOS comparator is a _____ .
   - (a) compensated CMOS OPAMP
   - (b) uncompensated CMOS OPAMP
   - (c) partially compensated CMOS OPAMP
   - (d) none of these
4. The equivalent resistance of a switched capacitor is _____ .
   - (a) proportional to clock frequency
   - (b) inversely proportional to clock frequency
   - (c) proportional to the square of the clock frequency
   - (d) inversely proportional to the square of the clock frequency
5. The equivalent resistance of a switched capacitor is _____ .
   - (a) proportional to capacitance
   - (b) proportional to the square of the capacitance
   - (c) inversely proportional to the capacitance
   - (d) inversely proportional to the square of the capacitance

Multiple Choice Questions

1. An ideal current source is a two-terminal element whose current
   - (a) is constant for any voltage across the source
   - (b) is monotonically decreased with the increase of voltage across the source
   - (c) is monotonically increased with the increase of voltage across the source
   - (d) none of these
2. Which one effect does not cause any deviation of a current mirror circuit from the ideal situation?
   - (a) channel length modulation
   - (b) threshold offset between the two transistors
   - (c) imperfect geometrical matching
   - (d) DIBL effects
3. The sensitivity of the BJT voltage reference is
   - (a) smaller than the MOS voltage reference
   - (b) equal to the MOS voltage reference
   - (c) greater than the MOS voltage reference
   - (d) none of these
4. Diffusion current dominates at
   - (a) strong inversion
   - (b) weak inversion
   - (c) strong and weak inversion both
   - (d) cannot be determined
5. Drift current dominates at
   - (a) strong inversion
   - (b) weak inversion
   - (c) strong and weak inversion both
   - (d) cannot be determined

True or False

1. A current mirror circuit can be used as a current amplifier by increasing the (*W/L*) ratios of the mirrored and source MOSFET.
2. After pinch off, current is saturated as drift velocity is saturated.
3. In cascode current mirror, the output resistance is increased.
4. In switched capacitor-based resistor realization, the resistor value is inversely proportional to the clock frequency.
5. Compensation is not required in a comparator circuit.

Short-answer Type Questions

1. What are the effects that cause a practical current mirror to behave differently from an ideal one? Discuss any one of them.
2. Explain how the combination of switches and a capacitor can emulate a resistor.
3. Explain with appropriate diagram, the operation of a Flash ADC.
4. Prove that for a bilinear-switched capacitor realization of the resistor, the equivalent resistance is $T/(4C)$, where T is the clock period and C is the capacitance of the circuit.
5. Write short notes on current reference.
6. What are the basic advantages and limitations of a switched capacitor?
7. Draw the circuit of MOS voltage level shifter and explain its working principle.
8. Explain the working of switched capacitor first-order high-pass filter with circuit diagram.
9. What are lock and capture range of PLL and when are they equal?
10. What is a MOS switch?
11. What are current sources and current sinks?
12. What are voltage and current reference circuits?
13. Define CMRR of CMOS OPAMP.
14. What is ICMR of CMOS OPAMP?
15. What is Miller effect?
16. What is Miller compensation in CMOS OPAMP circuit?
17. What is FPAA?
18. What is an oversampling ADC?
19. Derive the expression for MOS resistance and explain why very high resistance cannot be emulated using MOS resistor with the help of MOS resistance characteristics.
20. What is a CMOS switch? How is it different from the MOS switch? Draw the resistance characteristics of the CMOS switch.
21. What is a MOS diode/active resistor? Draw the small signal equivalent circuit and find out the expression for output resistance.
22. Explain with circuit diagram and necessary expressions how voltage division is achieved using MOS circuits.
23. Explain how MOSFET can be used as current source or sink.
24. What are the advantages of switched capacitor circuits?
25. Explain the operation of the MOS voltage reference circuit with circuit diagram.
26. Explain the operation of the MOS current reference circuit with circuit diagram.
27. Draw the CMOS differential amplifier circuit and explain how it works as a differential amplifier.
28. What are the purposes of an output amplifier and what are its implementation schemes?
29. Draw the circuit of class A amplifier using CMOS. Explain the operation.
30. Draw the circuit of source follower using CMOS. Explain the operation.

31. Draw the circuit of push–pull amplifier using CMOS. Explain the operation.
32. What are the basic building blocks of CMOS OPAMP? Draw them.
33. What is compensation of CMOS OPAMP and why is it used?
34. Explain the working of a switched capacitor first-order low-pass filter with circuit diagram.
35. Explain the working of a switched capacitor first-order high-pass filter with circuit diagram.
36. What are lock and capture range of PLL and when are they equal?

Long-answer Type Questions

1. Define a current sink/source. Obtain the expressions for small signal output resistance of an n-channel MOSFET. Show how the output resistance of a current sink can be increased. Explain how a current mirror can be used as a current amplifier.
2. Explain with a circuit diagram the operation of a differential amplifier and draw its voltage characteristics. What does CMMR mean for a differential amplifier? Draw different configuration of differential amplifier depending on the active load configuration.
3. What is an over-sampling converter? Explain the working principle with necessary circuit diagram of an over-sampling ADC.
4. Design a two-stage CMOS OPAMP that meets the following specifications: $A_v > 5000$, $V_{DD} = 2.5$ V, $V_{SS} = -2.5$ V, GB = 5 MHz, CL = 10 pF, SR > 10 V/μs. $V_{outrange} = \pm 2$ V, ICMR = -1 to 2 V, $P_{diss} \le 2$ mW.

	V_{t0} (V)	$\mu_n C_{ox}$ (=A/V^2)
nMOS	0.6	60
pMOS	-0.7	25

5. Draw the CMOS differential amplifier circuit. Find out the expression for transconductance gain of the CMOS differential amplifier. Draw the transconductance characteristics.
6. Draw the two-stage CMOS OPAMP circuit and explain its working principle.
7. What is compensation of CMOS OPAMP and why is it used? Draw the two-stage CMOS OPAMP circuit with compensation.
8. Show that the relationship between the *W/L* ratios of two-stage CMOS OPAMP, which guarantees that $V_{SG4} = V_{SG6}$, is given by

$$S_6/S_4 = 2 \times S_7/S_5$$

where $S_i = W_i/L_i$.
9. What is a comparator circuit? What are different design approaches for a comparator? What is resolution of a comparator? Draw the CMOS comparator circuit and find out the expression of voltage gain.
10. Draw the regenerative comparator circuit. Explain the working principle. Draw the small signal equivalent model and find out the expression for latch time constant.
11. Draw the circuit diagram of parallel/flash-type ADC and explain the working principle. What are the pros and cons of this type of ADC?
12. What is an over-sampling converter? Explain the working principle with necessary circuit diagram of an over-sampling ADC.
13. What is a PLL and what are basic building blocks of PLL? What are lock range and capture range of PLL? Discuss the applications of PLL.

14. What is FPAA? Draw the architecture of FPAA. What are the applications of FPAA?
15. Calculate the small signal output resistance and transconductance for the nMOS current sink shown in Fig. 4.7. Given: $I_{out} = 80$ μA, $\lambda = 0.04$ V^{-1} ($L = 1$ μm), $W/L = 2$, $K' = 100$ μA/V^2.
16. Calculate the output resistance of a cascode current sink as shown in Fig. 5.13. Given: $I_{out} = 80$ μA, $\lambda = 0.04$ V$=^1$ ($L = 1$ μm), $(W/L)_1 = 2$, $(W/L)_2 = 1$, $K' = 100$ μA/V^2.
17. Design a first-order low-pass switched capacitor filter to have a 3 dB frequency at 1 kHz. Assume that the clock frequency is 20 kHz.
18. Design a current sink of 30 μA assuming the mirror transistor current is 10 μA. Given $V_{DD} = -V_{SS} = 2.5$ V, $V_{GS} = 1.2$ V and W/L ratio for the mirror transistor as 15/5.
19. Determine the CMRR of the following circuit:

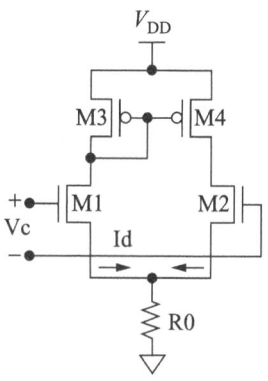

Digital CMOS Logic Design

6.1 Introduction

In this chapter, we introduce the digital logic circuit design using CMOS transistors, which are most popular because of low power dissipation and less area requirement compared to any other logic circuits. It is worthwhile mentioning that almost 90% of the total semiconductor devices are fabricated using silicon CMOS technology. Silicon (Si) is the most suitable semiconductor material for VLSI circuit fabrication because of its native oxide SiO_2. CMOS logic is a combination of nMOS and pMOS logic. The nMOS and pMOS transistors are both functionally and structurally complement to each other. Hence, the combination of nMOS and pMOS is known as complementary MOS or CMOS.

6.2 Digital Logic Design

In digital logic design, there are three primary logic operations: (a) NOT, (b) OR, and (c) AND. Using these three primary logics, any other logic such as NAND, NOR, XOR, or XNOR can be derived.

In NOT logic, if the input (A) is TRUE, the output (F) is FALSE; and if the input (A) is FALSE, the output (F) is TRUE. The NOT logic can be realized using a simple circuit as shown in Fig. 6.1(a). When the switch S1 is ON, the bulb will not glow. When the switch S1 is OFF, the bulb will glow.

In OR logic, if both the inputs are FALSE, the output is FALSE. Otherwise, the output is TRUE. The OR logic can be realized using two switches S1 and S2 connected in parallel, as shown in Fig. 6.1(b). When both switches are OFF, the bulb will not glow; otherwise, the bulb will glow.

In AND logic, if both the inputs are TRUE, the output is TRUE. Otherwise, the output is FALSE. The AND logic can be realized using two switches S1 and S2 connected in series as shown in Fig. 6.1(c).

The mechanical switches shown in Fig. 6.1 can be replaced by MOS transistors as the MOS transistor behaves as a switch when it is operated between the cut-off and saturation regions.

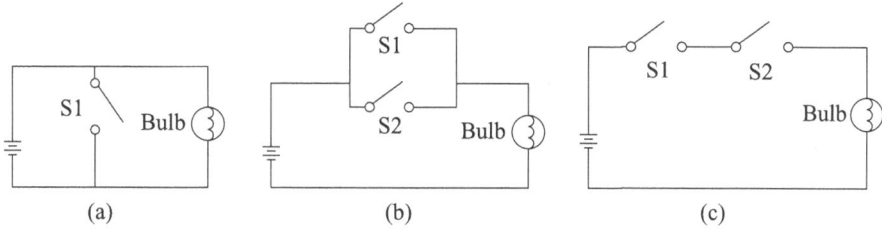

Fig. 6.1   (a) Realization of NOT logic using one switch; (b) realization of OR logic using two switches; (c) realization of AND logic using two switches

As shown in Fig. 6.2(a), an nMOS transistor can be modelled as a switch connected between the drain (D) and source (S) and the switch is controlled by gate (G). When the gate is logic high (H), the nMOS is ON and the switch is closed, and D is connected to S [see Fig. 6.2(b)]. When the gate is logic low (L), the nMOS is OFF and the switch is open, and D is disconnected from S [see Fig. 6.2(c)].

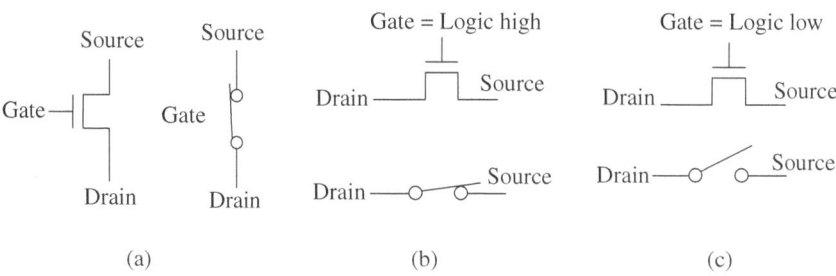

Fig. 6.2   (a) nMOS transistor and its switch model; (b) ON nMOS modelled as closed switch; (c) OFF nMOS modelled as open switch

Similarly, a pMOS can also be modelled as a switch as shown in Figs 6.3(a)–6.3(c).

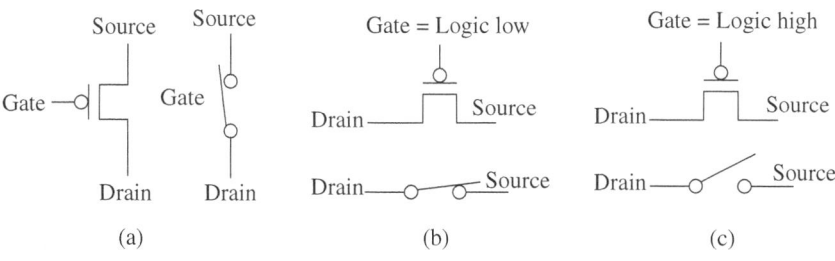

Fig. 6.3   (a) pMOS transistor and its switch model; (b) ON pMOS modelled as closed switch; (c) OFF pMOS modelled as open switch

In reality, an nMOS can pass logic low perfectly but cannot pass logic high perfectly. On the other hand, a pMOS can pass logic high perfectly but cannot pass logic low perfectly. This is illustrated in Figs 6.4(a) and (b).

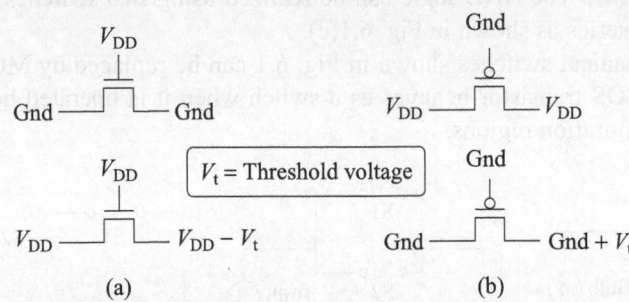

(a)   (b)

Fig. 6.4 (a) Logic level passing through nMOS; (b) logic level passing through pMOS

The logic degradation mechanism is explained in the following text with the help of a circuit shown in Fig. 6.5.

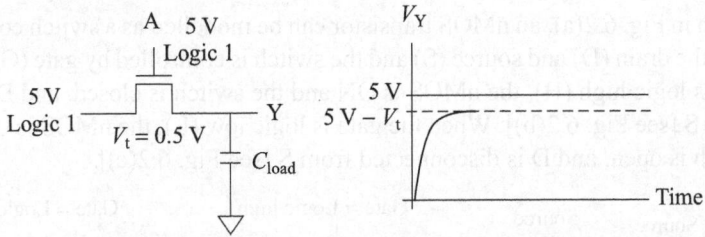

Fig. 6.5 Transfer of logic 1 through an nMOS transistor

Let us consider the case of transfer of logic 1 through an nMOS transistor having threshold voltage $V_t = 0.5$ V, as shown in Fig. 6.5. To make the nMOS transistor ON, we apply logic 1 (= 5 V) to its gate. At the input, we apply logic 1 to transfer it to the output. Assume that the load capacitor is initially at logic 0 (= 0 V). Hence, gate-to-source (V_{GS}) potential difference is 5 V and gate-to-drain (V_{GD}) potential difference is 0 V. Therefore, the ON condition for the nMOS transistor is satisfied. Current flows through the nMOS transistor from the drain to the source and the load capacitor slowly gets charged. As the load capacitor gets charged, the potential at node Y (V_Y) increases. As V_Y increases, V_{GS} decreases. When V_Y reaches (5 V$-V_t$), V_{GS} is just V_t. Under this condition, the nMOS transistor is still ON. But any further increase in V_Y causes V_{GS} to become less than V_t, and makes the nMOS transistor OFF. So no further increase in V_Y is possible. Hence, the maximum voltage at the output is 5 V$-V_t$, and, thus we get a degraded logic 1 at the output.

In the case of logic 0 transfer through an nMOS transistor, the gate and drain terminals are at 5 V and the source terminal is at 0 V. Note that in this case input node is source and the output node is drain that is initially charged to logic 1. Readers must keep in mind that the source and drain are mutually interchangeable due to symmetrical structure of the MOSFET. Now V_{GS} is 5 V and V_Y discharges through the nMOS transistor keeping V_{GS} at 5 V all the time. So the load capacitor can be fully discharged making V_Y to become 0 V. Hence, an nMOS transistor can pass logic 0 perfectly.

Similarly, we can explain that a pMOS transistor passes logic 1 perfectly but passes degraded logic 0. Hence, the bottom line is that the pMOS should be used for pulling a node to logic high, and nMOS should be used for pulling a node to logic low.

6.3 CMOS Logic Design

Any Boolean logic function (F) has possible values: either logic 0 or logic 1. For some of the input combinations, $F = 1$ and for all other input combinations, $F = 0$. So in general, any Boolean logic function can be realized using a structure as shown in Fig. 6.6. The switch S1 is closed and the switch S2 is open for input combinations that produce $F = 1$. The switch S1 is open and the switch S2 is closed for other input combinations that produce $F = 0$.

As shown in Fig. 6.6, the output (F) is either connected to V_{DD} or to the ground, where the logic 0 is represented by the ground and the logic 1 is represented by V_{DD}. So the basic requirement of digital logic design is to implement the pull-up switch (S1) and the pull-down switch (S2). As the pMOS transistors can pass logic 1 perfectly, they are used in pull-up switch realization. Similarly, as the nMOS transistors can pass logic 0 perfectly, they are used in pull-down switch realization.

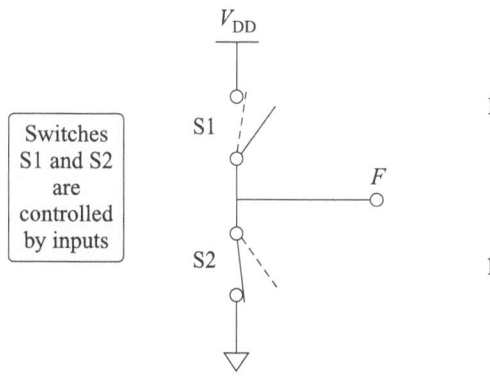
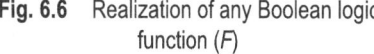

Fig. 6.6   Realization of any Boolean logic function (*F*)

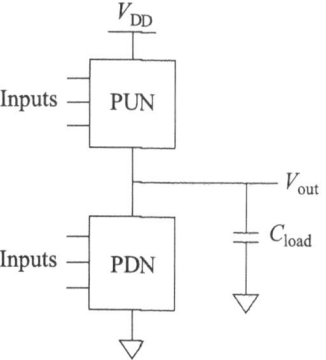

Fig. 6.7   A general CMOS logic circuit

A generalized CMOS logic circuit consists of a pull-up network (PUN) and a pull-down network (PDN), as shown in Fig. 6.7. The PUN comprises pMOSFETs and the PDN comprises nMOSFETs. Depending on the applied input logic, the PUN connects the output node to V_{DD}, and the PDN connects the output node to the ground.

6.4 CMOS Design Methodology

The basic CMOS design methodology involves three steps:

- Given the Boolean expression, take its complement
- Design PDN by realizing
  - AND terms using series-connected nMOSFETs
  - OR terms using parallel-connected nMOSFETs
- Design PUN just reverse (or dual) of the PDN

6.5 Design of CMOS Inverter (NOT) Gate

A CMOS inverter is the simplest logic circuit that uses one nMOS and one pMOS transistor. The nMOS is used in PDN and the pMOS is used in the PUN, as shown in Figs 6.8(a)–(c).

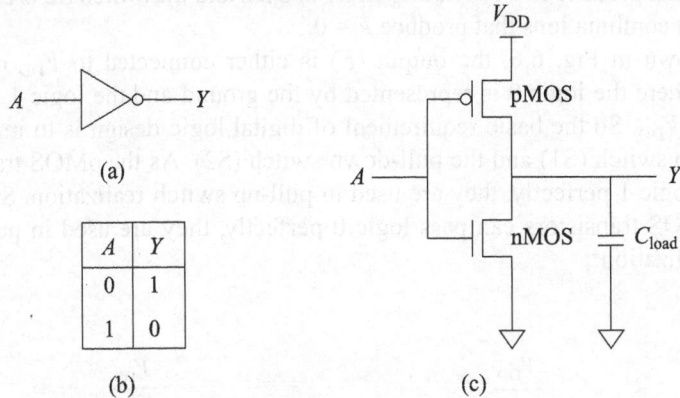

(a)

A	Y
0	1
1	0

(b)   (c)

Fig. 6.8 (a) Symbol for inverter; (b) truth table of inverter; (c) CMOS realization of inverter

Operation   When input is low, the nMOS is OFF and the pMOS is ON. Hence, the output is connected to V_{DD} through pMOS. When the input is high, the nMOS is ON and the pMOS is OFF. Hence, the output is connected to the ground through nMOS. We can connect a capacitor at the output node as shown in Fig. 6.8 to represent the load seen by the inverter. The load capacitor is charged to V_{DD} through pMOS when the input is low and is discharged to the ground through nMOS when the input is high.

6.6 Design of Two-input NAND Gate

To illustrate the design methodology, let us consider a simple example of a two-input NAND gate design. The two-input NAND function is expressed by

$$Y = \overline{A \cdot B} \tag{6.1}$$

Step 1: Take complement of Y

$$\overline{Y} = \overline{\overline{A \cdot B}} = A \cdot B \tag{6.2}$$

Step 2: Design the PDN
In this case, there is only one AND term. So there will be two nMOSFETs in series, as shown in Fig. 6.9.

Step 3: Design the PUN
In PUN, there will be two pMOSFETs in parallel, as shown in Fig. 6.10.

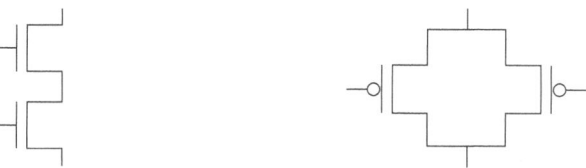

Fig. 6.9   Pull-down network   **Fig. 6.10**   Pull-up network comprising
comprising nMOSFETs   pMOSFETs

Now join the PUN and PDN as shown in Fig. 6.11(c). Note that we have realized $\overline{Y}$, rather than Y because the inversion is automatically provided by the nature of the CMOS circuit operation.

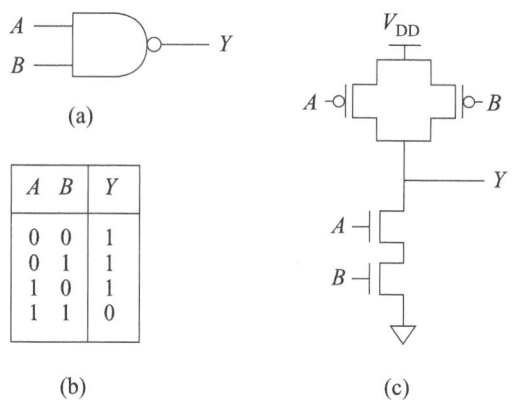

A	B	Y
0	0	1
0	1	1
1	0	1
1	1	0

(b)   (c)

Fig. 6.11   Two-input NAND gate: (a) symbol; (b) truth table; (c) CMOS realization

Operation   When $A = 0$ and $B = 0$, both the nMOS transistors are OFF and both pMOS transistors are ON. Hence, the output is connected to V_{DD} and we get logic high at the output.

When $A = 1$ and $B = 0$, the upper nMOS is ON and lower nMOS is OFF. So, output cannot be connected to the ground. Under this condition, left pMOS is OFF but right pMOS is ON. Hence, the output is connected to V_{DD}, and we get logic high at the output.

When $A = 0$ and $B = 1$, the upper nMOS is OFF and lower nMOS is ON. So, output cannot be connected to ground. Under this condition, left pMOS is ON but right pMOS is OFF. Hence, the output is connected to V_{DD}, and we get logic high at the output.

When $A = 1$ and $B = 1$, both nMOS transistors are ON and both pMOS transistors are OFF. Hence, the output is connected to the ground, and we get logic low at the output. This is illustrated in Figs 6.11(a) and (c). This proves by the truth table of NAND gate shown in Fig. 6.11(b).

6.7 Design of Two-Input NOR Gate

Let us consider another example of a two-input NOR gate. The two-input NOR function is expressed by

$$Y = \overline{A + B} \qquad (6.3)$$

Step 1: Take complement of Y

$$\overline{Y} = \overline{\overline{A + B}} = A + B \qquad (6.4)$$

Step 2: Design the PDN
Here, there is only one OR term. Hence, there will be two nMOSFETs connected in parallel, as shown in Fig. 6.12.

Step 3: Design the PUN
In the PUN, two pMOSFETs will be connected in series, as shown in Fig. 6.13.

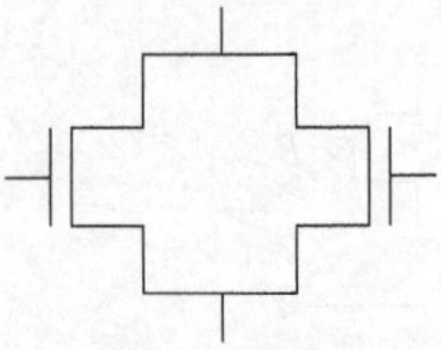

Fig. 6.12 Pull-down network comprising nMOSFETs

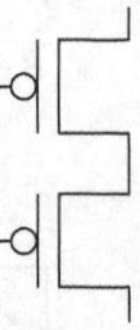

Fig. 6.13 Pull-up network comprising pMOSFETs

Now, join the PUN and PDN as shown in Fig. 6.14(c).

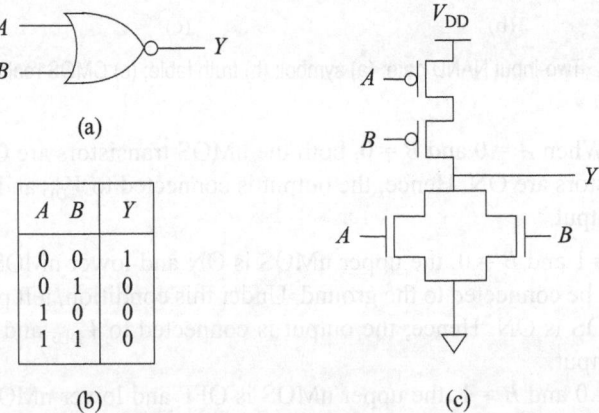

A	B	Y
0	0	1
0	1	0
1	0	0
1	1	0

Fig. 6.14 Two-input NOR gate: (a) symbol; (b) truth table; (c) CMOS realization

Operation When $A = 0$ and $B = 0$, both nMOS transistors are OFF and both pMOS transistors are ON. Hence, the output is connected to V_{DD} and we get logic high at the output.

When $A = 1$ and $B = 0$, the upper pMOS is OFF and lower pMOS is ON. So, output cannot be connected to the V_{DD}. Under this condition, left nMOS is ON and right nMOS is OFF. Hence, the output is connected to the ground and we get logic low at the output.

When $A = 0$ and $B = 1$, the upper pMOS is ON and lower pMOS is OFF. So, output cannot be connected to V_{DD}. Under this condition, left nMOS is OFF and right nMOS is ON. Hence, the output is connected to the ground and we get logic low at the output.

When $A = 1$ and $B = 1$, both nMOS transistors are ON and both pMOS transistors are OFF. Hence, the output is connected to V_{DD} and we get logic low at the output. This proves the truth table of NOR gate as shown in Fig. 6.14(b).

6.8 Classification of CMOS Digital Logic Circuit

CMOS logic circuits are mainly classified into two categories as follows:
- Combinational logic circuit
- Sequential logic circuit

In the combinational logic circuit, the output is determined by the present logic inputs. However, in the sequential logic circuit, the output is determined by the present inputs and past outputs. The examples of combinational logic circuits are Inverter, NAND gate, NOR gate, multiplexer, demultiplexer, decoder, encoder, half-adder, full-adder, etc. The examples of sequential logic circuits are flip-flops, latches, registers, counters, etc.

There are other CMOS design styles as given below:
- CMOS transmission logic
- Complementary pass-transistor logic
- Dynamic CMOS logic
- Domino CMOS logic
- NORA CMOS logic
- Zipper CMOS logic

6.9 Combinational Logic Circuit

In this section, we discuss how the combinational logic circuit is designed with a few examples. As we have discussed before, any CMOS logic circuit consists of a PUN and a PDN. The PUN comprises only pMOSFETs, and the PDN comprises of only nMOSFETs. There is a fundamental reason why pMOSFETs are used in PUN and the nMOSFETs in PDN (also explained before).

6.9.1 Design of Complex Logic Circuit

Let us now design a complex Boolean function given below:

$$Y = \overline{A(B+C)+DE} \tag{6.5}$$

Step 1: Take complement of Y

$$\bar{Y} = \overline{\overline{A(B+C)+DE}} = A(B+C)+DE \qquad (6.6)$$

Step 2: Design of PDN
Let $Z = B = C$, then Eqn (6.6) becomes

$$\bar{Y} = AZ + DE \qquad (6.7)$$

Now, $Z(=B=C)$ is realized by two nMOSFETs connected in parallel. Let us call this as sub-logic realizing Z. In Eqn (6.7), there are two AND terms which are ORed. AZ is realized by one nMOSFET connected in series with sub-logic realizing Z. The term DE is realized by two nMOSFETs connected in series. Finally, two sub-logics realizing AZ and DE are connected in parallel.

Step 3: Design of PUN
We just do the reverse connection of nMOSFETs to realize the PUN using pMOSFETs. The complete CMOS logic is shown in Fig. 6.15.

The Boolean function given in Eqn (6.5) can also be realized, as shown in Fig. 6.16. In this case, we expand Eqn (6.6) into:

$$\bar{Y} = AB + AC + DE \qquad (6.8)$$

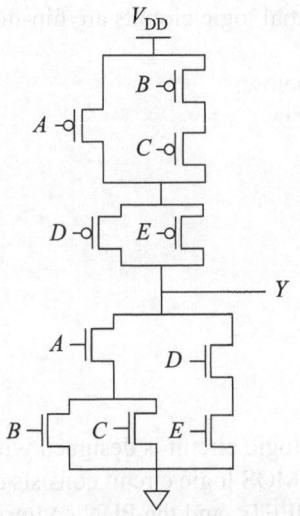

Fig. 6.15 CMOS logic realizing the Boolean function given in Eqn (6.5)

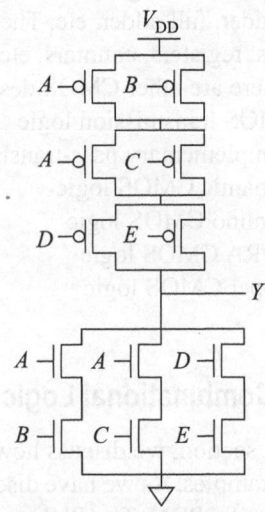

Fig. 6.16 CMOS logic realizing the Boolean function given in Eqn (6.8)

Comparing the realizations shown in Figs 6.15 and 6.16, we can see that in Fig. 6.16, we need more transistors than in Fig. 6.15. This means that the CMOS logic realization with the minimum number of transistors requires judicious function optimization.

6.9.2 Design of AND-OR-INVERT and OR-AND-INVERT Gate

Any Boolean function can be expressed either in the SOP (sum-of-products) or POS (product-of-sum) form. The AND-OR-INVERT (AOI) gate is suitable for realizing functions in the SOP form. Here a function is expressed as a summation (OR) of product (AND) terms. For example, a function X in the SOP form with a complement can be written as

$$X = \overline{AB + CD + EFG} \tag{6.9}$$

The AOI implementation of the function X is shown in Fig. 6.17. Each AND term is implemented by a series combination of nMOS transistors. The OR operation between the AND terms is realized by a parallel connection of each series-connected nMOS transistors subcircuit. The PUN is implemented as a dual of an nMOS network.

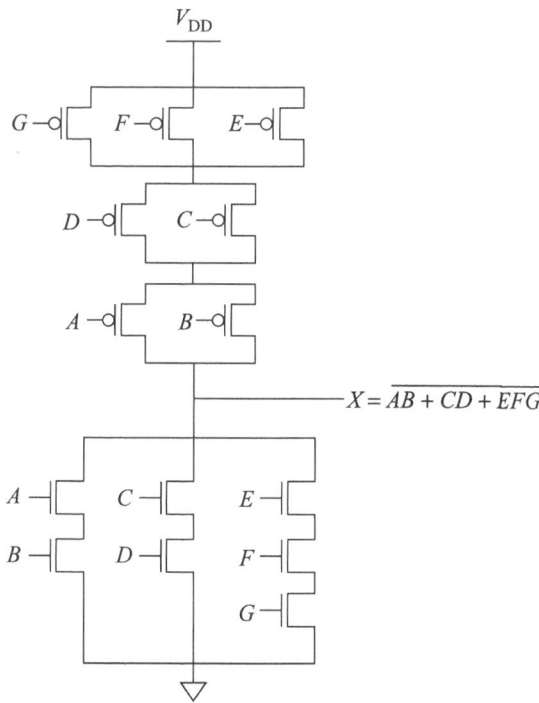

Fig. 6.17 AOI gate realization using CMOS logic

Similarly, a function expressed in the POS form with a complement can be written as

$$Y = \overline{(A + B)(C + D)(E + F + G)} \tag{6.10}$$

The OR-AND-INVERT (OAI) realization of the function Y is shown in Fig. 6.18. Each OR term is implemented by a parallel combination of nMOS transistors. The AND operation between the OR terms is realized by the series connection of each parallel-connected nMOS. The PUN is implemented as a dual of the nMOS network.

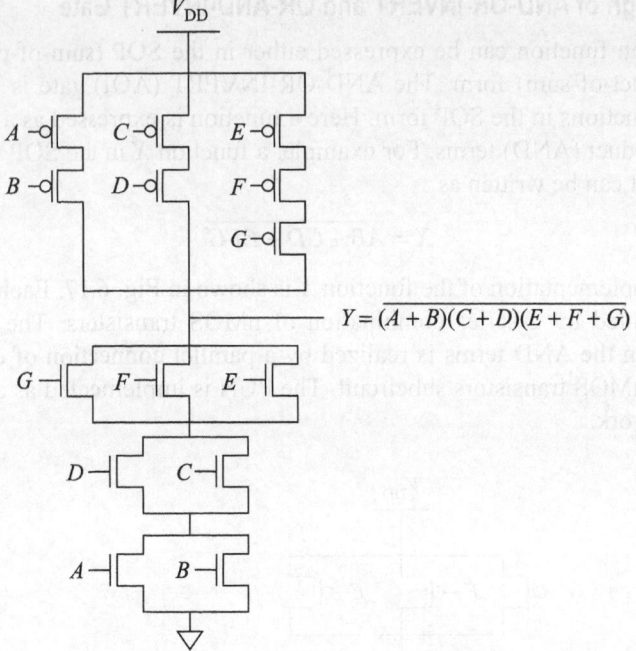

Fig. 6.18   OAI gate realization using CMOS logic

6.9.3 Design of XOR Gate

The Boolean expression for two-input XOR gate is given by

$$Y = A\overline{B} + \overline{A}B \tag{6.11}$$

Step 1: Take complement of Y

$$\overline{Y} = \overline{A\overline{B} + \overline{A}B} = \overline{A\overline{B}} \times \overline{\overline{A}B} = (\overline{A} + B) \times (A + \overline{B}) = \overline{A}\,\overline{B} + AB \tag{6.12}$$

Step 2: Design of PDN

Here, there are two AND terms and one OR term. Each of the AND term is realized by two series-connected nMOSFETs. The OR term is realized by parallel connection of two series-connected nMOSFETs realizing AND terms.

Step 3: Design of PUN

In PUN, we need four pMOSFETs connected in a reverse manner. The complete CMOS circuit realizing two-input XOR gate is shown in Fig. 6.19.

   Note: In realizing the two-input XOR gate, we have assumed that the inputs are available both in the normal and complemented form. This is a valid assumption because in a full-chip design, the inputs might come from the output of flip-flop or latch where both the normal and complemented forms of the output (Q and $\overline{Q}$) are available. In case the complemented input is not available, an inverter is to be used to generate the complemented input.

6.9.4 Design of Half-adder Circuit

A half-adder circuit has two inputs and two outputs: Sum and Carry. The truth table of a half-adder circuit is shown in Table 6.1.

Table 6.1   Truth table of half-adder

A	B	Sum	Carry
0	0	0	0
0	1	1	0
1	0	1	0
1	1	0	1

Boolean expressions for the sum and carry are given by

$$\text{Sum} = A\bar{B} + \bar{A}B \tag{6.13}$$

$$\text{Carry} = AB \tag{6.14}$$

CMOS realization of Sum is shown in Fig. 6.19 and the Carry is shown in Fig. 6.20.

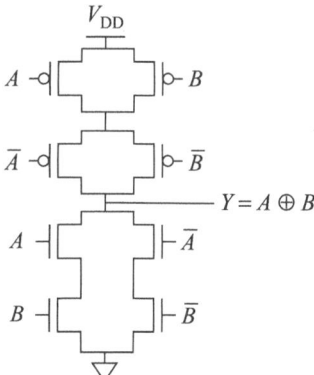

Fig. 6.19   CMOS realization of two-input XOR gate

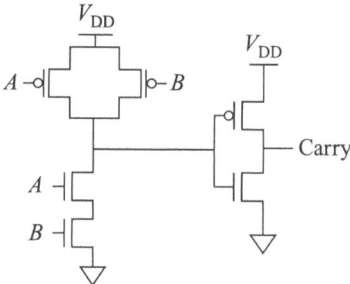

Fig. 6.20   CMOS realization of carry function of half-adder circuit

6.9.5 Design of Full-adder Circuit

A full-adder circuit has three inputs: A, B, and Carry-in (Cin) and two outputs: Sum and Carry-out (Cout). The truth table of a full-adder circuit is shown in Table 6.2.

Boolean expressions for the Sum and Cout are given by

$$\text{Sum} = \bar{A}\bar{B}C\text{in} + \bar{A}B\overline{C\text{in}} + A\bar{B}\overline{C\text{in}} + ABC\text{in} \tag{6.15}$$

$$C\text{out} = AB + AC\text{in} + BC\text{in} \tag{6.16}$$

As shown in Table 6.3, taking $\overline{C\text{out}}$ as input, we could simplify the expression for Sum as

$$\text{Sum} = \bar{A}\overline{C\text{out}} + \bar{B}\overline{C\text{out}} + C\text{in}\,\overline{C\text{out}} + ABC\text{in}$$

$$= (A + B + C\text{in})\,\overline{C\text{out}} + ABC\text{in} \tag{6.17}$$

Table 6.2   Truth table of full-adder

A	B	Cin	Sum	Cout
0	0	0	0	0
0	0	1	1	0
0	1	0	1	0
0	1	1	0	1
1	0	0	1	0
1	0	1	0	1
1	1	0	0	1
1	1	1	1	1

Table 6.3   Truth table of full-adder (modified to include $\overline{\text{Cout}}$)

A	B	Cin	Cout	$\overline{\text{Cout}}$	Sum
0	0	0	0	1	0
0	0	1	0	1	1
0	1	0	0	1	1
0	1	1	1	0	0
1	0	0	0	1	1
1	0	1	1	0	0
1	1	0	1	0	0
1	1	1	1	0	1

The complete CMOS realization of full-adder is shown in Fig. 6.21.

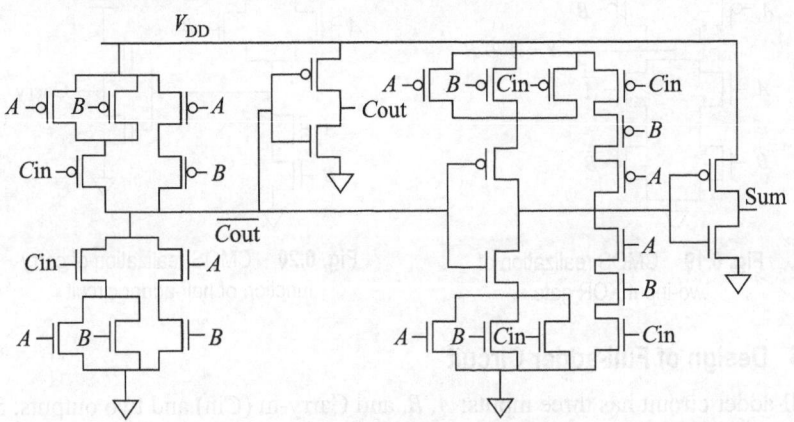

Fig. 6.21   CMOS realization of full-adder circuit

Example 6.1   Design a 2 × 2 input AND–OR gate to realize the following Boolean function:

$$Y = AB + CD$$

Solution   To realize the Boolean function, first we find out the complement of the function.

Complement of the function, $\overline{Y} = \overline{AB + CD} = \overline{AB} \times \overline{CD} = (\overline{A} + \overline{B}) \times (\overline{C} + \overline{D})$. We can see $\overline{Y}$ has two OR terms and one AND term. Realize each OR term by parallel connection of two nMOS. Connect them in series to realize the AND term as shown in Fig. 6.22.

Example 6.2 Realize a tri-state buffer.

Solution A tri-state buffer can be realized as shown in Fig. 6.23(c).

The circuit implementation as shown in Fig. 6.23(d) is not recommended as it suffers from dynamic charge sharing problem.

Example 6.3 Design a 2:1 multiplexer using static CMOS logic.

Solution A 2:1 multiplexer can be realized, as shown in Fig. 6.24(c).

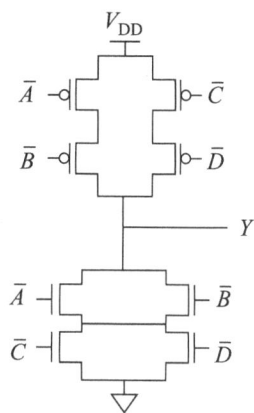

Fig. 6.22 Schematic to realize function $Y = AB + CD$

C	A	Y
0	0	0
0	1	1
1	0	Z
1	1	Z

(a)

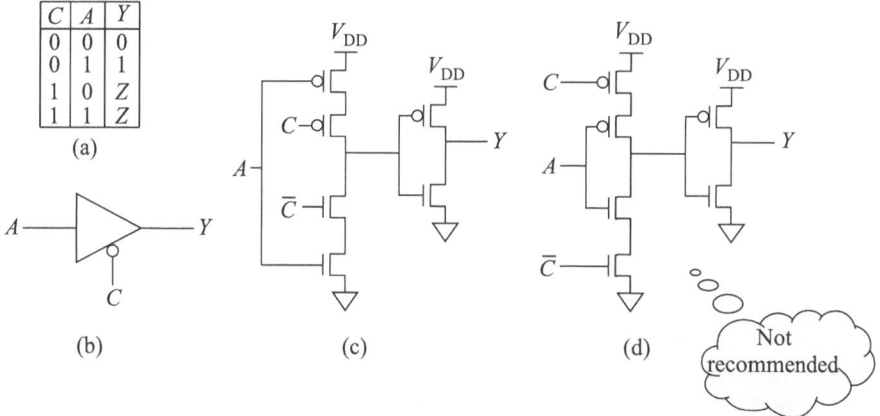

(b) (c) (d) Not recommended

Fig. 6.23 Tri-state buffer: (a) truth table; (b) symbol; (c) and (d) CMOS realization

S	A	B	Y
0	0	0	0
0	0	1	0
0	1	0	1
0	1	1	1
1	0	0	0
1	0	1	1
1	1	0	0
1	1	1	1

(a)

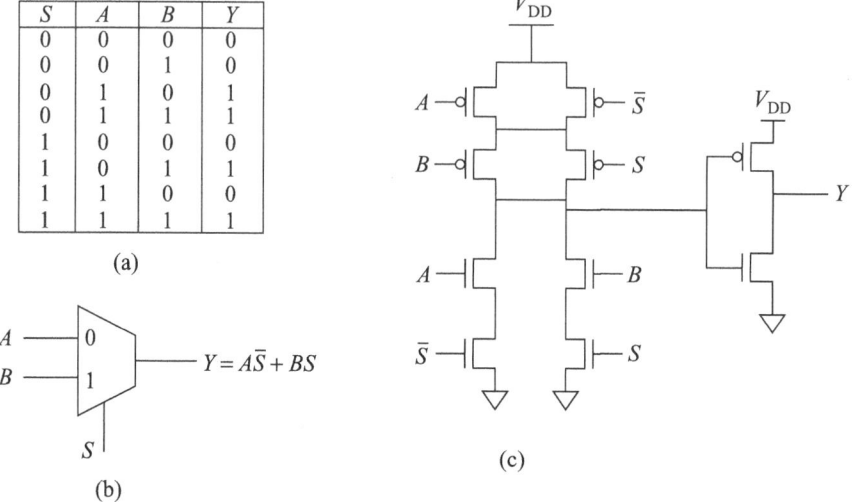

$Y = A\bar{S} + BS$

(b)

(c)

Fig. 6.24 2:1 Multiplexer: (a) truth table; (b) symbol; (c) CMOS realization

6.10 Sequential Logic Circuit

In the combinational logic circuits, the output is determined by the input logic levels. The combinational circuit lacks the storing capability of any previous events. In the sequential logic circuits, the output is determined by the present input logic level, as well as the past output logic level. In the sequential circuit, there is a feedback connection between the output and the input.

The generalized structure of a sequential circuit is shown in Fig. 6.25. It consists of a combinational logic block and a memory circuit in the feedback path.

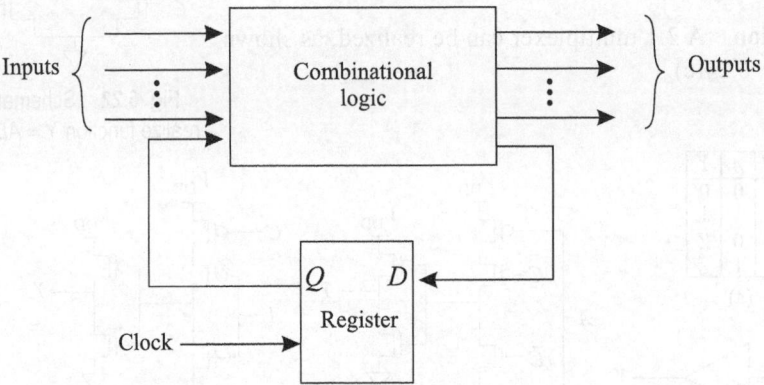

Fig. 6.25　Schematic of a sequential logic circuit

The feedback path contains the register that has the bit storing capability. The sequential logic circuit is also known as *regenerative* logic circuit. The regenerative circuits are formed with positive feedback. They are classified into three following types:

- Bistable circuits—have two stable states.
- Monostable circuits—have only one stable state.
- Astable circuits—have no stable state, rather they oscillate between two states.

Bistable circuits are the most commonly used circuits. Latches, flip-flops, registers, and memory are the examples of bistable circuits.

6.10.1　Working Principle of Bistable Circuits

Let us consider the basic bistable circuit consisting of two cross-coupled inverters (Inv1 and Inv2) as shown in Fig. 6.26.

The input and output of the first inverter (Inv1) are V_{in1} and V_{out1}. The voltage transfer characteristic (VTC) of the first inverter is shown in Fig. 6.27(a).

The input and output of the second inverter (Inv2) are V_{in2} and V_{out2}. The VTC of the second inverter is shown in Fig. 6.27(b). In this plot, the axes for input and output voltages are just interchanged.

When two inverters are connected back-to-back, we can write $V_{in2} = V_{out1}$ and $V_{in1} = V_{out2}$. In order to find out the operating points of the circuit shown in

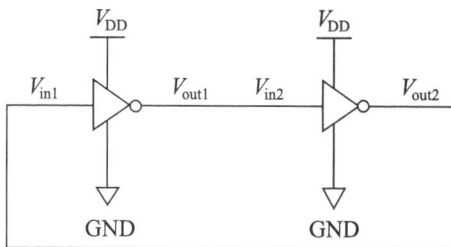

Fig. 6.26 Basic bistable circuit with two back-to-back connected inverters

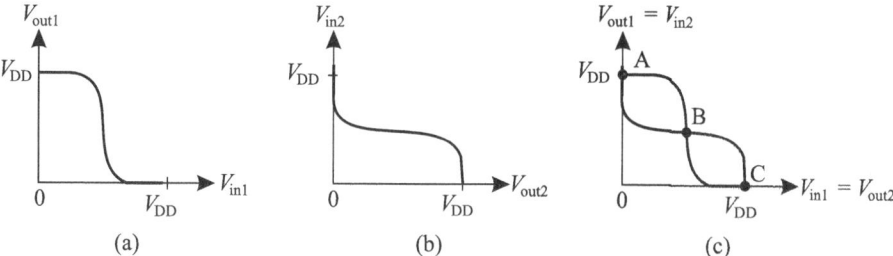

Fig. 6.27 Voltage transfer characteristics

Fig. 6.26, let us now combine the two VTCs as shown in Fig. 6.27(c). As shown in Fig. 6.27(c), there are three operating points: A, B, and C.

Now, let us suppose that the operating point is at A. That is, the input of the first inverter $V_{in1} = 0$, therefore, $V_{out1} = 1 = V_{in2}$ and consequently $V_{out2} = 0 = V_{in1}$. Hence, the circuit remains stable at this operating point. Any small deviation from the operating point does not cause any change in the outputs.

Similarly, let us suppose that the operating point is at C. That is, the input of the first inverter $V_{in1} = 1$, therefore, $V_{out1} = 0 = V_{in2}$ and consequently $V_{out2} = 1 = V_{in1}$. Hence, the circuit remains stable at this operating point. In this case also, any small deviation from the operating point does not cause any change in the outputs.

Now, let us suppose that the operating point is at B. Now, say the input V_{in1} is slightly reduced by an amount ΔV due to some noise. The output V_{out1} and the input V_{in2} will be increased which will cause V_{out2} to decrease and so to V_{in1}. This will continue till V_{in1} reaches 0 and V_{out1} reaches V_{DD}. Similarly, we can show that at the operating point B, a small increase in V_{in1} will continue to decrease V_{out1} till it reaches 0 and V_{in1} reaches V_{DD}. Therefore, at the operating point B, a small deviation in the voltage level will move the operating point to either A or C.

Hence, the circuit has two stable operating points A and C and B is an unstable operating point. Since the circuit has two stable operating points and hence two stable states (either 0 or V_{DD}), the circuit is called *bistable circuit*.

6.10.2 Design of SR Latch Circuit

In the SR latch logic circuit, there are two inputs: S (Set) and R (Reset) and two complementary outputs: Q and $\bar{Q}$. The input and output relationship is described in Table 6.4.

Table 6.4   Truth table of the NOR-based SR latch circuit

S	R	Q_{n+1}	$\overline{Q_{n+1}}$	Operation
0	0	Q_n	$\overline{Q_n}$	Hold
1	0	1	0	Set
0	1	0	1	Reset
1	1	0	0	Not allowed

The gate-level schematic of the SR latch circuit is shown in Fig. 6.28.

Figure 6.29 shows the CMOS realization of the NOR-based SR latch circuit.

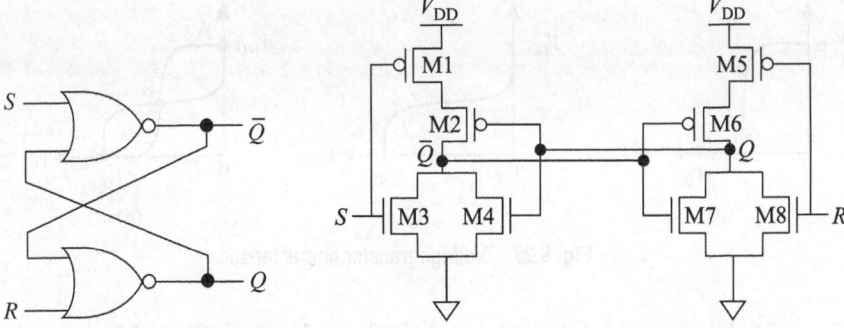

Fig. 6.28   Gate-level schematic of SR latch realized using NOR gate

Fig. 6.29   CMOS realizations of NOR-based SR latch circuit

The next state of SR latch can be expressed as

$$Q_{n+1} = S + \overline{R}Q_n \qquad (\text{for } SR = 0) \qquad (6.18)$$

The condition $SR = 0$ implies that both S and R cannot be 1 at the same time. An equation that expresses the next state value of a latch in terms of its present state and inputs is referred to as a 'next-state equation' or a 'characteristic equation'.

Operation   When $S = 0$ and $R = 0$, the nMOS transistors M3 and M8 are OFF. The pMOS transistors M1 and M5 are ON. This cannot determine the output logic. Let us assume $Q = 0$, then M2 is ON and M4 is OFF. Thus, the node Q is connected to V_{DD}, i.e., $\overline{Q} = 1$. As $\overline{Q} = 1$, M7 is ON and M6 is OFF. Thus, node Q is connected to the ground, i.e., Q remains at logic 0 state. If we assume $Q = 1$, then M4 is ON and M2 is OFF. Thus, node $\overline{Q}$ is connected to the ground, i.e., $\overline{Q} = 0$. As $\overline{Q} = 0$, M6 is ON and M7 is OFF. Thus, node Q is connected to V_{DD}, i.e., Q remains at logic 1 state. This proves the first row of the truth table of SR latch.

When $S = 0$, $R = 1$, the transistor M8 is ON. Thus, irrespective of other input and past output conditions, the node Q is connected to the ground, i.e., Q is in logic 0 state.

When $S = 1$ and $R = 0$, the transistor M3 is ON. Thus, irrespective of other input and past output conditions, the node $\overline{Q}$ is connected to the ground, i.e., $\overline{Q}$ is in logic 0 state. This makes M5 and M6 ON; thus, node Q is connected to V_{DD}, i.e., Q is in logic 1 state.

If S and R inputs are zero at the same time, both the outputs will be zero. Hence, this combination is not allowed.

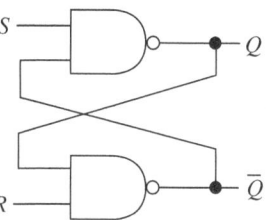

The SR latch circuit can also be realized using NAND gates as shown in Fig. 6.30.

Table 6.5 shows the truth table of NAND-based SR latch circuit.

Fig. 6.30 Gate-level schematic of SR latch realized using NAND gate

Figure 6.31 shows the CMOS realization of the NAND-based SR latch circuit.

Table 6.5 Truth table of the NAND-based SR latch circuit

S	R	Q_{n+1}	$\overline{Q}_{n+1}$	Operation
0	0	1	1	Not allowed
0	1	1	0	Set
1	0	0	1	Reset
1	1	Q_n	$\overline{Q}_n$	Hold

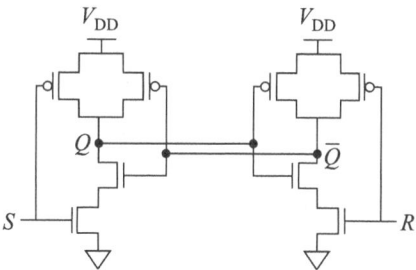

Fig. 6.31 CMOS realizations of NAND-based SR latch circuit

6.10.3 Design of Clocked Latch and Flip-Flop Circuit

The SR latch circuits discussed in Section 6.10.1 are asynchronous in nature in which the output changes as the input logic level changes. In a design, the inputs might come from different paths encountering different path delays. Hence, depending on which input comes first, the output might change accordingly, which might not be the desirable output. To avoid this problem, a clock input is added such that the output changes only in the active period of the clock signal. Generally, the clock signal is a periodic square waveform applied simultaneously to all the gates in the system.

Figure 6.32 shows the gate-level schematic of the clocked NOR-based SR latch.

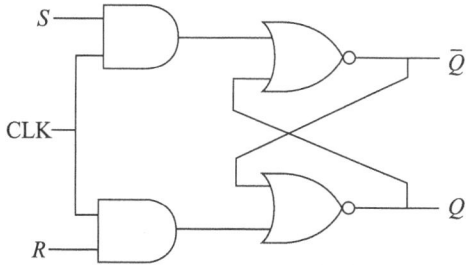

Fig. 6.32 Schematic of the clocked NOR-based SR latch circuit

Figure 6.33 shows the CMOS design of the NOR-based SR latch circuit.

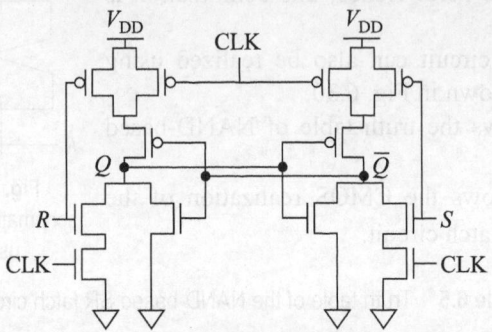

Fig. 6.33 CMOS realization of the clocked NOR-based SR latch circuit

The NAND-based SR latch can also be realized with clock input. This is left as an exercise to the readers.

6.10.4 Design of Clocked JK Latch

The JK latch is commonly known as JK flip-flop. It has three inputs: J, K, and CLK; and two complementary outputs: Q and $\overline{Q}$. JK flip-flop has no not-allowed inputs, unlike SR latches. It allows all possible input combinations. The truth table is shown in Table 6.6.

Table 6.6 Truth table of JK flip-flop

J	K	Q_n	$\overline{Q}_n$	Q_{n+1}	$\overline{Q}_{n+1}$	Operation
0	0	0	1	0	1	Hold
		1	0	1	0	
0	1	0	1	0	1	Reset
		1	0	0	1	
1	0	0	1	1	0	Set
		1	0	1	0	
1	1	0	1	1	0	Toggle
		1	0	0	1	

Figure 6.34 shows the gate-level design of the JK flip-flop.

From the gate-level design of JK flip-flop shown in Fig. 6.34, we can write the expressions for the output as

$$\overline{Q} = \overline{S+Q} = \overline{\overline{Q} \cdot J \cdot CLK + Q}$$

$$Q = \overline{R+\overline{Q}} = \overline{Q \cdot K \cdot CLK + \overline{Q}}$$

Therefore, to implement the PDN for implementing the logic for Q, we require three nMOS transistors connected in series with one nMOS connected in parallel. The inputs to the series connected nMOS transistors are Q, K, and CLK and the input to the parallel connected nMOS transistor is $\overline{Q}$.

Similarly, to implement the PDN for implementing the logic for $\overline{Q}$, we require three nMOS transistors connected in series with one nMOS connected in parallel. The inputs to the series connected nMOS transistors are $\overline{Q}$, J, and CLK and the input to the parallel connected nMOS transistor is Q.

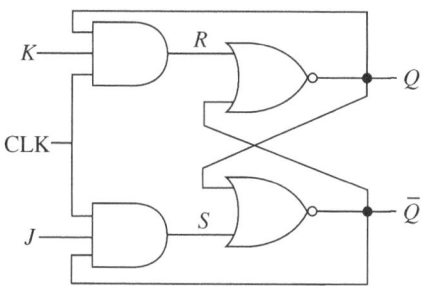

Fig. 6.34   Gate-level design of JK flip-flop

Figure 6.35 shows the CMOS design of the JK flip-flop.

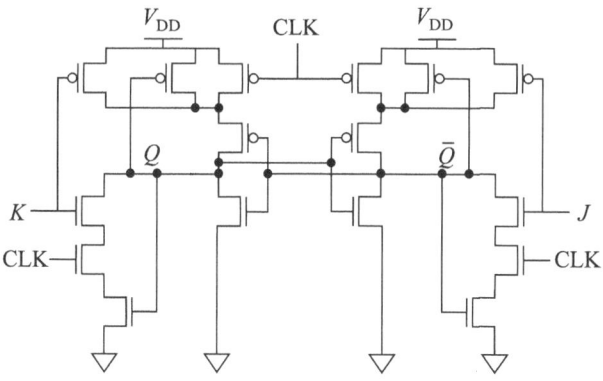

Fig. 6.35   CMOS design of JK flip-flop

6.10.5 Design of CMOS D Flip-Flop

D flip-flop can be designed from the JK flip-flop by connecting an inverter between the J and K inputs. The truth table of D flip-flop is shown in Table 6.7.

Table 6.7   Truth table of D flip-flop

D	Q	Q_n	Q_{n+1}	$\overline{Q}_{n+1}$	Operation
0	0	1	0	1	Hold
0	1	0	0	1	Reset
1	0	1	1	0	Set
1	1	0	1	0	Hold

When the CLK is at logic 1, the D input is propagated to Q output. When the CLK is at logic 0, the Q output is hold. The CMOS design of D flip-flop is shown in Fig. 6.36. When CLK input is at logic 0, the circuit simply works as two cross-coupled inverter circuits and therefore it holds its outputs. When the CLK input is set to logic 1, depending on the data input D, the output is either set or reset as illustrated in the truth table.

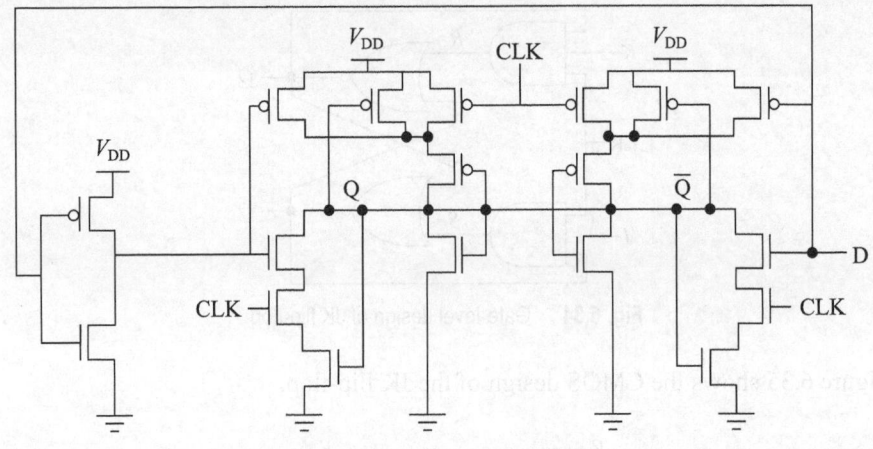

Fig. 6.36 CMOS D flip-flop

6.11 Pseudo-nMOS Logic

In the pseudo-nMOS logic, the PDN is realized by a single pMOS transistor. The gate terminal of the pMOS transistor is connected to the ground. It remains permanently in the ON state. Depending on the input combinations, output goes low through the PDN. An inverter realized using pseudo-nMOS logic is shown in Figs 6.37(a)–(c). The PDN is realized using the process described in Section 6.4.

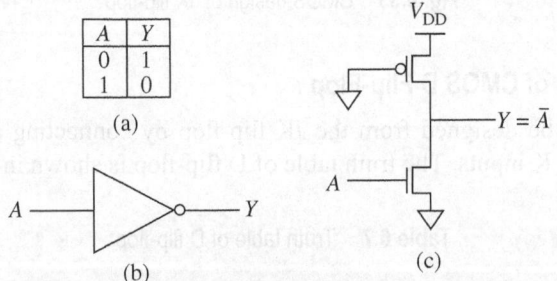

Fig. 6.37 Inverter: (a) truth table; (b) symbol; (c) pseudo-nMOS realization

---

Example 6.4 Design an XNOR gate using pseudo-nMOS logic.

Solution The pseudo-nMOS realization of two-input XNOR gate is shown in Fig. 6.38.

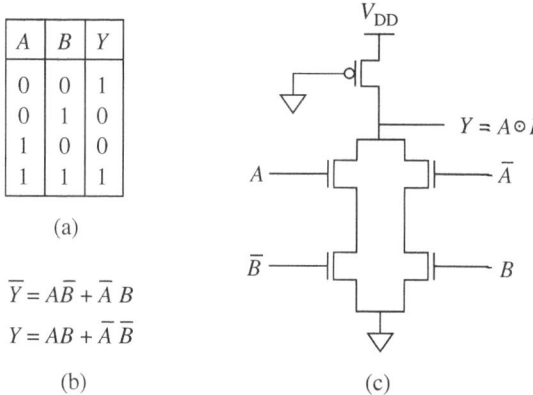

A	B	Y
0	0	1
0	1	0
1	0	0
1	1	1

(a)

$\overline{Y} = A\overline{B} + \overline{A}\,B$

$Y = AB + \overline{A}\,\overline{B}$

(b)   (c)

Fig. 6.38   XNOR gate: (a) truth table; (b) Boolean expression; (c) pseudo-nMOS realization

6.12 CMOS Transmission Gate

CMOS transmission gate (TG) is a parallel connection of nMOSFET and pMOSFET that realizes a simple switch. The inputs to the gate of the nMOSFET and pMOSFET are complementary to each other. It is also known as *pass gate*. Figure 6.39 shows a TG.

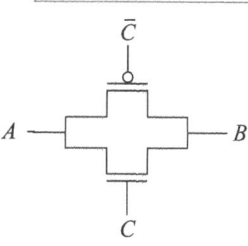

Fig. 6.39   CMOS transmission gate

When signal C is high, $\overline{C}$ is low, thus both nMOS and pMOS transistors are ON, and the nodes A and B are short-circuited. So the input logic is transferred to the output. When signal C is low, $\overline{C}$ is high, thus both nMOS and pMOS transistors are OFF, and the nodes A and B are open-circuited, this is called high-impedance state.

Using CMOS, TG logic circuits can be designed, and it might require lesser number of transistors as compared to the standard CMOS design. This will be clear as we go through some examples in Section 6.12.1.

6.12.1 Design of Combinational Logic Circuits Using CMOS TGs

Example 6.5   Design a 2:1 multiplexer.

Solution   Figure 6.40 shows the design of a two-input multiplexer circuit.

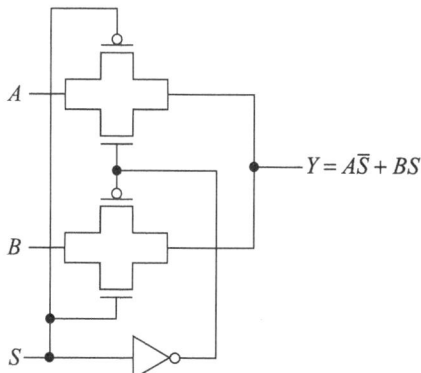

$Y = A\overline{S} + BS$

Fig. 6.40   Design of two-input MUX using CMOS TGs

Example 6.6   Design a two-input XOR gate.

Solution   Figure 6.41 shows the design of a two-input XOR gate.

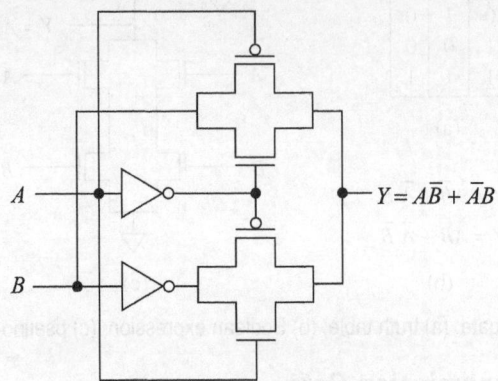

Fig. 6.41   Design of XOR gate using CMOS TGs

Example 6.7   Design of a Boolean function $F = ABC + \overline{A}C + B\overline{C}$.

Solution   Figure 6.42 shows the design of a Boolean function $F = ABC + \overline{A}C + B\overline{C}$.

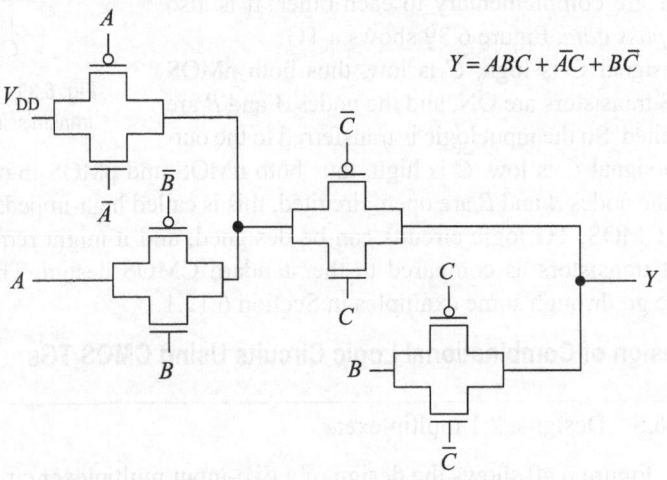

Fig. 6.42   Design of 3-variable Boolean function using CMOS TGs

6.12.2   D-Latch and Edge-triggered Flip-flop

D-latch stores a logic level. This is often called 1-bit register or delay flip-flop. It has single data input D and two outputs: Q and $\overline{Q}$. This is obtained simply by connecting an inverter between the J and K inputs of the JK flip-flop. D input is directly connected to the J input, and the inverted D input is connected to the K input of the JK flip-flop. Block diagram of the D-latch is shown in Fig. 6.43.

Truth table of the D-latch is shown in Table 6.8.

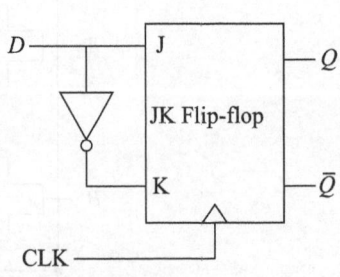

Fig. 6.43   Block diagram of the D-latch

Table 6.8 Truth table of D-latch

D	Q_{n+1}
0	0
1	1

Figure 6.44 shows the design of the D-latch using CMOS TGs.

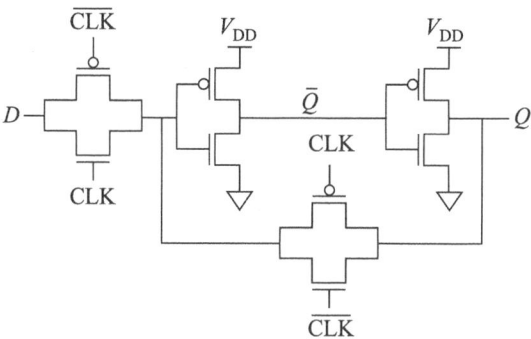

Fig. 6.44 Design of the D-latch using CMOS TGs

The D-latch discussed above is a level triggered flip-flop. This means that the output of the flip-flop changes as long as the clock signal remains at logic high. This has a serious problem in digital logic circuits where output might change during the ON-state of the clock pulse several times if data input changes. But we would like to have output that changes only once in a clock cycle. One way of solving this problem is to design the clock such that it remains at an enabling level for a very short time. In many situations, it may be found that the narrow clock pulses are too narrow to trigger the flip-flop reliably. A better solution is to design the flip-flop such that the output changes only when the clock input makes a transition, rather than at the enabling level of the clock input. One such flip-flop, where output changes only on the clock transitions is a master–slave flip-flop.

Figure 6.45 shows a master–slave flip-flop, which is designed by cascading two D-latches. The clock signal is applied to the first stage (master) and the inverted clock signal is applied to the second stage (slave).

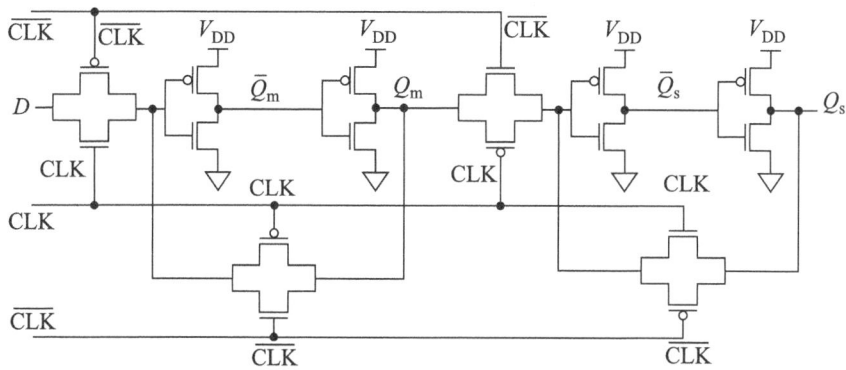

Fig. 6.45 CMOS edge-triggered master–slave D-flip-flop

6.13 Dynamic CMOS Logic

In the static CMOS logic the functionality of pull-up network is just redundant as the pull-down network alone is sufficient to realize the functionality. In order to reduce the number of transistors, pseudo-nMOS logic is used where pull-up network is replaced by a pMOS transistor with gate terminal connected to the ground. However, pseudo-nMOS logic suffers from large static power dissipation problem.

An alternate method of reducing transistor count is to use dynamic CMOS logic. It is very similar to the pseudo-nMOS logic except one additional nMOS transistor MN connected between the pull-down network and the ground. The pMOS transistor in the PUN and the extra nMOS transistors in the PDN are operated by a clock signal ϕ as shown in Fig. 6.46.

The dynamic logic circuit operates in two phases of a single clock pulse (ϕ). In the pre-charge phase ($\phi = 0$), the output is pre-charged to a logic high level. In the evaluation phase ($\phi = 1$), the output is evaluated based on the applied input logic. Figure 6.47 shows a dynamic CMOS circuit implementing the Boolean function $Y = AB + A\overline{B}C$.

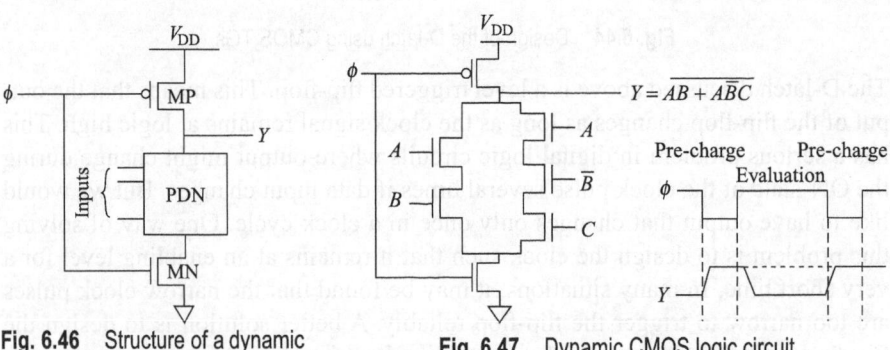

Fig. 6.46 Structure of a dynamic CMOS logic circuit

Fig. 6.47 Dynamic CMOS logic circuit

The dynamic logic circuits offer the following advantages over standard CMOS logic circuits:
- Low power dissipation
- Small area due to less number of transistors
- Large noise margin

However, the dynamic CMOS logic circuit has a serious problem when they are cascaded. In the pre-charge phase ($\phi = 0$), output of all the stages are pre-charged to logic high. In the evaluation phase ($\phi = 1$), the outputs of all the stages are evaluated simultaneously. Suppose in the first stage, the inputs are such that the output is logic low after the evaluation. In the second stage, the output of the first stage is one input, and there are other inputs. If the other inputs of the second stage are such that output of it discharges to logic low, then the evaluated output of the first stage can never make the output of the second stage logic high. This is because, by the time the first stage is being evaluated, output of the second stage is discharged, since evaluation happens simultaneously. Remember that the output cannot be charged to logic high in the evaluation phase ($\phi = 1$, pMOSFET in PUN is OFF), it can only be retained in the logic high depending on the inputs.

6.14 Domino CMOS Logic

In order to avoid the problem of dynamic CMOS logic discussed in the previous section, domino CMOS logic is proposed. Domino CMOS logic is a slightly modified version of the dynamic CMOS logic circuit. In this case, a static inverter is connected at the output of each dynamic CMOS logic blocks. The addition of the inverter solves the problem of cascading of dynamic CMOS logic circuits. Figure 6.48 shows a cascaded domino CMOS logic circuit.

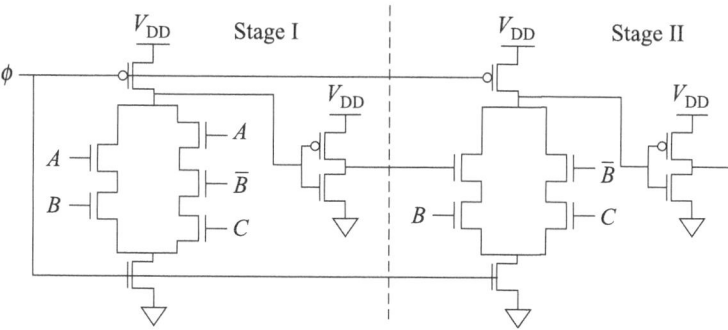

Fig. 6.48   Cascaded Domino CMOS logic circuit

In the pre-charge phase ($\phi = 0$), the outputs of the dynamic CMOS logic circuits are pre-charged to logic high and the output of the static inverter is logic low. In the evaluation phase ($\phi = 1$), outputs of the dynamic CMOS logic circuits can either go to logic low or remain at logic high. Consequently, the output of the static inverter can make only a 0→1 transition in the evaluation phase. So, irrespective of the input logic, the output of the static inverter cannot make a 1→0 transition in the evaluation phase.

Domino CMOS helps in reducing the number of transistors as compared to the static CMOS logic. In static CMOS logic we require $2N$ transistors to implement an N-input logic function. However, in dynamic CMOS logic we require only $N + 2$ transistors and for Domino CMOS logic two additional transistors are required. It also solves the cascading problem of dynamic CMOS logic. But it is suitable only for non-inverting logic (the expressions having no complement over whole expression). For inverting logic the expression must be reorganized (to remove the complement over whole expression) before it can be realized using domino CMOS logic.

Example 6.8   Design the following logic using domino CMOS logic and compare it with the static CMOS design.

$$Y = AB + CD + EFG$$

Solution   The static CMOS logic design for the given function will be same as shown in Fig. 6.17 except two additional transistors required for inverting the output. This requires $14 + 2 = 16$ transistors. The domino CMOS logic design is shown in Fig. 6.49 which requires only 11 transistors.

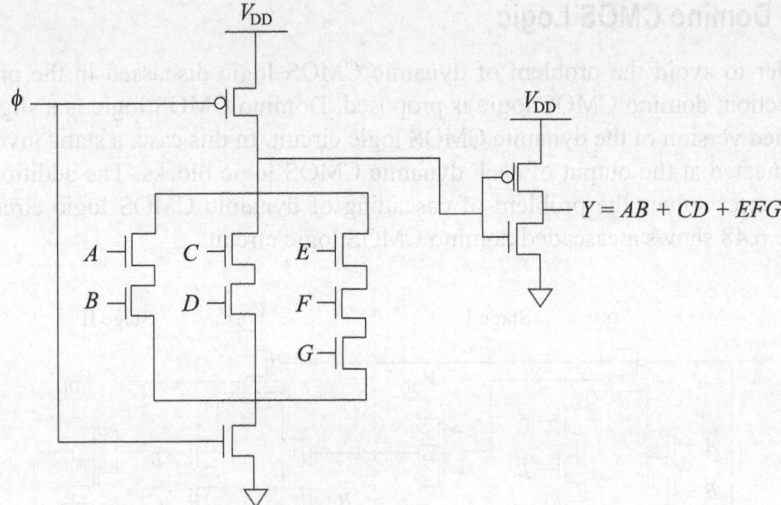

Fig. 6.49   Domino CMOS logic circuit realizing function $Y = AB + CD + EFG$

6.15 NORA CMOS Logic

The dynamic CMOS logic can also be realized using a pMOS transistor to implement the logic through PUN. The PDN is realized using a single nMOS transistor operated by inverted clock $(\bar{\phi})$. When $\phi = 0$, $\bar{\phi} = 1$, the nMOS is ON and the output is pre-discharge to 0 V.

NORA stands for NO Race. The NORA logic style is a modified version of the domino CMOS logic. In this logic style the static CMOS inverters are replaced by the dynamic logic circuits using pMOSFETs. The pMOSFET-based dynamic circuit is operated using inverted clock $(\bar{\phi})$. Unlike nMOSFET-based dynamic circuits, it is pre-charged when $(\phi = 1)$ and is evaluated when $\phi = 0$. An example of this logic is shown in Fig. 6.50.

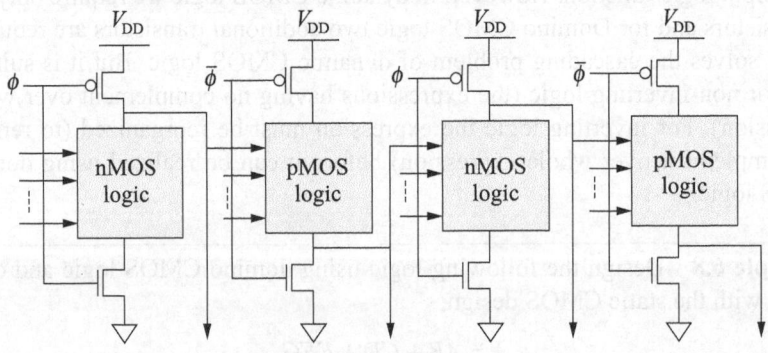

Fig. 6.50   NORA CMOS logic

When clock pulse $\phi = 0$, the output nodes of the nMOSFET logic circuits are pre-charged to V_{DD} through the pMOSFETs and the output nodes of the pMOSFET-based logic circuits are pre-discharged to 0 V through the nMOSFETs.

When clock pulse makes a low-to-high transition, all the nMOSFET-based and pMOSFET-based logic circuits evaluate their outputs depending on the input signals.

NORA logic has a number of advantages over the domino CMOS logic. Firstly, the static CMOS inverter is not required in this logic which saves the chip area. Secondly, NORA logic circuits can be connected in cascade with alternate ϕ-based and $\bar{\phi}$-based networks.

The NORA logic circuit has a major problem of low-noise margin. The floating output is susceptible to pick up noise due to coupling and charge sharing. That is why this logic is not very popular.

6.16 Zipper CMOS Logic

In a Zipper CMOS logic, a slightly different clocking scheme is used for the pre-charge (ψ_1) and pull-down (ϕ) transistors in nMOSFET-based logic circuits, and $\bar{\phi}$ and ψ_2 for the pull-up and pre-discharge transistors in pMOSFET-based logic circuits. This type of clocking scheme is used in order to keep the pMOSFET pre-charge and nMOS discharge transistors under the subthreshold region. Thus, the transistors operating under subthreshold region conduct weekly to compensate the charge leakage and charge sharing problems. An example of such logic style and the clocking scheme is shown in Figs 6.51(a) and (b).

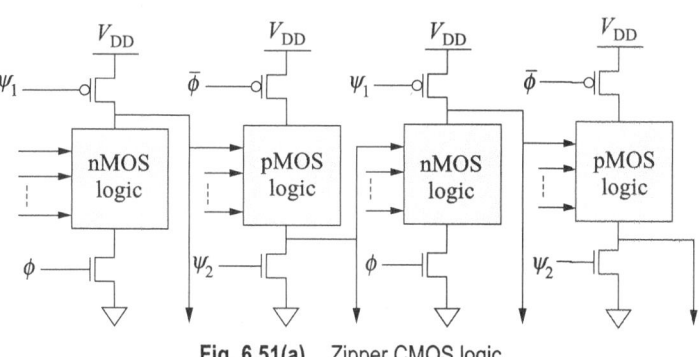

Fig. 6.51(a) Zipper CMOS logic

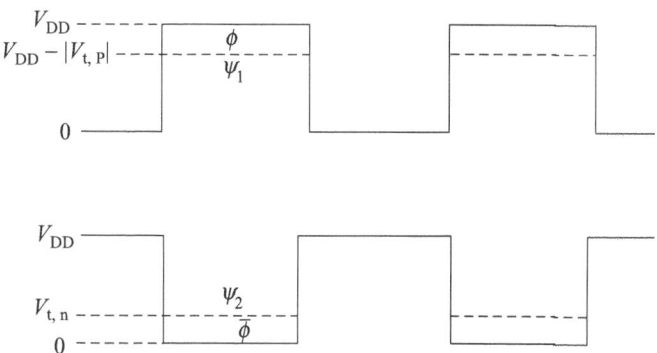

Fig. 6.51(b) Clock signals for zipper CMOS logic

6.17 True Single-phase Clock (TSPC) Dynamic CMOS Logic

The two-phase clocking logic circuits explained earlier require two clocks to be non-overlapping. Two clock signals must be routed carefully in order to reduce clock skew. The TSPC logic is proposed by Yuan et al., which eliminates the requirement of two-phase clocking. TSPC logic requires only single phase clocking, as shown in Fig. 6.52.

When CLK = 1, TSPC logic simply works as static CMOS logic. But when CLK = 0, only the PUN is activated and the PDN is deactivated. Hence, the circuit just holds its output logic level.

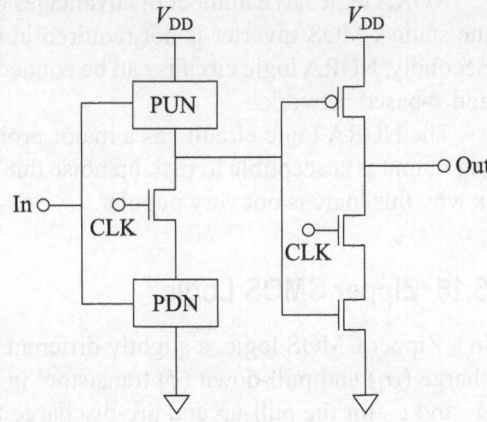

Fig. 6.52 CMOS logic using TSPC logic with latched output

Example 6.9 Design an AND latch circuit using TSPC logic.

Solution An AND latch using TSPC logic can be implemented as shown in Fig. 6.53. The PUN is implemented using two pMOS transistors in parallel, whereas the PDN is implemented using two nMOS transistors in series.

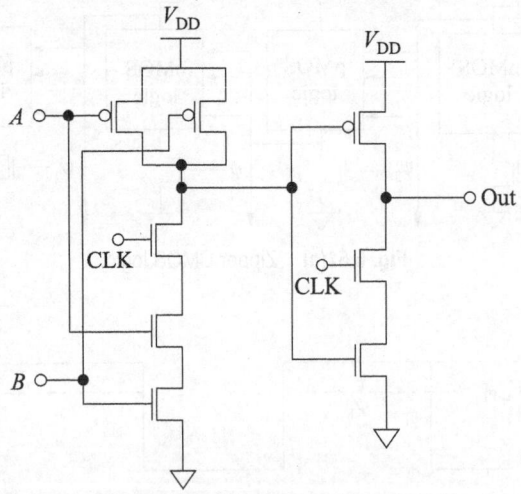

Fig. 6.53 AND latch using TSPC logic

6.18 Pass Transistor Logic

In pass transistor logic, only nMOS transistors are used to design the logic. The inputs signals are applied to both the gate and drain/source terminals. To illustrate the pass transistor logic, let us consider the design of a two-input AND gate. The truth table of two-input AND gate is shown in Fig. 6.54(a).

A B	F = AB	F
0 0	0	When A = 0,
0 1	0	F = 0
1 0	0	When A = 1,
1 1	1	F = B

(a)

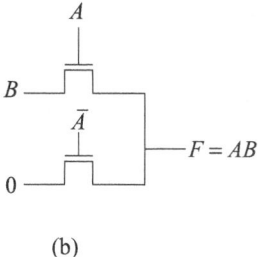

(b)

Fig. 6.54   (a) Truth table of two-input AND gate; (b) realization of two-input AND gate using pass transistor logic

As seen from the truth table, the function (F) is evaluated for two cases: (i) when $A = 0$, $F = 0$; (ii) when $A = 1$, $F = B$. This logic can be realized using two nMOS transistors connected in parallel as shown in Fig. 6.54(b). When $A = 1$, the upper nMOS is ON, so the output is B ($F = B$). When $A = 0$, the lower nMOS is ON, so the output is 0 ($F = 0$).

This logic is known as pass transistor logic (PTL). This logic uses lesser number of transistors as compared to conventional CMOS logic. The disadvantage of the logic is that the logic high output is degraded as nMOS cannot pass logic high perfectly.

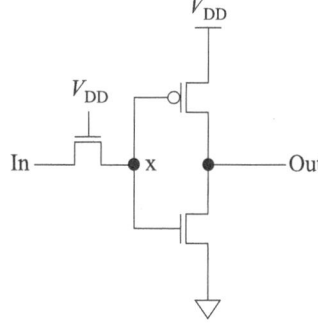

6.18.1 Degradation of Logic High Output Level

The nMOS transistor cannot pass the logic high level perfectly. To analyse this degradation, let us consider a PTL circuit as shown in Fig. 6.55.

Fig. 6.55   PTL circuit with CMOS inverter at the output

The voltage waveforms at nodes In, Out, and x are shown in Fig. 6.56. Clearly, it shows that the logic high output (at node x) of the PTL gate is degraded.

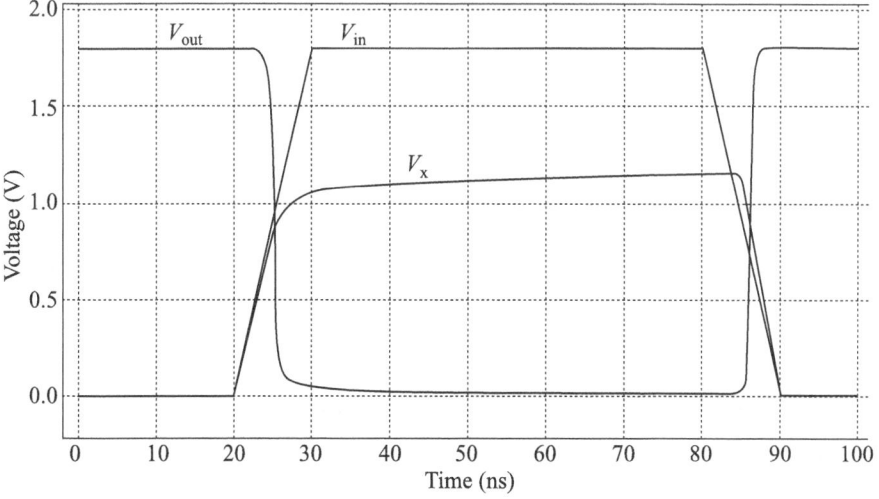

Fig. 6.56   Output waveform of a PTL gate

Due to the logic degradation, the PTL gates cannot be cascaded, as the logic degradation gets carried over several stages, and the final output logic becomes incorrect. To avoid the carry of degradation, a CMOS inverter is connected at the output of the PTL gate as shown in Fig. 6.55.

6.18.2 Design of XOR Gate Using PTL

An XOR gate can be realized using PTL, as shown in Fig. 6.57.

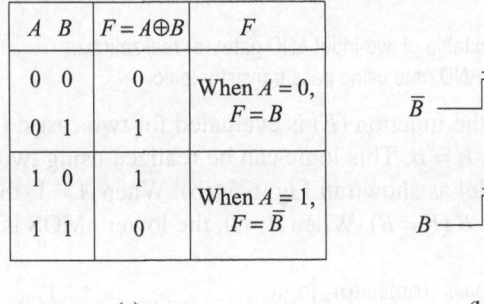
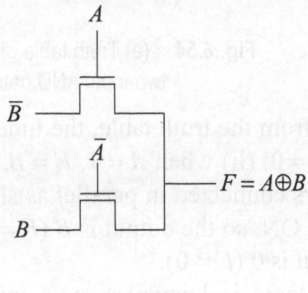

(a)   (b)

Fig. 6.57 (a) Truth table of XOR gate; (b) XOR gate implemented using PTL

The input and output waveforms of the XOR gate implemented using PTL are shown in Fig. 6.58. It shows that the output waveform is a function of the input transition. For example, when $A = 1$ and $B = 0 \rightarrow 1$, output waveform falls from high to low. Again, when $B = 1$ and $A = 0 \rightarrow 1$, output waveform falls from high to low. But in these two cases, the output waveform is different in shape. Thus, the PTL gate has another issue of having transition-dependent output waveform.

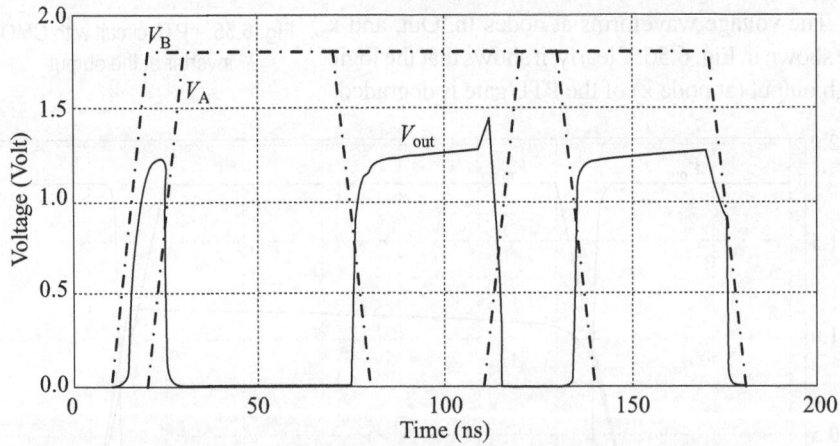

Fig. 6.58 Input–output waveforms of XOR gate using PTL

6.19 Complementary Pass Transistor Logic

In the complementary pass transistor logic (CPL), the inputs are applied both in the true and complement form, and the outputs are also evaluated in both true and complement form, using pass transistor logic. This logic is also known as differential pass transistor logic (DPL). An example of XOR/XNOR is illustrated in Fig. 6.59.

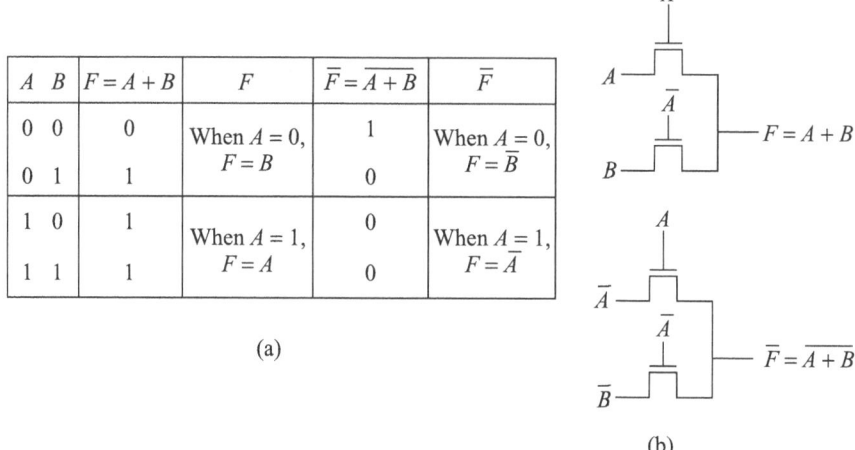

A B	$F = A \oplus B$	F	$\overline{F} = A \odot B$	$\overline{F}$
0 0	0	When $A = 0$,	1	When $A = 0$,
0 1	1	$F = B$	0	$F = \overline{B}$
1 0	1	When $A = 1$,	0	When $A = 1$,
1 1	0	$F = \overline{B}$	1	$F = B$

(a)

(b)

Fig. 6.59   (a) Truth table for XOR/XNOR gate; (b) realization of XOR/XNOR function using CPL

6.19.1 Design of OR/NOR Gate Using CPL

Using CPL, the OR and NOR functions can be implemented as shown in Fig. 6.60. The Boolean expression for an OR gate can be written as

$$F = A + B = A + \overline{A}B = A \times A + \overline{A} \times B$$

A B	$F = A + B$	F	$\overline{F} = \overline{A + B}$	$\overline{F}$
0 0	0	When $A = 0$,	1	When $A = 0$,
0 1	1	$F = B$	0	$F = \overline{B}$
1 0	1	When $A = 1$,	0	When $A = 1$,
1 1	1	$F = A$	0	$F = \overline{A}$

(a)

(b)

Fig. 6.60   (a) Truth table for OR/NOR gate; (b) realization of OR/NOR function using CPL

Similarly, the Boolean expression for a NOR gate can be written as

$$\overline{F} = \overline{A + B} = \overline{A} \times \overline{B} = A \times \overline{A} + \overline{A} \times \overline{B}$$

6.19.2 Design of AND/NAND Gate Using CPL

Using CPL, the AND and NAND functions can be implemented as shown in Fig. 6.61. The Boolean expression for an AND gate can be written as

$$F = AB = A \times B + A \times \overline{A}$$

A B	F = A B	F	$\overline{F} = \overline{A\,B}$	$\overline{F}$
0 0	0	When $A = 0$, $F = A$	1	When $A = 0$, $F = \overline{A}$
0 1	0		1	
1 0	0	When $A = 1$, $F = B$	1	When $A = 1$, $F = \overline{B}$
1 1	1		0	

(a)

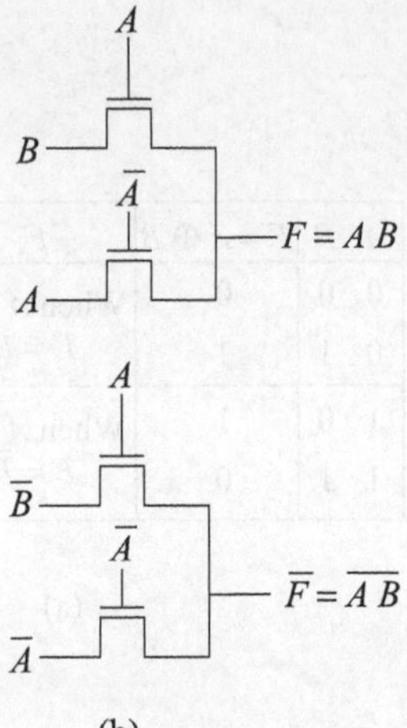

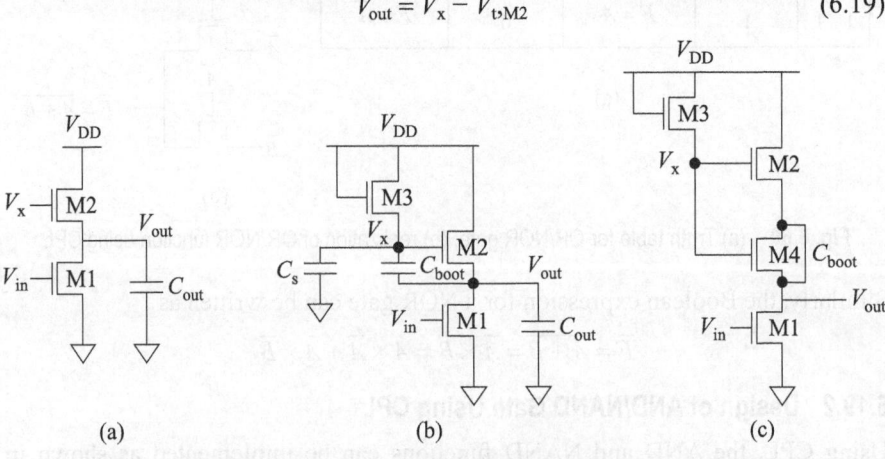

(b)

Fig. 6.61 (a) Truth table for AND/NAND gate; (b) realization of AND/NAND function using CPL

Similarly, the Boolean expression for an AND gate can be written as

$$\overline{F} = \overline{AB} = \overline{A} + \overline{B} = \overline{A} + \overline{B} \times (A + \overline{A})$$

$$= \overline{A} + \overline{A} \times \overline{A} + A \times \overline{B} = \overline{A} \times \overline{A} + A \times \overline{B}$$

6.20 Voltage Bootstrapping

It is a technique to mitigate the problem of threshold voltage drop in digital circuits. The logic degradation occurs in pass transistor logic and enhancement load inverter circuits. To explain the bootstrapping technique, let us consider an inverter circuit with an enhancement-type load as shown in Fig. 6.62(a). When the input voltage (V_{in}) is low, transistor M1 is turned Off. The output voltage (V_{out}) is logic high with a degradation of threshold voltage of the nMOS transistor load and is given by

$$V_{\text{out}} = V_{\text{x}} - V_{\text{t,M2}} \tag{6.19}$$

(a)   (b)   (c)

Fig. 6.62 (a) Inverter with enhancement-type nMOS load, (b) dynamic bootstrapping, (c) bootstrapping capacitor realized with a MOS capacitor.

The logic degradation is overcome by increasing the voltage V_x. This technique is illustrated in the circuit shown in Fig. 6.62(b) where two capacitors C_s and C_{boot} are added along with a diode-connected nMOS transistor M3. The node X is connected to ground by C_s whereas it is connected to output by C_{boot}. This configuration increases the voltage V_x as given by

$$V_x \geq V_{DD} + V_{t,M2} \tag{6.20}$$

When V_{in} is high, M1 is turned ON, the output voltage is low given by $V_{out} = V_{OL}$. At this point the voltage V_x is equal to $(V_{DD} - V_{t,M3})$. Total charge on node x can be written as

$$Q_x = C_S (V_{DD} - V_{t,M3}) + C_{boot} (V_{DD} - V_{t,M3} - V_{OL}) \tag{6.21}$$

When V_{in} is low, M1 is turned OFF, and the load capacitor starts charging towards logic high. Now let us consider that V_{out} is fully charged to V_{DD} and at this point the voltage at node x is V_x. So the total charge on node x can be written as

$$Q'_x = C_s V_x + C_{boot} (V_x - V_{DD}) \tag{6.22}$$

During the switching of input from high to low, the total charge at node x cannot change instantaneously. Hence, we can write

$$Q_x = Q'_x$$

$$C_s(V_{DD} - V_{t,M3}) + C_{boot} (V_{DD} - V_{t,M3} - V_{OL}) = C_s V_x + C_{boot} (V_x - V_{DD}) \tag{6.23}$$

Rearranging Eqn (6.23), we can write

$$V_x = (V_{DD} - V_{t,M3}) + \frac{C_{boot}}{C_s + C_{boot}} (V_{DD} - V_{OL}) \tag{6.24}$$

If $C_{boot} \gg C_s$, Eqn (6.24) becomes Eq. (6.25)

$$V_x = 2V_{DD} - V_{t,M3} - V_{OL} \tag{6.25}$$

Equation (6.25) shows that V_x can be increased significantly using the bootstrapping technique. The bootstrapped capacitance (C_{boot}) is realized using an nMOS transistor (M4) with its source and drain connected together as illustrated in Fig. 6.62(c). The capacitance C_s is due to the parasitic capacitances C_{GB} and C_{SB} of M2 and M3 transistors, respectively.

6.21 Differential CMOS Logic

In the differential CMOS logic, the inputs and outputs are to be given and derived both in the true and complement form, respectively connected in parallel. The PUN network is build using only two pMOS transistors. The structure of differential CMOS logic is shown in Fig. 6.63. This logic is also known as cascode voltage switch logic (CVSL) (Kan and Pulfrey1987).

The advantages of differential CMOS logic are as follows:

■ As the output is available in both the true and complement form, it eliminates the use of extra inverter circuits where the inputs are required in both true and complement form.

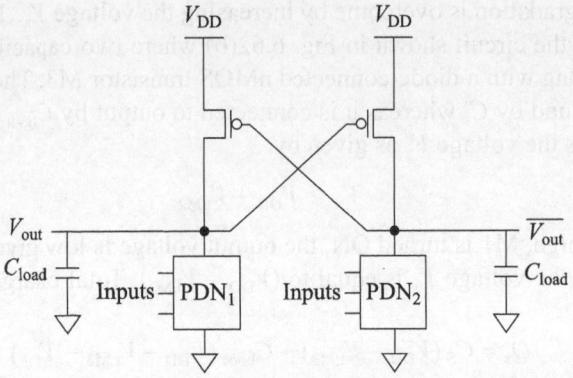

Fig. 6.63 Differential CMOS logic

- It uses just two pMOS transistors in the PUN, instead of using as many pMOS as nMOS transistors used in PDN. This saves a lot of area.

Operation When PDN_1 conducts for some of the input combinations, PDN_2 does not. So the outputs are complementary to each other. The structure of PDN_1 and PDN_2 are dual of each other.

Example 6.10 Design a AND/NAND circuit using differential CMOS logic.

Solution AND/NAND circuit can be realized using differential CMOS logic as shown in Fig. 6.64.

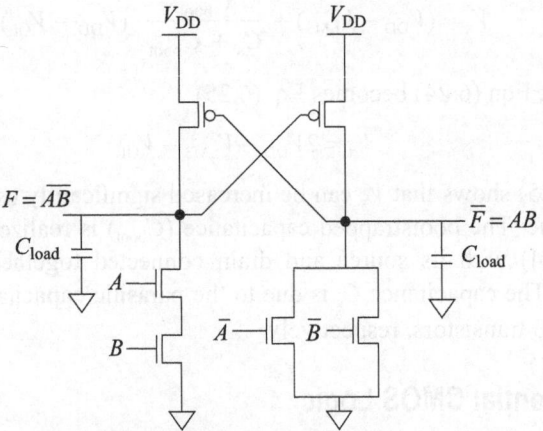

Fig. 6.64 AND/NAND gate realization using differential CMOS logic

6.22 Adiabatic Logic

Adiabatic logic is a special class of logic using the adiabatic concept of thermo-dynamics for low power VLSI circuits (Chandrakasan and Brdersen 2001). The term adiabatic means there is no heat exchange between a system and the environment. This concept is used for designing adiabatic logic circuits. In this case, the energy stored in the load capacitor during the charging operation is retuned to the power supply during the discharging operation of the load capacitor. To illustrate the logic, let us consider an adiabatic buffer/inverter circuit as shown in Fig. 6.65.

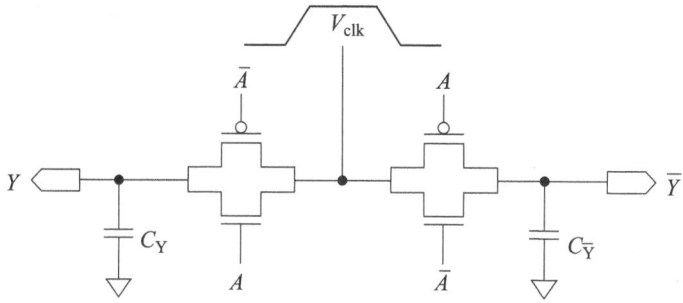

Fig. 6.65   Adiabatic buffer/inverter circuit

In this logic, the V_{DD} is replaced by a power clock which ramps up to V_{DD} with a slope, stays at V_{DD} for some time, and then ramps down to the ground with a slope. When input signal (A) is high, the left transmission gate is ON; hence, the capacitor C_Y charges according to the V_{clk} waveform. So, the output Y is high. The right transmission gate is OFF and the output $\bar{Y}$ is low. While V_{clk} ramps up to V_{DD}, the capacitor C_Y charges up to V_{DD}. After the completion of charging, the output node is at logic high. At this time, only the logic is evaluated and can be applied to the input of similar other gates. When the V_{clk} ramps down to the ground, the capacitor C_Y discharges and returns back the stored charge to the power supply clock. In this way the energy saving is done in the adiabatic logic circuits.

6.23 Dual-threshold CMOS Logic

Generally, a high-speed low-power design requires a reduction in the supply voltage (V_{DD}) and threshold voltage (V_t) of the MOS transistors. But reduction of V_t of the MOS transistor results in high subthreshold leakage problem. In order to avoid the leakage problem, the dual-threshold CMOS logic is used. In this logic, there are two modes of operation: the active mode when the circuit operates to function; and the stand-by mode when the circuit remains in the sleep mode. The general structure of dual-V_t CMOS logic is shown in Fig. 6.66. There are two types of MOS transistors used in this logic: (a) high V_t and (b) low V_t. The low V_t transistors are used to design the core logic that evaluates the function at high speed. When the device operates in the active mode, the stand-by signal is low; hence, the high V_t transistors are ON, and the circuit operates as normal CMOS logic at high speed. But when the circuit does not operate in the stand-by mode, the stand-by signal is high, and the high V_t transistors are OFF. So, they avoid any subthreshold leakage due to the low V_t transistors.

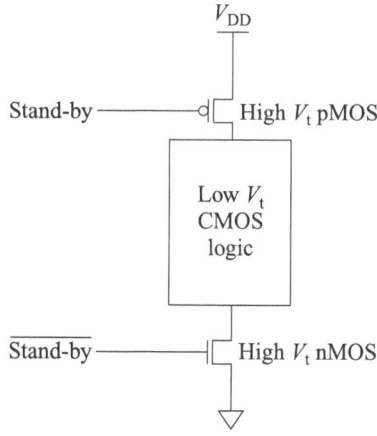

Fig. 6.66   Dual V_t CMOS logic circuit

SUMMARY

- In CMOS digital logic design, all MOSFETs act as switches operating either in cut-off or in saturation region.
- A pMOS passes logic 1 perfectly, but passes a degraded logic 0. On the other hand, nMOS passes logic 0 perfectly, but passes a degraded logic 1.
- In static CMOS logic structure, a pull-up network (PUN) is implemented with pMOS transistors and a pull-down network (PDN) is implemented with nMOS transistors.
- In general, product (AND) terms are implemented by a series-connected nMOS transistors, and sum (OR) terms are implemented with parallel-connected nMOS transistors. Inverse or dual of nMOS network is done in pMOS network.
- In pseudo-nMOS logic, the entire PUN is implemented by a single pMOS transistor with gate terminal connected to ground.
- The dynamic CMOS logic works in two-phases of a single clock. In pre-charge phase ($\phi = 0$), output is pre-charged to logic 1. In the evaluation phase ($\phi = 1$), output is evaluated based on the input signals.
- Domino CMOS logic is obtained by adding a static CMOS inverter at the output of the dynamic CMOS logic after each stage.
- In NORA logic, the static CMOS inverters of Domino logic are replaced by the pMOS dynamic logic, operated with a complemented clock ($\overline{\phi}$) waveform.
- In the differential CMOS logic, both inputs and outputs are in true and complemented form.
- Adiabatic logic and dual-threshold CMOS logic are targeted for low power applications.

SELECT REFERENCES

Aur, S., D.E. Hocevar, and P. Yang 1987, *Circuit Hot Electron Effect Simulation*, Electron Devices Meeting, 1987 International, Vol. 33, pp. 498–501.

Bhasker, J. 2003, *A VHDL Primer*, Pearson Education.

Brown, S. and Z. Vranesic 2002, *Fundamentals of Digital Logic with VHDL Design*, Tata McGraw-Hill, New Delhi.

Chandrakasan, A.P. and R.W. Brodersen 2001, *Minimizing Power Consumption in CMOS Circuits*, Department of EECS, University of California at Berkeley.

De Micheli, G. 2003, *Synthesis and Optimization of Digital Circuits*, Tata McGraw-Hill, New Delhi.

Gerez, S.H. 2007, *Algorithms for VLSI Design Automation*, Wiley.

Hayes, J.P. 1998, *Computer Architecture and Organization*, 3rd edn, McGraw-Hill.

Jacob, G.H. 2004, *Design of CMOS Cell Libraries for Minimal Leakage Currents*, Masters Thesis, Computer Science and Engineering Technical University of Denmark, 13 August.

Kan, M.C. and D.L. Pulfrey 1987, 'A Comparison of CMOS Circuit Techniques: Differential Cascode Voltage Switch Logic Versus Conventional Logic', *IEEE Journal of Solid-State Circuits*, Vol. SC-22, no. 4, August.

Kang, S.M. and Y. Leblebici 2003, *CMOS Digital Integrated Circuits: Analysis and Design*, 3rd edn, Tata McGraw-Hill, New Delhi.

Kittel, C. 1996, *Introduction to Solid State Physics*, 7th edn, Wiley.

Mano, M.M. 2001, *Computer System Architecture*, Prentice-Hall.

Martin, K. 2004, *Digital Integrated Circuit Design*, Oxford University Press.

May, G.S. and S.M. Sze 2004, *Fundamentals of Semiconductor Fabrication*, Wiley.

Rabaey, J.M., A. Chandrakasan, and B. Nikolic 2008, *Digital Integrated Circuits: A Design Perspective*, 2nd edn, Pearson Education.

Smith, M.J.S. 2002, *Application Specific Integrated Circuits*, Pearson Education.

Streetman, B.G. 1995, *Solid State Electronics Devices*, 3rd edn, Prentice-Hall of India, New Delhi.

Sutherland, I., R. Sproull, and D. Harris 1999, *Logical Effort: Designing Fast CMOS Circuits*, Elsevier.

Taub, H. and D. Schilling 1977, *Digital Integrated Electronics*, McGraw-Hill.

Weste, N.H.E., D. Harris, and A. Banerjee 2009, *CMOS VLSI Design: A Circuits and Systems Perspective*, 3rd edn, Pearson Education.

EXERCISES

Fill in the Blanks

1. _____ can pass a logic 1 perfectly, but cannot pass a logic 0 perfectly.
   - (a) nMOS transistor
   - (b) pMOS transistor
   - (c) CMOS transistor
   - (d) none of these
2. _____ can pass a logic 0 perfectly, but cannot pass a logic 1 perfectly.
   - (a) nMOS transistor
   - (b) pMOS transistor
   - (c) CMOS transistor
   - (d) none of these
3. Pull-up network (PUN) connects output node to _____ .
   - (a) V_{DD}
   - (b) ground
   - (c) input
   - (d) all of these
4. Pull-down network (PDN) connects output node to _____ .
   - (a) V_{DD}
   - (b) ground
   - (c) input
   - (d) all of these
5. AND terms are realized by _____ connections of nMOS in PDN.
   - (a) series
   - (b) parallel
   - (c) cascade
   - (d) anti-parallel

Multiple Choice Questions

1. Minimum number of transistors required to design a XOR gate is
   - (a) six
   - (b) eight
   - (c) twelve
   - (d) ten
2. Which of the following has minimum propagation delay?
   - (a) ECL
   - (b) TTL
   - (c) RTL
   - (d) DTL
3. Which of the following is/are advantages of CMOS?
   - (a) Wide range of supply voltage
   - (b) Greater noise margin
   - (c) Large packing density
   - (d) All of these
4. Compared to bipolar technology, output drive current in CMOS is
   - (a) lower
   - (b) higher
   - (c) same
   - (d) dependent upon use
5. Pseudo-nMOS logic provides which of the following advantages?
   - (a) Static power dissipation is less compared to CMOS logic.
   - (b) It is much faster compared to other logics.
   - (c) It requires less number of transistors compared to CMOS logic.
   - (d) It is more noise immune.

True or False

1. OR terms are realized by series connections of nMOS in PDN.
2. Two-input NAND gate requires six transistors in static CMOS logic.
3. The AND-OR-INVERT (AOI) gate is suitable for realizing functions in the SOP form.
4. Pseudo-nMOS logic gate is designed using only nMOS transistors.
5. Dynamic CMOS logic operates with two non-overlapping clock pulses.

Short-answer Type Questions

1. Design a CMOS half-adder circuit.
2. Realize a 2:1 MUX using CMOS transmission gate.
3. What is pseudo-nMOS gate and what is its advantage over CMOS gate?
4. Implement the Boolean function $F = A(B + CD)$ using static CMOS logic.
5. Design the following circuits using transmission gates.
   (a) D flip-flop
   (b) Two-input XOR gate
6. Explain why NOR gates are preferred for nMOS circuits, while NAND gates are preferred for static CMOS circuits.
7. Design a static CMOS circuit to implement the following Boolean function:

$$F = \overline{D} \times \overline{E} \times \overline{A} + \overline{B} \times \overline{C}$$

8. Draw the circuit of a CMOS full-adder circuit and explain its operation.
9. Design a clocked inverter using static CMOS logic.
10. Design a two-input XOR gate using pseudo-nMOS logic.
11. Why NOR-based design preferred in pseudo-nMOS logic compared to NAND-based design in CMOS logic?
12. Realize the circuit as shown in Fig. 6.66 using pseudo-nMOS logic.
13. Realize the logic circuit shown in Fig. 6.66 in static CMOS logic.

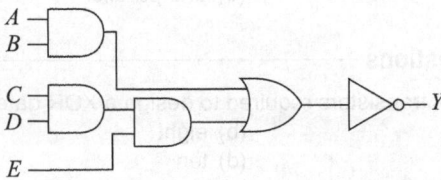

Fig. 6.66   Logic circuit for Questions 12 and 13.

14. Realize the following Boolean function using static CMOS logic:

$$Y = \overline{ABC} + \overline{B}D + A\overline{C}$$

15. Realize the following Boolean function using pseudo-nMOS logic:

$$Y = \overline{A + B(C + D)}$$

16. Design an even parity generator using static CMOS logic. Assume 3-bit input.
17. Realize the following Boolean function using differential CMOS logic:

$$Y = (A + \overline{B})C + \overline{A}D$$

18. Design a full-adder circuit using differential CMOS logic.

19. Design the pull down network (PDN) for the following Boolean expression:

$$F = \overline{W + X(Y + Z)}$$

20. Design the pull down network (PDN) for the following Boolean expression:

$$F = \overline{(W + X)(Y + Z)}$$

21. Design the pull up network (PUN) for the following Boolean expression:

$$F = \overline{WX + (Y + Z)}$$

22. Design the pull up network (PUN) for the following Boolean expression:

$$F = \overline{(W + XY + Z)}$$

Long-answer Type Questions

1. Explain the operation of a CMOS inverter with a proper circuit diagram. Draw the inverter characteristic curve and explain the various regions in the curve. What is CMOS noise margin?
2. Design a 3:2 decoder circuit using static CMOS logic.
3. Design a binary-to-grey code converter circuit using static CMOS logic.
4. Design a compressor circuit using static CMOS logic.
5. A logic gate has $V_{OH} = 5$ V, $V_{OL} = 0.2$ V, $V_{IH} = 2.5$ V, and $V_{IL} = 0.8$ V. Calculate the noise margins.
6. In a resistive load inverter, load resistor is 5 kΩ, and the nMOS is modelled by a 200 Ω resistor. Calculate the output voltage when input voltage is high.
7. For $\mu_n/\mu_p = 3.0$, $V_{tn} = 0.6$ V and $V_{tp} = -0.6$ V, find the inverter threshold voltage. Assume $V_{DD} = 1.5$ V.

Semiconductor Memories

7.1 Introduction

Semiconductor memories are storage devices intended for storing digital information. The following three criteria are important while designing semiconductor memories:

- *Area efficiency of the memory:* It determines the number of bits stored per unit area. Thus, the overall storage capacity of a memory is determined by this factor.
- *Speed of the memory:* It defines the access time of the memory. Access time is the time required to store and/or read binary data.
- *Power consumption:* It defines the power efficiency of the memory and is very important for low-power applications.

A simplified view of a memory array with row and column decoders is shown in Fig. 7.1. It has 2^M number of words with 2^N bits each.

The size of a memory determines its storage capacity. Generally, circuit designers specify the memory size in terms of *bits* which is equivalent to the number of storage cells. It has number of words with bits each. An 8-*bit* group is termed as 1-*byte*—chip designers prefer to specify memory size in *bytes*.

The memory array is divided into M number of rows and *N* number of columns. The group of bits in each row is termed as a *word*. Typically, system designers prefer to specify the memory size in words. Each word in the memory is addressed by a unique address. A memory with M number of address lines has 2^M number of words, each containing *N* number of bits. The content of each word is typically termed as *data*.

Storing data into the memory is called *memory read*, and the time required to read the memory is called the *read-access* time. Similarly, retrieving data from the memory is called *memory write*, and the time required to write from the memory is called the *write-access* time.

The *row decoder* is used to reduce the number of bits used for selecting a memory address, and the *column decoder* is used to select the number of bits to be retrieved as data.

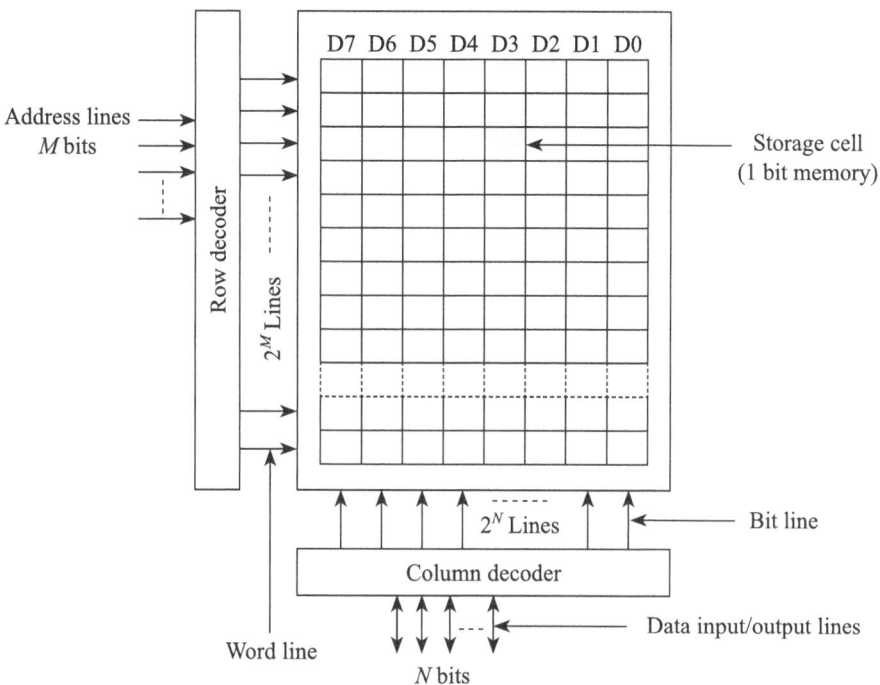

Fig. 7.1 Memory array with row and column decoders

Depending on their structure and access time, memories are classified into several categories. Table 7.1 displays the classification of semiconductor memories.

Table 7.1 Classification of semiconductor memories

Random Access Memory (RAM)		Read Only Memory (ROM)	Sequential Access Memory (SAM)		Content Addressable Memory (CAM)
Read/Write Memory (RAM)		Read Only Memory (ROM)	Shift Registers	Queues	Binary and Ternary
Static RAM (SRAM)	Dynamic RAM (DRAM)	Mask ROM PROM EPROM EEPROM Flash ROM	Serial In Parallel Out (SIPO) Parallel In Serial Out (PISO)	First In First Out (FIFO) Last In First Out (LIFO)	

7.2 Static RAM

Static RAM (SRAM) is also called *volatile memory*. It does not require any refreshing operation as the DRAM, which will be explained in the next section. However, the data stored in the SRAM is not lost as long as the power supply is ON.

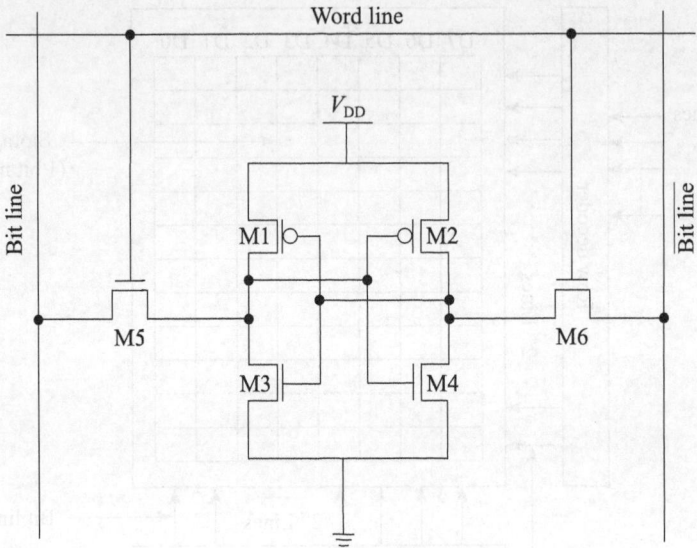

Fig. 7.2   6-transistor SRAM cell

A 1-bit memory cell with 6 transistors is shown in Fig. 7.2. We see that two CMOS inverters are connected back-to-back. Transistors M1 and M3 form one inverter, transistors M2 and M4 form another inverter, and the transistors M5 and M6 are used as a switch controlled by the word line.

7.2.1   Operations of SRAM

Following are the two operations of SRAM

Read Operation In order to read data from the cell, word line is enabled. This turns the transistors M5 and M6 ON. Hence, the stored data is available in both true and complemented form in the *Bit Line* and $\overline{Bit\ Line}$ bit lines, respectively.

Write Operation In order to write data into a cell, again word line enabled. The data to be written is made available in the bit lines, and this data is stored in the latch.

7.3   Dynamic RAM

In dynamic RAM (DRAM), the binary data is stored in a parasitic capacitance which discharges with time. In order to retain the stored data, the capacitor must be charged periodically. This operation is called *dynamic refreshing* of DRAM.

DRAM can be implemented using one, three, or four transistors. Figure 7.3 illustrates the different DRAM architectures.

7.3.1   Operation of DRAM

The 4-transistor DRAM structure shown in Fig. 7.3(a) has four nMOS transistors. The binary data is written in complemented form by enabling the word line and is stored in the parasitic storage capacitances. Since there is no restoring path from V_{DD} to these capacitances, the stored charge is lost. Therefore, in order to retain the

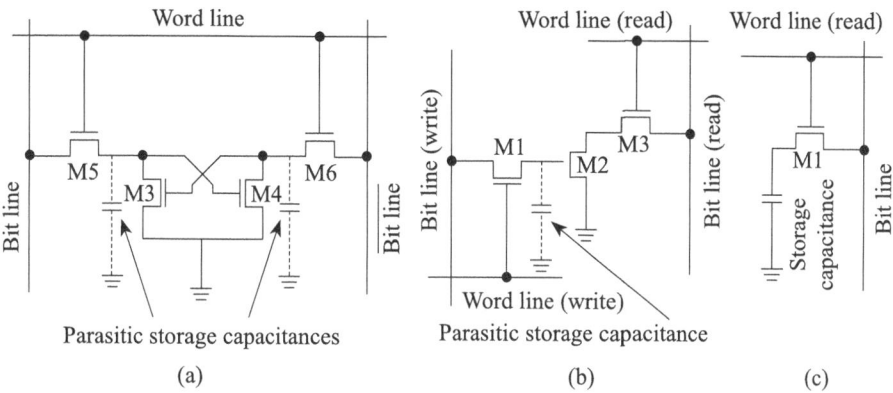

Fig. 7.3 DRAM cell architectures: (a) 4-transistor; (b) 3-transistor; (c) 1-transistor

logic level, the capacitors must be refreshed periodically. During the read operation, the word line is enabled and the stored data becomes available at the bit lines both in true and complemented form.

In the 3-transistor DRAM architecture, there are three nMOS transistors as shown in Fig. 7.3(b). Only M2 is used for storing the binary data in association with its parasitic gate capacitance. In the write mode, the write word line is enabled, and the logic from the write bit line is passed to the parasitic storage capacitance. In the read mode, the read word line is enabled, and the complement of the stored data becomes available in the read bit line.

The 1-transistor DRAM cell, which is mostly used in high-density DRAM architecture, consists of a dedicated storage capacitor as shown in Fig. 7.3(c). During the write operation, the word line is enabled and the data from the bit line is stored in the capacitor. During the read operation, the word line is enabled and the stored data becomes available at bit line.

7.4 Read Only Memory

Both SRAM and DRAM discussed in the previous sections are volatile memories, and the data stored in them is lost once the power supply is switched OFF. Therefore, in order to store data for indefinite time, various types of non-volatile memories have been proposed. We shall discuss the ROM and flash memory in this chapter.

A 4×4 ROM array architecture is shown in Fig. 7.4. The pMOS transistors act as pull-up devices for the bit lines, and the nMOS transistor between the word line and the bit line connects the bit line to the ground. When the word line is high, the nMOS transistor is turned ON, and hence the bit line gets connected to the ground. Therefore, the absence of an nMOS transistor at the cross-point of a word line and bit line indicates storage of logic 1, while its presence indicates storage of logic 0.

The ROM is permanently programmed by fabricating the required nMOS transistors at the cross-points for storing logic 0, while the cross-points without any nMOS transistors store logic 1.

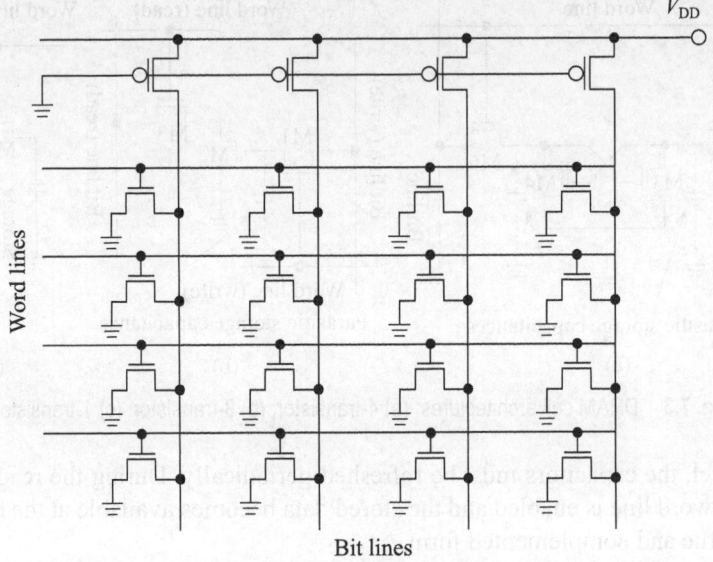

Fig. 7.4   A 4×4 ROM array

7.5 Flash Memory

A flash memory cell as shown in Fig. 7.5(a) uses two gates: one is the control gate and another is the floating gate. Under the normal mode of operation, there is no charge on the floating gate, and the transistor behaves like a normal transistor having a low threshold voltage. However, when a high voltage is applied to the control gate, the floating gate is charged, the threshold voltage is increased, and the transistor becomes permanently OFF.

The flash transistors are placed at the cross-points of the word line and bit line as shown in Fig. 7.5(b). When the flash transistor is programmed, its threshold voltage (V_t) becomes high; otherwise, its threshold voltage remains low. A cross-point with a high-V_t transistor stores logic 1, while a cross-point with low-V_t transistor stores logic 0.

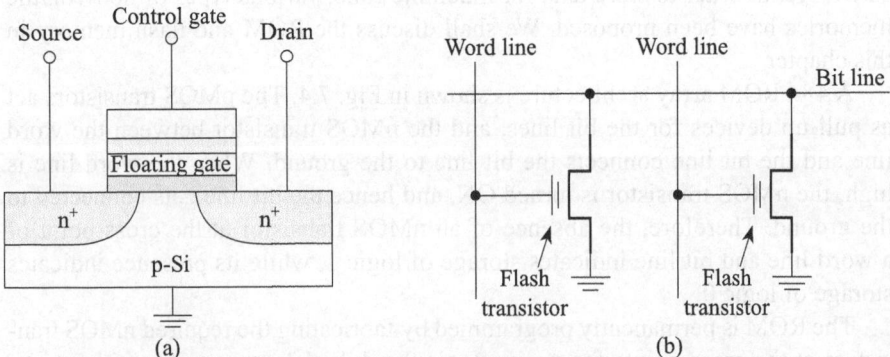

Fig. 7.5   (a) Flash transistor; (b) cross-point with flash transistor

7.6 Content Addressable Memory

Content addressable memory (CAM) is a special kind of memory, which has data searching capability. In a conventional memory, the input is the address and the output is the content of the memory location specified by the address bits. On the other hand, in the CAM, the input data is associated with something stored in the memory, and the output is the address of the memory location where the specified data is stored. CAM is used in applications that require high search speeds.

The architecture of a CAM cell is very similar to the standard 6-transistor SRAM cell. In addition to these six transistors, there are three transistors in a CAM cell to provide data search facility. The architecture of a 9-transistor CAM cell is shown in Fig. 7.6. The transistors M1–M6 form the basic SRAM cell architecture for data read and write operations, while the data search operation is performed by the three transistors M7–M9.

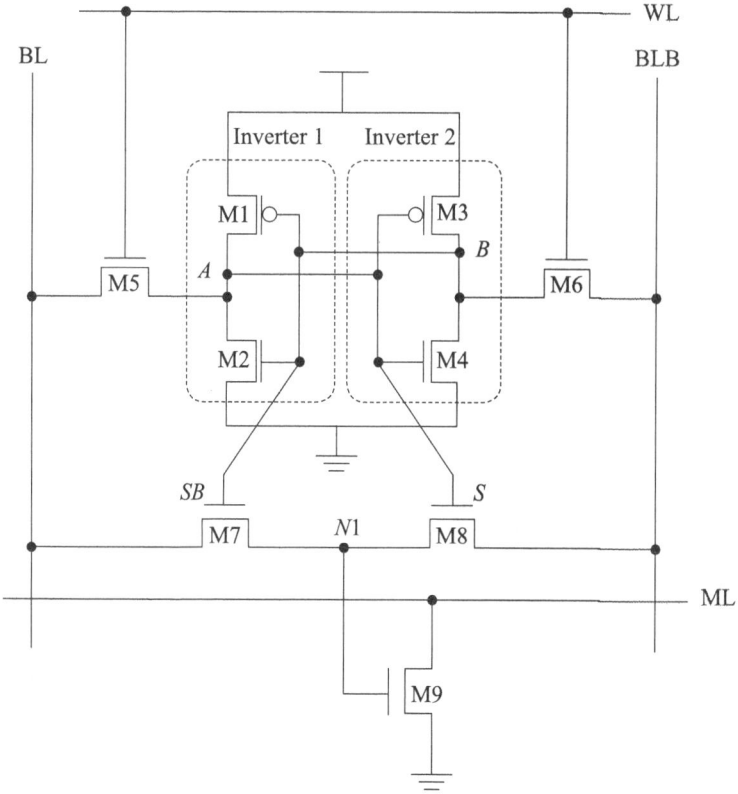

Fig. 7.6 Schematic of CAM cell

7.6.1 Read Operation

The complementary bit lines (BL and BLB) are initially pre-charged to V_{DD}. When the word line (WL) is made high, the transistors M5 and M6 are turned ON and the stored data (S and SB) become available at the bit lines. For example, if the stored data S = 1 and SB = 0, then BL remains high and BLB becomes low. During the read operation, the read-upset problem arises (Rabaey et al., 2003). The content of the CAM cell should not change during the read operation.

However, the voltage at node B increases by an amount ΔV, which is defined by the following equation:

$$\Delta V = \frac{V_{\text{DSATn}} + \text{CR}(V_{\text{DD}} - V_{\text{tn}}) - \sqrt{V_{\text{DSATn}}^2 (1 + \text{CR}) + \text{CR}^2 (V_{\text{DD}} - V_{\text{tn}})^2}}{\text{CR}} \qquad (7.1)$$

where CR is the cell ratio, and is defined by the following equation:

$$\text{CR} = \frac{(W/L)_4}{(W/L)_6} \qquad (7.2)$$

This voltage rise must be lower than the threshold voltage of transistor M2, in order keep it OFF; otherwise the stored data in the CAM cell will be lost. Further, to avoid the read-upset problem, the value of CR must be large.

7.6.2   Write Operation

In order to write into the cell, the complementary data is forced onto the bit lines and the word line is made high. The transistors M5 and M6 are turned ON and the logic levels of nodes A and B become equal to the logic level of the bit lines. The new data can only be stored in the cell, if the pull-up transistor is not as strong as the pass transistor.

During write operation, similar to the read operation, there is a constraint on the W/L ratio of the transistors M3 and M6. Let us assume that we have to write logic 0 into the cell where logic 1 is stored. For this, we set the bit lines BL = 0 and BLB = 1. Before the write operation, the stored data are S = 1 and SB = 0. Now, the voltage at node A must be pulled low so that M4 is turned OFF and M3 is turned ON, thus making the node B to go high.

The voltage at node A is a function of the pull-up ratio (PR) of the cell, which is defined by the following equation:

$$PR = \frac{(W/L)_1}{(W/L)_5} \qquad (7.3)$$

For the operation, the pull-up transistor M1 must be less conductive than the pass transistor M5 so that the node A goes low. Therefore, the value of PR must be small. Remember that for best read stability and write-ability, the pull-up transistor must be weak, the pass transistor must be medium, and the pull-down transistor must be strong.

7.6.3   Search Operation

During the search operation, the match line (ML) is initially pre-charged to V_{DD}. When data is to be searched, the search data is applied to the bit lines. The match line remains at logic high if the stored data matches with the search data; otherwise, it becomes low, indicating that the stored data do not match with the search data.

For example, when the stored data are S = 1 and SB = 0, and we place search data in the bit lines BL = 1 and BLB = 0, the match line should remain high indicating the match. In this case, the transistor M7 is turned OFF, M8 is turned ON, and the voltage on node $N1$ becomes low (logic level of BLB through M8) turning the transistor M9 OFF, so that the match line remains high. Alternately, if we place search data BL = 0 and BLB = 1, the match line goes low indicating a mismatch. In this case, the voltage on node $N1$ becomes high (logic level of BLB through M8), the transistor M9 is turned ON, and the match line becomes low.

The set of two-lobed curves, known as the butterfly curve, is used to determine SNM graphically. The SNM is determined by the length of the side of the largest square that can be fixed inside the lobes of the butterfly curve.

An alternate to the graphical method is the simulation method: two DC noise voltage sources V_n are connected in series with the cross-coupled inverters (Grossar et al. 2006). By increasing the magnitude of the noise voltage sources, SNM is measured when the stored data are flipped.

Though SNM is used as a metric for read stability, there are some drawbacks of using SNM as stated in Grossar et al. (2006). Therefore, as an alternative to SNM, the read stability is analyzed using the N-curve of the cell. The static current noise margin (SINM) and static voltage noise margin (SVNM) measured from the N-curve are used as a metric for read stability. In this work, we have measured SINM and SVNM from the N-curve of the cell.

7.7 Design of CAM Architecture

The architecture of a 4-bit CAM is illustrated in the schematic diagram shown in Fig. 7.7. A 4-bit search data is the input to the CAM. There are four memory locations at which four words are stored. Each stored word has a match line, and the match lines are connected to the encoder through the match-line-sense amplifiers (MLSAs). The match lines indicate whether or not the stored word matches with the search data. When the stored data matches with the search data, the corresponding match line becomes high, and the encoder generates the address of the stored data that is matched.

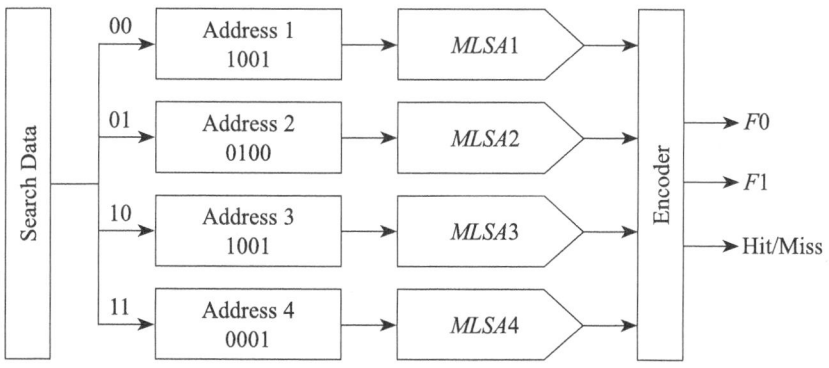

Fig. 7.7 Architecture of 4-bit CAM

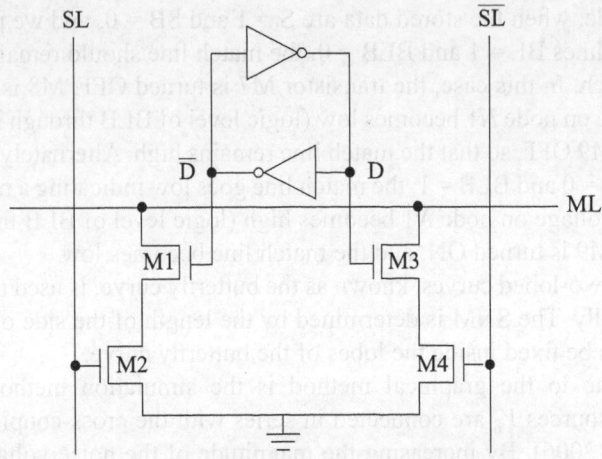

Fig. 7.8 A NOR-type CAM cell

The architecture of a NOR-type CAM cell is shown in Fig. 7.8. The following is a description of the working of a CAM cell:

The search data and its complement are applied to the search lines SL and $\overline{SL}$, respectively. Before the search operation, the match line (ML) is pre-charged to high. When the search data does not match with the stored data, the ML is pulled down to low. For example, if the stored data are D = 1 and $\overline{D}$, and the search data are SL = 1 and $\overline{SL}$ = 0, all the pull-down transistors (M1, M2, M3, and M4) are turned OFF. Therefore, the ML remains high, indicating the match condition. On the other hand, if the search data are SL = 0 and $\overline{SL}$ = 1, the pull-down transistors M3 and M4 are turned ON. Therefore, the ML discharges to ground, indicating a mismatch.

In the 4-bit CAM architecture, four CAM cells are connected in parallel as shown in Fig. 7.9. When any one of the four stored bits does not match with the search data, the ML is pulled down. Otherwise, when all four bits match with the stored data, the ML remains at logic high.

In this design, a current race MLSA is used, as shown in Fig. 7.10. The following is a description of the working of this CAM:

The ML is pre-charged to by setting signals MLPRE and $\overline{EN}$ to high. During this phase, the search data are also placed in the search lines SL and $\overline{SL}$. In the next phase (evaluation phase), the signals MLPRE and $\overline{EN}$ are both set to low. If the

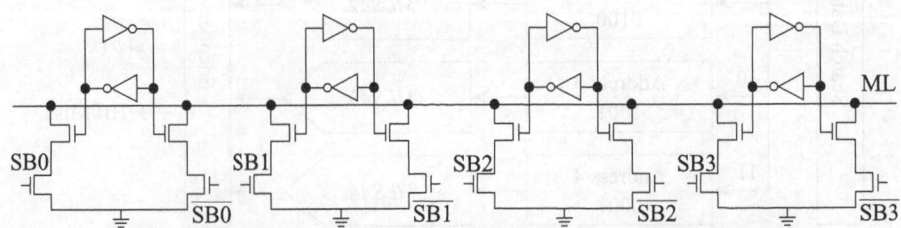

Fig. 7.9 Schematic of an address block with 4 CAM cells connected in parallel

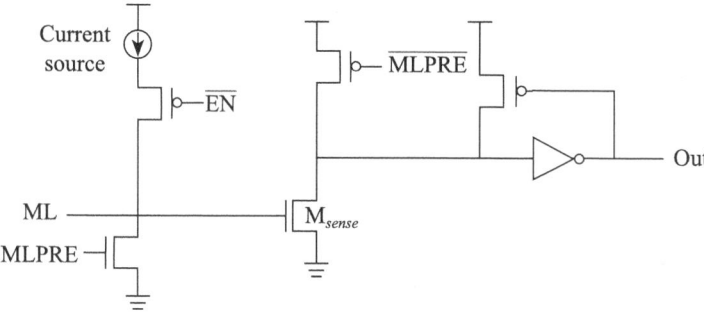

Fig. 7.10 Schematic of MLSA

ML is in the match state, then it is charged up to logic high through the pMOS and the current source. The ML remains at logic high, turning the match line sensing transistor (M_{SENSE}) ON, so that the output becomes high. In case of data mismatch, the ML is charged up to a very small voltage. Therefore, the M_{SENSE} transistor is turned OFF and the output becomes low.

When the search data match with the stored data, the Hit/Miss (H/M) output becomes high, indicating a match, and the encoder generates the address of the stored data. On the other hand, if there is a mismatch, the H/M signal goes low and the address bits are also set to low. In this design, there are two address bits, $F0$ and $F1$.

SUMMARY

- Read-write memory (RWM) is popularly known as RAM.
- DRAM has large packing density. It requires periodic refreshing of the stored data.
- CAM is a special class of memory, wherein a given data is searched and the address of the memory location, in case of a match, is output.

SELECT REFERENCES

Grossar, E., M. Stucchi, K. Maex, and W. Dehaene, 'Read Stability and Write-ability Analysis of SRAM Cells for Nanometer Technologies', *IEEE Journal of Solid-state Circuits,* 41(11), 2577–2588, 2006.

Pagiamtzis, K. and A. Sheikholeslami, 'Content Addressable Memory (CAM) Circuits and Architectures: A Tutorial and Survey', *IEEE Journal of Solid-state Circuits,* 41(3), 712–727, 2006.

Rabaey, C., and Nikolic, *Digital Integrated Circuits: A Design Perspective,* Pearson Education, Delhi, 2003.

EXERCISES

Fill in the Blanks

1. One byte is equal to _____ bits.
   (a) 4   (b) 6   (c) 8   (d) 16
2. A word in a memory refers to _____.
   (a) Address of a memory   (b) Content of a memory location
   (c) Size of the memory   (d) Number of bits in memory address
3. A content addressable memory (CAM) is a class of _____.
   (a) SRAM   (b) DRAM   (c) Flash memory   (d) ROM
4. Memory write operation means _____.
   (a) Storing data into the memory   (b) Retrieving data from the memory
   (c) Erasing data into the memory   (d) Copying data into the memory
5. A ROM can have _____.
   (a) Storage capacity   (b) Retrieving capacity
   (c) Store once and read many times capacity
   (d) Store many times and read once capacity

Multiple Choice Questions

1. Size of the memory means
   (a) Area of the memory   (b) Power of the memory
   (c) Speed of the memory   (d) Storage capacity of the memory
2. Row decoder is used to
   (a) Increase the number of address bits   (b) Reduce the number of address bits
   (c) Reduce the number of words   (d) Increase the number of words
3. In DRAM the data is stored by
   (a) The pass transistors   (b) The MOSFETs
   (c) The parasitic storage capacitors   (d) The cross-coupled inverters
4. In a memory, each location can have
   (a) Only one address   (b) Multiple addresses
   (c) Both a and b   (d) Two addresses
5. In a flash transistor, the control gate is used to
   (a) Control the gate current   (b) Turn the transistor ON
   (c) Charge the floating gate   (d) Control the drain current

True or False

1. SRAM requires dynamic refreshment of the memory.
   (a) True   (b) False
2. Flash memory has a single gate terminal.
   (a) True   (b) False
3. ROM can be used for read and write purpose.
   (a) True   (b) False
4. Read-upset problem arises during the writing of a memory.
   (a) True   (b) False
5. FIFO is a class of random access memory.
   (a) True   (b) False

Short-answer Type Questions

1. How is CAM different from a RAM?
2. Discuss the merits and demerits of a DRAM over SRAM.

3. What is the read-upset problem and how can it be solved?
4. Discuss the different types of memory.
5. What is the working principle of a flash memory?

Long-answer Type Questions

1. Discuss the working principle of the different types of DRAM cells.
2. Design the architecture of a 4-bit CAM. Explain its working principle.
3. What is the difference between RAM and SAM? What are the different types of SAM?
4. Discuss the 6-transistor SRAM architecture and its operational mechanism.
5. Why is flash memory used? Draw the structure of a flash transistor and explain its operation.

## CHAPTER 8

BiCMOS Technology and Circuits

KEY TOPICS

- CMOS vs BJT
- BiCMOS technology
- Bipolar logic
- BiCMOS logic circuits
- BiCMOS two-input NAND logic
- BiCMOS two-input NOR logic

- Complex logic using BiCMOS
- Applications of BiCMOS circuits
- Disadvantages of BiCMOS technology
- Silicon–Germanium BiCMOS technology

8.1 Introduction

The technology that enables fabrication of both bipolar junction transistor (BJT) and CMOS devices onto a same wafer is referred to as BiCMOS technology. The circuits designed together with BJT and CMOS transistors are known as BiCMOS circuits. In this chapter, we shall discuss the BiCMOS technology and circuits with examples.

Due to the large packaging density, low power dissipation, and high noise margin, CMOS has been the main technology for design and fabrication of ICs. But it suffers from timing issues as the technology advances. With the advancement of CMOS technology, the transistors become faster; however, the interconnection wires become slower. To drive the large capacitive load, the drive current capability of the CMOS logic circuits is increased using buffer circuits. However, the buffer circuits require a large silicon area. Another important fact about buffers is that the delay does not decrease monotonically as its size is increased. The high gate-capacitance of the buffer itself reduces the speed after a certain optimum buffer size. Hence, there is a speed bottleneck in the CMOS circuits.

On the other hand, BJTs are capable of driving a large load due to their high drive current capability as compared to the MOS devices using the same silicon area. But BJTs are not preferred for logic design due to more power consumption, poor packing density, and low noise margin.

BiCMOS technology, which was developed in the 1980s, basically combines BJT and CMOS transistors into a single IC for getting advantages of both. The BJTs are used for large current capability and the CMOS transistors are used for implementing the digital logic.

8.2 Comparison Between CMOS and BJTs

The main advantages of CMOS transistors over BJTs are as follows:
- Lower static power dissipation
- Higher noise margins
- Higher packing density, i.e., lower manufacturing cost per device
- High yield with large integrated complex functions

The other features of CMOS transistors are as follows:
- High input impedance (low drive current)
- Scalable threshold voltage
- High delay sensitivity to load (fan-out limitations)
- Low output drive current (issue in driving large capacitive loads)
- Low transconductance, where transconductance $g_m \propto V_{in}$
- Bi-directional capability (drain and source are interchangeable)
- A near-ideal switching device

Now, let us describe the advantages of BJTs over CMOS transistors:
- Higher switching speed
- Higher current drive per unit area, higher gain
- Generally better noise performance and better high frequency characteristics
- Better analog capability
- Improved I/O speed (particularly significant with the growing importance of package limitations in high-speed systems)

The other features of BJTs are as follows:
- High power dissipation
- Lower input impedance (high drive current)
- Low voltage swing logic
- Low packing density
- Low delay sensitivity to load
- High g_m ($g_m \propto V_{in}$)
- High unity gain bandwidth at low currents
- Essentially unidirectional

Combining the BJTs with CMOS transistors onto a same chip results in a number of advantages as described below:
- Improved speed over purely CMOS technology
- Lower power dissipation than purely bipolar technology (simplifying packaging and board requirements)
- Flexible I/Os (i.e., TTL, CMOS, or ECL)
  - BiCMOS technology is well suited for I/O intensive applications.
  - ECL, TTL, and CMOS input and output levels can easily be generated with no speed or tracking consequences
- High performance analog
- Latch-up immunity

8.3 BiCMOS Technology

BiCMOS technology enables fabrication of both CMOS transistors and BJTs on the same wafer. An NPN bipolar device structure is shown in Fig. 8.1. The process used to create the n-well region in CMOS can be used to create the *n*-type collector region in BJT. The source and drain diffusion steps in the CMOS process can be used to create the base contact and emitter regions. The emitter contact is formed by the polysilicon gate of the MOS transistor.

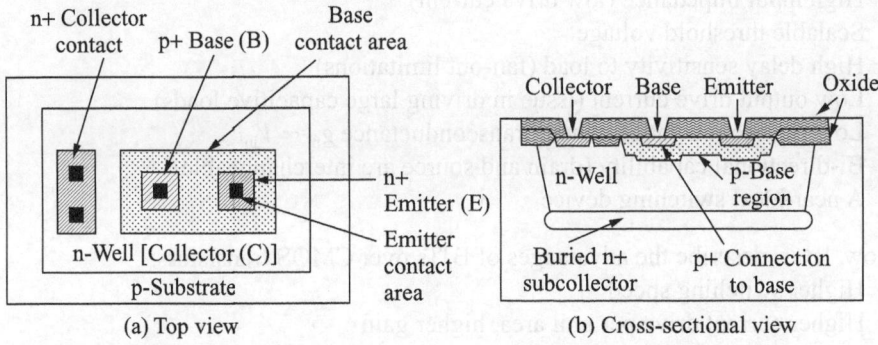

Fig. 8.1 Bipolar junction transistor: (a) top view; (b) cross-sectional view

The p-base region requires an additional process step for bipolar device fabrication. Typically, the BiCMOS fabrication requires additional 3–4 mask layers as compared to the CMOS fabrication process. A simplified BiCMOS structure with an npn bipolar and CMOS transistors is shown in Fig. 8.2.

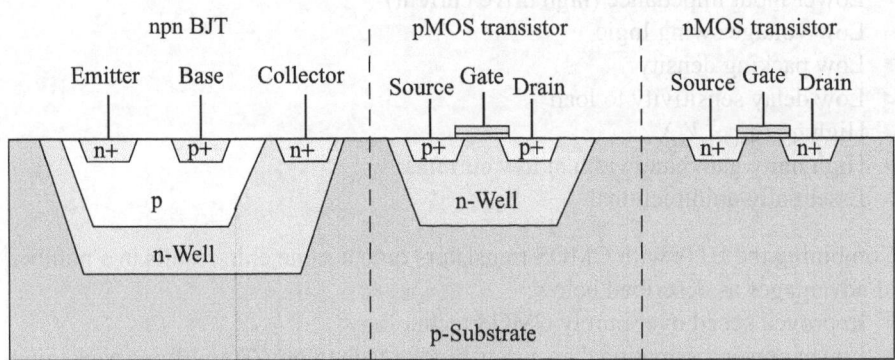

Fig. 8.2 BiCMOS device cross-sectional view

8.4 Bipolar Logic

The bipolar transistors can be used to implement digital logic. A simple circuit that implements an inverter logic using an npn bipolar transistor and a resistor is shown in Fig. 8.3(a).

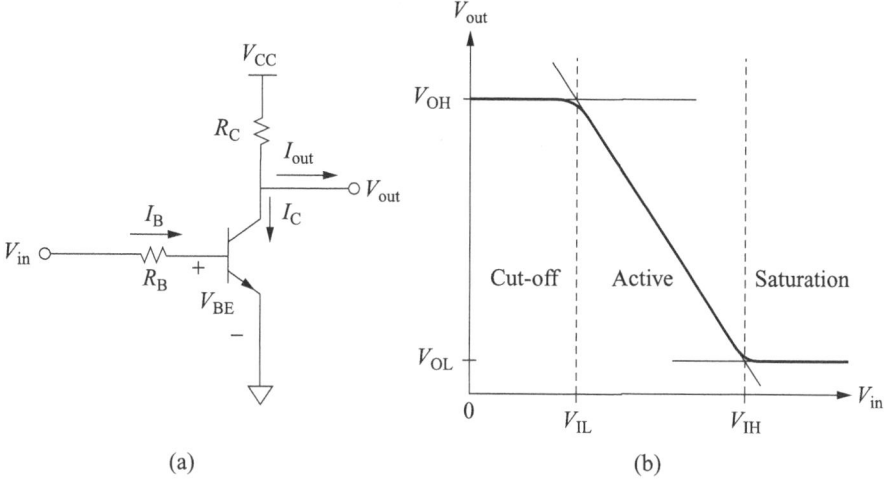

Fig. 8.3　(a) Inverter logic using BJT; (b) static characteristics

For an unloaded inverter, we can write the expression for output voltage as

$$V_{out} = V_{CC} - I_C R_C \tag{8.1}$$

The collector current can be expressed as

$$I_C = \alpha I_E - I_{CO}$$
$$= \alpha(I_C + I_B) + I_{CO}$$
$$= \frac{\alpha}{1-\alpha} I_B + \frac{I_{CO}}{1-\alpha}$$
$$= \beta I_B + (1+\beta) I_{CO} \approx \beta I_B \tag{8.2}$$

where I_E is the emitter current, I_B the base current, I_{CO} the reverse saturation current, α the common-base current gain, and β is the common-emitter current gain. Typically, α is less than but very close to unity. The parameter $\beta = \alpha/(1-\alpha)$ typically varies from 100 to 1000.

Operation　Let us now explain the operation of the circuit as an inverter. When the input voltage is low, the transistor operates in the cut-off region. There is no current flow through the device; hence, the output voltage is high (i.e., V_{CC}). When the input voltage is high, the transistor is ON. There is a large current flow through the device and the resistor R_C; hence, the output voltage is low (i.e., close to 0 V). Thus, the circuit behaves like an inverter.

Static Characteristics

The static characteristic of the BJT-based inverter circuit is shown in Fig. 8.3(b). The static characteristic can be divided into three parts for the three operating regions of the BJT.

1. *Cut-off region*　When the input voltage is zero, the transistor operates in the cut-off region. As no current flows through the device, output voltage is equal

to the power supply voltage. Hence, the *logic high output voltage* can be expressed as

$$V_{OH} = V_{CC} \tag{8.3}$$

The output voltage will be a voltage drop across R_C less than the power supply voltage, if there is a non-zero output current.

2. *Active region* As the input voltage is increased beyond the minimum voltage required to turn ON the npn transistor, the transistor starts conducting. There is a current through the resistor R_C, and the output voltage starts to decrease from V_{CC}. The *logic low input voltage* is defined as the input voltage at which the bipolar transistor enters into the forward active region and can be expressed as

$$V_{IL} = V_{BE,\text{turn-on}} \tag{8.4}$$

As the input voltage is increased further, the current through the transistor increases, and the output voltage continues to decrease until the transistor enters into the saturation region. The *logic high input voltage* is defined as the input voltage at which the bipolar transistor enters into the saturation region, and is expressed as

$$V_{IH} = V_{BE,\text{sat}} + R_B I_B$$
$$= V_{BE,\text{sat}} + \frac{R_B}{R_C} \times \frac{1}{\beta} (V_{CC} - V_{CE,\text{sat}}) \tag{8.5}$$

3. *Saturation region* When the input voltage is increased beyond V_{IH}, the output voltage is equal to the collector-to-emitter voltage under saturation ($V_{CE,\text{sat}}$), which is a relatively small constant value. The *logic low output voltage* is expressed as

$$V_{OL} = V_{CE,\text{sat}} \tag{8.6}$$

Dynamic Characteristics

When a pulse input is applied at the input of the BJT inverter, an inverted pulse is obtained at its output as shown in Fig. 8.4. The output waveform has certain delay with respect to the input waveform and also has some rise and fall times associated for low-to-high and high-to-low transitions.

When the input voltage makes $0 \to 1$ transition, the output voltage makes $1 \to 0$ transition. However, the output takes some time to respond to the input. This can be explained as follows. Initially, the BJT operates in the cut-off region when the input voltage is low. After the input makes $0 \to 1$ transition, the transistor still remains in the cut-off region for a duration t_1 as shown in Fig. 8.4. During t_1, the base–emitter and base–collector capacitances get charged and only after that the transistor is turned ON. The collector current increases and the voltage across the collector resistance R_C also increases. As a result the output voltage (collector-to-emitter voltage, V_{CE}) decreases. This continues for an interval t_2 till the transistor goes into saturation mode. The transistor works in the saturation region when the input voltage remains high.

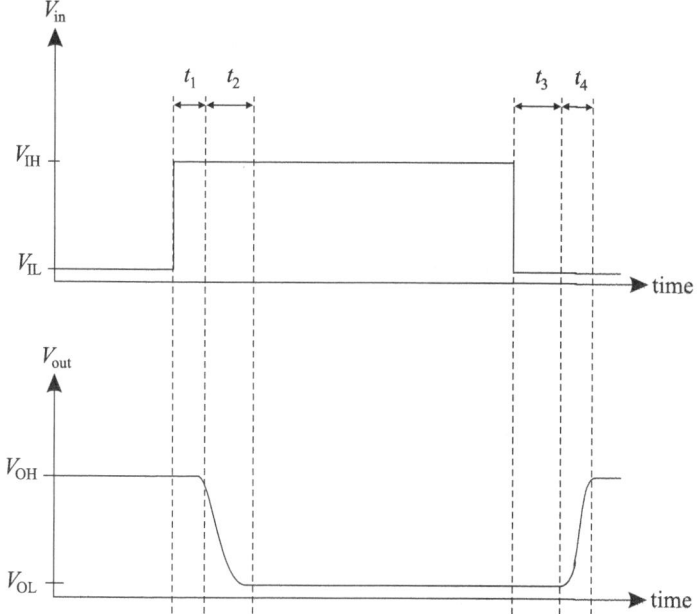

Fig. 8.4   Input and output waveforms of the BJT inverter

When the input voltage makes $1 \rightarrow 0$ transition, the output voltage makes $0 \rightarrow 1$ transition. However, the transistor takes some time (t_3) to recover from the saturation mode through the removal of over drive base charge. The transistor then goes into forward active mode during time t_4 when the output voltage goes from low to high.

The delays explained here are due to the intrinsic capacitance and carrier transit time of the transistor. If a load capacitor is connected then the delay will be increased further except the saturation recovery delay time t_3.

8.5 BiCMOS Logic Circuits

In this section, we shall discuss the BiCMOS logic circuits. Let us first consider the BiCMOS inverter logic as shown in Fig. 8.5.

The MOS transistors, M1 and M2, perform the inverter logic operation, whereas the BJTs Q1 and Q2 are used to drive the output load.

Operation

1. When the input voltage (V_{in}) is low, the transistor M1 is ON, and therefore, the transistor Q1 conducts. The transistor M2 is OFF, and hence, the transistor Q2 is non-conducting. The transistor M1 supplies base current for Q1 and keeps the base voltage of

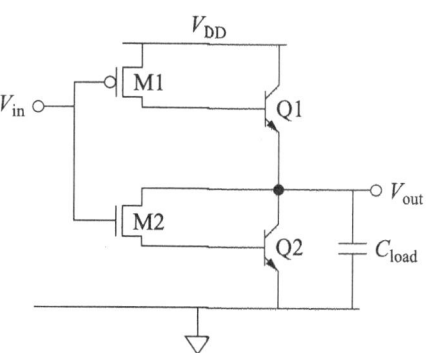

Fig. 8.5   BiCMOS inverter

Q1 at V_{DD}. Hence, the load capacitor charges through Q1 until the output voltage becomes $V_{DD} - V_{BE,Q1}$. So, we get logic high at the output.

2. When V_{in} is high, M1 is OFF and hence, Q1 is non-conducting. The transistor M2 is ON, and hence, Q2 is also ON. The load capacitor discharges through Q2 until the output voltage reaches $V_{CE,sat,Q2}$. Hence, we get logic low at the output.

The advantages that we gain using this BiCMOS logic are as following:

- The transistors Q1 and Q2 present low impedances when turned ON into saturation and the load capacitance will be charged or discharged rapidly.
- Output logic levels will be close to rail voltages. As $V_{CE,sat}$ is quite small and $V_{BE} \approx 0.7 \, \text{V}$, the BiCMOS inverter noise margins will also be good.
- The inverter has high input impedance due to the MOS gate input.
- The inverter has low output impedance.
- The inverter has high drive capability but occupies a relatively small area.

However, this simple arrangement to implement the inverter logic is not suitable for high speed operation. There is no discharge path for current from the base of either bipolar transistor when it is being turned OFF, i.e., when $V_{in} = V_{DD}$, M1 is OFF, and no conducting path to the base of Q1 exists. Similarly, when $V_{in} = 0$, M2 is OFF and no conducting path to the base of Q2 exists. This slows down the operation of the circuit.

In order to provide the discharge path for two bipolar transistors, two MOS devices are added, as shown in Fig. 8.6.

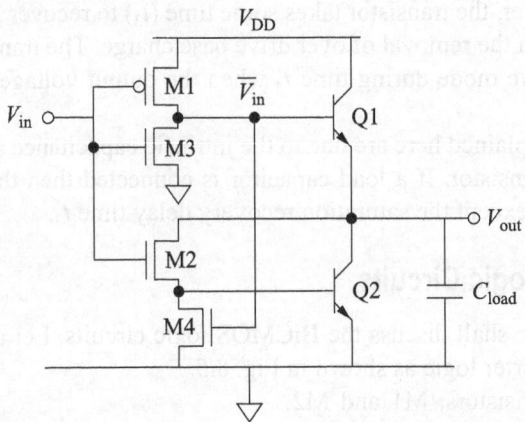

Fig. 8.6 BiCMOS inverter with base pull-down transistor

The transistor M3, which is ON when $V_{in} = V_{DD}$, provides the discharge path for the base of Q1. Similarly, the transistor M4 provides the discharge path for the base of Q2. M4 is ON when $V_{in} = 0$ or $\overline{V}_{in} = V_{DD}$.

8.6 BiCMOS Two-Input NAND Logic

The two-input NAND logic can be implemented as shown in Fig. 8.7. The MOS transistors M1 and M2 form the pull-up logic, and M3 and M4 form the pull-down logic. The two BJTs Q1 and Q2 are used to drive the output load. The transistors

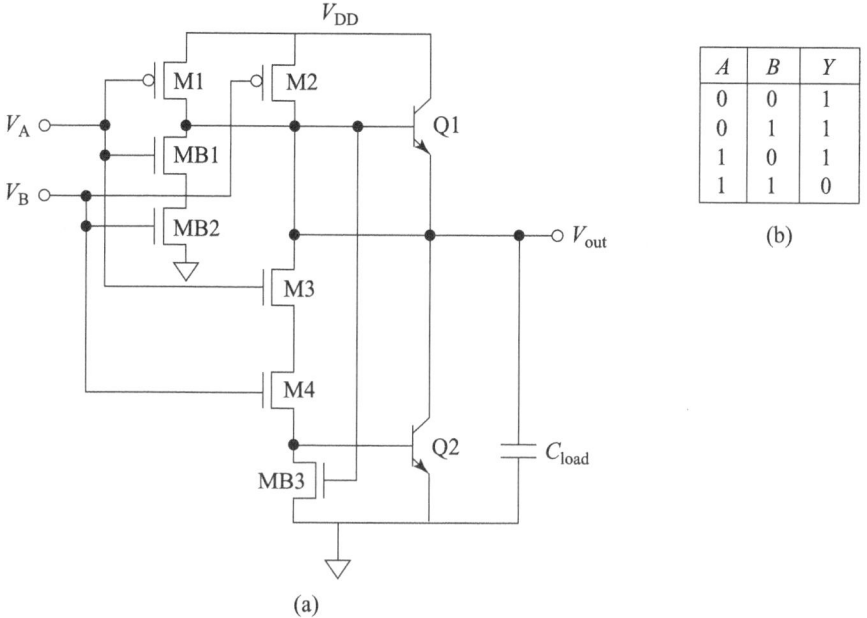

A	B	Y
0	0	1
0	1	1
1	0	1
1	1	0

(b)

(a)

Fig. 8.7 (a) BiCMOS two-input NAND logic; (b) truth table

MB1 and MB2 are used to discharge base of Q1. The transistor MB3 is used to discharge the base of Q2.

Operation
1. When either of inputs or both inputs are at logic 0, either of the transistors M1 and M2 or both is/are ON. Therefore, Q1 is conducting. The load capacitor charges through Q1. In this condition, both M3 and M4 are OFF; hence, there is no discharge path for the load capacitor. Hence, we get logic 1 at the output.
2. When both the inputs are at logic 1, the transistors M1 and M2 are OFF. Therefore, the transistor Q1 is OFF. At this condition, the transistors M3 and M4 are ON. The transistor Q2 is ON, and the load capacitor discharges through Q2.
3. When both inputs are high, the transistor Q1 is OFF and its base has to be discharged. This is done through the discharge path formed by the transistors MB1 and MB2. MB1 and MB2 are both ON when both inputs are logic high.

Similarly, when any one of the inputs or both is/are at logic 0, the transistor Q2 is OFF. The base of Q2 is discharged through the transistor MB3. MB3 is ON when either of the inputs or both is/are at logic low through either M1 or M2.

8.7 BiCMOS Two-input NOR Logic

The two-input NOR logic can be implemented as shown in Fig. 8.8. The MOS transistors M1, M2 form the pull-up logic and M3, M4 form the pull-down logic. The two BJTs Q1 and Q2 are used to drive the output load. The transistors MB1 and MB2 are used to discharge the base of Q1 and the transistor MB3 is used to discharge the base of Q2.

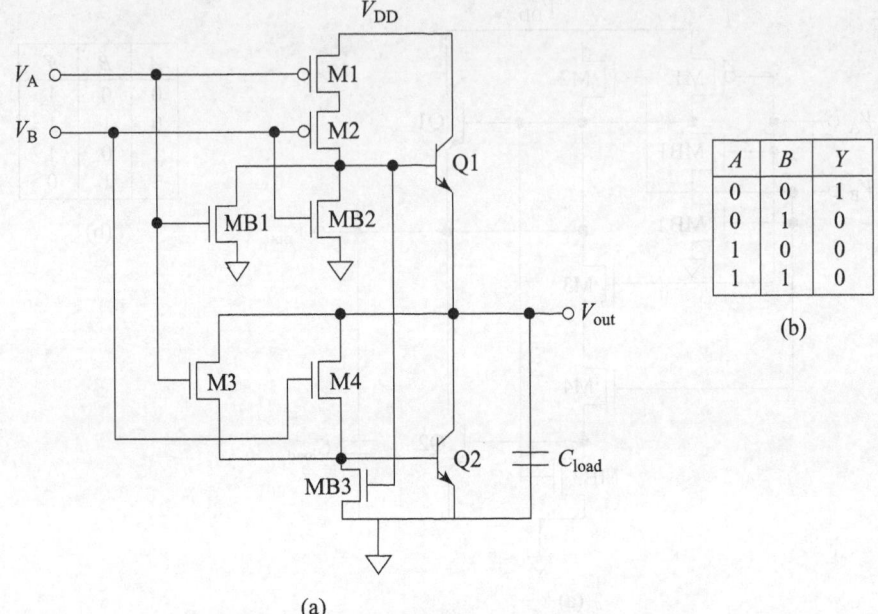

A	B	Y
0	0	1
0	1	0
1	0	0
1	1	0

(b)

(a)

Fig. 8.8   (a) BiCMOS two-input NOR logic; (b) truth table

Operation

1. When both the inputs are at logic 0, the transistors M1 and M2 are ON. Therefore, the Q1 is conducting. The load capacitor charges through Q1. In this condition, both M3 and M4 are OFF; hence, there is no discharge path for the load capacitor. Consequently, we get logic 1 at the output.

2. When either of the inputs or both the inputs are at logic 1, either of the transistors M3 and M4 or both is/are ON. Therefore, the Q2 is conducting. The load capacitor discharges through Q2. At this condition, the transistors M1 and M2 are OFF. The transistor Q2 is OFF, and the load capacitor cannot charge through Q1. Hence, we get logic 0 at the output.

3. When any one of the inputs or both is/are at logic high, the transistor Q1 is OFF and its base has to be discharged. This is done through the parallel discharge path formed by the transistors MB1 and MB2. MB1 and MB2 are ON when inputs are logic high.

Similarly, when both the inputs are at logic low, the transistor Q2 is OFF. The base of Q2 is discharged through the transistor MB3. MB3 is ON when both the inputs are at logic low through a series combination of M1 and M2.

8.8 Complex Logic Using BiCMOS

Any complex binary logic can be implemented using BiCMOS logic as shown in Fig. 8.9. The pull-up network (PUN) is constructed using pMOS transistors, whereas the pull-down network (PDN) is constructed using nMOS transistors. The PUN and PDN are constructed following the steps discussed in Section 6.3 in Chapter 6. In

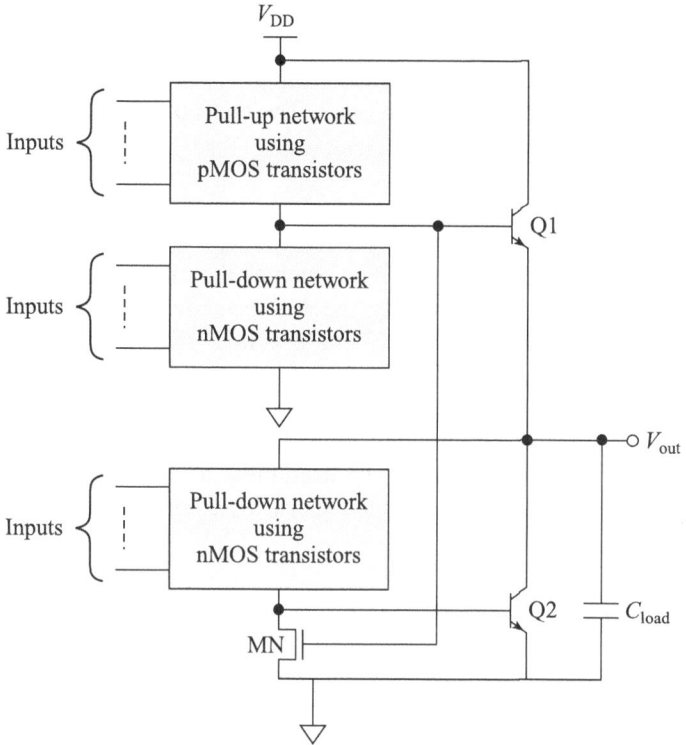

Fig. 8.9 General BiCMOS architecture to implement any complex binary logic

the PDN, the AND and OR terms are implemented by a series and parallel combination of nMOS transistors, respectively. The dual is applied for implementing PUN.

Two BJTs Q1 and Q2, are used to provide low resistive charging and discharging path for the load capacitor. The upper PDN is used to provide the discharge path for base of Q1, and the lower PDN is used to implement the logic. The transistor MN is used to provide the discharge path for the base of Q2.

8.9 Applications of BiCMOS Circuits

Following are some applications of BiCMOS circuits:
- Mixed analog/digital systems
  - Bipolar is used for high performance analog design
  - CMOS is used for high-density and low-power digital design
- High density, high speed RAMs
  - MOS transistors are used to design logic cells
  - BiCMOS is used for sense amplifiers and peripheral circuits
- High performance microprocessor/microcontroller
- Gate array
- Flash A/D converters
  - Bipolar is used for high speed, low offset, and low power comparators
  - CMOS is used for high density and low power encoding logic

In summary, the BiCMOS technology significantly enhances the speed performance while incurring negligible power and area penalty. It can provide applications with CMOS power and densities at high speed, which were previously the exclusive domain of bipolar transistors. The concept of a 'system on a chip' becomes a reality with BiCMOS.

8.10 Disadvantages of BiCMOS

The BiCMOS technology has few issues, such as the extra mask layers increase the manufacturing process complexity, there is speed degradation due to scaling, and the cost of processing is high. There is almost 1.25–1.4 times increase in die cost over conventional CMOS. Considering the packaging and testing costs, the total manufacturing cost of BiCMOS chips ranges from 1.1–1.3 times that of CMOS. The extra cost incurred in developing a BiCMOS technology is balanced by the enhanced chip performance. The usefulness of the manufacturing equipments and clean rooms can be extended for at least one technology generation, thus lowering the capital cost.

8.11 Silicon–Germanium BiCMOS Technology

The silicon–germanium heterojunction bipolar transistor (SiGe HBT) is the first practical bandgap-engineered device to be realized in silicon. The performance of SiGe HBT is competitive with group III–V technologies. It can be fabricated using conventional silicon IC manufacturing process technology at a lower cost.

In the recent past, the SiGe technology has become a very important technology for many wireless applications due to the continuous increase in consumer demand for low power portable products. SiGe technology offers the opportunity to integrate a high performance HBT with CMOS analog and digital circuits on a single chip. These transistors offer higher performance at lower currents, as well as high gain and low noise figure at frequencies up to 10 GHz; hence, these are extremely useful for RF applications. Several literature have reported frequencies up to 200 GHz range for SiGe HBT fabricated using standard CMOS process technology and tools. SiGe HBTs exhibit frequency response above 300 GHz at 300 K, and above one-half of THz (500 GHz) at cryogenic temperatures.

SUMMARY

- BiCMOS technology facilitates the fabrication of bipolar and CMOS devices on to the same substrate.
- Bipolar devices have more current driving capability than CMOS devices for a given device area.
- CMOS devices have a large packing density due to a small area, low power dissipation, and large noise margin compared to bipolar devices.
- Combination of bipolar and CMOS devices in BiCMOS circuits offers high speed and low power dissipation.

■ In BiCMOS circuits, the logic functionality is implemented using CMOS transistors, whereas the output stage that drives the load is designed using bipolar devices.

■ BiCMOS technology requires a few additional process steps which makes it costlier than a purely CMOS process.

■ SiGe technology has become very important. Devices fabricated using SiGe technology can operate up to hundreds of GHz frequency range.

SELECT REFERENCES

Alvarez, A.R., et al. 1990, 'An Overview of BiCMOS Technology and Application', *IEEE International Symposium on Circuits and Systems*, 1–3 May.

Baker, R.J., H.W. Li, and D.E. Boyce 2004, *CMOS Circuit Design, Layout, and Simulation*, Prentice-Hall of India, New Delhi.

Bellaouar, A., M.I. Elmasry, and H.K. Sherif 1995, 'Bootstrapped Full-Swing BiCMOS/BiNMOS Logic Circuits for 1.2–3.3 V Supply Voltage Regime', *IEEE Journal of Solid-State Circuits*, vol. 30, no. 6.

Cressler, D. 2008, 'Silicon–Germanium as an Enabling IC Technology for Extreme Environment Electronics', *Aerospace Conference, 2008 IEEE*, 1–8 March, pp. 1–7.

Gamal, A.E., J.L. Kouloheris, D. How, and M. Morf 1989, 'BiNMOS: A Basic Cell for BiCMOS Sea-of-Gates', *IEEE Custom Integrated Circuits Conference*.

Gonzalez, E.A. 2004, 'BiCMOS Processes, Trends and Applications', *DLSU ECE, Technical Report*, 29 November.

Harame, D.L. et al. 2001, 'Current Status and Future Trends of SiGe BiCMOS Technology', *IEEE Transactions on Electron Devices*, vol. 48, no. 11.

Harame, D.L. et al. 2002, 'The Emerging Role of SiGe BiCMOS Technology in Wired and Wireless Communications', *Fourth IEEE International Caracas Conference on Devices, Circuits and Systems*, Aruba, 17–19 April.

John, J.P. et al. 2002, 'Optimization of a SiGe: C HBT in a BiCMOS Technology for Low Power Wireless Applications', *Bipolar/BiCMOS Circuits and Technology Meeting*, pp. 193–6.

Kang, S.M. and Y. Leblebici 2003, *CMOS Digital Integrated Circuits: Analysis and Design*, 3rd ed., Tata McGraw-Hill, New Delhi.

Kempf, P. 2002, 'Silicon Germanium BiCMOS Technology', *Gallium Arsenide Integrated Circuit (GaAs IC) Symposium, 2002, 24th Annual Technical Digest*, pp. 3–6.

Klose, H. et al. 1989, 'BiCMOS: A Technology for High-speed/High-density ICs', *IEEE International Conference on Computer Design: VLSI in Computers and Processors*, 2–4 October.

Larry, W. and E.L. Gould 1992, 'Optimal Usage of CMOS within a BiCMOS Technology', *IEEE Journal of Solid-State Circuits*, vol. 27, no. 3, March.

Norman, P.J., S. Menon, and S. Sidiropoulos 1994, 'Circuit and Process Directions for Low-Voltage Swing Submicron BiCMOS', *Western Research Laboratory Technical Note*, TN-45, March.

Paul, G.Y.T. et al. 1993, 'Study of BiCMOS Logic Gate Configurations for Improved Low-Voltage Performance', *IEEE Journal of Solid-State Circuits*, vol. 28, no. 3, March.

Sakurai, T. 1992, 'A Review on Low-Voltage BiCMOS Circuits and a BiCMOS vs CMOS Speed Comparison', *Proceedings of the 35th Midwest Symposium on Circuits and Systems*, vol. 1, 9–12 August.

Soyuer, M. 2000, 'The Impact of SiGe BiCMOS Technology on Microwave Circuits and Systems', *Proceedings of the 30th European Solid-State Device Research Conference*, 11–13 September, pp. 34–41.

EXERCISES

Fill in the Blanks

1. The fastest logic family is _____.
   (a) TTL   (b) CMOS
   (c) ECL   (d) IIL
2. The logic family which consumes the least amount of power is _____ .
   (a) DTL   (b) RCTL
   (c) CMOS   (d) none of these
3. BiCMOS technology means _____ .
   (a) fabrication of BJT and CMOS in the same IC
   (b) fabrication of BJT and CMOS in a different IC
   (c) interfacing of CMOS and TTL ICs
   (d) interfacing of CMOS and ECL ICs
4. The common-base current gain is typically _____ .
   (a) equal to unity   (b) close to unity
   (c) less but close to unity   (d) greater but close to unity
5. The common-emitter current gain is typically between _____ .
   (a) 1–10   (b) 1000–10,000
   (c) 100–1000   (d) 0.1–1.0

Multiple Choice Questions

1. Conventional BiCMOS inverter logic requires
   (a) two pMOS transistors and two nMOS transistors
   (b) one pMOS transistor and three nMOS transistors
   (c) three pMOS transistor and one nMOS transistors
   (d) three pMOS transistor and three nMOS transistors
2. The two-input NAND and NOR logic requires
   (a) seven nMOS transistors
   (b) seven pMOS transistors
   (c) two pMOS and five nMOS transistors
   (d) five pMOS and two nMOS transistors
3. In the BiCMOS logic, two pull-down networks are used
   (a) to implement the logic
   (b) one to implement the logic and the other for base discharge
   (c) both for base discharge
   (d) to increase the speed
4. The BJTs are used in BiCMOS to achieve
   (a) higher speed   (b) low power
   (c) large noise margin   (d) less area
5. In BiCMOS circuits,
   (a) CMOS is used for implementing logic and BJT is used for high drive current
   (b) BJT is used for implementing logic and CMOS is used for high drive current
   (c) CMOS is used for implementing logic and BJT is used for low power
   (d) CMOS is used for high speed and BJT is used for high drive current

True or False

1. BJTs have more current driving capability than CMOS using the same silicon area
2. CMOS has more power dissipation than BJTs
3. Fabrication of bipolar devices require extra mask layers than CMOS

4. BiCMOS fabrication is costlier than CMOS fabrication
5. BiCMOS circuits offer higher speed than CMOS

Short-answer Type Questions

1. Explain how BJT can be used to implement digital logic with the help of a circuit diagram.
2. Draw the BiCMOS device structure and explain the fabrication in brief.
3. Implement a two-input NAND logic using BiCMOS.
4. Implement a two-input NOR logic using BiCMOS.
5. Implement a two-input AND logic using BiCMOS.
6. Implement a two-input OR logic using BiCMOS.
7. Implement a two-input XOR logic using BiCMOS.
8. Implement a two-input XNOR logic using BiCMOS.
9. Discuss the pros and cons of BiCMOS technology.
10. Discuss the applications of BiCMOS circuits.

Long-answer Type Questions

1. Draw and explain the operation inverter using a BJT and a resistor. For the BJT inverter, calculate the noise margins with the following circuit parameters:

$$V_{CC} = 5 \text{ V}, \qquad R_B = 10 \text{ k}\Omega, \qquad R_C = 2 \text{ k}\Omega, \qquad \beta = 100,$$

$$V_{BE(on)} = 0.6 \text{ V}, \qquad V_{BE(sat)} = 0.7 \text{ V}, \qquad V_{CE(sat)} = 0.2 \text{ V}$$

2. Implement the following Boolean expression using BiCMOS logic:

$$F = \overline{AB + CD}$$

3. Implement the following Boolean expression using BiCMOS logic:

$$F = \overline{A(B + CD)}$$

4. Design the 2:1 multiplexer using BiCMOS logic.
5. Draw and explain the general structure of BiCMOS logic. Discuss why the base discharge path is required and how is it implemented using a suitable example.

Logic Synthesis

9.1 Introduction

In this chapter, we have discussed the logic synthesis which is the very first step of the VLSI design flow. Once designers come up with new idea or new algorithm, they describe the algorithm in a formal hardware description language (HDL) explaining the behaviour of the system or block. Then the system or block needs to be designed structurally using digital logic blocks to achieve the desired behaviour. Converting the behavioural design into the structural design is simply the process of logic synthesis. Synthesis can be done at different levels of the integrated circuit (IC) design. First, we explain synthesis approaches with suitable examples at different levels. We then discuss the basic logic synthesis steps, followed by different design styles, logic synthesis goals, synthesis tools, basic algorithms for logic synthesis, and sequential logic optimization techniques. At the end of the chapter, we have given examples of using multiplexers to design logic circuits.

9.2 Introduction to Synthesis

The circuit designers very often use two terms: (a) synthesis and (b) analysis. *Synthesis* is a process of designing the circuit, given the input and output. In other words, it is the process of implementing the hardware for known input and output. *Analysis* is a process of finding out the output of a given circuit and its input. Figure 9.1 illustrates the difference between analysis and synthesis. In the VLSI design process there are several synthesis steps. When the designers design a circuit with transistors, it is called transistor level synthesis. At the logic design step, logic synthesis is used, which is a process of designing the logic circuit for a given input and output. Similarly, there is block level synthesis, and at the highest level, top-level synthesis.

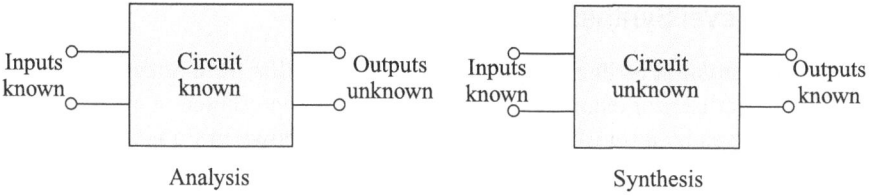

Fig. 9.1 Analysis vs synthesis

9.3 Transistor-level or Circuit-level Synthesis

Let us begin our discussion with a very simple design—an inverter circuit. An inverter is a logic circuit which inverts the input logic. Given the logic 0 at the input, it produces logic 1 at the output and vice versa. We will implement the inverter circuit following different circuit configurations. All the implementations should produce the same output.

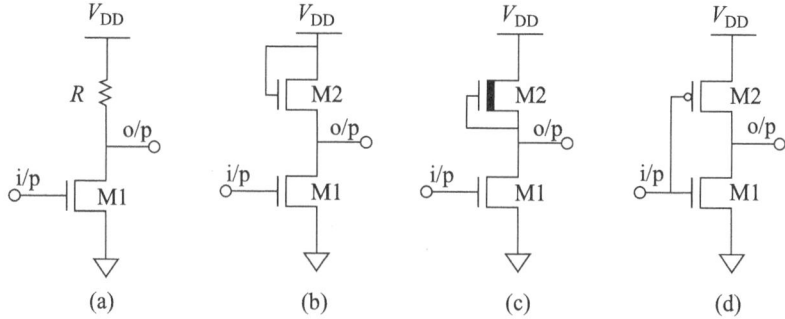

Fig. 9.2 Different implementations of an inverter circuit: (a) resistive load inverter; (b) enhancement-type nMOS load inverter; (c) depletion-type nMOS load inverter; (d) CMOS inverter

Figure 9.2 shows different implementations of an inverter circuit. When input is low, the transistor M1 is OFF. Thus, the output is connected to V_{DD} through pull-up network, and we get logic high at the output.

- In Fig. 9.2(a), the pull-up network is the resistor R.
- In Fig. 9.2(b), the pull-up network is the enhancement-type nMOS transistor M2, which is always ON as its gate is connected to V_{DD}.
- In Fig. 9.2(c), the pull-up network is the depletion-type nMOS transistor M2, which is always ON as its gate-to-source voltage is 0.
- In Fig. 9.2(d), the pull-up network is the pMOS transistor M2, which is ON when input is low.

When input is high, the nMOS M1 is ON in all the four circuits. Thus, the output is connected to the ground, and we get logic low at the output.

There could be some more implementations of the inverter circuit. All the implementations have the same functionality but they would have different speed, area, and power capabilities. So depending on the speed, area, and power requirements, a designer needs to try different circuit configurations to achieve the best performance. This example depicts the basic idea behind the transistor-level synthesis.

9.4 Logic-level Synthesis

Logic-level synthesis means designing a gate-level netlist from a register transfer level (RTL) netlist. For example, a half-adder can be represented in many ways. It can be expressed by a set of Boolean expressions as shown in Eqn (9.1):

$$\text{Sum} = A\bar{B} + \bar{A}B = A \oplus B$$
$$\text{Carry} = AB$$

(9.1)

It can also be expressed using a truth table, as shown in Table 9.1.

Table 9.1 Truth table of half-adder

A	B	Carry	Sum
0	0	0	0
0	1	0	1
1	0	0	1
1	1	1	0

Both these two representations do not specify any architecture of the half-adder. Although we know the inputs and outputs of the half-adder, we do not know what will be the architecture of a half-adder. So the designers come up with several structures at the logic-level to satisfy the input–output relationship. That means the designers perform logic-level synthesis. Some of the possible implementations of a half-adder at the logic-level are described in Figs 9.3(a) and (b).

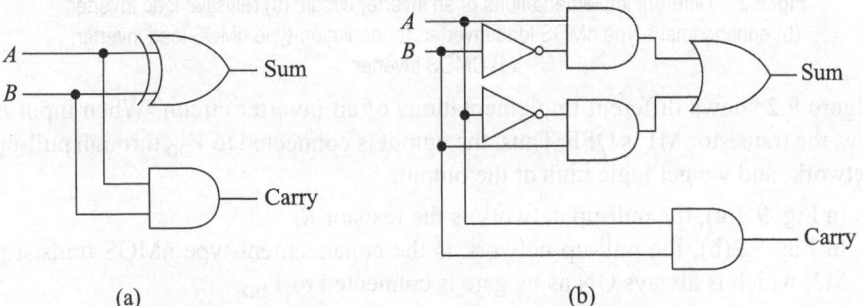

Fig. 9.3 Half-adder: (a) one-level logic; (b) three-level logic

9.5 Block-level Synthesis or Datapath Synthesis

Consider an *n*-bit adder design using serial and parallel architecture. The basic building block of an *n*-bit adder is a 1-bit adder or full-adder circuit. First, consider the serial architecture, as shown in Fig. 9.4.

A serial adder adds two *n*-bit binary numbers bit-by-bit. To compute the *n*-bit sum, it requires *n* clock cycles. In every clock cycle, one sum and carry bit is generated. The carry bit is stored in a D flip-flop and it is used in the next clock cycle. This is the least expensive circuit in terms of area for the addition of two

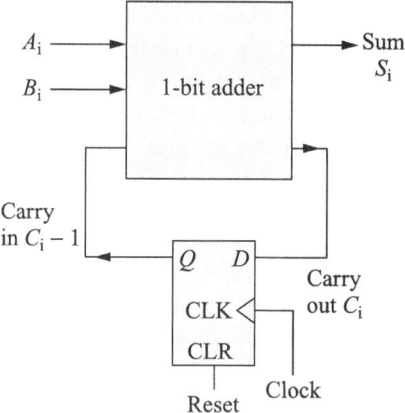

Fig. 9.4 A bit-serial adder

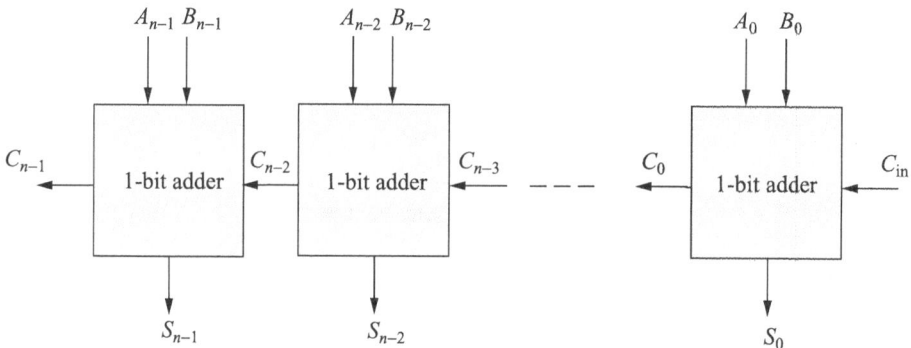

Fig. 9.5 An n-bit parallel adder (ripple-carry-adder)

n-bit numbers. The disadvantage of this design is that it is slow as it adds bit-by-bit serially.

Let us now consider the parallel adder for addition of two n-bit numbers. The structure of the parallel adder is shown in Fig. 9.5. The parallel adder is formed by cascading n full-adders. Each full-adder adds two bits along with a previous carry bit, and produces a sum and carry bit. Since the carry propagates through the full-adder chain, it is also known as ripple-carry-adder. In this adder, all the n sum bits and carry bit are available in a single clock cycle. The maximum propagation delay is $n\tau$, where τ is the delay of a full-adder. This design is fast compared to the serial adder, but it requires much more area.

9.6 Logic Synthesis

As discussed in the last section, the synthesis is possible at various levels in the VLSI design cycle. But if a chip designer needs to perform the synthesis tasks in the design cycles, the design cycle time would increase tremendously, and it may not always lead to an optimized circuit design. A VLSI chip typically contains hundreds of thousands of gates. Hence, it is impractical to synthesize the design manually at all levels. So a chip designer needs automated software which would do the tasks and come up with an optimum circuit from the area, speed, and power point of view. This automated tool is called logic synthesizer which basically works on logic minimization techniques. The logic minimization technique is mainly based on some advanced algorithms and some set of rules. Figure 9.6 illustrates the typical logic synthesis process.

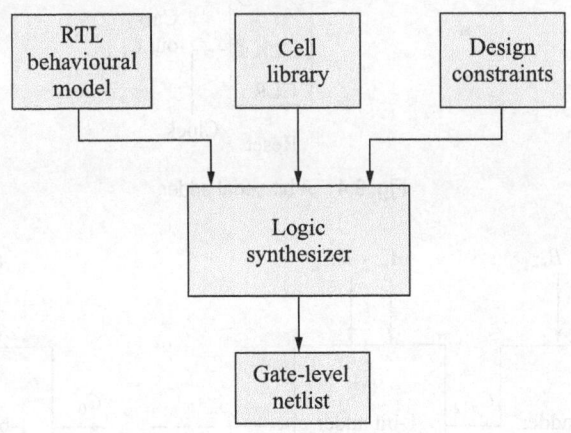

Fig. 9.6 A logic synthesis flow

Definition   It is a process of converting a high-level description of the design into an optimized gate-level representation, given a standard cell library and certain design constraints.

Standard Cell Library   It is a database containing logic gates such as AND, OR, NOT, etc. It can have macro cells such as adders, multiplexers, and flip-flops.

Design Constraints   A set of constraints such as maximum chip area, operating speed in terms of minimum clock frequency, maximum power dissipation, etc.

Logic synthesis translates an RTL netlist into gate-level netlist. The RTL or gate-level netlist is written in a HDL. There are two HDLs: (a) Verilog and (b) VHDL. Both the languages were developed in the early 1980s. A designer first writes an RTL to put his or her idea of a new product or algorithm in HDL using behavioural modelling style, and simulates to check if it produces correct results. If not, the RTL is corrected till it meets the functionality. Once the behavioural description of a design is ready, the designer can use the synthesis tool to synthesize his or her design. He or she requires some more inputs, such as the design constraints and the cell library. By far, the readers have the familiarity about the cell library. Let us introduce the design constraints here. A typical design constraints file would have the area, power, and the speed requirements. The logic synthesizer generates a gate-level netlist, a structural model, which contains the logic cells from the cell library. The synthesizer works in such a way that the constraints are met as specified in the constraints file. The gate-level netlist is simulated once again and results are compared against the behavioural simulation.

9.6.1 Steps

The logic synthesizer first parses the RTL netlist and stores the information in the form of a data structure. Typically, a graph is built using linked lists. Next, the logic minimization is performed by identifying the minimum cover in the Karnaugh map and the output is expressed in the sum-of-products (SOP) or product-of-sums (POS) form. The logic optimization step uses several factoring, substitution, and elimination steps, and simplifies the Boolean expressions. Finally, in the technology mapping step, it performs a mapping from the abstract optimized representation to a netlist of cells that are available in the technology-dependent cell library. Both combinational and sequential RTL descriptions are synthesized.

9.6.2 Design Styles

Building a large digital system requires hierarchical decomposition of the full IC. In the decomposition process, the whole circuit is partitioned into several functional blocks:

- datapath logic
- control logic
- memory block

The *datapath logic* performs basic data processing operations: addition, subtraction, multiplication, division under arithmetic operations, logical AND, OR, NOT, shift, etc., under logical operations. The *control logic* generates control signals for the datapath logic to select different functions to be performed at different times, for example, when to add two numbers or to multiply two numbers. The purpose of the *memory block* is to store the data or program, or both.

It is often found that different methodologies are used for designing different blocks. For example, the design of the datapath is done in a different way than the design of controllers. The datapath design may adopt a serial or parallel architecture based on the area and speed trade-offs. The controller, on the other hand, is designed at a higher level as a finite state machine (FSM). A finite state machine is a sequential circuit which undergoes a finite number of states depending on the inputs and present state of the machine. While designing FSMs, state assignment, encoding, and minimization techniques are used.

9.6.3 Tools

Logic synthesis tool is an electronic design automation (EDA) software which helps the VLSI chip designers in two ways: (a) automatically translates the behavioural design into gate-level design and (b) optimizes the chip area, speed, power, and testability. This helps the designer to reduce the design cycle time tremendously, reduce the chances of design error, and produce good quality designs. The most popular synthesis tools are listed in Table 9.2.

9.6.4 Goals

The primary goal of the logic synthesis process is to meet the area, speed, and power constraints. The main aspect of the automated logic synthesis is to achieve

Table 9.2   Synthesis tools

Sl. No.	Vendor	Tool
1.	Synopsys	Design compiler
2.	Cadence	Encounter RTL compiler
3.	Magma	Talus design

high quality designs from the initial specifications. The optimization problem is to minimize:

- chip area occupied by the logic gates and the interconnects between them.
- critical path delay; a *critical path* is defined as the longest path through the logic.
- power dissipated by the logic gates.

Another goal of the logic synthesis process is to achieve designs with a good degree of testability. Testing a manufactured chip is very crucial to determine if a fabricated chip really works as per the specifications.

9.7 Algorithms

The size of the VLSI circuits enforces innovation of newer algorithms, models, and methodologies. The automated software internally uses graph theory, models, Boolean algebra, optimization theories, theory of finite state machine, automate theory, etc., to produce high quality designs in a reduced cycle time.

9.7.1 Boolean Algebra

The operations of logic circuits are described by Boolean algebra. Table 9.3 shows the laws of Boolean algebra for the SOP and POS forms.

Table 9.3   Laws of Boolean algebra

Sl. No.	Laws	SOP form	POS form
1.	Idempotent	$x + x = x$	$x \cdot x = x$
2.	Commutative	$x + y = y + x$	$x \cdot y = y \cdot x$
3.	Associative	$x + (y + z) = (x + y) + z$	$x \cdot (y \cdot z) = (x \cdot y) \cdot z$
4.	Absorptive	$x \cdot (x + y) = x$	$x + (x \cdot y) = x$
5.	Distributive	$x \cdot (y + z) = (x \cdot y) = (x \cdot z)$	$x + (y \cdot z) = (x + y) \cdot (x + z)$
6.	Complementarity	$x + x' = 1$	$x \cdot x' = 1$
7.	Involution	$(x')' = x$	
8.	Operations with 0 and 1	$x + 0 = x$ $x + 1 = x$	$x \cdot 0 = x$ $x \cdot 1 = x$

A set of values $B = \{0, 1\}$ and the operators '+' and '·' are used in Boolean expressions. The '+' operator is called 'OR' operator, and the operator '·' is called 'AND' operator.

Theorems of Boolean Algebra

Theorem 1   Complementation in a Boolean algebra is unique.

Proof   Suppose x' and y are the complements of x. We can show that $y = x'$. If y is a complement of x, then $x + y = 1$ and $xy = 0$.

Theorem 2   (Involution) In a Boolean algebra:

$$(x')' = x$$

Theorem 3   In a Boolean algebra:

$$x + x'y = x + y$$
$$x(x' + y) = xy$$

Theorem 4   (DeMorgan's laws) In a Boolean algebra:

$$(x + y)' = x'y'$$
$$(xy)' = x' + y'$$

Theorem 5   (Consensus theorem) In a Boolean algebra:

$$xy + x'z + yz = xy + x'z$$
$$(x + y)(x' + z)(y + z) = (x + y)(x' + z)$$

Table 9.4   Boolean expressions for 16 functions of two variables

Boolean function	Operator symbol	Name	Comments
F_0		Null	Binary constant 0
$F_1 = xy$	$x \cdot y$	AND	x and y
$F_2 = x\bar{y}$	x / y	Inhibition	x but not y
$F_3 = x$		Transfer	x
$F_4 = \bar{x}y$	y / x	Inhibition	y but not x
$F_5 = y$		Transfer	y
$F_6 = x\bar{y} + \bar{x}y$	$x \oplus y$	Exclusive-OR	x or y but not both
$F_7 = x + y$	$x + y$	OR	x or y
$F_8 = \overline{x + y}$	$x \downarrow y$	NOR	Not OR
$F_9 = xy + \bar{x}\bar{y}$	$x \odot y$	Equivalence	x equals y
$F_{10} = \bar{y}$	$\bar{y}$	Complement	Not y
$F_{11} = x + \bar{y}$	$x \subset y$	Implication	If y, then x
$F_{12} = \bar{x}$	$\bar{x}$	Complement	Not x
$F_{13} = \bar{x} + y$	$x \supset y$	Implication	If x, then y
$F_{14} = \overline{xy}$	$x \uparrow y$	NAND	Not AND
$F_{15} = 1$		Identity	Binary constant 1

9.7.2 Boole's Expansion Theorem

In designing a large digital logic, we have to deal with a large number of input variables. In such cases, we split the large function into smaller ones with the help of Boole's expansion theorem. If f is a Boolean function of n input variables $(x_1, x_2, ..., x_n)$, then according to this theorem, we can write the following:

$$f(x_1, x_2, ..., x_n) = x_1' \cdot f(0, x_2, ..., x_n) + x_1 \cdot f(1, x_2, ..., x_n)$$
$$= [x_1' + f(1, x_2, ..., x_n)] \cdot [x_1 + f(0, x_2, ..., x_n)]$$

where $f(0, x_2, ..., x_n)$ and $f(1, x_2, ..., x_n)$ are the co-factors of the function f. $f(0, x_2, ..., x_n)$ and $f(1, x_2, ..., x_n)$ are the function f evaluated for $x = 0$ and $x = 1$, respectively. This theorem is also known as *Shannon's expansion theorem*.

9.7.3 Minterm Canonical Form

A Boolean function can be expressed by an infinite number of formulae. The minimization of a Boolean function is basically identifying if two or more representations of a function is same or not. There must be a form of each function which is unique, and this form is known as *canonical form*. Let us introduce here the minterm canonical form. We express a function f by recursively applying the Boole's expansion theorem as:

$$f(x_1, x_2, ..., x_{n-1}, x_n) = f(0, ..., 0, 0)x_1'...x_{n-1}'x_n'$$
$$+ f(0, ..., 0, 1)x_1'...x_{n-1}'x_n$$
$$\vdots$$
$$+ f(1, ..., 1, 1)x_1...x_{n-1}x_n$$

where $f(0, ..., 0, 0), f(0, ..., 0, 1), ..., f(1, ..., 1, 1)$ are the discriminants of the function f and the product terms $(x_1'...x_{n-1}'x_n'), (x_1...x_{n-1}'x_n), ..., (x_1...x_{n-1}x_n)$ are called the minterms.

9.7.4 Maxterm Canonical Form

Similarly, the maxterm canonical form is expressed as

$$f(x_1, ..., x_{n-1}, x_n) = [f(0, ..., 0, 0) + x_1 + \cdots + x_{n-1} + x_n]$$
$$\cdot [f(0, ..., 0, 1) + x_1 + \cdots + x_{n-1} + x_n']$$
$$\vdots$$
$$\cdot [f(1, ..., 1, 1) + x_1' + \cdots + x_{n-1}' + x_n']$$

where the sums $(x_1 + \cdots + x_{n-1} + x_n), (x_1 + \cdots + x_{n-1} + x_n'), ..., (x_1' + \cdots + x_{n-1}' + x_n')$ are called the maxterms.

9.7.5 Sum of Products and Product of Sums

Any Boolean function can be either expressed by a sum of products (SOP or $\Sigma\Pi$) or product of sums (POS or $\Pi\Sigma$). The SOP and POS are also known as disjunctive normal form (DNF) and conjunctive normal form (CNF), respectively.

For example, an SOP and POS forms are given as

$$f = x_1 x_2 + x_2' x_3 + x_1 x_3' = (x_1 + x_2)(x_2' + x_3)(x_1 + x_3')$$

The cost of a function is determined by the number of literals and the number of terms.

9.7.6 Implicants and Prime Implicants

An implicant of a function is a product term p that is included in the function f. A prime implicant of function f is an implicant that is not included in any other implicant of f. A prime implicant is essential if it includes a minterm which is not included in any other prime implicant.

9.7.7 Quine's Prime Implicants Theorem

A minimal SOP must always consist of a sum of prime implicants, if any definition of cost is used in which the addition of a single literal to any formula increases the cost of the formula.

9.7.8 Simplification of Boolean Expressions

There are several methods for simplification of Boolean expressions. The most widely used simplification methods are given by

- *Karnaugh map method* In this method, a graphical map is formed and logical adjacency is considered between the squares in the map to eliminate input variables from the product or sum terms. This method cannot be used for functions having more than six input variables.
- *Quine–McCluskey minimization method* In this method also, grouping of minterms is done based on logical adjacency, but it can be implemented as a computer program for functions with a large number of input variables. This method is also known as tabular method.
- *Boolean minimization* In this technique, theorems of Boolean algebra are used to find the minimal expression. But this method is not well defined, and results depend on the experience of the designers.
- *Expresso-II and mis-II* (multilevel interactive synthesis) These computer programs are embedded in EDA tools to achieve efficient realization of logic circuits.

Tabular Method

Tabular method is suitable for automated software development to produce a simplified form of Boolean functions with many variables. This method has the following two main tasks:

1. To find out prime implicants.
2. Choose the prime implicants that give the simplest expression with minimum number of literals.

Step 1 Form group of minterms according to the number of 1's contained.

Step 2 Any two minterms that differ by 1-bit is combined and the unmatched bit is removed. A check mark ($\sqrt{}$) is placed to the right of binary numbers if they are same in every position except one. This is illustrated in Table 9.5.

Step 3 Repeat the searching and comparing process until the two variable terms are formed.

Step 4 The unchecked terms (Table 9.6) form the prime implicants.

Example 9.1 Determine the prime implicants of the function:

$$F(w, x, y, z) = \sum(1, 4, 6, 7, 8, 9, 10, 11, 15)$$

Table 9.5 Tabular method

(a) wxyz		(b) wxyz		(c) wxyz	
1	0001 $\sqrt{}$	1,9	–001	8,9,10,11	10––
4	0100 $\sqrt{}$	4,6	01–0	8,10,9,11	10––
8	1000 $\sqrt{}$	8,9	100– $\sqrt{}$		
		8,10	10–0 $\sqrt{}$		
6	0110 $\sqrt{}$	6,7	011–		
9	1001 $\sqrt{}$	9,11	10–1 $\sqrt{}$		
10	1010 $\sqrt{}$	10,11	101– $\sqrt{}$		
7	0111 $\sqrt{}$	7,15	–111		
11	1011 $\sqrt{}$	11,15	1–11		
15	1111 $\sqrt{}$				

Table 9.6 Prime implicants

Decimal	Binary	Term
	$w\,x\,y\,z$	
1, 9	–0 0 1	$\bar{x}\,\bar{y}\,z$
4, 6	0 1–0	$\bar{w}\,x\,\bar{z}$
6, 7	0 1 1–	$\bar{w}\,x\,y$
7, 15	–1 1 1	$x\,y\,x$
11, 15	1–1 1	$w\,y\,z$
8,9,10,11	1 0––	$w\,\bar{x}$

Selection of Prime Implicants

A prime implicant table is generated as shown in Table 9.7. The prime implicants are put in rows and the minterms in each column. A cross ($\times$) is placed in each row to identify the minterm that generates the prime implicants. A mimimum set of prime implicants are selected to represent the function.

Prime implicants that cover minterms with a single $\times$ in their column are called essential prime implicants. A check mark ($\sqrt{}$) is placed against them to indicate that they are selected.

Table 9.7 Selection of prime implicants

		1	4	6	7	8	9	10	11	15
$\bar{x}\,\bar{y}z$	1,9 √	×					×			
$\bar{w}x\bar{z}$	4,6 √		×	×						
$\bar{w}xy$	6,7			×	×					
xyz	7,15				×					×
wyz	11,15								×	×
$w\bar{x}$	8,9,10,11 √					×	×	×	×	
		√	√			√		√		

Now we inspect all the essential prime implicants to check if they cover all the minterms. In this case, they cover all the minterms, except 7 and 15. So the prime implicant xyz is to be selected. Hence, the function can be expressed as

$$F = \bar{x}\,\bar{y}z + \bar{w}x\bar{z} + w\bar{x} + xyz$$

9.7.9 Iterated Consensus

Distance-1 merging theorem states that $AB + AB' = A$. This is a special form of consensus theorem. This theorem is used in the tabular method. A complete sum is obtained which is the sum of all the prime implicants.

9.8 Boolean Space

In this section, we will discuss the basic concepts behind the combinational logic synthesis. The logic circuits, without any memory is called *combinational circuits*. The behaviour of the combinational circuit is entirely dependent on the inputs. Let us consider a combinational logic circuit with m inputs and n outputs. The behaviour of this circuit is expressed as

$$f : B^m \rightarrow B^n$$

where f is the Boolean function and $B = \{`0`, `1`\}$ is the set of Boolean values.

The Boolean functions determine the outputs depending on the input signals. But there are some cases where the behaviour of a circuit is not fully specified using a Boolean function. These types of circuits are called incompletely specified circuits. In incompletely specified circuits, few input combinations never occur, or the output can be anything for some of the input combinations. In such cases, we need a third state to represent the unknown state. This is called *don't-care* condition and is represented by '−' symbol. In this case, the Boolean function is given by

$$f : B^m \rightarrow Y^n$$

where $Y = \{`0`, `1`, `−`\}$.

In general, the space B^m is partitioned into three sets:
1. ON-set contains all points in B^m space for which the output is '1'.
2. OFF-set contains all points in B^m space for which the output is '0'.
3. DC-set contains all points in B^m space for which the output is '–'.

Having don't-care conditions often yields smaller circuits. For a fully specified combinational circuit, the DC-set is a null set. The B^m space is shown in Fig. 9.7 for $m = 3$.

The $2^3 = 8$ corners of the cube represent 8 possible input combinations. As shown in Fig. 9.7, the encircled corners are the ON-set and the rest of the corners are OFF-set. In this case, the DC-set is null.

Any Boolean variable x or its negation (complement) $\bar{x}$ is termed as *literal*. A product of literals is called a *product term*. A product of n literals, denoting a single point in Boolean space, is called a *minterm*. The sum of minterms is a canonical expression, which basically represents the truth table.

A cube whose points are in the ON-set or the DC-set of Boolean function f, is called implicant of f.

A set of implicants that covers all the minterms of function f is called *cover* of f. A function is always unique but many different covers are possible (refer to Fig. 9.8).

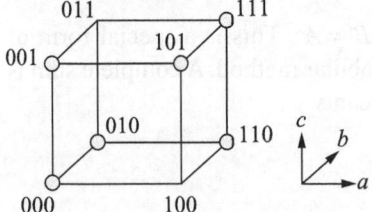

Fig. 9.7 B^3 Boolean space for input a, b, c

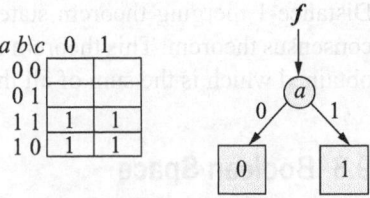

Fig. 9.8 A 3-input Boolean function

It has many covers as shown below:

$$F = a$$

$$F = ab + a\bar{b}$$

$$F = abc + ab\bar{c} + a\bar{b}c + a\bar{b}\bar{c}$$

Prime implicant is one which is not contained in any other implicant, and has at least one point in the ON-set. A cover by primes is known as *prime cover*. *Irredundant cover* is that cover which cannot delete any implicant and still cover. Some minterms are covered only by a prime implicant, which is called *essential primes*.

N.B. Quine's theorem: There is a minimum cover that is prime!

9.9 Binary Decision Diagram (BDD)

The Boolean space methodology becomes impractical to handle for a large number of input signals. The Boolean space B^m grows exponentially with m. An alternative methodology is followed to overcome this problem. Binary decision diagram (BDD) is the alternative method as shown in Fig. 9.9.

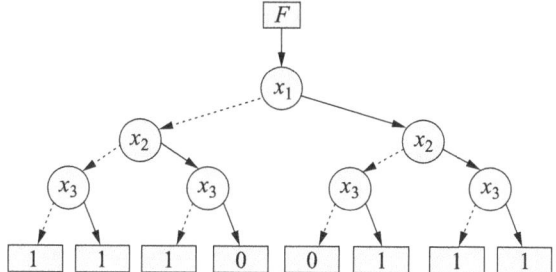

Fig. 9.9 Ordered binary decision diagram (OBDD) of the Boolean function *f*

9.9.1 Ordered Binary Decision Diagram (OBDD)

Let us consider a Boolean function as shown below:

$$f = \bar{x}_1 \bullet \bar{x}_2 \bullet \bar{x}_3 + \bar{x}_1 \bullet x_2 \bullet \bar{x}_3 + \bar{x}_1 \bullet \bar{x}_2 \bullet x_3 + x_1 \bullet \bar{x}_2 \bullet x_3 + x_1 \bullet x_2 \bullet \bar{x}_3 + x_1 \bullet x_2 \bullet x_3$$

Expanding the function *f* using the Boole's expansion theorem, we get

$$f = x_1 \bullet (x_2 \bullet (x_3 \bullet '1' + \bar{x}_3 \bullet '1') + \bar{x}_2 \bullet (x_3 \bullet '1' + \bar{x}_3 \bullet '0')) +$$

$$\bar{x}_1 \bullet (x_2 \bullet (x_3 \bullet '0' + \bar{x}_3 \bullet '1') + \bar{x}_2 \times (x_3 \bullet '1' + \bar{x}_3 \bullet '1'))$$

This fully expanded form has graph representation, as shown in Fig. 9.9.

This graph representation is called OBDD, which has a directed tree structure. Each vertex has two children. The two edges originated from a vertex are called *high* (positive co-factor) and *low* (negative co-factor). The high is represented by solid lines whereas the low is represented by dashed lines. Each leaf vertex has a value of '1' or '0'. An OBDD is a directed tree and is denoted by $G(V, E)$. Each vertex $v \in V$ is characterized by an associated variable $\phi(v)$, a *high* subtree $\eta(v)$, and a *low* subtree $\lambda(v)$.

Binary decision diagram (BDD) is an efficient way of representing and manipulating the large Boolean expression (Bryant). BDD is extensively used in the area of synthesis, and it is canonical. A form is canonical in that the representation of a Boolean expression is unique.

9.9.2 Reduced Ordered Binary Decision Diagram (ROBDD)

An OBDD is reduced to achieve a reduced ordered binary decision diagram (ROBDD). The steps to reduce the OBDD are as follows:

1. Merge all identical leaf vertices *v* with same $\phi(v)$, and appropriately redirect their incoming edges.
2. Proceeding from bottom to top, process all vertices: if two vertices *u* and *v* are found for which $\phi(u) = \phi(v)$, $\eta(u) = \eta(v)$, and $\lambda(u) = \lambda(v)$, merge *u* and *v* and redirect incoming edges.
3. For vertices *v* for which $\eta(v) = \lambda(v)$, remove *v* and redirect its incoming edges to $\eta(v)$.

Example 9.2 A function *F* and its OBDD are as shown in Fig. 9.10. Reduce the OBDD to obtain ROBDD.

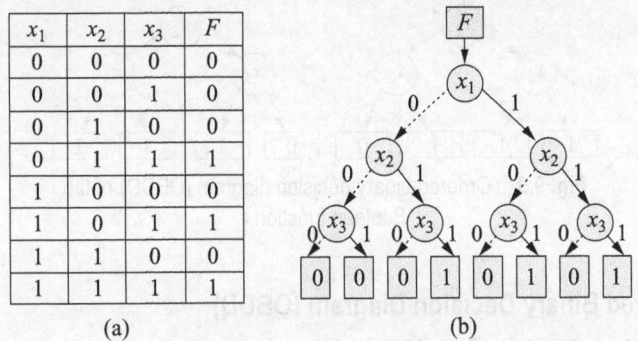

x_1	x_2	x_3	F
0	0	0	0
0	0	1	0
0	1	0	0
0	1	1	1
1	0	0	0
1	0	1	1
1	1	0	0
1	1	1	1

(a)   (b)

Fig. 9.10 Example of OBDD

Solution The solution is shown in Fig. 9.11.

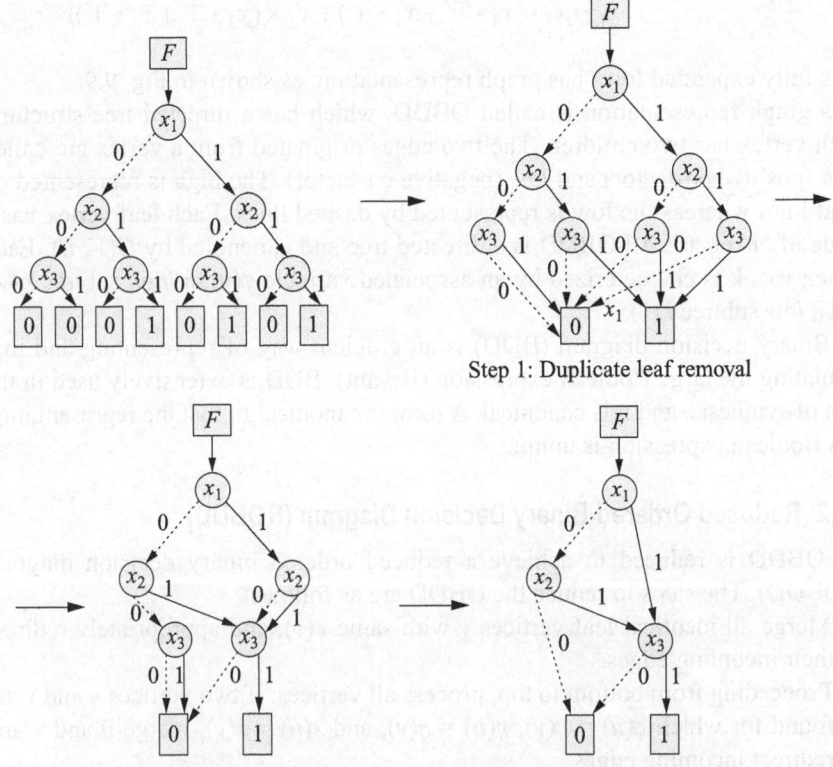

Step 1: Duplicate leaf removal

Step 2: Duplicate vertex removal

Step 3: Redundant test removal

Fig. 9.11 OBDD reduction procedure

Example 9.3 Let us consider the following Boolean expression, given in the SOP form as:

$$f = a\bar{b} + \bar{a}b$$

Construct ROBDD.

Solution
A BDD for this function is shown in Fig. 9.12. The root of the BDD is the function *f*. The first child node of the root is input variable *a*, and the subsequent child nodes are other input variables *b*. The leaf nodes are the possible values 1 and 0 of the function *f*. The edge labelled T indicates *if true then*. The edge labelled E indicates *else* (if false). To find out the value of the function *f* for *a* = 0 and *b* = 1, we start from the root. First we see variable *a* and its value is 0. We follow the edge labelled E. Then we come to a node labelled *b*. Now its value is 1, so we follow the edge labelled T. Finally, we reach the leaf node labelled 1. This is the value of the function *f*(0,1) = 1. This is desired value of the function *f* for *a* = 0 and *b* = 1.

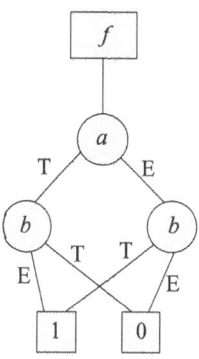

Fig. 9.12 A binary decision diagram

9.9.3 Binary Decision Diagram of Basic Logic Functions

The basic logic functions, such as AND, OR, NOT, are represented in the form BDD, as shown in Fig. 9.13.

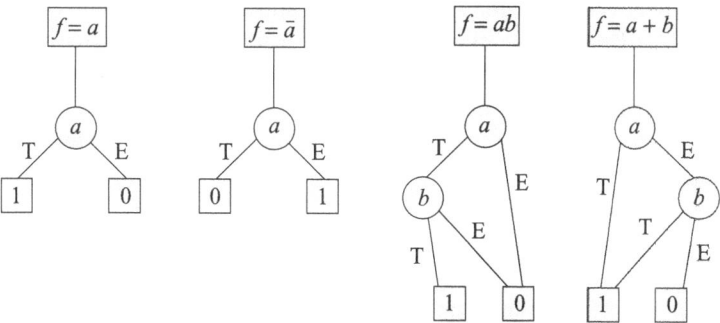

Fig. 9.13 BDDs of basic logic functions

9.9.4 Variable Ordering

BDD is basically a directed acyclic graph (DAG). The root of the BDD is the function, and the leaf nodes are the two logic values 1 and 0. The intermediate nodes of the BDD are the input variables of the function. The BDDs, as shown in Fig. 9.12 is an ordered BDD where ordering is $a \leq b$. The size and structure of the BDD depends on the ordering of variables. This is illustrated in Fig. 9.14. Let us consider a Boolean expression as: $f = abc + b'd + c'd$. The BDD as shown in Fig. 9.14(a) is a ROBDD for the ordering $a \leq b \leq c \leq d$. The BDD for optimal ordering of $a = d = b = c$ is shown in Fig. 9.14(b). The BDD as shown in Fig. 9.14(c) is obtained for the optimal ordering $b \leq c \leq a \leq d$, because this ordering yields the minimum number of nodes.

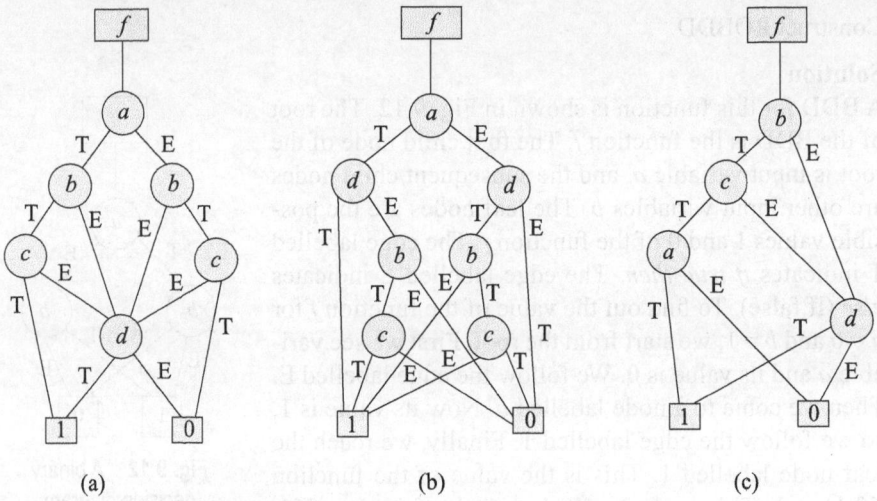

Fig. 9.14   BDD of the function $f = abc + b'd + c'd$ for the different ordering: (a) $a \leq b \leq c \leq d$, (b) $a \leq d \leq b \leq c$, (c) $b \leq c \leq a \leq d$

9.9.5 Applications of Binary Decision Diagram

The binary decision diagram is widely used in the VLSI design automation process. It is extensively used for logic synthesis and verification. The compact representation of a Boolean function using BDD leads to smaller circuits. It can be very efficiently used for logic verification. Verification is the process of checking a circuit, whether it is implemented as per the specifications of the circuit. Let us consider a specified function as F and its implementation as G. If the ROBDD of F and G are identical, we can say G is a correct implementation of F.

9.10   Logic Synthesis—Advantages and Disadvantages

The advantages of logic synthesis are as follows:
1. Automatically manages many details of the design process.
   - Fewer bugs
   - Improved productivity
2. Abstracts the design data (HDL description) from any particular implementation technology.
   - Designs can be resynthesized targeting different chip technologies; for example, first implement in FPGA, then in ASIC.
3. In some cases, leads to a more optimal design than could be achieved by manual means (e.g., logic optimization).

Disadvantage of logic synthesis is
   - It may lead to non-optimal designs in some cases

9.11 Logic Synthesis Techniques

There are a variety of general and ad hoc methods as given in the following text:

1. **Instantiation** It maintains a library of primitive modules (AND, OR, etc.) and adds to the user-defined modules.
2. **Macro expansion/substitution** A large set of language operators (+, −, Boolean operators, etc.) and constructs (if-else, case) expand into special circuits.
3. **Inference** Special patterns are detected in the language description and treated specially (e.g., inferring RAM blocks from variable declaration and read/write statements, FSM detection, and generation from always blocks).
4. **Logic optimization** Boolean operations are grouped and optimized with logic minimization techniques.
5. **Structural reorganization** Advanced techniques, including sharing of operators, and retiming of circuits, and others.

9.12 Sequential Logic Optimization

The behaviour of the sequential circuits is expressed either by state diagram or state table. A simple way to represent the sequential circuit behaviour is by finite state machine models. A block diagram of a finite state machine (FSM) is shown in Fig. 9.15.

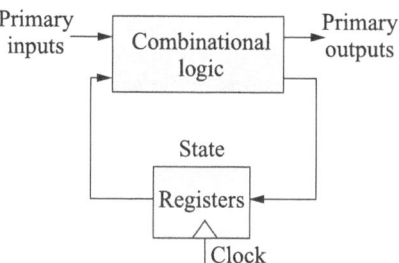

Fig. 9.15 Block diagram of a finite state machine

The state transition diagram and a state table representation are shown in Fig. 9.16. We shall discuss the two main techniques of sequential logic optimization: (a) state minimization and (b) state encoding.

9.12.1 State Minimization

It is the technique of reducing the number of states of an FSM without disturbing the behaviour of the FSM. The reduction of state will ultimately reduce the required number of logic gates.

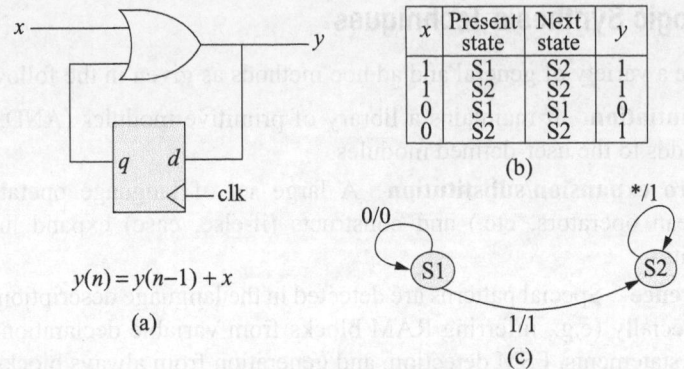

(b)

$$y(n) = y(n-1) + x$$

(a)

(c)

Fig. 9.16 FSM circuit models: (a) sequential logic circuit; (b) state table; (c) state transition diagram

Theorem *Two states of a finite state machine are equivalent if and only if, for any input, they have identical outputs, and the corresponding next states are equivalent.*

The state minimization process is described by the following steps.

Step 1 Compute the equivalent classes. This is done by iterative refinement of a partition of the state set. The collection of the states is called a *state set*. Based on the outputs, the state set is partitioned first. Let $\prod_i, i = 1, 2, ..., n_s$, be the partitions, where n_s is the number of states.

Step 2 Each partition in set $\prod_i$ contains states whose outputs are same for any input.

Step 3 Refine the partition blocks by further splittings so that all states in any block of $\prod_{i+1}$ have next states in the same block of $\prod_i$ for any possible input.

Step 4 Stop if the iteration converges. The iteration will converge if $\prod_{i+1} = \prod_i$ for some value of i.

The maximum number of iteration steps to converge is n_s.

Example 9.4 Let us consider the state diagram shown in Fig. 9.17. Reduce the number of states using the state minimization technique.

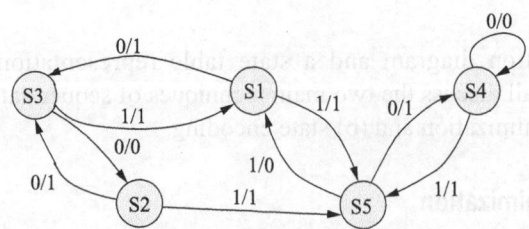

Fig. 9.17 State transition diagram

Solution The state table of the state diagram is shown in Table 9.8.

Table 9.8   State table

Input	Present state	Next state	Output
0	S1	S3	1
1	S1	S5	1
0	S2	S3	1
1	S2	S5	1
0	S3	S2	0
1	S3	S1	1
0	S4	S4	0
1	S4	S5	1
0	S5	S4	1
1	S5	S1	0

Step 1   The state set is partitioned first according to the outputs as

$$\Pi_1 = \{\{s_1, s_2\}, \{s_3, s_4\}, \{s_5\}\}$$

Step 2   Check each block of Π_1 to see if the corresponding next states are identical. In this case, the next states s_1 and s_2 match. But the next states of s_3 and s_4 are different. So, partition the $\{s_3, s_4\}$ block. Hence, the state is refined as

$$\Pi_2 = \{\{s_1, s_2\}, \{s_3\}, \{s_4\}, \{s_5\}\}$$

Step 3   Check the blocks again. In this case, no further refinement is possible. Hence, s_1 and s_2 are equivalent states and is combined as s_{12}. This is shown in Table 9.9.

Table 9.9   State table after minimization

Input	Present state	Next state	Output
0	S12	S3	1
1	S12	S5	1
0	S3	S12	0
1	S3	S12	1
0	S4	S4	0
1	S4	S5	1
0	S5	S4	1
1	S5	S12	0

The state diagram can be modified as shown in Fig. 9.18.

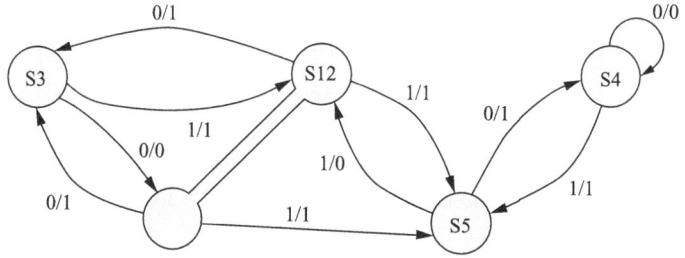

Fig. 9.18   State diagram after merging the states S1 and S2

Example 9.5 Consider the state diagram shown in Fig. 9.19. Reduce the number of states using the state minimization technique.

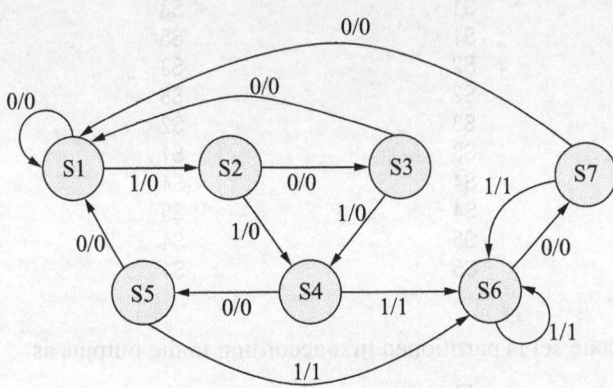

Fig. 9.19 State diagram

Solution For the given state diagram, the state table can be written as shown in Table 9.10.

Table 9.10 State table

Input	Present state	Next state	Output
0	S1	S1	0
1	S1	S2	0
0	S2	S3	0
1	S2	S4	0
0	S3	S1	0
1	S3	S4	0
0	S4	S5	0
1	S4	S6	1
0	S5	S1	0
1	S5	S6	1
0	S6	S7	0
1	S6	S6	1
0	S7	S1	0
1	S7	S6	1

According to the outputs, the first state partition set can be written as

$$\Pi_1 = \{\{S1, S2, S3\}, \{S4, S5, S6, S7\}\}$$

Now, checking each partition block to see if their next states are identical, we find that only states S5 and S7 have identical next states. Hence, partition set can be refined as

$$\Pi_2 = \{\{S1\}, \{S2\}, \{S3\}, \{S4\}, \{S5, S7\}, \{S6\}\}$$

Therefore, the states S5 and S7 are equivalent and can be combined to a single state S57. Substituting S5 and S7 by equivalent state S57, we can rewrite Table 9.10 as Table 9.11:

Table 9.11 Reduced state table

Input	Present state	Next state	Output
0	S1	S1	0
1	S1	S2	0
0	S2	S3	0
1	S2	S4	0
0	S3	S1	0 ·
1	S3	S4	0
0	S4	S57	0
1	S4	S6	1
0	S57	S1	0
1	S57	S6	1
0	S6	S57	0
1	S6	S6	1

Again, applying partition theory, we can write the first partition set as

$$\Pi_1 = \{\{S1, S2, S3\}, \{S4, S5, S6\}\}$$

Checking the next states of each of the partition block, we find that the states S4 and S6 have identical next states. Hence, partition set can be refined as

$$\Pi_2 = \{\{S1\}, \{S2\}, \{S3\}, \{S4, S6\}, \{S57\}\}$$

Therefore, the states S4 and S6 are equivalent and can be combined to a single state S46. Substituting S4 and S6 by equivalent state S46, we can rewrite Table 9.11 as Table 9.12:

Table 9.12 Reduced state table

Input	Present state	Next state	Output
0	S1	S1	0
1	S1	S2	0
0	S2	S3	0
1	S2	S46	0
0	S3	S1	0
1	S3	S46	0
0	S46	S57	0
1	S46	S46	1
0	S57	S1	0
1	S57	S46	1

Now, checking Table 9.12, we find there are no two states having identical next states as well as outputs. As no further state minimization is possible, the state diagram of Fig. 9.19 reduces to the state diagram as shown in Fig. 9.20.

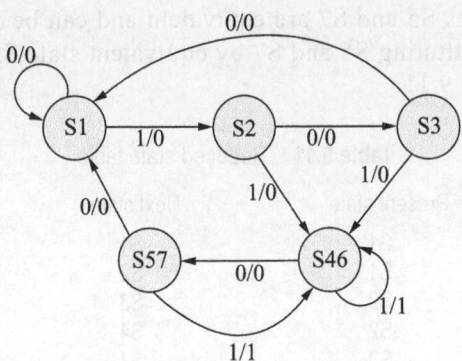

Fig. 9.20 Reduced state diagram

9.12.2 State Encoding

Determining the binary representations of the states of a finite state machine is known as the state encoding or state assignment problem. State encoding determines the size of the design and speed of the design. *Encoding length* is the number of bits required to represent the states.

The simplest encoding method is to encode a state by setting a corresponding bit to 1, and setting the remaining bits to 0. This is known as *1-hot* state encoding.

The minimum length codes uses $n_b = \log_2 n_s$ bits to represent each state, where n_s is the number of states. This code assigns states in the binary counting order.

Another encoding technique is to use the Gray code. Gray code has one advantage in that there is only one change required in going from one state to the next state.

The three different encoding techniques for state assignment of Example 9.5 are illustrated in Table 9.13.

Table 9.13 Different state encoding techniques

State	1-hot code	Binary code	Gray code
S1	00001	000	000
S2	00010	001	001
S3	00100	010	011
S46	01000	011	010
S57	10000	100	110

9.13 Building Blocks for Logic Design

Combinational logic circuits can be implemented using different basic building blocks, such as NAND and NOR structures, multiplexers, demultiplexers, decoders, and encoders. In the following subsections, we shall discuss logic design using multiplexers and decoders.

9.13.1 Design of Multilevel NAND Logic Circuits

The NAND gate is called universal gate as it can be used to implement any logic function. Figure 9.21 illustrates the implementation of NOT, AND, OR, and XOR gate by NAND gates.

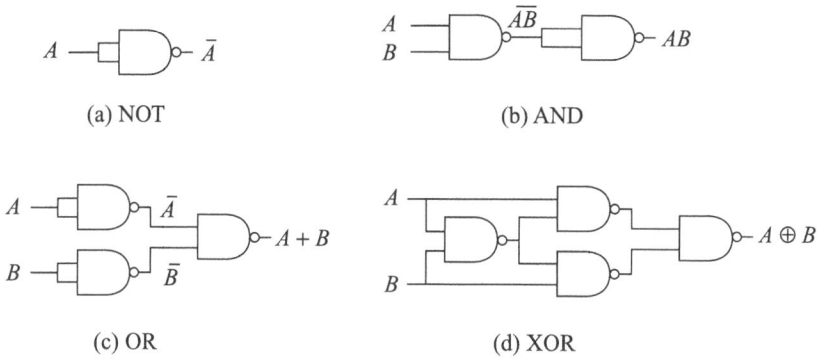

(a) NOT   (b) AND

(c) OR   (d) XOR

Fig. 9.21   Implementation of NOT, AND, OR, and XOR gate by NAND gates

9.13.2 Design of Multilevel NOR Logic Circuits

NOR gate is also called universal gate as it can be used to implement any logic function. Figure 9.22 illustrates the implementation of NOT, AND, OR, and XNOR gate by NOR gates.

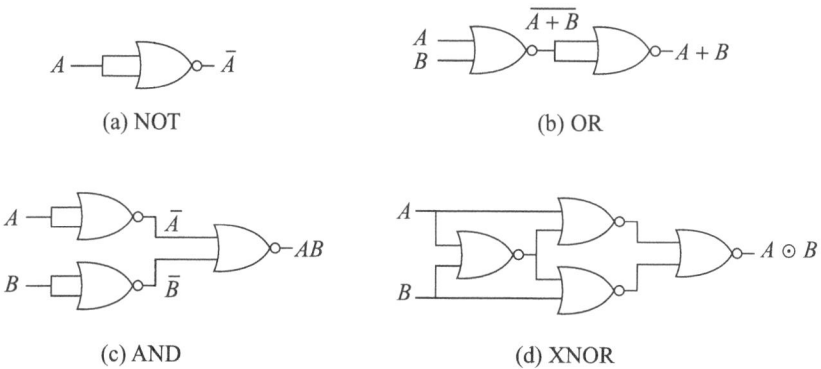

(a) NOT   (b) OR

(c) AND   (d) XNOR

Fig. 9.22   Implementation of NOT, AND, OR, and XNOR gate by NOR gates

9.13.3 Designing with Multiplexers

Any Boolean function can be realized using multiplexers. However, if a function is given in simplified form, it needs to be expanded to get all the minterms so that the expansion theorem can be applied. Depending on the number of input signals partitioned, there are several styles of design as follows:

1. *Type-0 design:* When no input signal is partitioned off or removed from the select lines.

2. *Type-1 design:* One input signal is removed from the selection lines.
3. *Type-2 design:* Two input signals are removed from the selection lines.
4. *Type-3 design:* Three input signals are removed from the selection lines.

Example 9.6 Design a circuit using a MUX to implement the following function by applying Shannon's expansion theorem with reference to variables A and B.

$$F(A,B,C) = A + B\overline{C}$$

Signal list: F, A, B, C.

Solution Applying Shannon's expansion theorem, we can write the function F as follows:

$$F(A,B,C) = F(0,0,C)\overline{A}\,\overline{B} + F(0,1,C)\overline{A}B + F(1,0,C)A\overline{B} + F(1,1,C)AB$$

Now, let us calculate values of the sub-functions.

$$F(0,0,C) = 0 \qquad\qquad F(0,1,C) = \overline{C}$$

$$F(1,0,C) = 1 \qquad\qquad F(1,1,C) = 1$$

The circuit is shown in Fig. 9.23.

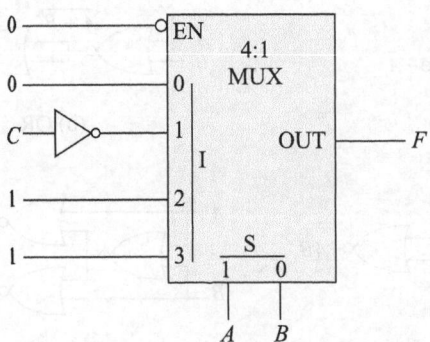

Fig. 9.23   Circuit diagram for the function
$$F(A,B,C) = A + B\overline{C}$$

Example 9.7 Design a circuit using a MUX to implement the following function by applying Shannon's expansion theorem:

$$F(A,B,C,D) = \sum m(4,5,6,7,10,14)$$

Signal list: F, A, B, C, D.

Solution

Type-0 Design

Let us generate the truth table for the given function as follows (Fig. 9.24):

Table 9.14 Truth table

A	B	C	D	F
0	0	0	0	0
0	0	0	1	0
0	0	1	0	0
0	0	1	1	0
0	1	0	0	1
0	1	0	1	1
0	1	1	0	1
0	1	1	1	1
1	0	0	0	0
1	0	0	1	0
1	0	1	0	1
1	0	1	1	0
1	1	0	0	0
1	1	0	1	0
1	1	1	0	1
1	1	1	1	0

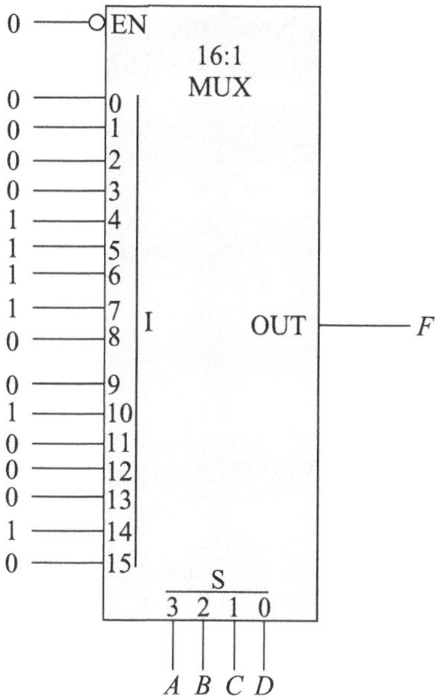

Fig. 9.24 Type-0 design of Example 9.7

Figure 9.24 shows the type-0 design of Example 9.7.

Type-1 Design

One signal is removed from the select lines (Fig. 9.25):

Table 9.15 Truth table

A	B	C	D	F	F(D)
0	0	0	0	0	0
0	0	0	1	0	
0	0	1	0	0	0
0	0	1	1	0	
0	1	0	0	1	1
0	1	0	1	1	
0	1	1	0	1	1
0	1	1	1	1	
1	0	0	0	0	0
1	0	0	1	0	
1	0	1	0	1	$\bar{D}$
1	0	1	1	0	
1	1	0	0	0	0
1	1	0	1	0	
1	1	1	0	1	$\bar{D}$
1	1	1	1	0	

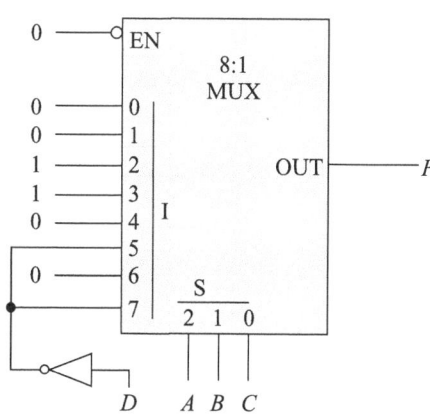

Fig. 9.25 Type-1 design of Example 9.7

Figure 9.25 shows the type-1 design of Example 9.7.

Type-1 Design: Alternative Approach

Let us choose C to be removed from the select lines. This is done by writing C in the last column, and the function values are rewritten according to the *ABDC* combination (see Fig. 9.26).

Table 9.16   Truth table

A	B	D	C	F	F(C)
0	0	0	0	0	0
0	0	0	1	0	
0	0	1	0	0	0
0	0	1	1	0	
0	1	0	0	1	1
0	1	0	1	1	
0	1	1	0	1	1
0	1	1	1	1	
1	0	0	0	0	C
1	0	0	1	1	
1	0	1	0	0	0
1	0	1	1	0	
1	1	0	0	0	C
1	1	0	1	1	
1	1	1	0	0	0
1	1	1	1	0	

The corresponding type-1 design is shown in Fig. 9.26.

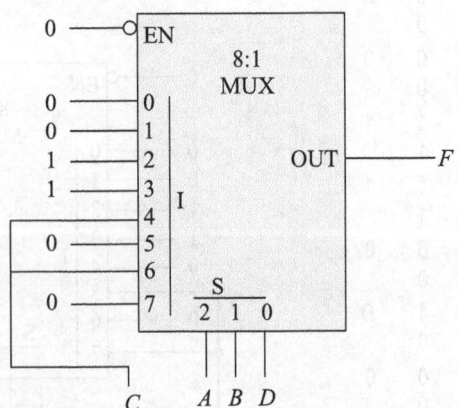

Fig. 9.26   Type-1 design (*C* is chosen for data line) of Example 9.7

Type-2 Design

Two signals are removed from the select lines (Fig. 9.27).

Table 9.17   Truth table

A	B	C	D	F	F(C,D)
0	0	0	0	0	
0	0	0	1	0	
0	0	1	0	0	0
0	0	1	1	0	
0	1	0	0	1	
0	1	0	1	1	
0	1	1	0	1	1
0	1	1	1	1	
1	0	0	0	0	
1	0	0	1	0	$C\bar{D}$
1	0	1	0	1	
1	0	1	1	0	
1	1	0	0	0	
1	1	0	1	0	$C\bar{D}$
1	1	1	0	1	
1	1	1	1	0	

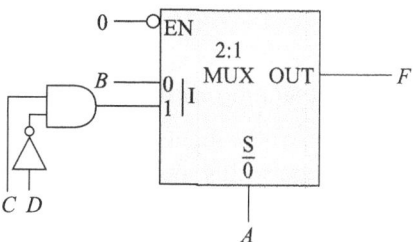

Fig. 9.27   Type-2 design of Example 9.7

Figure 9.27 shows the type-2 design of Example 9.7.

Type-3 Design

Three signals are removed from the select lines. Figure 9.28 shows the type-3 design of Example 9.7.

As more and more signals are partitioned off, the size of the multiplexer is reduced at the cost of some other logic gates. So, depending on the functions we

Table 9.18   Truth table

A	B	C	D	F	F(B,C,D)
0	0	0	0	0	
0	0	0	1	0	
0	0	1	0	0	
0	0	1	1	0	B
0	1	0	0	1	
0	1	0	1	1	
0	1	1	0	1	
0	1	1	1	1	
1	0	0	0	0	
1	0	0	1	0	
1	0	1	0	1	
1	0	1	1	0	$C\bar{D}$
1	1	0	0	0	
1	1	0	1	0	
1	1	1	0	1	
1	1	1	1	0	

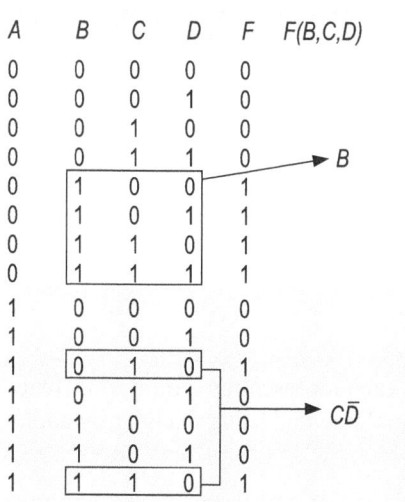

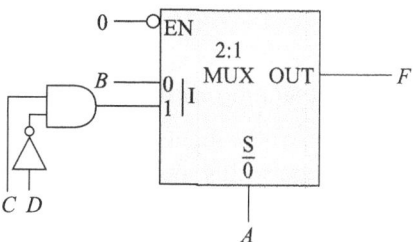

Fig. 9.28   Type-3 design of Example 9.7

are to choose the type of the design so that total area is reduced. If multiple output functions are to be designed, we require multiple multiplexers. Statistically, it is found that type-1 design is the most preferred one.

Example 9.8 Design a 3-bit odd-parity generator using 4:1 multiplexer.

Solution The truth table of a 3-bit odd-parity generator is shown in Table 9.19.

Table 9.19 Truth table of 3-bit odd-parity generator

A	B	C	P	P(c)
0	0	0	1	$\overline{C}$
0	0	1	0	
0	1	0	0	C
0	1	1	1	
1	0	0	0	C
1	0	1	1	
1	1	0	1	$\overline{C}$
1	1	1	0	

Considering A and B as select lines, we can write the following combinations:
- When $A = 0, B = 0, P = \overline{C}$
- When $A = 0, B = 1, P = C$
- When $A = 1, B = 0, P = C$
- When $A = 1, B = 1, P = \overline{C}$

Hence, the 3-bit odd-parity generator can be implemented using a 4:1 multiplexer, as shown in Fig. 9.29.

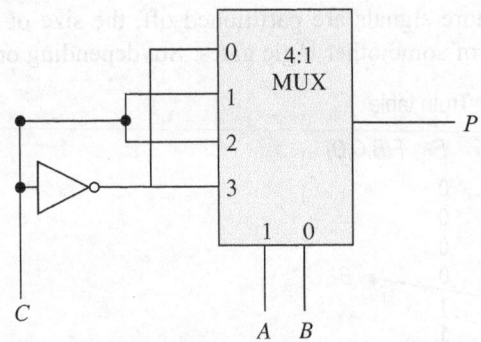

Fig. 9.29 Odd-parity generator using 4:1 multiplexer

9.13.4 Designing with Decoders

A decoder with n inputs produces 2^n outputs each representing a minterm. Hence, any function with n input variables can be implemented using decoder by adding OR gates to its outputs.

Example 9.9 Design a full subtractor circuit with decoder.

Solution   A full subtractor has three inputs x, y, and z and two outputs, difference (D) and borrow (B). The outputs are expressed in the SOP form as

$$D = \sum(1, 2, 4, 7)$$
$$B = \sum(1, 2, 3, 7)$$

The difference and borrow can be implemented by ORing the required minterms, as shown in Fig. 9.30.

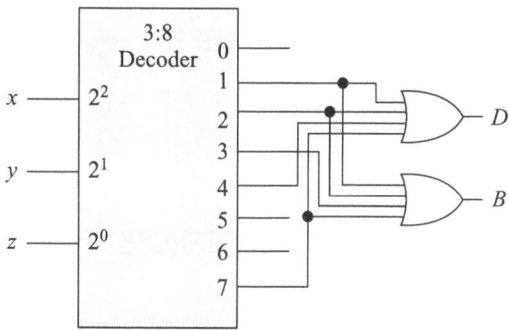

Fig. 9.30   Full subtractor designed with a decoder and two OR gates

Example 9.10   Implement the following Boolean functions using a decoder:

$$F_1(A, B, C) = \sum m(0, 1, 3)$$
$$F_2(A, B, C) = \sum m(1, 2, 3, 5, 7)$$
$$F_3(A, B, C) = \sum m(0, 2, 3, 6)$$

Solution   The solution is illustrated in Fig. 9.31. The function F_1 is implemented by ORing minterms m_0, m_1, and m_3 using the 3-input OR gate. The function F_2 is implemented by ORing minterms m_1, m_2, m_3, m_5, and m_7 using the 5-input OR gate. The function F_3 is implemented by ORing minterms m_0, m_2, m_3, and m_6.

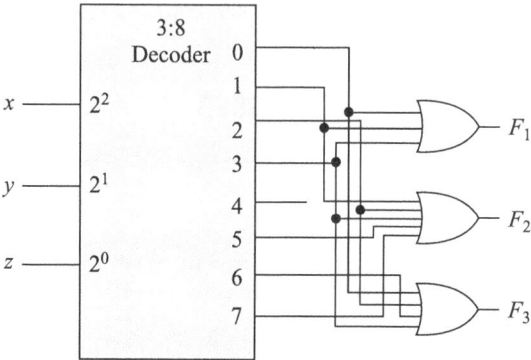

Fig. 9.31   Decoder implementing functions F_1, F_2, and F_3

SUMMARY

- Logic synthesis is a process of translating behavioural or RTL netlist into gate-level netlist using a set of design constraints, and standard library cells.
- Synthesis can be performed at different levels of abstraction such as, transistor-level, gate-level, logic-level, block-level, and architecture-level.
- Logic synthesis goals are to reduce chip area, power dissipation, and propagation delay.
- Binary decision diagrams are a useful way of representation of logic circuits, and are used in computer programs for describing netlists.
- State minimization and state encoding are the two main steps of sequential logic synthesis.
- A canonical form is a unique form of Boolean expression, either expressed using minterms or maxterms.

SELECT REFERENCES

Bhasker, J. 2003, *A VHDL Primer*, Pearson Education.

Brown, S. and Z. Vranesic 2002, *Fundamentals of Digital Logic with VHDL Design*, Tata McGraw-Hill, New Delhi.

DeMicheli, G.D. 2003, *Synthesis and Optimization of Digital Circuits*, Tata McGraw-Hill, New Delhi.

Hayes, J.P. 1998, *Computer Architecture and Organization*, 3rd ed., McGraw-Hill International Editions.

Gerez, S.H. 2007, *Algorithms for VLSI Design Automation*, Wiley.

Hatchel, G.D.H. 1998, *Logic Synthesis and Verification Algorithms*, Kluwer.

Kang, S.M. and Y. Leblebici 2003, *CMOS Digital Integrated Circuits: Analysis and Design*, 3rd ed., Tata McGraw-Hill, New Delhi.

Mano, M.M. 2001, *Computer System Architecture*, Prentice-Hall, New Delhi.

Mano, M.M. 2001, *Digital Logic and Computer Design*, Prentice-Hall, New Delhi.

Martin, K. 2004, *Digital Integrated Circuit Design*, Oxford University Press.

Rabaey, J.M., A. Chandrakasan, and B. Nikolic 2008, *Digital Integrated Circuits: A Design Perspective*, 2nd ed., Pearson Education.

Smith, M.J.S. 2002, *Application Specific Integrated Circuits*, Pearson Education.

Taub, H. and D. Schilling 1977, *Digital Integrated Electronics*, McGraw-Hill.

Weste, N.H.E., D. Harris, and A. Banerjee 2009, *CMOS VLSI Design: A Circuits and Systems Perspective*, 3rd ed., Pearson Education.

EXERCISES

Fill in the Blanks

1. For 1-hot state encoding _____ .
   (a) $n_b = n_s$ (b) $n_b = \log_2 n_s$
   (c) $n_b = n_s!$ (d) $n_b = (n_s - 1)!$
2. With n_b as the encoding length, and n_s as the number of states, the number of possible encodings are _____ .
   (a) $2^{nb}!/\left(2^{nb} - n_s\right)!$ (b) $(2^{nb} - 1)!/ (2^{nb} - n_s)!n_b!$
   (c) 2^{nb} (d) $2^{nb!}$

3. The complexity of the equivalency checking algorithm is _____ .
   - (a) $O(n_s^2)$
   - (b) $O(n_s \log n_s)$
   - (c) $O(n_s)$
   - (d) $O(n_s!)$
4. The number of transistors required to implement the function $(a \times b + c)$, using NAND, NOR, and inverter gates, is _____ .
   - (a) 12
   - (b) 14
   - (c) 16
   - (d) 10
5. Design constraints specifies _____ .
   - (a) maximum chip area
   - (b) minimum clock frequency
   - (c) maximum power dissipation
   - (d) all of these

Multiple Choice Questions

1. The goals of VLSI logic synthesis are
   - (a) Minimization of area, increase operating speed, and reduce power consumption
   - (b) Minimize power consumption, decrease propagation delay, and reduce area
   - (c) Minimize power consumption, increase propagation delay, and reduce area
   - (d) Minimization of area, reduce operating speed, and reduce power consumption
2. In the VLSI logic design process, we can
   - (a) minimize both area and delay
   - (b) minimize area at the cost of delay
   - (c) maximize speed by decreasing area
   - (d) minimize delay by reducing area
3. According to consensus theorem
   - (a) $xy + x'z + yz = xy + x'z$
   - (b) $(x + y)(x' + z)(y + z) = (x + y)(x' + z)$
   - (c) both (a) and (b)
   - (d) none of these
4. A Boolean space consists of
   - (a) ON-set
   - (b) OFF-set
   - (c) DC-set
   - (d) all of these
5. Two states of a finite state machine are equivalent
   - (a) if and only if, for any input, they have identical outputs, and the corresponding next states are equivalent.
   - (b) if and only if, for any input, they have identical outputs.
   - (c) if and only if, for any input, corresponding next states are equivalent.
   - (d) if and only if they have identical outputs, and the corresponding next states are equivalent.

True or False

1. A canonical form of Boolean expression is a unique form.
2. A prime implicant of function f is an implicant that is not included in any other implicant of f.
3. Binary decision diagram (BDD) is canonical.
4. In Gray code, there is only one change from one state to the next state.
5. In Type-2 MUX-based design, two input signals are removed from the selection lines.

Short-answer Type Questions

1. What is Shannon's expansion theorem? Design a 2-input OR-gate using 2:1 MUX with the help of this theorem.
2. What is logic synthesis and what are the main goals of the logic synthesis?
3. Explain the technology-independent cost model with suitable example.
4. What is technology mapping? Explain with suitable example.
5. What is a finite state machine (FSM)? How is it synthesized?
6. Draw the state diagram for the following circuit:

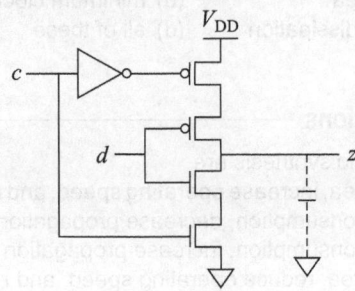

7. What is canonical form? Express the following Boolean function in minterm canonical form:

$$f = x_1x_2 + a\bar{x}_2$$

8. What do you mean by satisfiability don't care and observability don't care?
9. What is an implicant and what is a prime implicant? Give examples.
10. Find out the prime implicants of the following Boolean expression:

$$f = yz' + xy'z$$
$$d = x'z$$

11. What is Distance-1 merging theorem. Define a complete sum.
12. Find out the complete sum of the following expression:

$$x_1x_2 + \bar{x}_2x_3 + x_2x_3x_4$$

13. What is a binary decision diagram (BDD)? What are its properties?
14. Draw the BDD of the following circuit:

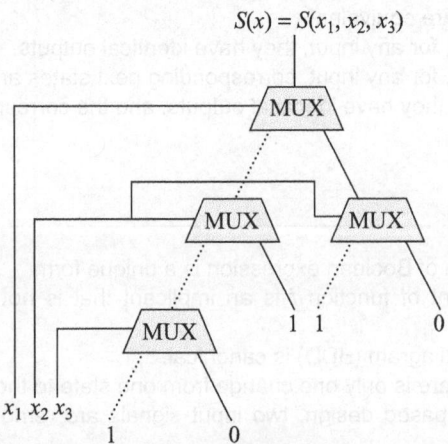

15. Draw the BDD of the following Boolean equation:

$$f = abc + \bar{b}d + \bar{c}d$$

16. What is a synchronous logic network? Explain with suitable circuit diagram.
17. Explain the state minimization technique with a suitable example.
18. Explain the state encoding problem with a suitable example.
19. Discuss the FSM design flow with necessary diagrams.
20. Prove or disprove the following identities:
    (a) $x' \oplus (x+y) = x+y'$
    (b) $(x+y)(x+yz) = x+yz$
    (c) $ab+b'cd'+acd' = ab+b'cd'$
    (d) $xy+x(y+z) = x(y+z)$
    (e) $ab+b'cd'+acd' = ab+cd'$
21. Using the consensus theorem, simplify the following:
    (a) $wxy+wx'z+wyz$
    (b) $wxy'+x'yz+wz$
    (c) $vw'y+vyz+wyz$
    (d) $xyz+wx'+wy'+wz$
22. Draw the BDD for the following function:
    $$f = abd'+ab'd+a'c+a'c'd$$

Long-answer Type Questions

1. Minimize the following Boolean expression using tabular method. Find out the prime implicants.
$$f = \bar{x} \times \bar{y} + wxy + \bar{x} \times y \times \bar{z} + w \times \bar{y} \times z$$

2. What is consensus theorem? Explain with examples. Simplify the following expressions using this theorem:
   (a) $wxy+wx'z+wyz$
   (b) $wxy'+x'yz+wz$
   (c) $vw'y+vyz+wyz$
   (d) $xyz+wx'+wy'+wz$

3. For the following formula:
$$xy+x'y'z+xy'z'$$
Do the following:
   (a) Write an equivalent POS formula.
   (b) Write a SOP formula for the complement of the function.
   (c) Compute a complete sum using the tabular method.
   (d) Compute a complete sum using the POS that you found.

4. Explain the state minimization for completely specified finite state machines. State the equivalency checking algorithm. Minimize the states for the following state machine using this algorithm.

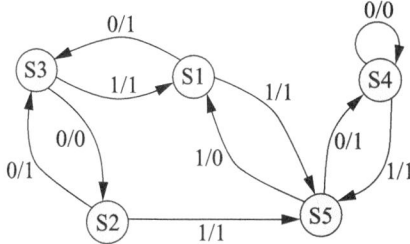

5. What is state encoding or assignment problem? Discuss the state encoding problem for a two-level circuit.

6. For the formula $xyz + x'y + y'z$, do the following:
   (a) Write an equivalent POS formula.
   (b) Write a SOP formula for the complement of the function.
   (c) Compute a complete sum using the tabular method.
   (d) Compute a complete sum using iterated consensus.
   (e) Compute a complete sum using the POS that you found.
   (f) Compute a complete sum using the recursive multiplication procedure.
7. Design a 3-bit even-parity generator using 4:1 multiplexer.
8. Design a full-adder using a 3:8 decoder.
9. Minimize the states for the following state machine.

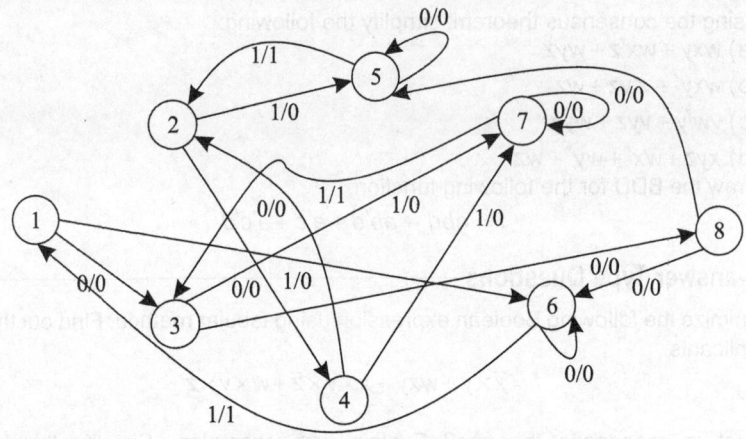

10. Implement the following expression using a single 4:1 multiplexer.

$$Y(A,B,C) = \sum m(0,1,4,6,7)$$

Timing Analysis

10.1 Introduction

Timing analysis is a significant part of the whole VLSI design process. If a designer is asked about the important things while designing a VLSI chip, the answer would be: (a) meeting the functionality and (b) meeting the timing specification. Meeting the functionality means that the chip should do what it is intended to do. For example, if a design is supposed to multiply two 4-bit numbers, it should perform the same without any error. But what does meeting the timing specification mean? This means how fast the multiplication job is done. In other words, if the design is supposed to multiply two 4-bit numbers in 10 ps, it should exactly do the same. The time required to perform any operation is mainly determined by two factors: (a) the speed of the logic gates and interconnects and (b) the algorithm for doing the operation. The first part is mainly determined by the speed of the process technology, whereas in the second part, chip designer has to come up with design techniques that enable speedy operation.

10.2 Delay in Any System

Any system, when it operates, has a finite delay in producing output for a given input. The delay is inherent to all systems, whether analog or digital. The inherent delay is generally called the time constant of the system. For example, consider a person who suddenly touching a hot body, will withdraw his hand quickly. This is called 'reflex action' in terms of biology. But the time to withdraw the hand will vary from person to person. This response time is basically the time constant of the person. Similarly, for any mechanical system, the machine needs a finite time to respond to a given input. A large response time indicates a slow system. Conversely, a small response time indicates a fast system. The time constant cannot be made zero, it can only be reduced to a small value by changing the system parameters. For any electrical circuit, there is also a response time. For example, an *RC* circuit will take a finite amount of time to charge the capacitor, and this time is the *RC* time constant of the circuit.

10.3 Delay in VLSI Circuits

VLSI circuits also suffer from the delay problem. A typical VLSI circuit is intended to perform certain basic operations, such as addition, subtraction, multiplication, division, or even more complex operations like computation of discrete Fourier transform (DFT), and convolution, etc. Now these operations will take a finite amount of time. The delay of the circuit is the time delay between the input and the output. The delay is again very much algorithm-dependent and implementation-dependent. For example, direct computation of DFT is much slower than the DFT computation using FFT algorithm. This is an example of how an algorithm can reduce the processing time. Again, consider addition of two binary numbers. A bit-serial adder is much slower than the bit-parallel adder. This is an example of how implementation can reduce delay in operations. In this chapter, we discuss the delay of VLSI circuits from the architecture or implementation point of view.

In all VLSI circuits, the input signals pass through the circuit and reach the output. While passing through the circuit, the signals traverse through two components: (a) devices and (b) interconnects.

The devices have a finite amount of delay in passing the input signals to the output. In CMOS circuits, the devices are the pMOS and nMOS transistors. An nMOS or pMOS transistor takes some time to respond because of its parasitic resistances and capacitances. The response time of a logic gate is called the *gate delay*.

The interconnects are nothing but wires to transmit signals from one point to another point in a circuit. But an interconnect is not a simple wire, where we cannot ignore the transmission delay. The interconnects have parasitic resistances and capacitances which cause the signals to take certain amount of time to reach the destination from the source. This time is called the *interconnect delay*. Hence, we can write delay of a circuit as follows:

$$\text{Delay} = \sum \text{Gate delay} + \sum \text{Interconnect delay}$$

10.4 Delay in CMOS Inverter

Let us consider a CMOS inverter circuit to analyse delay. Figure 10.1(a) shows a CMOS inverter circuit.

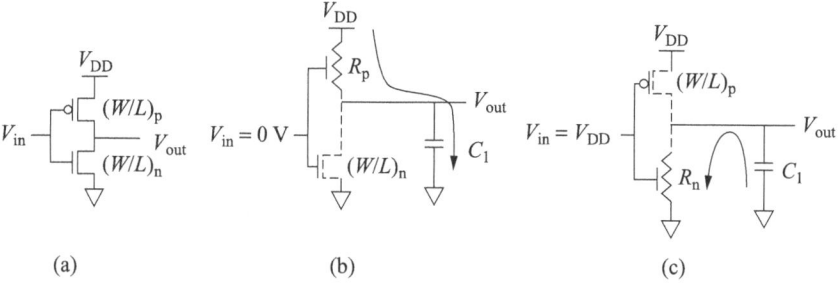

(a) (b) (c)

Fig. 10.1   (a) CMOS inverter circuit; (b) equivalent circuit of inverter at low-to-high transition; (c) equivalent circuit of inverter at high-to-low transition

As shown in Fig. 10.2, in the input and output waveforms of the CMOS inverter, there are time delays between the input and output waveforms. A very simple model for this delay is analysed assuming an equivalent circuit of CMOS inverter as shown in Figs 10.1(b) and (c). During the input high-to-low transition, the nMOS transistor becomes OFF, and the pMOS transistor becomes ON. We model the ON pMOS transistor by a voltage-controlled resistor R_p, which is given by

$$R_p = \frac{1}{\mu_p C_{ox}(V_{GS} - |V_{tp}| - V_{DS})} \times \left(\frac{L}{W}\right)_p \qquad (10.1)$$

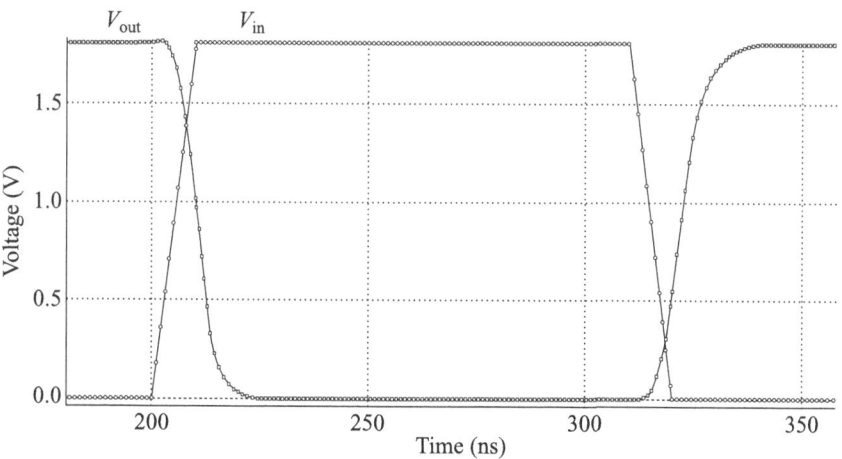

Fig. 10.2   CMOS inverter—input and output waveforms

Similarly, during input low-to-high transition, the nMOS transistor becomes ON, and the PMOS transistor becomes OFF. We model the ON nMOS transistor by a voltage-controlled resistor R_n, which is given by

$$R_n = \frac{1}{\mu_n C_{ox}(V_{GS} - V_{tn} - V_{DS})} \times \left(\frac{L}{W}\right)_n \qquad (10.2)$$

Case 1 **Input makes high-to-low transition**

When the input makes high-to-low transition, the load capacitor C_1 charges through the pMOS transistor as shown in Fig. 10.1(b).

During the charging of the load capacitor, the output voltage is related to the input voltage according to the relation

$$V_{\text{out}} = V_{\text{in}} \times (1 - e^{-t/R_p C_1}) \tag{10.3}$$

Let us now calculate the delay between the input and the output waveforms considering a 50% trip point. Let t_1 be the time at which $V_{\text{out}} = V_{\text{in}}/2$, then we can write

$$V_{\text{out}} = V_{\text{in}}/2 = V_{\text{in}} \times (1 - e^{-t_1/R_p C_1}) \tag{10.4}$$

or

$$\frac{1}{2} = 1 - e^{-t_1/R_p C_1}$$

or

$$e^{t_1/R_p C_1} = 2$$

or

$$t_1 = \ln(2) \times R_p C_1 = 0.69 \times R_p C_1 \tag{10.5}$$

Equation (10.5) gives the expression for time delay for the output low-to-high (TPLH) transition.

Let us now calculate the rise time of the output waveform which is defined as time difference between 20% and 80% trip points.

Let t_2 and t_3 be the times at which output waveform is at 20% and 80% trip points, respectively; then we can write

$$V_{\text{out}} = V_{\text{in}} \times 0.2 = V_{\text{in}} \times (1 - e^{-t_2/R_p C_1}) \tag{10.6}$$

and

$$V_{\text{out}} = V_{\text{in}} \times 0.8 = V_{\text{in}} \times (1 - e^{-t_3/R_p C_1}) \tag{10.7}$$

Simplifying Eqns (10.6) and (10.7), we get

$$t_{\text{rise}} = t_3 - t_2 = R_p C_1 \times \ln\left[\frac{V_{\text{in}} - 0.2 \times V_{\text{in}}}{V_{\text{in}} - 0.8 \times V_{\text{in}}}\right] = R_p C_1 \times \ln(4) = 1.3863 \times R_p C_1 \tag{10.8}$$

Equation (10.8) gives the expression for the rise time of the output waveform.

Case 2 **Input makes low-to-high transition**

When the input makes low-to-high transition, the load capacitor C_1 discharges through the nMOS transistor as shown in Fig. 10.1(c).

During the discharging of the load capacitor, the output voltage is related to the input voltage according to the relation

$$V_{\text{out}} = V_{\text{in}} \times e^{-t/R_n C_1} \tag{10.9}$$

Let us now calculate the delay between the input and the output waveforms considering a 50% trip point. Let t_4 be the time at which $V_{\text{out}} = V_{\text{in}}/2$, then we can write

$$V_{\text{out}} = V_{\text{in}}/2 = V_{\text{in}} \times e^{-t_4/R_n C_1} \tag{10.10}$$

or

$$e^{t_4/R_n C_1} = 2$$

or

$$t_4 = \ln(2) \times R_n C_1 = 0.69 \times R_n C_1 \tag{10.11}$$

Equation (10.11) gives the expression for time delay for the output high-to-low (TPLH) transition.

Let us now calculate the fall time of the output waveform which is defined as the time difference between 80% and 20% trip points.

Let t_5 and t_6 be the times at which output waveform is at 80% and 20% trip points, respectively; then we can write

$$V_{\text{out}} = V_{\text{in}} \times 0.8 = V_{\text{in}} \times e^{-t_6/R_n C_1} \qquad (10.12)$$

and

$$V_{\text{out}} = V_{\text{in}} \times 0.2 = V_{\text{in}} \times e^{-t_6/R_n C_1} \qquad (10.13)$$

Simplifying Eqns (10.12) and (10.13), we get

$$t_{\text{fall}} = t_6 - t_5 = R_n C_1 \times \ln\left[\frac{0.8}{0.2}\right] = R_n C_1 \times \ln(4) = 1.3863 \times R_n C_1 \qquad (10.14)$$

Equation (10.14) gives the expression for the fall time of the output waveform.

10.5 Slew Balancing

The rise and fall times of the input/output waveforms are often called *slew*. The rise time is called rise slew, and the fall time is called fall slew. According to Eqns (10.8) and (10.14), the rise and fall times are functions of the load capacitor and the pMOS and nMOS transistor's equivalent resistances, respectively. Making the rise and fall times equal is known as *slew balancing*. For perfect slew balancing, the equivalent resistances of the pMOS and nMOS transistors should be equal. Equating $R_p = R_n$, we get from Eqns (10.1) and (10.2),

$$R_p = \frac{1}{\mu_p C_{\text{ox}}(V_{\text{GS}} - |V_{\text{tp}}| - V_{\text{DS}})} \times \left(\frac{L}{W}\right)_p = R_n = \frac{1}{\mu_n C_{\text{ox}}(V_{\text{GS}} - V_{\text{tn}} - V_{\text{DS}})} \times \left(\frac{L}{W}\right)_n$$

or

$$\frac{\mu_n C_{\text{ox}}(V_{\text{GS}} - V_{\text{tn}} - V_{\text{DS}})}{\mu_p C_{\text{ox}}(V_{\text{GS}} - |V_{\text{tp}}| - V_{\text{DS}})} = \frac{(W/L)_p}{(W/L)_n} \qquad (10.15)$$

In CMOS inverter

$$\left.\begin{array}{l} V_{\text{GS}} = V_{\text{in}} - V_{\text{DD}} \\ V_{\text{DS}} = V_{\text{out}} - V_{\text{DD}} \end{array}\right\} \text{ for pMOS transistor}$$

and

$$\left.\begin{array}{l} V_{\text{GS}} = V_{\text{in}} \\ V_{\text{DS}} = V_{\text{out}} \end{array}\right\} \text{ for nMOS transistor}$$

The threshold voltage of pMOS and nMOS transistors are equal, i.e., $V_{\text{tn}} = |V_{\text{tp}}|$. Hence, we get from Eqn (10.15)

$$\frac{(W/L)_p}{(W/L)_n} = \frac{\mu_n}{\mu_p} \qquad (10.16)$$

The ratio of (*W/L*) of pMOS and nMOS transistors is the ratio of electron and hole mobility. Typically, $(\mu_n/\mu_p) \approx 2.5$ and hence, the (*W/L*) ratio of pMOS transistor is 2.5 times the (*W/L*) ratio of nMOS transistor in a CMOS inverter.

For a particular technology, the channel length is same for both pMOS and nMOS transistors. The designers must not change the channel length of the PMOS and nMOS transistors. In order to make (*W/L*) ratio of pMOS transistor 2.5 times than that of nMOS transistor, the channel width of the pMOS transistor is kept 2.5 times larger than the channel width of the nMOS transistor.

10.6 Transistor Equivalency

Two or more transistors can be represented by a single transistor which is equivalent to the combinations of transistors connected in series or parallel.

Case 1 **Two transistors connected in parallel**

Let us consider two transistors of identical channel length (*L*) connected in parallel, as shown in Fig. 10.3. Let the widths of the transistors be W_1 and W_2.

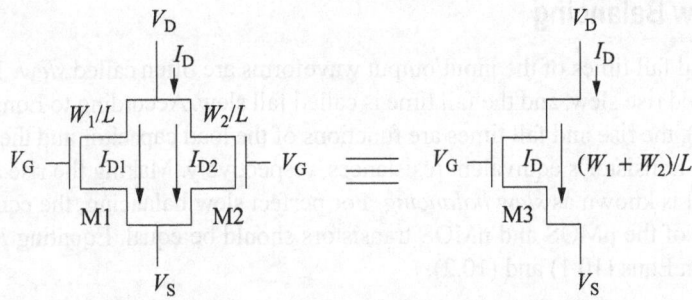

Fig. 10.3   Two nMOS transistors connected in parallel and their equivalent transistor

Let I_{D1} and I_{D2} be the currents through the nMOS transistors M1 and M2. Then we can write

$$I_{D1} = \frac{\mu_n C_{ox}}{2} \times \frac{W_1}{L} \left[2(V_{GS} - V_{tn})V_{DS} - V_{DS}^2 \right] \tag{10.17}$$

and

$$I_{D2} = \frac{\mu_n C_{ox}}{2} \times \frac{W_2}{L} \left[2(V_{GS} - V_{tn})V_{DS} - V_{DS}^2 \right] \tag{10.18}$$

The total current I_D is the sum of I_{D1} and I_{D2}, and is written by

$$I_D = I_{D1} + I_{D2} = \frac{\mu_n C_{ox}}{2} \times \frac{W_1}{L} \left[2(V_{GS} - V_{tn})V_{DS} - V_{DS}^2 \right]$$

$$+ \frac{\mu_n C_{ox}}{2} \times \frac{W_1}{L} \left[2(V_{GS} - V_{tn})V_{DS} - V_{DS}^2 \right]$$

$$= \frac{\mu_n C_{ox}}{2} \times \frac{(W_1 + W_2)}{L} \left[2(V_{GS} - V_{tn})V_{DS} - V_{DS}^2 \right]$$

$$= \frac{\mu_n C_{ox}}{2} \times \frac{W_{eq}}{L} \left[2(V_{GS} - V_{tn})V_{DS} - V_{DS}^2 \right] \tag{10.19}$$

We can see from Eqn (10.19) that the current I_D, which is the sum of currents I_{D1} and I_{D2} can be obtained by an equivalent nMOS transistor having channel width

$$W_{eq} = W_1 + W_2 \tag{10.20}$$

Hence, two transistors having identical channel length connected in parallel are equivalent to a single transistor of channel width equal to the sum of the widths of the transistors.

Case 2 **Two transistors connected in series**

Let us consider two transistors of identical channel length (L) connected in series, as shown in Fig. 10.4. Let the widths of the transistors be W_1 and W_2.

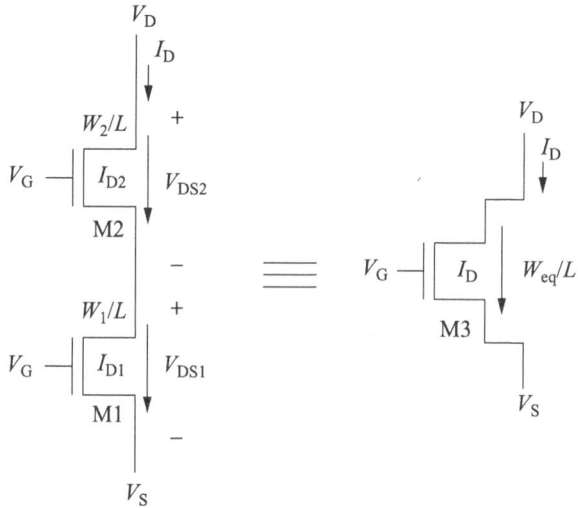

Fig. 10.4   Two nMOS transistors connected in series and their equivalent transistor

If I_{D1} and I_{D2} be the currents through the nMOS transistors M1 and M2, we can write

$$I_{D1} = \frac{\mu_n C_{ox}}{2} \times \frac{W_1}{L}\left[2(V_{GS1} - V_{tn})V_{DS1} - V_{DS1}^2\right] \tag{10.21}$$

and

$$I_{D2} = \frac{\mu_n C_{ox}}{2} \times \frac{W_2}{L}\left[2(V_{GS2} - V_{tn})V_{DS2} - V_{DS2}^2\right] \tag{10.22}$$

But the transistors M1 and M2 are carrying the same current as they are in series. So,

$$I_{D1} = I_{D2} = I_D \tag{10.23}$$

We can write

$$V_{GS2} = V_{GS1} - V_{DS1} \tag{10.24}$$

and

$$V_{DS2} = V_{DS} - V_{DS1} \tag{10.25}$$

where $V_{DS} = V_D - V_S$.

Substituting Eqns (10.21) and (10.22) in Eqn (10.23), we get

$$\frac{\mu_n C_{ox}}{2} \times \frac{W_1}{L} \left[2(V_{GS1} - V_{tn}) V_{DS1} - V_{DS1}^2 \right]$$

$$= \frac{\mu_n C_{ox}}{2} \times \frac{W_2}{L} \left[2(V_{GS2} - V_{tn}) V_{DS2} - V_{DS2}^2 \right] \quad (10.26)$$

Now, substituting Eqns (10.24) and (10.25) in Eqn (10.26), we get

$$\frac{\mu_n C_{ox}}{2} \times \frac{W_1}{L} \left[2(V_{GS1} - V_{tn}) V_{DS1} - V_{DS1}^2 \right]$$

$$= \frac{\mu_n C_{ox}}{2} \times \frac{W_2}{L} \left[2(V_{GS1} - V_{DS1} - V_{tn})(V_{DS} - V_{DS1}) - (V_{DS} - V_{DS1})^2 \right] \quad (10.27)$$

Simplifying Eqn (10.27), we can write

$$\frac{\mu_n C_{ox}}{2} \times \frac{W_1}{L} \left[2(V_{GS1} - V_{tn}) V_{DS1} - V_{DS1}^2 \right]$$

$$= \frac{\mu_n C_{ox}}{2} \times \frac{W_2}{L} \left[2(V_{GS1} - V_{tn}) V_{DS} - 2(V_{GS1} - V_{tn}) V_{DS1} - V_{DS}^2 + V_{DS1}^2 \right] \quad (10.28)$$

Rearranging Eqn (10.28), we get

$$\frac{\mu_n C_{ox}}{2} \times \frac{(W_1 + W_2)}{L} \left[2(V_{GS1} - V_{tn}) V_{DS1} - V_{DS1}^2 \right]$$

$$= \frac{\mu_n C_{ox}}{2} \times \frac{W_2}{L} \left[2(V_{GS1} - V_{tn}) V_{DS} - V_{DS}^2 \right] \quad (10.29)$$

Let the channel width of the equivalent transistor M3 be W_{eq}; then we can write the current through M3 as

$$I_D = \frac{\mu_n C_{ox}}{2} \times \frac{W_{eq}}{L} \left[2(V_{GS} - V_{tn}) V_{DS} - V_{DS}^2 \right] \quad (10.30)$$

Substituting Eqns (10.21) and (10.30) in Eqn (10.29), we get

$$\frac{(W_1 + W_2)}{W_1} I_{D1} = \frac{W_2}{W_{eq}} I_D \quad (10.31)$$

Combining Eqns (10.23) and (10.31), we get

$$W_{eq} = \frac{W_1 W_2}{W_1 + W_2} = \frac{1}{\dfrac{1}{W_1} + \dfrac{1}{W_2}} \quad (10.32)$$

Hence, two transistors having identical channel length connected in series are equivalent to a single transistor of channel width (W_{eq}) and equal to the reciprocal of sum of the reciprocal of the widths of the transistors.

10.7 Case Study: Effect of Transistor Size on Propagation Delay

Let us now consider a CMOS inverter circuit to study the effect of the transistor size on the propagation delay of the circuit. To study this effect, we create a SPICE netlist as shown below:

```
*SPICE netlist to study the effect of Transistor size on
*propagation delay.

.include "E:\CMOS-CELLS\SPICE MODELS\model.txt"
.option scale=0.18u

.param CL=1p
.param N=2

C1 y Gnd 'CL'
M2 y a Gnd Gnd NMOS L=1 W='2*N' AD='5*2*N' PD='2*2*N+10'
+ AS='5*2*N' PS='2*2*N+10'
M3 y a Vdd Vdd PMOS L=1 W='5*N' AD='5*5*N' PD='2*5*N+10'
+ AS='5*5*N' PS='2*5*N+10'
v4 Vdd Gnd 1.8
v5 a Gnd pulse(0.0 1.8 0 10n 10n 100n 200n)
.tran 1n 500n

.plot tran v(a) v(y)

.measure tran TPHL trig v(a) val=0.9 cross=1
+ targ v(y) val=0.9 cross=1
.measure tran TPLH trig v(a) val=0.9 cross=2
+ targ v(y) val=0.9 cross=2

.measure tran Trise trig v(y) val=1.62 cross=1
+ targ v(y) val=0.18 cross=1
.measure tran Tfall trig v(y) val=0.18 cross=2
+ targ v(y) val=1.62 cross=2
.step lin param CL 1p 5p 1p

.end
```

First, to study the effect of the load capacitor on the propagation delay, we vary the load capacitor from 1 pF to 5 pF with step 1 pF. Figure 10.5 illustrates the effect of load capacitance on the delay.

As we can see from Fig. 10.5, the output waveform is getting delayed as the load capacitance is increased. This indicates that the same inverter circuit becomes slower as the load increases. Now this increase in delay can be compensated by reducing the effective resistance of the transistors, i.e., increasing the channel width of the transistors. We obtain the delay and slew values for different transistor widths as depicted in Table 10.1.

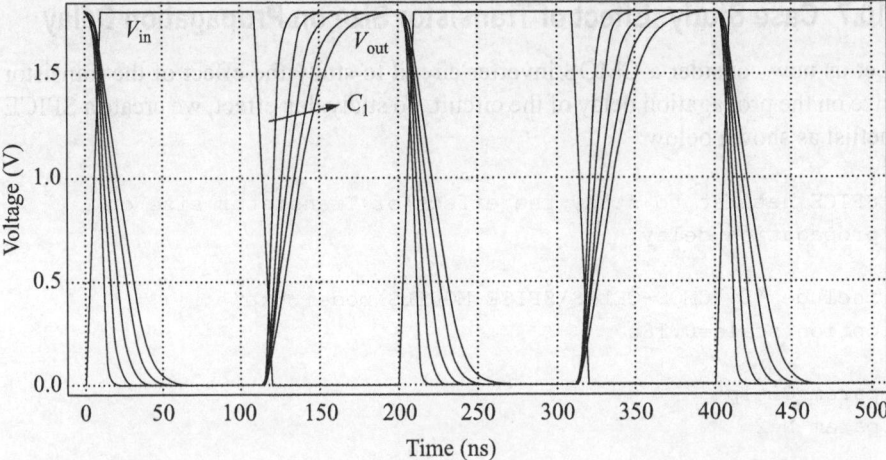

Fig. 10.5   Input and output waveforms for different values of load capacitance

Table 10.1   Typical delay and slew values of the CMOS inverter for different *W/L* ratio and load capacitance for 0.18 μm technology

W/L ratio	C_1 (F)	TPHL (s)	TPLH (s)	Fall time (s)	Rise time (s)
$(W/L)_n = 2$ $(W/L)_p = 5$	1.0e–12	**4.9564e–09**	5.8481e–09	10.3167e–09	8.5038e–09
	2.0e–12	8.1697e–09	**9.4286e–09**	1.2911e–08	1.5580e–08
	3.0e–12	1.1376e–08	1.3020e–08	**1.8748e–08**	2.2978e–08
	4.0e–12	1.4607e–08	1.6621e–08	2.4755e–08	**3.0535e–08**
	5.0e–12	1.7835e–08	2.0232e–08	3.0877e–08	3.8166e–08
$(W/L)_n = 4$ $(W/L)_p = 10$	1.0e–12	**3.5863e–09**	4.1126e–09	5.0699e–09	5.3488e–09
	2.0e–12	5.6378e–09	**6.0968e–09**	8.1584e–09	8.9350e–09
	3.0e–12	10.5613e–09	8.0287e–09	**1.1445e–08**	1.2729e–08
	4.0e–12	9.4986e–09	9.9423e–09	1.4848e–08	**1.6598e–08**
	5.0e–12	1.1432e–08	1.1876e–08	1.8322e–08	2.0547e–08
$(W/L)_n = 6$ $(W/L)_p = 12$	1.0e–12	**2.8726e–09**	3.3307e–09	4.1981e–09	4.2528e–09
	2.0e–12	4.5431e–09	**4.8866e–09**	6.3349e–09	6.6215e–09
	3.0e–12	5.9258e–09	6.1954e–09	**8.6238e–09**	9.0952e–09
	4.0e–12	10.3001e–09	10.5091e–09	1.0969e–08	**1.1670e–08**
	5.0e–12	8.6712e–09	8.8206e–09	1.3356e–08	1.4287e–08

As we can see in Table 10.1, the delay value increases as the load capacitance increases. Again, the delay value decreases as the (*W/L*) ratio increases (look at the numbers column-wise in **bold**).

10.8   Design of Two-input NAND Gate for Equal Rise and Fall Slew

The two-input NAND gate is realized in CMOS logic as shown in Fig. 10.6(b). The procedure of design is discussed in Chapter 6. In this section, we discuss the procedure of selecting the transistor dimensions.

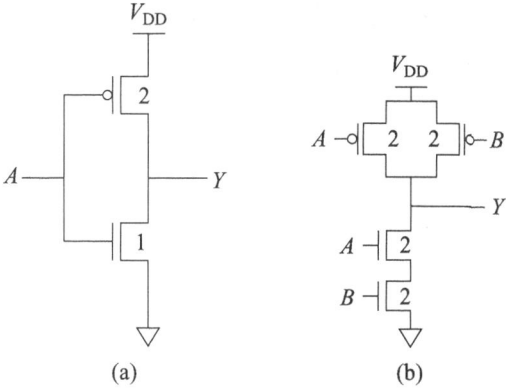

Fig. 10.6 (a) CMOS inverter; (b) CMOS two-input NAND gate

To design any complex logic CMOS circuit, we must make the effective resistance of the PUN and the PDN equal. For example, in the case of the CMOS inverter as shown in Fig. 10.6(a), the ratio of the (W/L) of pMOS and nMOS transistors is taken as 2:1. For simplicity, we have assumed mobility ratio of electron and hole as 2.0. We also assume that the channel lengths of all the transistors in the CMOS design are equal.

Compared to the inverter circuit, the PDN of the NAND circuit is to have W/L ratio as 1. Similarly, the PUN of the NAND circuit is to have W/L ratio as 2. In the PDN, the nMOS transistors are in series, hence their equivalent W/L ratio must be 1. Let us assume that W/L ratio of each of the nMOS transistor in PDN is x, then we can write from Eqn (10.32),

$$\frac{1}{\dfrac{1}{x} + \dfrac{1}{x}} = 1 \tag{10.33}$$

Simplifying Eqn (10.33), we get

$$x = 2 \tag{10.34}$$

Hence, the W/L ratio of the nMOS transistors in the PDN is 2, as shown in Fig. 10.6(b).

In the PUN of the NAND gate, two pMOS transistors are in parallel. Hence, their equivalent W/L ratio can be obtained from Eqn (10.20). As compared to the inverter circuit, the PUN should have W/L ratio as 2. Hence, if the width of each pMOS transistor is y, using Eqn (10.20), we can write

$$y + y = 2 \quad \text{or} \quad y = 1 \tag{10.35}$$

Hence, the W/L ratio of the pMOS transistors would be 1. But two pMOS transistors do not always operate simultaneously. So if we choose the W/L ratio to be 1, then the PUN resistance would be twice than the desired equivalent resistance when only one pMOS operates. Hence, for parallel transistors, we consider each transistor separately for selecting the W/L ratio. In the NAND circuit the W/L ratio of the PMOS transistors will be 2, as shown in Fig. 10.6(b).

10.9 Design of Two-input NOR Gate for Equal Rise and Fall Slew

The two-input NOR gate is realized in CMOS logic as shown in Figs 10.7(a) and (b). In the NOR gate, the PUN has two pMOS transistors in series. Hence, their W/L ratio must be double than that of the pMOS in the CMOS inverter. In PDN, the nMOS transistors are in parallel, so their W/L ratio would be same as that of nMOS in the CMOS inverter. Therefore, the W/L ratio of the pMOS transistors will be 4 and nMOS transistors will be 1, as illustrated in Fig. 10.7(b).

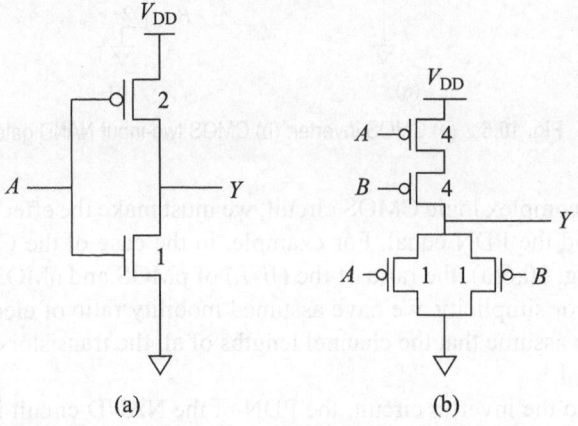

(a)   (b)

Fig. 10.7   (a) CMOS inverter, (b) CMOS two-input NOR gate

Example 10.1   Design a three-input NAND gate with transistor widths such that the effective rise and fall resistance are equal to that of an inverter.

Solution The three-input NAND gate is expressed by the Boolean expression

$$Y = \overline{ABC} \tag{10.36}$$

The PDN will consists of three nMOS transistors in series, and the PUN will consist of three pMOS transistors in parallel as shown in Figs 10.8(a) and (b).

As the three nMOS transistors are connected in series, their W/L ratio will be 3 obtained from Eqn (10.32).

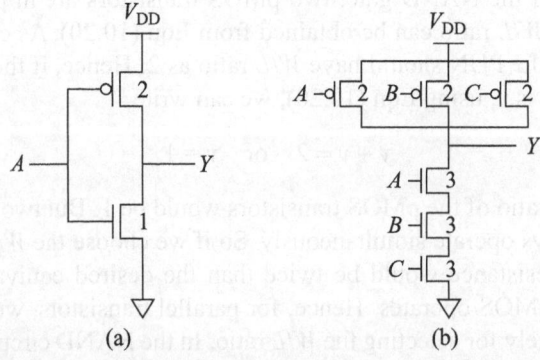

(a)   (b)

Fig. 10.8   (a) CMOS inverter; (b) CMOS three-input NAND gate

Example 10.2 Design an AOI gate with transistor widths, such that the effective rise and fall resistances are equal to that of an inverter

$$Y = \overline{AB + C} \qquad (10.37)$$

Solution The AOI gate is realized as shown in Figs 10.9(a) and (b).

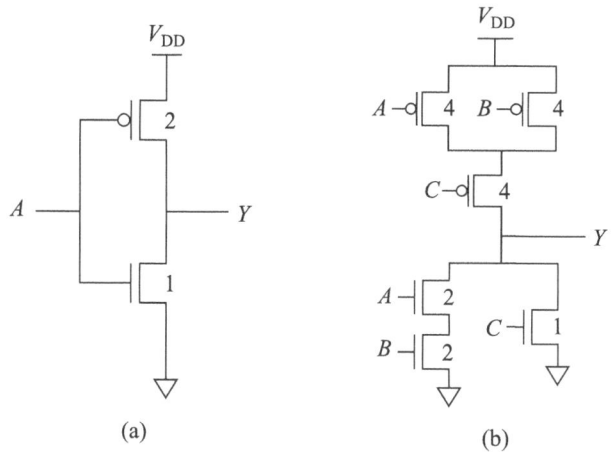

(a)   (b)

Fig. 10.9 (a) CMOS inverter; (b) CMOS AOI gate realizing function $Y = \overline{AB + C}$

10.10 MOS Capacitances

The propagation delay of a gate is analysed using a simple but fairly accurate RC delay model, where R is the effective resistance of the MOS transistors, and C is the load capacitance. In Section 10.9, we have analysed the effective resistance of the MOS transistors to design the logic functions with appropriate transistor W/L ratios. Let us now analyse the capacitances in the circuit. There are many components of the load capacitance that contribute to the propagation delay. The main three components of the load capacitance are as follows:

- Gate capacitance
- Junction capacitance
- Interconnect capacitance

10.10.1 Gate Capacitance

There are three components of the gate capacitance with respect to the other three terminals of the MOSFET:

- Gate-to-source capacitance (C_{gs})
- Gate-to-drain capacitance (C_{gd})
- Gate-to-bulk capacitance (C_{gb})

There are also gate-source and gate-drain overlap capacitances. As shown in Figs 10.10(a) and (b), the gate terminal has some overlap at the source and drain end. This overlap is required for the device to be fabricated properly considering the fabrication system mask alignment tolerance. But because of the overlap, there

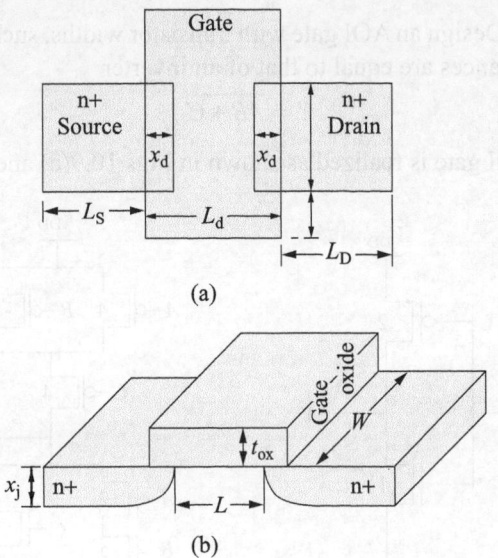

Fig. 10.10 MOSFET: (a) top view; (b) cross-sectional view

are two capacitances: gate-source overlap capacitance (C_{GSO}) and gate-drain overlap capacitance (C_{GDO}). The value of the overlap capacitance is given by

$$C_{GSO} = C_{GDO} = C_{ox} \times W \times x_d \quad (10.38)$$

where

$$C_{ox} = \frac{\varepsilon_{ox}}{t_{ox}} \quad (10.39)$$

C_{ox} = gate oxide capacitance per unit area; ε_{ox} = dielectric constant of the gate oxide material; t_{ox} = gate oxide thickness; W = channel width; and x_d = gate-source/drain overlap length.

The overlap capacitances are independent of the terminal voltages and are fixed for transistor dimensions.

When the MOS transistor operates under *cut-off region*, there is no channel region. Hence, the gate-to-source and gate-to-drain capacitances are zero, i.e., $C_{gs} = C_{gd} = 0$. Hence, the gate capacitance is entirely determined by the gate-to-bulk capacitance as given by

$$C_{gb} = C_{ox} \times W \times L \quad (10.40)$$

In the *linear region* of operation, the channel is formed and it shields the bulk from the gate. Hence the gate-to-bulk capacitance is zero, i.e., $C_{gb} = 0$. The gate-to-channel capacitance is shared equally by the gate-to-source and gate-to-drain capacitances. Therefore, we can write

$$C_{gs} = C_{gd} \cong \frac{1}{2} W \times L \times C_{ox} \quad (10.41)$$

When the MOS transistor operates under the *saturation region*, the channel is pinched off at the drain end. Hence, the gate-to-drain capacitance is zero ($C_{gd} = 0$). The gate-to-bulk capacitance is also zero ($C_{gb} = 0$) under this condition. The gate capacitance is entirely determined by the gate-to-source capacitance which is given by

$$C_{gs} \cong \frac{2}{3} W \times L \times C_{ox} \tag{10.42}$$

Hence, the gate capacitance and its components can be summarized as shown in Table 10.2.

Table 10.2 Gate capacitances of the MOS transistor

Region of operation	Cut-off	Linear	Saturation
C_{gb}	$C_{ox} \times W \times L$	0	0
C_{gs}	0	$\frac{1}{2} W \times L \times C_{ox}$	$\frac{2}{3} W \times L \times C_{ox}$
C_{gd}	0	$\frac{1}{2} W \times L \times C_{ox}$	0
C_G (overlap)	$C_{ox} \times W \times x_d$	$C_{ox} \times W \times x_d$	$C_{ox} \times W \times x_d$
C_G (total)	$C_{ox} \times W \times L + 2 \times C_{ox} \times W \times x_d$	$C_{ox} \times W \times L + 2 \times C_{ox} \times W \times x_d$	$\frac{2}{3} C_{ox} \times W \times L + 2 \times C_{ox} \times W \times x_d$

10.10.2 Junction Capacitance

The junction capacitances are also called *diffusion capacitances*. In the MOS transistor, there are two PN junctions between the source and bulk, and the drain and bulk. These PN junctions have depletion layer capacitances. The depletion layer capacitance again has two components. One is due to the bottom wall, and other is due to the side wall, as illustrated in Fig. 10.11.

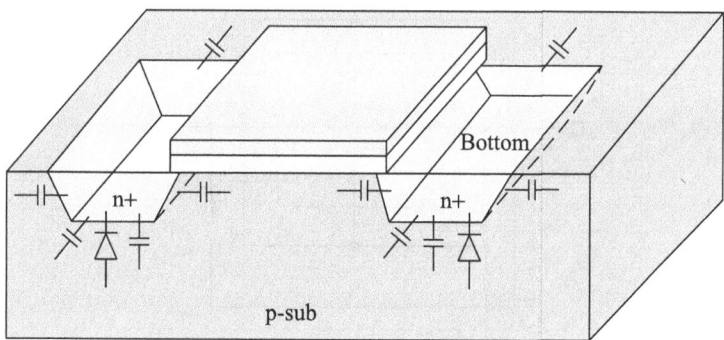

Fig. 10.11 MOSFET junction capacitances

The vertical junctions form capacitances with the bottom plate of the source and the drain regions which is given by

$$C_{jb} = C_j \times W \times L_S = C_j \times W \times L_D \tag{10.43}$$

where C_j is the vertical junction capacitance per unit area.

The lateral junctions also form capacitances with the side walls of the source and drain regions which is given by

$$C_{jsw} = C_{js}x_j \times (W + 2 \times L_S) = C_{js}x_j \times (W + 2 \times L_D) \tag{10.44}$$

where C_{js} is the side-wall junction capacitance per unit length.

Hence, the total junction capacitance is the sum of the vertical junction capacitance and the side-wall junction capacitance. We can write

$$
\begin{aligned}
C_{S\text{-junc}} = C_{S\text{-diff}} &= C_{jb} \times \text{Area} + C_{jsw} \times \text{Perimeter} \\
&= C_{jb} \times W \times L_S + C_{jsw} \times (2 \times L_S + W)
\end{aligned}
\tag{10.45}
$$

Equation (10.45) gives expression for the junction capacitance at the source end. A similar expression can be written for the junction capacitance at the drain end just by replacing L_S with L_D. Note that the side wall at the gate side is not considered for side-wall junction capacitance as this side is not a PN junction, but rather a conducting channel.

10.10.3 MOS Transistor Capacitance

The four-terminal MOS transistor capacitance model is shown in Fig. 10.12.

The gate capacitances and the junction capacitances are combined to form the lumped capacitances between the four terminals of the MOS transistors. These capacitances are given by

$$
\begin{aligned}
C_{GS} &= C_{gs} + C_{GSO} \\
C_{GD} &= C_{gd} + C_{GDO} \\
C_{GB} &= C_{gb} \\
C_{SB} &= C_{S\text{-diff}} \\
C_{DB} &= C_{D\text{-diff}}
\end{aligned}
\tag{10.46}
$$

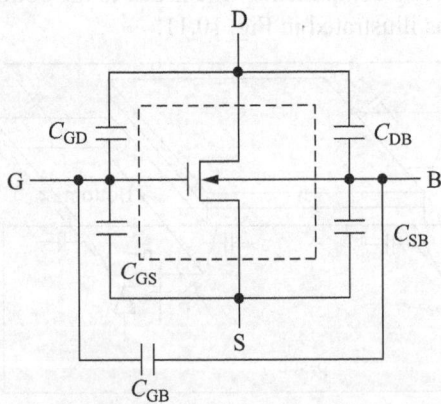

Fig. 10.12 MOS transistor capacitances

10.10.4 Effective Load Capacitance

The propagation delay of a logic gate is determined by the two factors: (1) the effective resistance in the charging or discharging path and (2) the effective value of load capacitance. Let us now consider a CMOS inverter driving another CMOS logic gate, as shown in Fig. 10.13.

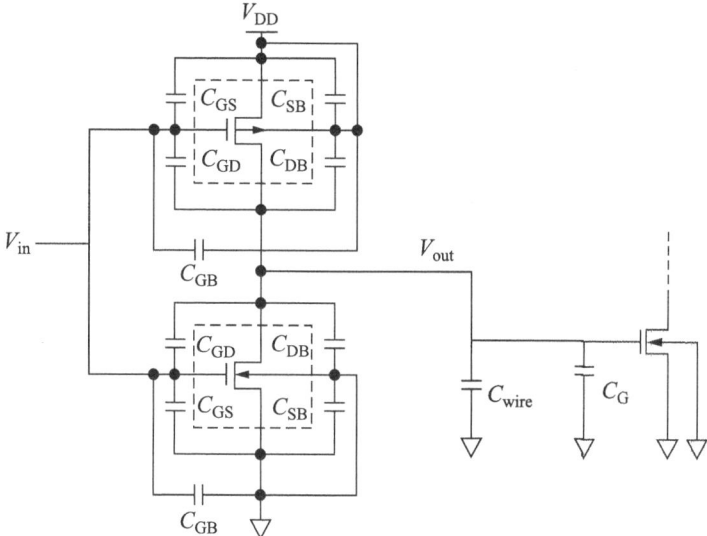

Fig. 10.13   A CMOS inverter driving another CMOS logic

The effective load capacitance at the output terminal of the CMOS inverter is given by

$$C_l = C_{GD,NMOS} + C_{GD,PMOS} + C_{DB,NMOS} + C_{DB,PMOS} + C_{wire} + C_G \qquad (10.47)$$

where C_{wire} and C_G are interconnect capacitance and the gate capacitance of the nMOS connected at the output of the CMOS inverter.

Equation (10.47) shows that the effective load capacitance at the output of a CMOS logic gate depends on the gate-to-drain and the drain-to-bulk capacitances. These capacitances in turn depend on the width of the transistors according to Eqns (10.41) and (10.45). So, when we increase the width of the transistor to reduce the propagation delay through the gate, it increases the effective load capacitance. Hence, continuously increasing the width of the transistors does not necessarily reduce the propagation delay through the gate.

10.11 Dependency of Delay and Slew on Power Supply Voltage

The propagation delay and slew values depend on the effective resistances and capacitances. But the effective resistance has a dependency on the voltage according to Eqn (10.1). To study the effect of V_{DD} on the propagation delay and slew, we create a SPICE netlist and vary V_{DD}. The SPICE netlist is given below:

```
* SPICE netlist to check propagation delay and slew
* variation with VDD

.include "E:\CMOS-CELLS\SPICE MODELS\model.txt"
.option scale=0.18u

.param CL=1p
.param N=1
```

```
.param VDD=2.0
C1 y Gnd 'CL'
M2 y a Gnd Gnd NMOS L=1 W='2*N' AD='5*2*N'
+ PD='2*2*N+10' AS='5*2*N' PS='2*2*N+10'
M3 y a Vdd Vdd PMOS L=1 W='6*N' AD='5*6*N'
+ PD='2*6*N+10' AS='5*6*N' PS='2*6*N+10'
v4 Vdd Gnd 'VDD'
v5 a Gnd pulse(0.0 'VDD' 0 10n 10n 100n 200n)

.tran 1n 250n
.plot tran v(a) v(y)

.measure tran TPHL trig v(a) val='VDD/2'
+ cross=1 targ v(y) val='VDD/2' cross=1
.measure tran TPLH trig v(a) val='VDD/2'
+ cross=2 targ v(y) val='VDD/2' cross=2

.measure tran Tfall trig v(y) val='0.9*VDD'
+ cross=1 targ v(y) val='0.1*VDD' cross=1
.measure tran Trise trig v(y) val='0.1*VDD'
+ cross=2 targ v(y) val='0.9*VDD' cross=2

.step lin param VDD 1.0 2.8 0.1

.end
```

The propagation delay and slew values are shown in Table 10.3.

The variation of propagation delay and slew with V_{DD} is shown in Figs 10.14 and 10.15.

Table 10.3   Propagation delay and slew values for different values of V_{DD}

V_{DD} (V)	TPHL (s)	TPLH (s)	Fall time (s)	Rise time (s)
1.00	9.0201e−09	1.1653e−08	1.2024e−08	1.7124e−08
1.10	10.9081e−09	9.7116e−09	1.0469e−08	1.3909e−08
1.20	10.1253e−09	8.4669e−09	9.4729e−09	1.1939e−08
1.30	6.5333e−09	10.5510e−09	8.7531e−09	1.0599e−08
1.40	6.0756e−09	6.8577e−09	8.2444e−09	9.6230e−09
1.50	5.7363e−09	6.3585e−09	10.9983e−09	8.9378e−09
1.60	5.4283e−09	5.9277e−09	10.7450e−09	8.3464e−09
1.70	5.1787e−09	5.5756e−09	10.5133e−09	10.8425e−09
1.80	4.9660e−09	5.2995e−09	10.3061e−09	10.4471e−09
1.90	4.7946e−09	5.0620e−09	10.1130e−09	10.1436e−09
2.00	4.6423e−09	4.8604e−09	6.9395e−09	6.8745e−09
2.10	4.5118e−09	4.7436e−09	6.7834e−09	6.7592e−09
2.20	4.3983e−09	4.5874e−09	6.6450e−09	6.5966e−09
2.30	4.2963e−09	4.4477e−09	6.5474e−09	6.4495e−09
2.40	4.2643e−09	4.3204e−09	6.5142e−09	6.3219e−09
2.50	4.1863e−09	4.2064e−09	6.4269e−09	6.2119e−09
2.60	4.1143e−09	4.1002e−09	6.3529e−09	6.1026e−09
2.70	4.0529e−09	4.0026e−09	6.2865e−09	5.9965e−09
2.80	3.9997e−09	3.9130e−09	6.2628e−09	5.9165e−09

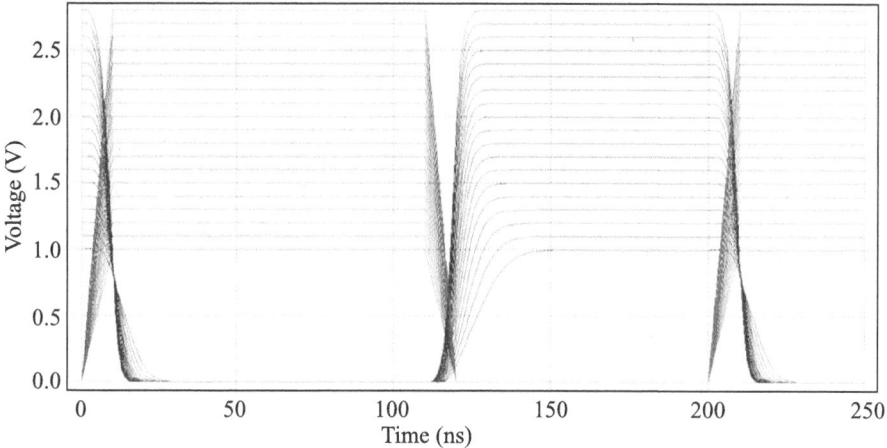

Fig. 10.14   Variation of delay and slew values for different values of V_{DD}

10.12 Design Techniques for Delay Reduction

Designers have different choices to reduce the propagation delay of the logic gates. The main approaches are:

- *Reduce the effective load capacitance*   As the effective load capacitance partly depends on the dimension of the MOS transistors, designers can design the MOS transistors such that the effective load capacitance reduces. Another important method is to reduce interconnect capacitance to reduce effective load capacitance. This is normally done at the technology level by using low-k dielectric material for interconnect separation.

- *Increase the W/L ratio of the MOS transistors*   Increasing *W/L* ratio is one useful technique to reduce the propagation delay of the MOS transistors. However, arbitrarily increasing *W/L* ratio causes the effective load capacitance to increase, and hence increase the propagation delay.

- *Increase the value of power supply voltage*   As explained in Fig. 10.15, the delay is reduced by increasing the power supply voltage. But again, there are

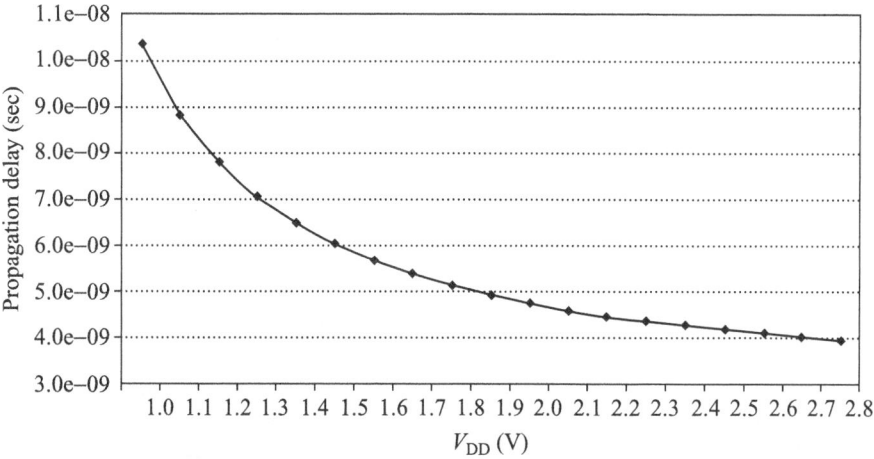

Fig. 10.15   Variation of propagation delay values for different values of V_{DD}

limitations in increasing the power supply voltage from the reliability point of view. The gate oxide breakdown and channel hot electron effects limit the upper limit of the power supply voltage.

An example of delay dependency on power supply voltage is to fix the IR drop problem. If a manufactured chip malfunctions, then one suspect is the IR drop problem. It is often checked by increasing the power supply voltage. After increasing the power supply voltage, if the chip works properly, then it is concluded that the chip suffers from IR drop problem.

10.13 Intrinsic Delay of Inverter

Let us consider a CMOS inverter circuit with equal rise and fall time. This is achieved by choosing the proper W/L ratio of the pMOS and nMOS transistors. The load capacitance of the inverter (C_{load}) can be assumed to be the sum of two capacitances: (a) the internal capacitance ($C_{internal}$) which is due to the gate-to-drain (Miller) capacitance and the diffusion capacitance; and (b) the external capacitance ($C_{external}$) which is due to the interconnect capacitance and input capacitance of the next stage. Hence, we can write

$$C_{load} = C_{internal} + C_{external} \tag{10.48}$$

Let us assume that the effective resistance for charging and discharging of this load capacitor as R_{eq}; we can then write the propagation delay through the gate as

$$\tau_p = 0.69 R_{eq} C_{load} = 0.69 R_{eq} (C_{internal} + C_{external}) \tag{10.49}$$

Rearranging Eqn (10.49), we can write

$$\tau_p = \tau_{p0} \left(1 + \frac{C_{external}}{C_{internal}} \right) \tag{10.50}$$

where $\tau_{p0} = 0.69 R_{eq} C_{internal}$ is called the intrinsic delay or delay without any load. The intrinsic delay is due to the self loading of the inverter.

10.14 Inverter Sizing Effect on Propagation Delay

In this section we will investigate the sizing effect of the inverter on the propagation delay. Let us assume that an 1x drive inverter or minimum size inverter has equivalent resistance R_{1x} and internal capacitance C_{1x}. Then we can write the propagation delay of the 1x drive inverter as

$$\tau_{p1x} = 0.69 R_{1x} (C_{1x} + C_{external}) = \tau_{p0} \left(1 + \frac{C_{external}}{C_{1x}} \right) \tag{10.51}$$

Let us scale up the transistors in the 1x drive inverter by 2 to design a 2x drive inverter. Scaling up a transistor basically means the width is scaled up a factor. The length of the transistors is not changed. Then the 2x drive inverter will have equivalent resistance $R_{2x} = 0.5 R_{1x}$ according to Eqn (10.2). The internal capacitance of the 2x drive inverter will be $C_{2x} = 2 \times C_{1x}$ as capacitance is directly proportional to the width of the transistor. If the 2x drive inverter drives the same external load then we can write the propagation delay as

$$\tau_{p1x} = 0.69\, R_{2x}(C_{2x} + C_{external})$$
$$= 0.69 \times 0.5\, R_{1x} \times (2 \times C_{1x} + C_{external})$$
$$= \tau_{p0}\left(1 + \frac{C_{external}}{2 \times C_{1x}}\right) \qquad (10.52)$$

It is observed from Eqns (10.51) and (10.52) that the intrinsic delay remains constant, and the extrinsic delay is scaled down by a factor of 2 between the 1x drive and 2x drive inverters. In general, if the scaling factor is S, then we can write the propagation delay as

$$\tau_{p} = \tau_{p0}\left(1 + \frac{C_{external}}{S \times C_{internal-1x}}\right) \qquad (10.53)$$

The normalized delay (τ_{p}/τ_{p0}) is shown in Fig. 10.16. It is clear from Fig. 10.16 that the extrinsic delay is reduced with increasing the transistors widths. But the reduction is not very sharp for large transistor widths. For example, at $S = 10$, the normalized delay is 1.2, whereas at $S = 40$, the normalized delay is 1.05. It indicates for huge increase in transistor width (4 times), the delay reduction is only a small fraction. Hence, increasing the transistor widths arbitrarily does not reduce the propagation delay. Designers needs to choose an optimum value of transistor width to trade off between the delay and chip area.

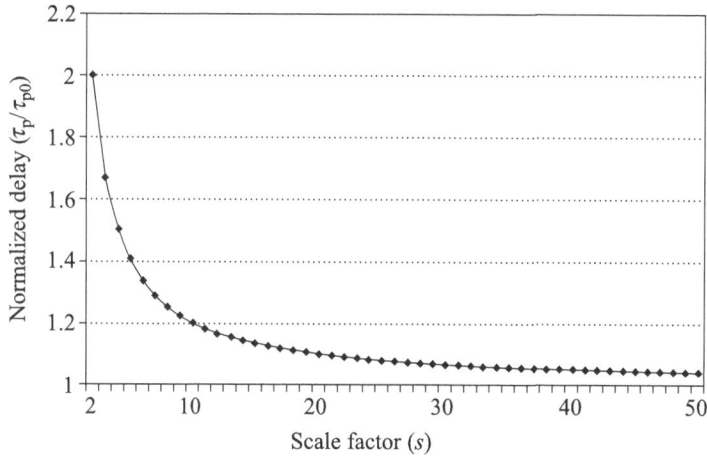

Fig. 10.16 Normalized delay of the inverter with increasing size with scale factor S

10.15 Inverter Chain Design

In a full chip design, it is often found that there is a critical path through which the delay is maximum. To reduce the critical path delay, the common practice is to insert a high speed buffer. A high speed buffer is a logic gate which has a large drive current capability, so that it can drive a large capacitive load with less propagation delay. The large drive current can be achieved by having large transistor widths. But this in turn causes large input gate capacitance which loads the previous stage of the buffer and hence, the situation becomes worse. To avoid this problem, instead of having a large sized inverter, a chain of inverters with increasing size is used for a high speed buffer design. Figure 10.17 shows a chain of $(n + 1)$ inverters. Let us assume that the first inverter is a minimum size inverter with input capacitance C_{in} and output capacitance C_{out}. The next inverters are designed to have widths x times larger than the previous ones.

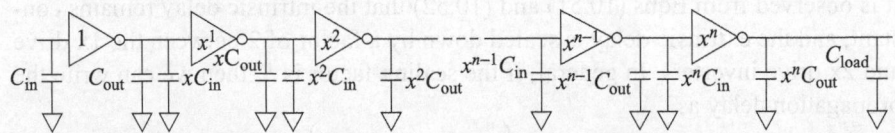

Fig. 10.17 Inverter chain of $(n + 1)$ inverters with increasing size

The last inverter has transistors' widths as n times the first one. We can write the input and output capacitance of the ith stage inverter as $C_{in,i} = x^i C_{in}$ and $C_{out,i} = x^i C_{out}$ The load capacitance of the last inverter is given by

$$C_{load} = x^{(n+1)} C_{in} \tag{10.54}$$

We can write the propagation delay of ith stage inverter using Eqn (10.50) as

$$\tau_{p,i} = \tau_{p0}\left(1 + \frac{C_{external}}{C_{internal}}\right) = \tau_{p0}\left(1 + \frac{x^{i+1}C_{in}}{x^i C_{out}}\right) = \tau_{p0}\left(1 + \frac{xC_{in}}{C_{out}}\right) \tag{10.55}$$

Hence, the total propagation delay through the $(n + 1)$ stages can be written as

$$\tau_{total} = \sum_{i=0}^{n} \tau_{p,i} = (n+1) \times \tau_{p0}\left(1 + \frac{xC_{in}}{C_{out}}\right) \tag{10.56}$$

From Eqn (10.54), we can write

$$(n+1) = \frac{\ln\left(C_{load}/C_{in}\right)}{\ln x} \tag{10.57}$$

Substituting Eqn (10.57) in Eqn (10.56), we get

$$\tau_{total} = \frac{\ln\left(C_{load}/C_{in}\right)}{\ln x} \times \tau_{p0}\left(1 + \frac{xC_{in}}{C_{out}}\right) \tag{10.58}$$

For minimum value of the total propagation delay, the derivative of τ_{total} with respect to x must be zero. Hence, we can write

$$\frac{\partial \tau_{total}}{\partial x} = \tau_{p0} \times \ln\left(\frac{C_{load}}{C_{in}}\right)\left[\frac{1}{\ln x} \times \frac{C_{in}}{C_{out}} - \left(\frac{1}{\ln x}\right)^2 \times \frac{1}{x} \times \left(1 + \frac{xC_{in}}{C_{out}}\right)\right] = 0 \tag{10.59}$$

or
$$\frac{1}{\ln x} \times \frac{C_{\text{in}}}{C_{\text{out}}} - \left(\frac{1}{\ln x}\right)^2 \times \frac{1}{x} - \left(\frac{1}{\ln x}\right)^2 \times \frac{C_{\text{in}}}{C_{\text{out}}} = 0$$

or
$$\frac{C_{\text{in}}}{C_{\text{out}}} \left(1 - \frac{1}{\ln x}\right) = \frac{1}{x \ln x}$$

or
$$\frac{C_{\text{out}}}{C_{\text{in}}} = x(\ln x - 1) \tag{10.60}$$

Equation (10.60) must be solved for x to find out the optimum scale factor for the inverter chain to design the high speed buffer. If we consider the output capacitance of the inverter to be zero, Eqn (10.60) yields $x = e$, the natural number 2.718.

10.16 Effect of Input Slew on Propagation Delay

So far we have analysed delay assuming the input signal has a fixed rise and fall times. However, the input signals for different gates in the design will experience different input rise and fall times. Hence, we need to study the effect of the variation of input slew on the propagation delay of the gate, and on the output slew as well.

To study this effect, we can create a SPICE netlist and simulate it by changing the input slew values.

```
* SPICE netlist to study variation of propagation delay and
* Slew with input Slew

.include "E:\CMOS-CELLS\SPICE MODELS\model.txt"

.option scale=0.18u
.param CL=1p
.param N=1
.param VDD=1.8
.param TR=8n
C1 y Gnd 'CL'
M2 y a Gnd Gnd NMOS L=1 W='2*N' AD='5*2*N' PD='2*2*N+10' AS='5*2*N'
PS='2*2*N+10'
M3 y a Vdd Vdd PMOS L=1 W='6*N' AD='5*6*N' PD='2*6*N+10' AS='5*6*N'
PS='2*6*N+10'
v4 Vdd Gnd 'VDD'
v5 a Gnd pulse(0.0 'VDD' 0 'TR' 'TR' 100n 200n)

.tran 1n 250n
.plot tran v(a) v(y)

.measure tran TPHL trig v(a) val='VDD/2' cross=1
+ targ v(y) val='VDD/2' cross=1
.measure tran TPLH trig v(a) val='VDD/2' cross=2
+ targ v(y) val='VDD/2' cross=2

.measure tran Trise trig v(y) val='0.9*VDD' cross=1
+ targ v(y) val='0.1*VDD' cross=1
```

```
.measure tran Tfall trig v(y) val='0.1*VDD' cross=2
+ targ v(y) val='0.9*VDD' cross=2

.step lin param TR 8n 12n 1n

.end
```

Figure 10.18 shows the input and output waveforms for a CMOS inverter with different propagation delay and slew values.

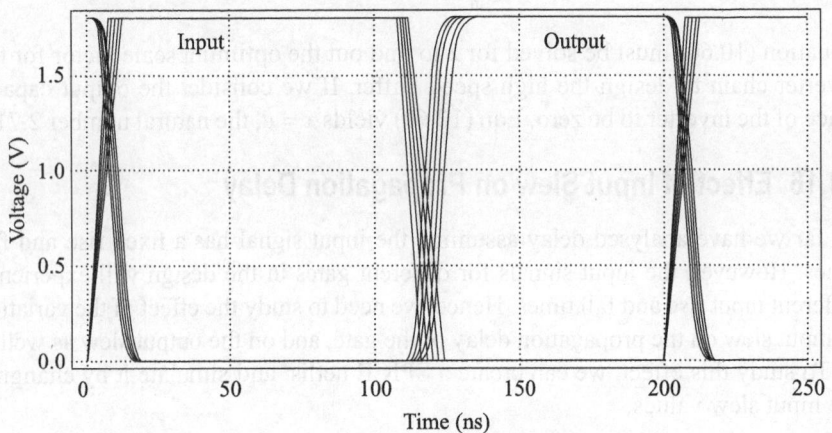

Fig. 10.18   Input and output waveforms with different slew values

The variation of propagation delay in input slew values linearly increases with the input slew, as shown in Fig. 10.19.

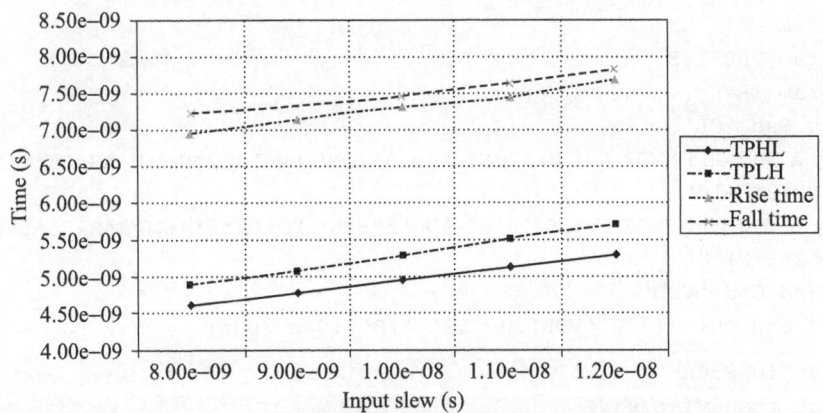

Fig. 10.19   Variation of propagation delay and slew values with input slew

10.17 Delay Dependence on Input Patterns

We have analysed the propagation delay for the single-input CMOS inverter. In this section, we will analyse the propagation delay for a multi-input gate. Let us consider a two-input NAND as shown in Fig. 10.20 for this purpose.

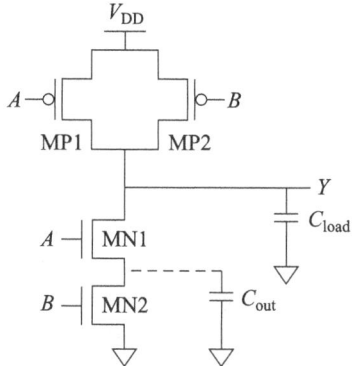

Fig. 10.20 Two-input NAND gate

We can create a SPICE netlist for the two-inputs NAND gate as shown below:

```
* SPICE netlist for 2-input NAND gate

.include "E:\CMOS-CELLS\SPICE MODELS\model.txt"
.option scale=0.18u

.param CL=1p
.param N=1
.param VDD=1.8
.param TR=10n
C1 y Gnd 'CL'
M2 y a vdd vdd PMOS L=1 W='6*N'
+ AD='5*6*N' PD='2*6*N+10' AS='5*6*N' PS='2*6*N+10'
M3 y b Vdd Vdd PMOS L=1 W='6*N'
+ AD='5*6*N' PD='2*6*N+10' AS='5*6*N' PS='2*6*N+10'
M4 y a n1 gnd NMOS L=1 W='4*N'
+ AD='5*4*N' PD='2*4*N+10' AS='5*4*N' PS='2*4*N+10'
M5 n1 b gnd gnd NMOS L=1 W='4*N'
+ AD='5*4*N' PD='2*4*N+10' AS='5*4*N' PS='2*4*N+10'
v4 Vdd Gnd 'VDD'
va a gnd BIT ({1110111101} pw=50n on=1.8 rt=10n ft=10n)
vb b gnd BIT ({1011111101} pw=50n on=1.8 rt=10n ft=10n)

.tran 1n 500n
.plot tran v(a) v(b) v(y)

.measure tran TPLHb trig v(b) val='VDD/2'
+ cross=1 targ v(y) val='VDD/2' cross=1
.measure tran TPHLb trig v(b) val='VDD/2'
+ cross=2 targ v(y) val='VDD/2' cross=2
.measure tran TPLHa trig v(a) val='VDD/2'
+ cross=1 targ v(y) val='VDD/2' cross=3
.measure tran TPHLa trig v(a) val='VDD/2'
+ cross=2 targ v(y) val='VDD/2' cross=4
```

```
.measure tran TPLHab trig v(a) val='VDD/2'
+ cross=3 targ v(y) val='VDD/2' cross=5
.measure tran TPHLab trig v(a) val='VDD/2'
+ cross=4 targ v(y) val='VDD/2' cross=6
.end
```

The input and output waveforms are shown in Fig. 10.21. Careful observation indicates that the delay for low-to-high and high-to-low transition of the output waveform is different for different input combinations. The delay values are shown in Table 10.4. The largest high-to-low delay occurs when both inputs A and B make 0 to 1 transition. The smallest delay for low-to-high transition occurs when both inputs go to low. When both the inputs go to low, the output goes low-to-high through two parallel paths to V_{DD}. Two ON pMOS transistors in parallel have effective value of low resistance. Hence, this case produces the smallest delay for low-to-high transition of the output.

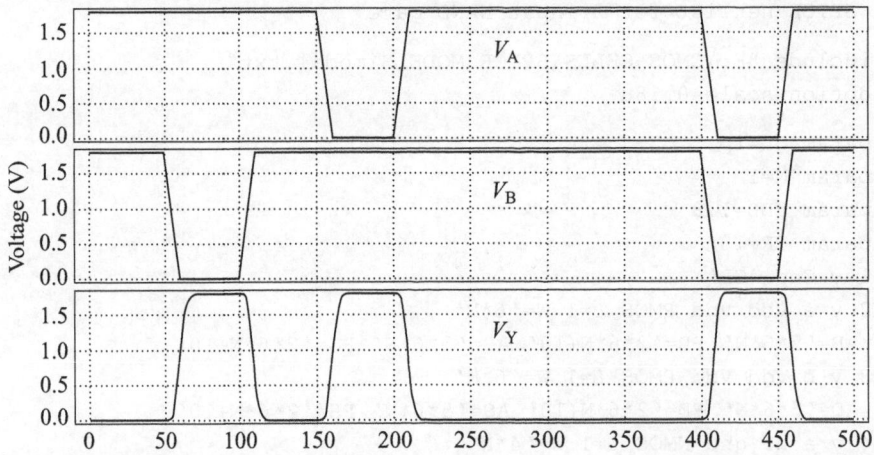

Fig. 10.21 Delay dependence on input patterns for two-input NAND gate

Table 10.4 Propagation delay values for different input patterns

Input patterns	Delay (ns)	Input patterns	Delay (ns)
TPLH for $A = 1$ and $B = 1 \to 0$	5.3123	TPHL for $A = 1$ and $B = 0 \to 1$	3.9910
TPLH for $B = 1$ and $A = 1 \to 0$	5.2944	TPHL for $B = 1$ and $A = 0 \to 1$	4.6385
TPLH for $A, B = 1 \to 0$	3.5365	TPHL for $A, B = 0 \to 1$	4.9938

The worst case delay for low-to-high transition of the output again depends on which input goes low first. In this case, when $A = 1$ MN1 is ON, and B goes low, the pMOS MP2 has to charge both C_{load} and the parasitic capacitance at the drain node of MN2 (C_{out}). But when $B = 1$, A goes low, since MN1 is OFF, the pMOS MP1 has to charge only C_{load}. Hence, the delay is less.

For high-to-low transition of the output, the delay depends on the initial conditions of the internal nodes.

The most important point is that the delay values are a function of the input patterns. This makes the delay characterization and the delay analysis much more complicated.

10.18 Logical Effort

Sutherland et al. (1999) first introduced the concept of logical effort in 1999. Logical effort is a method for estimating delay in the CMOS circuit. The best design is selected by comparing delays of different CMOS logic structures. It is known that NAND and NOR gates are two universal gates, i.e., using either of the gates, we can implement any Boolean function. The very next question comes to our mind is that which one to use, NAND or NOR? What is better? The answer becomes easy if we analyse these gates based on logical effort. Logical effort can be utilized for selecting the optimum transistor sizes of the CMOS circuits.

Let us now write the propagation delay of any gate using Eqn (10.49) as

$$\tau_p = 0.69 R_{eq} C_{load} = 0.69 R_{eq} (C_{p,out} + C_{out}) \tag{10.61}$$

where C_{out} is the external load capacitance connected to the gate and $C_{p,out}$ is the parasitic output capacitance of the gate.

Let us now consider a minimum size inverter driving an identical inverter. Its propagation delay is written as

$$\tau_{p,inv} = 0.69 R_{inv} C_{inv} \tag{10.62}$$

where R_{inv} and C_{inv} are the effective resistance and input capacitance of the minimum size inverter.

Normalizing Eqn (10.61) with respect to the delay of a minimum size inverter using Eqn (10.62), we get

$$d = \frac{\tau_p}{\tau_{p,inv}} = \frac{R_{eq}(C_{p,out} + C_{out})}{R_{inv} C_{inv}}$$

(Assuming effective resistance of the gate and minimum size inverter are identical, i.e., $R_{eq} = R_{inv}$)

$$= \frac{R_{eq} C_{p,out}}{R_{inv} C_{inv}} + \frac{C_{out}}{C_{inv}}$$

$$= p + \frac{C_{out}}{C_{in}} \times \frac{C_{in}}{C_{inv}}$$

$$= p + hg$$

or
$$d = p + g \times h \tag{10.63}$$

where p is known as the *parasitic delay* and it is given by

$$p = \frac{R_{eq} C_{p,out}}{R_{inv} C_{inv}} = \frac{\text{Intrinsic delay of the gate}}{\text{Intrinsic delay of minimum size inverter}} \tag{10.64}$$

h is known as the *electrical effort* and it is given by

$$h = \frac{C_{out}}{C_{in}} \tag{10.65}$$

g is known as the *logical effort* and it is given by

$$g = \frac{C_{in}}{C_{inv}} \tag{10.66}$$

The product of logical effort and electrical effort (gh) is known as *effort delay*. The electrical effort is basically the effective fan-out of the gate which is defined as the ratio between the output load capacitance connected to the output of the gate and the input capacitance of the gate.

The *logical effort* is defined as the ratio between the input capacitance of the gate and the input capacitance of a minimum size inverter. It basically represents how much worse a gate is than a minimum size inverter in driving the same output current. Logical effort is very important in designing logic cells as it depends only on the circuit topology. Hence, designers can select alternative topologies for implementing the same logic functions, and check for minimum logical effort to produce minimum propagation delay.

Let us now find out the logical effort for the basic CMOS logic gates. For this purpose, we consider the CMOS inverter, two-input NAND, and two-input NOR gate, as shown in Fig. 10.22.

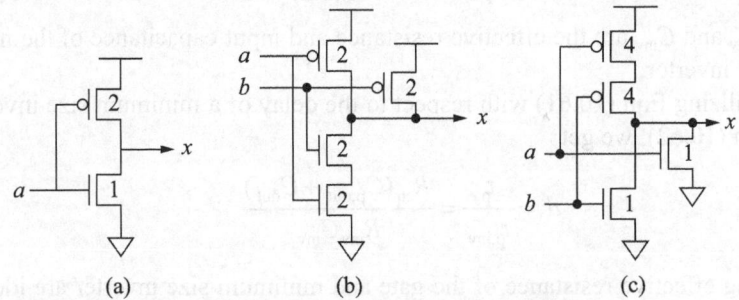

Fig. 10.22   (a) CMOS inverter; (b) two-input NAND gate; (c) two-input NOR gate
(Numbers indicate the relative width of the transistors)

Table 10.5 shows the values of logical effort for basic gates.

Table 10.5   Logical effort for inputs of static CMOS gates, assuming $\gamma = 2$

	Number of inputs				
Logic gate	1	2	3	4	N
Inverter	1	–	–	–	–
NAND	–	4/3	5/3	6/3	$(n+2)/3$
NOR	–	5/3	7/3	9/3	$(2n+1)/3$
Multiplexer	–	2	2	2	2
XOR	–	4	12	32	–

γ is the ratio of an inverter's pull-up transistor width to the pull-down transistor width

$$\text{Logical effort} = \frac{\text{Sum of transistor widths per input of a gate}}{\text{Sum of transistor widths per input of an inverter}}$$

The parasitic delay of a logic gate is independent of the load capacitance it drives. This delay is a form of overload that accompanies any gate.

The main contributor to the parasitic delay is the capacitance of the source/drain regions of the transistors that drive the output of the gate.

Table 10.6 presents the estimates of parasitic delay for basic logic gates. The parasitic delays are given as multiples of the parasitic delay of a minimum size inverter, denoted as p_{inv}, which is given by

$$p_{inv} = \frac{C_{p,out}}{C_{inv}} \tag{10.67}$$

Table 10.6 Estimates of parasitic delay of various logic gate types

Logic gate	Parasitic delay
Inverter	p_{inv}
n-input NAND	np_{inv}
n-input NOR	np_{inv}
n-way multiplexer	$2np_{inv}$
Two-input XOR, XNOR	$4p_{inv}$

Figure 10.23 shows the plot of normalized delay as a function of electrical effort. This is a plot of Eqn (10.63). The slope of the curve represents the logical effort.

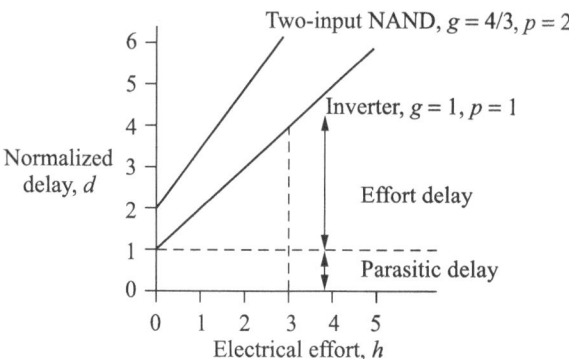

Fig. 10.23 Normalized delay vs electrical effort for an inverter and a two-input NAND gate

10.19 Classification of Digital Systems

Digital systems are broadly classified into two main categories as follows:
■ Combinational or asynchronous circuits
■ Synchronous circuits

10.19.1 Asynchronous Circuits

In *asynchronous circuits,* the signal can make a transition at any time. There is no dependency between the input signals in asynchronous circuits, whereas in *synchronous circuits,* the components are operated at a single clock cycle. There is dependency between the input data and the clock signal in synchronous circuits. A very simple example is a normal inverter and a clocked inverter as shown in Fig. 10.24.

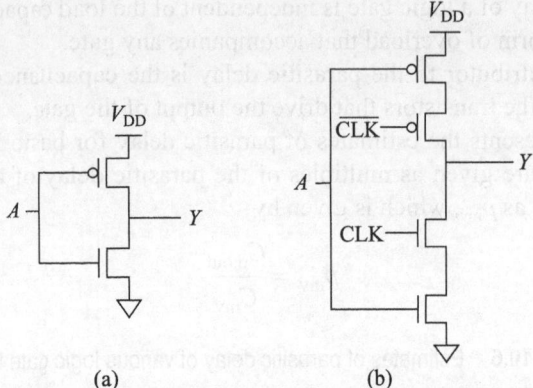

Fig. 10.24 (a) Normal CMOS inverter; (b) clocked CMOS inverter

In normal inverter, the input A can arrive at any time and correspondingly, the output Y is evaluated after the propagation delay of the gate. But in the case of a clocked inverter, if the CLK = 1, only output Y is evaluated depending on the input signal at A. Even if the input arrives at A at any time, the output will not change until the CLK signal becomes logic 1.

Now there is a problem of asynchronous circuits. As the output changes for any change in the output, the correct output data must be checked at the proper instant of time. Otherwise, a wrong output will be taken. That is why, mostly all digital systems are built using synchronous circuits.

10.19.2 Synchronous Digital Design

Let us consider a basic synchronous circuit as shown in Fig. 10.25. It is a combination of synchronous and asynchronous logic blocks. The synchronous blocks are operated at a single clock. The minimum clock period is limited by the propagation delay through the combinational logic block. In every clock cycle, new values propagate in the circuit. We only care about the final, steady state output at the end of each cycle, which will be captured in the flip-flops. For correct data to be captured in the flip-flops, the following two very important timing constraints must be satisfied:

- Set-up time
- Hold time

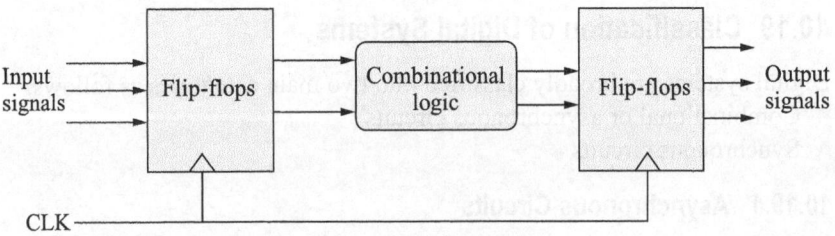

Fig. 10.25 Synchronous design

The *set-up time* of a flip-flop is defined as the minimum time before the clocking transition, the data must arrive at the input of the flip-flop. The *hold time* of a

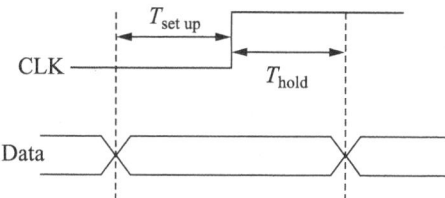

Fig. 10.26 Set-up time and hold time of a flip-flop

flip-flop is defined as the minimum time after the clocking transition the data must not change. The set-up time and hold time are shown in Fig. 10.26.

If an input data changes within the hold time of a flop-flop, there is a hold time violation. Similarly, when a signal arrives late than the set-up time of the flip-flop, there is a set-up time violation. No flip-flop in a design should have set-up time and hold time violations.

10.20 Definitions of Timing Terms

Critical path   It is the path between an input and an output having the maximum delay. Any further delay in this path slows down the entire operation of the design. After the timing delay of the design is calculated, the critical path is identified by a backtracking method.

Arrival time   It is the time taken for the signal to arrive at a certain point. The arrival time of the clock signal is generally taken as the reference time. After delays of all paths are calculated, the arrival time of any signal is calculated. Arrival times are specified by timing windows with a pair of values—the earliest and latest possible time at which a signal can arrive.

Required time   It is the latest time by which a signal must arrive at a certain point in the circuit. The RTL designers normally specify the timing constraints along with the VHDL or Verilog netlist. From the timing constraints, the required time at the primary outputs are calculated. Then, a backward traversal is followed to find the required time at the input of every gate.

Slack   It is the difference between the required time and the arrival time. A positive slack at a node indicates that the signal can be slowed down at that node till the slack is zero. Conversely, negative slack at a node indicates that the signal must be speeded up to meet the arrival time.

Clock skew   It is the variation in clock signals due to different clock traversal path to the clock inputs of different synchronous elements.

Clock jitter   It is the timing variation in clock signals.

10.21 Timing Analysis

Timing analysis basically means
- Measuring the speed of the circuit, e.g., clock speed
- Estimation of critical delay
- Slack determination
- Critical path identification

First, a clocking scheme is defined to set a timing measurement reference. A simple analogy is Greenwich Mean Time (GMT) which is used as a reference for measuring time worldwide. After the reference has been set for every node, the signal delay is calculated (*x* picoseconds) after or before the rise or fall transition of the clock. The timing attributes are attached to objects in the connectivity structure of the netlist. The timing attributes are

- Valid time
- Required time (earliest and latest)
- Slack
- Slope of the input/output waveforms
- Arrival time
- Departure time
- Set-up time
- Hold time
- Propagation delay
- Path delay

10.22 Timing Models

There are different timing models that are used for timing analysis, as given below:

(a) *Unit delay model*   In the unit delay model, the delay through any gate is considered to be one time unit. Interconnects have zero delay.

(b) *Fixed delay model*   In the fixed delay model, each gate has a fixed delay independent of circuit structure. Interconnects have zero delay.

(c) *Complex delay model*   In the complex delay model, every gate in the design has its own delay depending on input and output conditions. Nets have non-zero parasitic delay. The gate delays are analysed considering the following conditions:

- Load capacitance of the gate
- Non-ideal input waveform
- Delays have any value between some minimum and maximum value
- Different delays from different gate inputs to output
- Crosstalk induced delay

Example 10.3   Using the unit delay model, find the path with the longest delay in the following circuit of a full-adder shown in Fig. 10.27.

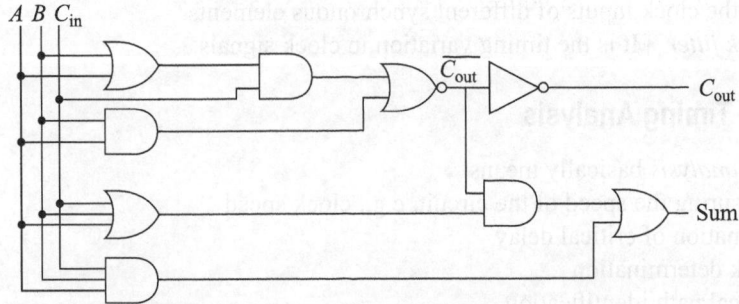

Fig. 10.27   Gate level realization of full-adder circuit

Solution In the unit delay model, each gate has a fixed delay of one unit and the interconnect has zero delay. Hence, we can write the delay values of each gate and propagate it to the next stages, as shown in Fig. 10.28.

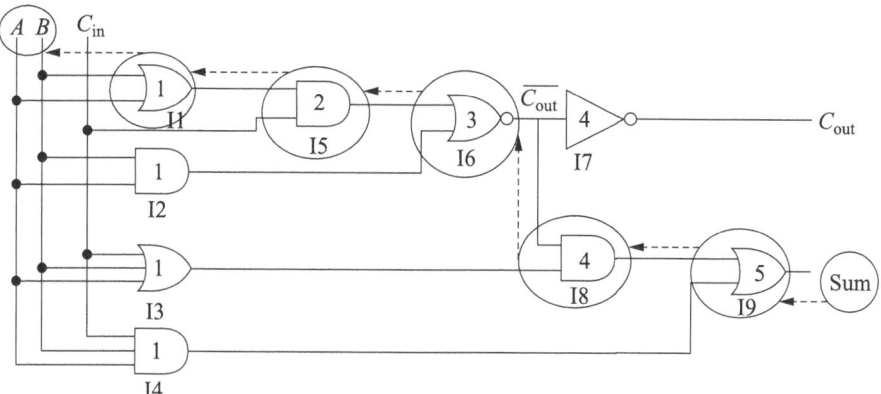

Fig. 10.28 Full-adder circuit with gate delays and critical path

The *Sum* output of full-adder needs five logic stages, and the C_{out} output needs four logic stages in this particular implementation. The Sum is implemented using the relationship shown in Eqn (10.68):

$$Sum = A\overline{C}_{out} + B\overline{C}_{out} + C_{in}\overline{C}_{out} + ABC_{in} \qquad (10.68)$$

In each logic stage, the delay values are incremented by a one-time unit.

To find out the longest path, we need to do a backtracking from the output having maximum delay. In this case, *Sum* output is having maximum delay. So, we start from *Sum* output. Select the gate that produces *Sum* output (Instance: I9 in Fig. 10.28). Next, we select the gate having maximum delay (Instance: I8) amongst the gates that are driving the instance I9. Then, we select instance I6 and so on, until the primary inputs are reached. By doing a backtracking in this way, we get the path having longest delay. In this example, the longest path is

$$B \rightarrow I1 \rightarrow I5 \rightarrow I6 \rightarrow I8 \rightarrow I9 \rightarrow Sum$$

The following path is also having the longest delay:

$$A \rightarrow I1 \rightarrow I5 \rightarrow I6 \rightarrow I8 \rightarrow I9 \rightarrow Sum$$

10.23 Timing Analysis Goals

The main aspects of timing analysis are to ensure that
- the circuit meets the timing specifications
- there are no functional violations due to timing error (e.g., no races)

10.24 Timing Analysis at the Chip Level

A circuit can be checked for functionality and timing by simulating it with circuit simulators, like SPICE. However, doing a SPICE simulation at the chip level is impractical, as at the chip level, the number of circuit elements is too huge. Hence, performing timing analysis at the chip level has become a major area of research. There are two main methodologies for timing analysis at the chip level as given below:

- Static timing analysis (STA)
- Dynamic timing analysis (DTA)

In *static timing analysis*, the circuit is not simulated. The delays for the gates and interconnects are analysed from the pre-characterized timing characterization data generated at the cell level. Static timing verification has become an essential part in VLSI design flow. Static timing verification is fairly accurate and a very fast timing analysis methodology. This concept is utilized in the timing driven logic synthesis techniques.

In *dynamic timing analysis*, the circuit is simulated with the input bit patterns or input vectors. This method is most accurate, but slow. The simulation time is more for more number of input vectors. The quality of timing results is also a function of input vectors.

10.25 Static vs Dynamic Timing Verification

Problems with both approaches have resulted in the formation of a new tool category named *hybrid timing verification*. It selectively combines both static and dynamic timing in an attempt to create the best of both techniques. A comparison between the static and dynamic timing analysis is given in Table 10.7.

Table 10.7   Comparison between the static and dynamic timing analysis

Static	Dynamic
Reasonable accuracy	Good accuracy
Fast analysis time	Slow analysis time
High capacity	Limited capacity
Requires no test patterns	Requires test patterns
High coverage	Limited coverage
No logic behaviour is comprehended	Comprehends logic functionality
Not suitable for synchronous design	Suitable for both synchronous and asynchronous designs

10.26 Factors Impacting Timing Delay

The primary factors that impact the timing delay are as follows:

- Power supply voltage (V)
- Operating temperature (T)
- Process variation (P)

Secondary factors are as follows:

- Pin-to-pin delay
- Load capacitance
- Shape of the input waveform (non-ideal slope)
- Signal threshold for delay measurement (50%)
- Interconnect model

10.27 Static Timing Analysis—Case Study

Let us now consider a circuit shown in Fig. 10.29. First, we will analyse the circuit using unit delay model.

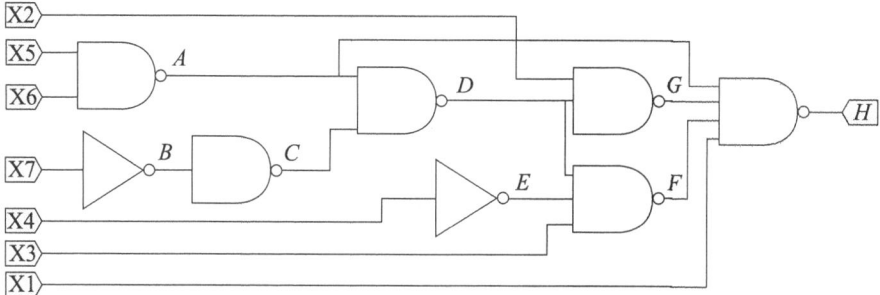

Fig. 10.29 A combinational circuit

To calculate the signal arrival time at each node of the circuit, we make the following assumptions:

- Every gate has a propagation delay of 1 time unit
- Same delay from every gate-input-pin
- Same delay for up and down transitions
- Same delay for all loads, regardless of fan-out
- Nets are ideal wires: same value everywhere on the net
- Logic is combinational (acyclic: no logic loops allowed)
- All primary inputs are ready at $t = 0$

10.27.1 Delay Graph

Delay graph is a graph constructed for the circuit to find out the signal arrival time at each node. We can build a delay graph for the circuit as shown in Fig. 10.30. To construct the graph we assume, graph vertices = nets and graph edges = gate delays.

It is a direct acyclic graph (DAG) showing causality. In practice, it is implemented on the circuit (hyper graph) data structure.

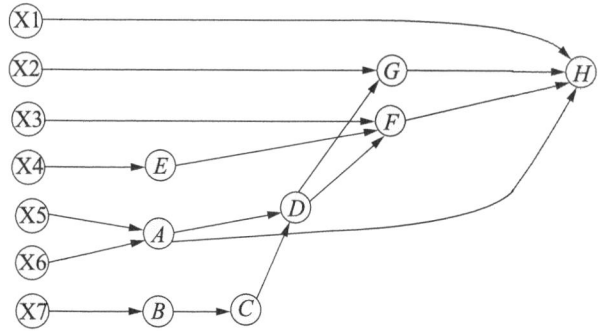

Fig. 10.30 Direct acyclic graph (DAG) of circuit shown in Fig. 10.29

10.27.2 Computing Signal Timing

The signal arrival time can be calculated using an algorithm, which is called levellizing algorithm. The algorithm is expressed in pseudo code as

```
Levellizing algorithm:
Initialize level, L = 0 ;
While (any vertex has not been assigned a level) {
  For each vertex V {
  If (all predecessors of V have an assigned level < L) then
    level_of_V = L ;
    }
  L=L+1;
}
```

The levellizing algorithm is illustrated in Figs 10.31(a) and (b)

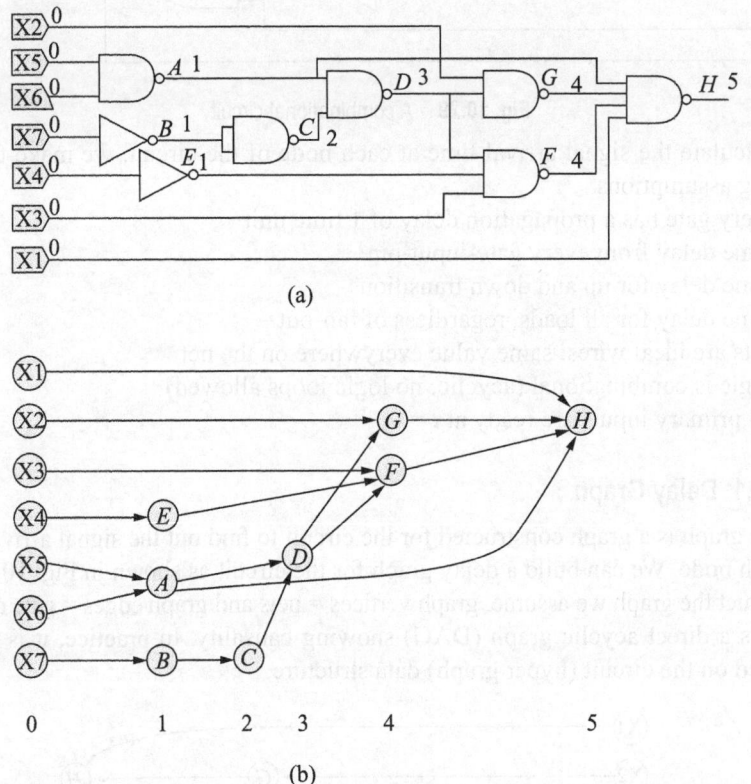

(a)

(b)

Fig. 10.31 (a) Fig. 10.30 redrawn with timing levels; (b) DAG with timing levels

The timing level count is a crude estimate of latest signal arrival time. Sometimes, this is used to detect long multistage logic paths in RTL models.

To find the critical path, a methodology is used which is called *backtracking*. We know the longest delay from primary inputs to each node in the circuit after setting the timing levels. The longest delay to output H is 5 time units in this example. By *tracking back,* we can find the *critical path.*

Steps to find out the critical path are given below:

■ Start at the output node
■ Recursively, go to the predecessor with biggest level

The signal takes maximum time to traverse this path. Hence, this path is the limiter of circuit speed.

In this example, the critical path is $(X7 \rightarrow B \rightarrow C \rightarrow D \rightarrow G/F \rightarrow H)$. The critical path is shown in Fig. 10.32.

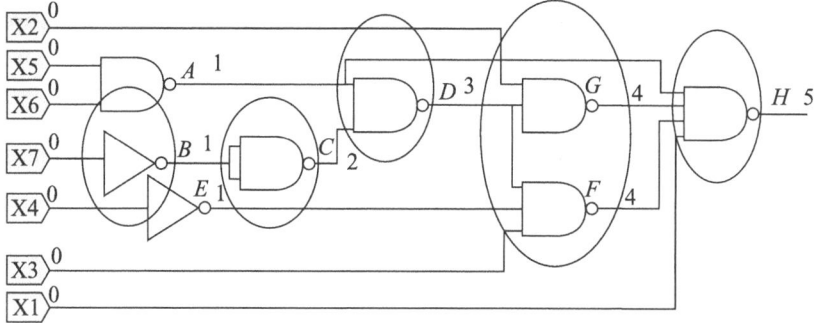

Fig. 10.32   Circuit showing critical path (path connecting the encircled gates)

10.28   Fixed Delay Model

We have so far analysed only unit delay model. Let us now consider the fixed delay model. In this model, every gate has a fixed delay. To build the DAG, we assume vertices = nets; and edge weights = gate delays. Consider the circuit shown in Fig. 10.33 to construct the DAG.

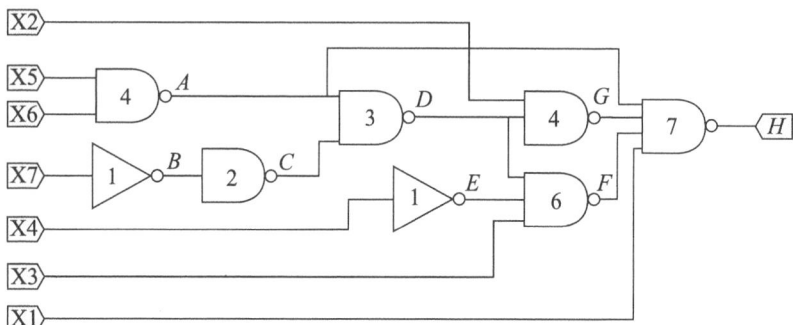

Fig. 10.33   Fixed gate delay of each gate

The DAG of the circuit shown in Fig. 10.33 is shown in Fig. 10.34.

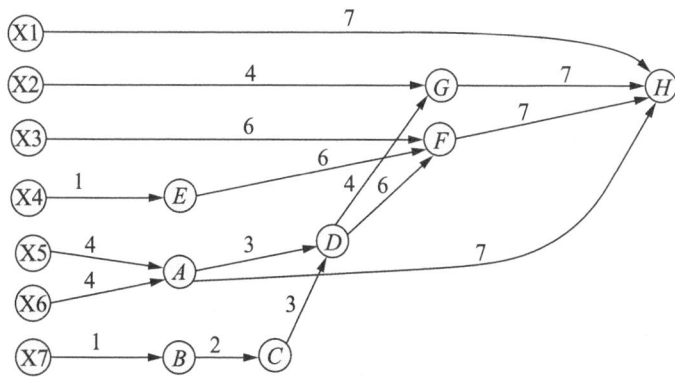

Fig. 10.34   Gate delays are shown as weights on edges of the graph

10.29 Checking Timing Constraints

Timing constraints are the required times for each circuit node for a desired clock speed. If the timing constraints are met, then only a circuit can operate at the specified clock speed. Otherwise, there will be timing violations and the circuit cannot produce correct output. To check if the timing constraints are met, the signal arrival time at each circuit must be calculated and checked against the required time. If a signal cannot meet the required time, it means the delay in this path must be reduced.

10.29.1 Computing Latest Signal Arrival Times

Signal arrival time can be calculated using the levellizing algorithm with a little modification as shown below:
- Start by assigning 0 to PIs
- Take the latest input transition
- Use same algorithm as in levellizing, but add up the delays labelled on the graph edges, instead of just incrementing the level
- Use a FIFO queue to keep track of all gates whose inputs arrival times are ready
- Place first-level gates on the queue
- Process gates from the queue

A pseudo code for calculating the signal arrival time can be written as

```
For each primary input V, Time (V) = 0;
For each remaining vertex in levellized order do {
   Time(V) = MAX ( time of predecessors to V +
      delay from predecessor) over all predecessors of V

}
```

10.29.2 Straightforward Extensions

This algorithm can be extended to consider the following for more accurate timing analysis:
- Dealing with different pin-to-pin delays within each gate
  - Same delay graph can be used, but edges corresponding to a gate can have different weights for different pin-to-pin delays.
- Finding the earliest signal arrival time
  - Same algorithm can be used, but min delay to be considered instead of max delay.
- Finding the 'shortest' path
  - Backtrack on the delay graph using minimum arrival times
- Assigning non-zero arrival times to primary inputs
  - To identify a critical, late-arriving input signal

10.29.3 Computing Latest Signal Required Times

We have so far computed the latest (or earliest) *arrival time*, also called *valid time*, of a signal to a node in the circuit. These are determined by the max/min delay from all inputs into the node. We can also talk about the latest/earliest time when a signal is *required to arrive* at each node, for correct circuit operation. This is called *required time.* Required time at node V is determined by max/min delay from the node V to all the outputs.

To calculate the required time, we start from the sink and assign the required time of arrival at the sink. Now, we walk the graph backwards: against the direction of all arrows. Calculate the required time at the previous node by subtracting the path delay from the required time at the sink. This is very much the same algorithms as we used for arrival times. Proceeding thus, we get the required times at all nodes and finally, at the source (see Fig. 10.35).

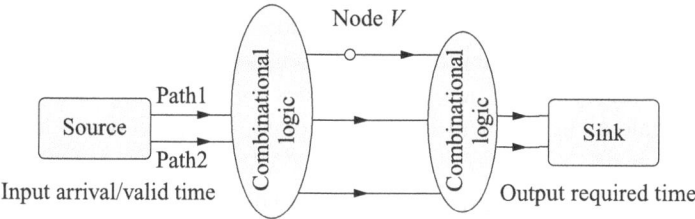

Fig. 10.35   Illustration of valid time and required time at node *V*

We can express valid time and required time as

```
Valid time at node V = (Valid time at source) + (Path delay to V)

Required time at V = (Required time at sink) - (Path delay from V)
```

10.29.4 Computing Slack at Node V

Slack is defined as the time difference between the earliest required time and the latest arrival time at any node. It is expressed as

```
Slack at V = (Earliest required time at V) - (Latest arrival time at V)
```

If all paths start and finish at flip-flops:

```
Slack at V = (Critical delay) - (Max delay to V) - (Max delay from V)
           = (Critical delay) - (Delay of longest path through V)
```

Slack indicates if the circuit is sensitive to the timing of signal V. If there is positive slack (T), the signal at V can be delayed by time T without increasing the longest path delay through the block. This can help reduce power, area, or noise. If the slack is 0, V is on the critical path. Negative slack means that the signal will arrive too late at the sink, and indicates the desirable speed-up at V.

10.29.5 Detecting False Paths

In the static delay analysis, the logic is not considered. For example, let us assume unit gate delay for the circuit shown in Fig. 10.36. In this circuit, the longest *topological path* is *a-c-d-y-z* with a delay of 4 time units. But logically, the AND gate requires $e = 1$, while the OR gate requires $e = 0$. Hence, no event can propagate from *a* to *z*, i.e., the path cannot be *sensitized.* The longest true path is *a–c–d–y*, with a delay of 3 time units. A path which cannot be sensitized is called a *false path.* The main drawback of the static delay analysis is the need to eliminate false paths. It does not use any test inputs vectors, i.e., it is a data-independent analysis.

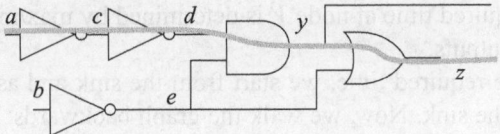

Fig. 10.36 Circuit showing a false path

10.29.6 When Can a Path be False?

- *Case 1* A gate along the path is controlled, the path provides a non-controlling value, a side-input provides a controlling value.
- *Case 2* A gate along the path is controlled, both the path and a side-input provide controlling values, but the side-input provides the controlling value first.
- *Case 3* A gate along the path is not controlled, but a side-input had the last controlling value.

Figure 10.37 illustrates the above three cases of false path.

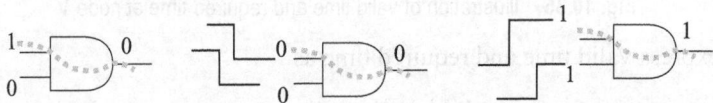

Fig. 10.37 Different cases for false paths

10.29.7 Intentional False Paths

The classical example for intentional false-path is carry-bypass adder circuit as shown in Fig. 10.38. Topologically, the longest path is from c0 to c2 via the carry chain. But the conditions for such carry propagation make the MUX select the '1' input, which sets the carry faster. Circuit timing problems can be solved by creating intentional bypass logic like this, to make the long critical path 'false'.

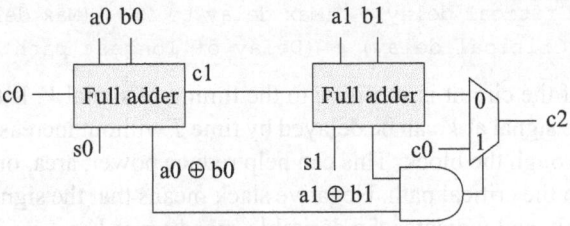

Fig. 10.38 Carry-bypass adder

10.30 Timing Verification in Sequential Synchronous Circuits

In sequential circuits, the registers are used to store the logic values. Generally, the registers are D flip-flops (DFFs). The roles of DFFs are:

- To stop propagation of signals that are too fast
- To store stable values of state variables

DFF outputs are treated as primary inputs (PIs), and DFF inputs act like primary outputs (POs). The timing constraints are derived from the requirements for correct operation of the DFFs.

10.30.1 Timing Constraints at Flip-flops

DFF operates on the edge of the clock, the logic value is the transferred value from D to Q. The Q output will change after a delay through the DFF (clock-to-Q). Q is kept stable at all other times.

For correct operation of DFF, the D input must stabilize before the edge of clock by some time. This time is known as set-up time ($T_{set\text{-}up}$) of the DFF. Hence, the latest required time at D input imposes the max delay constraint. Similarly, D input must stay stable until some time after the edge of clock, which is known as the hold time (T_{hold}) of the DFF. Thus, the earliest required time at the D input imposes the min delay constraint. This is illustrated in Fig. 10.39.

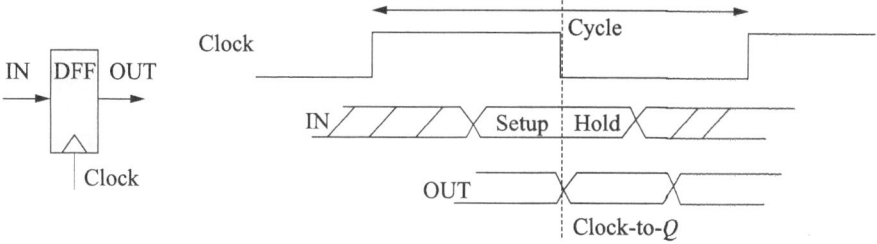

Fig. 10.39   Set-up and hold time of the DFF

10.30.2 Flip-flop Constraints with Clock Skew

Skew is defined as the difference in time of arrival of clock signal to different FFs. Hence, we can write the max delay constraint as

$$(\text{Longest\_path\_delay} + T_{clock\_to\text{-}Q}) < (T_{cycle} - T_{set\text{-}up} - T_{skew})$$

Set-up time violation indicates that the signal arrives after its latest required time. Similarly, the min delay constraint can be written as

$$(\text{Shortest\_path\_delay} + T_{clock\_to\text{-}Q}) > (T_{hold} + T_{skew})$$

Hold time violation indicates that the (new) signal arrives before its earliest required time (while the old value should still hold stable). Figures 10.40(a) and (b) illustrate the clock skew. In a chip, the clock signal generated from the clock source is distributed through the chip along different paths to the sequential elements. The clock signals traversing different paths encounter different delays, and arrive at different times at the clock input of the DFF causing clock skew.

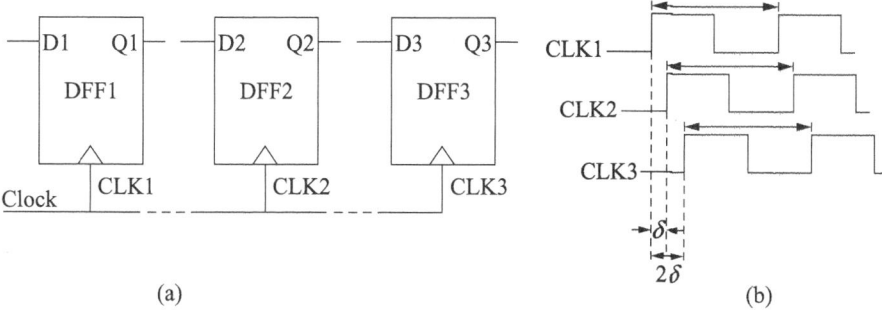

(a)   (b)

Fig. 10.40   (a) Synchronous D flip-flops; (b) clock signals appearing at different times having clock skew

10.30.3 Clock Jitter

Clock jitter is defined as the variation of clock time period at a given point in the chip. Due to clock jitter, clock period is either increased or decreased from the nominal value. Clock has a direct impact on the timing of sequential circuits. Normally, a clock jitter is generated by a voltage controlled oscillator (VCO). VCO is an analog circuit which is very sensitive to the intrinsic device noise and power supply variations. Hence, the output of VCO varies, causing the clock jitter.

10.31　Hierarchical Timing Specification and Verification

A large system cannot be analysed at the gate or transistor level for timing. Hence, a large design is analysed hierarchically. The design is divided into higher-level blocks. Then analyse the design with timing models of higher-level blocks. There are two important types of models for a block:

- *Timing behaviour* abstract timing description of the block as implemented: pin-to-pin delays, output valid times, set-up requirements propagated to block inputs. Composition of such models for all blocks produces a full system model.
- *Timing specifications* imposed on the block by its environment: arrival times of valid inputs, latest required arrival times of outputs. Generating a consistent set of specifications for all the blocks is called *time budgeting*. If behaviour does not meet specification, optimization or rebudgeting is required.

10.32　Issues with Static Delay Modelling

We have assumed:

- Gates have fixed delays
- Up and down transitions are equal
- Gate inputs are all equivalent
- Input waveform does not matter
- Fan-out and loading do not matter
- No delay along wires

But in practice, the assumptions are invalid due to the following reasons:

- Gate delay depend on many parameters (e.g., temperature, V_{DD}, process)
- Different up and down delays
- Pin-to-pin delays should be characterized
- Faster input slope $\rightarrow$ shorter delay
- Bigger capacitive load $\rightarrow$ longer delay
- Wires on chips are *RC* networks.
- Interconnect delay becomes up and down delays
- Different transistors pull up and down
- Pull-up path through p-channel device is usually weaker
- Analysis must be done twice
- In the end, take the worst case of up and down transitions
- Timing analyser must comprehend inverting logic

10.33 First-order Gate Delay Model

The gate delay is modelled as

$$\text{Gate delay} = T_0 + (A \times R_{\text{out}} \times C_{\text{load}}) + (B \times T_{\text{rise}}) \qquad (10.69)$$

where R_{out} = Effective output resistance of gate (a linearized Thevenin model assumed)

C_{load} = Capacitive load (*a lumped capacitive load is assumed*: this is the total capacitance of wires + input capacitance of fan-out gates)

T_{rise} = Rise time of the signal at the gate's input (slope is inversely proportional to T_{rise})

The gate output rise/fall time is modelled as

$$\text{Gate rise/fall time} = (A_1 \times R_{\text{out}} \times C_{\text{load}}) + (B_1 \times T_{\text{rise}}) \qquad (10.70)$$

10.34 Parasitic Extraction

The interconnect wire delay is becoming comparable to the device delay in DSM designs. Even the interconnect delay is dominating over the device delay. Hence, interconnect parasitic elements (R, L, and C) are extracted with greater accuracy and are considered for accurate timing analysis. The parasitic elements are extracted in the post-layout phase after the detailed route step. So far, the parasitic R and C are affecting the delay significantly compared to the inductance of the wire. Hence, the RC extraction is mainly performed for timing analysis. Inductance extraction is done for IC package pins, but not for on-chip interconnects. There are standard RC extraction tools available from the industry, as well as from the academic institutes. Synopsys StarRC is one such popular RC extraction tool. The extracted RC values are specified in a format called standard parasitic exchange format (SPEF), which is approved by IEEE.

10.35 Timing Convergence Problem

Meeting the timing constraints are the most time-consuming part in the VLSI design flow. The total delay in a path is the sum of the gate delay and the interconnect delay. After the synthesis, the gate level netlist is ready and the gate delays can be used from the standard cell library characterization database for timing analysis. But the accurate timing analysis considering both gate delay and interconnect delay can only be done after the detailed routing step. However, the timing analysis must be done ahead of the layout design flow to come up with a gate level netlist for meeting the timing specifications early in the flow, so that the layout design flow can start from there. Hence, meeting timing requirements always needs a number of iterations starting from synthesis to layout. More the number of iterations, more is the design cycle time and hence, the cost of the design increases. So, it is always tried to reduce the VLSI design cycle time to reduce the time-to-market and hence, catch the market early.

To reduce the number of timing iterations following guidelines are considered:

- Synthesis must predict the delays somehow
- Wire load models are used for pre-layout RC estimation
- The DSM effects on timing must be considered early in the flow

With good prediction, the loop converges quickly. A typical timing analysis flow is depicted in Fig. 10.41.

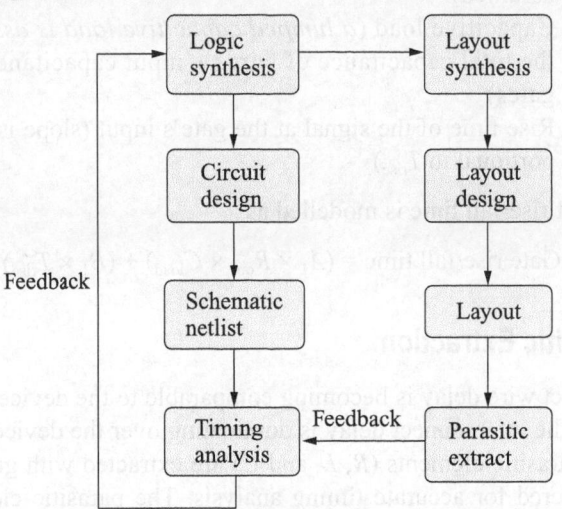

Fig. 10.41 Timing analysis in the VLSI design flow

10.35.1 Approaches to Timing Convergence

- Have a placement-aware synthesis, which is a merger of placement and logic synthesis
- Careful delay budgeting among blocks such that synthesis can converge well within the blocks
- Perform synthesis with accurate wire load model for the interconnects
- Insert buffers on critical nets
- Solve the timing problem by gate sizing
- Distribute the slack such that all stages have equal electrical efforts
- Pay in area and power to reduce delay (but assume that sizing does not change the layout significantly)
- Have better wire planning
- Insert buffers on all global nets (optimally) and compute interconnection delays
- Distribute remaining slacks for 'constant delay synthesis' within blocks

10.36 Timing-driven Logic Synthesis

In logic synthesis, the behavioural level netlist is transformed into a gate level netlist. For example, the following behaviour of a full-adder may be converted into a gate level design as shown in Fig. 10.42.

$$\text{Sum} = A \oplus B \oplus C_{\text{in}}$$
$$C_{\text{out}} = AB + AC_{\text{in}} + BC_{\text{in}}$$

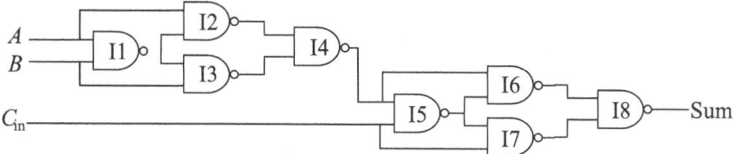

Fig. 10.42   Realization of sum function of a full-adder

Both the realization of sum and carry function use only NAND gates. The instances of NAND gates used to realize the sum function look the same. However,
the instances I1, I4, and I5 are a bit different than the instances I2, I3, I6, I7, and
I8. The fan-out of the three instances I1, I4, and I5 is two while the fan-out of the
other instances is only one. Thus, if all instances are having identical area, these
three instances would have more delay than others. In order to compensate this,
the instances can be chosen from the standard cell library having more drive capability. Hence, the NAND gate with $2x$ drive capability to be used for the instances
I1, I4, and I5 and the NAND gate with $1x$ drive capability to be used for the instances I2, I3, I6, I7, and I8.

As shown in Fig. 10.43, all the instances of NAND gate used to implement
carry function have single fan-out. Therefore, $1x$ drive NAND gate can be used to
implement carry function.

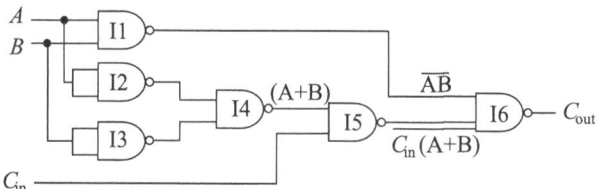

Fig. 10.43   Realization of carry function of a full-adder

A timing-driven synthesis is a methodology that performs synthesis considering
the timing constraints and creates netlist which has less chance of having timing
violations.

The inputs to this flow are given by
- a netlist which contains functional circuit description
- timing specifications (arrival times and required times for signals)
- cell library (with cell timing models)

The outputs of the flow are given by
- logic network of library cells implementing the circuit
- meet all timing constraints
- minimize circuit area

10.37 Gate and Device Sizing

Up-sizing of a gate basically means that the channel width of the transistors used
to realize the gate is increased. On the contrary, the reduction of the channel width
of the transistors used to realize the gate is known as down-sizing. Initially, the

increase in channel width causes delay though to gate to decrease. But the increasing channel width also increases the gate's input capacitance, creating a bigger load on its predecessors. Therefore, the delay starts increasing even with the increase in transistor width after a certain point. There is an optimum point for which the delay is minimum as illustrated in Fig. 10.44. In a simple inverter chain, an optimal tapering ratio is about 4. If the most of the capacitance is due to gates (not wires) then the optimum delay is shared almost equally among gates on the path. Sizing depends on the *logical effort* of the gates (complicated gates have bigger logical effort and should drive smaller capacitances). The real circuits where all gate sizes are parameters show a complex optimization problem.

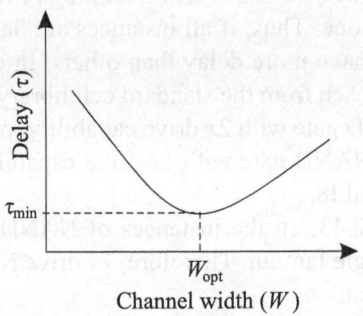

Fig. 10.44 Delay vs MOS transistor channel width (*W*)

10.38 Typical Delay-area Trade-off

The chip area versus delay plot is shown in Fig. 10.45. As the device dimension is increased, the delay is reduced at a faster rate in Region 2, compared to Region 1. As shown in Region 2, delay can be reduced significantly adding a small area. Whereas in Region 1, large area is a required area for a small delay improvement.

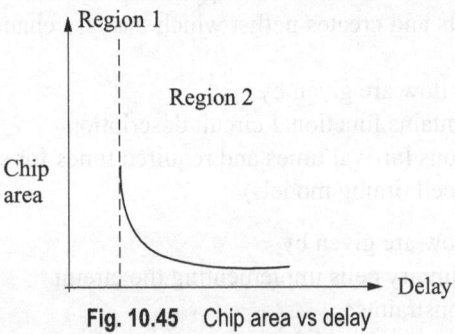

Fig. 10.45 Chip area vs delay

10.39 Timing-driven Layout Synthesis

In timing-driven synthesis, timing is used as *cost function,* instead of wire length in placement and routing tools. In this approach, the delay modelling must be incorporated into the tools. While routing, it must use estimation for interconnect capacitance and resistance. Interconnect capacitance affects timing and is proportional to wire length; so, placement for minimum wire length is most important. Timing is used to identify nets which are more critical than others, and give them priority (timing-driven placement). In submicron CMOS technology, interconnect resistance grows and becomes comparable with driver resistance. Hence, timing-driven routing becomes important.

SUMMARY

- In VLSI circuits, total delay is the sum of the device delay and the interconnect delay. With the advancement of VLSI technology, the interconnect delay dominates over the gate delay.
- Aspect ratio of pMOS transistors is typically 2.5 times the aspect ratio of the nMOS transistors. This is due to the fact that mobility of holes is 2.5 times lesser than that of electrons.
- Two transistors having identical channel length connected in parallel are equivalent to a single transistor of channel width equal to the sum of the widths of the transistors.
- Two transistors having identical channel length connected in series are equivalent to a single transistor of channel width equal to the reciprocal of the sum of the reciprocal of the widths of the transistors.
- With the increase in load capacitance, the delay through the logic circuits increases.
- With the increase in power supply voltage, the delay through the logic circuits decreases.
- With the increase in input rise/fall slew, the delay through the logic circuits increases.
- Propagation delay through a logic circuit is a function of input bit patterns.
- Logical effort of NAND is lesser than that of a NOR gate. This makes NAND a better choice for logic design.
- For correct operation of a flip-flop, the data must arrive before the clock arrives by an amount equal to the set-up time of the flip-flop. Similarly, data must not change after clocking transition within the specified hold time of the flip-flop.
- Propagation delay through a gate depends on process (P), temperature (T), and power supply voltage (V).

SELECT REFERENCES

Agarwal, A., D. Blaauw, V. Zolotov, S. Sundareswaran, K.M.Z. Gala, and R. Panda 2003, *Statistically Delay Computation Considering Spatial Correlations*, Proceedings of the ASP-DAC 2003, January, pp. 271–6.

Arif, I.A.-S., B. Nowak, and C. Chu 2004, *Fitted Elmore Delay: A Simple and Accurate Interconnect Delay Model*, IEEE Transactions on VLSI Systems.

Blaauw, D., K. Chopra, A. Srivastava, and L. Scheffer, 'Statistical Timing Analysis: From Basic Principles to State-of-the-Art', *Transactions on Computer-Aided Design of Integrated Circuits and Systems (T-CAD), IEEE.*

Chadha, R. and J. Bhasker 2009, *Static Timing Analysis for Nanometer Designs,* Springer.

Kang, S.M. and Y. Leblebici 2003, *CMOS Digital Integrated Circuits: Analysis and Design,* 3rd ed., Tata McGraw-Hill, New Delhi.

Martin, K. 2004, *Digital Integrated Circuit Design,* Oxford University Press.

Martin, G., L. Scheffer, and L. Lavagno 2005, *Electronic Design Automation for Integrated Circuits Handbook*, 2 vols, CRC Press, Florida.

Mehrotra, V. and D. Boning 2001, *Technology Scaling Impact of Variation on Clock Skew and Interconnect Delay*, International Interconnect Technology Conference (IITC), June 2001, San Francisco, CA.

Rabaey, J.M., A. Chandrakasan, and B. Nikolic 2008, *Digital Integrated Circuits: A Design Perspective,* 2nd ed., Pearson Education.

Smith, M.J.S. 2002, *Application Specific Integrated Circuits,* Pearson Education.

Sutherland, I., R. Sproull, and D. Harris 1999, *Logical Effort: Designing Fast CMOS Circuits,* Elsevier.

Vygen, J. 2006, 'Slack in Static Timing Analysis,' *Computer-Aided Design of Integrated Circuits and Systems', IEEE Transactions*, September 2006, vol. 25, no. 9, pp. 1876–85.

Weste, N.H.E., D. Harris, and A. Banerjee 2009, *CMOS VLSI Design: A Circuits and Systems Perspective,* 3rd ed., Pearson Education.

EXERCISES

Fill in the Blanks

1. The equivalent (W/L) of two MOS transistors with (W_1/L) and (W_2/L) connected in parallel is _____ .
   (a) $(W_1/L) + (W_2/L)$
   (b) $(W_1/L) . (W_2/L)$
   (c) $1/(L/W_1 + L/W_2)$
   (d) $(W_1/L)/(W_2/L)$

2. The equivalent (W/L) of two MOS transistors with (W_1/L) and (W_2/L) connected in series is _____ .
   (a) $[(W_1/L) (W_2/L)/(W_1/L + W_2/L)]$
   (b) $(W_1/L) (W_2/L)$
   (c) $(W_1/L) + (W_2/L)$
   (d) $1/(L/W_1 + L/W_2)$

3. Critical path delay is the _____ .
   (a) longest path delay
   (b) smallest path delay
   (c) optimum path delay
   (d) none of these

4. *PVT* corner in the context of VLSI design indicates _____ .
   (a) parameter, voltage, temperature
   (b) parameter, voltage, time
   (c) process, voltage, time
   (d) process, voltage, temperature

5. A timing analyser finds out _____ .
   (a) shortest delay in a circuit
   (b) longest delay in a circuit
   (c) nominal delay in circuit
   (d) all of the above

Multiple Choice Questions

1. Lowest propagation delay through a gate is due to
   (a) strong transistor, low temperature, high voltage
   (b) weak transistor, high temperature, high voltage
   (c) strong transistor, high temperature, high voltage
   (d) weak transistor, low temperature, low voltage

2. What is the fix mechanism for slower circuit operation than predicted?
   (a) slow clock
   (b) raise V_{DD}
   (c) either (a) or (b) or both
   (d) none of these

3. The critical path in a design refers to
   (a) the path having maximum delay
   (b) a path with minimum delay
   (c) the path having optimum delay
   (d) a path with no delay

4. A pin-to-pin delay refers to
   (a) delay through the logic cell including net delay
   (b) delay through the logic cell excluding net delay
   (c) net delay excluding the logic cell delay
   (d) net delay including the logic cell delay
5. Slack is defined as the time difference between
   (a) the earliest required time and the latest arrival time at any node
   (b) the latest required time and the earliest arrival time at any node
   (c) the latest required time and the latest arrival time at any node
   (d) the earliest required time and the earliest arrival time at any node

True or False

1. In fixed delay model, every gate has fixed delay.
2. The latest/earliest time when a signal is required to arrive at each node, for correct circuit operation is called required time.
3. The valid time is arrival time of signal at a node.
4. A path which can be sensitized is called a false path.
5. Clock skew is the offset in arrival time of clock signals at the clock input of the synchronous elements in a sequential circuit.

Short-answer Type Questions

1. Draw the CMOS clocked inverter circuit and explain the operation. What is its advantage?
2. Why is CMOS NAND logic better compared to CMOS NAND logic? Explain with the help of transistor sizes.
3. Design a 3-input NAND gate with transistor widths chosen to achieve effective rise and fall resistance equal to that of a unit inverter (R).
4. Consider a 5 mm long, 0.32 μm wide metal2 wire in a 180 nm process. The sheet resistance is 0.05 Ω square and the capacitance is 0.2 fF/μm. Construct a 3-segment π-model for the wire.
5. What is crosstalk delay in VLSI? Discuss the origin of crosstalk delay and how does it affect the performance of an IC.
6. What is crosstalk noise in VLSI? Discuss the origin of crosstalk delay and how does it affect the performance of an IC.
7. What do you mean by design margin and design corners in VLSI? Discuss each of them in brief.
8. Find out the interconnect delay using the Elmore delay model for the following network:

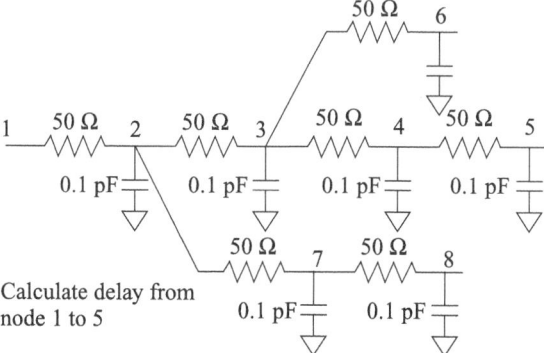

Fig. 10.46 *RC* network

9. Discuss the static timing analysis methodology.
10. What is wire-load model? How is it used in the timing analysis?
11. What are the different delay models? Discuss them in brief.
12. What do you mean by set-up time and hold time of a flip-flop? How can these problems occur in a design?
13. Show that the equivalent (W/L) ratio of n series connected transistors with identical (W/L) ratio is

$$\left(\frac{W}{L}\right)_{\text{equivalent}} = \frac{1}{n}\left(\frac{W}{L}\right)$$

14. Realize the following function in CMOS logic with transistors W/L ratio such that the rise and fall resistances are same as that of CMOS inverter.

$$Y = \overline{A \cdot B + C \cdot D}$$

15. Realize the following function in CMOS logic with transistors W/L ratio such that the rise and fall resistances are the same as that of the CMOS inverter.

$$Y = \overline{A \cdot (B + C) + D \cdot E}$$

Long-answer Type Questions

1. Explain the process, temperature, and voltage dependency of propagation delay.
2. Derive the expression for logical effort. Explain why NAND is preferred over NOR gate using the concept of logical effort.
3. Discuss the static timing analysis (STA) with suitable example.
4. Explain the timing models. Define critical path. What is meant by timing driven logic synthesis. What is inverter buffer chain?
5. Derive the expressions for rise and fall time of an inverter circuit.
6. Design a logic circuit to implement following Boolean function. The output must have equal rise and fall times.

$$Y = AB + C(D + E)$$

7. Discuss the origins of the MOS capacitances, and explain how they affect the circuit performance.
8. Calculate the delay of a FO4 (fan-out-of-4) inverter. Assume that the minimum size inverter with no load is 15 ps.

## CHAPTER 11

Physical Design—Floorplanning, Placement, and Routing

11.1 Introduction

Physical design is the process of generating the physical layout of the VLSI circuit from the schematic or gate level netlist. The physical layout contains all the mask layers with their exact locations and dimensions. The physical design phase typically starts after the logic design and verification steps, as illustrated in Fig. 11.1. The physical design phase mainly consists of four steps, namely, partitioning, floorplanning, placement, and routing. At the end of the physical design step, the design is verified for its correctness. Then the mask layers are generated from the layout. The mask design is sent to the foundry for integrated circuit (IC) manufacturing.

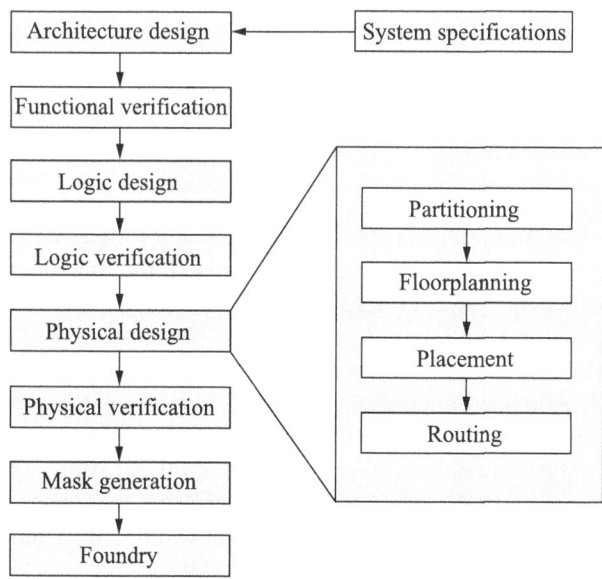

Fig. 11.1   VLSI design flow with detailed physical design steps

11.2 Partitioning

Partitioning is a process by which the entire VLSI circuit is divided into a smaller number of sub-circuits. The partitioning is done such that the number of interconnections between the sub-circuits are minimum. Typically, the entire circuit is partitioned into a number of blocks based on the functionality. Partitioning algorithms are discussed later in this chapter.

11.3 Floorplanning

Floorplanning is the starting step of the physical design in the VLSI design flow. In this step, it is planned to accommodate all the design components and their interconnects within a minimum area. The components are the functional blocks such as datapath logic, control logic, memory elements, etc. These components are of rectangular shape with different aspect ratios. The blocks need to be properly placed and connected within a minimum area. Floorplanning step comes up with a two-dimensional layout with a plan of each block. In electronic design automation (EDA), a floorplan of an IC is a schematic representation of tentative placement of its major functional blocks.

11.3.1 Goals

The goals of floorplanning step are as follows:
- To arrange the blocks on the chip
- To decide the I/O pad locations
- To decide the number of power pads and their locations
- To decide the power distribution style
- To decide the clock distribution and their locations

11.3.2 Inputs

The main inputs to the floorplanning step are as follows:
- A hierarchical netlist of the design which contains all blocks along with their area and connectivity information
- Pin location of each block
- Information whether the blocks are fixed or flexible

A netlist is a representation of the actual design in any formal hardware description language (e.g., VHDL or Verilog). It contains all the functional and control logic cells and blocks, soft and hard macros, memory blocks, registers, etc. It also describes the input and output ports of each block and the connectivity between the blocks. For example, let us consider a multifunctional circuit as shown in Fig. 11.2. The circuit contains six major blocks, three registers (A, B, and C), one functional unit (ALU), one control unit (CU), and one multiplexer unit (MUX). The control unit selects the operands and the function to be performed by applying proper control signals. The ALU performs the selected function on the operand B and another operand selected from C, or A, or 0, and the result is stored in register C.

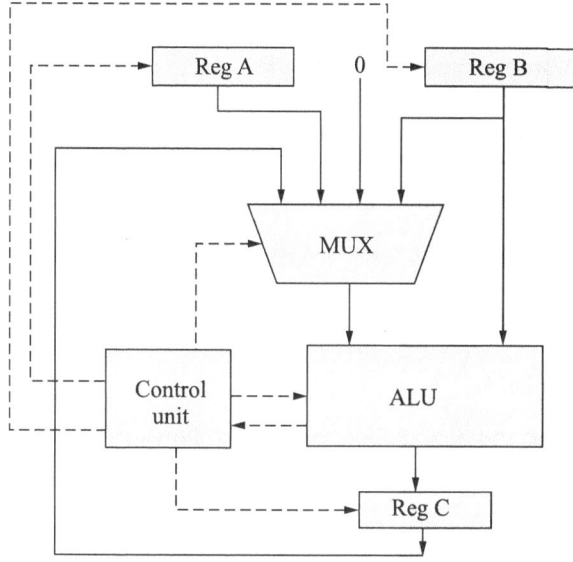

Fig. 11.2 Multifunctional circuit

The whole circuit is partitioned into six major blocks. Each block is represented by a rectangle which has a fixed amount of area requirement. The blocks are then planned in an estimated area by an initial floorplan, as shown in Fig. 11.3.

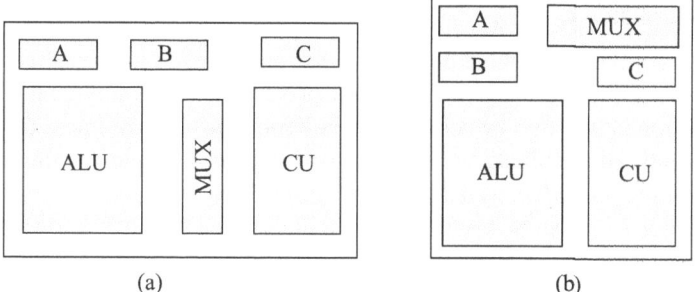

Fig. 11.3 Two different floorplans

The pin locations of each block needs to be known so that the blocks are considered to be placed adjacent to each other if they exchange signals frequently. During the floorplanning, it is also required to know if any block is flexible or fixed. A *flexible block* is normally made of standard cells; hence, their size is fixed, but they can be rearranged so that the block shape is flexible, whereas *fixed blocks* are not alterable. So, after placing the fixed blocks, the flexible blocks can be placed at the vacant areas.

Channel Definition

Although the blocks are planned, it is very important to allocate space for interconnects between the blocks. Channel definition is the process of space allocation for interconnects.

11.3.3 Objectives

The objectives of floorplanning are as follows:
- Minimize the die area
- Reduce the interconnect length (especially for critical nets)
- Maximize routability
- Determine the shapes of flexible blocks

11.3.4 Classes

Two types of floorplanning classes are normally used:
- Slicing floorplan
- Non-slicing floorplan

An example of slicing and non-slicing floorplan is shown in Fig. 11.4.

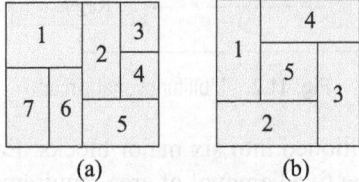

Fig. 11.4   Floorplan: (a) slicing; (b) non-slicing

Slicing Floorplan

In slicing floorplan, the whole die area is first partitioned into two slices of equal or unequal sizes using either a horizontal or vertical line. Individual blocks are again partitioned into two by using either horizontal or vertical lines. This process continues until all the blocks are separated. This process of partitioning a die area is called *slicing floorplan*, as depicted in Fig. 11.5(a).

The slicing floorplan is represented by a binary tree structure known as slicing tree, as shown in Fig. 11.5(b). The leaf nodes of the slicing tree are the blocks of the design. The other nodes are either represented by V or H. H signifies a horizontal partition, whereas V signifies a vertical partition.

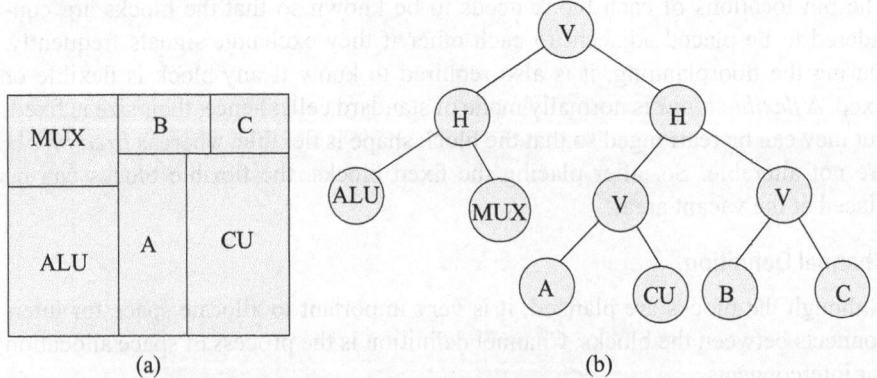

Fig. 11.5   (a) Slicing floorplan; (b) slicing tree

A skewed slicing tree is one in which no node and its right child are the same. A slicing floorplan and its skewed slicing tree are shown in Figs 11.6(a) and (b). The non-skewed slicing tree is shown in Fig. 11.6(c).

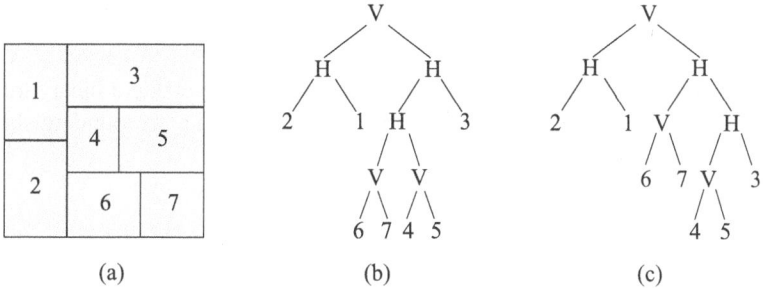

(a)   (b)   (c)

Fig. 11.6 (a) Slicing floorplan; (b) skewed slicing tree; (c) non-skewed slicing tree

Non-slicing Floorplan

If a floorplan is obtained with no recursive through cuts, it is called *non-slicing floorplan*. Figure 11.4(b) illustrates the non-slicing floorplan. It is also known as wheel or spiral floorplan. In this floorplan, the blocks are not obtained by bisections; it must have at least five blocks.

Abutment

It is a process of placing two blocks adjacent to each other without leaving any space between them for interconnections, as illustrated in Fig. 11.7.

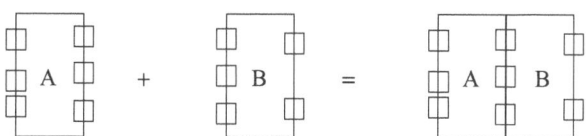

Fig. 11.7 Two blocks are joined by abutment

11.3.5 Hierarchical Floorplanning

Floorplanning can also be classified into two classes: (a) hierarchical floor planning and (b) non-hierarchical floorplanning. A floorplan is *hierarchical* if it is constructed recursively either by vertical slices or horizontal slices. Otherwise, the floorplan is *non-hierarchical*. By default, all slicing floorplans are hierarchical.

A floorplan is called hierarchical of order n, if it is obtained by recursive partitioning a rectangle into at most n parts.

11.3.6 Floorplanning Algorithms

There are several floorplanning algorithms. Some of them are discussed in this subsection.

Sizing Algorithm for Slicing Floorplans

Sizing algorithm for slicing floorplans are as follows:
1. The shape functions of all leaf cells are given as piecewise linear functions.
2. Traverse the slicing tree in order to compute the shape functions of all composite cells (bottom–up composition).

3. Choose the desired shape of the top-level cell; as the shape function is piece-wise linear, only the break points of the function need to be evaluated, when looking for the minimal area.
4. Propagate the consequences of the choice down to the leaf cells (top–down propagation).

Polish Expression It is a string of symbols obtained by traversing a binary tree in the post-order. Figure 11.8 illustrates the slicing floorplan, and corresponding slicing tree and polish expression.

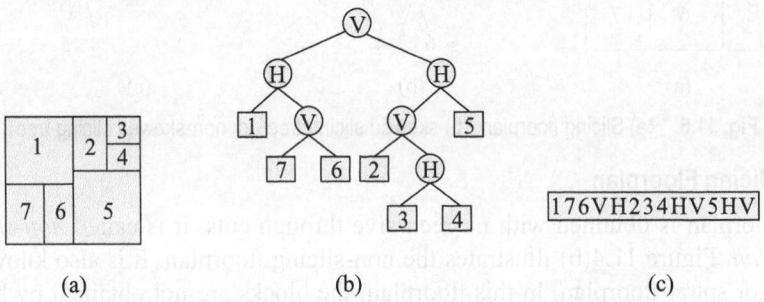

(a)   (b)   (c)

Fig. 11.8 (a) Slicing floorplan; (b) slicing tree; (c) polish expression

Issues of Polish Expression Issues of polish expression are as given below:
- Multiple representations for some slicing trees—when more than one cut in one direction cut a floorplan; this is illustrated in Fig. 11.9.

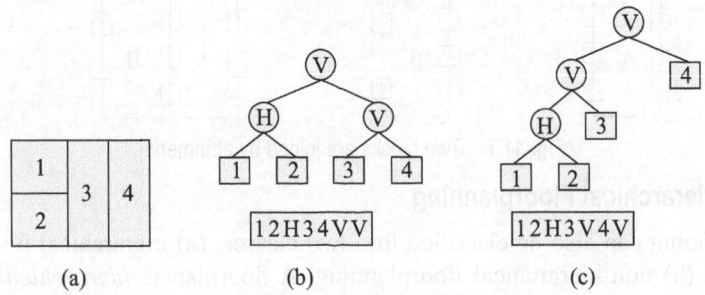

(a)   (b)   (c)

Fig. 11.9 (a) Slicing floorplan; (b) slicing tree; (c) alternate slicing tree

- Larger solution space
- A stochastic algorithm (e.g., simulated annealing) will be more biased towards floorplans with multiple representations—more likely to be visited

Solutions to resolve the issues with polish expression are as follows:
1. Assign priorities to the cuts
2. In a top–down tree construction
    - pick the rightmost cut
    - pick the lowest cut

This will ensure that no two same operators are adjacent in the polish expression, i.e., no 'VV' or 'HH' in the polish expression.

Simulated Annealing

Simulated annealing (SA) is an optimization technique first introduced by Kirkpatrick et al. (1983). It is formulated on the basis of the statistical mechanics of annealing of solids. A low energy state is always a highly ordered state. For example, let us consider the crystal growth technique that is used to grow defect-free single crystal semiconductor for use in IC fabrication. In this crystal growth technique, the poly-crystalline semiconductor is heated to a high temperature, then cooled slowly to form the single crystal. This controlled and slow cooling is called *annealing*. Annealing results in a highly ordered arrangement of atoms from a highly disordered state. The simulated annealing technique is also a similarly controlled operation that results in highly optimized desirable solutions from the set of poor unordered solutions.

Simulated annealing is a popular technique that is used for floorplanning. Although it is excellent in a local search processing, it is inadequate for global search. It repeatedly generates succeeding solutions using the local search procedure. Some of the solutions are accepted and some are rejected, according to a predefined acceptance rule. The acceptance rule is analogous to the annealing processes in metallurgy, as shown in Fig. 11.10. It is an optimization scheme with non-zero probability for accepting inferior (uphill) solutions. The probability depends on the difference of the solution quality and the temperature. The probability is typically defined by $\min\{1, \exp(-\Delta C/T)\}$, where ΔC is the difference in the cost of the neighbouring state and that of the current state, and T is the current temperature. In the classical annealing schedule, the temperature is reduced by a fixed ratio λ (normally 0.85) for each iteration of annealing.

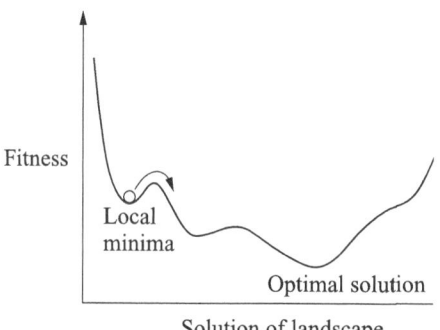

Fig. 11.10 Simulated annealing

Simulated annealing components that are used for floorplanning can be written as follows:

- Solution space → Slicing floorplan
- Cost function → Area of a floorplan (determine how good a solution is)
- Perturbation rules → Transforming a floorplan to a new one
- Simulated annealing engine
  - A variable T, analogous to temperature
  - An initial temperature T_0 (e.g., $T_0 = 40{,}000$)
  - A freezing temperature T_{freez} (e.g., $T_{\text{freez}} = 0.1$)
  - A cooling schedule (e.g., $T = 0.85 \times T$)

The pseudo-code for the simulated annealing algorithm can be written as follows:

```
Procedure simulated annealing
  curSoln = random initial solution;
  T = T0; // initial temperature
  While (T > Tfreez) do
  For I = 1 to NUM_MOVES_PER_TEMP_STEP do
  nextSol = perturb (curSoln);
  deltaCost = cost (nextSol)- cost (curSoln);
  if (acceptMove (deltaCost, T) then
  curSoln = nextSoln;
  T = coolDown (T);

Procedure acceptMove (delataCost, T)
  If delataCost < 0 then return TRUE;
  Else
  Boltz = exp (-deltaCost/kT);
  R = random (0,1);
  If R < Boltz then return TRUE
  Else return FALSE;
```

Summary of the algorithm:

(a) Good moves are always accepted
(b) Bad moves are accepted only
  ■ when $T = T_0$, bad moves acceptance probability = 1
  ■ when $T = T_{freez}$, bad move acceptance probability = 0
(c) Boltzmann probability function, Boltz = $\exp(-\text{deltaCost}/kT)$

where k = Boltzmann constant, chosen so that all moves at the initial temperature are accepted.

Wong–Liu Algorithm

Before explaining the algorithm let us first formulate the floorplanning problem. Consider a module M of rectangular shape with height h, width w, and area A. The aspect ratio of M is defined as h/w. A soft module is a module whose shape can be changed within a given range of aspect ratio and area. A floorplan having N number of such modules is a rectangle R which is subdivided by horizontal and vertical lines into N non-overlapping rectangles, such that each rectangle can accommodate the module assigned to it.

The floorplanning cost is defined by the following expression:

$$\text{Cost} = A + \lambda \times L \qquad (11.1)$$

where A = area of the floorplan = $h \times w$
  λ = user defined constant which defines the relative importance of A and L.
  L = interconnect length = $\sum c_{m,n} \times d_{m,n}$
where $c_{m,n}$ = connectivity between blocks m and n.
  $d_{m,n}$ = Manhattan distance between the centers of rectangles of blocks m and n.

The objective of floorplanning is to construct a floorplan R to minimize the cost defined by Eqn (11.1).

A slicing floorplan is represented by a slicing tree. Each internal node of the slicing tree is labelled by V or H, corresponding to a vertical or a horizontal cut, respectively. Each leaf corresponds to a basic module and is labelled by a number from 1 to N. A polish expression is obtained if the slicing tree is traversed in post-order. A polish expression is called *normalized* if there is no consecutive V's or consecutive H's in the sequence.

According to Wong and Liu (1986), there is a one-to-one mapping between the set of normalized polish expressions of length $(2N - 1)$ and the set of slicing floorplans with N modules. The solution space is constructed using the set of all normalized polish expressions for the simulated annealing method. They have defined three types of moves (M1, M2, and M3) to transform a polish expression into another for searching the solution space efficiently. The flexibility of the soft modules is used to select the best floorplan among all the equivalent ones represented by the same polish expression. To examine a polish expression they have used an efficient shape function (explained in the next section). The cost function given by Eqn (11.1) is used. This algorithm is very efficient and the performance is very good, but this method does not consider any placement constraint.

To consider the placement constraints, Young et al. (1999) have modified the cost function by introducing a cost component in Eqn (11.1) as

$$Cost = A + \lambda \times L + \zeta \times D \tag{11.2}$$

where D = penalty cost due to boundary constraint and ζ is another constant.

Shape Function Evaluation   It is the process of optimizing area and delay by determining flexible blocks shapes and hence, the floorplan. The flexible cells are those cells whose aspect ratio can be changed keeping the area constant. If w is the width and h is the height of the cell, the area of the cell A ($= w \times h$) is constant. Hence, the plot of h vs w is a hyperbola. This is called shape function 1.

As very thin and tall cells are not feasible, the shape function will be a combination of hyperbola, and one vertical and one horizontal straight line.

As the cells are built using transistors, their width and height cannot change continuously. So the shape function cannot be a hyperbola.

Generally, a piecewise linear function is a better choice to represent any shape function. The points, where the function changes its direction, are called the break points of the piecewise linear function.

The different shape functions are shown in Fig. 11.11.

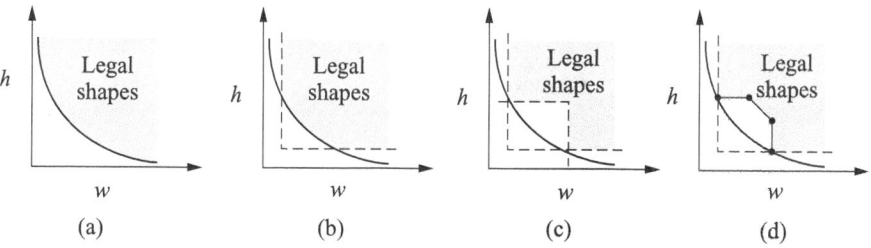

Fig. 11.11   Shape functions: (a) hyperbola; (b) hyperbola except tall and thin cells; (c) discrete set of (h, w) values; (d) piecewise linear

Genetic Algorithm for Floorplanning

The area-optimized floorplanning techniques—slicing and non-slicing floor-planning methods—can achieve well area minimization. Genetic algorithm (GA) is also used for floorplanning, and it produces a good compact floorplan. *Genetic algorithm* is a class of search and optimization method that mimics the biological evolution process, which is used for solving problems in a wide domain. It was proposed by Professor Holland in 1975. Figure 11.12 shows a flowchart of the GA.

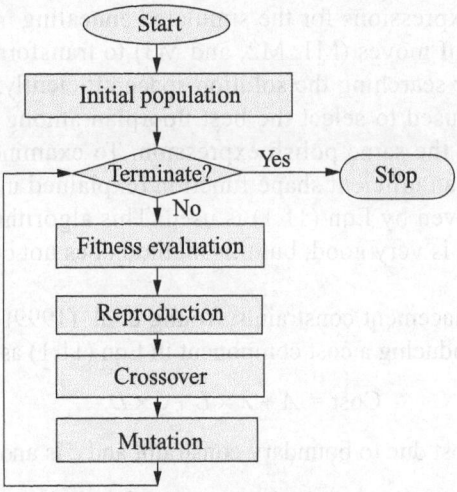

Fig. 11.12   Flowchart of the genetic algorithm

The GA operates through a simple cycle of stages as given below:
- Creation of initial population with binary strings called chromosomes
- Evaluation of fitness score by a fitness function
- Selection of best strings based on that fitness function that is further based on the objectives of the optimization problem
- A new set of chromosomes is generated by reproduction, crossover, and mutation—similar to natural selection operators

In the reproduction process, good chromosomes are selected, based on the fitness score and duplicated. In crossover (Fig. 11.13), randomly selected two chromosomes are picked up and some portions of the chromosomes are exchanged with a probability P_c. Finally, in the mutation process, one or more randomly selected bits are altered with a small mutation probability P_m.

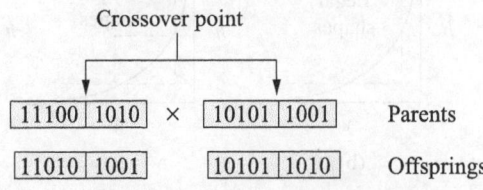

Fig. 11.13   Illustration of crossover

Each cycle in the GA produces a new set of chromosomes (possible solutions) for a given problem until a termination criterion is satisfied. It can very effectively search a large solution space while ignoring regions of the space that are not useful. Genetic algorithm is very popular for its very time-efficient searches.

There are many fields in a chromosome. Each field has several bits. Some of the bits are fixed either 1 or 0, and some other bits are don't-care (*). A bit string with '*' is called a schema. There must be many schema in an initial population pool.

The sample size of schema H in generation $(t + 1)$ is given by

$$m_H(t+1) = m_H(t)\frac{f_H}{f_{av}}\left[1 - p_c\frac{\delta_{(H)}}{(L-1)}\right] \times \left[1 - p_m\right]^{O_{(H)}} \tag{11.3}$$

where $m_H(t) =$ sample of schema H at generation t

$f_H =$ fitness of schema H

$f_{av} = \dfrac{1}{N}\displaystyle\sum_{i=1}^{N} f_i =$ population average fitness at time t

$N =$ population size

$p_c =$ probability of crossover

$L =$ word length of the chromosome

$\delta_{(H)} =$ defining length

$p_m =$ mutation probability

$O_{(H)} =$ order of schema H

A GA-based floorplanning method begins with a randomly generated *initial population*, which consists of many randomly generated chromosomes (floorplans). The floorplan design is encoded into integer and character (H/V) mixed strings to form the chromosome. Different slicing trees are obtained from different polish expressions, which represent the different orientations of the blocks obtained by rotations. The area estimation function is used to calculate the fitness score (dead space) of each population. The chromosomes with better fitness score will survive at each generation. The reproduction, crossover, and mutation operations generate new sets of chromosomes—or new floorplans. The iteration stops when the termination criterion is satisfied.

As discussed before, the floorplan can be represented by two different classes: (a) slicing and (b) non-slicing representations. It is found empirically that a slicing floorplan achieves a better area optimization compared to the non-slicing floorplan (Young et al., 1999).

Hence, the GA is explained with the slicing tree representations of floorplans.

A slicing floorplan in a rectangular area is obtained by slicing it recursively through horizontal or vertical cuts into a set of rectangular regions to accommodate a set of functional modules. A slicing floorplan is also represented by the polish expression. Let us consider an example of slicing tree, the polish expression, and its corresponding floorplan, as shown in Fig. 11.14. The slicing tree is a binary tree and is constructed in bottom–up approach.

Let us consider a slicing floorplan with four blocks as shown in Fig. 11.15.

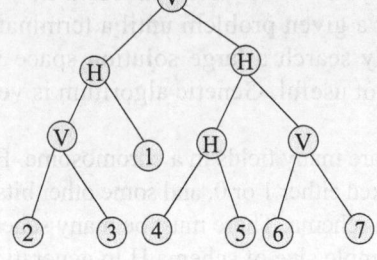

H : Horizontal cut
V : Vertical cut

Polish expression = 2 3 V 1 H 4 5 H 6 7 V H V

(a)   (b)

Fig. 11.14   (a) Slicing floorplan; (b) slicing tree

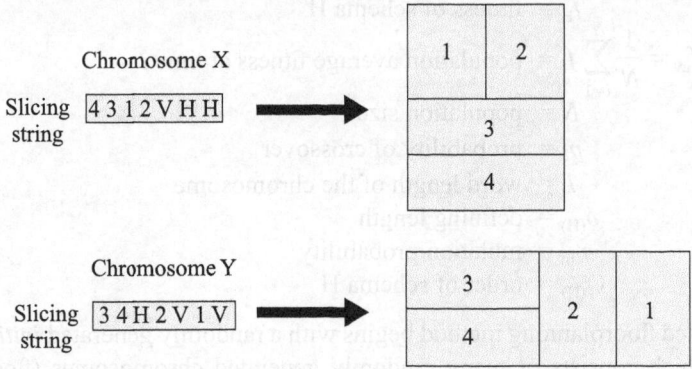

Fig. 11.15   Chromosome encoding for floorplanning

The slicing string is generated by four unique integers, which represent four functional blocks and two slicing operators, V and H. The V operator represents a vertical cut while the H operator represents a horizontal cut. The slicing string is used to find good solutions (minimal area) as the evolution progress.

Two chromosomes and their floorplans are shown in Fig. 11.15. This representation is used to obtain the exact location and orientation of each block on a chip. A fitness function is then used to find out the suitability of a chromosome, based on the optimization objectives.

This optimization goal is to achieve a minimum total chip area to accommodate a set of modules. The fitness of a chromosome is defined by Eqn (11.4).

$$\text{Fitness} = \frac{A_{\text{blocks}}}{A_{\text{chip}} - A_{\text{blocks}}} \tag{11.4}$$

where A_{blocks} = total area of the blocks and A_{chip} = total floorplan area.

The difference of A_{chip} and A_{blocks} is called the *dead space*. The fitness is the ratio of the total area of the blocks and the dead space of a floorplan. Higher fitness implies less dead space. The chromosomes having higher fitness have a chance to survive to the next generation.

The initial population is generated by setting the appropriate population size, the crossover probability p_c, and the mutation probability p_m. The initial population must be large, so that, the genetic diversity within the population can increase for many generations. But too many populations increase the computation time for each generation. However, it takes fewer generations to find the best solution.

11.3.7 Input/Output and Power Planning

A chip contains the input/output (I/O) pins for receiving input signals from the outside world, and for transmitting output signals to the outside world. It also contains the power and ground pins for getting the power supply. During the floorplanning stage, proper planning for the power and I/O pads is required.

A chip die area is estimated on two factors:

- Core size
- Number of pads

A die is called *core-limited* if its area is decided on the core size. A die is called *pad-limited* if its area is decided on the number of pads.

The die has bond pads at the periphery which are the I/O and power/ground ports of the internal circuitry. The IC package pins are provided to connect the IC to the outside world. The package pins are metallic in structure and often called the lead frame. The bond pads are connected to the lead frames via thin metallic wires called bond wires. A typical IC chip layout with the die and package is shown in Fig. 11.16.

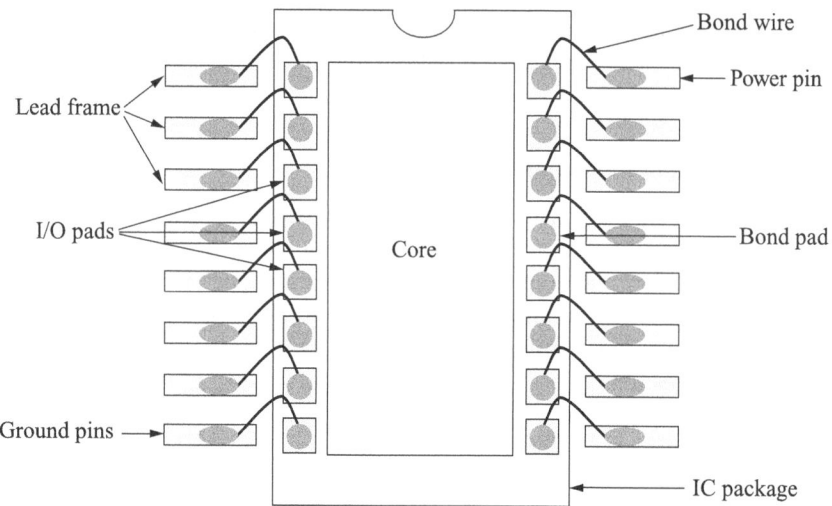

Fig. 11.16 Typical IC chip layout with the core (die) and the package

Normally, a chip would contain two power supply pins: (a) for the core logic which is V_{DD} and (b) for the I/O pads which is $V_{DD}IO$. The power supply connected to I/O pads experiences high transient overshoot and undershoots due to high switching currents. The I/O pads are also connected to electrostatic discharge (ESD) protection circuits, which are specially designed to withstand very high voltage for a short duration. To avoid the injection of the noise into the core logic through power supplies, the core supply is separated from the I/O supply.

11.3.8 Clock Planning

An IC chip contains a clock input pin which is used to receive the clock pulse from the outside world. The clock is used to synchronize the operations in the chip. The clock signals need to be distributed to all the flip-flops in the chip. The flip-flops could be placed anywhere in the chip. So the clock signals arrive at the clock input of the flip-flops after traversing different interconnect lengths. This will cause the clock signals to appear at the clock input of the flip-flops at slightly mismatched timing, i.e., with the clock skew. Hence, the synchronous operation cannot be performed. This is taken care of by the proper clock planning in the floorplanning stage and during the clock routing stage. The clock pin should be placed in such a way that the clock distribution network has an equal path to traverse from the master clock pin to the destinations.

11.3.9 Delay Estimation During Floorplanning

In the layout design cycle, it is always important to check that the timing requirements are met. The accurate delay analysis depends on the delay associated with the interconnect. The delay of the interconnect depends on the parasitic resistance and capacitance (RC). But during the floorplanning stage, the parasitic RC values of the interconnect cannot be determined. However, during the floorplanning, the interconnect length can be estimated on heuristics. Based on the statistics, a wire-load table is generated to estimate the delay of the interconnect. At the floorplanning stage, the fan-out (i.e., the number of gates driven by output of a gate) of each gate is known. Hence, delay of the circuit can be estimated on the pre-characterized delay table of standard cells and wire-load table for interconnects.

11.3.10 Floorplanning Guidelines

Floorplanning is the most critical part in physical design which forms the foundation of the chip layout. The accurate circuit timing and performance is dependent on the quality of floorplanning. Quality of floorplanning determines the quality of final designs. Poor floorplanning often leads to various problems, as given below:

- Design failing to meet timing specifications
- Routing congestion
- High power dissipation
- Large chip area
- Huge IR-drops
- Signal integrity issues

Floorplanning involves certain decisions at the beginning of the physical design flow. They are given by

- Pin/pad locations
- Hard macro placement
- Placement and routing blockage
- Location and area of the soft macros, and their pin locations
- Number of power pads and its location

Some of the tips that help in making correct floorplanning decisions are given below:

1. Fix the location of the pin or pad always considering the surrounding environment to which the block or chip is interacting. This can avoid routing congestion and also benefits in effective circuit timing. For example, if a chip/block containing the RX_DATA bus is going to sit on right-hand side, the TX_DATA bus should also be placed on the right-hand side of the chip/block.

2. Provide sufficient number of power/ground pads on each side of the chip for effective power distribution. Decide the number of power/ground pads based on the power report and IR-drop in the design.

3. Proper macro placement is crucial for the performance of any design. A design typically contains a number of hard macros varying from memory, PLL, and processors. Fly-line analysis must be done while placing the macros. This analysis gives a clear view of the interconnection with other macros and IO pins. Orientation of these macros affects floorplanning in a great deal. Figure 11.17 compares two approaches of macro placement, where the approach shown in Fig. 11.17(b) is the more desirable approach.

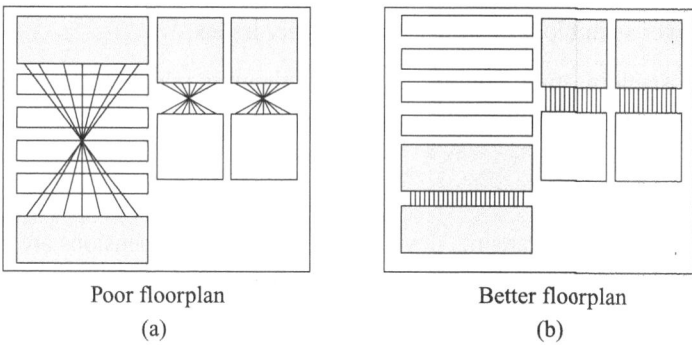

<table>
<tr><td>Poor floorplan</td><td>Better floorplan</td></tr>
<tr><td>(a)</td><td>(b)</td></tr>
</table>

Fig. 11.17 Two styles of macro placement

4. Do not spread standard cells throughout the chip area. This will create small placement traps. Figure 11.18(a) shows a floorplan with many pockets and isolated regions between the macros that can trap a standard cell, and limit the routing access. A layout design engineer must create layout having homogeneous standard cell area with aligned macros, as shown in Fig. 11.18(b).

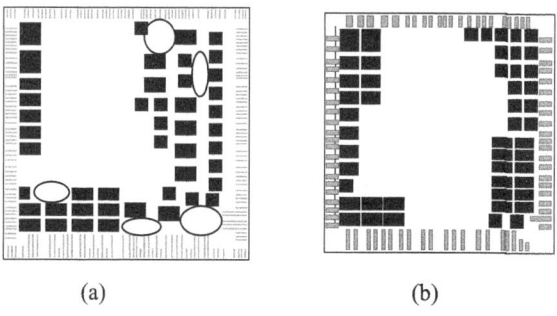

<table>
<tr><td>(a)</td><td>(b)</td></tr>
</table>

Fig. 11.18 Standard cells placed: (a) across the chip; (b) homogeneously

5. Creating standard cell placement blockage at the corner of the macro helps to reduce the routing congestion. Also, create standard cell placement blockage in long thin channel between macros.
6. Avoid uneven routing resources in the design by using the proper aspect ratio (width/height) of the chip. For example, consider a library with four routing layers, and the standard-cell rails are in layer 1. This arrangement reduces the horizontal routing resources to layer 3, whereas vertical routing resources are available on layers 2 and 4. Lack of horizontal routing resources can lead to congestion on layers 1 and 3. Greater width of the chip than its height compounds the problem.
7. For designs that have horizontal overflow, to increase utilization, cell row separation can be increased to increase the horizontal routing resources.
8. In a hierarchical design, cluster-based implementation enables placing the standard cells of the given module in a predefined region. This ensures that the interacting cells will be sitting in close proximity and save routing resource. The length of the routes will also be small, which in turn reduce the delay of the path.
9. If the design contains an analog block and the routing blockage is not defined in the physical library, then this creates routing blockage for all layers over the analog block. Analog blocks are more susceptible to noise, and signal routes going over such blocks cause signal integrity issues.

It is worth spending time and efforts in floorplanning which will save iterations and make the design cycle faster.

11.4 Placement

As the technology is advancing, device and interconnect dimensions are shrinking. But it has been found that circuit delay, power dissipation, and area are dominated by the interconnections, as compared to the devices. Hence, placement is becoming very critical in today's high performance VLSI design. It is the second step in VLSI physical design flow after the floorplanning step. Floorplanning step generates an initial layout of a design with the location of the fixed block, and location and shape of the flexible blocks. In the placement step, the logic cells are placed within the flexible blocks. It determines the physical layout of a chip. The routability of the layout is mainly determined by the placement. In this section, we will discuss the goals and objectives of placement followed by the placement algorithms.

An example circuit and its corresponding layout are shown in Fig. 11.19.

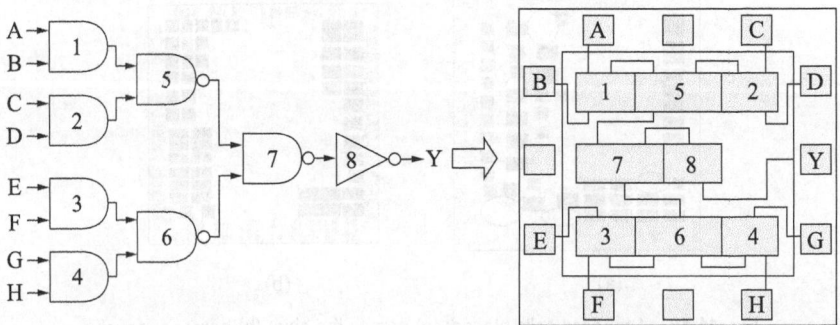

Fig. 11.19   A logic circuit and its corresponding layout

11.4.1 Goals and Objectives

The main purpose of placement is to place all the standard cells within the flexible blocks in a chip. The objectives of the placement step are as follows:
1. Minimize the chip area.
2. Minimize the delay on critical paths.
3. Ensure that layout is routable.
4. Minimize power dissipation of the design.
5. Minimize the crosstalk between the signals.

The placement goals are as follows:
1. Minimize the total estimated wire length of all the nets.
2. Minimize the interconnect congestion.

11.4.2 Placement Algorithms

Let us now discuss the various placement algorithms. There are mainly two placement algorithms as shown below:
- Constructive placement
- Iterative placement

In the constructive placement method, a set of rules is used. In this method, once a cell is placed, it is fixed. An already placed cell is not moved further. However, in the iterative placement, the cells are assigned to some locations, but they are moved around to obtain a better configuration.

Constructive Placement

The most widely used constructive placement method follows two algorithms:
- Min-cut algorithm
- Clustering method

Both of these methods basically partition the whole circuit into subcircuits, ensuring the number of connections between the subcircuits is minimum.

In the min-cut algorithm, the circuit is divided into two subcircuits of similar size while minimizing the number of connections between the subcircuits. The algorithm is called *min-cut*, as the minimum number of connections between the subcircuits is cut while partitioning. The subcircuits obtained from bipartition can be placed in separate halves of the layout, either in upper and lower halves, or in the left and right halves. This bipartitioning (see Fig. 11.20) is used recursively until the subcircuits become basic cells. The min-cut algorithm works in the top–down approach as it starts with the complete circuit and ends with basic logic gates.

The clustering method works just opposite to the min-cut approach. It follows the bottom–up approach. In

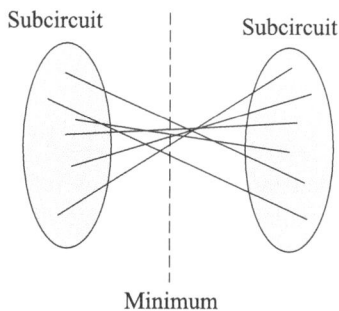

Fig. 11.20 Illustration of circuit bipartitioning

this method, a cell is first picked up and then finds out other cells that are sharing the common nets. These cells form a cluster to which more cells are appended sharing the same nets. This clustering method continues until the cluster becomes the whole circuit.

Iterative Placement

In the iterative placement, an already placed layout is taken, and the placed cells are perturbed or moved around to generate a new placement. If the new placement is better than the old one, the old one is replaced by the new one. This way the placement can be improved; hence, the method is also called *iterative placement improvement*.

The algorithm consists of the following basic steps:

- Initial configuration
- Cost computation
- Perturb to generate new placement
- Checking the acceptance criteria
- Stop if the number of iterations reach the limit

Figure 11.21 shows the flowchart of the iterative placement improvement algorithm.

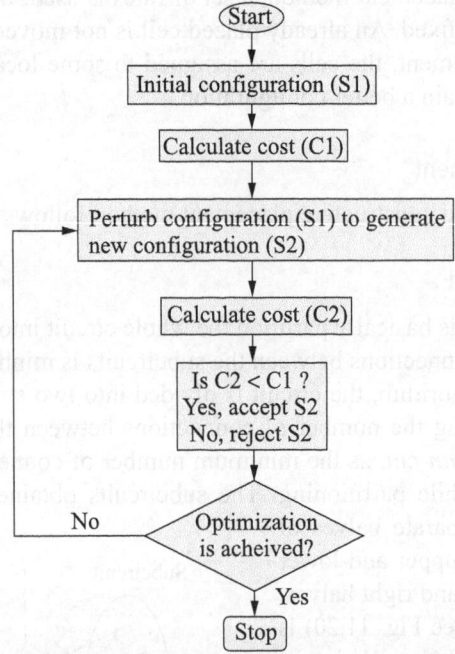

Fig. 11.21 Flowchart for the iterative placement

Simulated annealing can be used for iterative placement improvement. According to the simulated annealing algorithm, the following tasks are performed:

1. Select logic cells for trial perturbation (usually random).
2. Perturb the initial placement.
3. Calculate the cost function (E).

4. If $\Delta E \leq 0$ then exchange the logic cells. If $\Delta E > 0$, then exchange logic with a probability $\exp(-\Delta E / T)$.
5. Go back to step 2 if number of iterations has not reached the limit. Reduce the temperature, for example, $T_{n+1} = 0.85 T_n$.

11.4.3 Partitioning Algorithms

We have seen that the min-cut algorithm is based on the circuit partitioning technique. In this section, we will focus on the two types of partitioning techniques given below:

- Constructive partitioning
- Iterative partitioning improvement

The constructive partitioning uses seed growth or cluster technique. In this partitioning, a basic logic cell called seed is taken. Other cells, which are not yet partitioned, are added to the seed, based on the number of connections. This process is repeated until all cells are considered for partitioning.

The iterative partitioning improvement is based on the interchange and group migration process. In the interchange process, the logic cells are interchanged or swapped to improve the partitioning. In the group migration process, the groups of logic cells are interchanged. The group migration process is based on a useful algorithm proposed by Kernighan and Lin 1970.

Kernighan–Lin Algorithm

To describe the Kernighan–Lin algorithm, let us define some models. Let us consider bipartition of a circuit, as shown in Fig. 11.22. The nets inside the subcircuits are called the internal edges, and the nets connected between the subcircuits are called the external edges. Assigning a cost to each edge of the circuit, we can define a cost matrix as

$$C = c_{ij} \tag{11.5}$$

where $c_{ij} = c_{ji}$ and $c_{ii} = 0$.

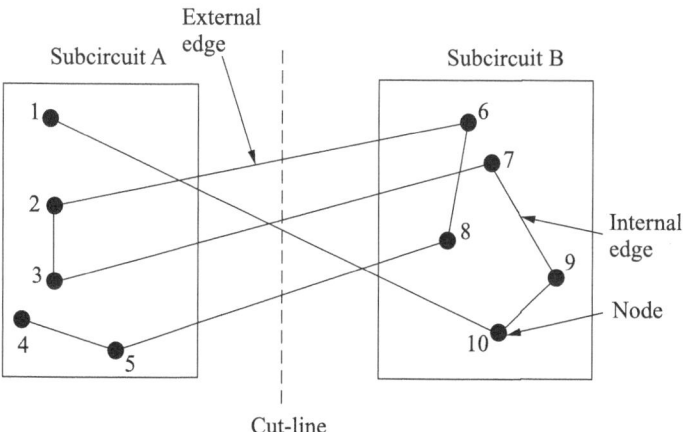

Fig. 11.22 Bipartitioned circuit

Let the two partitions be A and B having m number of nodes each. The main goal here is to swap nodes between A and B, such that the number of external edge is minimum. The cut cost is written as

$$W = \sum_{a \in A, b \in B} c_{ab} \qquad (11.6)$$

For any node a in subcircuit A, the external edge cost is defined as

$$E_a = \sum_{y \in B} c_{ay} \qquad (11.7)$$

For example, in Fig. 11.22, the external edge costs are $E_1 = 1, E_2 = 1, E_9 = 0$. The internal edge cost is defined as

$$I_a = \sum_{z \in A} c_{az} \qquad (11.8)$$

For example, in Fig. 11.22, the internal edge costs are $I_1 = 0, I_2 = 1, I_9 = 2$.
Similarly, we can define the edge cost for subcircuit B. The difference cost is given by

$$D_x = E_x - I_x \qquad (11.9)$$

For example, in Fig. 11.22, the difference costs are $D_1 = 1, D_2 = 0, D_9 = -2$.
If the nodes a and b are swapped, the reduction in cut-cost can be measured as

$$g_j = D_a + D_b - 2c_{ab} \qquad (11.10)$$

For example, if we swap the nodes 1 and 9, the cut-cost will be

$$g = D_1 + D_9 - 2c_{19} = 1 - 2 - 2 \times 0 = -1$$

According to the Kernighan–Lin (K–L) algorithm, a group of node pairs are swapped for which the gain will increase.

The steps of the K–L algorithm are:
1. Compute the difference cost for all the nodes a_i and b_i, where $a_i \in A, b_i \in B$.
2. Find the nodes a_i and b_i for which gain g_i is maximum.
3. Move a_i to B and b_i to A.
4. Do not consider swapping a_i and b_i in the next pass.
5. Find k for which $g_{max} = \sum_{i=1}^{k} g_i$.
6. If $g_{max} > 0$, swap the nodes $a_1, a_2, ..., a_k$ with the nodes $b_1, b_2, ..., b_k$.
7. Repeat steps 1 through 6 until $g_{max} = 0$.

11.4.4 Timing-driven Placement

The interconnect delay dominates the gate delay in modern IC technologies, as shown in Fig. 11.23. Hence, placement becomes a major factor in determining circuit timing. The timing-driven placement implies that the placement should be

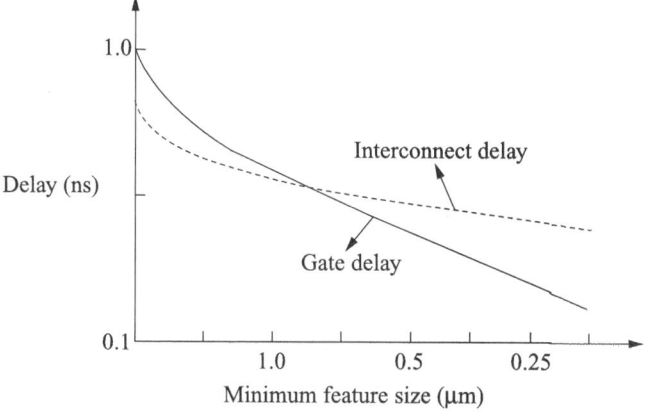

Fig. 11.23 Gate delay and interconnect delay vs minimum feature size

done in such a way that the timing constraints are met. There are two methods for doing timing-driven placement:

- Net-based approach
- Path-based approach

In the net-based approach, the net weights are calculated on the number of times they appear in the critical paths of the circuit. There is another method for calculating net weight. This is known as zero-slack algorithm, in which the delays are added to the nets until the slacks are zero. The net delays are then translated as the net weights.

In the path-based approach, the critical paths are considered during the placement. But this process is more complex than the net-based approach. Very few commercial tools have this feature.

11.4.5 Congestion-driven Placement

If too many wires are placed in a routing channel, the router cannot finish routing. This region of overcrowded wires in a local region is called *congested region*. Hence, in addition to wire length optimization, congestion reduction is another important aspect of the placement step. The congestion is estimated as the overflow on each edge as

$$\text{Overflow on each edge} = \text{Routing demand} - \text{Routing supply}$$

The total overflow is estimated as

$$\text{Total overflow} = \sum_{\text{all edges}} \text{Overflow} \tag{11.11}$$

Figure 11.24 shows a congestion estimation example for a chip. The chip is divided into a number of global bins. The edges are passed through the centre of the global bins. The edge that connects the centres of the global bins is called a global edge. For a global edge e, the routing supply s_e is a function of length of the edge and the technology parameters. The routing demand d_e is number of nets crossing the edge e. If $d_e > s_e$, then the edge is said to be congested.

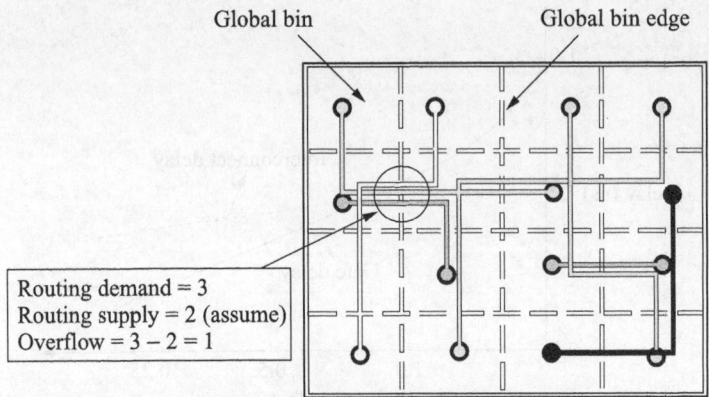

Fig. 11.24 Congestion estimation

Congestion optimization involves the following basic steps:

- Estimation of overflow
- Identification of congested region
- Expansion of congested region
- Reduction of congestion

As the number components in a circuit increases, the congestion problem becomes more and more important. The multiple routing layers are provided for more routing resources for routing, but even then, the routing congestion cannot be prevented in local regions. Excessive congestion in a region can result in a local shortage of the routing resource, or the router may fail to complete the routing.

11.4.6 Wire Length Estimation

The quality of a particular placement is evaluated based on the wire lengths. The length of each net is estimated and summed up to get the total wire length. Assuming certain wire spacing and wire width, the total wiring area is estimated. There are several ways to estimate the wire lengths.

Half Perimeter Wire Length (HPWL)

In this method, a smallest rectangle is considered that covers all the terminals of a net. The sum of the width and the height of the rectangle is an estimation of the wire length.

Minimum Rectilinear Spanning/Steiner Tree (MRST)

In this method, the wire length is estimated either by Manhattan distance or Euclidean distance. The Manhattan distance is given as $|x_1 - x_2| + |y_1 - y_2|$, whereas the Euclidean distance is given as $\sqrt{(x_1 - x_2)^2 + (y_1 - y_2)^2}$.

Squared Euclidean Distance

In this method, the placement cost is calculated by summing over all the cells rather than over the nets. The placement cost is defined as

$$\text{Placement cost} = \frac{1}{2} \sum_{i=1}^{n} \sum_{j=1}^{n} c_{ij} \left[(x_i - x_j)^2 + (y_i - y_j)^2 \right] \qquad (11.12)$$

where n is the number of cells, c_{ij} is the edge weight, and (x_i, y_i) and (x_j, y_j) are the coordinates of the cells.

Minimum Chain

In this method, wire length is estimated starting from one vertex connect to the closest one, then to the next closest, and so on.

Source-to-sink Connection

In this method, wire length is estimated by connecting one pin to all the other pins of the net. But this method is not accurate for uncongested chips.

Complete Graph

In this method, the wire length is estimated as

$$\text{Wire length} = \frac{2}{n} \sum_{i, j \in \text{nets}} \text{Distance}(i, j) \tag{11.13}$$

The different wire length estimation is depicted in Fig. 11.25.

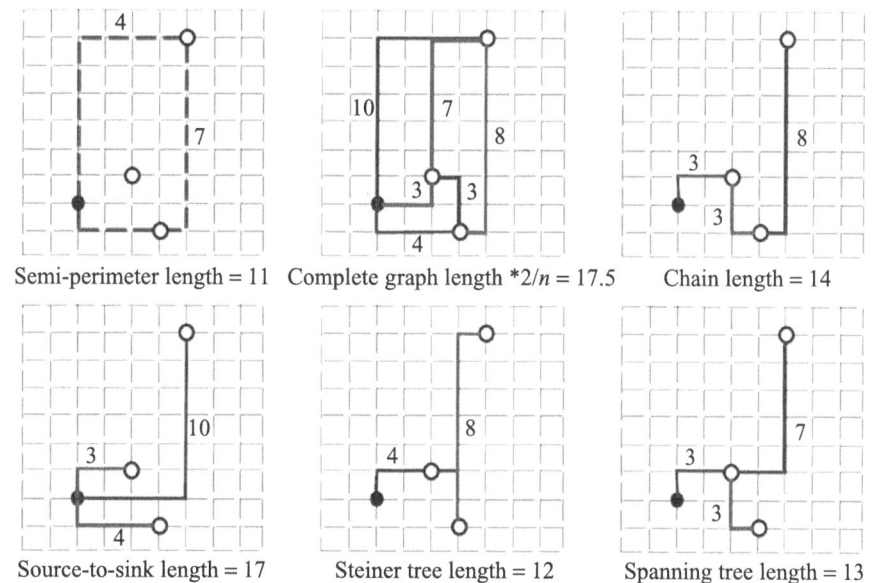

Semi-perimeter length = 11   Complete graph length *2/n = 17.5   Chain length = 14

Source-to-sink length = 17   Steiner tree length = 12   Spanning tree length = 13

Fig. 11.25   Different wire length estimation methods

11.4.7 Standard Cell Placement

The placement problem is mainly of two types: (a) standard cell placement and (b) building block placement. The standard cells are pre-designed, pre-tested, and pre-characterized logic gates such as inverter, NAND, NOR, XOR, flip-flops, etc. In a standard cell-based IC, the standard cells are interconnected to form the complete chip. The standard cells are of equal height. They are placed in rows separated by routing channels. An example of standard cell-based IC layout is shown in Fig. 11.26. The standard cell-based design is well suited for automated physical design flow. Its advantages are high productivity and efficient utilization of the chip area.

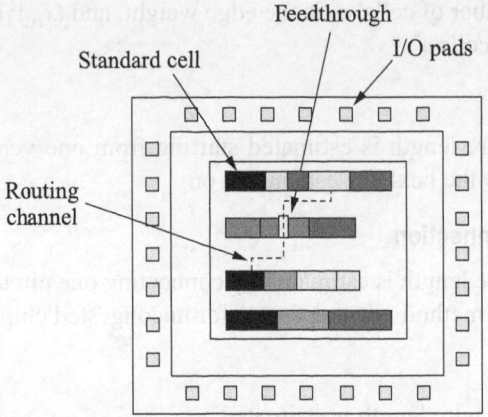

Fig. 11.26 Standard cell layout

11.5 Routing

Routing simply means drawing paths from the source to the destination in a layout. In the physical design flow, after the floorplan and placement step, the routing is done. In this step, the interconnection lines are drawn for the signal, power, ground, and clock nets. Although it seems a pretty simple process, it is a very hard problem to solve when the number of connections is very large.

To understand the routing problem, we consider a city design example. When a city planner designs the layout of a city, a very important job for him is to draw roads. The roads must be designed in such way that the traffic should be smooth. There should not be any congestion. Once he/she decides the floorplan of the city, he/she will first draw highways for long haul traffic movement followed by main roads connecting different parts of the city, and the entry and exit points on the highways. Next he/she will then draw the narrow roads for each block in the city. And finally, he will draw the narrow lanes for each house. He/she also needs to take care that every house gets power, telephone/cable/Internet connections.

In case of VLSI chip design also, when a chip designer starts designing the layout of the chip, after the floorplan and placement, he/she needs to draw interconnects. The city's power connection problem can be compared as supplying V_{DD} to each cell in the design. The telephone connection problem can be thought of as supplying the clock to the sequential blocks in the chip. The laying of roads in the city can be compared to the signal route between the blocks in the design. Whenever there is heavy congestion in a city, we must build flyover for the crossroads. Similarly, to avoid the routing congestion problem, we must have a multilayer interconnection methodology.

In this section, we discuss the goals of routing, the basic algorithms, and a few examples of the routing problem.

11.5.1 Goals and Objectives

At the end of placement phase, the exact location of the circuit blocks and pins are determined, and a netlist is generated which specifies the required interconnections, and region for interconnections (routing region) is also generated.

The main goal of routing is to complete the interconnections among the blocks according to the specified netlist. While interconnecting, the nets or interconnecting wires must satisfy constraints coming from the manufacturing and design style.

The objectives of routing are to

- Achieve 100% routing and
- Minimize
  - area of the chip
  - total wire length in the chip
  - layer changes for interconnection
  - net delay on critical paths

11.5.2 Routing Constraints

Designers have certain constraints that need to be accounted for while routing. Some of the constraints are due to manufacturing and CAD tool limitations, and others are design constraints.

Fab Line and CAD Tool Constraints

Figure 11.27 shows the typical interconnecting metal layers. The horizontal (H) and vertical (V) layers along x- and y-directions, respectively, are alternately drawn on top of each other. The metal layers are named M1, M2, M3, and so on from the bottom. Between the metal layers there is interconnecting vias (not shown in the figure) in the z-direction.

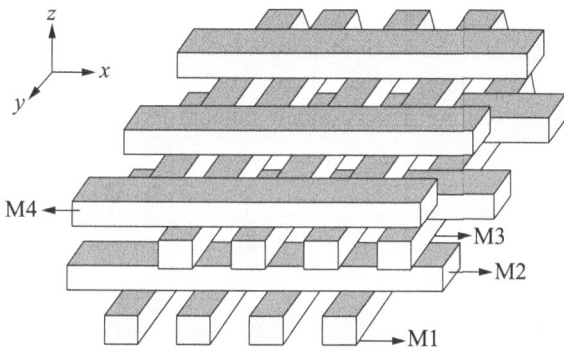

Fig. 11.27 Horizontal and vertical interconnecting metal layers

There are restrictions from the fabrication unit on the number of metal layers that can be drawn. There are design rules that specify the minimum width and spacing between the metal layers. A router must route the nets adhering to these design rules. The design rules are discussed in Section 6.4.2 in Chapter 6.

Following points are to be considered while routing:

- Presence of obstacles
- Wire orientation in a layer (H, V, H & V, D)
- Design rules on wire width and separation
- Grid vs gridless
- Number of terminals (usually two, but can be many for power and clock nets)
- Terminal positions (channel, switch box)

- Fixed vs floating terminals
- Net width (signal nets have less width but power/ground nets have greater width)
- Via restrictions
- Number of layers (number of metal layers)
- Net types (power, ground, clock, or signal net)
- Yield

The design constraints for routing are
- Heat and power dissipation
- Signal integrity (noise, crosstalk, power bounce)

11.6 Types of Routing

Routing can be classified into two main categories:
- Global routing
- Detailed or local routing

Global routing is the step where the actual routing is not done, rather, the routing is planned. It is done at the top level where routing regions are defined, the nets are assigned to routing regions, and all channel terminals are determined.

Detailed routing is the step where the actual routing is done. It is also known as local routing. Detailed routing is classified into two types:
- Area routing
- Channel routing

In area routing, the routing is carried out with obstacles. It is also called maze routing. This routing problem is to find a shortest path between two terminals on a grid with obstacles. In channel routing, dedicated space called routing channels are provided for routing. Channel routing is very much suitable for standard cell-based design (Fig. 11.28).

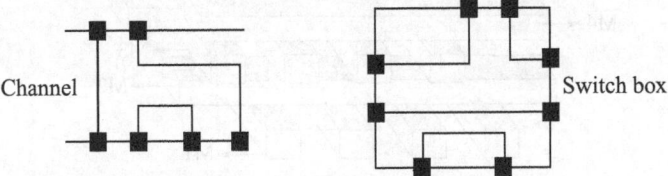

Fig. 11.28 Channel and switch box routing

11.6.1 Global Routing

Global routing is the planning for complete routing of the chip. It takes the netlist with the floorplan and placement information of the fixed and flexible blocks, and all the logic cells. It generates the routing information for every net that is used by the detailed router. The actual geometric layout is not done in the global routing. The main objective of the global routing step is to minimize the wire lengths, minimize the delay on critical paths, and increase the chance of complete routing by the router. The following are mainly the two approaches for global routing:
- Sequential approach
- Concurrent approach

Sequential Approach

In the sequential global routing approach, one net is taken at a time and the shortest path is calculated using tree or graph. There are two types of sequential routing: (a) order-independent routing and (b) order-dependent routing. In the order-independent routing, each net is taken and assigned channels, independent of whether it is taken first or last. After assigning all the nets, the nets from an overcrowded region are moved to a less-crowded region. In the order-dependent routing, the order of the consideration of nets impacts the result. The routing solutions are further improved using the simulated annealing algorithm or iterative improvement algorithm.

Concurrent Approach

In the concurrent approach, all the nets are considered at the same time. In this approach, the global routing problem is hierarchically divided into lower levels with a reduced problem size. Then the nets under a particular hierarchy level are considered at the same time. The hierarchy level can be traversed either by top–down approach or by bottom–up approach. For this reason, the concurrent approach is also known as *hierarchical routing*.

11.6.2 Detailed Routing

In the global routing, a tentative route is generated for each net. It assigns a list of routing regions for each net; but it does not specify the actual layout of wires. In the detailed routing step, the nets are assigned dimensions to form the actual geometric layout.

Area or Maze Routing

In this style of routing, terminals are allowed anywhere in the area available for routing. It finds out the shortest path between two terminals on a grid with obstacles. The first area routing algorithm was proposed by Lee in 1961. Later on, an improvement was made on Lee's algorithm in 1978 by Soukup. In 1977, Hadlock had proposed an alternate approach of routing using the A* searching algorithm.

Area Routing by Lee's Algorithm

This algorithm is used where the positions of the source and destination are throughout the entire area of the chip. This type of routing problem is known as the area routing problem. The aim of this algorithm is to draw a path from a source to a destination, avoiding any obstacles that may be present. The algorithm is depicted in Fig. 11.29.

The whole area is divided into a grid structure, where the separation between the grid points indicates the minimum wire width and their spacing. The minimum width and spacing rule is derived from the process technology. The routing blockage is modelled as obstacles and are marked as 'X' as shown in Fig. 11.29(a). The grid points are shown by dots, and the source and destination are represented by 'S' and 'D', as shown in Fig. 11.29(a). The algorithm tries to find the shortest path between the source and destination.

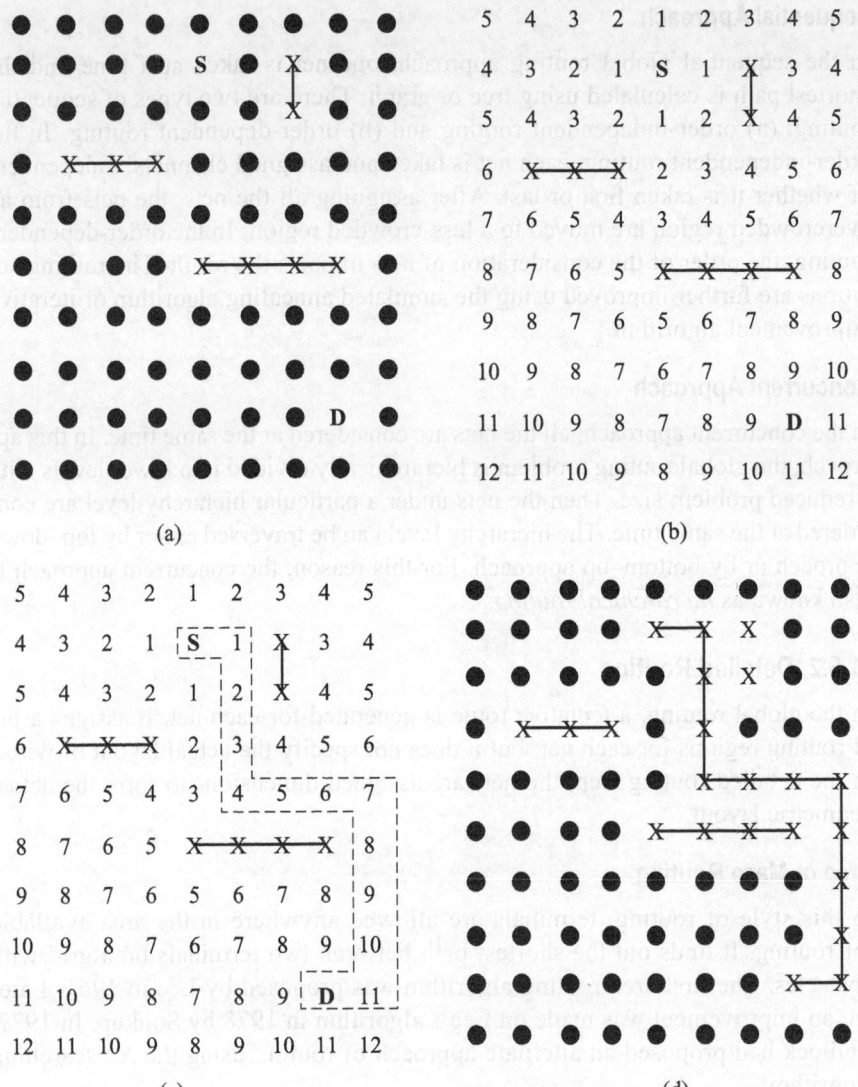

Fig. 11.29   Lee's routing algorithm: (a) source (S), destination (D), and obstacles (X);
(b) wave propagation from S to D; (c) connected path from S to D; (d) found path becomes
obstacle for next problem

Wave Propagation

A path between two points is realized by propagating a wavefront from the source
node outwards until the destination node is reached. This methodology is termed
as *wave propagation*. Starting from the node 'S', the grid points are labelled. The
labelling is done by putting integer '1' to the grid points belonging to first wavefront. The grids belonging to second wavefront are labelled by integer '2'. This
process continues until the destination node 'D' is reached. This is illustrated in
Fig. 11.29(b).

Backtracing

In the second step, starting from the destination node 'D', the shortest path is identified by backtracing the grid points with decreasing labels. This is illustrated in Fig. 11.29(c).

Clean-up

The path found using the wave propagation and backtracing methods become obstacles for the next routing problem. So the labels of all the grid points are removed so that next routing task can be taken up. This step is known as clean-up, and it is illustrated in Fig. 11.29(d).

Lee's algorithm is not suitable for the routing problem where more than two terminals are to be connected. This algorithm is also known as *Lee–Maze routing algorithm*.

Hightower Algorithm

Hightower algorithm is a modified version of Lee's algorithm. It is more efficient than the latter. It is also known as line-search algorithm. Figure 11.30 illustrates the algorithm.

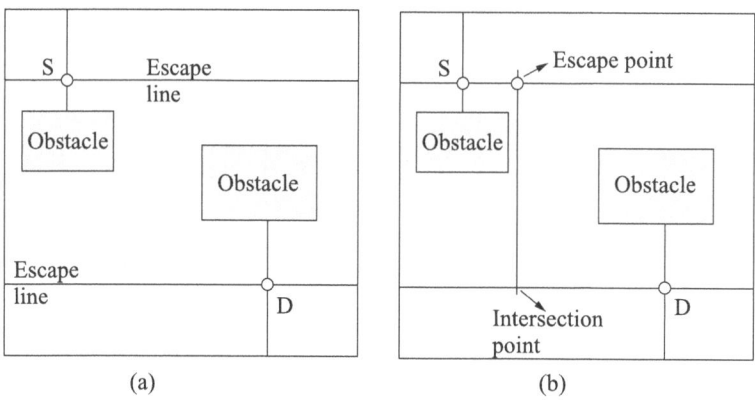

Fig. 11.30 Line-search algorithm: (a) escape lines are drawn from source 'S' and destination 'D'; (b) escape point and intersection point are found

In the first step, from the source and the destination nodes, one horizontal and vertical line are drawn towards each other. The lines are extended until they meet the obstacles. These lines are called *escape lines*. When the escape lines meet the obstacle, find out a point from which a line, which is right-angled to the present line, can be drawn avoiding the obstacle. This point is called *escape point*. Continue this process till the escape lines originating from the source and destination intersect each other. When the intersection point is found, the escape lines drawn from the source to the destination form the routing path.

The method is faster as compared to the Lee–Maze routing algorithm.

Channel Routing Algorithm

In a standard cell-based design, the standard cells are placed in rows separated by routing regions called *routing channels*. The nets have fixed terminals at the top and bottom. However, the nets have floating terminals on both the left and right ends. The main goal of channel routing is to reduce the net heights. The other goals of channel routing are minimization of total wire length and number of vias.

Left-edge Algorithm

In 1971, Hashimoto et al. had proposed the left-edge algorithm (LEA) (Fig. 11.31). It considers two layers for channel routing. For example, metal1 (m1) layer is used for routing in the horizontal direction, whereas metal2 (m2) layer is used for routing in the vertical direction. The steps of the LEA can be written so as to

1. Identify the net's horizontal segments in a routing channel.
2. Arrange the nets according to the leftmost edges.
3. Assign the first net to the first free track.
4. Assign the next net which fits in the same track.
5. Repeat Step 4 till first track is exhausted.
6. Consider next track and repeat steps from 3 to 5 until all net segments are assigned tracks.
7. Make the vertical connections to the top and bottom of the channel.

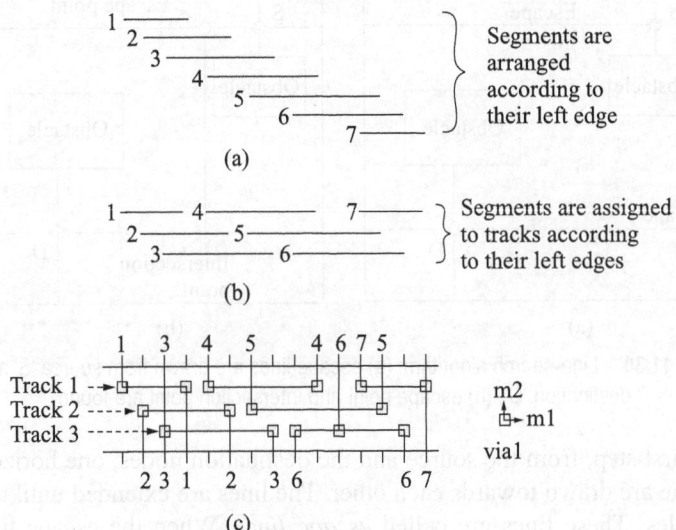

Fig. 11.31   Channel routing using left-edge algorithm: (a) segments are arranged according to their left edges; (b) segments are assigned to tracks based on their left edges; (c) channel route completed

11.7   Special Routing

The clock nets and the power nets are routed specially before the signal nets are routed. In this section, we will discuss the clock routing and power routing separately.

11.7.1 Clock Routing

Clock nets routed specially to avoid the clock skew problem. The clock skew problem arises due to the fact that clock signals traversing different net lengths arrive at the clock inputs of the synchronous elements at different times. This impacts the synchronous operation.

A clock routing is shown in Fig. 11.32(a), where the net lengths to the leaf nodes (sequential element) are kept same to avoid the clock skew. Another technique that is used to avoid the clock skew problem is called clock-buffer insertion. Clock buffers are inserted in the paths to equalize the net delays as shown in Fig. 11.32(b).

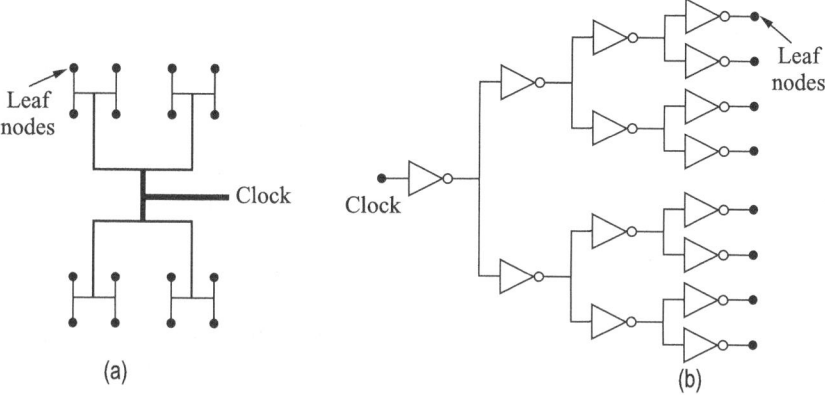

(a)　　　　　　　　　　　　　　(b)

Fig. 11.32　(a) Clock routing as a tree structure (b) Clock buffering technique

11.7.2 Power Routing

The cross-sectional area of the power nets is limited by the electromigration and voltage drop phenomena. Depending on the maximum current that a power net must carry, the width of the power net is decided by a power simulation tool or by hand calculation. The power distribution network is typically a grid structure for gate array-based design, whereas in standard cell-based design, it is composed of two horizontal layers: one for supplying V_{DD} and the other for the ground as illustrated in Fig. 11.33.

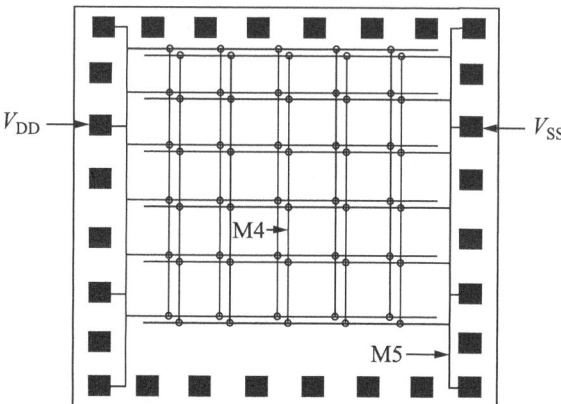

Fig. 11.33　Power and ground routing

11.8 Multilevel Routing

As the technology advances, the circuit size increases according to Moore's law, but the chip area is not increased proportionally. Even though the device sizes are reduced, the available routing area is not enough for the routing to be completed using one or two metal layers. Hence, there is a requirement to have multilevel routing, and fabrication technology must support this. Recent semiconductor fabrication technology allows at least five metal layers for routing. The two models that are followed in multilevel routing are given below:

- Reserved layer routing
- Unreserved layer routing

In the reserved layer routing, all interconnections are drawn in one direction, either horizontal or vertical, whereas in the unreserved layer routing, the interconnections are drawn in either direction. In the two-level routing, two types of routing models are followed: (a) HV routing and (b) VH routing. In the HV routing, metal1 is run in the horizontal direction and metal2 is run in the vertical direction. In the VH routing, metal1 is run in the vertical direction and metal2 is run in the horizontal direction. In the three-level routing, two routing models are followed: (a) VHV and (b) HVH routing. In VHV routing, metal1 and metal3 are run in the vertical direction and metal2 is run in the horizontal direction, and the opposite is followed for HVH routing. Generally, the higher level metal layers have a coarser pitch, and are used for power and clock lines rather than signal lines. The multilevel routing layers are shown in Fig. 11.34.

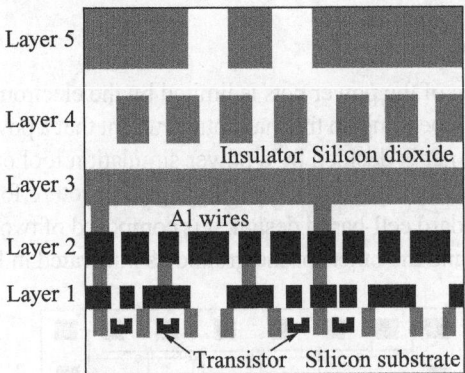

Fig. 11.34 Multilayer metallization

11.9 Physical Design EDA Tools

The most widely used EDA tools for physical design are shown in Table 11.1.

Table 11.1 Physical design tools

Vendor	Tool
Synopsys	Apollo
Cadence	SoC Encounter
Magma	Talus Vortex

11.10 Layout Database

There are different file formats for storing the layout information. The most widely used file formats are given below:

- *Library exchange format (LEF)* This is in ASCII text format used to store standard cell library information. It contains three sections: library, site, and macro.
- *Design exchange format (DEF)* It is an open specification for representing the physical layout of an IC in an ASCII format. It is used to represent a netlist, component placements, and routing information.
- *Graphic data system II (GDSII)* It is a layout database file in binary format. It is the de facto industry standard for data exchange of IC layout information.

SUMMARY

- Physical design is the process of generating IC layout information from the synthesized netlist.
- Floorplanning step plans the chip area with an estimation of die area, decides the location of I/O pads, number of power pads, and locations of power and clock pads.
- Placement step places the blocks and standard cells such that the chip area is reduced, routing length is minimized, and routing congestions are avoided.
- Routing step makes the interconnections between the blocks or cells in the chip.
- Via is the vertical interconnect between the different metal layers.

SELECT REFERENCES

Agrawal, B. et al. 1993, 'Timing-Driven Placement', *US Patent 5218551*, June.

Borah, M., R.M. Owens, and M.J. Irwin 1994, 'An Edge-Based Heuristic for Steiner Routing', *IEEE Transactions on Computer-Aided Design of Integrated Circuits and Systems*, vol. 13, no. 12, December.

Carl, S. and S.V. Alberto 1985, 'The TimberWolf Placement and Routing Package', *IEEE Journal of Solid-State Circuits*, vol. sc-20, no. 2, April.

Chang, Y.W. and S.P. Lin 2004, 'MR: A New Framework for Multilevel Full-Chip Routing', *IEEE Transactions on Computer-Aided Design of Integrated Circuits and Systems*, vol. 23, no. 5, May.

Cohoon, J.P., S.U. Hegde, W.N. Martin, and D. Richards 1988, 'Floorplan Design Using Distributed Genetic Algorithms', *IEEE*.

Cohoon, J.P., Shailesh U. Hegde, Worthy N. Martin, and Dana S. Richards 1991, 'Distributed Genetic Algorithms for the Floorplan Design Problem', *IEEE Transactions on Computer-Aided Design*, vol. 10, no. 4, April.

Cong, J. and C.L. Liu 1988, 'Over-the-Cell Channel Routing', *ICCAD*.

Cong, J., J. Fang, M. Xie, and Y. Zhang 2005, 'MARS: A Multilevel Full-Chip Gridless Routing System', *IEEE Transactions on Computer-Aided Design of Integrated Circuits and Systems*, vol. 24, no. 3, March.

Dai, W.M. and E.S. Kuh 1987, 'Simultaneous Floor Planning and Global Routing for Hierarchical Building-Block Layout', *IEEE Transactions on Computer-Aided Design of Integrated Circuits and Systems*, vol. 6, no. 5, pp. 828–37, September.

Dai, W.W.M. 1989, 'Hierarchical Placement and Floorplanning in BEAR', *IEEE Transactions on Computer-Aided Design of Integrated Circuits and Systems*, vol. 8, no. 12, pp. 1335–49, December.

Donath, W. 1979, 'Placement and Average Interconnection Lengths of Computer Logic', *IEEE Transactions on Circuits and Systems*, vol. 26, no. 4, pp. 272–7, April.

Dunlop and Kernighan 1985, 'A Procedure for Placement of Standard Cell VLSI Circuits', *IEEE Transactions on Computer-Aided Design*, January.

Funabiki, N. and Y. Takefuji 1992, 'A Parallel Algorithm for Channel Routing Problems', *IEEE Transactions on Computer-Aided Design*, vol. 11, no. 4, April.

Ganley, J.L. and J.P. Cohon 1994, 'Routing a Multi-Terminal Critical Net: Steiner Tree Construction in the Presence of Obstacles', *IEEE International Symposium on Circuits and Systems*, pp. 113–16.

Gerez, H. 2007, *Algorithms for VLSI Design Automation*, Wiley.

Ho, T.T., S.S. Iyengar, and S.Q. Zheng 1991, 'A General Greedy Channel Routing Algorithm', *IEEE Transactions on Computer-Aided Design*, vol. 10, no. 2, February.

Hong, Y. and H. Xianlong 2000, 'A New Timing-Driven Placement Algorithm Based on Table-Lookup Delay Model', *Chinese Journal of Semiconductors*, vol. 21, no. 11, pp. 1129–38.

Hou, W.T., H. Yu, and X.L. Hong 2001, 'A New Congestion-Driven Placement Algorithm Based on Cell Inflation', *Proceedings of the IEEE ASP-DAC2001* (Japan), pp. 605–8.

Hung, W.L., Y. Xie, N. Vijaykrishnan, C. Addo-Quaye, T. Theocharides, and M.J. Irwin 2005, 'Thermal-Aware Floorplanning Using Genetic Algorithms', *IEEE*.

Jens, L. 1997, 'A Parallel Genetic Algorithm for Performance-Driven VLSI Routing', *IEEE Transactions on Evolutionary Computation*, vol. 1, no. 1, pp. 29–39.

Jhang, Kyoung-Son, Soonhoi Ha, and Chu Shik Jhon 1998, 'Simulated Annealing Approach to Crosstalk Minimization in Gridded Channel Routing', *VLSI Design*, vol. 7, no. 1, pp. 85–95.

Kahng, A.B. and G. Robins 1002, 'A New Class of Iterative Steiner Tree Heuristics with Good Performance', *IEEE Transactions on Computer-Aided Design*, vol. 11, no. 7, July.

Kirkpatrick, S., C.D. Gelatt, and M.P. Vecchi 1983, 'Optimization by Simulated Annealing', *Science, New Series*, vol. 220, no. 4598, pp. 671–80, May.

Kleinhans, J., G. Sigl, F.M. Johannes, and K.J. Antreich 1991, 'GORDIAN: VLSI Placement by Quadratic Programming and Slicing Optimization', *IEEE Transactions on Computer-Aided Design*, vol. 10, no. 3, March.

La Potin, D.P. and S.W. Director 1986, 'Mason: A Global Floorplanning Approach for VLSI Design', *IEEE Transactions on Computer-Aided Design of Integrated Circuits and Systems*, October vol. 5, no. 4, pp. 477–89.

Lienig, J. and K. Thulasiraman 1994, 'A New Genetic Algorithm for the Channel Routing Problem', *7th International Conference on VLSI Design*, January.

Murata, H., K. Fujiyoshi, S. Nakatake, and Y. Kajitani 1996, 'VLSI Module Placement Based on Rectangle-Packing by the Sequence-Pair', *IEEE Transactions on Computer-Aided Design of Integrated Circuits and Systems*, vol. 15, no. 12, December.

Parakh, P.N., R.B. Brown, and D.A. Sakallah 1998, 'Congestion-Driven Quadratic Placement', *Proceedings of ACM/IEEE DAC, New Jersey*, IEEE Computer Society Press, pp. 275–8.

Peichen, P. and C.L. Liu 1995, 'Area Minimization for Floorplans', *IEEE Transactions on Computer-Aided Design of Integrated Circuits and Systems*, vol. 14, no. 1, pp. 123–32, January.

Rebaudengo, M. and M. Reorda 1998, 'GALLO: A Genetic Algorithm for Floorplan Area Optimization', *IEEE Transaction on CAD*, vol. 15, no. 8, August.

Rutenbar, R.A. 1989, 'Simulated Annealing Algorithms: An Overview', *IEEE Circuits and Device Magazine*.

Shi, W. 1996, 'A Fast Algorithm for Area Minimization of Slicing Floorplans', *IEEE Transactions on Computer-Aided Design of Integrated Circuits and Systems*, vol. 15, no. 12, December.

Smith, M.J. S0 2002, *Application Specific Integrated Circuits*, Pearson Education.

Srinivasan, A., K. Chaudhary, E.S. Kuh 1991, 'RITUAL: An Algorithm for Performance-Driven Placement of Cell-Based ICs', in: *Proceedings of the 3rd Physical Design Workshop*, Nemacolin Woodlands, Pennsylvania, May.

Szkaliczki, T., 'Routing with Minimum Wire Length in the Dogleg-Free Manhattan Model Is NP-Complete', *Journal in Computer Society for Industrial and Applied Mathematics*, vol. 29, no. 1, pp. 274–87.

Ting-Hai, C., Y. Hsu, J. Ho, K. Boese, and A. Kahng 1002, 'Zero Skew Clock Routing with Minimum Wirelength', *IEEE Transactions on Circuits and Systems—II: Analog and Digital Signal Processing*, vol. 39, no. 11, November.

Tsay, R.S., E.S. Kuh, and C.P. Hsu 1988, 'PROUD: A Sea-of-Gates Placement Algorithm', *IEEE Design Test Compute*, pp. 44–56, December.

Wang, M. and M. Sarrafzadeh 1999, 'On the Behaviour of Congestion Minimization during Placement, International Symposium of Physical Design', *ACM*, Pullman, pp. 145–50, April.

Wimer, Sh., K. Israel, and C. Israel 1988, 'Floorplans, Planar Graphs, and Layouts', *IEEE Transactions on Circuits and Systems*, vol. 35, no. 3, March.

Wong, D.F. and C.L. Liu 1986, 'A New Algorithm for Floorplan Design', in: *Proceedings of the 23rd ACM/IEEE Design Automation Conference*, 1986, pp. 101–7.

Xing, Z. and R. Kao 2002, 'Shortest Path Search Using Tiles and Piecewise Linear Cost Propagation', *IEEE Transactions on Computer-Aided Design of Integrated Circuits and Systems*, vol. 21, no. 2, p. 145, February.

Yongkui, H. and I. Koren 2007, 'Simulated Annealing Based Temperature Aware Floor Planning', *Journal of Low Power Electronics*, vol. 3, pp. 1–15.

Young, F.Y., D.F. Wong, and Hannah H. Yang 1999, 'Slicing Floorplans with Boundary Constraints', *IEEE Transactions on Computer-Aided Design of Integrated Circuits and Systems*, vol. 18, no. 9, September.

Young, F.Y., C.C.N. Chu, W.S. Luk, and Y.C. Wong 2001, 'Handling Soft Modules in General Nonslicing Floorplan Using Lagrangian Relaxation', *IEEE Transactions on Computer-Aided Design of Integrated Circuits and Systems*, vol. 20, no. 5, May.

Zhang, C.X., A. Vogt, and D.A. Mlynski 1991, 'Floorplan Design Using a Hierarchical Neural Learning Algorithm', *IEEE International Symposium on Circuits and Systems*, vol. 4, 11–14 June, pp. 2060–3.

EXERCISES

Fill in the Blanks

1. A hard macro refers to a _____ .
   - (a) flexible block
   - (b) fixed block
   - (c) flexible block with fixed aspect ratio
   - (d) flexible block with fixed pin locations
2. MRST stands for _____ .
   - (a) minimum rectilinear Steiner tree
   - (b) maximum rectilinear Steiner tree
   - (c) minimum race setting time
   - (d) maximum race set-up time
3. The physical interconnection happens in the _____ .
   - (a) global routing step
   - (b) detailed routing step
   - (c) both in the global and detailed routing step
   - (d) floorplanning step
4. Left-edge algorithm (LEA) is used for _____ .
   - (a) floorplanning
   - (b) placement
   - (c) routing
   - (d) partitioning

5. Via is the _____ between the different metal layers.
   (a) horizontal interconnect   (b) vertical interconnect
   (c) gap   (d) dummy metal

Multiple Choice Questions

1. Algorithm used for floorplanning is
   (a) genetic algorithm   (b) simulated annealing
   (c) floorplan sizing   (d) all of these
2. A feed-through cell is an empty cell (with no logic) that is used for
   (a) vertical interconnections   (b) horizontal interconnections
   (c) both (a) and (b)   (d) feeding power to a cell
3. With the increase in metal levels, the thickness of metal layers
   (a) increases   (b) decreases
   (c) remains same   (d) none of these
4. Hightower algorithm is used for
   (a) floorplanning   (b) placement
   (c) routing   (d) partitioning
5. As compared to the Euclidean distance, Manhattan distance is
   (a) always greater   (b) always lesser
   (c) can be equal or greater   (d) can be equal or lesser

True or False

1. Distance calculated using Manhattan model is more than Euclidean model.
2. Slicing tree is a binary tree.
3. A graph is an exact model of a network for partitioning purpose.
4. Slicing tree is always unique.
5. A non-slicing floorplan is called wheel or spiral floorplan.

Short-answer Type Questions

1. Write the floorplanning goals and objectives. What is logical partitioning and physical partitioning? What are partitioning algorithms?
2. Discuss the slicing and non-slicing floorplanning with necessary diagrams.
3. What is the mathematical model of the slicing floorplan? Explain with suitable example.
4. What is skewed and non-skewed slicing tree?
5. What are the floorplanning algorithms?
6. What do you mean by pad-limited and core-limited die?
7. Explain the slicing floorplanning with suitable example. Draw its corresponding slicing tree and write the polish expression.
8. What are goals and objectives of placement? What are the placement algorithms?
9. Discuss how wire length is estimated using Euclidean and Manhattan distance model.
10. What are the goals and objectives of routing? Classify the routing categories and discuss them in brief.
11. What is meant by global routing? Discuss global routing methodologies in brief.
12. What is meant by detailed routing? What are the routing constraints? What are the objectives of detailed routing?

Long-answer Type Questions

1. What is channel definition? What are goals and objectives of placement? What are the placement algorithms? Discuss how wire length is estimated using the Euclidean and Manhattan distance model.
2. Explain how genetic algorithm is used for floorplanning.
3. Explain the floorplanning using simulated annealing algorithm.

4. Discuss the floorplanning guidelines.
5. What do you mean by shape function evaluation for achieving optimal floorplanning?
6. Discuss the Wong–Liu algorithm for floorplanning.
7. Explain how I/O, power, and clock planning are done during floorplanning.
8. Explain how delay analysis can be done during the floorplanning.
9. Explain why placement is an important step in VLSI physical design flow in meeting circuit timing specifications.
10. Discuss the placement algorithms in detail with suitable examples.
11. Discuss the Kernighan–Lin partitioning algorithm with a suitable example.
12. Explain the timing-driven and congestion-driven placement.
13. Discuss the various wire length estimation methods.
14. Explain how standard cell placement is carried out.
15. What are the goals and objectives of routing? Classify the routing categories. Discuss the different area routing procedures with suitable examples.
16. What is meant by detailed routing? What are the routing constraints? What are the algorithms for detailed routing?
17. Write short notes on the following:
    (a) Left-edge algorithm (LEA)
    (b) Lee-Maze routing
    (c) Multilevel routing
18. Write short notes on the following:
    (a) Power routing
    (b) Clock routing
    (c) Global routing and detailed routing
19. What is meant by global routing? Discuss global routing methodologies.
20. Find out the shortest path between the source (S) and the destination (D) as shown in the following figure using Lee's routing algorithm. The symbol × indicates obstacles.

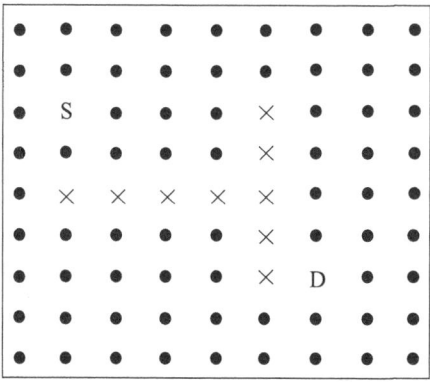

21. Represent the equivalent graph of the following circuit and using Kernighan–Lin algorithm, find out the optimum partitioning for the circuit.

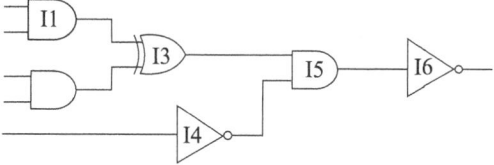

## CHAPTER 12

Verification and Reliability Analysis

KEY TOPICS

- Verification methodologies in VLSI design flow
- Logic verification
- Physical verification
- Design rule check
- Layout vs schematic check
- Electrical rule check
- Antenna check
- Metal density check
- Layout extraction
- Crosstalk analysis
- Introduction to reliability
- IC reliability—definitions

- Reliability issues in ICs
- Electromigration
- Time-dependent dielectric breakdown
- Channel hot carrier
- Negative bias temperature instability
- Power and ground bounce
- IR drop
- Latch-up
- ESD and EOS
- Soft error

12.1 Introduction

An integrated circuit (IC) chip is designed with a desired functionality and specifications. The specifications and the functionality are at a very high level. The functional description is generally written in the form of register transfer level (RTL) netlist using hardware description languages such as VHDL or Verilog. The specifications (e.g., power, speed, area) are also described in a formal language, which is usually an industry standard. Taking these two inputs, the synthesis flow generates a gate level netlist by mapping the circuit behaviour in terms of logic gates of a target technology library. It is followed by the timing and power analysis to check if the timing and power requirements are met at the top level. Then the physical design flow starts, which actually implements the physical layout of the chip that needs to be fabricated using the specified technology. The physical design starts with the floorplan, followed by placement and routing. VLSI design flow is a sequential process (Fig. 12.1) with feedback loops for corrections after each step, if required.

Both the logical and physical design steps must ensure that the implemented chip meets the desired functionality and target specifications. *Verification* is the process of checking the design against the given functionality and specifications. There are mainly two types of verifications performed: one is at the end of logical design phase, which is known as logic verification; and the other is at the end of the physical design phase, which is known as the physical verification.

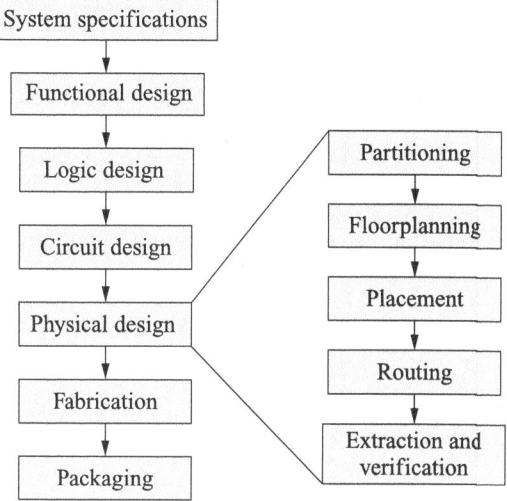

Fig. 12.1 VLSI design flow

In this chapter, we will discuss both the logic verification and physical verification.

12.2 Verification Methodologies in VLSI Design Flow

Figure 12.2 illustrates the verification methodologies at different parts of the VLSI design flow.

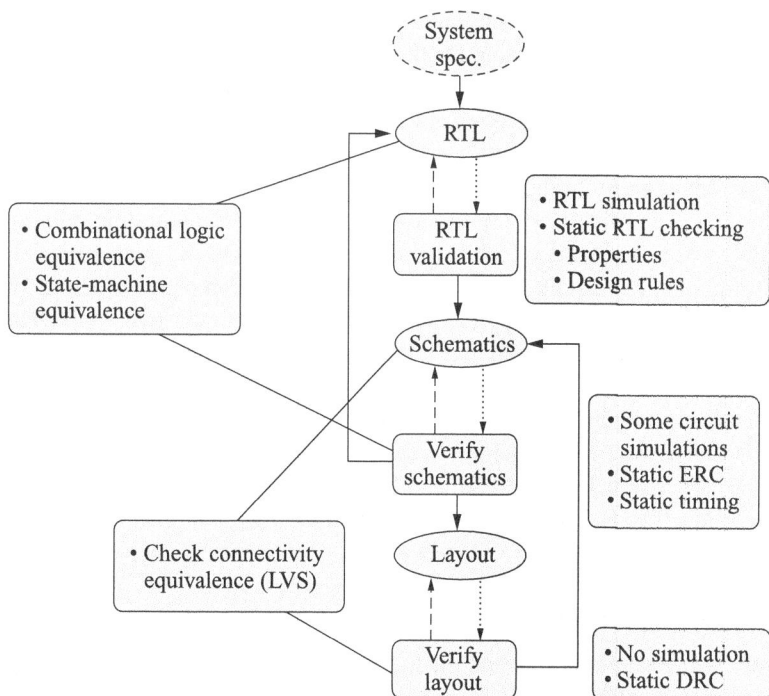

Fig. 12.2 Verification methodologies in VLSI design flow

Logic Functional Equivalence Checking This step checks for RTL versus gates, gates versus gates, and state-machine equivalence.

RTL Checking It checks for RTL coding style, design rules, and properties.

Static Timing Analysis It checks for timing delays. In this step, the noise analysis, power analysis, and electromigration analysis is also done at a higher level.

Electrical Rule Checking It checks for circuit configurations, composition rules, and reliability rules.

Layout Versus Schematics—Connectivity Verification It checks for graph isomorphism.

Layout Design Rule Checking It checks minimum sizes, minimum spacing area, and overlap rules.

12.3 Logic Verification

Logic verification is the process of checking the functionality of the circuit. It can be classified into the following two categories:

- Static verification
- Dynamic verification (simulation-based)

In the static verification process, no simulation is performed on the circuit. It checks against some rules, such as electrical rules, design rules, schematic check, etc. It is fast as compared to the dynamic verification process. In the dynamic verification process, the circuit is simulated to check its correctness. It is very slow but most accurate, and not applicable for the whole big system.

The reduced ordered binary decision diagram (ROBDD) discussed in Chapter 9 is widely used for logic verification. In this section, we will discuss how ROBDD can be used to check if the logic circuit implemented by the logic synthesis step meets the functionality of the circuit.

For a fully specified Boolean function f, the simplest methodology is to build a ROBDD with root vertex f, and to build another ROBDD with root vertex g, for the implemented circuit. Then, two ROBDDs are compared to check if f and g correspond to each other. For the same input patterns, the two ROBDDs must produce the same output. Hence, verification is just checking if $f = g$ for the same input combinations.

For an incompletely specified function f, the verification process requires some modifications in the BDD structure. In case of incompletely specified function, the output variable has three set of values {'1', '0', '–'}, where '–' represents 'don't' care condition. The BDD needs to be modified to include a third leaf node for the 'don't' care condition.

12.3.1 Formal Approaches

There are two formal approaches for logic verification: (a) equivalence checking and (b) model checking.

Equivalence Checking

Equivalence checking verifies that the implementation is equivalent to the formal specification as given below:

1. For combinational logic, if both functions can be expressed in ROBDD, then equivalence checking is very simple. It just checks if two ROBDDs are the same.
2. For sequential circuits, Coudert et al. (1997) developed an equivalent checking method between two deterministic finite automata.

Model Checking

Using a formal model of the circuit, model checking verifies certain desirable properties required for a correct implementation, e.g., liveness, fairness (absence of starvation or lockout), no bus conflicts, etc.

12.4 Physical Verification

Physical verification is the process of checking the layout for its correctness. At the end of the physical design phase, the layout is ready for creation of mask layers (pattern generation). But before sending it to the pattern generator, the layout must be checked for the following rules:

- Design rule check (DRC)
- Layout versus schematic (LVS) check
- Electrical rule check (ERC)
- Antenna check

12.5 Design Rule Check

Design rules are a set of rules or guidelines for the layout of an IC. These rules are derived from the IC fabrication process technology. Some examples of the design rules are metal-to-metal spacing, minimum metal width, minimum poly width, poly-to-metal spacing, minimum size of the contacts, etc. These rules are the outcome of the limitations of the fabrication process. For example, if two metal lines are fabricated very close to each other violating the minimum spacing rule, the metal lines may be shorted to each other. On the other hand, if a metal line is fabricated with a width that is less than the minimum specified width rule, the metal line might have a break causing an open circuit. If the layout is created adhering to the design rules, there is less chance of defects in the fabricated chip, and yield of the process is increased.

12.5.1 Design Rules

A layout is a combination of several layers which represent different parts of the devices and circuits. For example, a layout of the MOS device contains polylayer over active layer. There are a number of layers in a layout, as shown in Fig. 12.3.

There are rules for all the layers used to create the layout. The rules are mainly of three types as given below:

- Width rule
- Spacing rule
- Enclosure rule

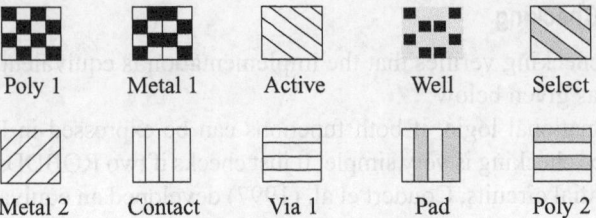

Fig. 12.3   Different layers in layout

Figure 12.4 shows the typical DRC checking process and three main DRC checks—width, spacing, and enclosure. The design rules are expressed in two different ways:

- Micron rules
- Lambda rules

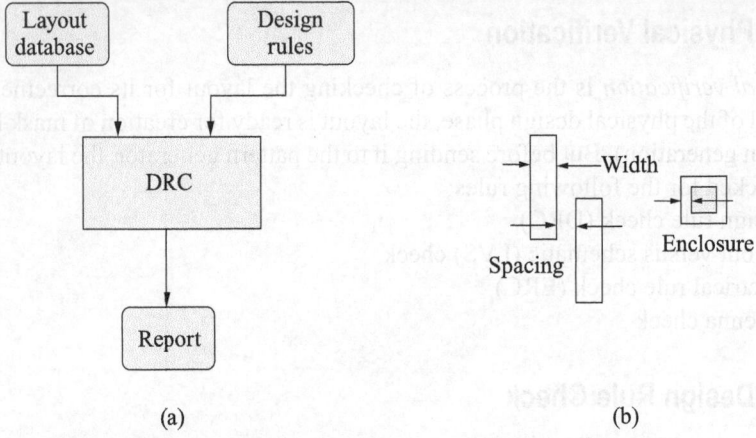

Fig. 12.4   (a) DRC checking process; (b) three basic DRC checks

Micron rules

In the micron rules, all the rules (i.e., the allowed feature sizes) are expressed in microns, whereas in the lambda rules, the features are expressed in terms of a single parameter called lambda (λ). The lambda-based rules are more suitable for scaling of process technology, as just replacing λ with a new value changes all the rules accordingly.

Lambda rules

The λ-rules were first introduced by Mead and Conway in 1980 (Mead 1980). Using the concept of scalable λ-rules, MOSIS (MOS Implementation Service, Information Sciences Institute, University of Southern California, that collects designs from academic, government, and commercial customers, forms one mask to share overhead cost) has developed a set of rules that cater for a variety of manufacturing processes.

12.5.2 MOSIS Design Rules (Mead Conway Design Rules)

MOSIS is an organization under Information Sciences Institute at the University of Southern California. IC chips can be fabricated using MOSIS IC fabrication facility. Table 12.1 shows the layout design rules for the MOSIS fabrication technology (Baker 2004).

Table 12.1 MOSIS design rules

Layer	Rule	Dimension	Lambda
Well	1.1	Width	10
	1.2	Spacing (wells at different potential)	9
	1.3	Spacing (wells at same potential)	0 or 6
	1.4	Spacing (wells of different type)	0
Active	2.1	Minimum width	3
	2.2	Minimum spacing	3
	2.3	Source/drain active to well edge	5
	2.4	Substrate/well contact active to well edge	3
	2.5	Minimum spacing between active to different implant	0 or 4
Poly1	3.1	Minimum width	2
	3.2	Minimum spacing	2
	3.3	Minimum gate extension of active	2
	3.4	Minimum active extension of poly1	3
	3.5	Minimum field poly to active	1
Select	4.1	Minimum select spacing to channel of transistor to ensure adequate source/drain width	3
	4.2	Minimum select overlap to active	2
	4.3	Minimum select overlap to contact	1
	4.4	Minimum select width and spacing	0 or 2
Contact to poly1	5.1	Exact contact size	2 × 2
	5.2	Minimum poly1 overlap	1
	5.3	Minimum contact spacing	2
Contact to active	6.1	Exact contact size	2 × 2
	6.2	Minimum active overlap	1
	6.3	Minimum contact spacing	2
	6.4	Minimum spacing to gate of transistor	2
Metal 1	7.1	Minimum width	3
	7.2	Minimum spacing	3
	7.3	Minimum overlap of poly contact	1
	7.4	Minimum overlap of active contact	1
Via 1	8.1	Exact size	2 × 2
	8.2	Minimum via1 spacing	3
	8.3	Minimum overlap by metal1	1
	8.4	Minimum spacing to contact	2
	8.5	Minimum spacing to poly or active edge	2
Metal 2	9.1	Minimum width	3
	9.2	Minimum spacing	4
	9.3	Minimum overlap of via1	1
Overglass	10.1	Minimum bonding pad width	100 × 100
	10.2	Minimum probe to pad width	75 × 75
	10.3	Pad overlap of glass opening	6
	10.4	Minimum pad spacing to unrelated metal2	30
	10.5	Minimum pad spacing to unrelated metal1, poly electrode or active	15

(Contd)

Table 12.1 *(Contd)*

Layer	Rule	Dimension	Lambda
Poly2	11.1	Minimum width	3
	11.2	Minimum spacing	3
	11.3	Minimum poly1 overlap	2
	11.4	Minimum spacing to active or well edge	2
	11.5	Minimum spacing to poly1 contact	3
Electrode for	12.1	Minimum width	2
transistor	12.2	Minimum spacing	3
	12.3	Minimum poly2 gate overlap of active	2
	12.4	Minimum spacing to active	1
	12.5	Minimum spacing or overlap of poly1	2
	12.6	Minimum spacing to poly1 or active contact	3
Electrode con-	13.1	Exact contact size	2 × 2
tact, analog	13.2	Minimum contact spacing	2
option	13.3	Minimum electrode overlap (on capacitor)	3
	13.4	Minimum electrode overlap (not on capacitor)	2
	13.5	Minimum spacing to poly1 or active	3
Via 2	14.1	Exact size	2 × 2
	14.2	Minimum spacing	3
	14.3	Minimum overlap by metal2	1
	14.4	Minimum spacing to via1	2
Metal 3	15.1	Minimum width	6
	15.2	Minimum spacing to metal3	4
	15.3	Minimum overlap of via2	2
NPN bipolar	16.1	All active contact	2 × 2
transistor	16.2	Minimum select overlap of emitter contact	4
	16.3	Minimum pbase overlap of emitter select	2
	16.4	Minimum spacing between emitter select and base select	4
	16.5	Minimum pbase overlap of pbase select	2
	16.6	Minimum select overlap of base contact	2
	16.7	Minimum nwell overlap of pbase	6
	16.8	Minimum spacing between pbase and collector active	4
	16.9	Minimum active overlap of collector contact	2
	16.10	Minimum nwell overlap of collector active	3
	16.11	Minimum select overlap of collector active	2
Capacitor well	17.1	Minimum width	10
	17.2	Minimum spacing	9
	17.3	Minimum spacing to external active	5
	17.4	Minimum overlap of active	3

12.5.3 Micron Rules

Table 12.2 describes the micron-based design rules for Orbit Semiconductors' 2 μm technology with double-metal, double-poly, and n-well layers.

Table 12.2 Micron rules

Layer	Rule	Dimension	Micron
n-well	1.1	Width	3.0
	1.2	Spacing	9.0
Active	2.1	Width	3.0
	2.2	Active to active	3.0
	2.3	n^+ active to n-well	7.0
	2.4	p^+ substrate contact to n-well	4.0
	2.5	n-well to n^+ well tie down	0.0
	2.6	n-well overlap of p^+ active	3.0
Poly1	3.1	Width	2.0
	3.2	Space	3.0
	3.3	Gate overlap of active	2.0
	3.4	Active overlap of gate	3.0
	3.5	Field poly1 to active	1.0
Poly2	4.1	Width	3.0
	4.2	Space	3.0
	4.3	Poly1 overlap of poly2	2.0
	4.4	Space to active or well edge	2.0
	4.5	Space to poly1 contact	3.0
Contact	5.1	Contact size	2.0 × 2.0
	5.2	Spacing	2.0
	5.3	Poly overlap	2.0
	5.4	Active overlap	2.0
	5.5	Poly contact to active edge	3.0
	5.6	Active contact to gate	3.0
Metal1	6.1	Width	3.0
	6.2	Spacing	3.0
	6.3	Overlap to contact	1.0
	6.4	Overlap of via	2.0
Via	7.1	Space to contact	2.0
	7.2	Size (except for pads)	2.0 × 2.0
	7.3	Spacing	3.0
Metal2	8.1	Width	3.0
	8.2	Space	3.0
	8.3	Metal2 overlap of via	2.0
Pad	9.1	Maximum pad opening	90 × 90
	9.2	Pad size	100 × 100
	9.3	Separation	75
p-base	10.1	p-base active to n-well	5.0
	10.2	Collector n^+ active to p-base active	4.0
	10.3	p-base active overlap of n^+/p^+ active	4.0
	10.4	p^+ active to n^+ active	7.0

12.5.4 DRC Checker

The design rule checker (DRC) finds the deviations from design constraints. The DRC program works with any technology, and handles all layout methodologies, including full custom, structured custom, standard cell, macro cell, gate array, and automated layout. DRC is used to check geometric spacing, width, enclosure, and overlap. Its hierarchical operation, incremental checking, and unique pattern-recognition capabilities reduce run times. DRC displays design rule violations graphically as an additional graphics layer on the layout, and lists them in the text files for debugging. If there are DRC errors in a layout, it must be corrected or modified so that there are no more DRC errors before it can be send out for pattern generation (PG). Some DRC tools are displayed in Table 12.3.

Table 12.3   Some industry standard DRC tools

EDA Vendor	Tool Name
Tanner EDA	HiPer verify
Mentor graphics	Calibre
Synopsys	Hercules
Cadence design systems	Dracula and Assura
Magma design automation	Quartz

12.6 Layout Versus Schematic (LVS) Check

DRC ensures that the layout is designed according to the design rules for the specified process technology. However, it does not guarantee that the layout really represents the circuit that the designer intended to fabricate. For example, let us consider a DRC clean layout of the inverter which has no substrate connections. A DRC checker will not catch this as an error. But this is a potential problem as the substrate will be floating, and will impact the functionality of the inverter. Hence, a checking must be done to ensure that the connections in the layout are the same as that of the inverter. Typically, this check is performed by generating a netlist from the layout and comparing it with the netlist for the schematic. This process of checking a layout with respect to a schematic is called layout versus schematic (LVS) check.

Typical LVS errors are given by

- *Shorts*   Two or more wires that should not be connected are connected together.
- *Opens*   Wires or components that should be connected are left dangling or only partially connected.
- *Component mismatches*   Components of an incorrect type have been laid out (e.g., a pMOS device instead of an nMOS device).
- *Missing components*   A component has been left out in the layout.
- *Property errors*   A component is laid out with wrong dimensions compared to the schematic.

12.6.1 LVS Checker

A typical LVS checker tool extracts the devices and their connection from the layout based on the specified extraction rules, and generates a netlist using the extracted devices and their connectivity. Then it compares the layout netlist to the schematic netlist, and displays mismatches between the layout and the schematic, both textually and graphically.

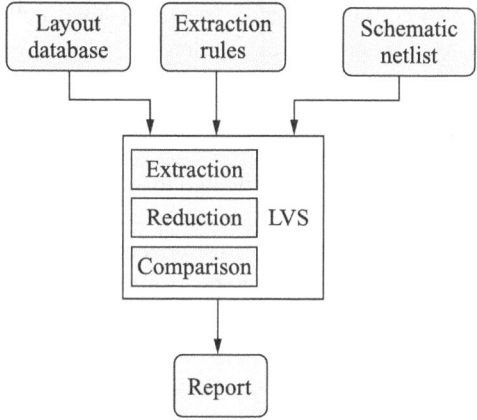

Fig. 12.5 LVS checking process

An LVS checker performs following three basic steps, as illustrated in Fig. 12.5.

1. *Extraction* The software program takes the layout database as input which contains all the layers, drawn to represent the circuit. It then performs logic operations between the layers to determine the device components and their connections, based on extraction rules.

2. *Reduction* In the reduction process, the software combines the extracted components into series and parallel combinations, if possible, and generates a netlist.

3. *Comparison* Then it compares the extracted layout netlist with the schematic netlist. If the two netlists match, it reports no LVS error, otherwise it reports the errors in a file.

Some LVS checker tools are displayed in Table 12.4.

Table 12.4 Some industry standard LVS checker tools

EDA Vendor	Tool Name
Tanner EDA	L-Edit LVS
Mentor graphics	Calibre
Synopsys	Hercules
Cadence design systems	PVS, Dracula and Assura
Magma design automation	Quartz

12.7 Electrical Rule Check

Another important check in the physical verification phase is the electrical rule check (ERC). In this checking, the connectivity is checked mainly to find the floating nodes that are not connected to the ground or power. It is also possible to check some types of connections which are not allowed, and are considered as errors.

The electrical rule checker checks for the network connectivity. The ERC program reports the electrical connectivity issues, such as floating interconnect and devices, and abnormal connections in the physical or schematic designs. It operates on the network generated from either the layout or the schematic. ERC performs conventional checks, such as verifying pull-up/pull-down and isolating inactive devices. It also converts a MOS transistor-level netlist into a gate-level netlist which can be used by gate-level simulators.

12.8 Antenna Check

The metal interconnects in an IC are fabricated in a plasma environment. Plasma is a mixture of ionized (partially or fully) gas composed of equal number of positive and negative charges, and a number of un-ionized molecules. Plasma is created when a large electric field is applied to a gas, causing the gas molecules to break down and become ionized.

In a multilevel metallization process technology, it is often found that some nets are routed through several metal layers as shown in Fig. 12.6. For multilevel metallization process, the bottom metal layers are fabricated first. The bottom metal layers are connected to either the gate or the active (source/drain) of the MOS transistors. The net is fully connected if all the metal layers have been fabricated. After one or two metal layers are fabricated, the metal leads are not yet fully connected if the connection is made using even higher metal layers. These partially processed interconnects are called *antenna*. During the processing of metal interconnects, these antennas collect large static charges in the plasma environment. Now, if they are connected to the gate of the MOS transistors, these large static charges discharge through the gate electrode causing the underlying thin gate oxide layer to breakdown, and hence damage to the MOS device.

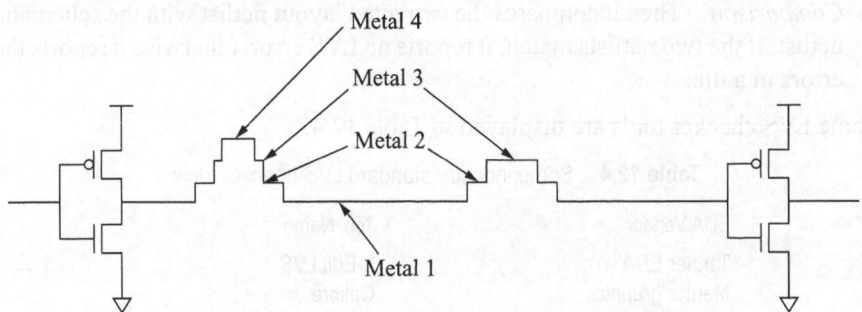

Fig. 12.6 Antenna effect in integrated circuit fabrication

The partially processed leads, if connected to the drain or source of the MOS device, can discharge easily without any damage to the oxide. The antenna rules specify the maximum area of metal that is allowed to connect to the gate of the MOS devices without any active (drain/source) discharge path. The antenna rule is generally expressed as the ratio of metal area to the gate area, called *antenna ratio*.

The antenna check refers to checking a layout for the antenna rules. If the metal area is connected to a gate area which is not within the allowed limit, the checker flags antenna violations. Antenna violations must be fixed by modifying the layout before it can be sent for PG. The typical fix mechanism of the antenna violations are given by

- Breaking the long metal lines by inserting a jumper in the metal interconnects.
- Adding the antenna protection diode which acts as a discharge path.

12.8.1 Jumper Insertion

It is process of breaking a long interconnect in the lower metal layer, and the broken parts are routed through the higher metal levels. This ensures that when the bottom metal layer is processed, it collects less static charge as its length is reduced. Figure 12.7 illustrates the jumper insertion process to fix the antenna violation. In Fig. 12.7(a), the length of the net is L which is only connected to the gate of the MOS device at the right end of the net when the metal1 layer is processed. If L is large enough, the antenna ratio will be violated. As shown in Fig. 12.7(b), a jumper is inserted by breaking the net at metal1 layer, and routed through the metal2, metal3, and metal4 layers. Hence, during the metal1 layer processing, the net length connected to the gate of the MOS device is much reduced ($L3 \ll L$), and satisfies the antenna ratio.

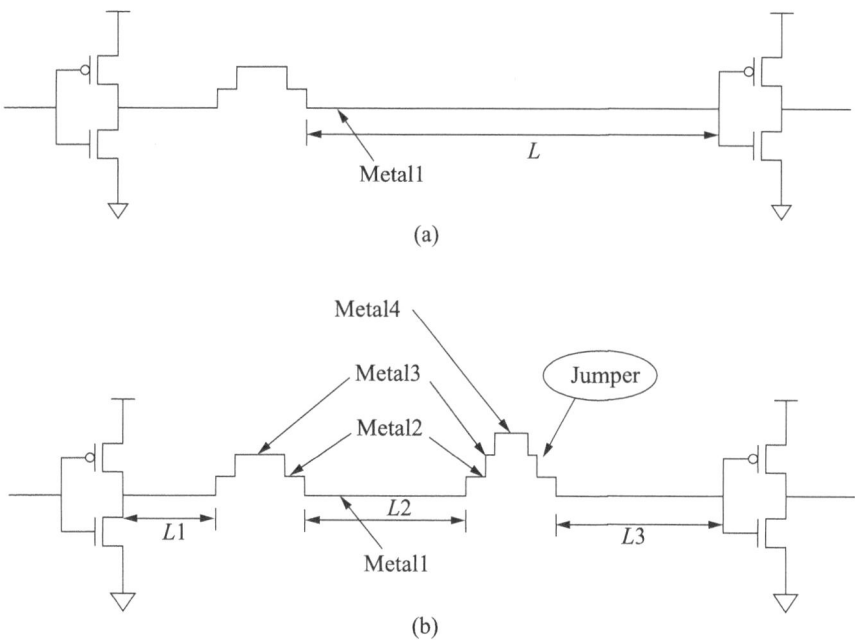

Fig. 12.7   (a) Antenna violation; (b) fix mechanism using jumper insertion

12.8.2 Antenna Protection Diode

Another antenna violation fix mechanism is to add an antenna protection diode to the antenna that violates the antenna ratio. The antenna protection diode is a MOS diode obtained by connecting the gate and source electrodes of an nMOS device, as shown in Fig. 12.8. The antenna length L shown in Fig. 12.8(a) violates the antenna ratio. Now, if the antenna diode is connected, the length of the net is not reduced; but, it is connected to a low resistance discharge path. The static charge collected by the net in plasma environment finds a discharge path through this diode. Hence, the gate of the MOS device is protected by the antenna protection diode.

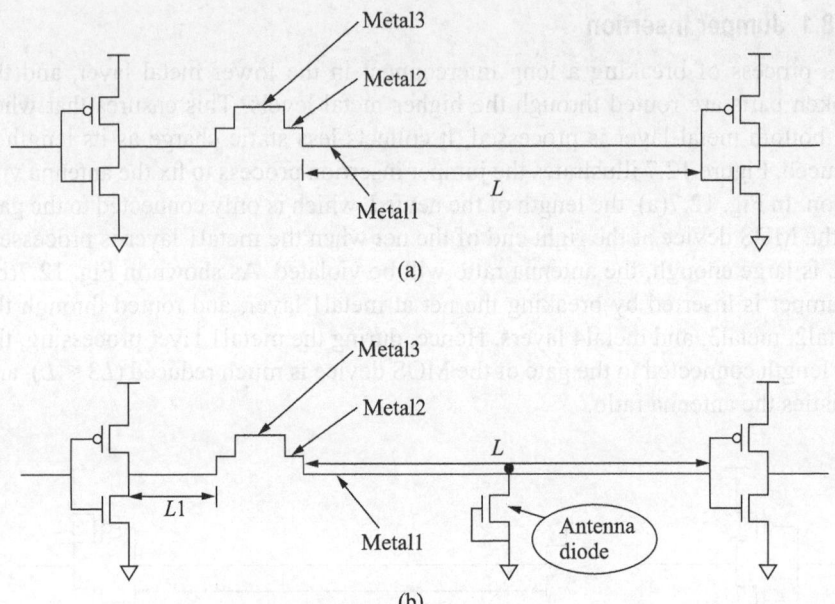

Fig. 12.8 Antenna protection diode addition

12.9 Metal Density Check

There is a rule that specifies the minimum and maximum number of metal layers within a specific area for advanced process technology. These rules are due to the chemical mechanical polishing (CMP) process, which is used for multilevel Cu metallization, in which patterns are created in the dielectric first, and then Cu is filled in the trenches, followed by CMP to remove the excess metal from the top surface of the dielectric, and leave Cu in the holes and trenches.

The metal density check is required to check if a given area violates the minimum or maximum metal layers.

12.10 Layout Extraction

Layout extraction is a process of taking out circuit information from the physical drawing or layout of the IC. There are mainly two types of information extracted from the layout: (a) information related to the devices and (b) information related to the interconnections.

12.10.1 Device Extraction

Device extraction is a methodology of extracting device information from an IC layout. It works on simple logic operations on the drawn layers in the layout. For example, a MOS device is identified if there is an overlap between the active and polylayer. The overlap can be identified by doing logic AND operation between two layers: active and poly.

12.10.2 Parasitic Extraction

Another extraction process is the parasitic extraction. In this process, all the parasitic circuit elements such as resistance, capacitance, and inductance are obtained from the geometry information and material properties. For example, resistances associated with interconnections can be obtained from the length (l), cross-sectional area (A), and resistivity of the material (ρ), given by the following equation:

$$R = \rho \frac{l}{A} = \rho \frac{l}{w \times t} \qquad (12.1)$$

The expression of capacitance per unit length for a conductor over a ground plane is given by

$$C = \varepsilon \times \left[1.13 \frac{w}{h} + 1.44 \left(\frac{w}{h} \right)^{0.11} + 1.46 \left(\frac{t}{h} \right)^{0.42} \right] \qquad (12.2)$$

where w = width, h = height from ground plane, and t = thickness of the conductor.

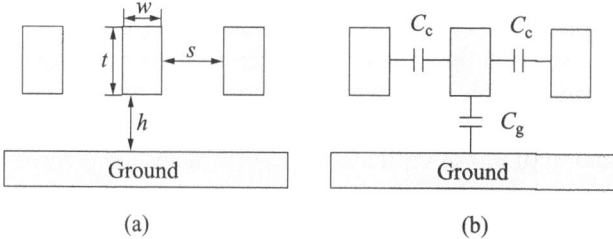

Fig. 12.9   (a) Parallel conductors above ground plane; (b) self and coupling capacitances

For two conductors running parallel over a ground plane, capacitance per unit length of the conductors with respect to ground is given by (Fig. 12.9):

$$C_g = \varepsilon \times \left[1.10 \frac{w}{h} + 0.79 \left(\frac{w}{h} \right)^{0.1} + 0.46 \left(\frac{t}{h} \right)^{0.17} \left(1 - 0.87 h^{\left(\frac{-s}{h} \right)} \right) \right] \qquad (12.3)$$

and coupling capacitance per unit length between the conductors is given by

$$C_c = \varepsilon \left[\frac{t}{s} + 1.2 \left(\frac{s}{h} \right)^{0.1} \left(\frac{s}{h} + 1.15 \right)^{-2.22} \right.$$

$$\left. + 0.253 \ln \left(1 + 7.17 \frac{w}{s} \right) \left(\frac{s}{h} + 0.54 \right)^{-0.64} \right] \qquad (12.4)$$

where w = width, h = height from ground plane, t = thickness, and s = spacing of the conductor, ε = dielectric constant = $\varepsilon_o \varepsilon_r$ where ε_o = permittivity of free space = 8.85×10^{-12} F/m and ε_r = relative dielectric constant.

The self and mutual inductances are given by

$$L_s = \frac{\mu_0 \times l}{2\pi} \left[\ln \left(\frac{2l}{w+t} \right) + \frac{1}{2} + \frac{0.22(w+t)}{l} \right] \qquad (12.5)$$

$$M = \frac{\mu_0 \times l}{2\pi} \left[\ln \left(\frac{2l}{d} \right) - 1 + \frac{d}{l} \right] \qquad (12.6)$$

where $\mu_0 = 4\pi \times 10^{-7}$ H/m and d is the centre-to-centre distance between the two wires. Some extraction tools are given in Table 12.5.

Table 12.5   Extraction tools

Vendor	Tool
Magma	QuickCap
Synopsys	StarRCX
Cadence	QRC

12.11 Crosstalk Analysis

Crosstalk is a phenomena that occurs in two physically adjacent wires in which a transition in one wire introduces a noise or overshoot/undershoot in another wire. The wire that affects the other nets is called the *aggressor net,* and the wire that is affected by the aggressor net is called the *victim net.* As the wires are physically separated by a dielectric material, there is parasitic capacitance between the wires, which is known as the coupling capacitance. Figure 12.10 shows how the parasitic coupling capacitance introduces a crosstalk glitch. The effect of the crosstalk noise on the circuit performance is discussed in the following text.

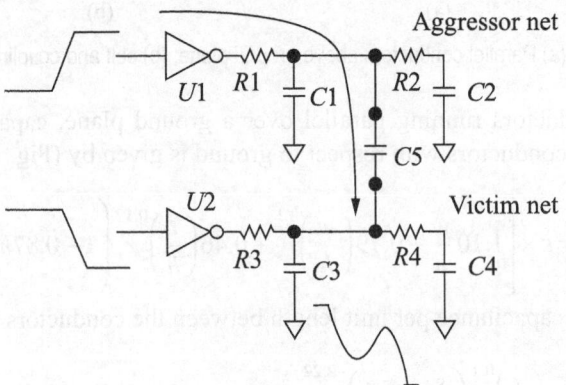

Fig. 12.10   Schematic of crosstalk in a circuit

1. When both the *aggressor net* and the *victim net* switches together, each one affects the rise and fall time of the other.
   (a) If the aggressor net rises from logic 0 to logic 1, and the victim net also rises from logic 0 to logic 1, the rise time of the victim net reduces.
   (b) If the aggressor net rises from logic 0 to logic 1, and the victim net falls from logic 1 to logic 0, the fall time of the victim net increases.
   (c) If the aggressor net falls from logic 1 to logic 0, and the victim net rises from logic 0 to logic 1, the rise time of the victim net increases.
   (d) If the aggressor net falls from logic 1 to logic 0, and the victim net also falls from logic 1 to logic 0, the fall time of the victim net decreases.
2. When the *aggressor net switches* and the *victim net is quiet*, the aggressor net introduces noise or overshoot/undershoot on the victim net.

(a) If the aggressor net rises from logic 0 to logic 1, and the victim net is held at logic 1, the victim net experiences an overshoot.
(b) If the aggressor net rises from logic 0 to logic 1, and the victim net is held at logic 0, the victim net experiences a rise glitch.
(c) If the aggressor net falls from logic 1 to logic 0, and the victim net is held at logic 1, the victim net experiences a fall glitch.
(d) If the aggressor net falls from logic 1 to logic 0, and the victim net is held at logic 0, the victim net experiences an undershoot.

Table 12.6 summarizes the above mentioned crosstalk effects.

Table 12.6 Effects of crosstalk on the circuit performance

Sl. No.	Aggressor	Victim	Results	Impacts
1.	Rises from logic 0 to logic 1	Rises from logic 0 to logic 1	Decrease in rise time	Timing
2.	Rises from logic 0 to logic 1	Falls from logic 1 to logic 0	Increase in fall time	Timing
3.	Falls from logic 1 to logic 0	Rises from logic 0 to logic 1	Increase in rise time	Timing
4.	Falls from logic 1 to logic 0	Falls from logic 1 to logic 0	Decrease in fall time	Timing
5.	Rises from logic 0 to logic 1	Held at logic 1	Overshoot	Reliability
6.	Rises from logic 0 to logic 1	Held at logic 0	Rise glitch	Functionality
7.	Falls from logic 1 to logic 0	Held at logic 1	Fall glitch	Functionality
8.	Falls from logic 1 to logic 0	Held at logic 0	Undershoot	Reliability

12.11.1 Crosstalk Glitch

When aggressor nets switch from low to high and victim net is at low logic level, there is a crosstalk-induced noise or glitch on the victim net. If the glitch area (= peak × width) is above a noise threshold, it can cause a functional violation. Therefore, the crosstalk glitch analysis is very much important in DSM designs. Glitch analysis flow extracts the *RC* parasitics from the layout, generates an equivalent circuit, and performs glitch analysis using circuit simulators.

12.11.2 Crosstalk Delay

Crosstalk has an impact on the timing of the circuit. When both the aggressor and victim nets switch together, depending on relative switching strength, the signal on victim nets can be either delayed or advanced. If the crosstalk induced delay is significant enough, it can cause set-up and hold-time violation for a sequential element like the D flip-flop in the circuit. Crosstalk delay analysis flow also extracts *RC* parasitics from the layout, generates equivalent circuit, and performs delay analysis using circuit simulators.

12.12 Introduction to Reliability

Reliability of an IC chip defines how long the chip can operate without any performance degradation in the field. It is not acceptable to have IC chips that work as per the specification just after fabrication, but do not sustain over device operating lifetime. Hence, lifetime reliability is an important design factor for VLSI circuits. There are several deep-sub-micron (DSM) issues that degrade both the MOS device and the interconnect wire. Under various operating conditions, MOS devices suffer from reliability degradation from a number of sources, including, channel hot electron (CHE), negative-bias temperature instability (NBTI), time-dependent dielectric breakdown (TDDB), radiation-induced damage, etc. Electromigration (EM) is another issue that affects the interconnect wires. These degradations can cause performance degradations (e.g., timing or power) or even functional problems in fabricated chips. Hence, a reliability aware circuit design must be adopted to improve the product lifetime. In this chapter, first we describe the reliability issues in detail. Then, we discuss their effects on the circuit behaviour and their remedies.

12.13 IC Reliability—Definitions

The *reliability of an IC* can be defined as the probability that the chip will function according to its specifications for a period of time. The reliability of a VLSI circuit is measured in terms of *mean time to failure* (MTTF). If the failures are assumed to occur at a constant rate, the failure rate λ can be written as

$$\lambda = (\text{MTTF})^{-1} \tag{12.7}$$

Failure rate, if plotted with respect to the operating lifetime of the device, shows a decreasing, then constant, and finally increasing nature, as shown in Fig. 12.11. This curve is known as the reliability bathtub curve.

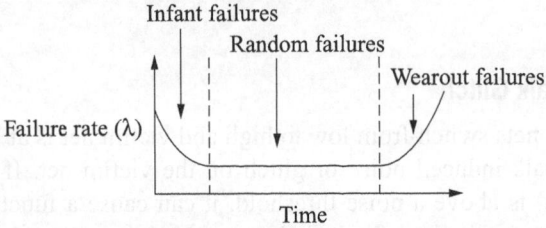

Fig. 12.11 Typical time dependence of system failure rate (bathtub curve)

The constant part of the bathtub curve signifies failures due to accidental overstress or overloads, occurring randomly over the operating lifetime of the product. This part of the curve is known as *random failures*. The constant failure rates are mainly due to the external factors.

In the end, the devices wear out. This is called ageing effect, and due to this, the failure rate starts increasing exponentially.

Infant mortality Early in the life of a product there are some chips which are failed during the test. The defects early in the life of a product, is termed as *infant mortality*. Almost all chips follow the same trend for failures as a function of operating lifetime. The failure rate decreases rapidly at the beginning of the lifetime to a low value, then remains constant till the end of lifetime, and finally the failure rate increases again rapidly.

The early failures are detected by a methodology called burn-in. The failure rate has an exponential dependency on the absolute temperature (T) as given by

$$\text{Failure rate} \propto \exp(-E_a / kT) \tag{12.8}$$

where E_a is known as activation energy, k the Boltzmann's constant ($= 8.62 \times 10^{-5}$ eV K^{-1}), and T is the absolute temperature. Equation (12.8) is known as the Arrhenius equation.

In the burn-in process, the chip is operated at an elevated temperature. This high temperature accelerates the early failure mechanisms, and the failures are detected early. Depending on the failure mechanisms, excess electrical stress, such as elevated voltage or current, also accelerates the failure mechanisms.

Some Terminologies

Mean Time Between Failures (MTBF)   MTBF is used to measure the reliability for a repairable product.

Mean Time to Failure (MTTF)   MTTF is used to measure the reliability for a fatal failure.

Failure in Time (FIT)   FIT is used to measure the reliability for VLSI chips. 1 FIT indicates one device failure in 10^9 device-hours. The total failure rate of a product is calculated by summing up the FIT rates of the individual components of the chip. For example, let us consider a product with the following components:
- Microprocessor — 5 FIT
- 10 RAM chips — 6 FIT each
- 100 TTL parts — 1 FIT each

The total failure rate of the product is $5 + 10 \times 6 + 100 \times 1 = 165$ FIT.

While designing a chip, a FIT rate budgeting is done. Suppose the chip would have total failure rate of 50 FIT. Now all the failure mechanisms are considered separately and each one is allocated a FIT rate. For example, the EM is allocated 10 FIT, CHE is allocated 10 FIT, TDDB is allocated 10 FIT, NBTI is allocated 10 FIT, and the remaining 10 FIT may be for some other failure mechanism. The FIT rate budgeting may or may not be equal. Depending on the operating conditions and the failure mechanism, unequal FIT rate budgeting may be adopted.

12.14 Reliability Issues in ICs

The IC product lifetime is determined by several reliability mechanisms. Some of the most important mechanisms are as follows:
- Electromigration (EM)
- Time-dependent dielectric breakdown (TDDB)
- Channel-hot-electron (CHE)
- Negative-bias temperature instability (NBTI)
- Electrostatic discharge (ESD)
- CMOS latch-up

12.15 Electromigration

Electromigration (EM) is a phenomenon which causes migration of atoms of metal interconnects in the VLSI chips due to electron wind. When the current density through a metal wire exceeds a threshold level, the electrons travel at a greater velocity. This is called *electron wind*. The high-speed electrons knock the metal atoms and displace them. The displacement of the metal atoms can cause void or hillock in the metal wire, and can cause either open-circuit or short-circuit failure. Electromigration occurs when the device is in operation for some time. It also depends on the operating temperature. EM can occur both for signal and power lines.

12.15.1 Power EM

For power lines, the direct current (DC) causes the EM. The MTTF due power EM is given as

$$\text{MTTF} = AJ^{-2} \exp\frac{-E}{kT} \tag{12.9}$$

where J is the current density, E is the activation energy ~ 0.5 eV, k is the Boltzmann's constant $(8.62 \times 10^{-5} \text{ eV K}^{-1})$, and T is the absolute temperature in Kelvin.

12.15.2 Signal EM

For signal lines, the alternate current (AC) cause the EM. The MTTF due to signal EM can be expressed as

$$\text{MTTF} = \frac{A \exp\dfrac{-E}{kT}}{\overline{J}\left|\overline{J}\right| + k_{AC/_{DC}}\left|\overline{J}\right|^2} \tag{12.10}$$

where $\overline{J}$ is the average of $J(t)$ and $\left|\overline{J}\right|$ is the average of $\left|\overline{J}\right|$. The constant $k_{AC/_{DC}}$ is to consider the relative effects of AC and DC. It is typically between 0.01 and 0.00001. EM is a very serious problem when MTTF is less than 10^5 hours or failure rate exceeds 10 FIT.

The only solution to the EM problem is to reduce the current density or increase the metal widths. Most of today's technology therefore includes maximum metal-width rules (fat metal rules) in the design rules file to avoid EM.

To find out the optimum metal widths, we need to know the current density. But the average current density depends on the actual input vectors applied when the device is actually used. So, during EM analysis, the average current density is estimated based on the statistical input vectors. Most often, a pessimistic approach is adopted, so that designers put enough margins in EM reliability, so that the device does not fail due to the EM problem. Figure 12.12 illustrates EM failures in the VLSI circuits.

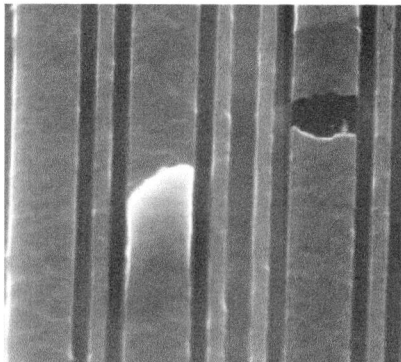

Fig. 12.12   Hillock and void created by electromigration

12.16 Time-dependent Dielectric Breakdown

Time-dependent dielectric breakdown (TDDB) is a very important issue in deep-sub-micron (DSM) designs. The gate oxide thickness of the MOS devices is extremely small in DSM designs. So the vertical electric field ($E_{ox} = V_{DD}/t_{ox}$) puts excessive electrical stress on the ultra-thin gate oxides. When this thin oxide is operated at high electric field over a period of time, depending on how long the device is operated, there is degradation in the gate oxide insulator. Eventually, the gate oxide may breakdown and the MOS device may get completely damaged. This problem is known as TDDB; it is also known as gate oxide integrity (GOI) problem in various literature. The failure rate due to TDDB is exponentially dependent on the device operating temperature and the gate oxide thickness.

As the technology advances, the gate oxide thickness is scaled down at a greater rate than the supply voltage. So, there is a continuous increase in the gate oxide stress. But manufacturers often ensure that the gate oxide is engineered in such a way that over the device operating lifetime, TDDB must not happen when the device is operated under normal operating condition (V_{DD}). However, due to circuit design issues, there could be over voltages in the signal or in the power lines that may cause the TDDB problem. Some of the causes of over voltages are as follows:

- Overshoot/undershoot due to crosstalk noise
- Overshoot/undershoot due to Miller effect
- Overshoot/undershoot due to *Ldi/dt* effect
- Power and ground bounce

12.16.1 Overshoot/Undershoot Due to Crosstalk

In VLSI chips, crosstalk noise occur when two or more metal lines (net)˙run parallel over a length. The field oxide (FOX) between the metal lines forms a coupling capacitor. Due to this coupling capacitor, when a signal makes a transition in one net, the adjacent net experiences a glitch. The net that causes the glitch is termed as *aggressor net,* and the affected line is termed as *victim net*. Now, if the victim net is already at logic high (V_{DD}) level, and the aggressor net makes a low-to-high transition, it causes *overshoot* on the victim net. Similarly, if the victim net is already at logic low (ground) level, and the aggressor net makes a high-to-low transition, it causes *undershoot* on the victim net. This is depicted in Figs 12.13 and 12.14.

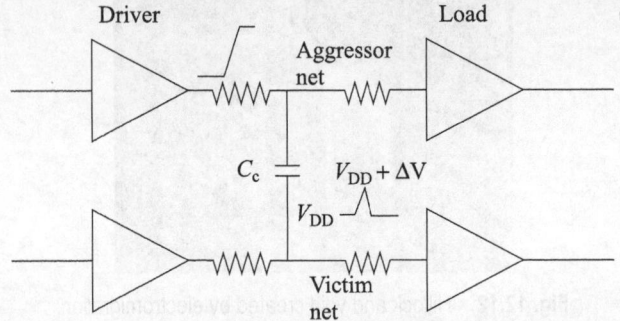

Fig. 12.13 Circuit schematic for crosstalk overshoot/undershoot analysis

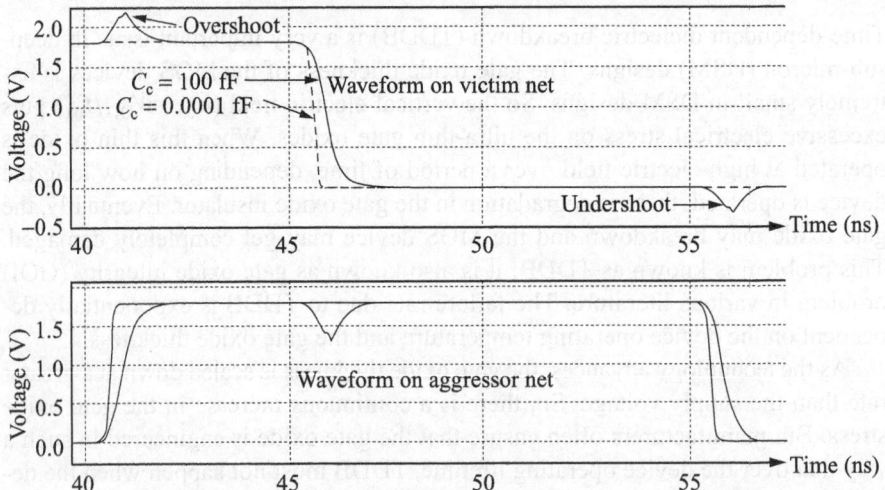

Fig. 12.14 Overshoot and undershoot waveforms due to crosstalk

12.16.2 Miller Effect

The capacitance between the input and output of a circuit is known as Miller capacitance. In CMOS circuits, the Miller capacitance is present in every MOS transistor due to the drain-gate capacitance (C_{DG}). For example, in a CMOS inverter, the Miller capacitance is shown in Fig. 12.15.

12.16.3 Overshoot/Undershoot Due to *Ldi/dt* Effect

The metal lines have parasitic resistance, inductance, and capacitance which form an *RLC* circuit. When there is a transition through the nets, due to the inductance, there will be overshoot or undershoot in the metal lines.

The power and ground metal lines have significant parasitic resistance, inductance, and capacitance which form an *RLC* circuit. When there is switching activity in the logic circuits, it will cause power and ground bounce, as shown in Fig. 12.16.

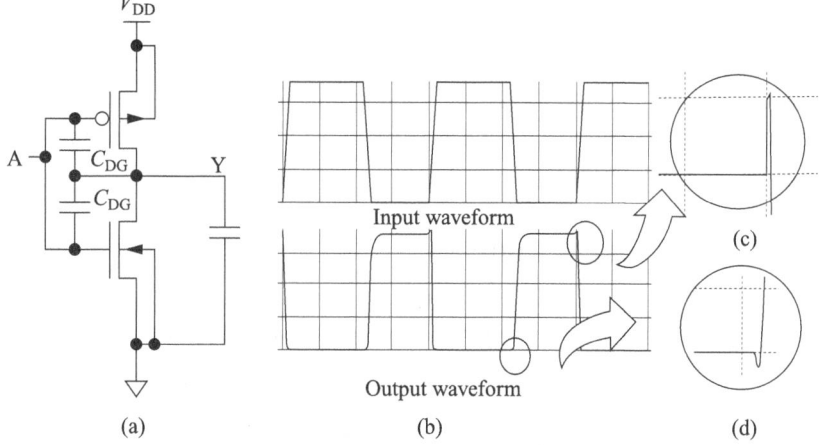

Fig. 12.15 (a) CMOS inverter with C_{DG}; (b) input–output waveform; (c) magnified view of overshoot; (d) magnified view of undershoot

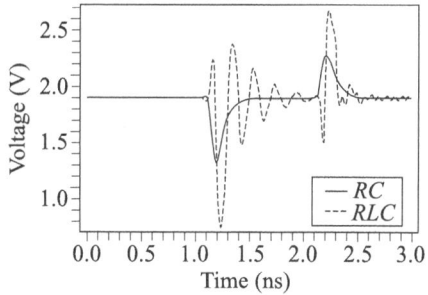

Fig. 12.16 Power and ground bounce

12.16.4 Power and Ground Bounce

The interconnect wires have parasitic resistance, capacitance, and inductance. It has been found that the parasitic resistance and capacitance (RC) of interconnect wires are the most dominating factors over inductance, and parasitic RC is extensively used for post-detailed route timing analysis. The on-chip parasitic inductance is still not critical. However, the parasitic inductance of the IC package lead-frames and bond-wires are significant. The package RLC has transmission line effects which can cause voltage drop, ringing, and overshoot in the power and ground lines when fast switching occurs. The fast switching introduces Ldi/dt drop in the power and ground lines, which is known as power and ground bounce. This can cause

- noise margin to be reduced
- the logic levels to be changed
- introduction of large spikes which can create functional violation

Some of the design techniques that are used to resolve the *Ldi/dt* issue are as follows:

- Separate power pins for core cells and input/output buffers
- Multiple power and ground pins
- Proper positioning of the power and ground pins in the package
- Placing large decoupling capacitors between the power and ground lines
- Suitable package design

12.17 Channel Hot Electron

As the technology advances, the MOS transistor length is scaled at a faster rate compared to the supply voltage. As a result, the lateral electric field in the channel region increases. At increased horizontal and vertical electric field, electrons gaining high kinetic energy (*hot electrons*) are injected into the gate oxide, and cause permanent changes in the oxide-interface charge distribution, degrading the *I–V* characteristics of the MOSFET. This effect is known as channel hot electron (CHE) effect. It is found only in MOS devices.

When the hot electrons are trapped in the oxide layer, the threshold voltage of nMOS device is increased. Threshold voltage on the gate of the nMOS device pulls the electrons up below the oxide layer. But the electrons trapped inside the oxide layer repel them down. To compensate this repulsion, additional voltage must be applied to the gate. Hence, the threshold voltage of nMOS increases. This increase in threshold reduces the current through the nMOS device; hence, the nMOS device becomes slow. This will not only reduce the speed of the design, but it could lead to functional violation as well.

Typically, when a technology is defined, it is ensured that the hot-carrier induced gate oxide degradation is acceptable over a period of device operating lifetime under normal operating conditions. But in circuits, there are several causes by which the drain of the nMOS devices can experience overshoots. These overshoots can accelerate the hot electron induced degradation. Again, just only overshoots at the drain of nMOS cannot cause hot-carrier induced degradation. The degradation occurs when the nMOS is ON and it experiences a drain overshoot. CHE analysis is very much challenging in VLSI circuits. The analysis must consider the relative switching of the nMOS gate and drain simultaneously for realistic CHE verification.

12.18 Negative-bias Temperature Instability

Similar to the hot electron effect in nMOS device, pMOS devices also suffer a problem which is known as negative bias temperature instability (NBTI). NBTI has become a very important reliability concern in digital as well as analog CMOS VLSI circuits. Due to the scaling of gate oxide thickness below 20 Å without corresponding scaling of the supply voltage, vertical electric across the gate oxide is increased. Under high vertical electric field and high operating temperature, NBTI effect became significant. The threshold voltage of pMOS transistors increase with time under NBTI.

NBTI arises due to the continuous generation of traps in the $Si–SiO_2$ interface of the pMOS transistors. When the MOS device structure is fabricated, undesirable Si dangling bonds exist due to structural mismatch at the $Si–SiO_2$ interface. These dangling bonds act as charged traps at the $Si–SiO_2$ interface. After the oxidation process, these dangling bonds are passivated by hydrogen by forming Si–H bonds. However, the Si–H bonds are very weak, and with time, can easily break during device operation (i.e., when negative gate bias is applied to the pMOS). These broken bonds act as interfacial traps, illustrated in Fig. 12.17. When the holes are trapped in the oxide interface, they start repelling holes in the pMOS channel region or inversion layer region. Then more negative voltage is required to be applied at the gate, which means increase in the threshold voltage (V_t) of the device. If the threshold is increased, the current through the pMOS decreases and the device becomes slow; hence it affects the performance of the circuit. NBTI impact becomes worse in ultra-deep sub-micron (UDSM) technology due to the high operating temperature and the usage of ultra-thin oxide (i.e., higher oxide field).

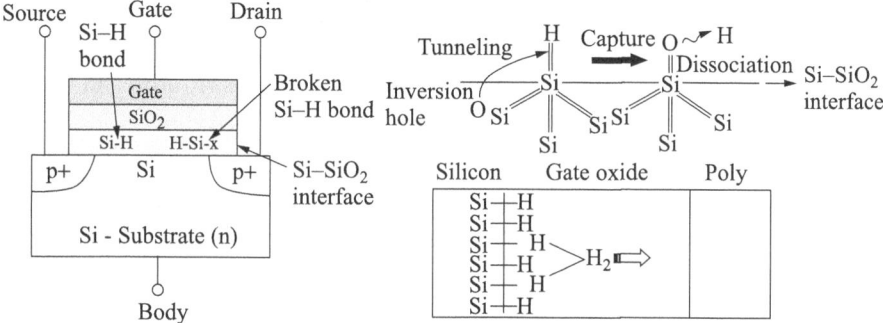

Fig. 12.17 Schematic of pMOS showing traps at the Si–SiO$_2$ interface

The effect of NBTI is a gradual increase of threshold voltage with time due to the continuous generation of traps at the Si–SiO$_2$ interface in negatively biased pMOS transistors at increased temperature. It is found experimentally that the increase in threshold voltage (V_t) due to NBTI under DC stress increases nonlinearly with time, as per the following relationship:

$$\Delta V_t(t) = K_{DC} \times t^n \tag{12.11}$$

where K_{DC} is a technology-dependent constant (e.g., process, temperature, V_{DD}, device geometry), and n is the time exponent of degradation ($n \geq 1/6$).

As in real circuit operation, the pMOS devices are not always under DC stress, but they are operated in pulsed mode. When the gate input of pMOS is high, the device is OFF, and it tries to recover the increase in threshold voltage. Hence, the DC degradation model is overly estimated. In order to correct this, a correction factor is introduced in Eqn (12.11) as shown in Eqn (12.12),

$$\Delta V_t(t) = \alpha \times K_{DC} \times t^n \tag{12.12}$$

where α is the correction factor which depends on the frequency of operation and the signal transition probability.

NBTI is also a strong function of operating temperature of the device. The temperature dependence can be expressed as

$$\Delta V_t(t) \approx e^{\frac{-E_a}{kT}} \tag{12.13}$$

where E_a is the activation energy (0.1–0.2 eV), k is the Boltzmann's constant, T is the operating temperature in Kelvin.

In summary, at high temperature and negative gate bias, the current through the pMOS device reduces. This phenomenon is known as NBTI. The causes of NBTI are
- Creation of interface states
- Trapping of holes in the oxide layer

NBTI results in decrease in device speed, and hence causes both timing and functional violations.

12.19 IR Drop

IR drop is a phenomenon, which is very important for power and ground lines in a VLSI circuit. The power/ground nets have finite distributed parasitic resistance. When current flows through the power/ground net, there is a finite voltage drop in the net or wire itself—this voltage drop is known as *IR* drop and can be expressed as

$$\Delta V = I \times R \tag{12.14}$$

where I is the current flowing through the power/ground net and R is the resistance of the power/ground net. Hence, effective power supply value is reduced by the amount of *IR* drop as given by

$$V'_{DD} = V_{DD} - \Delta V \tag{12.15}$$

The reduced power supply voltage has the following effects on the circuit behaviour:
1. The noise margin is reduced.
2. The logic levels are changed.
3. The logic level becomes a function of distance of a cell from the power/ground pin.
4. The cells at a far distance from the power/ground pins suffer more IR drop problems.
5. The IR drop can cause static power dissipation.
6. The reduced power supply voltage increases the propagation delay through a gate and hence, degrades the speed. The timing degradation can also cause the functional violation.

The solution to the IR drop issues is to come up with an effective power distribution network. Generally, a power distribution network follows a grid structure. The top-level metal layers, which are thick and having high pitch, are used for power routing. Another approach is to use multiple power/ground pins across the periphery of the chip instead of a single power supply.

The IR drop analysis as explained looks very simple but in reality, it is a hard problem. The IR drop phenomenon is a dynamic problem, i.e., it occurs when the gates make transitions. It is maximum, when, simultaneous switching event occurs. For accurate IR drop analysis, the timing information needs to be accounted for. Also the IR drop analysis must be done for the entire chip at a time. It can only be done after the detailed routing step when the interconnect layer and geometry information are available.

12.20 Latch-up

In CMOS circuits, the parasitic NPN and PNP bipolar junction transistors (BJT) form a thyristor or PNPN structure. When the thyristor is triggered by noise/glitch, the BJTs become ON, and it leads to a formation of short circuit between the power and ground lines, known as CMOS latch-up. Once the device is triggered, the current through the device continues to increase due to regenerative feedback action, then it cannot be stopped. The only way to stop it is to switch the power supply. To illustrate this, let us consider the structure of a CMOS inverter as shown in Fig. 12.18.

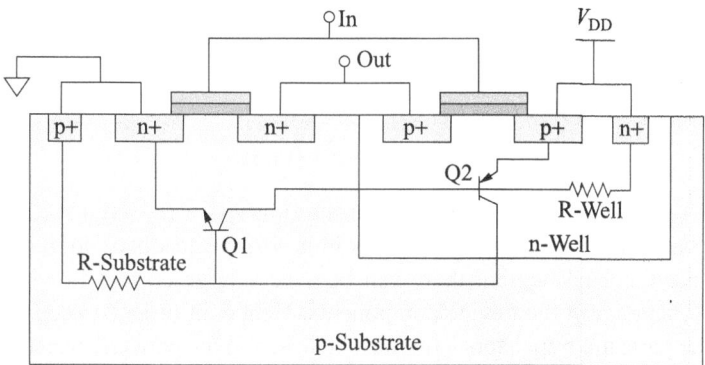

Fig. 12.18   Cross-sectional view of CMOS inverter with parasitic PNP–NPN structure

The equivalent circuit of the PNPN structure is shown in Fig. 12.19.

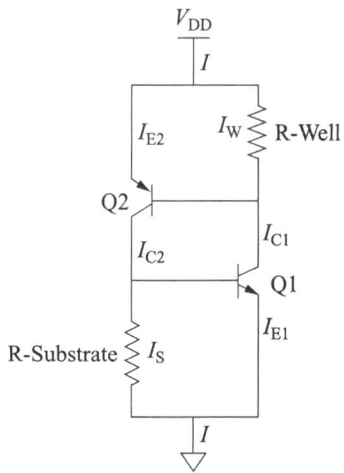

Fig. 12.19   Equivalent circuit of PNPN structure

If α_1 and α_2 be the gain of the NPN and PNP transistors, we can write

$$I_{C1} = \alpha_1 I_{E1} + I_{CBO1} \tag{12.16}$$

$$I_{C2} = \alpha_2 I_{E2} + I_{CBO2} \tag{12.17}$$

Again,

$$I = I_{E2} + I_w \tag{12.18}$$

$$I = I_{E1} + I_s \tag{12.19}$$

The total current can be written as

$$I = I_{C1} + I_{C2} = \alpha_1 I_{E1} + I_{CBO1} + \alpha_2 I_{E2} + I_{CBO2}$$

$$= \alpha_1(I - I_s) + \alpha_2(I - I_w) + I_{CBO1} + I_{CBO2}$$

$$I(1 - \alpha_1 - \alpha_2) = I_{CBO1} + I_{CBO2} - (\alpha_1 I_s + \alpha_2 I_w)$$

$$I = \frac{I_{CBO} - (\alpha_1 I_s + \alpha_2 I_w)}{1 - (\alpha_1 + \alpha_2)} \tag{12.20}$$

where

$$I_{CBO} = I_{CBO1} + I_{CBO2} \tag{12.21}$$

When the sum of the current gains approaches unity, i.e., $(\alpha_1 + \alpha_2) \approx 1$, the current becomes very large. The circuit operates in a positive feedback loop. Due to the regenerative feedback action, there can be a very large current flowing through the device, damaging the device components. Hence, in order to keep the current limited, the parasitic transistors gain must be low, and the parasitic resistances must be small.

Latch-up prevention techniques include the following:
1. Use of p$^+$ and n$^+$ guard rings around nMOS and pMOS connected to the ground and V_{DD}, respectively.
2. Reduction of R-well and R-substrate as much as possible by placing substrate and well contacts as close as possible to the source contact.
3. Keeping sufficient spacing between the nMOS and pMOS transistors.

System-level approaches to avoid latch-up are as follows:
1. Power supplies must be off before plugging a board. A plug-in in power supply ON condition can cause a surge voltage to appear on the signal pins which could trigger the latch-up.
2. The electrostatic discharge (ESD) can trigger latch-up. ESD protection circuits need special care such that it does not lead to triggering of latch-up.
3. The electron–hole pairs are generated when radiation, such as X-rays, cosmic, or alpha-rays, penetrate into the chip. These carriers can contribute to well or substrate currents, leading to latch-up.
4. Simultaneous switching of a large number of transistors can cause noise in the power/ground lines, which can drive the circuit into latch-up. This must be checked through simulation.

12.21 Electrical Overstress and Electrostatic Discharge

Electrical overstress (EOS) is a failure mechanism of VLSI circuits where the device is subjected to excessive voltage, current, or power. *Electrostatic discharge* (ESD) is the rapid transfer of large electrostatic charge into VLSI circuits through the I/O pads.

Both EOS and ESD are vulnerable for semiconductor devices, resulting in damages in the device. The mechanisms that damage the device are
- Dielectric or oxide punch-through
- Fusing of a conductor or resistor
- Junction damage or burn-out

Electrostatic discharge occurs when a charged body releases its static charge to the ground in a very short span of time. Typically, the ESD happens in a fraction of micro seconds. In an IC chip, ESD happens in several ways. Some of them are as follows:
1. IC package pin is touched by human hands
2. IC package pin is held with metallic tweezers
3. IC package pin is in contact with the ground

On the other hand, the EOS event lasts for a comparatively larger time than the ESD. Typically, an EOS event lasts for several microseconds or even milliseconds. General EOS events are either over-voltages or transient spikes.

12.21.1 Electrostatic Discharge Models

The most commonly used ESD models are as follows:
- Human body model (HBM)
- Machine model (MM)
- Charged-device model (CDM)

Human body model is an equivalent circuit to represent the ESD when the human body is in direct contact with the IC. Figure 12.20 illustrates this model. A high voltage is applied to charge the 100 pF capacitor, and then discharged through the 1.5 kΩ resistor on to the pin under test. In HBM, the ESD event occurs typically at 2–4 kV in the field.

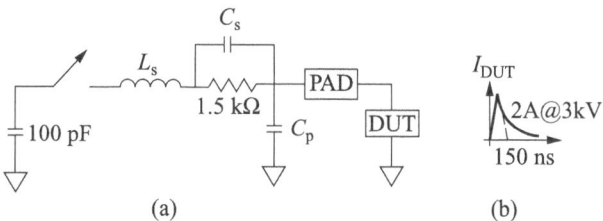

Fig. 12.20   Equivalent circuit of human body model (HBM)

The ESD *machine model* (MM) represents the ESD events when a charged body is in direct contact with the IC pin during device bonding, assembly, and testing. The ESD MM is tested with the 200 pF capacitor which is pre-charged to 400 V with an initial peak current of 7 A. Then a damped, oscillatory 16 MHz sinusoidal current is applied to test the device under test (DUT). An equivalent circuit for MM is shown in Fig. 12.21.

The ESD *charged-device model* (CDM) represents the discharge of a packaged IC. To test the CDM, first the device under test (DUT) is charged, and then discharged to the ground. A high-current short-duration (5 ns) pulse is applied to the DUT. An equivalent circuit is shown in Fig. 12.22 for the CDM.

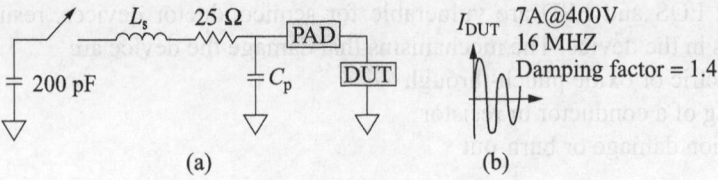

(a)   (b)

Fig. 12.21   Equivalent circuit of a machine model

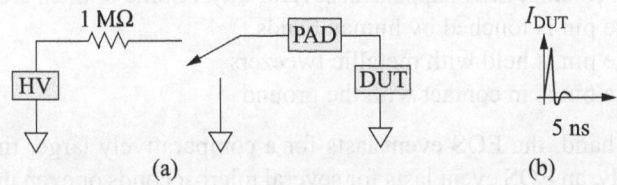

(a)   (b)

Fig. 12.22   Equivalent circuit of a charged-device model

12.21.2 Electrostatic Discharge Failures

Electrostatic discharge can cause several damages to microelectronic devices, including gate oxide breakdown, junction spiking, and latch-up.

1. Gate oxide failure is the breakdown of the dielectric (oxide) layer between the transistor gate and the channel, which leads to excessive leakage current or a functional failure.
2. Junction spiking means the migration of the metal through the source/drain junction of MOS transistors. This can cause high leakage or a functional failure.
3. Electrostatic discharge can trigger the latch-up failure (discussed in the previous section), which can cause an internal feedback mechanism to start that gives rise to temporary or permanent damage of the device.

12.21.3 Electrostatic Discharge Protection Circuit

Electrostatic discharge issue is taken care by designing ESD protection circuit, and connecting it to all the input and output pins. The ESD stress can be both positive and negative. So, the ESD protection circuit is designed to bypass the ESD currents to V_{DD} or V_{SS} pins. Figure 12.23 illustrates the simple bypass path to protect the gate of the core logic transistor.

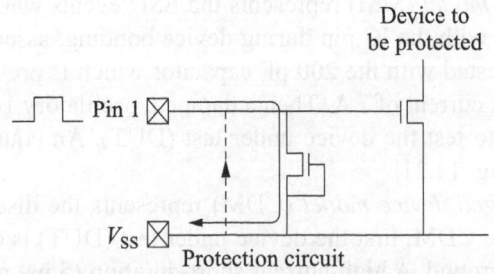

Fig. 12.23   Electrostatic discharge protection circuit

Electrostatic discharge protections at the input and output pads are shown in Fig. 12.24.

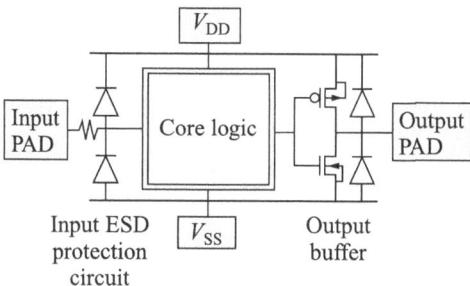

Fig. 12.24 Electrostatic discharge protection mechanism at the chip level

12.22 Soft Error

A *soft error* is a noise or glitch that appears in a circuit generated from outside the chip. It is caused by the charged particle striking the semiconductor devices. These charged particles come from radioactive materials (alpha particles) and cosmic rays (protons and neutrons). The physical phenomenon is often called soft error or *soft error rate* (*SER*), which only affects the memory elements by corrupting the stored data. The similar error in FPGA due to radiation is called *firm error* as it changes the configuration of the FPGA, thereby changing the functionality of the device.

Designers cannot control the sources of soft or firm errors; only careful design techniques can mitigate the effects of soft errors by utilizing error resistant programmable products. A chip designer can try to minimize the soft error rate by judicious device design, choosing the right material, package and substrate materials, and the right device geometry. One technique that is commonly used to reduce the soft error rate in digital circuits is called *radiation hardening*. Radiation hardening is often achieved by increasing the size of transistors which share a drain/source region at the node that has the highest probability of resulting in soft errors, if struck.

Soft error rate (SER) is the rate at which a device or system encounters or is predicted to encounter soft errors. It is commonly expressed as either number of failures-in-time (FIT), or mean-time-between-failures (MTBF). One FIT is equivalent to 1 error per billion hours of device operation. MTBF is usually given in years of device operation. For example, 1 year MTBF is equivalent to approximately 114,077 FIT.

Error checking and correction (ECC) is generally used to check and correct the soft errors. ECC is implemented by adding an extra chip to each multi-chip memory module (to generate the extra bits), and incorporating the ECC logic and circuitry at the system or board level.

Soft errors are becoming a great concern as memory size is increased, and the memory technology is shrunk. For example, using a relatively conservative soft error rate of 500 FIT/megabit, a system with 1 gigabyte of RAM is expected to have an error in every two weeks; a terabyte system would experience a soft error every few minutes. Soft errors can be disastrous for systems with large memories, critical applications, or high altitude locations. An error detection/correction mechanism is must in these cases, in spite of the higher cost in price and/or performance.

SUMMARY

- Verification is the process of checking the design for its correctness, both from the functionality and the specifications.
- Design rules are a set of rules that specify the geometric specifications between the layers used for drawing a layout. Design rules come from the manufacturing technology.
- LVS check is the equivalence checking between the circuit schematic netlist and the netlist extracted from layout.
- Electrical rule checker checks for network connectivity, such as short-circuit and open-circuit checks.
- The partially processed metal leads collect static charges from the plasma environment. These leads are called antennas, if connected to gate of MOS devices, the static charge discharges through the gate oxide and damages the gate oxide.
- Extraction is a process of pulling out information related to devices and interconnects from the drawn layout.
- Crosstalk can affect both functionality and timing of the VLSI circuits.
- Reliability of an integrated circuit ensures that the circuit operates with correct functionality and specifications for a specified period of operating lifetime.
- One FIT indicates one device failure in 10^9 device-operating-hours.
- Electromigration happens due to electron wind that can create open or short circuits.
- TDDB is an oxide breakdown phenomenon which depends on the electric field, operating lifetime, and has exponential dependency on the operating temperature.
- Hot electron effect is predominant in nMOS transistors, whereas NBTI is severe in pMOS transistors.
- IR drop slows down the circuit operating speed.
- ESD protection circuits are must at the I/O pads for an integrated circuit that protects the internal core from the external electrical hazards.
- Soft error is critical for memory circuits and FPGAs where stored information is corrupted due to soft errors.

SELECT REFERENCES

Abadeer, W. and W. Ellis 2003, 'Behaviour of NBTI under AC Dynamic Circuit Conditions', in: *Proceedings of the IEEE International Reliability Physics Symposium*, pp. 17–22, August.

Alam, M.A. 2003, 'A Critical Examination of the Mechanics of Dynamic NBTI for pMOSFETs', *IEEE International Electronic Devices Meeting*, pp. 14.4.1–14.4.4, December.

Alam, M.A. 2005, 'On the Reliability of Microelectronic Devices: An Introductory Lecture on Negative Bias Temperature Instability', *Nanotechnology 501 Lecture Series*, September.

Alam, M.A. and S. Mahapatra 2004, 'A Comprehensive Model of PMOS NBTI Degradation', *Journal of Microelectronics Reliability*, vol. 45, pp. 71–81, August.

Amerasekera, A. and C. Duvvury 2002, *ESD in Silicon Integrated Circuits,* 2nd ed., Wiley, London.

Chakravarthi, S., A.T. Krishnan, V. Reddy, C. Machala, and S. Krishnan 2004, 'A Comprehensive Framework for Predictive Modeling of Negative Bias Temperature Instability', in: *Proceedings of the IEEE International Reliability Physics Symposium*, pp. 273–82, April.

Chen, G., M.F. Li, C.H. Ang, J.Z. Zheng, and D.L. Kwong 2002, 'Dynamic NBTI of p-MOS Transistors and Its Impact on MOSFET Scaling', *IEEE Electron Device Letters*, pp. 734–6, December.

Chen, J.Z., A. Amerasekera, and C. Duvvury 1997, 'Design Methodofology for Optimizing Gate Driven ESD Protection Circuits in Submicron CMOS Processes', *EOS/ESD Symposium*.

Ershov, M., S. Saxena, H. Karbasi, S. Winters, S. Minehane, J. Babcock, R. Lindley, P. Clifton, M. Redford, and A. Shibkov 2003, 'Dynamic Recovery of Negative Bias Temperature Instability in p-type Metal Oxide Semiconductor Field-Effect Transistors', *Appl. Phys. Lett.*, vol. 83, p. 1647.

Diaz, C.H. 1994, 'Automation of Electrical Overstress Characterization for Semiconductor Devices', *Hewlett-Packard Journal*.

Johnston, A.H. 2000, 'Scaling and Technology Issues for Soft Error Rates', *4th Annual Research Conference on Reliability,* Stanford University, October.

Kamon, M., S. McCormick, and K. Shepard 1999, 'Interconnect Parasitic Extraction in the Digital IC Design Methodology', *IEEE*.

Ker, M.D. 1999, 'Whole-Chip ESD Protection Design with Efficient V_{DD}-to-V_{SS} ESD Clamp Circuits for Submicron CMOS VLSI', *IEEE Transactions on Electron Devices*, vol. 46, no. 1, p. 173, January.

Krishan, A.T., V. Reddy, S. Chakravarthi, J. Rodriguez, S. John, and S. Krishnan 2003, 'NBTI Impact on Transistor and Circuit: Models, Mechanisms and Scaling Effects', *IEEE International Electronic Devices Meeting*, pp. 14.5.1–14.5.4, December.

Kumar, S.V., C.H. Kim, and S.S. Sapatnekar 2006, 'An Analytical Model for Negative Bias Temperature Instability', *International Conference on Computer-Aided Design*, IEEE.

Mahapatra, S., P.B. Kumar, and M.A. Alam 2004, 'Investigation and Modeling of Interface and Bulk Trap Generation during Negative Bias Temperature Instability of p-MOSFETs', *IEEE Transactions on Electron Devices*, vol. 51, no. 9, 1377–9.

Mahapatra, S., P.B. Kumar, and M.A. Alam 2004, 'Investigation and Modeling of Interface and Bulk Trap Generation During Negative Bias Temperature Instability of p-MOSFETs', *IEEE Transactions on Electronic Devices*, pp. 1371–9, September.

Paul, B.C., K. Kang, H. Kufluoglu, M.A. Alam, and K. Roy 2003, 'Impact of NBTI on the Temporal Performance Degradation of Digital Circuits', *IEEE Electron Device Letters*, vol. 26, pp. 560–2, August.

Ramaswamy, S., C. Duvvury, S.-M. Kang 1995, 'EOS/ESD Reliability of Deep Sub-Micron nMOS Protection Devices', *IEEE International Symposium on Reliability Physics*, pp. 284–91, 4–6 April.

Rangan, S., N. Mielke, and E.C.C. Yeh 2003, 'Universal Recovery Behaviour of Negative Bias Temperature Instability [PMOSFETs]', *Electron Devices Meeting, 2003, IEDM 2003 Technical Digest*, IEEE.

Reddy, V., A.T. Krishnan, A. Marshall, J. Rodriguez, S. Natarajan, T. Rost, and S. Krishnan 2004, 'Impact of Negative Bias Temperature Instability on Digital Circuit Reliability', 21 July.

Stathis, J.H. and S. Zafar 2005, 'The Negative Bias Temperature Instability in MOS Devices: A Review', *IBM Semiconductor Research and Development Center (SRDC), Research Division,* TJ Watson Research Center, Yorktown Heights, New York.

Tsetseris, L., X.J. Zhou, D.M. Fleetwood, R.D. Schrimpf, and S.T. Pantelides 2005, 'Physical Mechanisms of Negative-Bias Temperature Instability', *Appl. Phys. Lett.,* vol. 86, p. 103–142.

Young, D. and A. Christou 1994, 'Failure Mechanism Models for Electromigration', *IEEE Transactions on Reliability,* June.

EXERCISES

Fill in the Blanks

1. TDDB has _____ dependency on the operating temperature.
   (a) exponential
   (b) linear
   (c) square-law
   (d) proportional
2. An aggressor net is _____
   (a) a net that affects other nets
   (b) a net that is affected by other nets
   (c) a power net
   (d) a ground net
3. Crosstalk affects _____.
   (a) timing
   (b) functionality
   (c) reliability
   (d) all of these
4. Crosstalk delay arises due to _____.
   (a) simultaneous switching of aggressor and victim nets
   (b) switching of only aggressor nets
   (c) switching of only victim nets
   (d) switching of aggressor and victim nets
5. One FIT indicates one device failure in _____ device-hour.
   (a) 10^9
   (b) 10^6
   (c) 10^{12}
   (d) 10^3

Multiple Choice Questions

1. Compared to static verification, dynamic verification is
   (a) accurate but slow
   (b) accurate and fast
   (c) fast but inaccurate
   (d) none of these
2. Electromigration can be reduced by
   (a) increasing metal cross-sectional area
   (b) reducing metal cross-sectional area
   (c) decreasing current density
   (d) using strong buffers
3. Mostly used ESD models are
   (a) human body models
   (b) machine models
   (c) charged device models
   (d) all of these
4. Design rules are mostly
   (a) minimum width rules
   (b) minimum spacing rules
   (c) enclosure rules
   (d) all of these
5. Antenna effect arises due to the
   (a) design defect
   (b) manufacturing process issues
   (c) radiation of signals
   (d) layout defect

True or False

1. Micron-based rules are scalable.
2. LVS checks RTL netlist versus gate-level netlist.
3. Hot electron effect is severe in pMOS transistors.
4. NBTI is severe in pMOS transistors.
5. Extraction is used only for parasitic.
6. Antenna diode acts as ESD protection circuit.

Short-answer Type Questions

1. What are the reliability issues in VLSI? Discuss them in brief.
2. What is reliability bathtub curve? Explain the different regions of this curve.
3. What is FIT rate? If nine devices fail after operating for 10 years, what is the FIT rate of this device?
4. What is antenna effect in VLSI? What is the fixation mechanism for this? Discuss with necessary diagrams.
5. What is crosstalk delay in VLSI? Discuss the origin of crosstalk delay and how it effects the performance of an IC.
6. What is crosstalk noise in VLSI? Discuss the origin of crosstalk delay and how it effects the performance of an IC.
7. Define FIT rate, MTTF, and MTBF. What are the uses of these metrics?
8. Draw the VLSI design flow with indicating verification at different stages of the design flow.
9. Discuss the parasitic extraction methodology.

Long-answer Type Questions

1. What do you mean by physical verification? Discuss several physical verification steps.
2. What are the different methods of logic verification? Discuss them in brief. With the help of ROBDD, show how LVS checking can be done.
3. What do you mean by design rules? Discuss different design rules with examples.
4. What do you mean by crosstalk analysis in VLSI circuits? Discuss with the help of schematic diagram how overshoot/undershoot analysis can be done.
5. (a) What are the reliability issues in VLSI? Discuss them in brief.
   (b) What is reliability bathtub curve? Explain different regions of this curve.
   (c) What is FIT rate? If 9 devices fail after operating for 10 years, what is FIT rate of the device?
6. Write short notes on the following:
   (a) Electromigration
   (b) Electrostatic discharge (ESD)
   (c) Hot carriers
   (d) CMOS latch-up
   (e) Gate oxide integrity or TDDB
   (f) Soft error

IC Packaging

KEY TOPICS

- Types of IC packages
- Package modelling
- Electrical package modelling
- IC package thermal modelling
- IC package stress modelling
- Package models
- Package simulation
- Flip-chip package

13.1 Introduction

The integrated circuits (ICs) are fabricated on a wafer in a batch. The bunch of wafers, processed simultaneously, is called a *lot*. From the processed wafer individual ICs called die are separated out by cutting the wafers. A die is then packaged in a plastic or ceramic compound structure to form an IC chip. Figure 13.1 illustrates a schematic view of the lot, wafer, die, and IC chip. The IC package is a cover to the internal core die which supports the die, and protects the die from damage. In addition to protection, the package serves the following important functions: it provides electrical signal exchange between the core die and the outside environment; it acts as a medium for heat dissipation; it provides power to the die. But with circuits approaching higher frequencies, the function of a package is not just limited to the above-mentioned functions.

Initially, the electrical performance of the packaged IC chip was limited by the die itself and very little because of the package. But with rise in the operating range of frequency, the parasitics associated with the package started affecting the performance of the IC. A variety of requirements for smaller, lighter, faster, and less expensive electronic products have not only led to better fabrication techniques, but also to better packaging techniques. With increase in the frequency range of operation, the electrical performance of different packages decides the packaging used for a particular application.

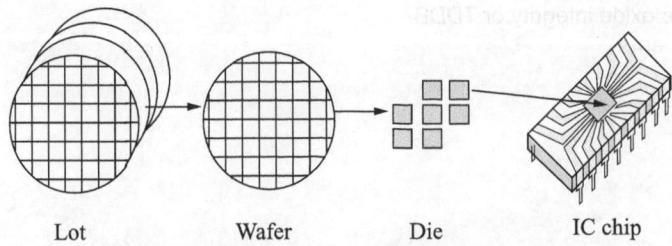

Lot	Wafer	Die	IC chip

Fig. 13.1   A schematic of lot, wafer, die, and IC chip

13.2 Types of IC Packages

The IC packages are mainly classified into two types:

- *Pin-through-hole* (*PTH*)   package pins are extended in the vertical direction so that they can be inserted through holes in the circuit board.
- *Surface-mount technology* (*SMT*)   package pins are extended in the horizontal direction so that they can be mounted on the surface of the circuit board.

Figure 13.2 illustrates the types of IC packages.

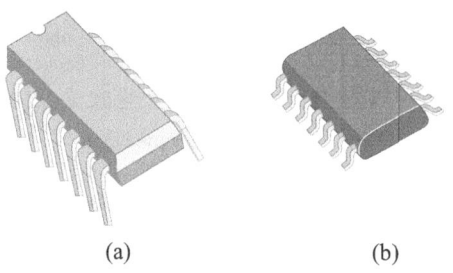

(a)   (b)

Fig. 13.2   IC package: (a) PTH; (b) SMT

13.2.1 Pin-through-hole Package

In the PTH package, the pins are inserted into holes in the printed circuit board (PCB) and soldered in place from the opposite side of the board. It can also be inserted in the socket installed on the PCB. The PTH packages can be classified based on the location of the pins as follows:

- *Single side*   Single side package has pins over a single side. The pins can be positioned in-line or zig-zag. So, again depending on pin positions they are classified as
  - SIP (single in-line package)
  - ZIP (zig-zag in-line package)
- *Dual side*   In the dual side package, the pins are located on both sides of the package in in-line pattern, hence the name DIP (dual in-line package).
- *Full surface*   Full surface package has pins located in arrays over the four sides of the package. This type of package is called PGA (pin grid array).

13.2.2 Surface-mount Technology Package

The surface-mount technology (SMT) packages have pins that are soldered directly onto the surface of the circuit board. These packages do not require any hole in the circuit board. SMT packages are preferred over PTH packages due to many of their advantages, such as

- SMT package can be mounted on both sides of the circuit board.
- They are small in dimensions.
- Their package parasitics are reduced over PTH packages.
- They increase circuit board wiring density.
- Manufacturing is easier than PTH.

SMT packages are classified into several types based on the pin locations. They are shown in Fig. 13.3.

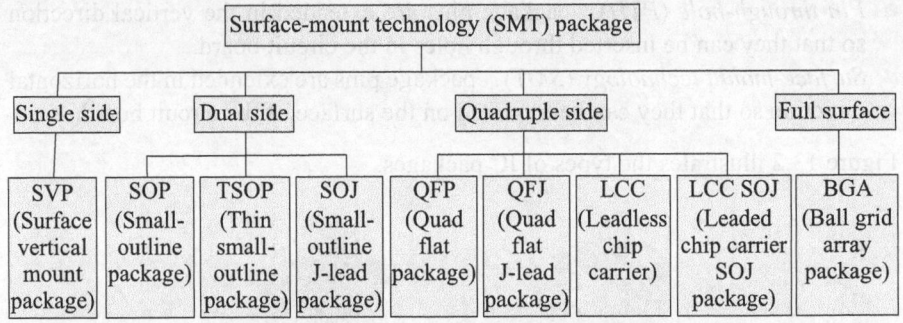

Fig. 13.3   SMT package types

13.2.3 Types of Packages Based on Package Material

Depending on the material used for packaging, the packages are classified into the following two types:

- *Plastic package*   Die-bonding and wire-bonding are used to connect the die bond pad to a metal lead-frame. The encapsulation is done by an injection of moulded plastic. These packages are inexpensive, but they have very high thermal resistance and absorb moisture. They are prone to early device failure due to package cracking, and hence are used only for low-cost applications.
- *Ceramic package*   In the ceramic packages there is no lead-frame. The metal leads are directly soldered on the package, and separated by ceramic (Al_2O_3) layers. These packages are costlier than the plastic packages, and hence used for high-cost applications.

13.3 Package Modelling

A cross-sectional view of a typical IC package is shown in Fig. 13.4. The metallic lead-frame structure is made available outside the chip for external electrical connections. The lead-frame structure is connected to the bond pads of the die through thin wires called bond-wires. The die is mounted on a heat spreader. Then the complete system is moulded using either plastic or ceramic compound material.

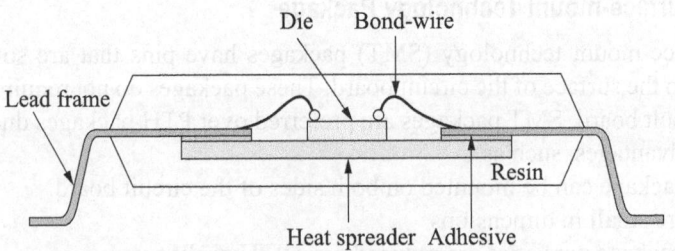

Fig. 13.4   Cross-sectional view of an IC package

Apart from protecting the internal core die, the package performs three main functions as given below:

- Provides electrical connections to the internal core die from the outside world.
- Dissipates the heat generated by the devices during circuit operation.
- Provides mechanical stability for the complete IC.

Hence, it is important to characterize the packages based on their electrical, thermal, and mechanical properties, by which the IC package can be selected for a particular application.

In this section, we shall explain the modelling of packages and their parameters.

13.4 Electrical Package Modelling

As the frequency of operation of VLSI chips have increased and power supply voltages have decreased, the function of the IC package has changed significantly. In earlier technologies, the main purpose the package was to protect the die from the outside environment. But now the package is no more a protective component, its electrical characteristics affect the circuit performance. Hence, the electrical characteristics of the IC package must be properly understood. To understand the electrical characteristics of the package, the package needs to be characterized and modelled electrically.

A package electrical model represents the physical properties of a package by its electrical characteristics. The electrical circuit characteristics of a package are determined by the following model parameters:

- Resistance (R)
- Inductance (L)
- Capacitance (C)
- Characteristic impedance (Z_0)

The above package model parameters are functions of the material of package and the structure/geometry of the package. The *RLC* parameters are combined to form an equivalent circuit model for the package electrical characterization.

The package model is of two types, both of which are necessary for fully understanding the electrical performance of the package.

The first is an input/output lead model which represents the signal path from the die to the board. There are several input/output lead model depending upon the accuracy of simulation. These models are given by

- Simple lumped circuit model
- Distributed lumped circuit model
- Single-conductor transmission-line model
- Multiple-conductor transmission-line model

Simple lumped models can be used for estimating simple effect, such as DC resistive voltage drop. But to analyse the effects such as time delay and crosstalk, more sophisticated models like the multiple-conductor transmission-line model must be used.

The second type of a package model is a power-distribution network that describes the power scheme of the package. Similar to the input/output lead model, the complexity of the power-distribution network can vary from a simple distributed lumped model to a complex circuit network called a partial-element equivalent circuit (PEEC) network. The simpler model is used to describe gross electrical characteristics of the power distribution network, such as DC resistive drop for the entire package. The complex models are required for the analysis of the effects of the power distribution topology.

In this section, we shall describe the basic electrical package modelling terminologies and derive an equivalent circuit using the package parasitics.

13.4.1 Package Resistance

The resistance of the package leads depends on the material and the lead dimensions. The package resistance cause a voltage drop ($V = IR$) across the package leads, and hence, is very important for the power and ground paths. The resistance of the package leads is given by

$$R = \rho \frac{l}{A} \tag{13.1}$$

where ρ is the resistivity of the material used for package leads, l is the length of the package lead, and A is the cross-sectional area. Typically, ceramic packages have a higher resistance as they use tungsten alloy metallization, as compared to plastic packages which use Cu or Cu alloy metallization.

13.4.2 Package Capacitance

The capacitance of the package pins depends on the lead geometry and the dielectric between the leads. The capacitance calculation for IC packages is complex as the simple parallel-plate capacitance formula does not hold well. The self and coupling capacitance calculation must take into account fringing effects, and the bends and corners of the lead geometry. Depending on the accuracy needed, two-dimension (2D), 2.5D, and 3D capacitance extraction methodologies are used.

There are number of package extraction software available by different EDA vendors. The EM solver FastCap (FastCap-web) software developed by Massachusetts Institute of Technology (MIT) also can be used to calculate the package capacitance with reasonable accuracy.

13.4.3 Package Inductance

The package inductance is another very important parameter that decides the electrical characteristic of the IC packages. The lead inductance can cause overshoot/undershoot in the signal lines, ground bounce due to the simultaneous switching of the circuits, and crosstalk between the signal lines due to mutual inductances between the leads. The lead inductance depends on the permeability of the lead material, and the lead geometry. The package inductance calculation is also done using three different extraction (2D, 2.5D, and 3D) methodologies depending on the accuracy needed. The EM solver FastHenry software developed by MIT can be used to calculate package inductance with good accuracy. In addition to the inductance calculation, FastHenry can calculate frequency-dependent lead resistance, taking care of the skin effect.

13.4.4 Characteristic Impedance

The characteristic impedance is a parameter that is used to characterize a transmission line. The package electrical equivalent model can be either lumped, or distributed, or transmission line depending on the frequency of operation. The characteristic impedance can be expressed as

$$Z_o = \sqrt{\frac{L}{C}} \qquad (13.2)$$

where L and C are the inductance and capacitance per unit length of the package pins.

The characteristic impedance decides the amount of signal reflection from the die to the outside world or vice versa, based on the mismatch between the load impedance and Z_o.

13.4.5 Electrical Equivalent Model

The electrical equivalent circuit model of a package can be lumped, distributed, or transmission line as illustrated in Fig. 13.5.

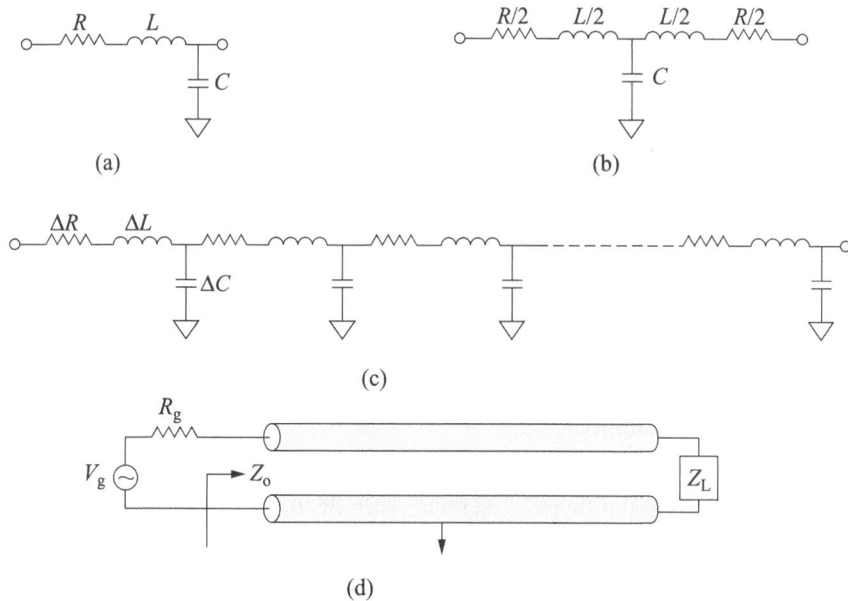

Fig. 13.5   Electrical model IC package pin: (a) lumped L-network; (b) lumped T-network; (c) distributed RLC network; (d) transmission line model

13.5 IC Package Thermal Modelling

Every IC dissipates some amount of power in the form of heat. The heat is generated when there is a current flow through a resistor. This is a well-known phenomenon, called Joule heating. The amount of power dissipated is simply expressed as I^2R, where I is the amount of current that flows through the resistor R. In an IC, there are many sources of power dissipation—dynamic power, short-circuit power, static power, leakage power, etc. These are discussed in Chapter 3. However, the heat that is produced by the devices must be carried away from the device, and transferred to the environment. The IC package which acts as the interface between the internal core to the outside environment is very important in deciding how fast the heat can be transferred from the core of the IC to the outside world. There are two important reasons why a device generated heat must be carried away from the device:

1. The first reason is to maintain the device temperature within the specified limit. Otherwise, the device performance cannot be guaranteed. For example, if the temperature is increased, the propagation delay through the gates is increased. The increase in delay causes the circuit to operate at a lower speed. In extreme cases, if the delay is increased beyond the timing specifications (e.g., set-up and hold time), then there would be functional violations as well. Hence, the devices must be operated within its allowed temperature range.

2. The second reason is to maintain the reliability criteria of each device. All the reliability phenomena (e.g., electromigration, TDDB, hot electron effect, NBTI) degrade the devices over a period of time. More importantly, all the reliability degradations increase exponentially with temperature. Typically, the normalized failure rate follows the Arrhenius equation as

$$\eta = e^{\frac{E_a}{k}\left(\frac{1}{T_{REF}} - \frac{1}{T}\right)} \tag{13.3}$$

where

E_a = activation energy (eV)

T = absolute junction temperature (K)

T_{REF} = reference temperature (K)

k = Boltzmann's constant: 8.616×10^{-5} (eV/K)

For example, for $E_a = 0.6$ eV, a 25°C increase in the junction temperature (325 K) above reference temperature (300 K) causes failure rate to increase almost 6 times above the reference failure rate. Hence, maintaining the junction temperature within the allowed limit is very important in ensuring the device reliability.

The generated heat transferred from the device to the outside environment is mainly controlled by the thermal characteristic of the IC package. Hence, thermal modelling of the IC package is very important in deciding the ICs thermal behaviour.

13.5.1 Mode of Heat Transfer

There are three different processes by which heat is transferred from one point to another point. These processes are as follows:

- Conduction
- Convection
- Radiation

Conduction In the conduction process, the heat flows from a high temperature region to a low temperature region through a medium—either solid, liquid, or gaseous. The one-dimensional heat flow equation can be written as:

$$Q = -K \frac{A(T_1 - T_2)t}{d} \tag{13.4}$$

where
  K = coefficient of thermal conductivity (J s^{-1}m^{-1}K^{-1})
  A = cross-sectional area through which heat flows
  t = time for which heat flows
  d = distance by which heat flows
  T_1 = temperature of hot end
  T_2 = temperature of cold end
The heat flow rate can be expressed as

$$\frac{dQ}{dt} = -\frac{KA\Delta T}{d} = -\frac{\Delta T}{(d\,/\,KA)} \tag{13.5}$$

Using the electrical analogy, if Q and T are analogous to charge and voltage, respectively, (d/KA) is analogous to electrical resistance. According to Eqn (13.5), thermal resistance is a function of thermal conductivity of the material and dimensions, and is independent of the amount of power dissipation.

Convection   In this mode of heat transfer, heat flows from a solid surface to a fluid, and occurs due to the bulk motion of the fluid. The heat transfer equation by the process of convection is referred to as Newton's law of cooling, and is expressed as

$$Q_c = hA(T_s - T_{amb})\,t \tag{13.6}$$

where
  Q_c = amount of heat flows by convection process
  T_s = surface temperature (K)
  T_{amb} = ambient temperature (K)
  A = surface area (m^2)
  h = average convective heat transfer coefficient (W/m^{2}C)
The convective thermal resistance can be expressed as (1/hA).
  Convection can be either forced or natural. In forced convection, fluid flow is created by an external factor, such as a fan. In case of natural convection, fluid motion takes place due to the density variations resulting from temperature gradients in the fluid. The continuous movement of the fluid transfers heat from the surface of the fluid to the ambient.

Radiation   In the radiation process, heat transfer occurs as a result of energy emission from a hot body by virtue of its temperature. Heat transfer by radiation does not require any medium. Radiant energy can also be transported by electromagnetic waves, or by photons. The amount of heat transferred by radiation, between two surfaces at temperatures T_1 and T_2, respectively, can be expressed as

$$Q_r = \varepsilon\sigma A(T_1^4 - T_2^4)F_{12} \tag{13.7}$$

where
  Q_r = amount of heat transfer by radiation (W)
  ε = emissivity ($0 < \varepsilon < 1$)
  σ = Stefan–Boltzmann constant, 5.67×10^{-8} (W/m^2 K^4)
  A = area (m^2)

F_{12} = shape factor between surfaces 1 and 2 (fraction of surface 1 radiation seen by surface 2)

T_1, T_2 = surface temperatures (K)

Normally, the radiation takes place when the source temperature is relatively much higher than the sink temperature. If the difference between the junction temperature and the ambient temperature in the IC is not small, heat transfer by radiation does not take place. However, for power ICs, the heat transfer by radiation must be considered.

13.5.2 Thermal Resistance

Thermal resistance is the resistance of the package to heat dissipation, and is inversely related to the thermal conductivity (K^{-1}) of the package. The thermal performance of IC packages is typically measured using the junction-to-ambient and junction-to-case thermal resistance values (Fig. 13.6). These parameters are defined by the following relations:

$$\theta_{jc} = \frac{T_j - T_c}{P} \tag{13.8}$$

$$\theta_{ca} = \frac{T_c - T_a}{P} \tag{13.9}$$

$$\theta_{ja} = \theta_{jc} + \theta_{ca} \tag{13.10}$$

where

θ_{jc} = junction-to-case thermal resistance (°C/W)
θ_{ca} = case-to-ambient thermal resistance (°C/W)
θ_{ja} = junction-to-ambient thermal resistance (°C/W)
T_j = average junction temperature (°C)
T_c = case temperature at a predefined location (°C)
P = power dissipation of the device (W)
T_a = ambient temperature (°C)

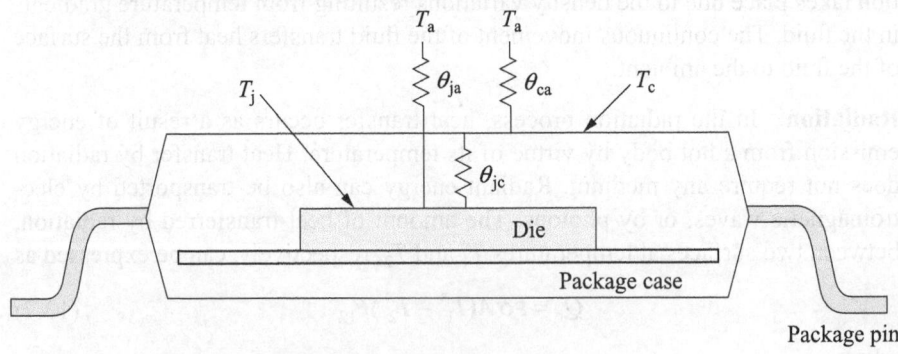

Fig. 13.6 IC package thermal model

The junction-to-case thermal resistance (θ_{jc}) is a measure of the internal thermal resistance of the package from the silicon die to the package surface. It depends on the thermal conductivity of the packaging materials, and on the package geometry.

The junction-to-ambient thermal resistance (θ_{ja}) is a combination of the package internal thermal resistance, and the conductive and convective thermal resistance from the package surface to the ambient. The value of θ_{ja} depends on material thermal conductivity, package geometry, and the ambient conditions such as coolant flow rates and the thermo-physical properties of the coolant.

In order to ensure the component functionality and long-term reliability, the maximum device operating temperature is limited by setting constraints on either the ambient temperature, or the package surface temperature measured at predefined locations. Depending on the environment (ambient and board temperatures), different cooling arrangements, such as fans or forced air cooling are used to keep the package case temperature within a specified limit.

13.5.3 Factors Impacting Thermal Resistance

The thermal resistance of the IC package depends on several factors. The factors are as follows:

- Size of the IC package—thermal resistance is inversely proportional to area. Hence, bigger the size of the package, lesser is the thermal resistance. Also, packages having more pins have less thermal resistance.
- Thermal conductivity of the packaging material—thermal resistance is inversely proportional to the thermal conductivity. Material having higher thermal conductivity (e.g., aluminium oxide used in ceramic package) has lesser thermal resistance. Hence, a ceramic package has less thermal resistance than plastic package.
- Heat sink—by using heat sink, the thermal resistance of the IC package is reduced. It is a metallic structure usually made of copper or aluminium attached to the bottom of the lead-frame.
- Die size—as die size increases, the thermal resistance decreases. The die with a larger dimension has low power density and large area for heat transfer.
- Power dissipation of the devices—the junction-to-case thermal resistance is independent of the device power dissipation; however, the junction to ambient temperature decreases with increase of device power dissipation.
- Air flow rate—the case-to-ambient thermal resistance decreases with the increase in air flow rate, but the junction-to-case thermal resistance does not change with air flow rate.

13.6 IC Package Stress Modelling

The IC package consists of different materials, which are used to form the complete package. But the different materials have different coefficient of thermal expansion (CTE). When the device operates, it generates heat and is dissipated in the IC package. Hence, different materials in the package expand differently due to this generated heat. The dissimilar thermal expansion between these different materials introduces mechanical stresses in the attached components during manufacture, and in operation. If the stress exceeds the strength of the material, it can initiate a crack at the weakest point. Once the crack is initiated, it propagates until complete failure occurs. Typical causes of failure of electronic packages are briefly discussed below.

Thermal Stress The thermal loading of the IC package and the resulting stress is defined as the stress that is associated with the change in temperature, and depends on the thermo-mechanical properties of the employed materials. Thermal stresses occur during fabrication, testing, and storage of the equipment, and more importantly during its normal operation. Thermal stresses and strains are due to dissimilar materials that expand and contract at different rates during temperature excursions, or can derive from a non-uniform temperature distribution.

The thermal stresses can cause the mechanical or structural failures (e.g., ductile rupture, brittle fracture, fatigue, creep, thermal relaxation, thermal shock, stress corrosion, and excessive deformation). In addition to the mechanical failures, the thermal stresses and strains can also lead to the functional failures due to the loss of the electrical performance of the component or device. For example, transistor junction failure can occur because of the elevated thermal stress in the IC if the heat produced by the chip cannot quickly escape.

Moisture-induced Stress The package is susceptible to the moisture absorption due to the porosity of the material itself. The moisture absorbed by the material is in the mixed liquid–vapour phase, and it is transported from the ambient to the material inside by the diffusion process. The moisture resides in the micropores or free spaces in material, and it contributes to the failures mainly from two aspects: (a) the evaporation of the moisture during temperature rise, generating high internal vapour pressure; and (b) the reduction of the interface strength with the moisture absorption. The combination of these two effects makes this problem prominent and severe during the surface mounting of electronic packages onto the printed circuit board.

Finite element (FE) modelling has been used since the mid-1950s, as the major technique for theoretical evaluations in mechanical and structural engineering, including the area of microelectronics. This technique is used by the powerful and flexible computer programs, which enable the package designers to obtain a solution to almost any stress–strain-related problem within a reasonable time.

Another modelling approach that is used for stress modelling is the probabilistic approach. The predictive modelling is used by the package designers for modelling of thermal stresses and other thermal phenomena, at all the stages of the analysis, design, testing, manufacturing, operation, and maintenance of the product or system. The traditional approach in predictive modelling is referred to as deterministic, i.e., it does not pay sufficient attention to the variability of parameters and criteria used. This approach is acceptable when the deviations (fluctuations) of package parameters from the mean values are small, and when the design parameters are known or can be predicted with reasonable accuracy.

13.7 Package Models

The IC packages are designed and a model is created for use in system level simulation. A typical model creation flow is depicted in Fig. 13.7.

Generally, package models are described either in the SPICE or in IBIS format. These models are described in the following sections.

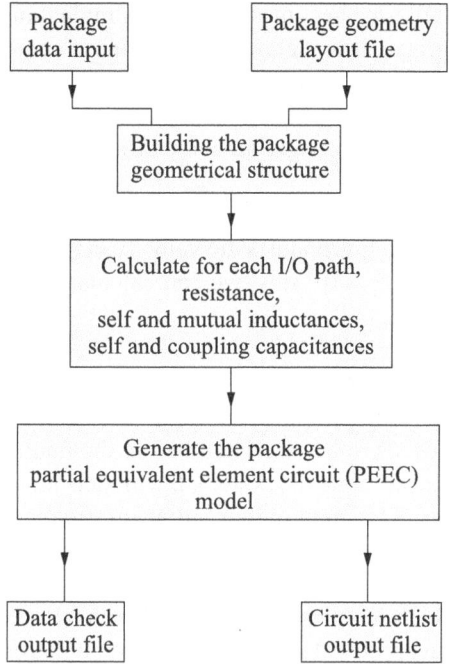

Figure 13.7   Flowchart for generating package model

13.7.1 SPICE Model of Package

The package *RLC* values, extracted using the package geometry and the material properties, can be described in a SPICE format. A typical package model in a SPICE format can be described as

```
*Package RLC data
*for pin 1
 R1 PI1 PM1 100Ohm
 L1 PM1 PO1 2.0mH
 C1 PO1 Gnd 3.0pF

*for pin 2
 R2 PI2 PM2 200Ohm
 L2 PM2 PO2 3.0mH
 C2 PO2 Gnd 1.0pF

*for pin 3
 R3 PI3 PM3 300Ohm
 L3 PM3 PO3 4.0mH
 C3 PO3 Gnd 2.0pF

*End of package data
```

The entire package *RLC* data can be expressed in a SPICE format with input and output external nodes. These nodes are to be connected to the appropriate circuit components while doing package simulation. The entire SPICE deck for a package can be expressed in the form of a SPICE SUBCKT also.

As the signal rise/fall time becomes sharp, the simple lumped RLC model does not hold good. If the propagation delay through a line is comparable to the quarter of the signal rise/fall time, the transmission line model must be used.

A line is electrically long when the propagation delay, $t_p > t_r / 4$, where t_r is the signal rise/fall time. The propagation delay through a line is calculated as

$$t_p = L / v \qquad (13.11)$$

where L is the length of the line and v is the velocity of signal propagation through the line. Assuming, $v = 1.5 \times 10^8$ m/s, we can calculate length L as 8.75 cm for 2.5 ns signal rise/fall time.

13.7.2 IBIS Model of Package

As the designs become more complex and the speed is increased, the performance of the overall system needs to be simulated before the final physical design steps (routing and layout). To simulate the system, the models must be developed for all the system components. The input/output buffer information specification (IBIS) has been the standard for device models. IBIS model is a behavioural model and it has less simulation runtime. Moreover, it does not reveal the proprietary information, unlike SPICE which requires SPICE model parameters (depends on process technology). IBIS is an approved standard within the Electronic Industry Alliance (EIA), and is also known as ANSI/EIA-656.

Table 13.1　Typical contents of an IBIS file

Header Information	Component, Package, and Pin Information	Model
IBIS version	Component	Model type
File name	Manufacturer	Temperature range
File revision	Package	Voltage range
Date	Pin	Pull-down, Pull-up, GND Clamp, power clamp reference
Source	Pin mapping	Ramp rate
Notes		Rising/Falling waveform
Disclaimer		
Copyright		

IBIS models are most commonly used for signal integrity (SI) analysis of high-speed boards or systems. An IBIS model contains $I–V$ and $V–t$ data of input and output buffers of a device in the ASCII-text format. There are some keywords in the IBIS format. The format of the IBIS model is shown in Table 13.1.

A standard IBIS model file consists of three sections as shown below:

- *Header information*　This section contains information about the IBIS file.
- *Component, package, and pin information*　This section contains information regarding the targeted device package, pin lists, pin operating conditions, and pin-to-buffer mapping.
- *V–I behavioural model*　This section contains $I–V$ data, as well as data for $V–t$ transition waveforms, which describe the switching properties of the particular buffer.

The syntax of the IBIS file is shown in Table 13.2.

Table 13.2 IBIS format

Format	Description
`[IBIS ver] <string>`	IBIS version
`[File name] <string>`	Name of the file.
`[File Rev] <string>`	File version
`[Date] <string>`	Date of creation
`[Source] <string>`	Source information
`[Notes] <string>`	Any notes
`[Disclaimer] <string>`	Disclaimer information
`[Copyright] <string>`	Copyright information
`| <string>`	Comment line
`[Component] <string>`	Component name
`[Manufacturer] <string>`	Manufacturer name
`[Package]`	Package data, lumped R, L, C values

`| variable typ min max`
 	
`|`
`|R_pkg <R1> <R2> <R3>`
`|L_pkg <L1> <L2> <L3>`
`|C_pkg <C1> <C2> <C3>`
`|`	

`[Pin] signal_name model_name`
`R_pin L_pin C_pin`
`<pinname> <signalname>`
`<modelname>`
`<s1> <s2> <s3>`
`|`
	Pin information

`[Pin Mapping] pulldown_ref`
`pullup_ref gnd_clamp_ref`
`power_clamp_ref`
`|`
`<s1> <s2> <s3> <s4> <s5>`
`|`
	Pin mapping information

`Model] <string>`	Name of the model
`Model_type <string>`	Type of model (i.e., input, output, bidirectional)
`Polarity <string>`	Type of polarity, e.g., inverting or non-inverting
`Enable <string>`	Type of enable, e.g., active-high, active-low
`Cref=<value>`	Reference capacitance of test load
`Vref=<value>`	Reference voltage
`Vinh=<value>`	Minimum upper threshold voltage. This is V_{IH} from the datasheet for a receiver input. For a differential input, V_{diff} defined under the differential pin mapping section will override V_{IH}
`Vinl=<value>`	Maximum lower threshold voltage. This is V_{IL} from the datasheet for a receiver input. For a differential input, V_{diff} defined under the differential pin mapping section will override V_{IL}
`C_comp <value1> <value2>` `<value3>`	Input die capacitance. This includes parasitic capacitance from the transistor and circuit elements, capacitance due to metallization, and pad capacitance. This does not include package capacitance. value1 = typical, value2 = min, and value3 = max.

(Contd)

Table 13.2 *(Contd)*

Format	Description
`[Temperature Range] <value1>` `<value2> <value3>`	Specified operating temperature range of device
`[Pullup reference] <value1>` `<value2> <value3>`	Pull up reference voltage
`[Pulldown reference]` `<value1> <value2> <value3>`	Pull down reference voltage
`[POWER Clamp reference]` `<value1> <value2> <value3>`	Reference voltage for the power clamp circuit
`[GND Clamp reference]` `<value1> <value2> <value3>`	Reference voltage for the ground clamp circuit

`[Pulldown]`
`|`
`|Vtable I(typ) I(min) I(max)`
`|`
`<val1> <val2> <val3> <val4>`
	Current–voltage characteristics of pull-down circuit

`[Pullup]`
`|`
`|Vtable I(typ) I(min) I(max)`
`|`
`<val1> <val2> <val3> <val4>`
	Current–voltage characteristics of pull-up circuit

`[GND_clamp]`
`|`
`|Vtable I(typ) I(min) I(max)`
`|`
`<val1> <val2> <val3> <val4>`
	Current–voltage characteristics of ground clamp circuit

`[POWER_clamp]`
`|`
`|Vtable I(typ) I(min) I(max)`
`|`
`<val1> <val2> <val3> <val4>`
	Current–voltage characteristics of the power clamp circuit

`[Ramp]`
`|`
`|variable typ min max`
`dV/dt_r <v1>/<t1> <v2>/<t2>`
`<v3>/<t3>`
`dV/dt_f <v1>/<t1> <v2>/<t2>`
`<v3>/<t3>`
	Ramp data, *dV/dt* for rise and fall transitions The slew rate is measured at the 20% and 80% points. R_load is the resistive load used to generate the *dV/dt* data. If R_load is not specified, the default value is 50 Ω

`R_load = <val>`
`R_fixture = <val>`
`V_fixture = <val>`
`V_fixture_min = <val>`
`V_fixture_max = <val>`
`| time V(typ) V(min) V(max)`
`[Rising Waveform]`
	Waveform data for rise transition

(Contd)

Table 13.2   *(Contd)*

Format	Description

`<t1> <v1> <v2> <v3>`
`<t2> <v1> <v2> <v3>`
`<t3> <v1> <v2> <v3>`	

`[Falling Waveform]`
`R_fixture = <val>`
`V_fixture = <val>`
`V_fixture_min = <val>`
`V_fixture_max = <val>`
`
	time V(typ) V(min) V(max)`
	| Waveform data for fall transition

`<t1> <v1> <v2> <v3>`
`<t2> <v1> <v2> <v3>`
`<t3> <v1> <v2> <v3>`	

`End [Model] ALL_IO_PINS`	End of model, end of component, and end of file
`End [Component] <string>`	
`[End]`	

The latest IBIS version 5.0 can be obtained from the following link:
http://eda.org/pub/ibis/ver5.0/ver5_0.txt

IBIS Input Model
The input model can be used for the receiver input, control input, or driver input. Figure 13.8 shows a typical input model.

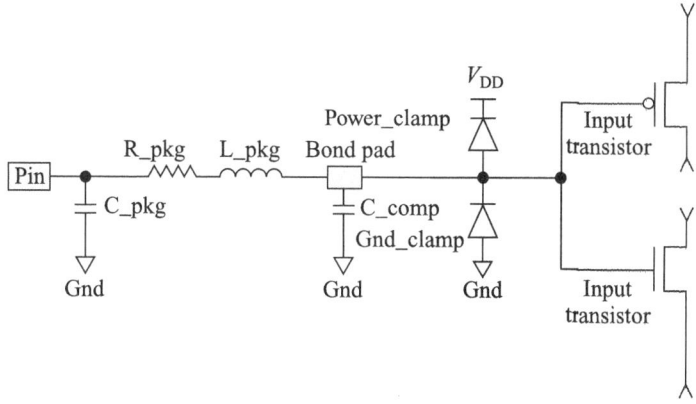

Fig. 13.8   IBIS input model

The input model contains the following elements:
- Package elements resistance, inductance, and capacitance (C_pkg, R_pkg, L_pkg).
- A clamp (Power_Clamp) that is active when the input voltage is above V_{DD}.
- A clamp (GND_Clamp) that is active when the input voltage is below ground.
- Die capacitance of the input (C_comp).

The power and ground clamp $I–V$ curves define the ESD structure of the input model. When the input voltage is between V_{DD} and the ground, the circuit operates in normal operating conditions. For input voltages higher than the supply V_{DD} or lower than ground, one of the diodes turns on to prevent the excess voltage at the circuit input. For typical input structures, the power clamp diode is forward-biased when the input is approximately 0.6 V above V_{DD}, and the ground clamp diode is forward-biased when the input is approximately 0.6 V less than the ground. An example of an IBIS LVDS input model is shown in Table 13.3.

Table 13.3 Example of an IBIS input model

```
[Model] LVDS_INPUT
Model_type Input
Vinh=1.30 | Vth
Vinl=1.10 | Vtl
|
| TYP MIN MAX
|
C_comp 1p 2p 3p
|
[Temperature Range] 27 105 -40
|
[Voltage Range] 1.8 1.62 1.98
|
[GND Clamp]
|
|Vtable I(typ) I(min) I(max)
|
1.1E+00 5.20E-01 4.32E-01 1.12E-01
...
...
2.41E+00 7.20E-07 6.43E-07 7.11E-07
|
[POWER Clamp]
|
|Vtable I(typ) I(min) I(max)
|
1.00E+00 2.11E-10 3.49E-10 2.85E-10
...
...
-1.48E+00 2.19E-09 3.34E-09 4.40E-09
|
| End LVDS_INPUT
```

IBIS Output Model

The example of an output model is shown in Fig. 13.9. It contains the power and ground clamp circuit, the lumped package *RLC* circuit, die input capacitance, and voltage, temperature, and reference voltage ranges.

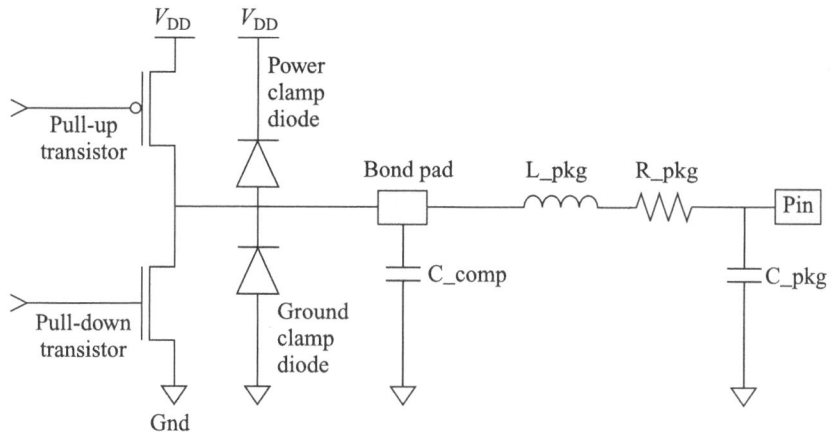

Fig. 13.9   IBIS output model

Following are the elements of an output model:

- Package parasitic resistance, inductance, and capacitance (C_pkg, R_pkg, L_pkg)
- A clamp (Power_Clamp) that is active when the output voltage is above V_{DD}
- A clamp (GND_Clamp) that is active when the output voltage is below the ground
- Die capacitance of the output (C_comp)
- A circuit (pull-up) that is active when the output is high
- A circuit (pull-down) that is active when the output is low
- A ramp rate that describes the output slew rate
- Rising and falling waveform data that describes the transient output

The example of an IBIS output model is shown in Table 13.4.

Table 13.4   Example of an IBIS output model

```
|
|  IBIS file myDesign.ibs created by Prof. D. Das
|  VLSI Design Laboratory 2010
|
|
[IBIS ver] 2.1
[File name] myDesign.ibs
[File Rev] 2.0
[Date] Jan 21 2010
[Source] V/I curve data extracted from silicon lab measure-
ments. Ramp data extracted from SPICE netlist. [Notes] V/I
max min curve data was measured in the lab under max and
min VDD and Temp conditions.
```

Table 13.4   *(Contd)*

```
[Disclaimer] This information is for demonstration
purposes only, and should not be used in actual design.
[Copyright] Copyright 2010, --- Corporation, All Rights
Reserved.
|
|
| Component myDesign_0.18um_18v
|
|
[Component] myDesign_0.18um_18v
[Manufacturer] --- Corporation
[Package]
|variable typ min max
|Un-comment the appropriate package
|
|R_pkg 10m 20m 30m | TQFP 144
|L_pkg 1.1nH 2.2nH 3.3nH | TQFP 144
|C_pkg 2.2pF 4.4pF 5.5pF | TQFP 144
.
.
.
|
|
[Pin] signal_name model_name R_pin L_pin C_pin
Y1     IO1_out      ALL_IO_PINS
| A1   IO1_in       INPUT1
| A2   IO1_en       ENABLE1
GNDP   GND          GND
VDD    VDD          POWER
|
[Pin Mapping] pulldown_ref pullup_ref gnd_clamp_ref power_
clamp_ref
|
ALLIO    GND    VDD    GND    VDDA
| I1     NC     NC     GND    VDDA
| E1     NC     NC     GND    VDDA
GND      GND    NC
VDD      NC     VDD
VDDA     NC     VDDA
|
|
| Model ALL_IO_PINS
|
|
[Model] ALL_IO_PINS
Model_type   I/O
Polarity     Inverting
Enable       Active-Low
Cref = 15.0pF
Vref = 1.0V
C_comp       1.4pF    2.4pF    3.4pF
```

(Contd)

Table 13.4 *(Contd)*

```
Vinl = 0.5V
Vinh = 1.0V
|
|
[Temperature Range]        27.0      105.0     -40.0
[Pullup Reference]         1.80V     1.62V     1.98V
[Pulldown Reference]       0.00V     0.00V     0.00V
[POWER Clamp Reference]    1.80V     1.62V     1.98V
[GND Clamp Reference]      0.00V     0.00V     0.00V
[Pulldown]
| voltage I(typ) I(min) I(max)
|
-1.30E+00 2.83E-02 4.00E-03 2.49E-02
-2.15E+00 2.67E-02 4.82E-03 2.47E-02
-4.00E+00 2.51E-02 4.64E-03 2.44E-02
.
.
|
[Pullup]
| voltage I(typ) I(min) I(max)
|
-1.30E+00 3.26E-03 3.44E-03 2.58E-03
-1.15E+00 3.28E-03 3.42E-03 2.52E-03
-1.00E+00 3.22E-03 3.46E-03 2.80E-03
.
.

.
|
[GND_clamp]
| voltage I(typ) I(min) I(max)
|
-1.30E+00 -8.21E-01 -3.31E-01 -4.80E-01
-1.15E+00 -7.27E-01 -3.79E-01 -4.23E-01
-1.00E+00 -7.24E-01 -3.28E-01 -4.67E-01
.
.

.
|
[POWER_clamp]
| voltage I(typ) I(min) I(max)
|
-1.30E+00 3.06E-01 4.21E-01 5.81E-01
-1.15E+00 3.93E-01 4.09E-01 5.68E-01
-1.00E+00 3.81E-01 4.98E-01 5.55E-01
.
.

.
|
|
[Ramp]
| variable typ min max
```

(Contd)

Table 13.4 *(Contd)*

```
dV/dt_r 2.98/0.33n 2.80/0.36n 2.16/0.34n
dV/dt_f 2.98/0.38n 2.80/0.25n 2.16/0.35n
R_load = 1.50M
|
[Rising Waveform]
R_fixture = 0.70k
V_fixture = 0.0
V_fixture_min = 0.0
V_fixture_max = 0.0
| time V(typ) V(min) V(max)
|
0.000S 0.000V 0.000V 0.000V
0.20nS -17.96mV -12.51mV -26.17mV
0.40nS -27.82mV -27.72mV 12.14mV
|......cont'd
|
[Falling Waveform]
R_fixture = 0.50k
V_fixture = 3.30
V_fixture_min = 3.00
V_fixture_max = 3.60
| time V(typ) V(min) V(max)
|
0.000S 1.30V 1.00V 1.60V
0.10nS 1.32V 1.01V 1.63V
0.20nS 1.33V 1.04V 1.42V
.

.

.
|
|
| End [Model] ALL_IO_PINS
|
| End [Component] myDesign_0.18um_18v
|
[End]
```

13.8 Package Simulation

As the frequency of operation is increased from MHz to GHz range, the parasitic effects of the package is becoming very important. The high frequency performance of the package limits the performance of the IC and the system. For instance, the self and the mutual inductance along with the coupling capacitance seriously affect the signal quality. They can introduce signal distortion and cross-talk problems. In order to estimate the systems performance, the full 3D field simulation is required.

In the package simulation process, the complete package is simulated by considering the metallic parts of the package as perfectly conducting medium and other parts are defined by their dielectric constant and conductivity. The entire structure is divided into a large number of small meshes. Different types of signals are applied to the input ports, and the resulting transmitted and reflected signals are estimated. The crosstalks on other ports are also estimated. A typical simulation result is illustrated in Fig. 13.10. It shows how the signal quality is affected during transmission.

Alternatively, a standard broadband Gaussian pulse is used for the excitation to obtain the S-parameters of the package. All input/output ports were excited successively to obtain a full S-parameter matrix for the entire package.

Some of the widely used software for IC package design is shown in Table 13.5.

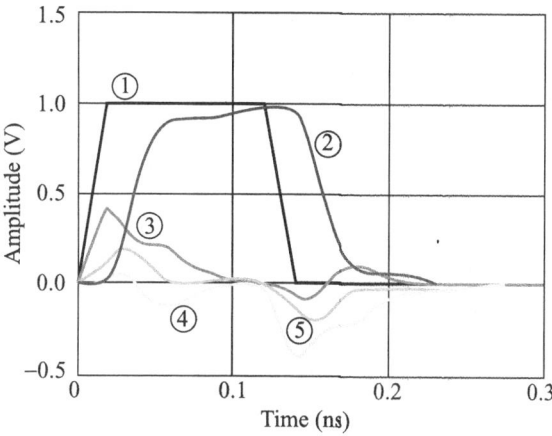

1 - Incident signal
2 - Transmitted signal
3 - Reflected signal
4 & 5 - Crosstalk on adjacent ports

Fig. 13.10   Typical package simulation results

Table 13.5   Some of the widely used software for IC package design

Sl. No.	Tool	Capabilities	Vendor
1.	XtractIM	IC package modelling	Sigrity
2.	FloTherm, FloMCAD	Thermal and mechanical analysis	Mentor graphics
3.	Pro/Engineer	Mechanical engineering, design and manufacturing	
4.	SolidWorks	3D mechanical CAD tool	
5.	CATIA	Computer-aided three-dimensional interactive application	
6.	Allegro Package Designer SpectreRF	Package/IC do-design, simulation.	Cadence
7.	ANSYS Mechanical	Mechanical design	Ansys
8.	PATRAN	Finite element analysis for solid modelling, meshing, and analysis	MSC software
9.	ABAQUS	Software package for finite element analysis	
10.	CoolIT	Thermal analysis software	CoolIT
11.	TPA	*RLC* extraction	Ansoft
	Slwave	Power and SI analysis	
12.	Package Interface Planner Encore	Die/package input/output planning and feasibility Package implementation	Synopsys

13.9 Flip-chip Package

In the flip-chip package, there is no bond-wire or lead-frame structure. Instead, the die is attached to the substrate of the package directly. The die is flipped onto the substrate for direct electrical connection using conductive bumps on the chip bond pads. In contrast to wire-bonding technology, the interconnection between the die and the carrier in flip-chip packaging is made using a *conductive bump* placed directly on the die surface. The bumped die is then flipped and placed face down so that the bumps directly connect to the carrier, as illustrated in Fig. 13.11.

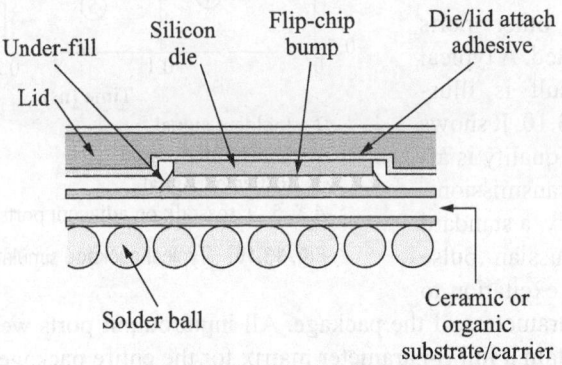

Fig. 13.11 Cross-sectional view of a typical flip-chip package

Flip-chip package can have a large number of pin-count within a small die and package area. It can be mounted using the standard PCB assembly technique and replaced using standard repair techniques.

The solder bumps are located on the active side of the silicon in a bumped die. Bumped die products are manufactured using a standard wafer fabrication process, deposition of solder bumps on the input/output pads, back-lapping, testing the wafer, and laser marking on the wafer backside. A bumped die product requires underfill material and fine-pitch interconnect layout. The lid structure is used for cooling the die. The lid is fabricated using a Kovar ring brazed to a sheet of tungsten copper.

The reliability of flip-chip contacts is determined by the difference in the coefficient of thermal expansion (CTE) between the chip and the ceramic substrate, or the organic printed circuit board (PCB). The CTE mismatch between the chip and the carrier induces a high thermal and mechanical stresses and strain at the contact bumps. The thermo-mechanical stress and strain introduces cracks in the joints. When these cracks become large, the contact resistance increases, and the flow of current is stopped. This ultimately leads to the electrical failure of the chip.

The reliability of the bump joints is improved if a bead of encapsulating epoxy resin is dispensed near the chip, and drawn by capillary action into the space between the chip and the carrier. The epoxy is then cured to provide the final flip-chip assembly. The epoxy-resin under-fill mechanically couples the chip and the carrier, and locally constrains the CTE mismatch, thus improving the reliability of the joints. The encapsulant also acts as a protective layer on the active surface of the chip. Good adhesion between the under-fill material, the carrier, and the chip surface is needed for stress compensation.

SUMMARY

- IC packages provide electrical connections between the internal circuitry and also protect the core die.
- IC packages are of two types: pin-through-hole and surface-mount based on the package pin structure.
- Based on the material of IC packages, they are either plastic or ceramic types.
- Package electrical modelling extracts the package parasitics, generates electrical equivalent circuit, and performs simulation. This is used to ensure that the signal transmission through the package pins do not cause too much of distortion.
- Thermal modelling of a package is required to find out the thermal distribution of the package and design proper heat dissipating system, so that the device junction temperature is within allowed limit.
- Package stress modelling analyses the different types stress arising in the IC package, and ensures that package is not cracked due to any mechanical stress.
- SPICE and IBIS models are the two most popular package models.
- Flip-chip package has no bond-wire and lead-frame structure. It can have a large number of package pins.

SELECT REFERENCES

Bansal, A., S. Yoon, J. Xie, Y. Li, and V. Mahadev 2005, 'Comparison of Substrate Finishes for Flip-Chip Packages', *2005 Electronic Components and Technology Conference*, vol. 1, pp. 30–7.

Bash, C.E. and R.L. Blanco 1997, 'Improving Heat Transfer from a Flip-Chip Package', *Hewlett-Packard Journal, August*.

Edwards, D.H., K. Groothuis, and S. Martinez 1987, 'Shear Stress Evaluation of Plastic Packages', *IEEE Transactions on Components, Hybrids, and Manufacturing Technology*, vol. 10, no. 4, pp. 618–27, December.

Fan, X.J., G.Q. Zhang, and L.J. Ernst 2002, 'A Micro-mechanics Approach in Polymeric Material Failures in Microelectronic Packaging', in: *Proc. EuroSimE*, pp. 154–64.

Fan, X.J. and T.B. Lim 1999, 'Mechanism Analysis for Moisture-induced Failure in IC Packages', *ASME International Mechanical Engineering Congress and Exposition, 11th Symposium on Mechanics of Surface Mount Assemblies*, IMECE/EPE-14, Nashville, Tennessee, November 14–19.

Giesler, J., G. O'Malley, M. Williams, and S. Machuga 1994, 'Flip-Chip on Board Connection Technology: Process Characterization and Reliability', *IEEE Trans. Components, Packaging, and Manufacturing Technology*, Part B, vol. 17, no. 3, pp. 256–63, August.

Hua, Y., C. Basaran, and D.C. Hopkins 2003, 'Numerical Simulation of Stress Evolution during Electromigration in IC Interconnect Lines', *IEEE Transactions on Components and Packaging Technologies*, vol. 26, no. 3, September.

Melamud, R., M. Hopcroft, C. Jha, B. Kim, S. Chandorkar, R. Candler, and T.W. Kenny 2005, 'Effects of Stress on the Temperature Coefficient of Frequency in Double Clamped Resonators', *Solid-State Sensors, Actuators and Microsystems*, 2005, Digest of Technical Papers. Transducers apos. The 13th International Conference, 1, pp. 392–5, 5–9 June.

Ricky, L., S.W., S.R. Hon, X.D. Zhang, and C.K. Wong, '3D Stacked Flip-Chip Packaging with Through Silicon Vias and Copper Plating or Conductive Adhesive Filling', *IEEE Electronic Components and Technology Conference*.

Shaneyfelt, M.R., J.R. Schwank, S.C. Witczak, D.M. Fleetwood, R.L. Pease, P.S. Winokur, L.C. Riewe, and G.L. Hash 2000, 'Thermal-Stress Effects and Enhanced Low Dose Rate Sensitivity in Linear Bipolar ICs', *IEEE Transactions on Nuclear Science*, vol. 47, no. 6, December.

Suhir, E. 2000, 'Thermal Stress Modeling in Microelectronics and Photonics Packaging, and the Application of the Probabilistic Approach: Review and Extension', *IMAPS International Journal of Microcircuits and Electronic Packaging*, vol. 23, no. 2.

Tong, Y.T., X.J. Fan, and T.B. Lim 1999, 'Modeling of Whole Field Vapor Pressure during Reflow for Flip-Chip BGA and Wire Bond PBGA Packages', *1st International Workshop on Electronic Materials & Packaging, Singapore*, 29 September–1 October.

Unchwaniwala, K.B. and M.F. Caggiano 2001, 'Electrical Analysis of IC Packaging with Emphasis on Different Ball Grid Array Packages', *2001 Electronic Components and Technology Conference, IEEE*.

Wang, Y., and H.N. Tan 1999, 'The Development of Analog SPICE Behavioral Model Based on IBIS Model', IEEE, VLSI, 1999. *Proceedings of the Ninth Great Lakes Symposium*, 4–6 March.

Witting, T., T. Weiland, F. Hirtenfelder, and W. Eurskens 2001, 'Efficient Parameter Extraction of High Speed IC Interconnects Based on 3D Field Simulation Using Fit', *Proceedings of the EMC 2001*, Zürich, Switzerland, pp. 281–6.

FastCap-web, http://www.rle.mit.edu/cpg/research_codes.htm, last accessed on 30 Dec 2010.

IBIS-web, http://www.eda.org/ibis/, last accessed on 30 Dec 2010.

EXERCISES

Fill in the Blanks

1. Ceramic package is _____ than plastic package.
   (a) more costly and reliable
   (b) more costly and less reliable
   (c) less costly and more reliable
   (d) less costly and reliable
2. IBIS model takes _____ simulation time.
   (a) more
   (b) less
3. Ceramic packages have _____ thermal resistance than plastic packages.
   (a) higher
   (b) lower
   (c) equal
   (d) none of these
4. FastHenry is used for _____ calculation.
   (a) only inductance
   (b) resistance and inductance
   (c) only resistance
   (d) capacitance
5. FastCap is used for _____ calculation.
   (a) only inductance
   (b) resistance and inductance
   (c) only resistance
   (d) capacitance

Multiple Choice Questions

1. SMT package means
   (a) Surface mount technology
   (b) Surface mount technique
   (c) Surface model technology
   (d) Surface model technique
2. Characteristic impedance depends on
   (a) only inductance
   (b) only capacitance
   (c) both inductance and capacitance
   (d) ratio of inductance and capacitance
3. Thermal resistance increases as die size
   (a) increases
   (b) decreases

4. IBIS is an acronym for
   (a) input/output buffer information specification
   (b) input buffer information specification
   (c) information for buffer interface specification
   (d) input/output buffer interface specification
5. IC packages are mainly classified as
   (a) surface-mount technology (b) pin-through-hole
   (c) both (a) and (b) (d) none of these

True or False

1. IBIS model reveals the proprietary information.
2. Plastic packages are costlier than ceramic packages.
3. SMT package enhances PCB wiring than the PTH package.
4. SMT package does not require soldering.
5. In the flip-chip package, there is no bond-wire or lead-frame structure.

Short-answer Type Questions

1. Discuss the various types of IC packages based on the structure and materials used for packaging.
2. Discuss the structure of the flip-chip package.
3. What are different package modelling domains? Explain them in brief.
4. What do you mean by the IBIS model? Discuss with an example.
5. Discuss the package simulation methodology in brief.
6. What do you mean by electrical modelling of packages?
7. What do you mean by thermal modelling of packages?
8. What do you mean by stress modelling of packages?
9. Compare the SMT and PTH packages.
10. As the frequency increases, the importance of package modelling becomes crucial. Justify.

Long-answer Type Questions

1. Discuss the electrical modelling of IC packages in detail. Draw the equivalent circuit for IC package. What is PEEC?
2. Discuss the thermal modelling of IC packages in detail. What do you mean by thermal resistance? Discuss how it depends on package parameters.
3. Draw the input and output model in IBIS modelling. Discuss the pros and cons of IBIS modelling over SPICE modelling of IC packages.
4. Write short notes on
   (a) flip-chip package
   (b) surface mount package
   (c) stress modelling
5. Write a SPICE netlist for package simulation using input/output buffers. Assume 24-pin package with equal package parasitics ($R = 50$ ohm, $L = 5$ mH, and $C = 2$ pF).

## CHAPTER 14

VLSI Testing

- Importance of testing
- Fault models
- Fault simulation
- Design for testability
- Ad hoc testing
- Scan test
- Boundary scan test
- Built-in self test (BIST)

- Automatic test-pattern generation
- IDDQ test
- Design for manufacturability (DMF)
- Design economics
- Yield
- Probe test

14.1 Introduction

After the fabrication of the integrated circuit (IC), it must be verified. VLSI testing means checking of the manufactured IC to verify its correctness. If test results are satisfactory, then only the chip is qualified to be shipped to the customer. Testing is done at several stages. In the first stage, the die is tested at the *wafer level*. A *test program* is used to test the die. The test program applies a set of input bit patterns which are known as *test vectors*, and checks the *test response*. Any chip that fails this test is marked as a bad chip. This test is often called the *production test*. In the second stage, the die is separated out from the wafer and packaged, and it is again tested with the same test vectors. This is called the *final test* or *package level* test. The chips that are shipped to the customer must pass the production and final test. At the customer end, the chip is placed in the board and the board is tested. If any chip malfunctions at the *board level test*, it is sent back to the manufacturer. The manufacturer has a failure analysis team which looks into the problem to find out the failure mechanisms. In the *system level* test, the board is placed in the system and checked if the system works as specified. The last stage of testing is done at the *field level,* when an end-user tests the system while using a system.

14.2 Importance of Testing

In summary, testing is done at five different levels as shown in Table 14.1.

Table 14.1 Cost of testing at different levels

Level	Cost (in $)
Wafer	0.01–0.1
Package	0.1–1
Board	1–10
System	10–100
Field	100–1000

The product quality is measured in terms of *defects*. More the number of defective chips, poorer will be the quality of the chip. Hence, the defect level of the chips must be kept low so that the average quality level is maintained. The defect level is measured in parts per million (ppm). The cost of testing increases by orders of magnitude from the wafer level to the field level. So, any defect that is detected early is always better for the manufacturer.

Another important aspect of testing is the testing time. Let us consider a combinational circuit having n inputs. To test this circuit completely, we need 2^n number of possible input combinations (or test vectors). In case of sequential circuits, the required test vectors would be 2^{n+m} if there are m number of registers. To understand it better, let us consider an example.

For a VLSI chip, if there are 50 inputs and 50 registers inside, it requires 2^{100} ($\approx 1.27 \times 10^{30}$) test vectors to fully test it out. Now, if testing one pattern requires 1 ns time, testing all the test vectors would require 4×10^{12} years.

The above example indicates that there must be innovative ways of testing the chip without applying so many test vectors, but catch all possible defects or most of the defects.

14.3 Fault Models

The *fault* is the manifestation of manufacturing defects in an IC. During the IC fabrication, the MOS devices could be fabricated incorrectly, or the interconnect wires could have open-circuit or short-circuit fault. All these defects lead to malfunctioning of the IC. In order to differentiate good and bad chips, the faults must be identified in the bad chips.

The fault models are used to identify different types of faults. There are many fault models. A list of fault models is shown in Table 14.2.

Table 14.2   Fault models

Sl. No.	Fault Model
1.	Physical fault
2.	Logical fault
3.	Degradation fault
4.	Parametric fault
5.	Timing or delay fault
6.	Open-circuit fault
7.	Short-circuit fault
8.	Stuck at fault
9.	Bridging fault

A physical fault can be mapped to a logical fault. A fault can occur at two levels: (a) chip level and (b) device level. The logical faults can be classified into two main subclasses:

- Degradation fault—this degrades the performance of the chip
- Fatal fault—this causes the chip to malfunction

The open-circuit and short-circuit faults are often grouped under fatal faults. A delay fault can be classified as a degradation fault, as it may not cause any functionality failure, but causes the chip to operate at a slower speed.

The open-circuit fault is caused by several reasons. Some of the possible causes are as follows:

- Bad contact
- Over-etched metal
- Break in poly silicon line
- Void formed due to electromigration

The short-circuit fault is also caused by various reasons. Some of the possible causes are as follows:

- Under-etching of metal lines
- Hillock formed due to electromigration
- Junction spiking
- Pinholes or shorts through the gate oxide
- Diffusion shorts

Another important fault is the bridging fault that happens in interconnects. It occurs mainly due to metal coverage problems.

14.3.1 Stuck-at Fault

The most popular fault model is the stuck-at fault model. In this model, there are two types of logical faults:

- Stuck-at-1 (abbreviated as SA1 or S@1)
- Stuck-at-0 (abbreviated as SA0 or S@0)

The stuck-at-fault normally occurs due to the short circuit of the gate of the MOS device to either the V_{DD} or to the ground and metal-to-metal shorts.

The number of fault sites in a circuit is given by (Number of principal inputs + Number of gates + Number of fan-out branches). The number of single stuck-at-fault is equal to twice the number of fault sites in the circuit.

Let us consider the circuit shown in Fig. 14.1 to model the stuck-at-fault. The circuit contains eight signal lines which are potential fault sites. These lines are represented as p, q, r, s, t, u, v, and w. Each of these lines can have SA0 or SA1 faults, thereby 16 possible single stuck-at faults. Table 14.3 shows the fault-free and faulty outputs for all possible input combinations.

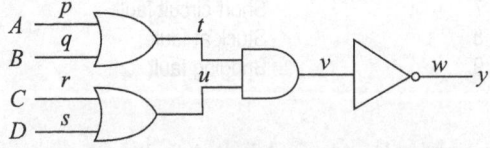

Fig. 14.1 Circuit for stuck-at-fault modelling

The highlighted boxes represent faulty outputs. The input patterns 0001, 0100, 0101, and 1010 cover all stuck-at-faults.

Table 14.3 Stuck-at-fault model

ABCD	0000	0001	0010	0011	0100	0101	0110	0111	1000	1001	1010	1011	1100	1101	1110	1111
Y	1	1	1	1	1	0	0	0	1	0	0	0	1	0	0	0
p SA0	1	1	1	1	1	0	0	0	1	1	1	1	1	0	0	0
p SA1	1	0	0	0	1	1	0	0	1	0	0	0	1	0	0	0
q SA0	1	1	1	1	1	1	1	1	1	0	0	0	1	0	0	0
q SA1	1	0	0	0	1	0	1	1	1	0	0	0	1	0	0	0
r SA0	1	1	1	1	1	0	1	0	1	0	1	0	0	0	1	0
r SA1	1	1	1	1	1	1	0	0	0	0	0	0	0	0	0	0
s SA0	1	1	1	1	0	1	0	0	0	0	0	0	0	1	0	0
s SA1	1	1	1	1	0	1	0	0	0	1	0	0	1	1	0	1
t SA0	1	1	1	1	1	0	1	1	1	1	1	1	1	1	0	1
t SA1	0	1	1	1	0	1	1	1	0	1	1	1	0	1	1	1
u SA0	1	1	1	1	1	1	1	1	1	1	1	1	1	1	1	1
u SA1	1	1	1	0	1	0	1	0	0	0	1	0	0	0	1	1
v SA0	1	0	0	0	0	0	0	0	0	0	0	0	0	0	0	0
v SA1	0	0	0	0	0	0	0	0	0	0	0	0	0	0	0	0
w SA0	0	0	0	0	0	0	0	0	0	0	0	0	0	0	0	0
w SA1	1	1	1	1	1	1	1	1	1	1	1	1	1	1	1	1

14.4 Fault Simulation

Fault simulation is the methodology used for testing a design after introducing a fault intentionally. The design is tested using a set of test vectors to detect any fault. There is a program that applies a set of test vectors to the primary inputs (PIs). This program is known as *test program*. The time required to apply a test vector or pattern, check the response at the primary outputs (POs), and verify the output response against the expected output response, is known as *test-cycle time*.

Fault simulation can measure the quality of the test program as it can verify if the test program can catch all the known faults that are introduced. It measures the *fault coverage,* which is defined as the ratio of the detected fault to the detectable faults.

14.4.1 Deterministic Fault Simulation

In the deterministic fault simulation technique, a set of test vectors are used to simulate a circuit and catch the faults. But if all the faults are not caught by the test vectors; they are modified and the fault simulation is repeated. There are mainly three types of deterministic fault simulations as given below:

- Serial fault simulation
- Parallel fault simulation
- Concurrent fault simulation

Serial Fault Simulation

In this process, two copies of the circuit are tested. The first copy is a good circuit and the second is generated from the first by inserting faults into it to make a faulty circuit.

In this process, each faulty circuit is simulated at a time. As it is done one after another, the process is inherently slow.

Parallel Fault Simulation

In the *parallel fault simulation*, the faulty circuits are simulated simultaneously. This process is very fast as compared to serial fault simulation.

Concurrent Fault Simulation

In the *concurrent fault simulation*, the whole circuit is not simulated, but only a part is simulated where the fault is introduced. In this process, many faults are simulated at the same time allowing a very fast simulation. But the process is more complex as it needs to follow the parts of the circuit that is affected after introduction of new faults.

14.4.2 Nondeterministic Fault Simulation

In the *nondeterministic fault simulation*, instead of testing every fault, a subset or sample of the faults is tested and extrapolate the fault coverage from the sample tested.

14.5 Design for Testability

A design is testable if it is controllable and observable. *Design for testability* (DFT) means the design must take into consideration the controllability and observability. The overhead of introducing the extra circuitry in the chip for DFT has a great impact on the cost of manufacturing test. It has a very good test coverage with a fewer number of test vectors. The built-in self test (BIST), scan test, and boundary-scan test (BST) to be discussed in this chapter are part of the DFT.

14.5.1 Controllability and Observability

While testing a circuit, it is often required to set a particular node to either logic 1 or logic 0. The nodes that are connected to the primary inputs are directly controllable. But it is not possible to set an internal node to a desired logic level directly. In such cases, it requires many test vectors to be applied to the primary inputs to set the node to a desired logic level. *Controllability* defines a measure of ease of setting an internal node logic level. It is the job of the chip designers to design the chip in such a way that all the nodes are easily controllable.

Similarly, any node that is connected directly to the primary output is easily observable. But for any internal node, it is not possible. The internal node logic level needs to be propagated to the primary outputs. *Observability* defines the degree at which the nodes are observed at the outputs. Again, good design practices are followed while doing the design to make a chip better observable.

14.6 Ad Hoc Testing

The *ad hoc testing* is basically a combination of different simple test strategies that are used to reduce the number of tests from a large set of test patterns. It is suitable for small designs where the systematic test methodologies, e.g., scan test, ATPG, or BIST are not available. Following are the basic methods of ad hoc testing:

- Large sequential circuits are partitioned into smaller ones.
- Extra test points are added.
- Multiplexers are added to multiplex the scan input and data input.
- Easy state resets are provided.

Though ad hoc testing is very effective, it depends on the architecture and requires expert knowledge. Hence, it is not very suitable for automation. The systematic and structured test methodologies are preferred over the ad hoc testing.

14.7 Scan Test

The automatic test generation methodology is well suited for combinational circuits but not very useful for sequential circuits. Therefore, for the sequential circuits, a different technique is used, which is known as *scan design*. In this process, all the flip-flops are replaced by a scan flip-flop (SFF). The SFF has two modes of operation: (a) normal mode in which the flip-flop is operated in the conventional mode; and (b) scan mode in which the flip-flops are connected serially to form a large chain of shift registers throughout the entire chip. A test compiler program automatically replaces the flip-flops by the SFF. This is known as *scan chain insertion*. By applying clock pulses, a large stream of data can be shifted in and out through the scan chain. Therefore, every sequential element can be thoroughly verified. A typical SFF is shown in Fig. 14.2.

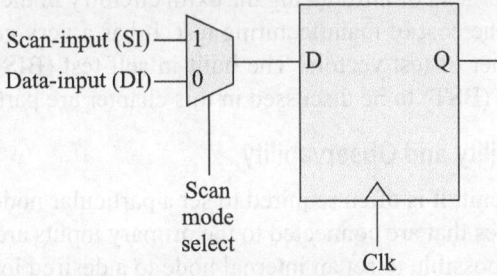

Fig. 14.2 A scan flip-flop

A 2:1 multiplexer is added to the input of a normal D-flip-flop to construct an SFF. When scan mode is selected to logic 1, the scan-input (SI) data goes to the D-input of SFF, and with a clock pulse, scan-input data shifts to the Q-output. The Q-output of SFF is connected to the SI input of next SFF, as well as to the input of the logic block. When scan mode is selected to logic 0, the data-input (DI) goes to the Q-output, the and normal operation proceeds.

14.7.1 Serial Scan Test

Figure 14.3 shows a schematic of the serial scan test. In the scan-mode, the scan-in data input flows through the chain of registers, as illustrated by the dotted line. In the normal mode, the normal input flows through the registers and the combinational logic blocks as shown by the solid lines.

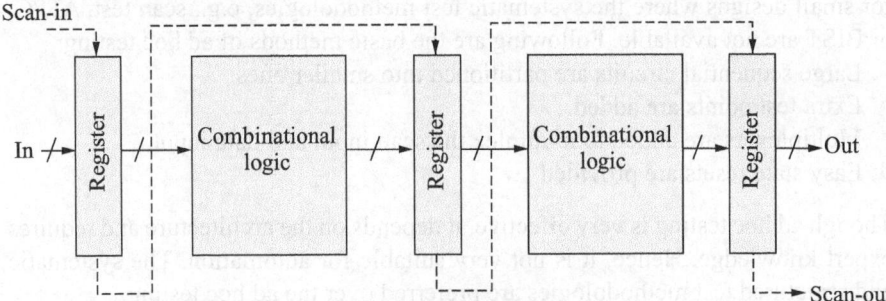

Fig. 14.3 Serial scan test

The most popular serial-scan test is called level-sensitive scan design (LSSD), which was developed by researchers from IBM in the 1970s. The LSSD is constructed using two latches, L1 and L2, as shown in Fig. 14.4. The first latch called the master latch is operated using two clocks CLK1 and CLK2. The second latch called the slave is operated using a third clock CLK3. The master latch has two data inputs: D1 (data) and D2 (scan-in).

In the normal circuit operation, the signals D1, CLK1, and Q act as latch input, clock, and the output. The test clocks CLK2 and CLK3 are kept low at this mode

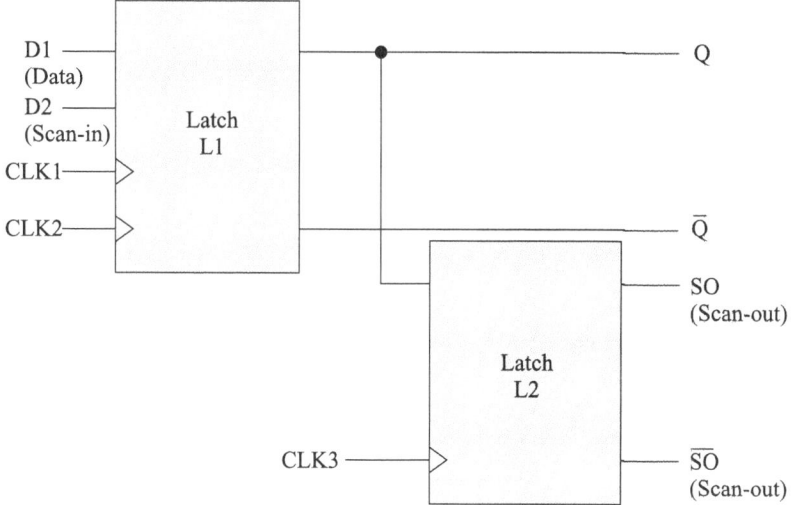

Fig. 14.4 LSSD reconfigurable D flip-flop

of operation. In the scan mode, the D2 and SO signals act as scan-in and scan-out. In scan mode, the clock CLK1 is kept low, and the clocks CLK2 and CLK3 are applied by two non-overlapping two-phase clock signals.

14.7.2 Parallel Scan

For large circuits, the size of the scan chain is too big, and there, the scan test requires a significant amount of time. To avoid this, the whole scan-chain is divided into smaller scan-blocks, and each block is scanned independently. This way it saves the overall scan test time.

14.8 Boundary Scan Test

Boundary scan test (BST) is the methodology used for board testing. A schematic for BST is shown in Fig. 14.5.

BST tests the connection of chips to the board. The *test-data input* (TDI) is sent at the input, and the *test-data output* (TDO) is checked against the data sent. BST operation is selected through the *test-mode select* (TMS) control line, and the test data is shifted using the *test clock* (TCLK).

A special logic cell, called boundary-scan register, is added to every I/O pad in the IC. These registers are connected in series to form a long chain of shift register called the *boundary-scan shift register*. The TDI signal is applied at the input of the boundary-scan shift register and its output is the TDO signal. The four signals TDI, TDO, TMS, and TCK are connected to a *test-access port* (TAP) controller, which performs the following operations:

- Selects the register
- Loads data into the registers
- Performs the test
- Shifts the data-out

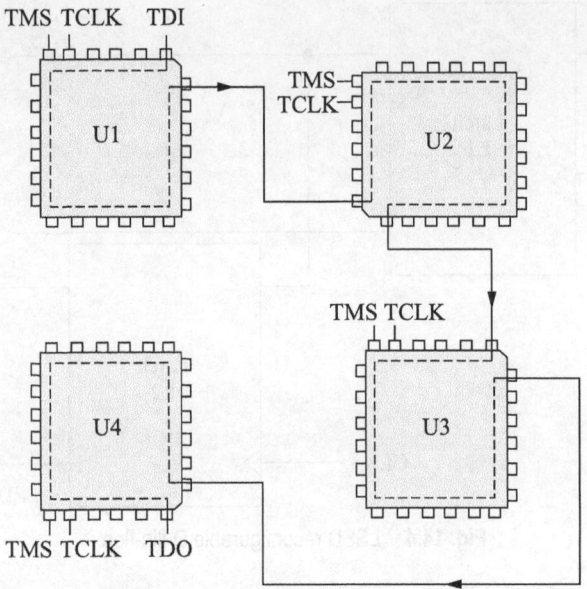

Fig. 14.5 Boundary-scan test architecture

There is an optional control signal called $\overline{\text{TRST}}$ (test reset signal) to asynchro-nously reset the TAP controller when the power-up signal is not automatically generated by the chip. In the normal mode of operation, the TCK and $\overline{\text{TRST}}$ sig-nals are kept low, and the boundary scan is disabled.

BST was first developed by the Joint Test Action Group (JTAG) which was formed in 1986 by a group of manufacturers in Europe. JTAG came up with test standards 2.0 for BST that were later approved by ANSI (American National Standards Institute) in 1990. The IEEE Standard 1149.1 is formed based on the JTAG 2.0 standard, which is followed for board testing.

14.9 Built-in Self Test

In built-in self test (BIST), a test logic circuit is incorporated in the chip. The extra circuit generates test patterns, applies them to the inputs, and tests the circuit. This extra circuit increases the chip size but reduces the test cycle time.

The components of the BIST module are: (a) pseudo-random-sequence-generator (PRSG) and (b) signature analyser.

14.9.1 Linear Feedback Shift Register

The *linear feedback shift register* (LFSR) is used to generate pseudo-random test vec-tors in the chip. The outputs at each stage of an LFSR are used as the input of the circuit. The LFSR is clocked for a large number of cycles, and the output is monitored.

A PRSG is a sequence of a particular length (n). The bit-pattern of the se-quence is random in nature but has a periodicity. That is why it is called pseudo-random sequence. The sequence is constructed using D-flip-flops and a XOR gate. An example of LFSR is shown in Fig. 14.6.

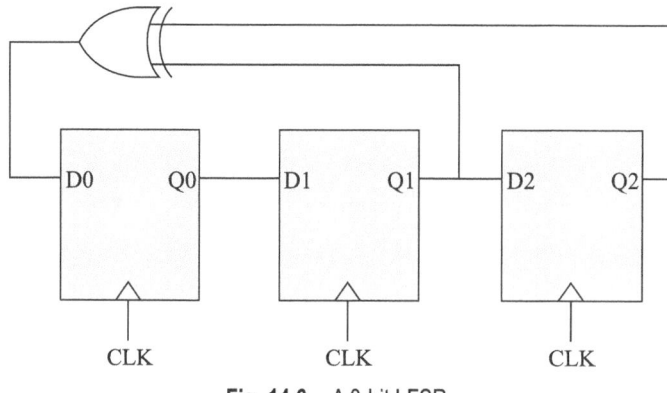

Fig. 14.6   A 3-bit LFSR

The length of the sequence is determined by the number of flip-flops (N) and is given by

$$n = 2^N - 1 \qquad (14.1)$$

The states of the circuit, as shown in Fig. 14.6 are shown in Table 14.4. It is assumed that initially, all the flip-flops are set to logic 1.

The sequence generated by the circuit as shown in Fig. 14.6 is 11100101. Note that the 000 state is not included in the LFSR states. If all zero states are included in a LFSR, it is known as complete feedback shift register (CFSR). CFSR is used in some special test situations.

Table 14.4   States of LFSR shown in Fig. 14.6

Clock pulse	Q0	Q1	Q2	Q0Q1Q2
1	1	1	1	111
2	0	1	1	011
3	0	0	1	001
4	1	0	0	100
5	0	1	0	010
6	1	0	1	101
7	1	1	0	110
8	1	1	1	111

14.9.2 Signature Analyser

A *signature analyser* is formed by adding an extra XOR gate at the input of the LFSR, as shown in Fig. 14.7. A binary input sequence is applied at the input IN. At the end of the input sequence, the shift register's output form a pattern. This pattern is known as a *signature*. A test input sequence is applied at the input, and the resulted output sequence is compared with the signature to determine a faulty circuit. If the length of the input sequence is long enough, it is unlikely that two different input sequences will produce the same signature. This test methodology is known as *signature analysis*.

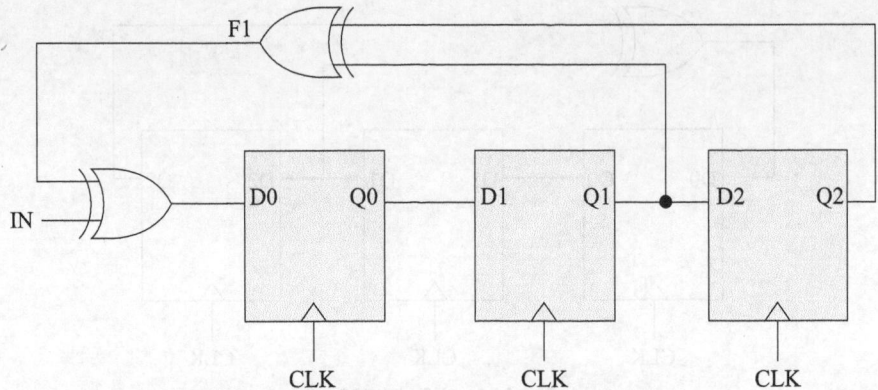

Fig. 14.7 Signature analyser of LFSR shown in Fig. 14.6

14.9.3 Built-in Logic Block Observer

A signature analyser and scan test circuit is combined to form a built-in logic block observer (BILBO) or BIST, as shown in Fig. 14.8.

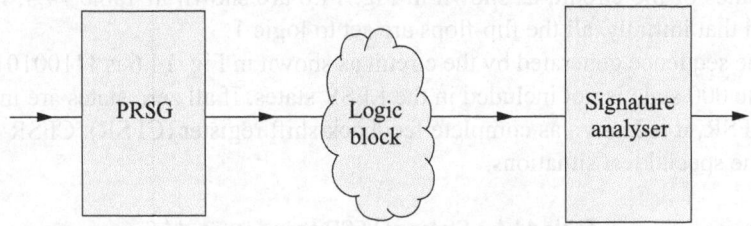

Fig. 14.8 Block diagram of BILBO

Many EDA tool vendors provide the tools that support BIST. There are two types of BIST: (a) LogicBIST and (b) MemBIST. The LogicBIST is used to test the core logic, whereas the MemBIST is used for memory testing.

14.10 Automatic Test-pattern Generation

It is a technique of automatic generation of test vectors to test a design. The excellence of the automatic test-pattern generation (ATPG) algorithm is determined by its fault coverage. As we have already discussed, testing a VLSI chip often requires a huge number of test vectors. So, there must be a methodology to automate this test pattern generation. There are several algorithms for ATPG. Some of them are: (a) D-calculus algorithm; (b) basic ATPG algorithm; (c) PODEM algorithm. The detailed discussion of these algorithms is beyond the scope of this book. But readers are advised to consult Smith (2002).

14.11 IDDQ Test

During the production test, a very simple test methodology is followed to test if any short is present between the V_{DD} and ground. The power supply current is measured and if it is found to exceed a certain value, it is concluded that the chip is defective. This test methodology is known as IDDQ test. IDDQ stands for quiescent supply current. IDDQ test quickly identifies bad chips, and saves the time for testing it in the tester.

14.12 Design for Manufacturability (DFM)

A design must be manufacturable. The circuit or the layout should be designed so that yield of the manufactured chip is high.

There are different levels at which the design can be optimized to increase the manufacturability, and hence improve the yield.

14.12.1 Optimizing Physical Layout

At the physical level, the following design rules, while drawing the layout of a chip, can improve the yield:
- Increase the metal-to-metal spacing to reduce the chance of a short circuit.
- Increase the overlapping of contacts and vias to reduce the chance of bad contact structure due to misalignment errors.
- Increase the number of vias at the wire intersections to reduce the chance of an open circuit.

14.12.2 Introducing Redundancy

Redundant hardware in chip can be utilized in case of fault in other parts of the chip. For example, an extra memory array in a chip can compensate a faulty memory array in a chip.

14.12.3 Minimizing Power Dissipation

The excess power dissipation in a chip can accelerate the degradations due to reliability issues. Hence, minimizing power dissipation is one of the design techniques which can improve the yield of a chip. Additionally, a suitable choice of the package and heat sink can also be exercised.

14.12.4 Wide Process Modelling

During the fabrication, there are always some process variations. So modelling the process variations and characterizing the circuits at the extreme process corners can reduce the chance of malfunctioning of a chip, which in turn can improve the yield.

14.12.5 Yield Analysis

If a chip is to be produced in high volume and the fabricated chips have a poor yield, a post-mortem analysis can be done to find out the defects. If any particular design has caused many failures that can be modified in the layout, the yield can be improved for the next batch of chips to be manufactured.

14.13 Design Economics

In this section, we will discuss the various components of costs associated with a chip. Let us consider an IC as a product. The total cost of the product is separated into two components:

■ Fixed costs
■ Variable costs

Fixed costs are independent of the number of chips sold. The variable cost includes the cost of assembly, cost of manufacturing, and the cost of parts used into the product. We can write the total cost as:

$$\text{Total cost} = \text{Fixed cost} + (\text{Variable cost} \times \text{Volume of chips}) \quad (14.2)$$

Fixed cost has many components such as

■ Training cost, which includes the learning cost to learn a new EDA tool
■ Hardware (wafer, equipment) cost
■ Software (EDA tool) cost
■ Design (salary of designers) cost
■ Cost for design of test
■ Non-recurring engineering (NRE) cost
  − Test program development cost
  − Masks cost
  − Simulation cost
■ Miscellaneous cost (insurance policy)

14.14 Yield

Yield of a process determines the profitability of a semiconductor company. *Yield* is defined as the ratio of number of good chips to the total number of fabricated chips on a wafer. Yield can be written as

$$\text{Yield} = \frac{\text{Number of good chips}}{\text{Total number of fabricated chips}} \quad (14.3)$$

Process yield is defined as the fraction of acceptable parts among all the parts fabricated. Defect level is the fraction of bad chips that pass the final package tests, and it is expressed in defects-per-million (DPM).

There are two types of defects in wafers: (a) randomly oriented (unclustered) defective dies on the wafer and (b) clustered defective dies on the wafer. Figure 14.9 illustrates the unclustered and clustered defects in the wafers.

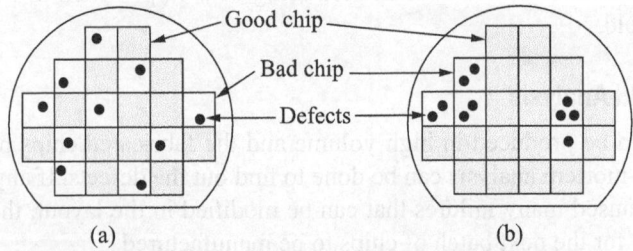

(a)   (b)

Fig. 14.9   (a) Unclustered defects; (b) clustered defects

The yield of the wafers shown in Figs 14.9(a) and (b) are $12/22 = 0.55$ and $17/22 = 0.77$, respectively.

The random defects are characterized by two factors: (a) the defect density, d and (b) the clustering parameter, α. The average number of defects on a chip of area A is $A \times d$. Yield can be expressed as

$$Y = \left(1 + \frac{Ad}{\alpha}\right)^{-\alpha}$$ (14.4)

For an unclustered model, Eqn (14.4) reduces to

$$Y = e^{-Ad}$$ (14.5)

Example 14.1   Consider a wafer with
- Defect density $d = 1.25$ defects/cm^2
- $\alpha = 0.5$
- Chip area, $A = 8$ mm $\times$ 8 mm $= 0.64$ cm^2

Find out the yield.

Solution   Yield calculated according to Eqn (14.4):

$$Y = \left(1 + \frac{0.64 \times 1.25}{0.5}\right)^{-0.5} = 0.62$$

Example 14.2   Given
- The process uses 8-inch wafers
- The cost of processing a wafer is $100
- Each wafer has 500 chips

Calculate the processing cost per chip. Calculate the processing cost if DFT is included, which increases the chip area by 10%.

Solution   (a) Processing cost per chip is

$$\text{Cost}_{chip} = \frac{\$100}{500 \times 0.62} = 32 \text{ cents}$$

Assume that the chip size is increased by 10% after DFT is included. Yield is then

$$Y = \left(1 + \frac{0.64 \times 1.1 \times 1.25}{0.5}\right)^{-0.5} = 0.60$$

Hence, there is a 2% reduction in yield after DFT is included.

With DFT, a wafer contains $500/1.1 = 454$ chips. Therefore, the processing cost is

$$\text{Cost}_{chip} = \frac{\$100}{454 \times 0.60} = 36 \text{ cents}$$

The cost is increased by 12.5% over no DFT.

14.15 Probe Test

Testing the individual die at the wafer level is called *probe*. Probing is the most direct approach for mass testing of unencapsulated semiconductor dies in the wafer form. Testing at the wafer level is done for two purposes: The first is to detect the bad die, and the second is to save on the cost of packaging faulty devices. The wafer-probe test costs of ICs are an order of magnitude less than the corresponding test costs of assembled packages.

In the probing process, the wafer is carefully mounted onto a movable plate. The wafer can be moved either manually or automatically by the machine in both vertical and horizontal directions. The electronic connection is made via a 'probe card'. A probe card is a printed circuit board (PCB) designed to match the bonding pad geometry of each die and connect it to the test equipment. The probe card has thin metal probes for making the connections between the card's circuit and the die bonding pads. These probes are lowered onto the pads of each die, the connections are made, and the test program is run to determine the pass/fail status of the die. After the test is finished, the probes are then lift. A bad die is marked with an ink dot in the centre. After a die is tested, the wafer is then moved into position for the next die to be tested. By this method, each die on the wafer is probed. The probe testing process leaves a mark on each pad to identify that the die has been tested.

A known good die (KGD) can be

- Packaged for the end user in some type of custom package
- Mounted directly on a substrate
- Combined with other dies in a multi-chip package (MCP)

The evolution of testing, moving from package level to wafer level is *merging*. The availability of the sophisticated probers, probe cards, and testers have increased the throughput and yield, thus reducing the cost of testing.

Summary

- Cost of testing increases exponentially as moved from wafer level to the field level.
- The faults in ICs are due to the manufacturing defects of ICs. These are modelled to identify different types of faults.
- The most popular fault models are stuck-at-1 and stuck-at-0 faults.
- Scan test uses the scan flip-flop as a scan-input other than the normal input. When ICs have to be scanned, scan input is selected.
- Boundary scan test is used to test the ICs at the board level.
- Built-in self-test introduces extra circuits in a chip for the testing purpose.
- IDDQ is a very simple way of checking any shorts between the power and ground.
- Yield is the ratio of defect-free chips to the total number of chips manufactured.

SELECT REFERENCES

Bhattacharya, S. and A. Chatterjee 2003, 'High Coverage Analog Wafer-Probe Test Design and Co-optimization with Assembled-Package Test to Minimize Overall Test Cost', *VLSI Test Symposium Proceedings 2003,* pp. 89–95, 27 April–1 May.

Kang, S.M. and Y. Leblebici 2003, *CMOS Digital Integrated Circuits: Analysis and Design,* 3rd ed., Tata McGraw-Hill, New Delhi.

Manna, W.R., F.L. Taberb, P.W. Seitzerc, and J.J. Broz 2004, 'The Leading Edge of Production Wafer Probe Test Technology', *IEEE International Test Conference.*

Mark, A. 2005, 'Wafer Probe Acquires a New Importance in Testing', *Chip Scale Review,* May/June.

Martin, K. 2004, *Digital Integrated Circuit Design,* Oxford University Press.

Rabaey, J.M., A. Chandrakasan, and B. Nikolic 2008, *Digital Integrated Circuits: A Design Perspective,* 2nd ed., Pearson Education.

Smith, M.J.S. 2002, *Application Specific Integrated Circuits,* Pearson Education.

Weste, N.H.E., D. Harris, and A. Banerjee 2009, *CMOS VLSI Design: A Circuits and Systems Perspective,* 3rd ed., Pearson Education.

EXERCISES

Fill in the Blanks

1. Testing cost is maximum at _____ level of testing.
   - (a) system
   - (b) field
   - (c) board
   - (d) wafer
2. BIST means _____ .
   - (a) board integrated system testing
   - (b) built-in system test
   - (c) built-in self test
   - (d) board-in self test
3. An n-bit LFSR will cycle through _____ .
   - (a) $2^n - 1$ states
   - (b) 2^n states
   - (c) 2^{n-1} states
   - (d) 2! states
4. JTAG means _____ .
   - (a) joint test action group
   - (b) joint telecom agency
   - (c) junior test activity guide
   - (d) joint test activity group
5. BST was originally developed by _____ .
   - (a) JTAG
   - (b) JETAG
   - (c) IEEE
   - (d) ANSI

Multiple Choice Questions

1. Cost of the die depends on
   - (a) wafer cost
   - (b) number of die per wafer
   - (c) yield
   - (d) all of these
2. What is the fix mechanism for the slower circuit operation as compare to predicted?
   - (a) slow clock
   - (b) raise V_{DD}
   - (c) either (a) or (b) or both
   - (d) none of these
3. The input test vectors used to test a module using a test bench are
   - (a) available in a file
   - (b) computed on the fly
   - (c) neither (a) nor (b)
   - (d) either (a) or (b)
4. Stuck-at-0 fault indicates that a node is shorted to
   - (a) V_{DD}
   - (b) ground
   - (c) floating
   - (d) open

5. Stuck-at-1 fault indicates that a node is shorted to
   (a) V_{DD} (b) ground
   (c) floating   (d) open

True or False

1. IDDQ is very simple way of checking any shorts between the power and ground.
2. Boundary scan test is used to test the ICs at the board level.
3. Yield increases as the chip area increases.
4. As defect density reduces, yield also reduces.
5. Cost per chip decreases as the volume of the chip increases.

Short-answer Type Questions

1. Draw the architecture of boundary scan test (BST) and explain the operation. Why is BST required?
2. Draw the BILBO/BIST architecture and explain the different modes of operation.
3. What is a signature analyser? Draw the architecture of a LFSR (linear feedback shift register) for the following characteristic polynomial, $1 + x + x^3$.
4. What is a CFSR (complete feedback shift register)? Explain how a LFSR can be converted to a CFSR.
5. What do you mean by BIST (built-in self test)? Explain it with necessary circuit diagrams.
6. Draw the parallel scan architecture and explain the principle of operation.
7. Draw the serial-scan architecture and explain the operating principle. What are the merits and demerits of the serial scan architecture?
8. What is a scan flip-flop? Explain the working of scan flip-flop with a necessary circuit diagram. What is a scan chain?
9. What do you mean by DFT (design for testability)? Discuss the different DFT methodologies.
10. What are the different level at which a die is tested? What do you mean by testing and verification? Discuss the various manufacturing defects and circuit maladies caused due to them.
11. What is a test program? What is a VCD file? Explain the testing mechanism.
12. What is a test bench? Discuss the test principle for combinational logic and sequential logic.
13. What are the different fault models? Discuss each of them.
14. Define the following terms:
    (a) observability   (b) controllability
    (c) fault coverage   (d) reliability
    (e) FIT (failure in time)
15. What is IDDQ testing? Why is it used?
16. What is fault simulation? What are the different fault simulation techniques? Discuss each of them.

Long-answer Questions

1. What is ATPG (automatic test pattern generation)? Name a few algorithms for ATPG.
2. (a) Draw the architecture of boundary scan test (BST) and explain the operation. Why is BST required?
   (b) Draw the BILBO (built-in logic block observation)/BIST architecture and explain the different modes of operation.
   (c) What is a signature analyser? Draw the architecture of a LFSR (linear feedback shift register) for the following characteristic polynomial, $1 + x + x^3$.

3. (a) What do you mean by BIST (built-in self test)? Explain it with necessary circuit diagrams.
   (b) Draw the parallel scan architecture and explain the principle of operation.
4. (a) Draw the serial-scan architecture and explain the operating principle. What are the merits and demerits of the serial scan architecture?
   (b) What is a scan flip-flop? Explain the working of scan flip-flop with a necessary circuit diagram. What is a scan chain?
   (c) What do you mean by DFT (design for testability)? Discuss the different DFT methodologies.
5. (a) What are the different level at which a die is tested? What do you mean by testing and verification? Discuss the various manufacturing defects and circuit maladies caused due to them.
   (b) What is a test program? What is a VCD file? Explain the testing mechanism.
   (c) What is a test bench? Discuss the test principle for combinational logic and sequential logic.
6. (a) What are the different fault models? Discuss each of them.
   (b) Define the following terms:
   - observability
   - controllability
   - fault coverage
   - reliability
   - FIT (failure in time)
7. (a) What is fault simulation? What are the different fault simulation techniques? Discuss each of them.
   (b) What is burn-in? Draw the reliability bathtub curve and explain the regions that they signify.
8. Find out the number of fault sites and the number of single stuck-at-fault for the circuit shown in Fig. 14.10. Determine the input patterns that cover all the single stuck-at-faults.

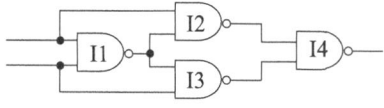

Fig. 14.10

Field Programmable Gate Array

15.1 Introduction

The ever-increasing demand of electronic systems has forced the semiconductor industry to push for more and more advanced VLSI technologies. As the VLSI technology advances, the device dimensions are shrunk, and millions of components are fabricated in a small area of silicon. This has enabled to put more functionality on the IC chip. But as the complexity of the designs increases, the cost of production and time to market have become a crucial factor in determining the design styles. Generally, the standard products (i.e., catalog ICs) are readily available and are cheap also. But for a specific application, standard products cannot achieve the expected functionality. Hence, the application-specific integrated circuit (ASIC) has evolved to meet a specific requirement. But very often, the ASIC product development is very costly and a time-consuming process. Hence, for a quick prototype development, a standard product with reconfigurable architecture has always been the demand. The programmable logic device (PLD) is such an option for the VLSI designers. There are different versions of PLDs—read only memory (ROM), programmable logic array (PLA), programmable array logic (PAL), complex PLD (CPLD), field programmable gate array (FPGA), and mask programmable gate array (MPGA). But typically, the low-density programmable ICs (ROM, PLA, and PAL) are known as PLDs. The complex PLD or CPLD is a high-density programmable device. In this chapter, we discuss these programmable logic devices with their structure and applications.

15.2 Programmable Logic Devices

Programmable logic devices (PLDs) are standard products, but can be programmed to function in a specific application. The programming can be done either by the end user or by the manufacturer. The PLDs, which are programmed by the manufacturer, are known as mask-programmable logic devices (MPLDs). The PLDs which are programmed by the end user are called field-programmable

logic devices (FPLDs). The architecture of PLDs is very regular and fixed. It cannot be changed by the end user. The PLDs have a wide range of applications, and have a low risk and cost in manufacturing large volume. Hence, the PLDs are cheaper. As the PLDs are premanufactured, tested, and placed in inventory in advance, the design cycle time is very short. The PLDs are classified into three categories based on the architecture and programmability as given below:
- Read only memory (ROM)
- Programmable logic array (PLA)
- Programmable array logic (PAL)

This section describes the above three PLD architecture and their applications.

15.2.1 Read Only Memory

Read only memory (ROM) is a storage device which can be programmed only once. As a result, the data remains intact and can be read as many times as possible. The stored data is not lost even if the power is removed, unlike random access memory (RAM). The structure of a ROM is shown in Fig. 15.1.

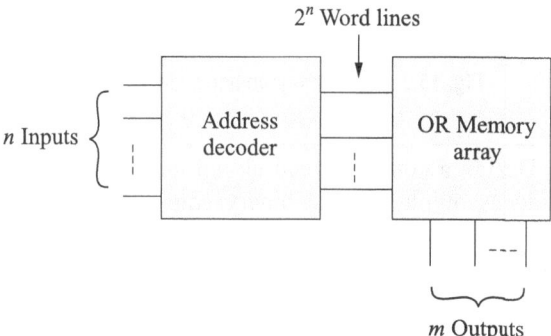

Fig. 15.1 A $2^n \times m$ ROM architecture

It consists of an address decoder with n input lines, and a programmable OR array with m output lines. The decoder produces 2^n minterms based on the n input lines. The minterms are ORed through programmable switches which can be made ON or OFF to select a particular minterm. The programmable switches can be implemented by either bipolar, CMOS, nMOS, or pMOS technologies. A ROM which is mask-programmed with the nMOS technology is shown in Fig. 15.2.

The nMOS transistors connect the decoder outputs to the output lines. The output line is normally pulled up to V_{DD}. When the decoder output goes high, the transistor gate connected to it, will turn ON, and the corresponding output line goes low. The presence of the nMOS transistor determines the pattern of 1s or 0s in the output lines. This pattern is determined by the mask layer which contains only those transistors where a connection is required.

Mask-programmed ROMs are used in the applications where the system requires data to be stored and not to be changed during the operation. For example, they are used to store monitor programs. ROMs are used in domestic appliances, industrial equipments, security systems, electronic terminals in retail stores, instrumentation, and so on.

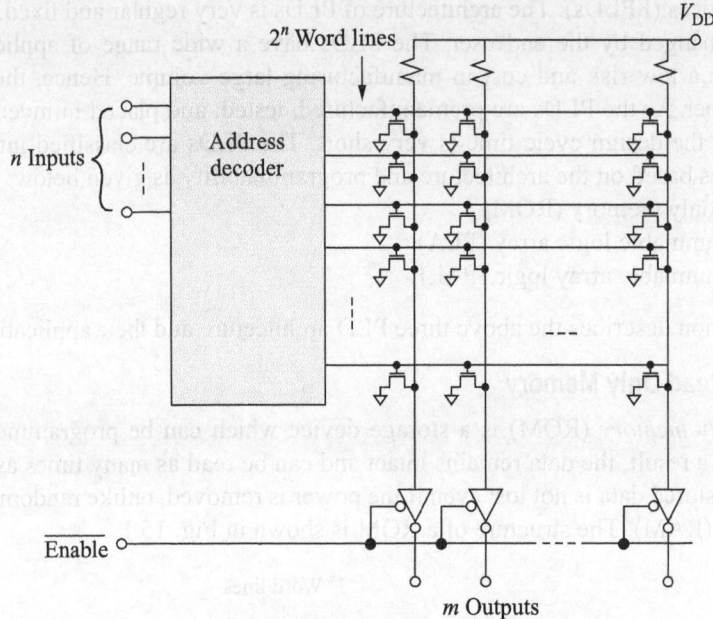

Fig. 15.2 Mask-programmable nMOS ROM

Example 15.1 Design a combinational circuit using ROM which takes a 3-bit number and produces outputs as the binary equivalent of the square of input numbers.

Solution Let us first derive the truth table of the combinational circuit that takes a 3-bit number and produces its square as the output.

The three input bits are A2, A1, and A0. The three input bits can have at the most eight combinations starting from 000 to 111. The maximum decimal equivalent value is 49 which requires six bits for representation. Hence, the combinational circuit would require at the most six output bits, which are represented as Y5, Y4, Y3, Y2, Y1, and Y0. The truth table of the circuit is shown in Table 15.1.

Table 15.1 Truth table of the circuit of Example 15.1

Inputs				Outputs						
A2	A1	A0	Decimal	Y5	Y4	Y3	Y2	Y1	Y0	Decimal
0	0	0	0	0	0	0	0	0	0	0
0	0	1	1	0	0	0	0	0	1	1
0	1	0	2	0	0	0	1	0	0	4
0	1	1	3	0	0	1	0	0	1	9
1	0	0	4	0	1	0	0	0	0	16
1	0	1	5	0	1	1	0	0	1	25
1	1	0	6	1	0	0	1	0	0	36
1	1	1	7	1	1	0	0	0	1	49

Out of six output bits, two bits—Y1 and Y0—can be implemented directly, as Y1 is always zero, and Y0 is the same as input A0. The remaining four bits Y5, Y4, Y3, and Y2 can be implemented using an 8 × 4 ROM, as shown in Fig. 15.3.

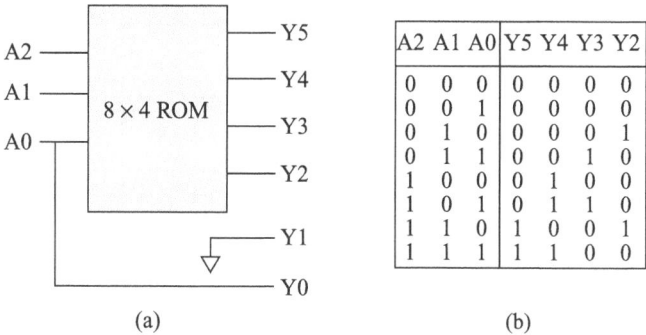

A2	A1	A0	Y5	Y4	Y3	Y2
0	0	0	0	0	0	0
0	0	1	0	0	0	0
0	1	0	0	0	0	1
0	1	1	0	0	1	0
1	0	0	0	1	0	0
1	0	1	0	1	1	0
1	1	0	1	0	0	1
1	1	1	1	1	0	0

(a)   (b)

Fig. 15.3 Implementation of the combinational circuit of Example 15.1:
(a) simplified form of the circuit using ROM;
(b) ROM truth table

15.2.2 Programmable Logic Array

Programmable logic array (PLA) is an IC chip used for two-level combinational logic circuits. It consists of an AND array followed by an OR array. Both the AND array and OR array are programmable. The architecture of PLA is shown in Fig. 15.4.

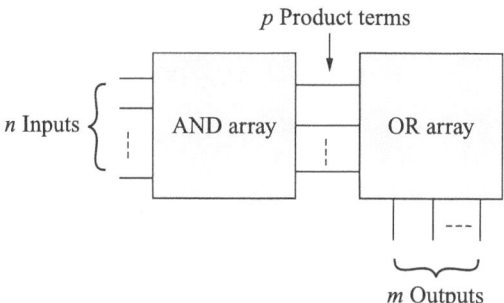

Fig. 15.4 PLA architecture in a block diagram

The AND array, also called the AND plane, implements the product terms, and the OR array, also called the OR plane, implements the sum of product (SOP) terms. In PLA, both the arrays are programmable. PLA has a limited number of product terms, not the minterms. Hence, to implement a logic using PLA, a minimal SOP form should be derived, so that the logic can be implemented using the available product terms. A PLA implemented using CMOS technology is shown in Fig. 15.5.

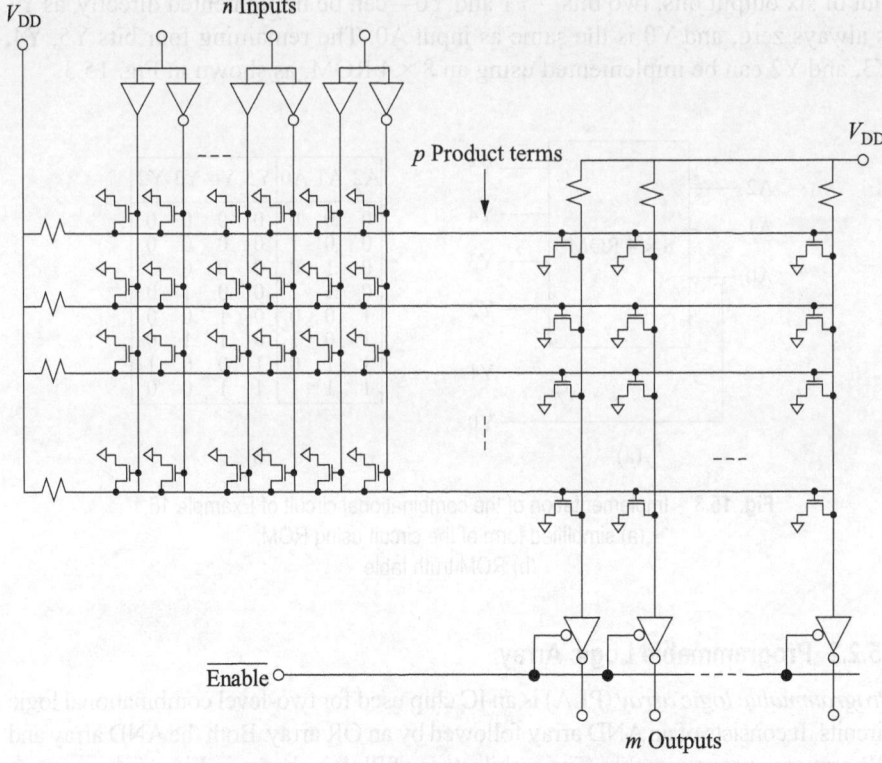

V_{DD} *n* Inputs

V_{DD}

p Product terms

Enable

m Outputs

Fig. 15.5   PLA architecture with CMOS technology

Example 15.2   Implement the following Boolean functions using a PLA:

$$F_1 = A\overline{B} + AC + BC$$

$$F_2 = \overline{A}B\overline{C} + AC$$

Solution   To implement the product terms of the functions F_1 and F_2, only those transistors are to be kept, which should form the product terms (Fig. 15.6).

For example, two transistors in the first row in the AND plane generates the product term $A\overline{B}$. The line is normally pulled to high, when $A = 0$ or $B = 1$. The line is pulled to the ground through the nMOS transistors. That is the line which implements the term W, where $W = A\overline{B}$. Similarly, the second, third, and fourth lines implement the terms X, Y, and Z as given by

$$X = AC$$

$$Y = BC$$

$$Z = \overline{A}B\overline{C}$$

Now, let us write function F_1 as

$$F_1 = A\overline{B} + AC + BC = W + X + Y = \left(\overline{\overline{W} \cdot \overline{X} \cdot \overline{Y}}\right)$$

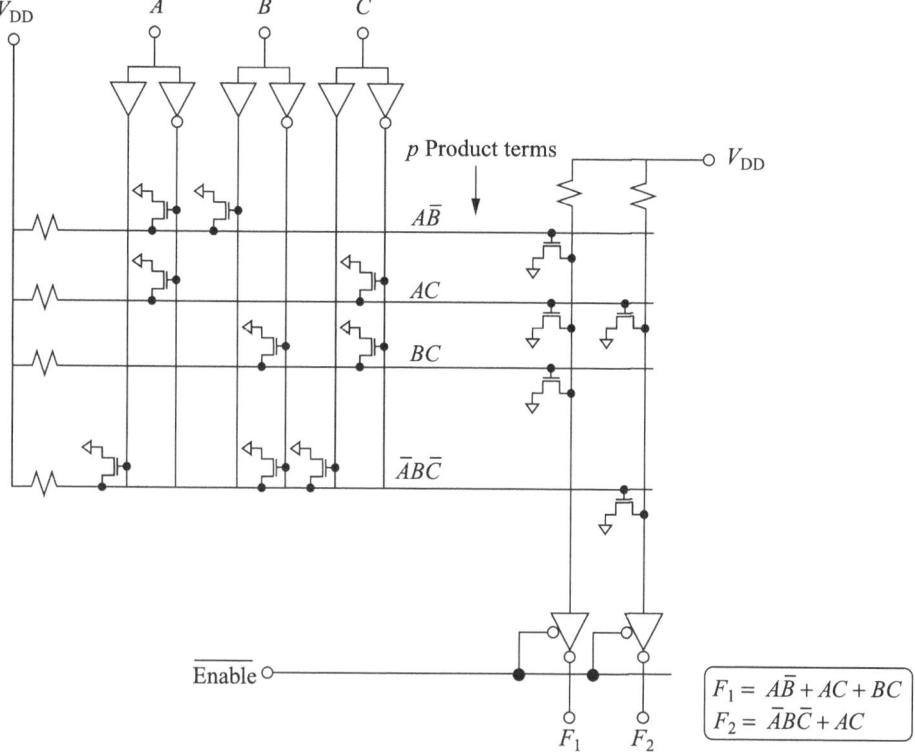

Fig. 15.6 PLA with three inputs and two outputs

Similarly, we can write the function F_2 as

$$F_2 = \overline{A}B\overline{C} + AC = Z + X = \left(\overline{\overline{Z} \cdot \overline{X}}\right)$$

In the OR plane, the nMOS transistors are kept to form the product terms $(\overline{W}\,\overline{X}\,\overline{Y})$ and $(\overline{Z}\,\overline{X})$. In the first column, the line is pulled to the ground if either of W, X, or Y is 1, otherwise the line is pulled to high. The inverting buffer at the output hence, implements the function F_1. Similarly, the transistors in the second column in the OR plane implement the complement of the function F_2, and the inverting buffer at the output implements the function F_2.

Example 15.3 Implement the following Boolean functions using PLA:

$$\text{Sum}(A, B, C_{in}) = \sum m(1, 2, 4, 7)$$

$$C_{out}(A, B, C_{in}) = \sum m(3, 5, 6, 7)$$

Solution The Boolean expressions for the given functions can be written as

$$\text{Sum} = \overline{A}\,\overline{B}\,C_{in} + \overline{A}\,B\,\overline{C}_{in} + A\,\overline{B}\,\overline{C}_{in} + ABC_{in}$$

$$C_{out} = AB + AC_{in} + BC_{in}$$

The functionality of a PLA to implement these functions can be represented as shown in Table 15.2.

Table 15.2 PLA implementation table

	Inputs			Outputs	
	A	B	C_{in}	Sum	C_{out}
$\bar{A}\,\bar{B}\,C_{in}$	0	0	1	1	0
$\bar{A}\,B\,\bar{C}_{in}$	0	1	0	1	0
$A\,\bar{B}\,\bar{C}_{in}$	1	0	0	1	0
ABC_{in}	1	1	1	1	0
AB	1	1	–	0	1
AC_{in}	1	–	1	0	1
BC_{in}	–	1	1	0	1

Inputs are represented by 1 for true form, 0 for complement form, and – for don't care. The outputs are represented by 1 if the term is present in the function, and 0 if the term is absent in the function. A simplified form of PLA is shown in Fig. 15.7.

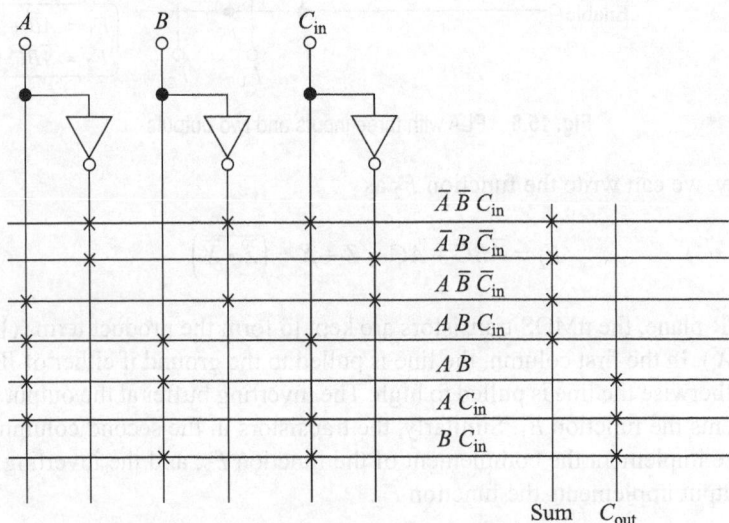

Fig. 15.7 Simplified form of PLA implementing Boolean functions for Sum and C_{out}

The connection between a vertical and horizontal line is represented by a crosspoint '×'.

15.2.3 Programmable Array Logic

The *programmable array logic* (PAL) is another class of programmable logic device with the AND array followed by the OR array, where the AND array is programmable but the OR array is fixed. The PAL architecture is shown in Fig. 15.8.

The OR array has permanently programmed connections as shown by dots in Fig. 15.8. The OR plane cannot be programmed. In this PAL architecture, each OR gate has two inputs; hence, the SOP must have two product terms. It may be noted that unlike PLA, the product terms cannot be shared between the OR gates. Each function must be simplified individually to reduce the product terms to maximum two. If the SOP expression contains more than two product terms, each OR gate can be used to implement the function partially, and then summed using the additional OR gate to implement the complete function. The following example illustrates the implementation of Boolean functions using PAL.

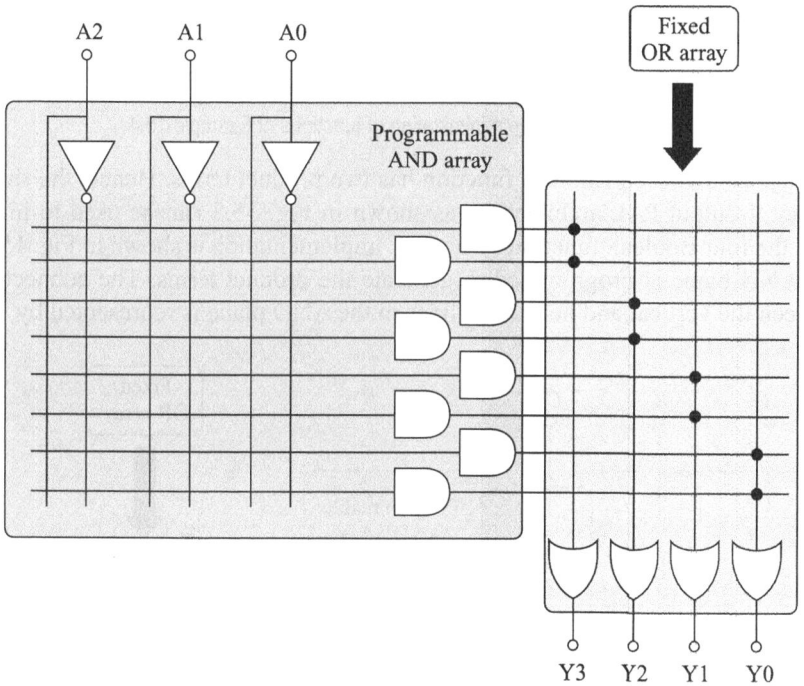

Fig. 15.8   PAL architecture

Example 15.4   Implement the following Boolean logic using PAL:

$$F_1(A,B,C) = \sum m(1,3,4,5,6,7)$$

$$F_2(A,B,C) = \sum m(0,1,4,5,6)$$

$$F_3(A,B,C) = \sum m(1,2,5)$$

$$F_4(A,B,C) = \sum m(0,1,3,7)$$

Solution   Let us first find out the minimum SOP form of the given function using the Karnaugh's map method. The K-maps and the corresponding minimum SOP forms are shown in Fig. 15.9.

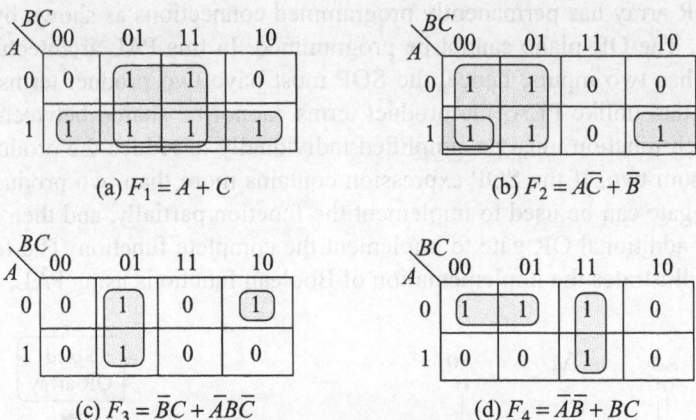

(a) $F_1 = A + C$

(b) $F_2 = A\overline{C} + \overline{B}$

(c) $F_3 = \overline{B}C + \overline{A}B\overline{C}$

(d) $F_4 = \overline{A}\overline{B} + BC$

Fig. 15.9   K-map minimization of functions of Example 15.4

We can see that each Boolean function has two product terms. Hence, the simple 3-input 4-output PAL architecture, as shown in Fig. 15.8 can be used to implement the four Boolean functions. The PAL implementation is shown in Fig. 15.10. The AND plane is programmed to generate the product terms. The connections between the vertical and horizontal lines in the AND plane is represented by '×'.

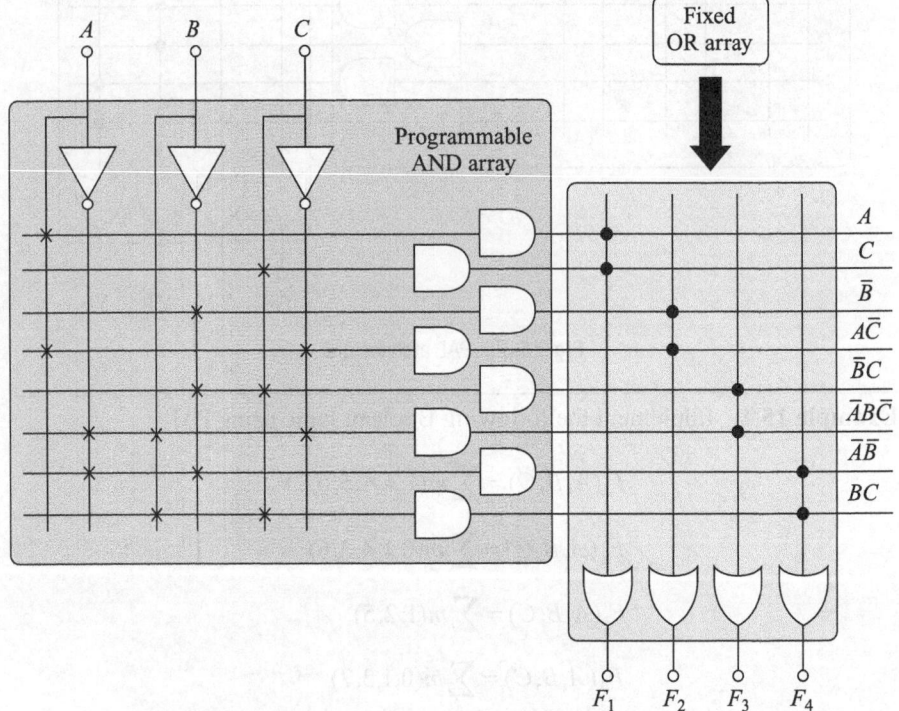

Fig. 15.10   PLA implementation of Boolean functions of Example 15.4

In the ROM-based design, the addition of an input signal increases the ROM size by two times. This in turn doubles the size of the AND and OR array. But in case of PLA or PAL, additional input can easily be accommodated without doubling the size. Commercially available PAL can have at the most 22 input lines.

A standard PAL (PAL16R8) with 16 inputs and 8 outputs is shown in Fig. 15.11. The IC has 20 pins, including the power and ground pins. All the inputs are available in both the true form and complemented form. There are eight 8-input OR gates connected to the word lines from the AND array. Any word line can be connected to any input, either in the true or complemented form. The outputs are taken from the Q-output of the D flip-flops through tri-state inverting buffers. Out of the 16 inputs, 8 inputs are the direct inputs, and the remaining 8 inputs are taken from the Q-outputs of the D flip-flops.

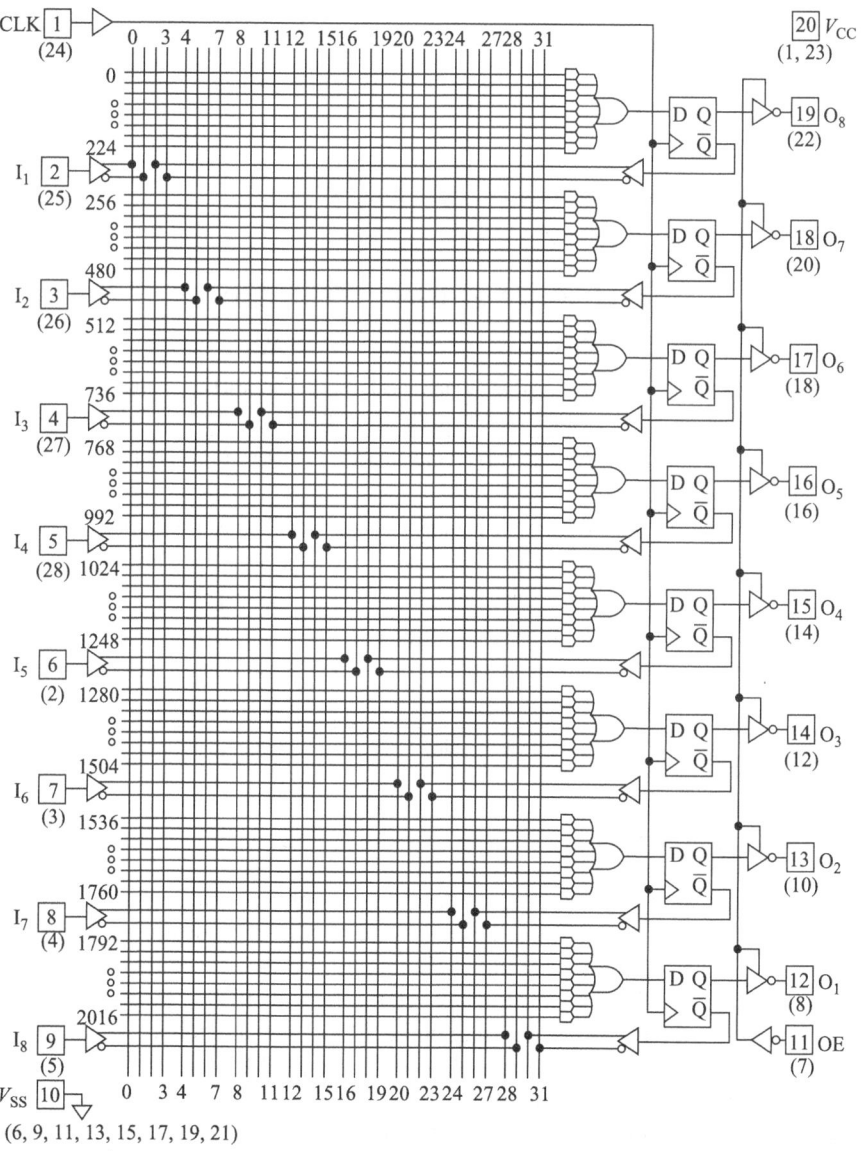

Fig. 15.11　The architecture of PAL16R8

The standard PAL chips are named following a nomenclature as explained in Table 15.3.

Table 15.3 PAL nomenclature

PAL ii t oo		
ii	Maximum number of inputs to the AND array	
t	Types of outputs	
	Combinational	Registered
	H-active high	R-registered
	L-active low	RP-registered with programmable polarity
	P-programmable polarity	V-versatile
	C-complementary	
oo	Maximum number of dedicated or programmed outputs	

15.2.4 Comparison between ROM, PLA, and PAL

Comparative studies are shown in Table 15.4 between the ROM, PLA, and PAL.

Table 15.4 Comparison between ROM, PLA, and PAL

ROM	PLA	PAL
The decoder (or AND array) implements all the minterms	The AND array implements a limited number of product terms	The AND array implements a limited number of product terms
AND array is not programmable	AND array is programmable	AND array is programmable
OR array is programmable	OR array is programmable	OR array is not programmable
Additional inputs double the size of AND and OR array	Additional input doesn't require doubling of size	Additional input doesn't require doubling of size
It can implement SOP with any number of terms	It can implement SOP with any number of terms	It can implement SOP with limited number of terms
	Costlier than PAL	Cheaper than PLA
Least flexible	Extremely flexible	Moderate flexible

15.3 Sequential PLD

The PLDs that we have discussed contain only combinational logic gates but no sequential elements or flip-flops. Digital systems are to be designed using both combinational and sequential circuits. Hence, to implement sequential programmable devices, flip-flops must be used externally with PLDs. In order to avoid the external use of flip-flops, the sequential PLDs are developed with D or JK flip-flops. The sequential PLD is also known as simple PLD or SPLD. The SPLD architecture is mostly based on combinational PAL and D flip-flops. The section of an SPLD which implements one SOP output through a register is known as a *macrocell*. A macrocell is shown in Fig. 15.12.

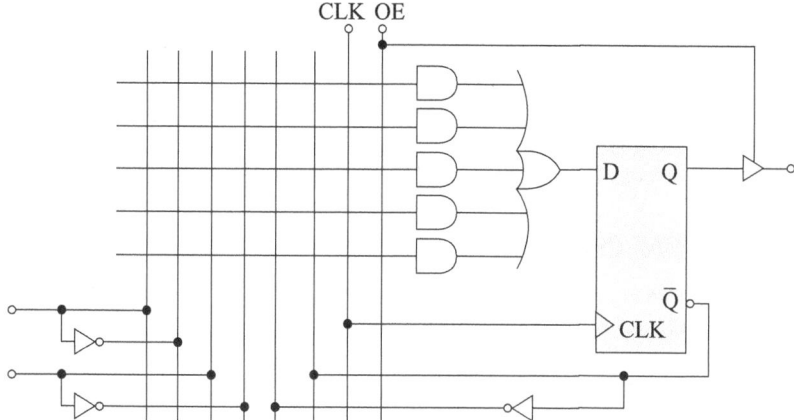

Fig. 15.12 A typical macrocell architecture

The AND-OR array is similar to the PAL architecture. The output of AND-OR array is passed through a D flip-flop triggered by a clock signal CLK. The final output is available through a tri-state buffer controlled by the output enable signal OE. The true and complemented form of the output signal is fed back to the input of the AND array. This provides the previous state of the output signal. A typical SPLD chip has 6–10 macrocells.

Some of the SPLD vendors are AMD, Altera, ICT, Lattice, Cypress, and Philips-Signetics.

15.4 Complex PLD

The complex PLD (CPLD) is an advanced programmable logic device which can implement large structures of field programmable combinational and sequential logic. The high level architecture of a typical CPLD is shown in Fig. 15.13. It contains array of PLDs and programmable interconnect fabric.

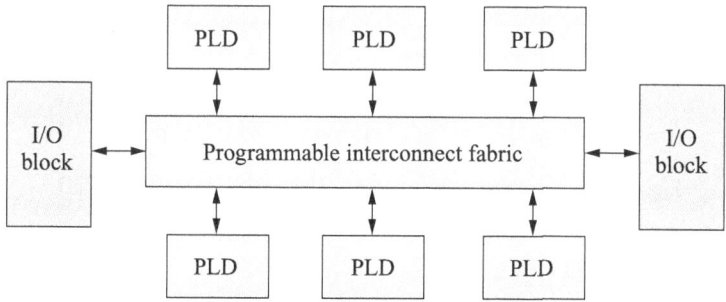

Fig. 15.13 Typical CPLD architecture

A CPLD can have a large number of inputs. The input/output block (IOB) provides connectivity to the pins of the IC. The input/output pins can be configured to act as either input or output pins. The interconnect fabric provides the connectivity between the PLD blocks, and also receives signals from, and sends to the IOBs. The individual PLD block contains typically 8–16 macrocells. In the large size CPLDs, the output of every macrocell is not connected to the output pins. However, all the macrocells are connected through the switch fabric.

The most commonly used CPLDs are: Altera MAX 7000 series and Xilinx XC9500 series. We shall discuss the architecture of these CPLDs in Sections 15.4.1 and 15.4.2.

15.4.1 Altera MAX CPLD Series

Altera CPLD series are MAX 5000, 7000, and 9000. We shall discuss the architecture of the MAX 7000 series CPLD. It consists of an array of logic array blocks (LABs), a programmable interconnect array (PIA), and an array of programmable IOBs. The schematic of MAX 7000 series PLD is shown in Fig. 15.14. The LAB contains 16 macrocells. The PIA connects the LABs, the IOBs, and the primary inputs.

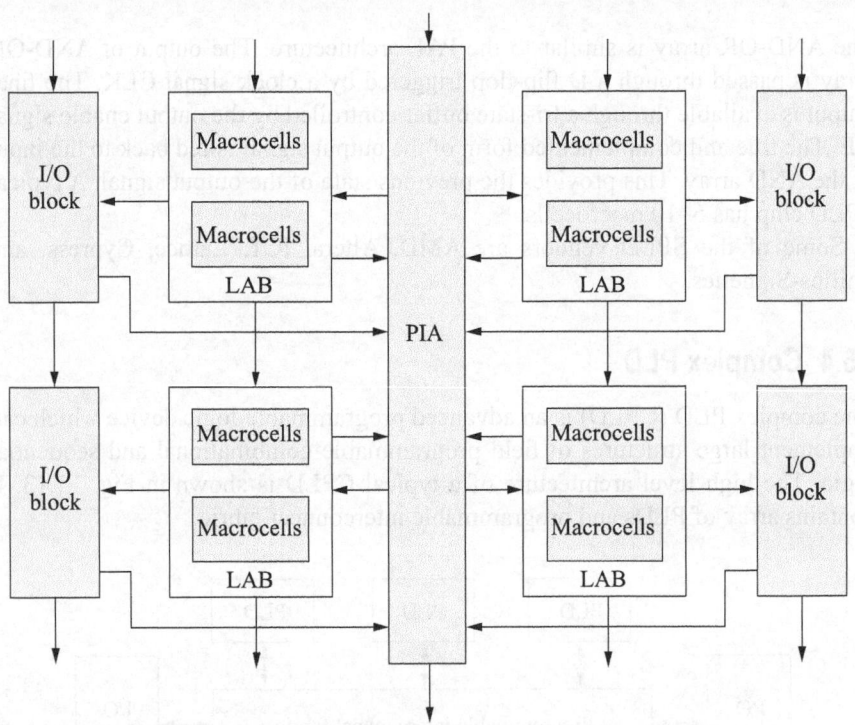

Fig. 15.14   Typical architecture of the Altera MAX 7000 series CPLD

15.4.2 Xilinx XC9500 CPLD series

The architecture of the Xilinx XC9500 series CPLD is shown in Fig. 15.15. It consists of an array of function blocks, a FastCONNECT switch matrix, an array of IOBs, a JTAG controller, and an in-system programming controller.

Each of the function blocks contains up to 18 macrocells, and 54 inputs and 18 outputs. The IOBs interface the input and output signals, and also the global clock and set/reset signals.

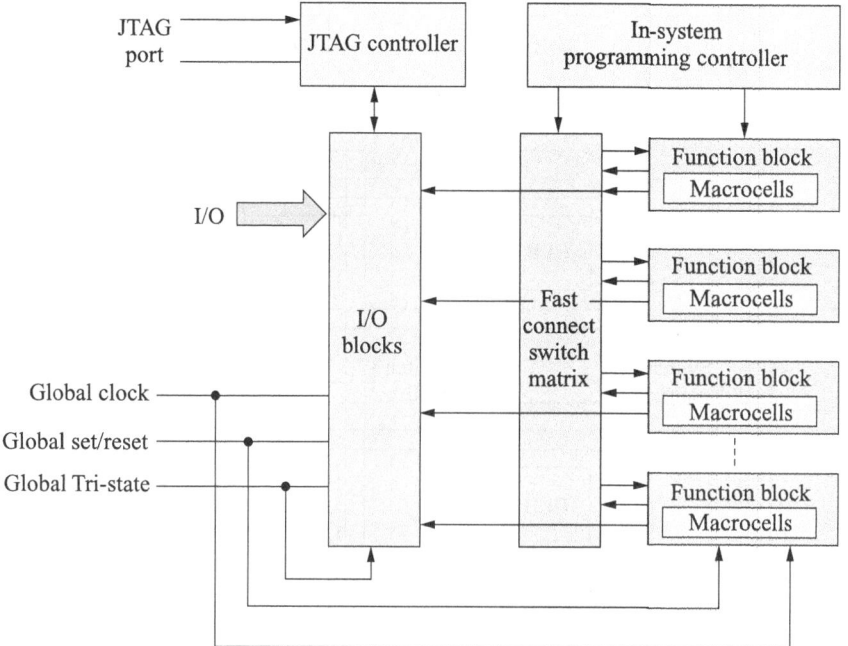

Fig. 15.15 Architecture of Xilinx XC9500 CPLD

15.5 Field Programmable Gate Array

Field programmable gate array (FPGA) is an IC which can be hardware-programmed to implement various logic functions. The end users of FPGA can program it to configure for any functionality—so it is called field programmable. FPGA is completely fabricated and standard parts are tested and available readily for use. FPGA can be used for prototyping of an idea into silicon in a very short time. In this section, we will discuss the basic architecture of FPGA, and FPGA-based VLSI design. We will also discuss the FPGA vendors with their FPGA compatible EDA tools.

15.5.1 Architecture

There are a number of different FPGA architecture. Basically, it contains three major components; the configurable logic block (CLB), switch matrix, and the IOB. A simple FPGA architecture is shown in Fig. 15.16.

There are three types of FPGA programming technology as shown in the following text:

- Antifuse-based
- EPROM-based
- SRAM-based

Antifuse-based FPGA

The antifuse FPGAs are programmed by applying high voltage between the two terminals of the fuse to break down the dielectric material of the fuse.

The antifuse switch used in FPGA is shown in Fig. 15.17. Antifuse structure is normally used in an open circuit condition. However, when they are programmed,

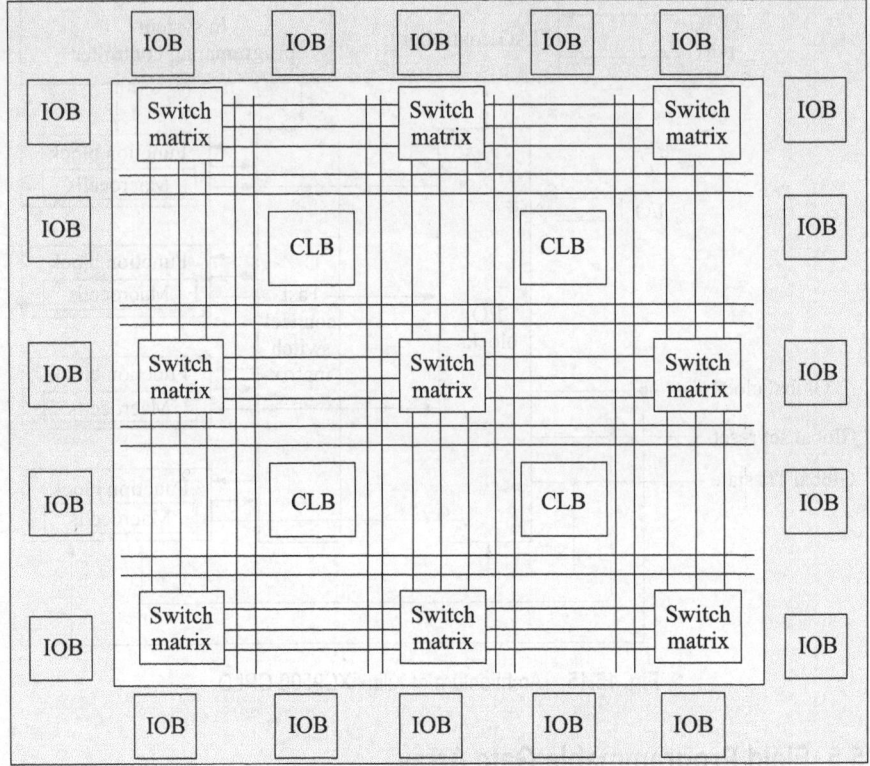

Fig. 15.16 Basic FPGA architecture

a low resistance path is established. As shown in Fig. 15.17, the top and bottom layers are conducting, and the middle layer is an insulator. In normal conditions, the insulating layer isolates the top and bottom layers. But when the antifuse is programmed, a low resistance path is established through the insulator. The antifuse switches have smaller on-resistance and parasitic capacitance than pass transistors and transmission gates. Hence, it supports higher switching speed. Antifuse switches are one-time programmable, so design changes are not possible.

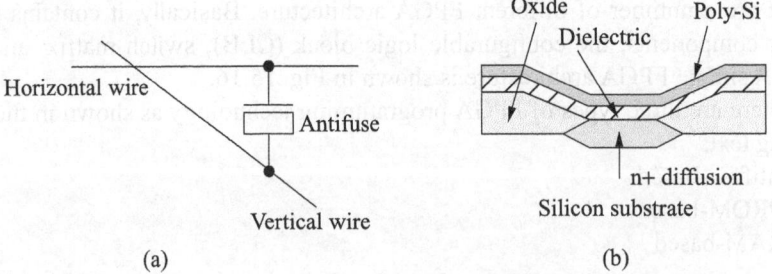

Fig. 15.17 Antifuse switch used in FPGA: (a) schematic, (b) structure

EPROM-based FPGA

The FPGAs use EPROM and EEPROM technology which are programmed using high voltages. The devices are reprogrammable and nonvolatile, and can be programmed while the devices are embedded in the system. The EPROM and EEPROM programming is based on the flash memory cell as shown in Fig. 15.18 which uses two gates, one is the control gate and another is the floating gate. Under normal mode of operation, there are no changes on the floating gate, and the transistor behaves like a normal transistor with low threshold voltage. When a high voltage is applied to the control gate, the floating gate is charged, and the threshold voltage is increased. The transistor becomes permanently OFF.

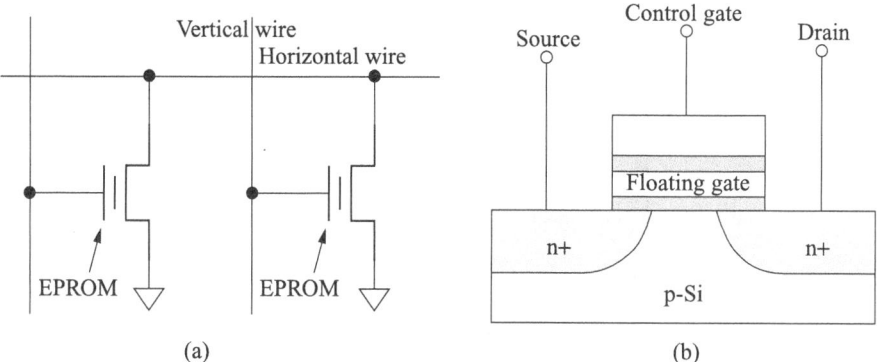

Fig. 15.18 (a) Switch interconnection; (b) EPROM (flash) memory cell

SRAM-based FPGA

In the SRAM-based FPGA, the logic functions are based on the stored bits in the SRAM. These devices use CMOS transmission gates for switching. The SRAM-based switch interconnect is explained in detail in Section 15.5.4.

There are a number of FPGA vendors in the market today. In the SRAM-based FPGA category, Xilinx and Altera are the leading manufacturers in terms of the number of users, with the major competitor being AT&T. For antifuse-based products, Actel, Quicklogic and Cypress, and Xilinx offer competing products.

15.5.2 Configurable Logic Block

The configurable logic blocks (CLBs) contain several modules such as lookup tables (LUT), multiplexers, gates, and flip-flops. LUT is a hardware that stores the truth table of a function in an SRAM to function as a combinational circuit. Figure 15.19 illustrates an SRAM-based LUT. The LUT-based structure is similar to the ROM-based architecture, as discussed in Chapter 1. As shown in Fig. 15.19, the 8×4 SRAM architecture implements the 3-input and 4-output combinational logic. The SRAM is used instead of ROM to have the capability of reprogramming the memory so that different functions can be implemented. But as the SRAM is volatile memory, the stored data gets erased once the power goes off.

15.5.3 Lookup Table

Combinational logic is stored in the form of a truth table, called a lookup table (LUT). The LUTs are also known as function generators. The capacity of a LUT is limited by the number of inputs. The advantage of the LUT is that the delay through it is constant.

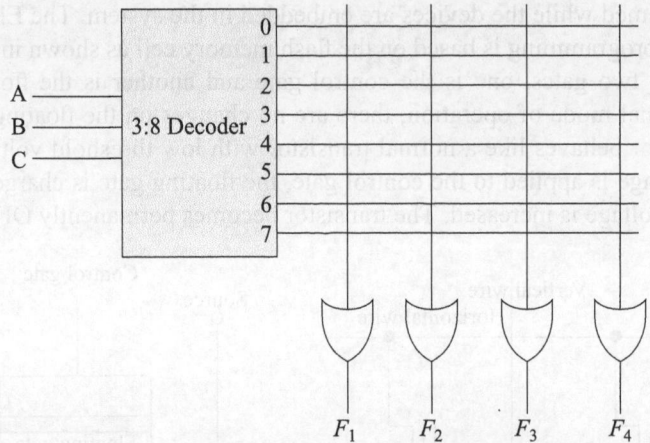

Fig. 15.19 An 8 × 4 SRAM architecture

A LUT is built using a configurable register array (SRAM) and a set of multiplexers. The register bits are programmed to implement the truth table of a function. The input signals are connected to the select lines of the multiplexers. An n-input LUT can implement any function of n number of inputs. To implement an n-input LUT, 2^n SRAM bits and 2^n:1 multiplexer are needed. Figure 15.20(a) shows a 4-input LUT with 4 inputs—A, B, C, and D. By setting the 16 bits in the SRAM, any function of 4 inputs can be implemented. This 4-input LUT can also be used as two 3-input LUTs as shown in Fig. 15.20(b).

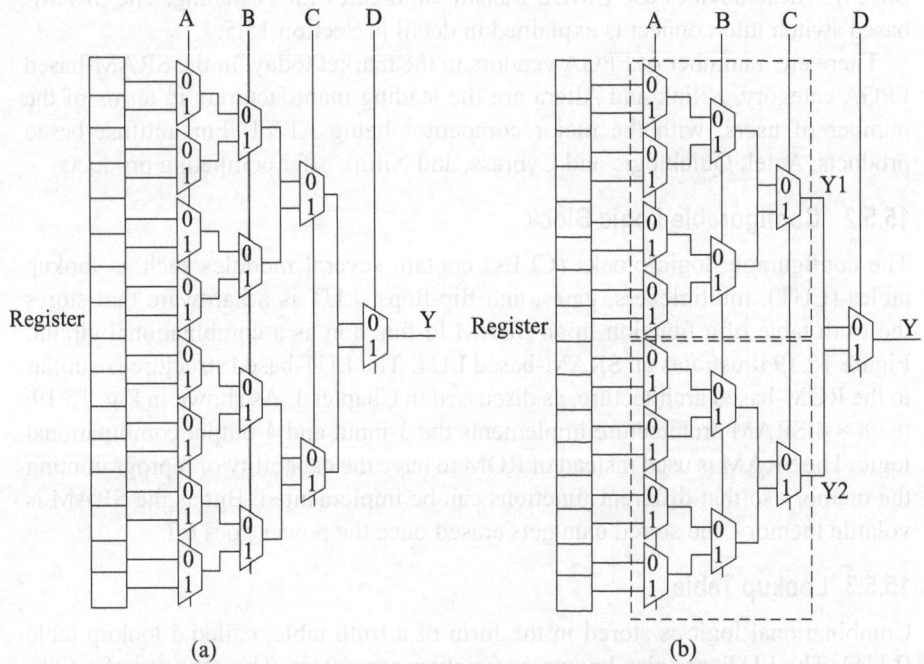

Fig. 15.20 (a) 4-input LUT; (b) 4-input LUT built using two 3-input LUTs and a 2:1 multiplexer

Similarly, bigger size LUTs can be built using small size LUTs, as shown in Fig. 15.21.

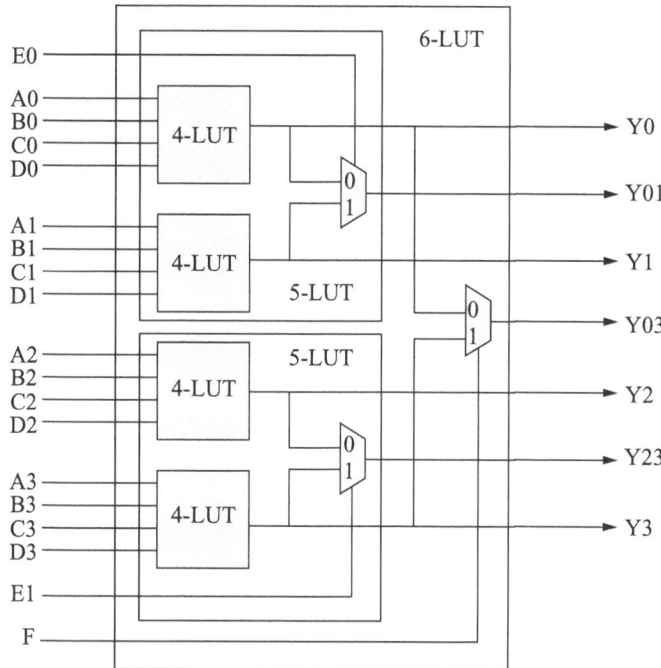

Fig. 15.21 Bigger size LUT implemented using 4-input LUTs

A 5-input LUT can be built using two 4-input LUTs and a 2:1 multiplexer. Similarly, a 6-input LUT can be built using two 5-input LUTs and a 2:1 multiplexer, and so on.

15.5.4 SRAM-based Crosspoint Switch Matrix

The interconnect switch fabric is shown in Fig. 15.22(a). Each crosspoint has six switches, which are controlled by the SRAM. Depending on the bit stored in the SRAM, the connection is established between the horizontal and the vertical interconnect wires. For example, in Fig. 15.22(a), the north-to-east (NE) connection is established by the SRAM containing a bit 1. This makes the nMOS transistor ON, and the connection between N and E is established. The internal circuit diagram of a SRAM cell is shown in Fig. 15.22(b).

15.6 Xilinx SRAM-based FPGA

The basic structure of Xilinx FPGAs comprises a two-dimensional array of logic blocks that can be interconnected via horizontal and vertical routing channels. Figure 15.23 illustrates typical Xilinx FPGA architecture. In 1985, Xilinx introduced the first FPGA family, called the XC2000 series. Now they have many versions such as XC3000, XC4000, XC5000, Spartan, and Virtex series. Xilinx has recently introduced an FPGA family based on antifuses, called the XC8100. The

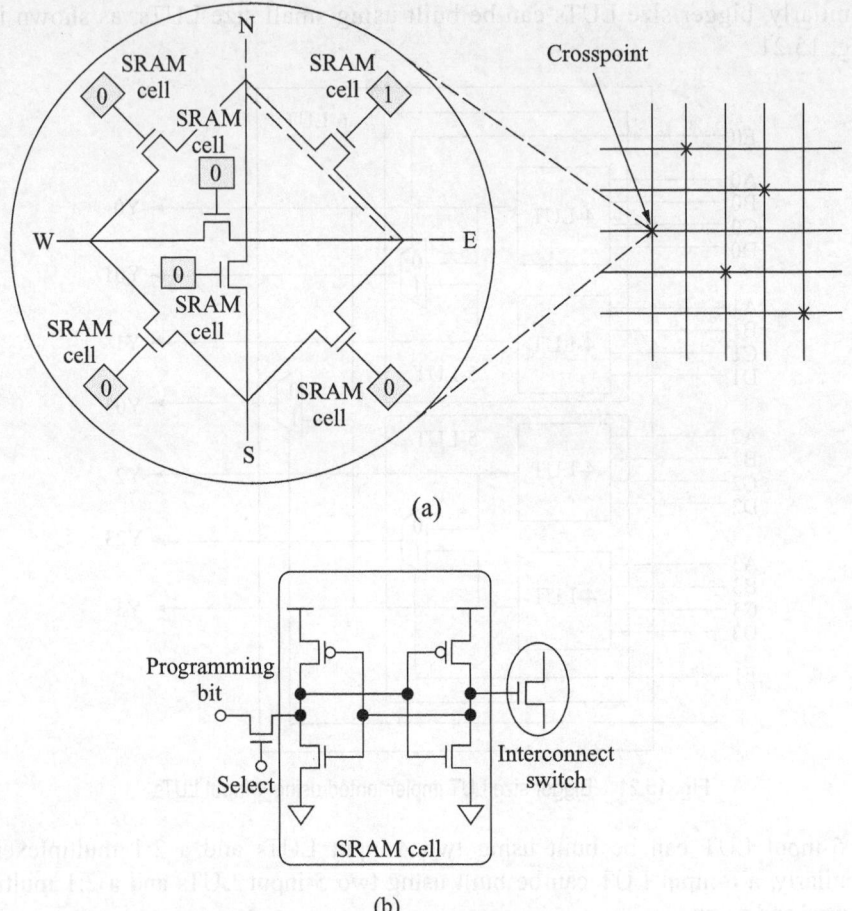

Fig. 15.22 SRAM-based crosspoint switch matrix;
(a) crosspoint switch, (b) SRAM cell

Xilinx 4000 family devices range in capacity from about 2000 to more than 15,000 equivalent gates. Generally, a 2-input NAND gate is used to represent the equivalent gate count in an IC.

The XC4000 consists of a CLB that is based on LUTs. A LUT with *n* inputs can realize any logic function with *n* inputs by programming the logic function's truth table directly into the memory. The XC4000 CLB contains three separate LUTs, as shown in Fig. 15.23. There are two 4-input LUTs that are fed by CLB inputs, and the third LUT is used in combination with the other two LUTs. This arrangement allows the CLB to implement a wide range of logic functions of up to nine inputs, two separate functions of four inputs, or other possibilities. Each CLB also contains two flip-flops.

Another key feature that characterizes an FPGA is its interconnect structure. The XC4000 interconnect is arranged in horizontal and vertical channels. Each channel contains some number of short, long, and very long wires. The short wire segments span a single CLB, the long segments span two CLBs, and very long segments span the entire length or width of the chip. Programmable

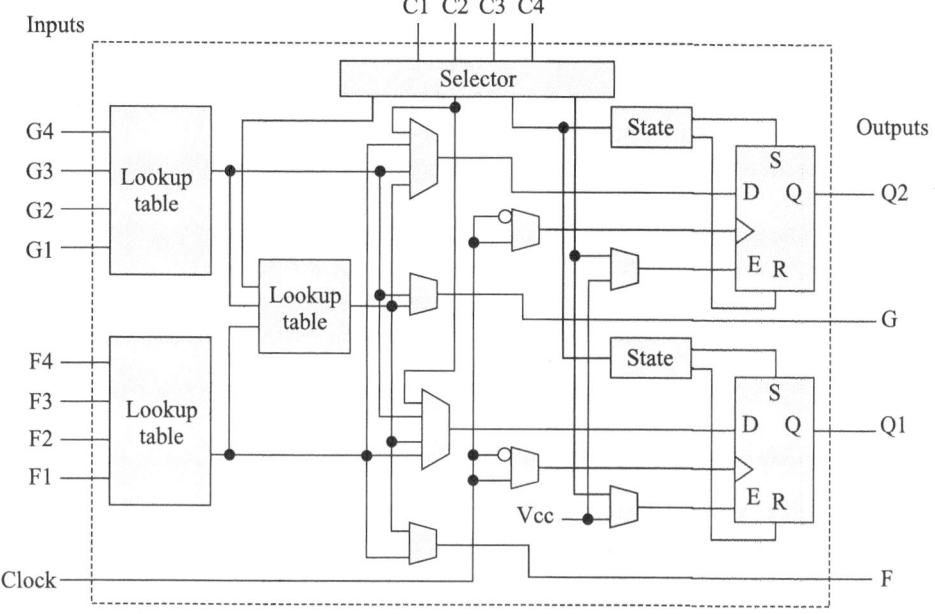

Fig. 15.23   Xilinx XC4000 CLB architecture

switches are available to connect the inputs and outputs of the CLBs to the wire segments, or to connect one wire segment to another. In Xilinx FPGA interconnect architecture, the signals pass through switches to reach from one CLB to another, and the total number of switches traversed depends on the particular set of wire segments used. Hence, speed/performance of an implemented circuit depends partly on how the wire segments are allocated to individual signals by the CAD tools.

15.7 Comparison between FPGA, ASIC, and CPLD

A comparative study between FPGA, ASIC, and CPLD is shown in Tables 15.5 and 15.6.

Table 15.5   FPGA vs ASIC

FPGA	ASIC (Application-specific integrated circuit)
Purchased from vendor as a standard part, then programmed by the user.	Made to customer specification by the vendor.
No production set-up costs (first unit costs the same as subsequent units).	High production set-up costs (often in the $100,000 range).
Fast turnaround time (can be programmed in a matter of minutes).	Slow turnaround time (often at least 6 weeks).
Relatively high per unit cost and low capability per chip.	Lower per unit cost—good for high volume production.
Design requires mostly writing HDL code (in VHDL or Verilog).	Design often requires knowledge of physical layout of silicon inside the IC.

Table 15.6 FPGA vs CPLD

FPGA	CPLD
Performance depends on the routing implemented for a particular application.	Predictable performance independent of internal placement and routing.
Functionality is implemented by lookup tables.	Functionality is implemented by PAL-like structures.
Suitable for medium to high density designs.	Suitable for low to medium density designs.
More complex and register-rich architecture.	Regular PAL-like architecture.
Channel-based interconnection fabric.	Crossbar type interconnection fabric.
Can be reprogrammed as many times as possible.	Can be reprogrammed a limited number of times.

15.8 FPGA-based System Design

A typical FPGA-based system design flow is shown in Fig. 15.24. The flow starts with the design specifications. The functional description of the system is written in a hardware description language (VHDL or Verilog) in the behavioural modelling style. The functionality is checked by performing behavioural simulation using a set of test vectors. The next step is to perform synthesis. The synthesis step translates the behavioural netlist into a gate level netlist. The synthesis step requires the behavioural netlist, the selected device family (e.g., Spartan, Virtex) name, and other synthesis directives. The gate level netlist is again checked for functionality. The user constraints are to be specified for timing, power, etc. Then using the user constraints and gate level netlist, the implementation step is performed. In

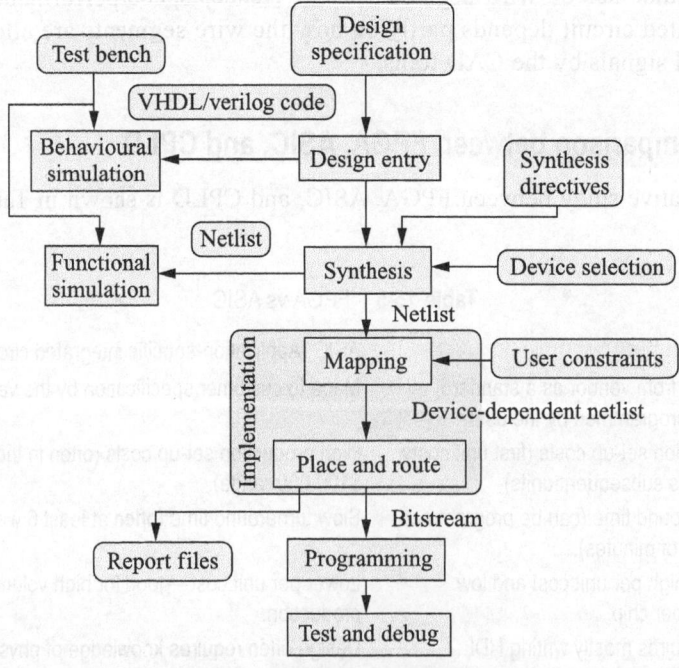

Fig. 15.24 FPGA-based system design flow

the implementation step, the mapping of the logic gates are done to the available functional blocks in the FPGA, and the placement and routing are done to complete the implementation. Next, the bitstream file is generated which contains the programming data. The bitstream file is downloaded through the JTAG cable into the FPGA device. Downloading the bitstream into the FPGA device is often referred to as FPGA programming. The final step is to test the FPGA device in the system, and debug for any problems in functionality.

15.9 IRSIM

IRSIM is a switch-level simulator used for simulating digital circuits. It models the transistors as switches, and parasitic resistance and capacitances are added to these switches for simulating timing delays. IRSIM was developed at Stanford. The details about IRSIM is found at http://opencircuitdesign.com/irsim/.

IRSIM considers the following three aspects of digital circuit behaviour:
1. Transistor state (ON or OFF)
2. Logic value (high or low)
3. Transition events (from logic 0 to logic 1, and logic 0 to logic 1)

15.9.1 Transistor State

IRSIM uses *linear model* for transistors with two states: high and low resistance between the source and drain. The high resistance indicates transistor's OFF state, whereas the low resistance indicates ON state.

15.9.2 Logic Value

IRSIM considers all voltages in a circuit as one of the three following values:

High–h, 1
Low–l, 0
Indeterminate or unknown–x

15.9.3 Transition Events

The change in the logic state of a signal is called an *event*. If there is an event on any path, IRSIM re-evaluates the path. In a well-constructed circuit, every path must be driven by a single value, high or low. However, in some circuits, a path may be driven by multiple values (bus contention), or it may be electrically isolated (tri-state). IRSIM can handle such conditions, but the behaviour is not well modelled.

15.10 Generalized Open Source Programmable Logic

Generalized open source programmable logic (GOSPL) is an open source platform for FPGA. It is targeted to serve as a research and development platform for the electronics and software community worldwide. The main components of GOSPL are as follows:

■ Source code for FPGA software tool suite, such as:

—Placer, router, synthesis, timing analyser, etc.
■ Graphical user interface (GUI)
■ Configuration bit generator that links with the FPGA device
■ Software user manual, release notes, coding guidelines, design documents, and test cases

Details on advanced FPGA architecture are available which can be used to understand the architecture, and to work on new architectural innovations and enhancements. It provides an opportunity for research scholars, professionals, and students to learn, develop, and optimize implementation flows, algorithms, and devices.

GOSPL is a platform where a software developer has to find solutions for very large NP hard problems, which is the most challenging part of any EDA tool development. A complete framework and flow is provided to experiment on different parts of the implementation flow, such as synthesis, placement, and routing. Details are found in the following link:

http://intranet.cs.man.ac.uk/apt/people/sfurber/uElecCV/VS25.php

SUMMARY

- Programmable devices offers flexibility in logic function design without changing the hardware components.
- PLDs are composed of AND array followed by OR array. In ROM, the AND array is fixed, whereas the OR array is programmable. In PLA, both the AND and the OR array are programmable. In case of PAL, the AND array is programmable but the OR array is fixed.
- The smallest block in a SPLD is a macrocell which contains the AND-OR array followed by a flip-flop.
- There are three types of FPGA programming—antifuse-based, EPROM-based, and SRAM-based.
- LUT is a hardware that stores the truth table of a combinational circuit in SRAM. By programming the SRAM bits the functionality can be changed.
- FPGAs are most suitable for prototype development in a very short time.

SELECT REFERENCES

Brown, S. and J. Rose 1996, *FPGA and CPLD Architecture: A Tutorial*, IEEE Design and Test of Computers.

Ciletti, M.D. 2005, *Advanced Digital Design with the Verilog HDL*, Pearson Education, New Delhi.

Mano, M.M. and M.D. Ciletti 2008, *Digital Design*, 4th ed., Pearson Prentice-Hall.

Streetman, B. and S. Banerjee 2006, *Solid State Electronic Devices*, 6th ed., Prentice-Hall.

Weste, N.H.E., D. Harris, and A. Banerjee 2009, *CMOS VLSI Design: A Circuits and Systems Perspective*, 3rd ed., Pearson Education.

EXERCISES

Fill in the Blanks

1. FPGA-based design has turn-around time _____ than ASIC-based design.
   - (a) less
   - (b) more
   - (c) equal
   - (d) more or less
2. Logic gate that is used to measure the gate equivalent/count in an IC is _____ .
   - (a) NOT gate
   - (b) 2-input NAND gate
   - (c) 2-input NOR gate
   - (d) 2-input XOR gate

3. Programmable logic array (PLA) has _____ .
   (a) fixed OR plane followed by a programmable AND plane
   (b) programmable AND plane followed by a fixed OR plane
   (c) fixed AND plane followed by a programmable OR plane
   (d) programmable AND plane followed by a programmable OR plane
4. Programmable array logic (PAL) has _____ .
   (a) fixed OR plane followed by a programmable AND plane
   (b) programmable AND plane followed by a fixed OR plane
   (c) fixed AND plane followed by a programmable OR plane
   (d) programmable AND plane followed by a programmable OR plane
5. ROM has _____ .
   (a) fixed OR plane followed by a programmable AND plane
   (b) programmable AND plane followed by a fixed OR plane
   (c) fixed AND plane followed by a programmable OR plane
   (d) programmable AND plane followed by a programmable OR plane

Multiple Choice Questions

1. In FPGA-based design, designers
   (a) design the layout and fabricate the IC
   (b) download the bit stream to program the device
   (c) both (a) and (b)
   (d) none of the above
2. Different FPGA programming technologies are based on
   (a) antifuse   (b) SRAM
   (c) EPROM   (d) all of these
3. LUT is used in
   (a) CPLD   (b) ASIC
   (c) FPGA   (d) SPLD
4. Which of the following is not a part of FPGA?
   (a) RTL   (b) I/O
   (c) PI   (d) CLB
5. SPLD is a macrocell which contains
   (a) AND-OR array followed by a flip-flop
   (b) a flip-flop
   (c) AND-OR array
   (d) AND array

True or False

1. SPLD contains sequential elements like a flip-flop.
2. CPLD is more complex than FPGA in terms of architecture.
3. FPGA can be programmed only one time.
4. GOSPL is an open source platform for FPGA.
5. IRSIM is a switch-level simulator used for simulating analog circuits.

Short-answer Type Questions

1. Explain the PLA architecture with necessary circuit diagrams and operating principle.
2. Explain the PAL architecture with necessary circuit diagrams and operating principle.
3. Design a combinational circuit using ROM which converts a 6-bit binary number to its corresponding 2-digit BCD.

4. Design the combinational circuit using ROM for a 7-segment display driver for the hexadecimal character generator of 4-bit binary inputs.
5. What is an FPGA? What are the main advantages of using an FPGA?
6. Discuss the architecture of SPLD.
7. Draw and explain the architecture of CPLD.
8. Compare FPGA with ASIC.
9. Compare FPGA with CPLD.
10. Draw and explain the FPGA-based system design flow diagram.
11. Distinguish between ROM, PLA, and PAL as elements realizing the Boolean function.
12. Explain the ROM-based design with a suitable example.
13. Write short note on the Xilinx FPGA architecture.

Long-answer Type Questions

1. What are the components of an FPGA? Discuss each of them in detail.
2. Draw the structure of SRAM-based FPGA and explain its operation.
3. What is a permanently programmed FPGA? Discuss in brief each component in it.
4. What is FPGA? Draw the architecture of FPGA and explain the operating principle. How does it differ from an ASIC?
5. Draw the architecture of PAL, PLA, and FPGA, and discuss the working principle.
6. Implement the following Boolean functions using PAL.

$$W(A,B,C,D) = \sum m(1,2,4,5)$$
$$X(A,B,C,D) = \sum m(0,1,2,7)$$
$$Y(A,B,C,D) = \sum m(3,5,6,7)$$
$$Z(A,B,C,D) = \sum m(0,4,6,7)$$

7. Why is FPGA preferred over CPLD? Explain the architecture of FPGA.
8. How is LUT used to program an FPGA? Explain with an example.
9. How is the logic capability of PLA measured? What are the differences between PAL and PLA? Implement the following Boolean functions using PLA:

$$f_1 = A\bar{B} + \bar{A}B$$
$$f_2 = A + (B + \bar{C})D$$

10. Explain the architecture of PLD.
11. Design a combinational circuit using an 8 × 4 ROM that accepts a 3-bit number and generates an output binary number equal to the square of the input number.
12. Implement the following logic using PAL.

$$F_1 = A\bar{B} + CD$$
$$F_2 = A + \bar{C}D$$

13. Implement the following logic using PLA.

$$F_1 = A\bar{C} + BD$$
$$F_2 = B + \bar{C}D$$

VLSI Process Technology

16.1 Introduction

Integrated circuit (IC) chips are designed and fabricated using very large scale integration (VLSI) technique. VLSI technique has two wings: (a) designing of circuits and layouts using EDA tools and algorithms is known as VLSI design; and (b) fabrication or manufacturing of VLSI circuits is known as VLSI technology or process technology. In this chapter, we discuss the basic steps for IC fabrication. First, we examine the different techniques for crystal growth followed by photolithography technique, oxidation, diffusion, ion implantation, etching techniques, and epitaxial growth techniques. The chapter concludes with metallization and packaging.

16.2 Crystal Growth

VLSI fabrication starts with a substrate material of either intrinsic or extrinsic type. A technique called lithography is used to create patterns on the substrate. Creating patterns on wafer means to fabricate the circuit components at the exact locations. For example, for a CMOS inverter circuit, the pMOS and nMOS will be fabricated physically. But to start the IC fabrication, we need the substrate or the wafer. In reality, the semiconductor material is not available in the elemental form. The elemental semiconductor needs to be extracted from its compounds. Then the crystalline form of the semiconductor needs to be grown. The techniques that grow the semiconductor crystals are known as crystal growth techniques. The grown crystal is often called the single crystal as it ensures that the growth of crystal follows a particular crystal plane.

16.2.1 Silicon Crystal Growth

Silicon (Si) and germanium (Ge) are the two elemental semiconductors which are of interest to the semiconductor device people because of their excellent properties. Although Ge was initially used to make semiconductor devices, eventually it was subdued by Si because of mainly the native oxide of Si, silicon dioxide

(SiO$_2$). More than 90% of semiconductor devices fabricated across the world, use Si, and that too is purely CMOS because of its low power and area requirements.

In this section, we discuss the single crystal Si growth technique. Si is available in nature abundantly in the form of sand (SiO$_2$). So, first the elemental Si is extracted from sand using a series of chemical reactions as follows:

1. $$SiC + SiO_2 \xrightarrow{\text{coal, coke, woodchips}} Si + SiO + CO \qquad (16.1)$$

Sand is taken with coal, coke, and woodchips in a furnace and is reacted at 1400°C to form metallurgical grade silicon (MGS). MGS is 98% pure, where impurities are in the order of parts per million (ppm).

2. $$Si + 3HCl \xrightarrow{300°C} SiHCl_3 \uparrow + H_2 \uparrow \qquad (16.2)$$

MGS is reacted with hydrogen chloride at 300°C to form trichlorosilane (SiHCl$_3$).

3. $$SiHCl_3 + H_2 \longrightarrow Si + 3HCl \qquad (16.3)$$

Next, fractional distillation of trichlorosilane is used to remove the impurities. It is then reacted with hydrogen gas at 1150°C (Otto, 1964) to reduce the hydrogen atom and produce electronic grade silicon (EGS). EGS has the highest purity and its impurity level is in the order of parts per billion (ppb).

The EGS that is obtained from the chemical process discussed above produces polycrystalline silicon which is not suitable for device fabrication. A single crystal is to be grown from EGS.

16.2.2 Czochralski Technique

In the Czochralski growth technique, the crystal growth takes place from the melt. The molten silicon is taken in a crucible. A seed crystal is placed just on top of the melt so that it touches the melt. Then the seed is pulled upwards slowly and rotated. The molten silicon is cooled slowly and solidifies underneath the seed in the desired crystal direction. The arrangement is shown in Fig. 16.1.

The grown crystal takes a cylindrical shape, which is known as an *ingot*. An ingot is shown in Fig. 16.2.

16.2.3 Dopant Distribution

During the process of crystal growth, dopants are introduced to make the grown crystal either *p*-type or *n*-type by adding boron or phosphorus into the melt, respectively. The doping concentration in the crystal and the melt differs from dopant to dopant. The ratio of dopant concentration in a solid (C_{solid}) to that in a liquid (C_{liquid}) is known as *segregation coefficient* or *distribution coefficient*,

$$k_{\text{d}} = \frac{C_{\text{solid}}}{C_{\text{liquid}}} \qquad (16.4)$$

If k_{d} is less than 1.0, then $C_{\text{liquid}} > C_{\text{solid}}$, which indicates dopants prefer to stay in the melt than in the solid. Hence, as the crystal grows, dopant concentration in the

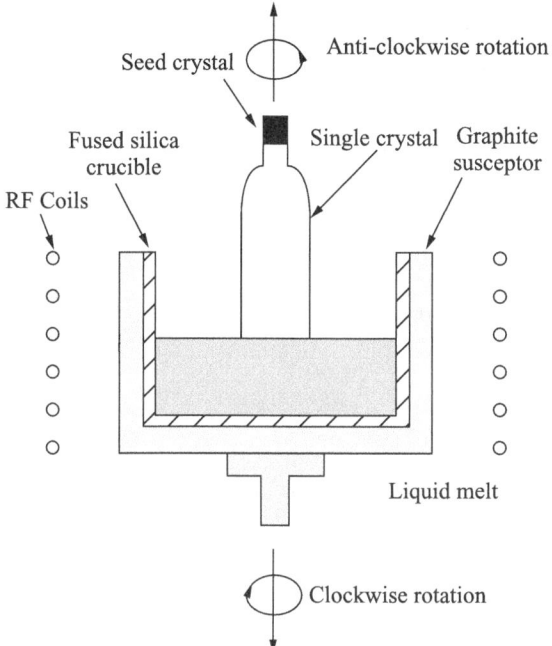

Fig. 16.1   Czochralski crystal growth technique   **Fig. 16.2**   Silicon ingot

melt increases. Table 16.1 lists segregation coefficients for some of the dopants for silicon.

Table 16.1   Typical segregation coefficients

Dopant	k_d	Type
Boron (B)	8×10^{-1}	P
Aluminium (Al)	2×10^{-3}	P
Gallium (Ga)	8×10^{-3}	P
Indium (In)	4×10^{-4}	P
Phosphorus (P)	3.5×10^{-1}	N
Arsenic (As)	3×10^{-1}	N
Antimony (Sb)	2.3×10^{-2}	N

Let us consider a crystal grown from its melt. The mass of the melt is m_0 and the initial doping concentration is C_0 in the melt (i.e., the amount of the dopant per gram of melt).

Assume that after a certain amount of crystal is grown, its mass is m_s. Hence, the remaining amount of melt is $(m_0 - m_s)$, and let us assume the amount of dopant remaining in the melt is m_d.

For a small amount of the crystal of mass dm_s, the reduction of the dopant $(-dm_s)$ from the melt is $C_{solid} \, dm_s$, where C_{solid} is the doping concentration in the crystal:

$$-dm_d = C_{solid} dm_s \tag{16.5}$$

Now, the mass of the melt left in the crucible is $(m_0 - m_s)$, and the doping concentration in the liquid, C_{liquid}, is given by

$$C_{liquid} = \frac{\text{Amount of dopant in melt}}{\text{Amount of melt}} = \frac{m_d}{m_0 - m_s} \tag{16.6}$$

Combining Eqns (16.5) and (16.6), and substituting $C_{solid}/C_{liquid} = k_d$ yields

$$\frac{dm_d}{m_d} = -k_d \left(\frac{dm_s}{m_0 - m_s} \right) \tag{16.7}$$

Initial condition: when $m_s = 0, m_d = C_0 m_0$
After some point: when $m_s = m_s, m_d = m_d$
Using these two limits, integrating Eqn (16.7), we get

$$\int_{C_0 m_0}^{m_d} \frac{dm_d}{m_d} = k_d \int_0^{m_s} \frac{-dm_s}{m_0 - m_s} \tag{16.8}$$

or

$$\ln m_d \Big|_{C_0 m_0}^{m_d} = k_d \ln(m_0 - m_s) \Big|_0^{m_s}$$

or

$$\ln \left(\frac{m_d}{C_0 m_0} \right) = k_d \ln \left(\frac{m_0 - m_s}{m_0} \right)$$

or

$$\frac{m_d}{C_0 m_0} = \left(\frac{m_0 - m_s}{m_0} \right)^{k_d}$$

or (substituting the value of m_d)

$$\frac{C_{liquid}(m_0 - m_s)}{C_0 m_0} = \left(\frac{m_0 - m_s}{m_0} \right)^{k_d}$$

or (substituting the value of C_{liquid})

$$\frac{C_{solid}}{k_d C_0} = \left(\frac{m_0 - m_s}{m_0} \right)^{k_d - 1}$$

or

$$C_{solid} = k_d C_0 \left(1 - \frac{m_s}{m_0} \right)^{k_d - 1} \tag{16.9}$$

where $\frac{m_s}{m_0}$ is called the *fraction solidified*.

As the crystal grows, the dopant concentration increases continually for $k_d < 1$, and decreases continually for $k_d > 1$. For $k_d \cong 1$, a uniform dopant distribution is obtained. Figure 16.3 shows the variation of C_{solid}/C_0 with respect to fraction solidified.

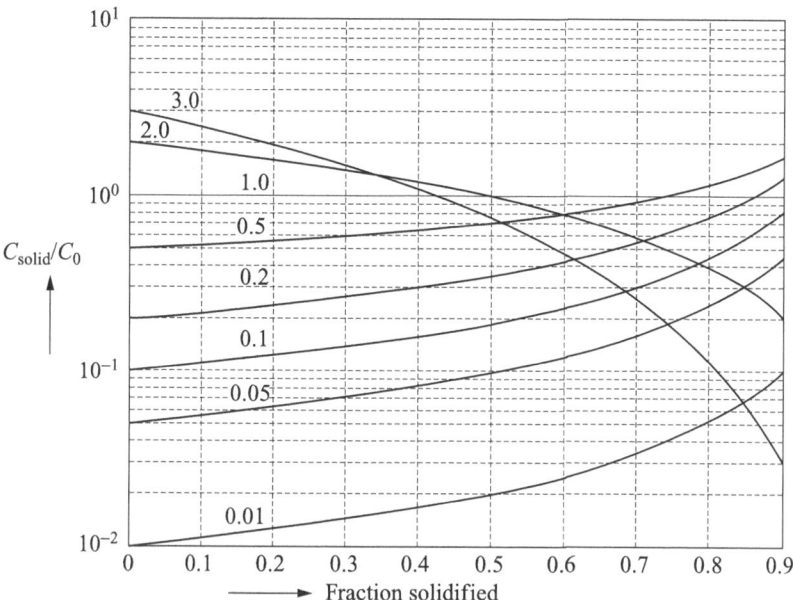

Fig. 16.3 Dopant distribution vs fraction solidified

Example 16.1   A Si crystal is grown from the melt to contain 10^{17} phosphorus atoms/cm^3. Assume $k_d = 0.35$, density of Si is 2.33 g/cm^3.
  (a) What should be the initial concentration of phosphorus atoms in the melt?
  (b) Assume the initial melt is of 6 kg. What is the amount of P to be added to the melt? Atomic weight of P is 31.

Solution

  (a) Initial concentration of P in the melt is

$$C_{liquid} = \frac{C_{solid}}{k_d} = \frac{10^{17}}{0.35} = 2.857 \times 10^{17}/cm^3$$

  (b) Volume of the melt $= \dfrac{\text{Mass}}{\text{Density}} = \dfrac{6000}{2.33} = 2.575 \times 10^3 \ cm^3$

Number of dopant atoms in the melt $=$ Volume of melt $\times$ Dopant concentration

$$= 2.857 \times 10^{17} \times 2.575 \times 10^3$$
$$= 7.357 \times 10^{20}$$

Therefore, the amount of $P = \dfrac{\text{Number of P atoms} \times \text{Atomic weight of P}}{\text{Avogadro's number}}$

$$= \frac{7.357 \times 10^{20} \times 31}{6.023 \times 10^{23}} = 0.0379 \text{ g}$$

16.2.4 Float-zone Technique

Float-zone technique is used to grow a single crystal silicon. The arrangement is shown in Fig. 16.4.

The polycrystalline silicon rod is kept vertical. A radio frequency (RF) coil is used to heat a small region and the region is melted. The impurities which prefer to stay in liquid than in solid come into the melt. As the molten zone is moved upward, it continues to become enriched with impurities and the crystal freezes from bottom with less or no impurity. The molten zone is moved along the length of the rod starting from the bottom seed and the crystal grows with the plane of the seed. This process achieves a high purity single crystal silicon.

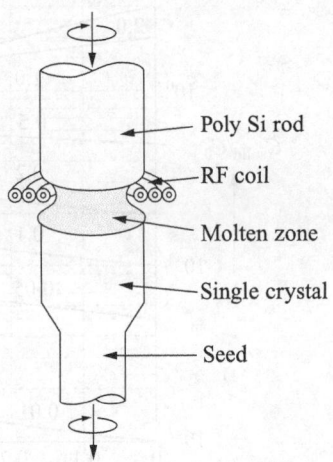

Fig. 16.4   Float-zone technique

16.2.5 Bridgman Technique

In this technique, the Ge and GaAs single crystal is grown. The molten Ge is taken in a crucible and a small seed crystal is placed at one end. The crucible is pulled slowly in the horizontal direction. As the molten zone moves out of the furnace, it is slowly cooled, and the Ge solidifies following the seed crystal. This technique is known as the horizontal Bridgman technique. The arrangement is shown in Fig. 16.5.

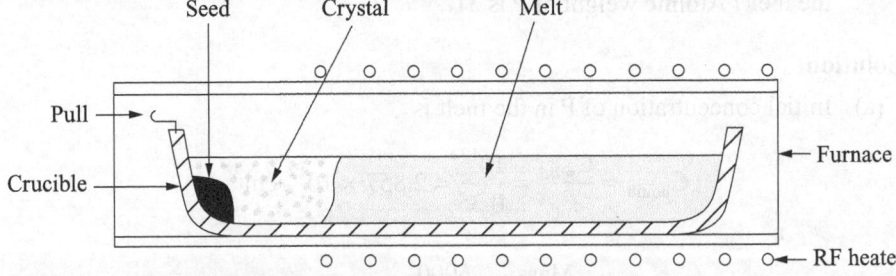

Fig. 16.5   Bridgman crystal growth technique

16.2.6 Wafer Preparation

An ingot is cut into thin slices which are known as wafers. Before the ingot is sliced, there are several steps. First, both the ends of the ingot are removed and its surface is ground so that any diameter variation is made constant. Then, the primary and secondary flat regions, as shown in Fig. 16.6 are cut along the entire length of the ingot. These flat regions are cut to identify the type (*p*-type or *n*-type) of the wafer and crystal orientation. The primary flat gives mechanical stability to the wafer when it undergoes further processing steps. The ingot is then sliced into wafers. A diamond tipped saw is used to slice the ingot. The sliced wafers are again lapped for better flatness. A mixture of glycerin and Al_2O_3 is used for the lapping purpose. The lapped wafers are finally polished once again to get a smooth surface so that the microelectronic devices can be fabricated using the photolithography technique. Typical wafer diameters are 5, 6, 8, and 12 inch. The wafer thickness is about 500–775 µm depending on the wafer diameter.

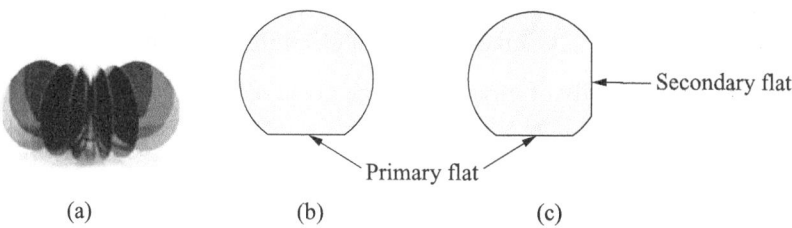

Fig. 16.6   Wafers: (a) with primary flat; (b) with both primary and secondary flats

16.2.7 Wafer Characterization

The electrical, optical, and mechanical characteristics of a crystal are strongly dependent on the crystal imperfections or defects. There are four different types of defects in a crystal:

- Point defects, such as
  - Substitutional
  - Interstitial
  - Vacancy
  - Frenkel
- Line defects or dislocations, such as
  - Edge
  - Screw
- Area defects, such as
  - Twins
  - Grain boundaries
- Volume defects

After the crystal growth, the wafers are cut and they are characterized for mobility, lifetime, and resistivity. Mobility is characterized using the Hall effect measurement. The lifetime is characterized using the 'Haynes–Shockley's experiment'. The resistivity is measured using a standard technique called four-probe technique.

16.2.8 Mobility

The mobility of a carrier in an electric field E, moving with drift velocity v_d is given by

$$\mu = \frac{v_d}{E}$$

(16.10)

The drift current density J is given by

$$J = nev_d = ne\mu E = \sigma E \qquad (16.11)$$

where n is the carrier concentration, e (=1.6×10^{-19}) the electronic charge, and σ is the conductivity. The resistivity is given by

$$\rho = \frac{1}{\sigma} = \frac{1}{ne\mu} \qquad (16.12)$$

Example 16.2 Calculate the resistivity of n-type Si having doped with 10^{17} phosphorous atoms/cm^3. Assume mobility of electron as 580 cm^2/V s.

Solution The resistivity of n-type Si for the given data is

$$\rho = \frac{1}{ne\mu} = \frac{1}{10^{17} \times 580 \times 1.6 \times 10^{-19}} = 0.10776 \ \Omega \ \text{cm}$$

16.2.9 Hall Effect

The carrier concentration n is measured by the Hall effect. The carrier concentration is given by

$$n = -\frac{I_x B_z}{et V_{AB}} \qquad (16.13)$$

where I_x is the applied current in the x-direction, B_z is the applied magnetic field in the z-direction, t is the thickness of the material, and V_{AB} is the voltage established due to the Hall effect. The Hall effect is illustrated in Fig. 16.7.

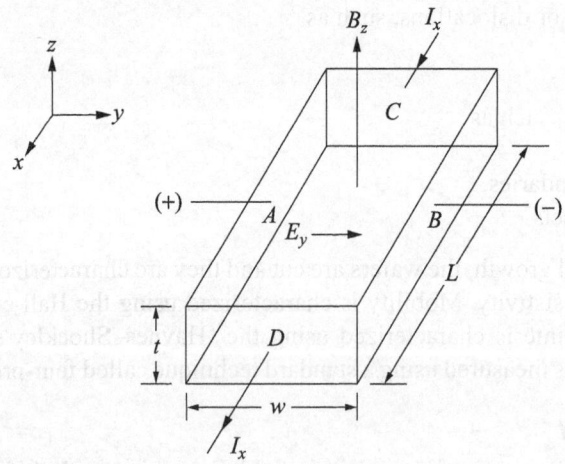

Fig. 16.7 The Hall effect

By measuring the Hall voltage V_{AB}, we can measure the carrier electron concentration according to Eqn (16.13).

The type of a sample, i.e., *n*-type or *p*-type is also determined by Hall's experiment by measuring the sign of the Hall voltage. For *p*-type material, the Hall voltage is positive, and for *n*-type material, the Hall voltage is negative.

The Hall coefficient R_H is given by

$$R_H = -\frac{1}{ne} \tag{16.14}$$

Combining Eqns (16.12) and (16.14), we obtain

$$\mu = \frac{R_H}{\rho} \tag{16.15}$$

Hence, the mobility of electrons can be measured by measuring the Hall coefficient and the resistivity of the material. For bulk semiconductor, Eqn (16.15) is used to measure the mobility. But for two-dimensional (2D) semiconductor (e.g., electrons in the inversion layer in a MOSFET), mobility of electrons is anisotropic. In the 2D systems, Haynes and Shockleys proposed methodology is used to directly measure mobility.

Example 16.3 A *p*-type Si sample with doping density 10^{16} atoms/cm^3 is using Hall experiment under the following conditions. Calculate the Hall voltage.

$$I_x = 2 \text{ mA}, \ B_z = 10^{-5} \text{ Weber/cm}^2, \text{ and thickness, } t = 100 \text{ μm}$$

Solution We calculate the Hall voltage developed in the sample as follows:

$$V = \frac{I_x B_z R_H}{t} = -\frac{I_x B_z}{e n_0 t} = -\frac{2 \times 10^{-3} \times 10^{-5}}{1.6 \times 10^{-19} \times 10^{16} \times 100 \times 10^{-4}} = -1.25 \times 10^{-3} \text{ V}$$

16.2.10 Haynes–Shockley Experiment

An electric field is applied along the length of the bar as shown in Fig. 16.8(a). Through the emitter contact, a pulse of holes (minority carrier) is injected into the bar. The hole pulse drifts towards the collector contact. While drifting, the pulse spreads out due to diffusion as shown in Fig. 16.8(b). At the collector end, the hole pulse is collected and observed using an oscilloscope.

If the hole pulse takes t amount of time to drift through the length d, we can write drift velocity as

$$v_d = \frac{d}{t} \tag{16.16}$$

Hence, mobility of the holes can be expressed as

$$\mu_p = \frac{v_d}{E} = \frac{d}{Et} = \frac{d}{(V/L)t} \tag{16.17}$$

where V is the applied sweep voltage and L is the length of the bar.

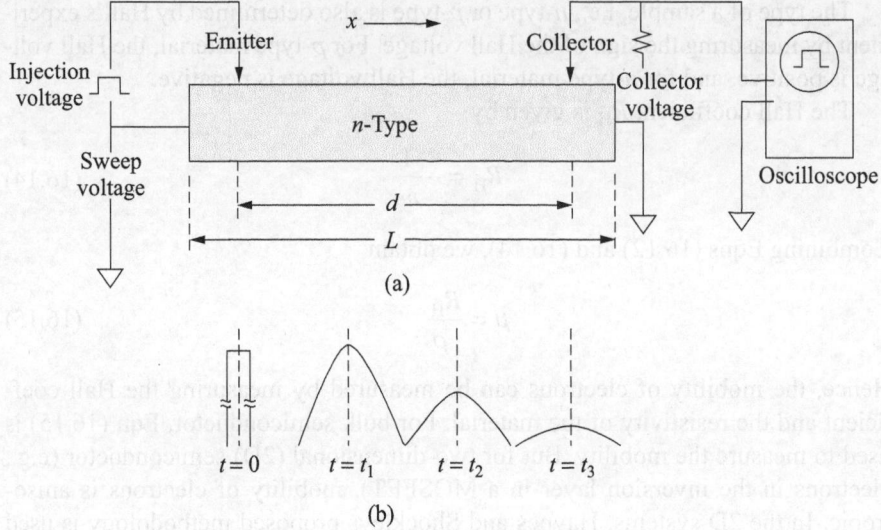

Fig. 16.8 (a) Experimental set-up of Haynes–Shockley experiment; (b) spreading of hole pulse

Measuring the drift time t, the mobility of the minority carrier (hole) can be measured using Eqn (16.17).

16.2.11 Lifetime Measurement

As shown in Fig. 16.8(b), it is observed that the pulse amplitude decreases and the pulse width increases as the pulse drifts towards the collector end. The amplitude decreases due to the recombination of holes with the majority carrier electrons present in the bar. The lifetime is defined as the time that a carrier survives before it recombines with the opposite polarity carrier. Hence, the pulse drift time can give a measure of the hole lifetime. The lifetime (τ_p) is given by

$$\mu_p = \frac{e\tau_p}{m_p^*} \tag{16.18}$$

where m_p^* is the effective mass of the hole. Using Eqn (16.18), the lifetime (τ_p) can also be measured knowing the mobility and effective mass of the carrier.

16.2.12 Resistivity of the Semiconductor

The resistance of a rectangular slab of semiconducting material, as shown in Fig. 16.9 is given by

$$R = \rho\frac{l}{A} = \rho\frac{l}{w \times t} = \frac{\rho}{t} \times \frac{l}{w} = \rho_s \times \frac{l}{w} \tag{16.19}$$

where ρ is the resistivity of the material and $\rho_s = \rho/t$ is the sheet resistance of the material. The unit of sheet resistance is ohm per square.

Four-probe Technique

Resistivity of a semiconducting material is measured by a technique called four-probe technique. The technique is depicted in Fig. 16.10.

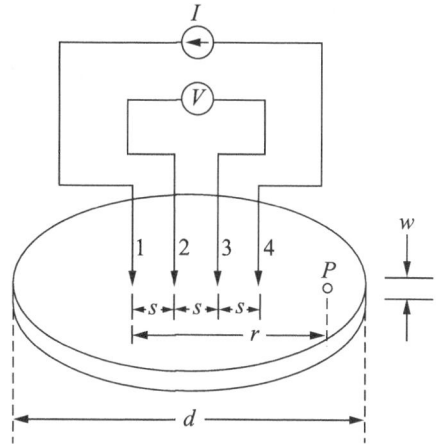

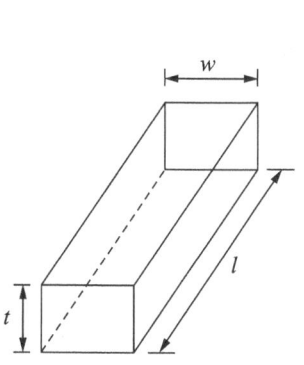

Fig. 16.9 A conductor of length *l*, width *w*, and thickness *t*

Fig. 16.10 Four-point probe technique to measure resistivity

In the four-probe technique, four-point probes are placed on the semiconducting substrate with a spacing *s*. A constant current source carrying a small current (*I*) is connected between the outer two probes. A voltmeter is connected between the inner two probes to measure the voltage. The resistivity of the sample can be obtained from the measured voltage and known current.

Let us consider a point *P* at a distance *r* from the first probe. The voltage at point *P* is given by

$$V = \frac{\rho I}{2\pi r} \tag{16.20}$$

where ρ is the resistivity of the semiconducting material. With respect to zero reference potential, the voltage measured is given by

$$V_0 = \frac{\rho I}{2\pi} \left(\frac{1}{r_1} - \frac{1}{r_4} \right) \tag{16.21}$$

where r_1 and r_4 are the distances from probes 1 and 4, respectively.
The voltage at probe 2 is given by

$$V_2 = \frac{\rho I}{2\pi} \left(\frac{1}{s} - \frac{1}{2s} \right) \tag{16.22}$$

The voltage at probe 3 is given by

$$V_3 = \frac{\rho I}{2\pi} \left(\frac{1}{2s} - \frac{1}{s} \right) \tag{16.23}$$

The measured voltage (*V*) between the probes 2 and 3 is given by

$$V = V_2 - V_3 = \frac{\rho I}{2\pi} \times \frac{1}{s} \tag{16.24}$$

Hence, the resistivity ρ is given by

$$\rho = 2\pi \times s \left(\frac{V}{I}\right) \tag{16.25}$$

The above derivation is valid for semi-infinite dimension. As the wafers are finite in both horizontal and vertical dimensions, Eqn (16.25) needs to be corrected. By introducing a correction factor (CF), we can rewrite Eqn (16.25) as

$$\rho = 2\pi \times s \times \text{CF} \times \left(\frac{V}{I}\right) \tag{16.26}$$

Typically CF = 4.532 for $d/s > 20$.

Example 16.4 In a four-probe technique, resistivity of n-type Si is measured with the following parameters. Find out the resistivity of the material.
Probe spacing, s = 0.5 mm
Probe radius = 30 μm
V = 10 mV
I = 0.4 mA

Solution Resistivity can be written from Eqn (16.26),

$$\rho = 2\pi s \times \text{CF} \times \left(\frac{V}{I}\right) = 2\pi \times 0.5 \times 10^{-3} \times \left(\frac{10 \times 10^{-3}}{0.4 \times 10^{-3}}\right) = 0.356 \quad \Omega \quad \text{m}$$

16.3 Photolithography

Photolithography is the technology to create a pattern on the silicon wafer using an ultraviolet (UV) ray of light. The steps are shown in Fig. 16.11.

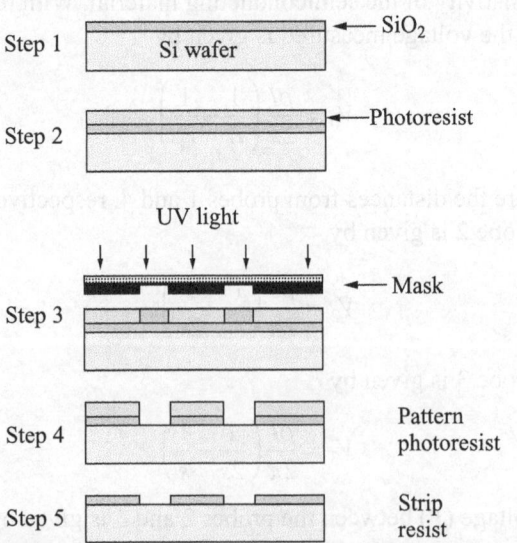

Fig. 16.11 Photolithography process: creating patterns on the photoresist, pattern transfer from the photoresist to the SiO$_2$ layer by etching (see Plate 7)

The components of the photolithography process are as follows:

- Si wafer
- Photoresist—a light-sensitive material
- Lens
- Mask (for each layer to be patterned) with the desired pattern
- UV light source and method of projecting the image of the mask onto the photoresist
- Developer solution—a method of 'developing' the photoresist, i.e., selectively removing it from the regions where it was exposed

The wafer is first cleaned and silicon dioxide (SiO_2) layer is deposited on the surface of the wafer. The wafer is then coated with the photoresist on the top. Then the UV light is projected on the wafer through the mask and a lens. The mask has certain regions transparent and other regions opaque. The transparent regions of the mask allow the UV light to pass through and fall on the photoresist. Depending on whether the photoresist is positive or negative, it undergoes some chemical changes and becomes more soluble or less soluble in an etchant solution. A pattern is formed on the photoresist. For positive photoresist, the pattern is the same as the mask, and for negative photoresist, the pattern is the inverse of the mask.

The wafer is dipped into a developer solution. The soluble part of the photoresist and the underneath SiO_2 layer are etched out. The photoresist is then stripped off and the replica of mask is formed on the SiO_2 layer.

16.3.1 Exposure Techniques

There are mainly two types of exposure techniques:

- Shadow printing
  - Contact printing
  - Proximity printing
- Projection printing

In the shadow printing, the wafer and the mask are either in direct contact or separated by a small gap. But in projection printing, the wafer and the mask are separated by a large distance, typically few centimetres away. Shadow printing is again of two types: *contact printing* and *proximity printing*. In contact printing, the mask and the wafer are in direct contact, whereas in proximity printing the mask and the wafer are separated by a small gap. The exposure mechanisms are illustrated in Fig. 16.12.

In shadow printing, the *minimum feature size* that can be patterned depends on the wavelength (λ) of the light used for exposure and the gap (g) between the mask and wafer. The minimum feature size is called *critical dimension* (CD) and is given by

$$CD = \sqrt{\lambda g} \tag{16.27}$$

For projection printing, the resolution of the projection system is given by

$$l_m = k_1 \frac{\lambda}{NA} \tag{16.28}$$

where k_1 is a constant (typically $k_1 = 0.6$) that depends on the process, and NA is the numerical aperture of the lens.

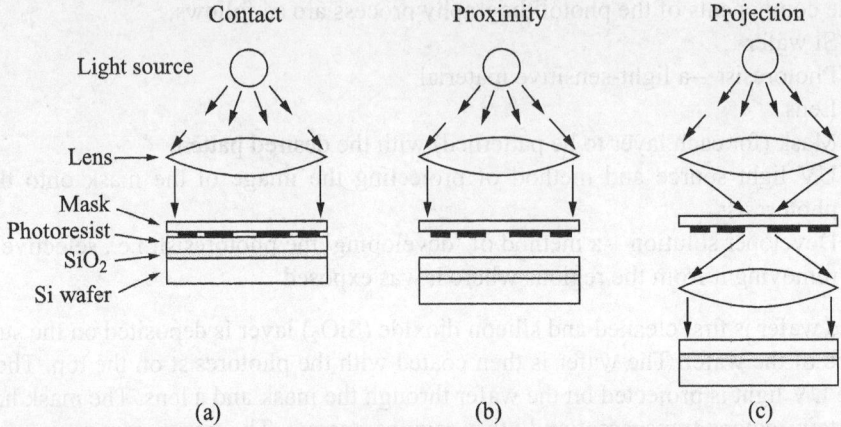

Fig. 16.12 Schematic diagram of (a) contact printing; (b) proximity printing; and (c) projection printing

The depth of focus (DOF) is given by

$$DOF = k_2 \frac{\lambda}{(NA)^2} \qquad (16.29)$$

where k_2 is another constant ($0.5 < k_2 < 1.0$) that depends on the process.

Figure 16.13 illustrates the numerical aperture of a lens and a projection imaging system

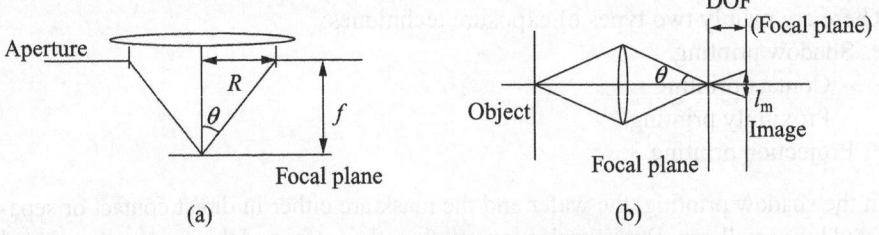

Fig. 16.13 (a) Numerical aperture of a lens; (b) projection image system

16.3.2 Comparison between Different Exposure Systems

- Limitations of contact printing are reduced mask life, high defect density in the printed pattern, and introduction of defects in the resist and the mask.
- In proximity printing, a small gap between mask and photoresist eliminates defects as there is no direct contact. But the disadvantages are increased diffraction effects and hence, a limit on the minimum feature size that can be printed.
- Projection printing is commonly used. The wafer is separated from the mask by several centimetres.
- Advantages of projection printing are an increased mask life and the features on the reticle do not have to be as small as the final image.

16.3.3 Clean Room

The IC fabrication requires a specially designed clean environment which is generally called a *clean room*. As the device dimension is very small, any tiny dust particle can cause disruption in the device processing. Therefore, a clean room environment is a must for IC fabrication. A clean room is defined by class: one in the British system and another in the metric system. In the British system, a class 10 clean room indicates there are 10 particles/ft^3 with particle diameter of 0.5 μm or larger. Similarly, a class M 2.544 clean room, in the metric system, indicates $10^{2.544} = 350$ particles/m^3 with particle diameter of 0.5 μm or larger.

Table 16.2 shows the typical particle size and number for the different classes of a clean room.

Table 16.2   For different classes of a clean room; number of particles/ft^3

Class	0.1 μm	0.2 μm	0.3 μm	0.5 μm	5 μm
1	3.5×10	7.5	3.0	1.0	
10	3.5×10^2	7.5×10	3.0×10	1.0×10^1	
100		7.5×10^2	3.0×10^2	1.0×10^2	
1000				1.0×10^3	7.0
10000				1.0×10^4	7.0×10

16.3.4 Mask

The mask is the photocopy of the layout generated at the end of the physical design step in the VLSI design flow. The layout information is transferred onto the mask using a pattern generator. The mask is made of fused silica covered with a chromium layer.

16.3.5 Resolution Enhancement Techniques

As the VLSI process technology advances, the device and interconnect dimensions scale down. In the sub-micron and nanometer regime, the printability and process window of the finer lithographic patterns are significantly reduced due to the fundamental limit of the micro/nano lithography systems and process variations. Till now, the 193 nm lithography systems are used to print sub-wavelength feature size (e.g., 65 nm or even 45 nm); with the aid of various resolution enhancement techniques (RET), as given by

- Optical proximity correction (OPC)
- Phase shift mask (PSM)
- Off-axis illumination (OAI)
- Sub-resolution assist feature (SRAF) insertion

These techniques modify illuminations, mask patterns, or transmissions.

16.3.6 Photoresist

It is a photosensitive polymer that is used to create patterns by the use of solvents after irradiation. It is of two types: positive photoresist (PPR) and negative photoresist (NPR).

Positive Photoresist

On exposure to light it becomes more soluble. It consists of a resin and a photoactive compound dissolved in an organic solvent. The unexposed regions are insoluble in the developer solution. Upon exposure, the chemical structure is changed and it becomes more soluble in the developer solution. The exposed regions are removed.

Negative Photoresist

On exposure to light it becomes less soluble. It consists of a chemically inert poly-isoprene rubber and a photosensitive compound. Its disadvantage is that the exposed resist has low molecular weight, so it swells as the unexposed part is dissolved in, developer solvent. The swelling distorts the pattern features which limits the resolution. Advantages are: (a) resistance to etching and (b) good adhesion to the substrate.

16.3.7 Pattern Generation

The mask is the design created using Layout Editor where the user specifies layout objects on different layers. The layout information is stored in a file called layout file. Generally, the GDSII format is used for layout file.

The pattern generator reads the layout file, and generates an enlarged master image of each mask layer, and the image printed on glass. The step and repeat camera is used for this purpose. It reduces the image and copies the image onto the mask, one copy for each die on a wafer. While overlapping masks of different layers, the alignment of masks is of great concern.

The most common layers used in the layout are given in Table 16.3.

Table 16.3 Mask layers

Sl. No.	Layer	Colour	Purpose	Note
1.	Metal1	Blue	First metal layer	
2.	Poly	Red	Poly-Si	
3.	Active	Green		
4.	n-select	Hashed blue		
5.	p-select	Dotted red		
6.	n-diff	Green	n-diffusion	Combination of active, n-select
7.	n-transistor	Green/red crosshatch	nMOS	Combined poly, n-diff
8.	p-diff	Brown	p-diffusion	Combination of active, p-select
9.	p-transistor	Brown/red crosshatch	pMOS	Combined poly, p-diff
10.	Polycontact	Black	polymetal contact	
11.	Active contact	Black	Metal—semiconductor contact	
12.	n-well	Hatched black	n-well	

16.3.8 Optical Proximity Correction

The small shapes that are nearly equal to the resolution limit of the photolithography system are modified when it is transferred. For example, a small square hole becomes a small circle. This is illustrated in Fig. 16.14.

To avoid this problem, the layout is modified near the bends and corners. The corners are over-shaped and the bends are under-shaped.

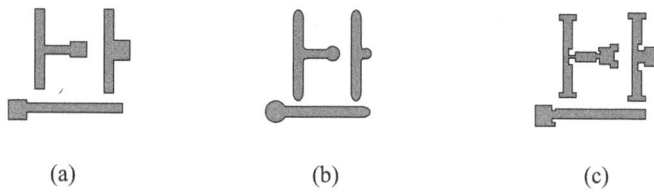

(a)   (b)   (c)

Fig. 16.14   (a) Layout without optical proximity correction (OPC); (b) image pattern of layout without OPC that is transferred; (c) layout with OPC

16.4 CMOS Technology: *n*-well and *p*-well Process

The CMOS process technology has the capability of fabricating both nMOS and pMOS on the same wafer. The substrate region of an MOS device is oppositely doped to that of the channel type. An nMOS device requires *p*-type substrate whereas a pMOS device requires *n*-type substrate. To accommodate both types of MOS devices on a single substrate (either *p*-type or *n*-type) special regions are created which are known as *wells*. If the substrate is of *p*-type then *n*-type well is created which is called *n-well*. If the substrate is of *n*-type then *p*-type well is created which is called *p-well*. The cross-sectional view of a CMOS inverter is shown in Fig. 16.15. It illustrates how an nMOS transistor is created on the *p*-type substrate and *n*-well is created on the *p*-type substrate to create pMOS transistor.

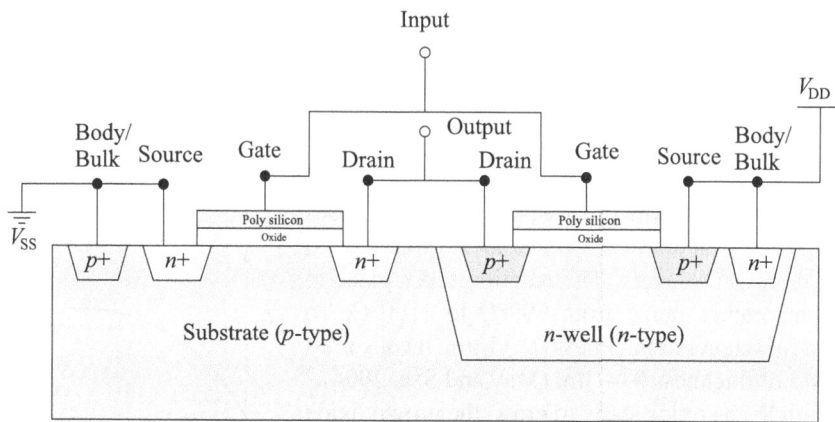

Fig. 16.15   Cross-sectional view of CMOS inverter

16.5 Oxidation

Silicon is the most popular semiconductor for IC fabrication because of its native oxide SiO_2. SiO_2 has excellent insulating properties. A very good quality oxide can be grown on the Si substrate. SiO_2 has a number of uses in the IC fabrication process. Some of its important uses are as follows:

- SiO_2 acts as a protecting buffer layer during device fabrication.
- It is used for device isolation.
- It is used in the MOSFET device (gate oxide).
- It is used for interconnected isolation (field oxide, FOX).

Oxidation is the process of growing the SiO_2 layer on top of the Si substrate. Two types of oxidation techniques are generally used:

■ Thermal growth (has excellent quality)
  - Dry oxidation
  - Wet oxidation
■ Chemical vapour deposition (required to put SiO_2 on materials other than Si)

The oxidation process starts from the top surface of the Si wafer and it slowly penetrates into the wafer. The separation line between the Si substrate and the grown SiO_2 layer is known as the Si–SiO_2 interface. The quality of oxide and the Si–SiO_2 interface greatly influences the behaviour of the MOSFET device. A layer called SiO_x $(0 < x < 2)$ at the interface between Si and SiO_2 is formed due to unsaturated Si bonds. These unsaturated bonds are called dangling bonds, and they behave like traps at the interface. The traps have energy states within the bandgap energy of Si, which are known as interface states.

16.5.1 Dry Oxidation

The chemical reaction that governs the dry oxidation is expressed as

$$Si + O_2\uparrow \rightarrow SiO_2 \tag{16.30}$$

The reaction takes place in the temperature range 900°C–1200°C.

16.5.2 Wet Oxidation

In the wet oxidation technique, water vapour is reacted with the Si layer and the chemical reaction is expressed as

$$Si + 2H_2O\uparrow \rightarrow SiO_2 + 2H_2\uparrow \tag{16.31}$$

In the thermal growth process, O_2 or H_2O diffuses through SiO_2, and reacts with Si at the interface to form more SiO_2 layers. The reaction takes place in the temperature range from 700°C to 1100°C. To form a SiO_2 layer of thickness (t_{ox}) 1 μm, it consumes Si layer of thickness 0.44 μm (May and Sze, 2004).

Initially, as oxide starts to grow, the grown oxide thickness (t_{ox}) linearly increases with time (t). After a certain thickness is grown, the growth becomes parabolic. This is illustrated in Fig. 16.16.

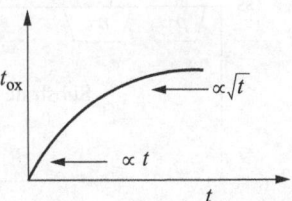

Fig. 16.16 Oxide thickness as a function of time

16.5.3 Chemical Vapour Deposition

Chemical vapour deposition (CVD) is another technique by which the SiO_2 layer can be grown. But it does not produce high quality oxide as in the thermal oxidation technique. The oxide grown using CVD technique is utilized for metal level isolation, as a mask during photolithography.
Temperature range:

■ 300°C–500°C for SiH_4 (silane)
■ 500°C–800°C for TEOS (tetraethylorthosilicate)

Process:
- Precursor gases dissociate at the wafer surface to form SiO_2
- No Si on the wafer surface is consumed
- Film thickness is controlled by the deposition time

The chemical reactions are given by

$$Si(C_2H_5O)_4 + 2H_2O \xrightarrow{700°C} SiO_2 + 4C_2H_6O \qquad (16.32)$$

$$SiH_4 + O_2 \xrightarrow{450°C} SiO_2 + 2H_2 \qquad (16.33)$$

In the CVD technique, oxide thickness grows linearly with time as indicated in Fig. 16.17.

16.5.4 Thin Oxide

An oxide having thickness 20 nm or lesser is called *thin oxide*. Thin oxides are used as a gate oxide layer in the MOSFET structure. Thin oxides are grown using the dry oxidation technique, as dry oxidation produces a very good quality oxide. But the dry oxidation process is very slow compared to the wet oxidation technique.

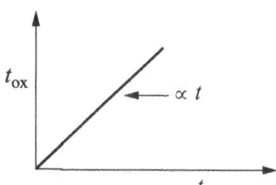

Fig. 16.17 Oxide thickness as a function of time in the CVD process

16.5.5 Thick Oxide

An oxide having thickness 20 nm or more is called *thick oxide*. Thick oxides are grown using the wet oxidation technique. It is fast as compared to the dry oxidation technique, but the quality of oxide is not as good as that produced in the dry oxidation technique. Thick oxides are used as field oxide.

16.6 Diffusion

The introduction of doping atoms into the semiconductor can be done by two techniques: (a) diffusion and (b) ion implantation. Diffusion is a process of displacement of particles or atoms from a high concentration region to a low concentration region. Using the diffusion process, dopants are introduced into the wafer. An ambient is created of the dopant atoms at the surface of the wafer and the temperature is raised to 800°C–1200°C.

For *n*-type material, arsenic (As) and phosphorus (P) are used as dopants, and for *p*-type material, boron (B) is used as a dopant.

16.6.1 Fick's Law

The basic diffusion mechanism is expressed by using the Fick's law of diffusion. It is given by

$$\frac{\partial C}{\partial t} = D \frac{\partial^2 C}{\partial x^2} \qquad (16.34)$$

where D is the diffusion coefficient and C is the dopant concentration.

16.6.2 Diffusion Profile

The diffusion profile is a function of the initial and boundary conditions of doping. Based on the boundary conditions, the diffusion process is classified into two processes:

- Constant surface concentration
- Constant total dopant

In the constant surface concentration diffusion process, the dopant supply is maintained so that dopant concentration at the surface of the substrate is kept constant, whereas in the constant total dopant diffusion process, the total number of dopant atoms is fixed. Applying the boundary conditions, for these cases to the Fick's diffusion law yields two solutions, and two profiles of dopants distribution. For the constant surface concentration diffusion, the profile follows an error function. For the constant total dopant diffusion, the profile follows the Gaussian distribution. The diffusion depth is dependent on the time and temperature used for diffusion.

Readers are suggested to refer May and Sze (2004) for details.

16.7 Ion Implantation

Ion implantation is another technique for introduction of impurities into the surface of the Si wafer or substrate. Individual dopant atoms are first ionized and they are accelerated. The accelerated ions impinge on the surface and penetrate into the substrate.

Advantages of ion implantation:

- Very precise control of impurity numbers is possible
- Low processing temperature
- Short processing time
- Low penetration depth of the implanted ions
- Implantation through thin layers (e.g. SiO_2, Si_3N_4) is possible.

16.7.1 Ion Stopping

When the projected ions travel inside the substrate, they undergo several collisions with the electrons and the nucleus of the host atoms. During the collision process, they lose their energy and finally come to rest. This mechanism is called *ion stopping*.

16.7.2 Ion Channelling

When the projected ions enter into the substrate, they get aligned with the gap between the host atoms, and they travel a large distance before finally coming to rest. This phenomenon is called *ion channelling*. Due to the ion channelling effect, the ions become uncontrollable and hence, the junction depth goes beyond control. There are several mechanisms to avoid the ion channelling effect. These are given as follows:

- Using an amorphous blocking layer
- Disorienting the wafer
- Creating a damaged layer at the wafer surface

16.7.3 Implantation Damage and Annealing

In the ion implantation process, when the highly energetic ions enter into the lattice, they suffer several collisions with the host atoms. Due to the collision mechanism, the host atoms are displaced from their original positions. If the number of displaced atoms per unit volume becomes comparable to the atomic density of the material, it loses its crystalline property and becomes amorphous material. This fact is known as *lattice damage*.

To rectify the damage caused by ion implantation, a methodology is used which is known as *annealing*. Annealing heals most of the damage. In this process, the wafer is treated alternately in very high and low temperature for a long time.

16.8 Etching

Etching is the process of selective removal of some portions of a layer. The photolithography technique is used to create patterns on the Si wafer. It involves the etching process which removes the selected portion of a layer. Etching is of two types:
- Isotropic—etch rate is same in all directions
- Anisotropic—etch rate is not same in all directions

Figure 16.18 illustrates the anisotropic and isotropic etching process.

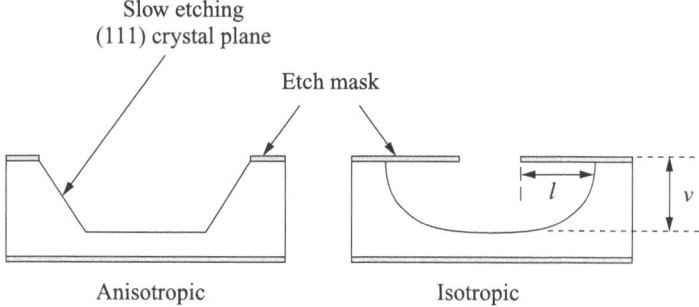

Slow etching
(111) crystal plane

Etch mask

Anisotropic   Isotropic

Fig. 16.18 Anisotropic and isotropic etching

The degree of anisotropy is defined as

$$A = 1 - \frac{l}{v} \tag{16.35}$$

where l is the lateral etch distance and v is the vertical etch distance.

16.8.1 Wet Chemical Etching

An etchant is a chemical solution that reacts with the layer that needs to be etched out. Depending on the material to be etched out, the chemical solution differs. The etchant solution is either sprayed on the wafer, or the wafer is dipped into the etchant solution. This type of etching is known as *wet chemical etching*. Wet chemical etching is isotropic in nature. In this process, the etch rate is the same in all directions.

An example of anisotropic etching is the etching of <111> crystal plane which results in V-shaped sidewalls when etched a hole in a <10 0> Si wafer in potassium hydroxide (KOH) solution. This is illustrated in Fig. 16.18.

16.8.2 Dry Etching

There are three types of dry etching process: (a) reactive ion etching (RIE), (b) sputter etching, and (c) vapour phase etching. Dry etching process is expensive compared to wet etching. Dry etching is used when vertical side walls are required, or when thin films with small feature resolution are to be obtained. Dry etching is also known as plasma etching. Plasma is a mixture of ionized and unionized gas molecules. It is produced by applying a very high electric field to a gas.

Reactive Ion Etching

In reactive ion etching (RIE), the substrate is placed inside a reactor in which several gases are injected. Plasma is created in the gas mixture using an RF power source, breaking the gas molecules into ions. The plasma ions move towards the surface and react with the material being etched to form another gaseous material. This is known as the chemical process of reactive ion etching. There is another part which is called the physical part, which is similar in nature to the sputtering deposition process. If the ions have high energy, they can knock out atoms of the material to be etched without a chemical reaction. The chemical process is isotropic, and the physical process is anisotropic in nature. Combining the physical and chemical processes, it is possible to control the anisotropy of the etching to form sidewalls that have shapes from rounded to vertical. A schematic of a reactive ion etching system is shown in Fig. 16.19.

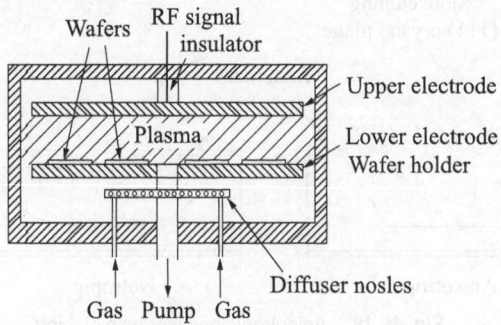

Fig. 16.19   Plasma etching

Sputter Etching

Sputter etching is basically RIE without reactive ions. The process is very similar to the sputtering deposition process. The only difference is that substrate is now subjected to the ion bombardment instead of the material target used in sputter deposition.

16.9 Epitaxial Growth

In this growth technique, thin layers of material can be grown on a substrate. A material with similar lattice structure other than the substrate material can also be grown using this technique. As thin layers are grown on a substrate wafer, this technique is known as *epitaxial growth* or simple epitaxy.

There are several methods of epitaxial growth as follows:

- Chemical vapour deposition (CVD)
- Liquid phase epitaxy (LPE)

- Physical vapour deposition (PVD)
- Molecular beam epitaxy (MBE)

For epitaxial growth, the important requirement is that the lattice constant of the substrate layer and the lattice constant of the layer to be grown should match. For example, the lattice constant of GaAs and AlAs are ~5.65 Å. Hence, the epitaxial layer of Al_xGa_{1-x} As can be grown on the GaAs substrate for any composition x ($0 \leq x \leq 1.0$).

Advanced epitaxial growth is used to grow very thin layers of lattice with slightly mismatched crystals. The grown layer is formed with a substrate lattice constant and hence, the atoms in the grown layer are either compressed or elongated. So they are called *strained layer*. Such a structure is called strained-layer super-lattice (SLS).

16.9.1 Chemical Vapour Deposition

Chemical vapour deposition (CVD) is also called vapour phase epitaxy (VPE). In this process, the chemical compounds of the material to be grown is reacted with other gas molecules and the released material is deposited on the substrate. For example, for Si epitaxial growth, four different compounds of Si can be used in the CVD process. They are silane (SiH_4), trichlorosilane ($SiHCl_3$), dichlorosilane (SiH_2Cl_2), and silicon tetrachloride ($SiCl_4$). A typical reaction is expressed as

$$SiCl_4\uparrow + 2H_2\uparrow \xleftrightarrow{1200°C} Si + 4HCl\uparrow \qquad (16.36)$$

Diborane (B_2H_6) is used for p-type dopant, and phosphine (PH_3) and arsine (AsH_3) are used for n-type dopant.

Advantages of CVD are as follows:
- Uniform step coverage
- Precise control of the composition and structure
- Fast deposition rate
- High throughput
- Low processing cost

16.9.2 Metalorganic CVD (MOCVD)

Metalorganic CVD (MOCVD) or organometalic CVD (OMCVD) is a technique to grow epitaxial layers from metalorganic compounds. For example, GaAs is grown from its chemical compound trimethylgallium [$(CH_3)_3Ga$].

$$(CH_3)_3Ga + AsH_3 \xrightarrow{700°C} GaAs + 3CH_4 \ [\qquad (16.37)$$

Trimethylgallium (TMGa) reacts with Arsine (AsH_3), and produces GaAs and methane. The epitaxial layer of GaAs is achieved of very high quality. Similarly, diethylzinc [DEZn, $(C_2H_5)_2Zn$] and diethylcadmium [DECd, $(C_2H_5)_2Cd$] are used to introduce for p-type dopants in GaAs, and trimethylaluminium [TMAl, $(CH_3)_3Al$] is used along with TMGa to grow AlGaAs.

Several thin layers of different material can be grown using this technique. The arrangement of the MOCVD system is shown in Fig. 16.20. MOCVD is used to grow hetero-structured and nano-structured devices.

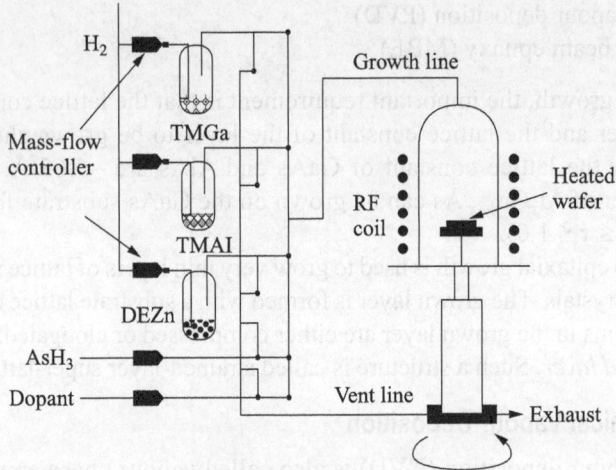

Fig. 16.20 MOCVD system

16.9.3 Molecular Beam Epitaxy

Molecular beam epitaxy (MBE) is a very useful technique to grow epitaxial layer on a substrate. It can be used to grow epitaxial layers composed of different materials. It is done in an evacuated chamber at very low pressure. The doping profile and the chemical compositions of the epitaxial layer can be precisely controlled. It is a very popular technique to grow hetero-structure and nano-structured devices. Figure 16.21 illustrates the arrangement of the MBE system.

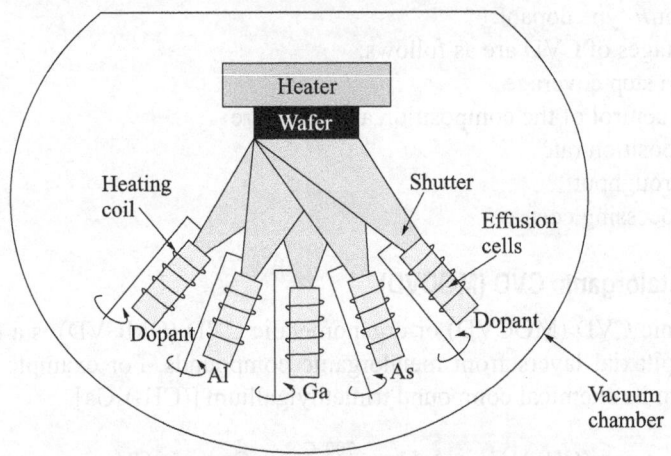

Fig. 16.21 MBE system

16.9.4 Physical Vapour Deposition

One of the PVD techniques is known as *sputtering*. It is used to deposit metal (Al) films required for metallization. Highly energetic Argon (Ar) ions hit the surface of a metal target, knocking atoms loose, which then land on the surface of the wafer. Sometimes the substrate is heated to ~300°C. Gas pressure is kept between 1 to 10 mTorr. The deposition rate is proportional to the ion current (I) and the sputtering yield (S). Figure 16.22 illustrates the sputtering technique.

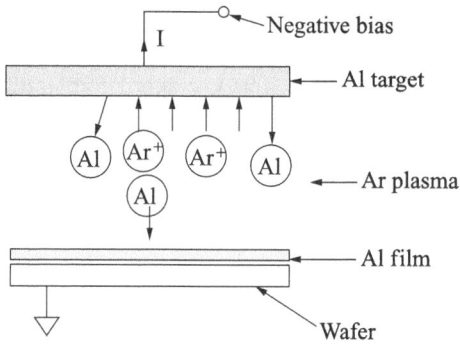

Fig. 16.22 PVD technique

16.10 Metallization

After completion of all the device fabrication, the next important step is the metallization process. In this process, the contacts between different layers and the interconnect layers are formed. Metallization is done using the CVD and PVD techniques. The photolithography technique is used to create patterns of interconnect layers.

The gate electrode of the MOSFET is often fabricated using poly-Si which is the polycrystalline form of Si. The CVD technique is used to grow poly-Si layers (see Fig. 16.23).

$$SiH_4 \xrightarrow{\;600°C\;} Si + 2H_2 \uparrow \tag{16.38}$$

The wafer is heated to ~600°C and Si-containing gas silane (SiH_4) is injected into the furnace.

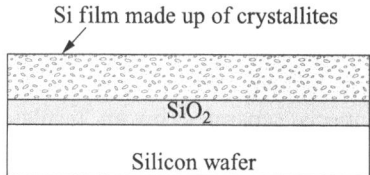

Fig. 16.23 Poly-Si deposition using the CVD technique

Poly-Si is used for gate electrode of MOSFET because of the following advantages:
- Easy to deposit poly-Si layer on SiO_2 layer as they have similar lattice constants.
- Similar thermal expansion coefficient gives rise to similar thermal expansion and hence, better mechanical stability.

Metal is used for interconnects because of high conductivity. Aluminium (Al) and copper (Cu) are mostly used materials for interconnects. Poly-Si is used for short interconnects. In VLSI circuits, interconnects cannot be completed in one or two metal levels. They use multilevel metallization, as shown in Fig. 16.24.

Metallization of interconnects imposes the following considerations for deep submicron technologies:
- Must not exceed maximum current density to be limited to avoid electromigration problem.

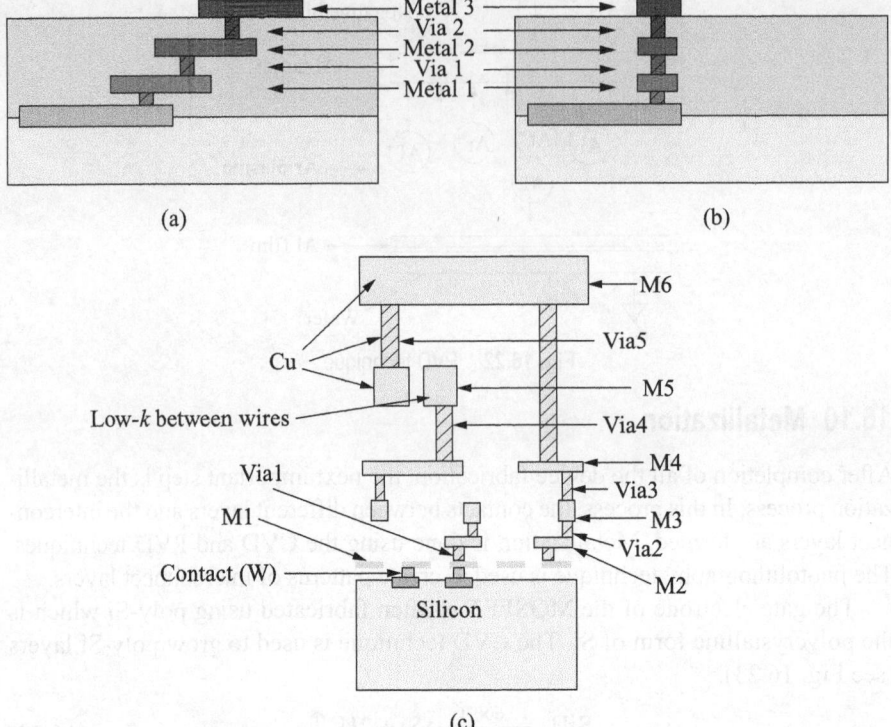

Fig. 16.24 Multilevel metallization: (a) 3-level metal connection to n-active without stacked vias;
(b) 3-level metal connection to n-active with stacked vias;
(c) 6-level metal interconnect (see Plate 8)

- However small, the interconnect has some resistance, giving rise to ohmic drop or IR drop which must be managed.
- Two metal interconnects separated by an insulator or dielectric introduce parasitic capacitances, which must be managed to avoid the crosstalk problem.
- Interconnects from high to low level metals require connections to each level of metal through vertical interconnects called via.
- Stacked vias are permissible in some processes.
- Silicides are used to reduce resistance, however, silicides are not used when interconnects are used as resistors.

16.11 Packaging

Packaging is the last step of the IC manufacturing process. The VLSI circuit (all the devices and their interconnections) is fabricated on a die. The die is packaged in a suitable compound material. A package provides
- Protection to the die from the environmental damage
- Interface to the outside world, i.e., system
- Power to the internal circuits
- Cooling of heat generated due to power dissipation

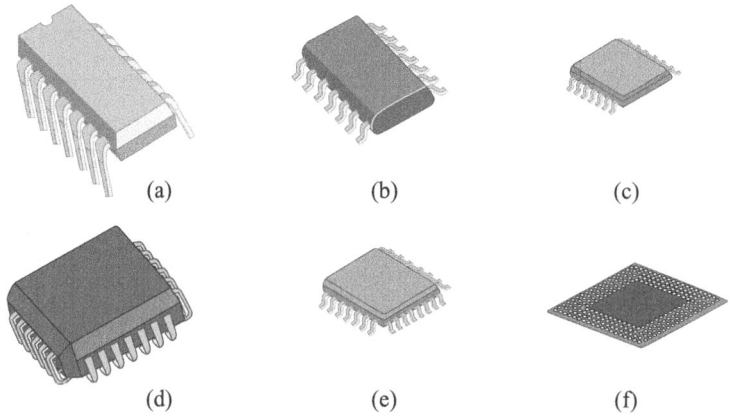

Fig. 16.25 IC packages: (a) PDIP—plastic dual in-line; (b) SOIC—small outline integrated circuit; (c) TSSOP—thin shrink small outline package; (d) PLCC—plastic lead chip carrier; (e) TQFP—thin quad flat package; (f) BGA—ball grid array

Figure 16.25 shows different IC package types.

Packaging involves two phases: (i) package design and (ii) package modelling. These are discussed in detail in Chapter 13.

16.11.1 Die Separation

ICs are manufactured in a batch. In a single wafer there are a number of dies. Each die has the complete VLSI circuit fabricated on it. After the die is fabricated, the dies are to be separated from each other. This is called *die separation*. Typically, a diamond tipped scriber is used to draw scribe lines along the boundary of the die. Modern technologies use a diamond saw to separate the dies from the wafer. The dies are then packaged.

16.11.2 Package Types

Depending on the material used for packaging and the structure of package pins, the IC packages are classified as follows:

- Pin-through-hole (PTH)—dual in-line package (DIP)
- Surface mount technology (SMT)—quad flat pack (QFP)
- Pin grid array (PGA)
- Ball grid array (BGA)
- Flip-chip

Figure 16.26 describes the package classifications in detail.

Most common IC packaging materials are plastic, ceramic, laminates (fiberglass, epoxy resin), metal, etc.

Chips are often 'pad-limited'. Chip area increases as the square of the number of pads. Hence, package design is the equally important as the circuit design.

The bond pads of the die are connected to the package pins by three well-defined methods as follows:

- Wire bonding
- Tape-automated bonding (TAB)
- Flip-chip bonding

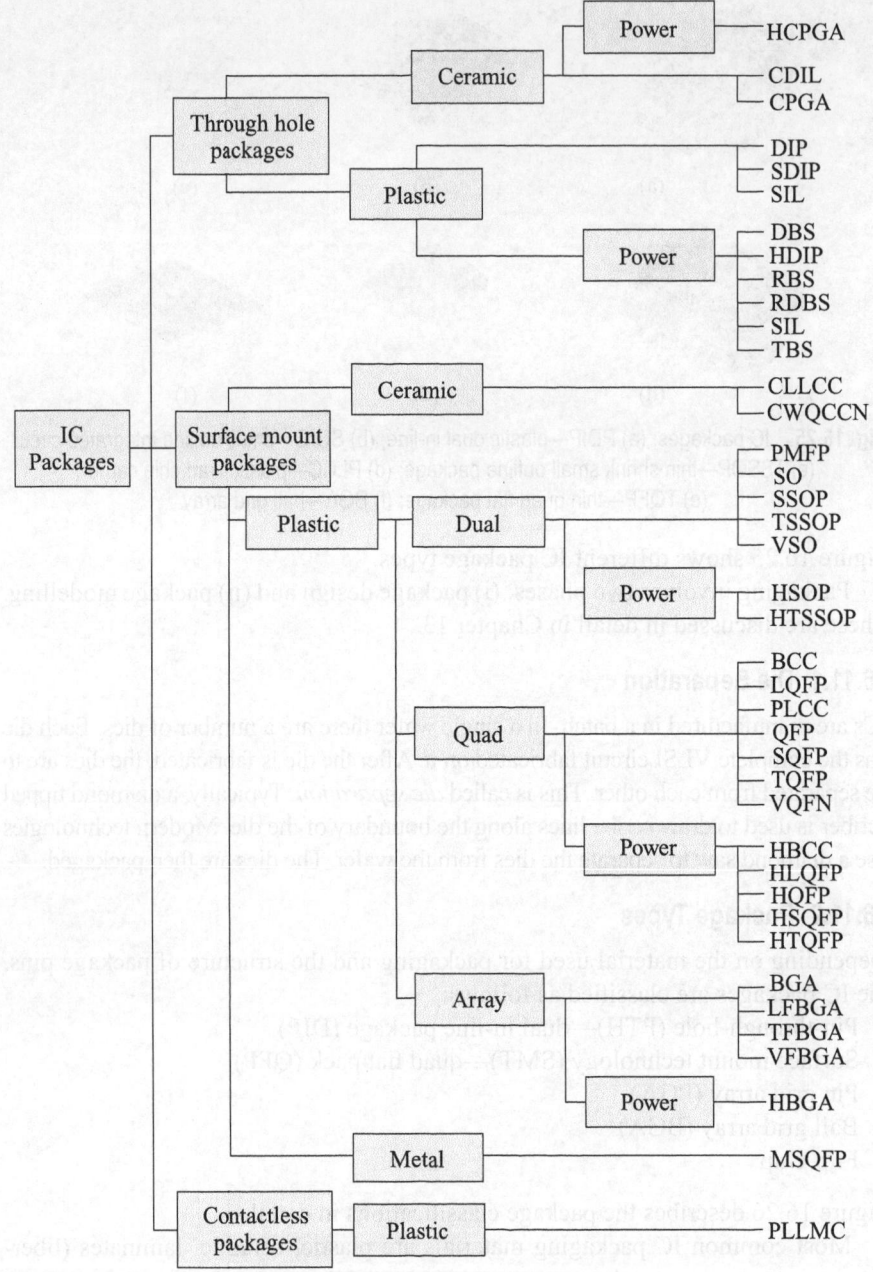

Fig. 16.26 IC package classifications

Figure 16.27 illustrates different bonding techniques.

16.11.3 Wire Bonding

In the wire bonding technology, a gold or aluminium wire called a bond wire is used to connect bond pads and the package pins.

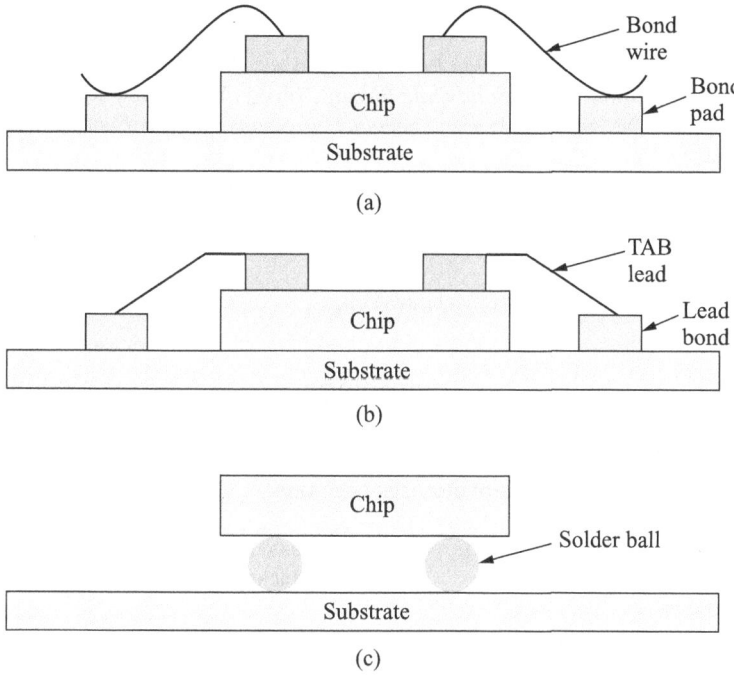

Fig. 16.27 Schematic of (a) wire bonding; (b) tape automated bonding; (c) flip-chip bonding (see Plate 8)

16.11.4 Tape-automated Bonding

In this case, all the bonding is done in one step. Hence, the throughput is large. A flexible polymer containing an interconnection pattern is used to mount the chip first. Then the bond pads are connected to the interconnection stripes using thermo compression technique.

16.11.5 Flip-chip Bonding

In this method, the chips are flipped on to the bumps of the package. No wire is used for bonding. The bumps are located anywhere in the chip. Hence, the package interconnect distance is reduced. While flipping the chip, the bond pads must be aligned to the bumps. In flip-chip bonding, throughput is large compared to wire bonding.

SUMMARY

- Silicon is used for fabricating almost 95% of the semiconductor devices worldwide.
- Electronic-grade silicon is the purest form of silicon and is used to prepare a single crystal silicon.
- Photolithography technique is used to create a pattern on the silicon substrate.
- Silicon-based IC fabrication technology is popular because of its native oxide, SiO_2.
- Diffusion and ion implantation are two techniques used to introduce dopants in a semiconductor substrate.
- MBE and MOCVD are two key technologies that can be used to create very low dimensional devices.

SELECT REFERENCES

Gandhi, S.K. 1994, *VLSI Fabrication Principles*, 2nd ed., Wiley, New York.

Kittel, C. 1996, *Introduction to Solid State Physics*, 7th ed., Wiley, New Jersey.

May, G.S. and S.M. Sze 2004, *Fundamentals of Semiconductor Fabrication*, Wiley, New Jersey.

Otto, S. et al. 1964, *Pyrolytic Method for Precipitating Silicon Semiconductor Material*, US Patent 3,120,451, February 1964.

Streetman, B.G. 1995, *Solid State Electronics Devices*, 3rd ed., Prentice-Hall of India, New Delhi.

Sze, S.M. 1988, *VLSI Technology*, McGraw-Hill.

EXERCISES

Fill in the Blanks

1. Upon exposure, the positive photoresist becomes _____ soluble in the developer solution.
   (a) more
   (b) less
2. Higher level metal layers have _____ thickness compared to lower level metal layers.
   (a) larger
   (b) equal
   (c) smaller
   (d) all of these
3. Unit of sheet resistance is _____ .
   (a) ohm/square
   (b) ohm
   (c) ohm m
   (d) ohm/m
4. Four probe technique is used to measure _____ .
   (a) resistivity of semiconducting material
   (b) mobility of carriers
   (c) carrier concentration
   (d) all of these
5. In photolithography, higher the radiation wavelength _____ .
   (a) smaller is the minimum feature size
   (b) larger is the minimum feature size
   (c) feature size is independent of it
   (d) none of these

Multiple Choice Questions

1. In metallurgical grade semiconductor, the impurity level is in the range of
   (a) ppm (parts per million)
   (b) ppb (parts per billion)
   (c) ppt (parts per trillion)
   (d) none of these
2. If the value of segregation coefficient is greater than one, it signifies that the
   (a) dopant concentration is more in solid than liquid
   (b) dopant concentration is less in solid than liquid
   (c) dopant concentration is equal in solid and liquid
   (d) dopant concentration is zero in solid than liquid
3. Higher mobility indicates
   (a) larger resistivity
   (b) smaller resistivity
4. Hall effect can be used to measure
   (a) mobility of carriers
   (b) type of semiconductor
   (c) carrier concentration
   (d) all of these
5. Higher effective mass of the carriers indicates
   (a) smaller mobility
   (b) larger mobility

True or False

1. Contact printing has higher resolution than proximity and projection printing.
2. Oxide quality is good if it is grown by the dry oxidation technique, as compared to wet oxidation technique.
3. Bridgman technique is used to grow a single crystal silicon.
4. OPC is a process of modifying the bends and corners in a layout.
5. To form a SiO_2 layer of thickness (t_{ox}) x μm, it consumes Si layer of thickness $0.24x$ μm.

Short-answer Type Questions

1. What is the suitable material for the VLSI interconnect?
2. What is the natural source of Si?
3. Explain the fabrication of SiO_2 using the wet oxidation technique. Discuss its relative merits and demerits over the dry one.
4. Write Fick's law of diffusion for a 3D isotropic medium. Explain its significance in integrated circuit processing.
5. Compare wet etching and dry etching. What do you mean by anisotropic etching? What is loading effect?
6. What is meant by a clean room class? Describe briefly how you can achieve the desired clean room condition necessary for IC fabrication.
7. Describe the RCA cleaning steps commonly used for removing various contaminants of a silicon wafer.
8. What do you mean by isotropic and anisotropic etching in IC technology? Name the commonly used isotropic and anisotropic etchants for Si.
9. What are the uses of poly-Si?
10. What are differences between the thick film and thin film technology?
11. List the processes for fabrication of VLSI circuits.
12. What is meant by impurity profile? How is the profile controlled during fabrication?
13. How is metallization done in VLSI fabrication?
14. Write down Fick's equations for one dimension for diffusion.
15. Discuss the chemical vapour deposition process for VLSI fabrication.

Long-answer Type Questions

1. Write short notes on
   (a) ion implantation process
   (b) clean room
   (c) sputtering
   (d) pattern generation
   (e) phase shift mask
2. What are integrated circuit resistors? What are the ways by which you can fabricate IC resistors? Explain them in brief. What is sheet resistance?
3. What is meant by etching? What are the different types of etching? Discuss the plasma etching process. Why is etching technique preferred over lift-off technique? What is antenna effect?
4. What are the purposes of metallization in IC fabrication? What are the desired properties of the metallization process integrated circuits? What are the materials used for metallization?
5. Describe different metallization techniques. What is electromigration in VLSI circuits and how is it solved? Draw a multilevel metallization structure.
6. Describe how a nMOS device is fabricated. Use diagrams to show the steps.
7. Draw the device structure of CMOS inverter. Show the different steps to fabricate a CMOS inverter.
8. Discuss the benefits of integrated circuits over discrete circuits. What are the types of integrated circuits? What is Moore's law in VLSI?

9. What do you mean by 'clean room'? What are the criteria of a clean room? Compare crystal growing using Bridgman and CZ method with their pros and cons.

10. Why is Si preferred over Ge for IC fabrication? What are the different quality grades of Si? Describe how a single crystal silicon ingot is prepared using the CZ method.

11. (i) Discuss the various steps of wafer preparation.

   (ii) An Si crystal is to be grown by the CZ method, and it is desired that the ingot contain 10^{16} boron atoms/cm^3.

   (a) What concentration of boron atoms should the melt contain to give this impurity concentration in the crystal during the initial growth? For B in Si, $k_d = 0.8$.

   (b) If the initial load of Si in the crucible is 50 kg, how many gram of boron should be added? The atomic weight of boron is 10.8 and density of silicon is 2.53 g/cm^3.

12. Describe the Si oxidation mechanisms. What are the uses of SiO_2 in VLSI circuits? Classify the SiO_2 layer formation techniques and discuss them in brief.

13. What are the effects of impurities and damage on the oxidation rate? What do you mean by thick and thin oxides, and discuss their properties.

14. Discuss the different methods used to introduce impurities to the Si with their pros and cons.

15. What are the different types of lithography processes? A proximity printer operates with a 10 µm mask-wafer gap, and a wavelength of 430 nm. Another printer uses a 40 µm gap with wavelength 250 nm. Which offers higher resolution? Describe the mask generation process.

16. Why are clean air and ultra pure water essential in IC fabrication? How should the wafer be treated before using it for IC fabrication? Describe the systems to be maintained in a clean room identified for VLSI fabrication. How is wafer chemically cleaned before diffusion?

17. Describe with a neat diagram the Czochralski's method of crystal growth. Derive an expression for maximum pull rate.

18. (a) Determine the kinetics of oxide growth on Si using Deal and Grove's model, and derive an expression for the oxide thickness.

   (b) Show that to grow an oxide layer of thickness x, a thickness of $0.44x$ of Si is consumed (Molecular weight of Si and SiO_2 are 28.09 g/cm^3 and 2.21 g/cm^3).

19. Why is metallization needed in the final step of IC fabrication? What materials are suitable for this purpose?

20. (a) Describe the process of etching with diagram and explain the effects of 'over-etching'.

   (b) Highlight the chemical vapour deposition (CVD) process as applicable to VLSI technology.

   (c) Why is SiO_2 very useful in IC fabrication?

   (d) What is the effect of pressure on etching?

21. The conducting region in Fig. 16.28 of thickness $t = 0.1$ µm and resistivity $\rho = 0.1 \ \Omega$ cm, is deposited on an insulating substrate. $L = 1$ mm, $W = 100$ µm. Determine the resistance between contacts A and B.

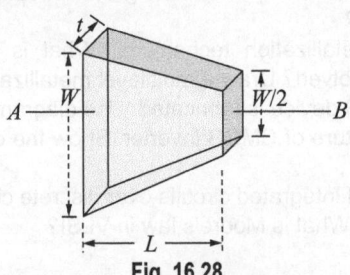

Fig. 16.28

Subsystem Design

17.1 Introduction

A digital system contains several functional blocks, which are known as subsystems. For example, an arithmetic logic unit (ALU) contains functional blocks, or subsystems such as adders, subtractors, multipliers, dividers, comparators, and shift registers. In this chapter, we shall learn how to design such subsystems.

17.2 Adders

Adders are the most fundamental subsystems of a digital circuit. A binary adder adds two binary inputs and produces the sum and carry as output. The simplest adder is a half-adder which adds two bits and produces two outputs: sum and carry. On the other hand, a full adder adds two bits along with one input carry. The process of adding two n-bit binary numbers is described in the following sub-section.

17.2.1 Carry Look-ahead Adder

A ripple carry adder is slow, as the final output requires all the full adder stages in order to generate the carry signals. If τ is the delay of each full adder block, it takes $n\tau$ units of time to generate the final output carry. On the other hand, a carry look-ahead adder (CLA) is a high-speed adder circuit that does not wait for each full adder to generate the carry signals. Instead, it finds the carry signal ahead by using a CLA generator circuit. This way, the CLA can add two n-bit binary numbers very fast.

Design Principle

The ith sum (S_i) and carry (C_i) of a full adder in the ripple-carry adder (RCA) are given by the following expressions:

$$S_i = A_i \oplus B_i \oplus C_{i-1} \tag{17.1}$$

$$C_i = A_i B_i + B_i C_{i-1} + A_i C_{i-1} \tag{17.2}$$

In order to generate the carry signal, we introduce two auxiliary signals, *generate* (G_i) and *propagate* (P_i) as defined by the following expressions:

$$G_i = A_i B_i \tag{17.3}$$

$$P_i = A_i + B_i \tag{17.4}$$

Now, Eqn (17.2) can be written as follows:

$$C_i = G_i + P_i C_{i-1} \tag{17.5}$$

or

$$C_{i-1} = G_{i-1} + P_{i-1} C_{i-2} \tag{17.6}$$

Substituting Eqn (17.6) in Eqn (17.5), we get the following expression:

$$C_i = G_i + P_i G_{i-1} + P_i P_{i-1} C_{i-2} \tag{17.7}$$

Therefore, for a 4-stage CLA, we have the following expressions:

$$C_0 = G_0 + P_0 C_{in} \quad \text{where } C_{in} \text{ is input carry} \tag{17.8}$$

$$C_1 = G_1 + P_1 C_0 = G_1 + P_1 G_0 + P_1 P_0 C_{in} \tag{17.9}$$

$$C_2 = G_2 + P_2 C_1 = G_2 + P_2 G_1 + P_2 P_1 G_0 + P_2 P_1 P_0 C_{in} \tag{17.10}$$

$$C_3 = G_3 + P_3 C_2 = G_3 + P_3 G_2 + P_3 P_2 G_1 + P_3 P_2 P_1 G_0 + P_3 P_2 P_1 P_0 C_{in} \tag{17.11}$$

Now, the sum in Eqn (17.1) can also be expressed using Eqns (17.3) and (17.4) as follows:

$$S_i = P_i \oplus G_i \oplus C_{i-1} \tag{17.12}$$

The gate level structure of a 4-bit CLA circuit is shown in Fig. 17.1. The sum and final carry bits are generated using four-level logic. If the average propagation delay of each logic level is τ, then the adder generates all the output bits after a time delay of 4τ, which is independent of the word size n.

17.2.2 Carry-save Adder

A carry-save adder (CSA) is very useful for adding more than two binary numbers. Hence, it is particularly useful in a multiplier circuit, where more than two partial products are added to obtain the final output.

Let us illustrate the concept of carry save addition with the help of the following example:

We add two binary numbers, $A = (20)_{10} = (10100)_2$ and $B = (5)_{10} = (00101)_2$.

The conventional method to add these two numbers is as follows:

Table 17.1

C5	C4	C3	C2	C1			←Carry
+	A4	A3	A2	A1	A0		←A
	B4	B3	B2	B1	B0		←B
	S4	S3	S2	S1	S0		←Sum

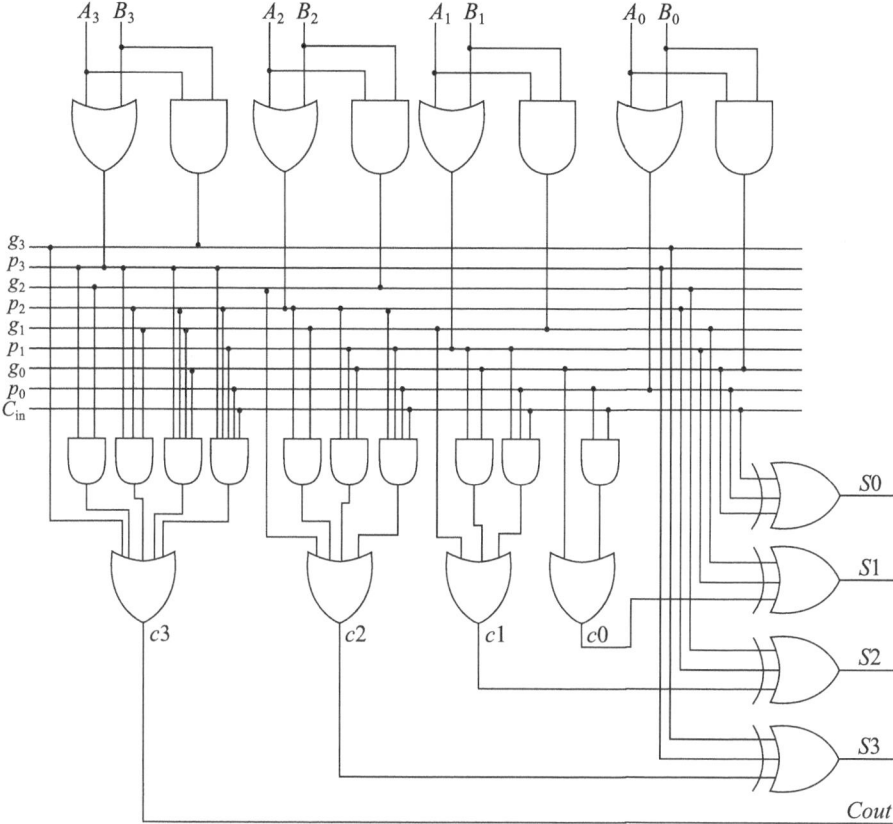

Fig. 17.1 A 4-bit CLA adder

We use this method to add the given numbers, as follows:

Table 17.2

0	0	1	0	0	←Carry
+ 1	0	1	0	0	←A (20)
0	0	1	0	1	←B (5)
1	1	0	0	1	←Sum
0 1	1	0	0	1	←Result (25)

The carry bits are added while adding the bits of A and B. However, it is also possible to add the carry bits to the sum bits, as shown in the following table:

Table 17.3

+ 1	0	1	0	0	←A (20)
0	0	1	0	1	←B (5)
+ 1	0	0	0	1	←Sum
0	0	1	0	0	←Carry
0 1	1	0	0	1	←Result (25)

The advantage of the second approach will be clear when we add more than two numbers, as shown in the following example:

We add three numbers, $A = (20)_{10} = (10100)_2$, $B = (5)_{10} = (00101)_2$, and $C = (15)_{10} = (01111)_2$.

Table 17.4

+	1	0	1	0	0	←A (20)
	0	0	1	0	1	←B (5)
	0	1	1	1	1	←C (15)
+		1	1	1	1	0	←Sum
	0	0	1	0	1		←Carry
	1	0	1	0	0	0	←Result (40)

We see that the final result can be obtained by adding the sum bits and carry bits using the ripple-carry addition technique.

Figure 17.2 illustrates the schematic of a 4-input CSA. There are four inputs, A, B, C, and D. The first three inputs, A, B, and C are added by the carry-save stage1, and then the carry-save stage2 adds the sum bits with the remaining input, D, along with the carry bits. The final result is obtained after the RCA adds the sum and the carry bits obtained through carry-save stage2.

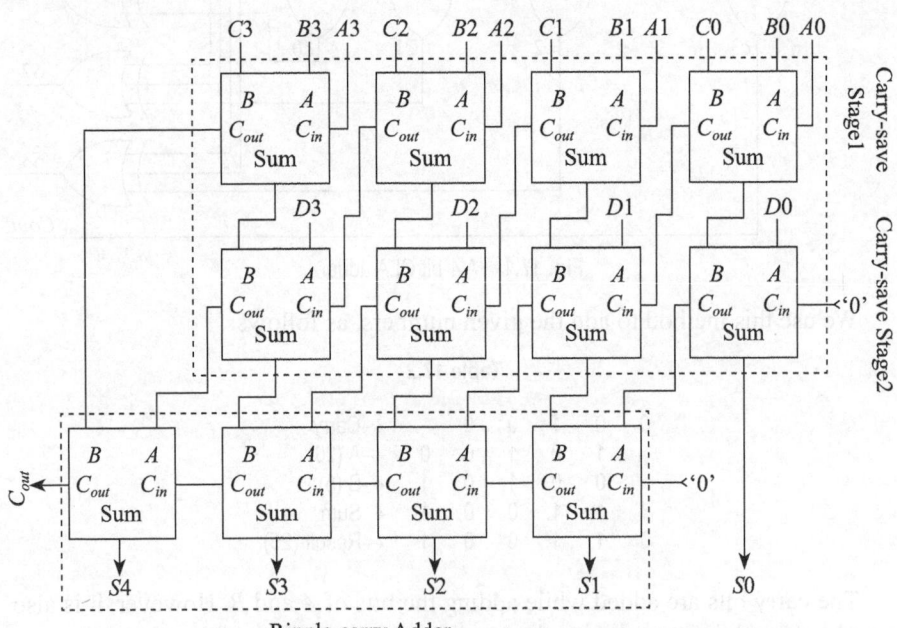

Fig. 17.2 Schematic of a 4-input CSA

A CSA is used to construct a tree multiplier which is known as the Wallace Tree multiplier.

17.2.3 Manchester Carry Chain Scheme

In the Manchester carry chain scheme, the carry is defined in terms of three control signals, generate (G), propagate (P), and kill (K). Table 17.5 illustrates truth table for a full-adder circuit:

Table 17.5   Truth table for a full-adder circuit

Inputs			Carry Output	Control Inputs		
C_{in}	A	B	C_{out}	$G = A \cdot B$	$P = A \oplus B$	$K = (A + B)'$
0	0	0	0	0	0	1
0	0	1	0	0	1	0
0	1	0	0	0	1	0
0	1	1	1	1	0	0
1	0	0	0	0	0	1
1	0	1	1	0	1	0
1	1	0	1	0	1	0
1	1	1	1	1	0	0

From the above truth table, we see that the control inputs, G and P are not active at the same time. Out of the eight possible input combinations, either G or P is active for six input combinations; and for the remaining two input combinations, an auxiliary control signal is generated which is termed as carry-kill (K).

The signal K is high when neither G nor P is active; and when K is high, the output carry (C_{out}) is low. Hence, the C_{out} can be defined using the three control signals, G, P, and K as follows:

- If $G = 1$, then $C_{out} = $ '1'
- If $P = 1$, then $C_{out} = C_{in}$
- If $K = 1$, the $C_{out} = $ '0'.

This logic can be implemented using switch-logic, as shown in Fig. 17.3.

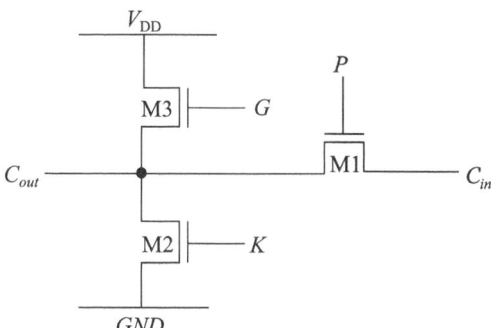

Fig. 17.3   Schematic for Manchester carry generator circuit

The signals, G, P, and K are active one at a time:

When signal P is active, the transistor M1 is ON and the input (C_{in}) passes to the output (C_{out}), i.e., $C_{out} = C_{in}$.

When signal G is active, the transistor M3 is ON and the output is pulled to logic high, i.e., $C_{out} = $ '1'.

When signal K is active, the transistor M2 is ON and the output is pulled to logic low, i.e., $C_{out} = $ '0'.

This logic can be implemented using static CMOS logic and without using the control signal K, as illustrated in Table 17.6. Note that the signals P and G are not active simultaneously.

Table 17.6

Inputs			Carry Output	Control Inputs			C'_{in}	C'_{out}	Control Logic
C_{in}	A	B	C_{out}	$G = A \cdot B$	$P = A \oplus B$	$K = (A+B)'$			
0	0	0	0	0	0	1	1	1	If $P = 0$, $C'_{out} = G'$
0	0	1	0	0	1	0	1	1	If $P = 1$,
0	1	0	0	0	1	0	1	1	$C'_{out} = C'_{in}$
0	1	1	1	1	0	0	1	0	If $P = 0$,
1	0	0	0	0	0	1	0	1	$C'_{out} = G'$
1	0	1	1	0	1	0	0	0	If $P = 1$,
1	1	0	1	0	1	0	0	0	$C'_{out} = C'_{in}$
1	1	1	1	1	0	0	0	0	If $P = 0$, $C'_{out} = G'$

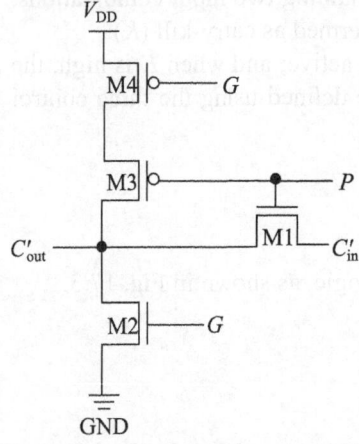

Fig. 17.4 Static CMOS realization of Manchester carry chain generation circuit

From the truth table, we see the following:

■ When $P = 1$, $C'_{out} = C'_{in}$
■ When $P = 0$, $C'_{out} = G'$

The Manchester carry chain generator circuit using the static CMOS logic is shown in Fig. 17.4.

The following is a description of the working of the circuit:

When the signal P is active, the transistor M1 is ON. Therefore, the input signal (C'_{in}) passes through the transistor M1 to the output (C'_{out}). When this happens, the other paths to the output signal are broken, as both the transistors M3 (as $P = 1$) and M2 (as $G = 0$) are OFF.

When the signal G is active, the transistor M1 is OFF and the transistor M3 is ON (as the signal P is active low). Now, when the signal G is low, the transistor M4 is ON and the output is high through M4. On the other hand, when the signal G is high, the transistor M2 is ON and the output is low through M2.

17.3 Multipliers

Multipliers are the most useful building blocks after the adders in digital signal processing or arithmetic computing systems. A multiplier has two binary inputs—one is called the *multiplicand* and the other is called the *multiplier*. It has one binary output, which is the *product* of the multiplicand and the multiplier.

Let A and B be two 4-bit numbers; their product can be written as follows:

$$P = \sum_{i=0}^{C'_{in}} a_i 2^i \times \sum_{j=0}^{3} b_j 2^j \qquad (17.13)$$

The 4×4 bit multiplication is illustrated in Fig. 17.5.

	a3	a2	a1	a0	
×	b3	b2	b1	b0	

= A (Multiplicand)
= B (Multiplicand)

			a3b0	a2b0	a1b0	a0b0	
		a3b1	a2b1	a1b1	a0b1		
	a3b2	a2b2	a1b1	a0b2			
a3b3	a2b3	a1b3	a0b3				

Partial
Product

p7	p6	p5	p4	p3	p2	p1	p0	= P (Product)

Fig. 17.5   Example of 4×4 bit multiplication

In general, the multiplier can be either unsigned type or signed type. An un-signed multiplier takes two binary inputs in unsigned format and outputs the product in the same format. The signed multiplier takes two binary inputs in two's complement format and outputs the result in the same format.

17.3.1   Array Multiplier

The design of an unsigned array multiplier is based on the simple pen and paper method of multiplication—taking one bit at a time from the multiplier, the partial products are generated by multiplying each bit to the multiplicand. These partial products are then written in rows, by shifting each row by one bit position to the right. These rows are then added column-wise. This multiplier has a very regular architecture and is most suitable for VLSI implementation. Figure 17.6 illustrates an unsigned array multiplier.

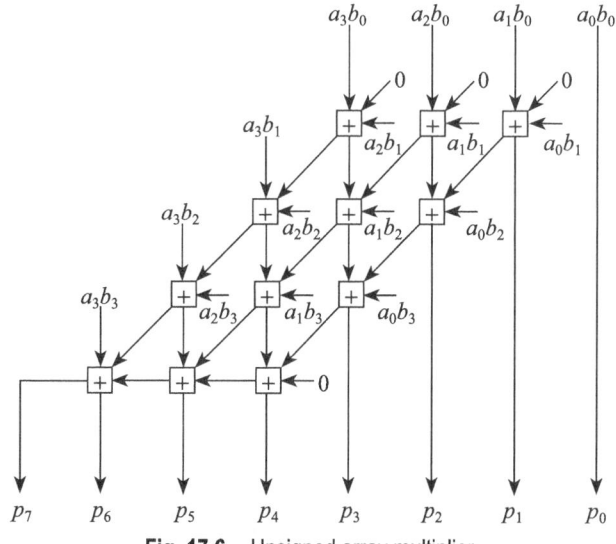

Fig. 17.6   Unsigned array multiplier

17.3.2 Signed Binary Multiplier

A signed binary number has the most significant bit (MSB) reserved as the sign bit. If the sign bit is '1', then the number is a negative number; otherwise, it is a positive number. A signed binary number having N number of bits can be represented as follows:

$$X = -2^{N-1}x_{N-1} + \sum_{k=0}^{N-2} 2^k x_k \qquad (17.14)$$

For a negative number, the MSB is one, i.e., $x_{N-1} = 1$ and for a positive number, the MSB is zero.

An N-bit signed number can also be represented in its two's complement form, as follows:

$$X = -2^N + X_1 \qquad (17.15)$$

where X_1 is the N-bit unsigned representation of the number X.

The 4-bit representation of binary numbers in both signed and two's complement format is shown in Table 17.7.

Table 17.7 Binary representation of 4-bit signed number

Decimal (X)	Signed Binary (two's complement)	Unsigned	X_1	2^n
7	0111	7		
6	0110	6		
5	0101	5		
4	0100	4		
3	0011	3		
2	0010	2		
1	0001	1		
0	0000	0		
−1	1111	15	15	16
−2	1110	14	14	16
−3	1101	13	13	16
−4	1100	12	12	16
−5	1011	11	11	16
−6	1010	10	10	16
−7	1001	9	9	16
−8	1000	8	8	16

Let us consider two N-bit signed numbers X and Y, which are represented as follows:

$$X = -2^{N-1}x_{N-1} + \sum_{k=0}^{N-2} 2^k x_k \qquad (17.16)$$

$$Y = -2^{N-1}y_{N-1} + \sum_{k=0}^{N-2} 2^k y_k \qquad (17.17)$$

The product ($P = X \times Y$) can be written as follows:

$$P = \left(-2^{N-1} x_{N-1} + \sum_{k=0}^{N-2} 2^k x_k \right) \left(-2^{N-1} y_{N-1} + \sum_{m=0}^{N-2} 2^m y_m \right)$$

$$\text{or } P = 2^{2N-2} x_{N-1} y_{N-1} + \sum_{k=0}^{N-2} x_k y_m 2^{k+m} - 2^{N-1} \sum_{k=0}^{N-2} x_k y_{N-1} 2^k - 2^{N-1} \sum_{m=0}^{N-2} x_{N-1} y_m 2^m$$

$$(17.18)$$

C.R. Baugh and B.A. Wooley proposed an architecture—similar to that of an unsigned array multiplier—for the signed multiplier, which is known as the Baugh–Wooley multiplier. The process of multiplication used by this multiplier is illustrated in Fig. 17.7.

				a_3	a_2	a_1	a_0
				b_3	b_2	b_1	b_0
			1	$\overline{pp_{30}}$	pp_{20}	pp_{10}	pp_{00}
			$\overline{pp_{31}}$	pp_{21}	pp_{11}	pp_{01}	
		$\overline{pp_{32}}$	pp_{22}	pp_{12}	pp_{02}		
	pp_{33}	$\overline{pp_{23}}$	$\overline{pp_{13}}$	$\overline{pp_{03}}$			
$\overline{P7}$	$P6$	$P5$	$P4$	$P3$	$P2$	$P1$	$P0$

Fig. 17.7 Illustration of 4-bit Baugh–Wooley multiplication process

Let us assume that the two signed 4-bit numbers are A ($= a_3 a_2 a_1 a_0$) and B ($= b_3 b_2 b_1 b_0$) and their product is P. The MSB of the first three partial product rows and all bits of the last partial product row except the MSB are inverted; '1' is added in the fourth column and the MSB of the final result is inverted.

The architecture of the Baugh–Wooley multiplier is shown in Fig. 17.8. As mentioned earlier, the architecture is very similar to that of an unsigned array multiplier with additional seven inverters.

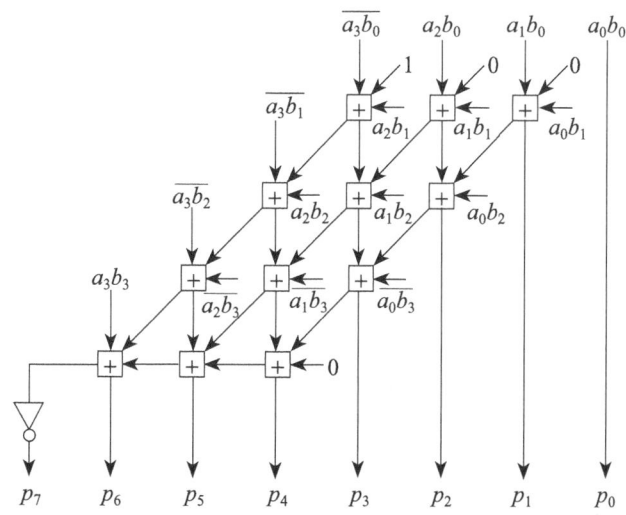

Fig. 17.8 Signed 4-bit Baugh–Wooley multiplier

17.3.3 Serial-parallel Multiplier

The serial-parallel method of multiplication follows the addition of partial products column-wise, as illustrated in Fig. 17.9. Each column is added in a one-clock cycle. Hence, for N-bit inputs, the serial-parallel multiplier requires $2N$ number of clock cycles to complete the operation.

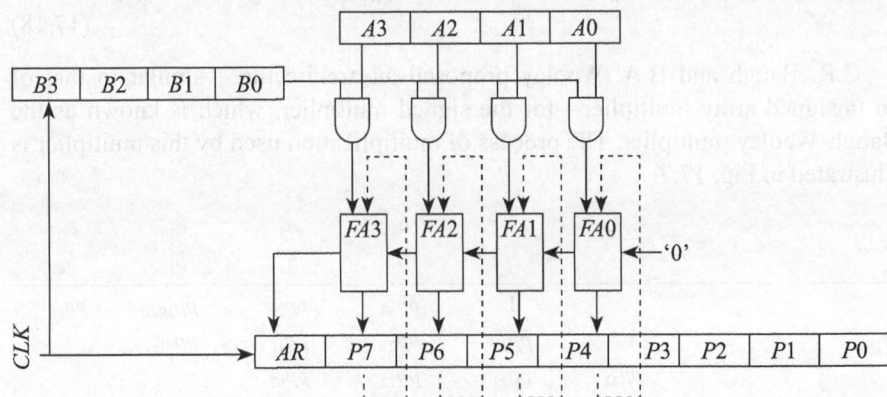

Fig. 17.9 Serial-parallel multiplier

The multiplicand is loaded into the register B and the multiplier is loaded into the register A. Table 17.8 illustrates the multiplication process using the serial-parallel method of multiplication for two numbers A ($= 13_{10} = 1101_2$) and B ($= 11_{10} = 1011_2$) that are loaded into the registers A and B.

Table 17.8

Clock Cycle		P7	P6	P5	P4	P3	P2	P1	P0
0th	Initialize	0	0	0	0	0	0	0	0
1st	Multiply & Add	1	1	0	1				
2nd	Shift	0	1	1	0	1	0	0	0
3rd	Multiply & Add	1	1	0	1				
4th	Shift	1	0	0	1	1	1		
5th	Multiply & Add	0	0	0	0				
6th	Shift	0	1	0	0	1	1	1	
7th	Multiply & Add	1	1	0	1				
8th	Shift	1	0	0	0	1	1	1	1

The 8-bit product register P is initialized to "00000000". In the first clock cycle, the first row of the partial product array is generated and added to the content of the register P. In the second clock cycle, the content of the register P is shifted by one bit position to the right. This process continues till the eighth clock cycle, at the end of which the register contains the product ($P = 13_{10} \times 11_{10} = 143_{10} = 10001111_2$) "10001111" in this example.

17.4 Drivers and Buffers

Drivers and buffers are special digital circuit components, which, unlike the other logic circuits, do not offer any logic functionality; they are just used to restore the signal swing levels. Normally, when the digital signals travel through long interconnects, their signal levels degrade after a certain length. In order to recover the signal levels back to their original values, buffers are used at regular intervals. Sometimes, for a logic cell, it also happens that the number of fan-out is more. In such cases, the logic cell cannot drive the required number of fan-out. Hence, we can say that the buffers are used to increase the drive capability of a logic cell.

Buffers can be of three types: inverting, non-inverting, and tri-state buffers.

17.4.1 Buffer Scaling and Design Issues

The propagation delay of a buffer can be reduced by increasing the (W/L) ratio of the transistors. However, increasing the transistor size also increases the output capacitance of the logic gate. Therefore, continuous increase of transistor size does not produce continuous decrease in propagation delay. In fact, at some point, increasing transistor size increases the propagation delay of the logic gate. The wide transistors also have larger gate capacitance, which leads to large input capacitance of the logic gate. Generally, a high speed buffer is not constructed using a single buffer stage; instead a chain of buffers with increasing sizes is used. This is known as a *tapered buffer*, as illustrated in Fig. 17.10.

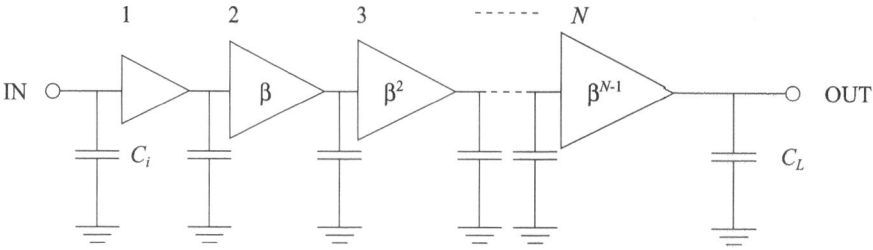

Fig. 17.10   Schematic of a tapered buffer

Tapered Buffer

The tapered buffer is constructed by using several buffers in cascade with increasing drive capability. The schematic of a tapered buffer is shown in Fig. 17.10. We see that each buffer is modelled using a capacitance and a conductance.

Let us assume that the logic level capacitance is C_i, conductance is g, and the logic level time-constant is $\tau_i = C_i/g$. Then, there are N stages in the buffer chain where the (W/L) ratio of $(k + 1)^{\text{th}}$ stage is β times the (W/L) ratio of the kth stage, as shown in the following expression:

$$\left(\frac{W}{L}\right)_{k+1} = \beta\left(\frac{W}{L}\right)_{k} \tag{17.19}$$

The capacitance, conductance, and time-constant of the kth stage can be expressed as follows:

$$C_k = \beta^{k+1} C_i \qquad (17.20)$$

$$g_k = \beta^k g \qquad (17.21)$$

$$\tau_k = \beta \tau_i \qquad (17.22)$$

The overall time-constant of the buffer, which is the sum of the time-constants of each buffer can be expressed as follows:

$$\tau_0 = \sum_{k=0}^{N-1} \tau_k = N\beta\tau_i \qquad (17.23)$$

The load capacitance of the output stage is given by the following expression:

$$C_L = \beta^N C_i \qquad (17.24)$$

From Eqn (17.23), we can write the number of stages (N) as follows:

$$N = \frac{\ln(C_L/C_i)}{\ln(\beta)} \qquad (17.25)$$

Substituting Eqn (17.24) in Eqn (17.22), we derive the following expression:

$$\tau_0 = \frac{\ln(C_L/C_i)}{\ln(\beta)} \times \beta\tau_i \qquad (17.26)$$

Hence, for optimum value of the overall time-constant τ_0, we can write the following expression:

$$\frac{\partial \tau_0}{\partial \beta} = \tau_i \ln(C_L/C_i) \left[\frac{1}{\ln(\beta)} - \beta \frac{1}{\{\ln(\beta)^2\}} \times \frac{1}{\beta} \right] = 0$$

or,

$$\frac{1}{\ln(\beta)} = \frac{1}{\{\ln(\beta)^2\}}$$

or,

$$\ln(\beta) = 1$$

which yields $\beta = e = 2.718$.

Therefore, the overall time-constant of the tapered buffer is given by the following expression:

$$\tau_0 = e \ln(C_L / C_i)\tau_i \qquad (17.27)$$

Hence, we see that the successive buffer design stages scale exponentially for optimum propagation delay.

SUMMARY

- Adders and multipliers are the most important building blocks of most digital logic hardware circuits.
- A ripple-carry adder has the smallest architecture amongst all other types of adders, but it is the slowest in terms of speed.
- The regular architecture of an unsigned array multiplier is very suitable for VLSI implementation of the circuit.
- Buffers are used to increase the drive capability of a logic cell. They are also used to restore the logic levels when the signals are transmitted over long interconnects.

SELECT REFERENCES

Baugh, C.R. and B.A. Wooley, "A Two's Complement Parallel Array Multiplication Algorithm", *IEEE Trans. Computers,* vol. 23, pp. 1045–1047, Dec. 1973.

Charles, H. Roth, *Digital Systems Design using VHDL*, PWS Publishing Company, Boston, 1998.

EXERCISES

Fill in the Blanks

1. A tapered buffer has optimum delay when the taper size is _____.
   (a) 2.718 (b) 3.414 (c) 2.5 (d) 4.3
2. An *N*-bit serial-parallel multiplier requires _____.
   (a) *N* clock cycles (b) 2*N* clock cycles
   (c) 2^N clock cycles (d) $\log_2 N$ clock cycles
3. The architecture of a Baugh-Wooley multiplier has additionally _____ as compared to unsigned array multiplier.
   (a) Four inverters (b) Six inverters
   (c) One inverter (d) Three inverters
4. Static CMOS realization of Manchester carry chain produces _____ output.
   (a) Normal (b) Complemented
   (c) Both a and b (d) None of these
5. Static CMOS realization of Manchester carry chain takes _____ input.
   (a) Normal (b) Complemented
   (c) Both a and b (d) None of these

Multiple Choice Questions

1. The expression for carry generate signal is
   (a) $G = AB$ (b) $G = A + B$ (c) $G = A \oplus B$ (d) $G = A \otimes B$
2. The expression for carry propagate signal is
   (a) $P = AB$ (b) $P = A + B$ (c) $P = A \oplus B$ (d) $P = A \otimes B$
3. The sum is generated in CLA using
   (a) Carry generate and propagate signals
   (b) Carry generate signal
   (c) Carry input, carry generate, and propagate signals
   (d) Carry input and carry generate signals

4. Carry-save adder is suitable for
   (a) Two inputs   (b) Three inputs   (c) Four inputs   (d) Any of these
5. Manchester carry chain has following input signals:
   (a) Generate, propagate, and kill signals
   (b) Generate and propagate signals
   (c) Only kill signal
   (d) Kill and propagate signals

True or False

1. A serial adder is the fastest adder.
   (a) True   (b) False
2. An array multiplier is a synchronous design.
   (a) True   (b) False
3. A carry look-ahead adder is best suited for large operand size.
   (a) True   (b) False
4. Baugh-Wooley multiplier is a signed multiplier.
   (a) True   (b) False
5. Tri-state buffer uses a control signal.
   (a) True   (b) False

Short Answer Type Questions

1. Design a tri-state inverting buffer using CMOS logic.
2. Design a Manchester carry chain adder.
3. What are different types of buffers? What are their uses?
4. Discuss the serial-parallel multiplication process with the help of a suitable example.
5. Design a serial adder.

Long Answer Type Questions

1. Design a 5-bit unsigned array multiplier.
2. Design an 8-bit signed Baugh-Wooley multiplier.
3. Design a tapered buffer and optimize its performance.
4. Compare the merits and demerits of RCA and CLA.
5. Design a Booth's multiplier.

## CHAPTER 18

Low Power Logic Circuits

KEY TOPICS

- Power dissipation
- Dynamic power
- Static power
- Leakage power
- Power reduction techniques
- Low power design

18.1 Introduction

Digital logic circuits, when operated, consume some power which is dissipated in the form of heat. In portable devices such as laptops, iPods, mobile phones, and tablets, the source of power is the battery which needs recharging after a certain period of operation. For the handheld devices, the users would like to have more battery life as one would not want to charge the battery too often. Therefore, in order to increase the battery lifetime, the amount of power that is dissipated by logic circuits must be minimized. Low power design is a technique especially targeted at reducing power dissipation in digital logic circuits. The low power logic circuits are specially designed with less power consumption.

18.2 Power Dissipation in Logic Circuits

Power dissipation in a digital logic circuit has a number of components. The average power dissipation can be expressed as

$$
\begin{aligned}
P_{\text{avg}} &= P_{\text{dynamic}} + P_{\text{short circuit}} + P_{\text{leakage}} \\
&= \alpha C_L V_{DD}^2 f_{\text{clk}} + I_{\text{sc}} V_{DD} + I_{\text{leakage}} V_{DD}
\end{aligned}
\tag{18.1}
$$

The first term in Eqn (18.1) represents a dynamic power dissipation which depends on the switching activity of logic circuits. The second term represents a short-circuit power dissipation, and the third term represents a leakage power dissipation. Here, f_{clk} is the clock frequency, α the activity factor, and C_L represents the load capacitance.

18.2.1 Dynamic Power Dissipation

CMOS logic circuits, depending on the input signals, produce either logic 1 or logic 0. Logic 1 is produced by charging the load capacitor C_L through the pull-up network (PUN), and logic 0 is produced by discharging the load capacitor C_L through the pull-down network (PDN). During the charging operation, energy is drawn from the power supply voltage source V_{DD}. The amount of energy drawn from the power supply is $C_L V_{DD}^2$. Half of this energy is stored in the capacitor and half is dissipated in PUN in the form of heat. During the discharging operation, no energy is drawn from the power supply voltage source. However, the stored energy in the capacitor is dissipated in PDN.

In a period of time T, if the output node switches n number of times, then the amount of energy drawn from the voltage source is $n C_L V_{DD}^2$. Therefore, for a period of time T, the amount of power dissipated for n number of charging and discharging operations is $n C_L V_{DD}^2 / T$. We can write the average power dissipation as

$$P_{\text{dynamic}} = \frac{n C_L V_{DD}^2}{T} \qquad (18.2)$$

Now, let us consider that the entire chip operates at a clock frequency of f_{clk}. The number of times a clock makes transitions in time T is

$$m = T \times f_{\text{clk}} \qquad (18.3)$$

or

$$T = \frac{m}{f_{\text{clk}}} \qquad (18.4)$$

From Eqs. (18.2) and (18.4), we can write

$$P_{\text{dynamic}} = \frac{n f_{\text{clk}} C_L V_{DD}^2}{m} = \alpha f_{\text{clk}} C_L V_{DD}^2 \qquad (18.5)$$

where $\alpha = n/m$. If the output node switches equal number of times as the clock then $n = m$ and $\alpha = 1$. But, in general, logic circuits do not switch equally with the clock rather at a reduced rate. Therefore, the average power dissipation in a logic circuit depends on the switching activity of the logic gate with respect to the clock frequency of the chip. The factor α is known as the switching activity factor, and its value lies between 0 and 1.

18.2.2 Static Power Dissipation

Static power dissipation or short-circuit power dissipation occurs when both pMOS and nMOS transistors are ON simultaneously. This situation arises when the rise and the fall times of the input pulse are sufficiently large. Figures 18.1 and 18.2 show the input and the output voltage waveforms for two cases: (i) when the rise and the fall times of the input pulse are large and (ii) when the rise and the fall times of the input pulse are small. In Fig. 18.1, we can see that the current flows through both pMOS and nMOS transistors simultaneously at each transition for a longer period of time. On the contrary, we can see in Fig. 18.2 that the current flows through the pMOS transistor only during the output $0 \to 1$ transitions and the current flows through the nMOS transistor only during the output $1 \to 0$ transitions.

When the rise and the fall times of the input pulse are small, only a dynamic current flows through the transistors for a small duration and this causes the dynamic power dissipation in CMOS logic gates. Whereas when the rise and the

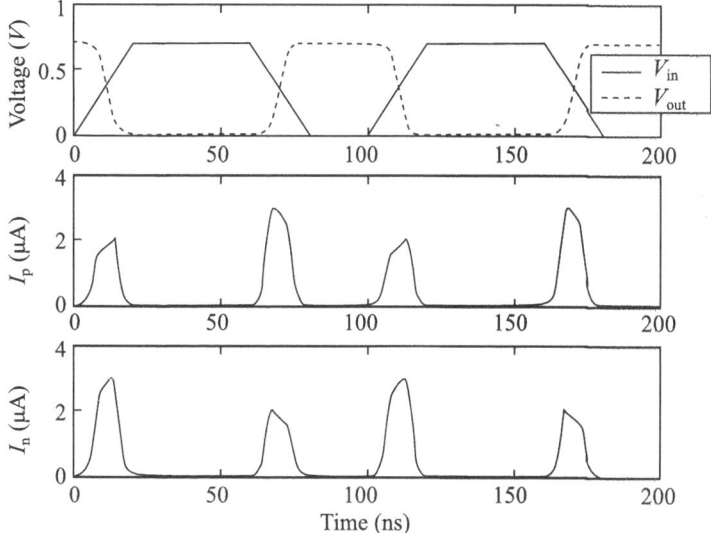

Fig. 18.1 Input and output voltage waveforms of a CMOS inverter, the waveform of current through the pMOS transistor, and the waveform of current through the nMOS transistor when the rise and the fall times of the input pulse are large.

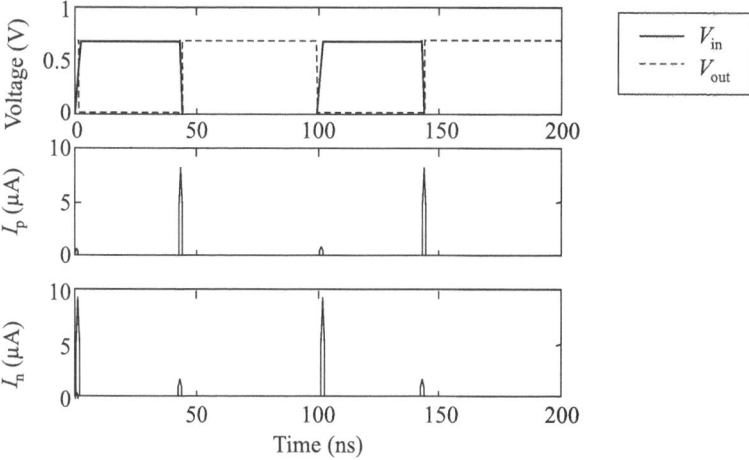

Fig. 18.2 Input and output voltage waveforms of a CMOS inverter, the waveform of current through the pMOS transistor, and the waveform of current through the nMOS transistor when the rise and the fall times of the input pulse are small.

fall times of the input pulse are large, a static current flows through the transistors for a long duration, forming a direct short-circuit path between the power supply voltage source (V_{DD}) and the ground. This causes the static power dissipation in CMOS logic gates.

Figure 18.3 illustrates how power dissipation increases with the increase in the rise/fall time of the input pulse for a constant load capacitance. From Eqn (18.5),

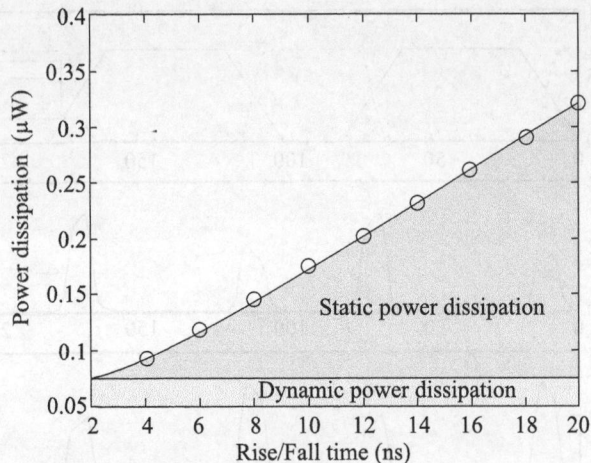

Fig. 18.3 Power dissipation as a function of the rise/fall time of the input pulse.

we can see that the dynamic power dissipation in a CMOS logic gate is independent of the rise/fall time of the input pulse. However, the static power dissipation increases with the increase in the rise/fall time of the input pulse. This is due to the static current flowing through pMOS and nMOS transistors at every input transition and for a longer period of time. The static power dissipation is predominant for a small load capacitance; however, when the load capacitance is large the dynamic power dissipation dominates the static power dissipation as illustrated in Fig. 18.4.

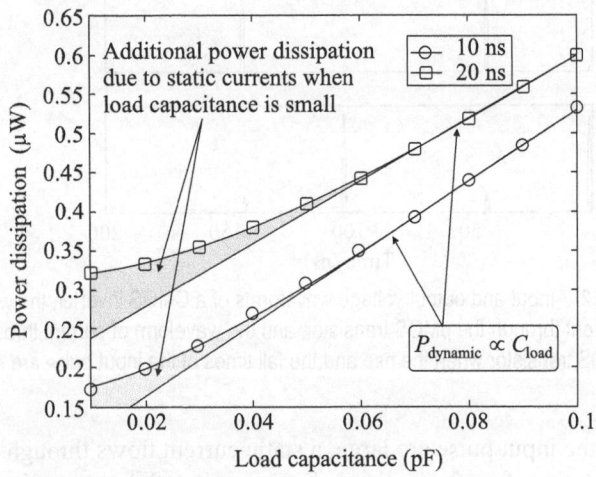

Fig. 18.4 Power dissipation as a function of load capacitance.

For a small load capacitance, the static power dissipation is significant as illustrated by the grey regions in Fig. 18.4.

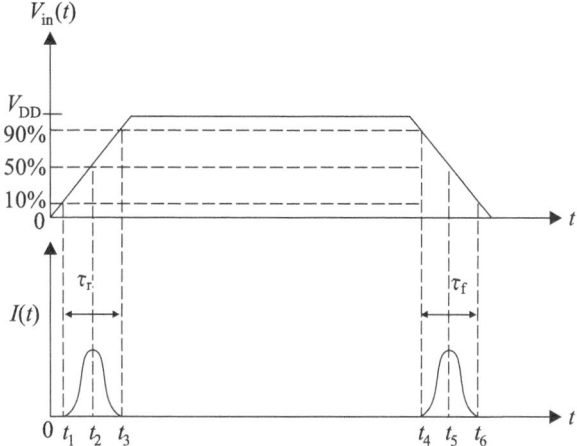

Fig. 18.5 The input pulse waveform and the corresponding static current waveform.

The expression for the static power dissipation can be derived as follows. Let us consider that the input waveform is a pulse to the inverter as shown in Fig. 18.5. The static current flows only during the transition of the input waveform as shown in Fig. 18.5.

The average static current can be expressed as

$$I_{mean} = 2 \times \frac{1}{T} \left[\int_{t_1}^{t_2} i(t)dt + \int_{t_2}^{t_3} i(t)dt \right] \qquad (18.6)$$

Factor 2 accounts for the static currents during the rising edge and the falling edge of the input pulse.

As the current waveform is symmetric about time t_2, we can write

$$I_{mean} = \frac{4}{T} \left[\int_{t_1}^{t_2} i(t)dt \right] \qquad (18.7)$$

When an nMOS transistor operates in the saturation region, we can write the current as

$$i(t) = \frac{K}{2} \left(V_{in}(t) - V_{tn} \right)^2 \qquad (18.8)$$

The rising and the falling edges of the input pulse can be written as

$$V_{in}(t) = \begin{cases} \dfrac{V_{DD} \times t}{\tau_r} & \text{for rising edge} \\[2em] V_{DD}\left(1 - \dfrac{t}{\tau_f}\right) & \text{for falling edge} \end{cases} \qquad (18.9)$$

Therefore, we can write

$$I_{\text{mean}} = \frac{4}{T} \int_{\tau \times V_{\text{tn}}/V_{\text{DD}}}^{\tau/2} \frac{K}{2}\left(\frac{V_{\text{DD}} \times t}{\tau} - V_{\text{tn}}\right)^2 dt \qquad (18.10)$$

Note that when $V_{\text{in}} = V_{\text{tn}}$, $t = t_1 = \tau$, so

$$t_1 = \frac{\tau \times V_{\text{tn}}}{V_{\text{DD}}}$$

Hence,
$$I_{\text{mean}} = \frac{2K}{T} \int_{\tau \times V_{\text{tn}}/V_{\text{DD}}}^{\tau/2} \left(\frac{V_{\text{DD}} \times t}{\tau} - V_{\text{tn}}\right)^2 dt \qquad (18.11)$$

Let us assume that

$$x = \frac{V_{\text{DD}} \times t}{\tau} - V_{\text{tn}} \Rightarrow dx = \frac{V_{\text{DD}}}{\tau} dt$$

Now, we can write Eqn (18.11) as

$$\begin{aligned}
I_{\text{mean}} &= \frac{2K}{T} \int_{\tau \times V_{\text{tn}}/V_{\text{DD}}}^{\tau/2} x^2 \frac{\tau}{V_{\text{DD}}} dt \\
&= \frac{2K}{3T} \times \frac{\tau}{V_{\text{DD}}} \left[x^3\right]_{\tau \times V_{\text{tn}}/V_{\text{DD}}}^{\tau/2} \\
&= \frac{2K}{3T} \times \frac{\tau}{V_{\text{DD}}} \left[\left(\frac{V_{\text{DD}} \times t}{\tau} - V_{\text{tn}}\right)^3\right]_{\tau \times V_{\text{tn}}/V_{\text{DD}}}^{\tau/2} \\
&= \frac{2K}{3T} \times \frac{\tau}{V_{\text{DD}}} \left[\left(\frac{V_{\text{DD}} \times \tau}{\tau \times 2} - V_{\text{tn}}\right)^3 - \left(\frac{V_{\text{DD}} \times \tau \times V_{\text{tn}}}{\tau \times V_{\text{DD}}} - V_{\text{tn}}\right)^3\right] \\
&= \frac{2K}{3T} \times \frac{\tau}{V_{\text{DD}}} \left[\left(\frac{V_{\text{DD}}}{2} - V_{\text{tn}}\right)^3 - 0\right] \\
&= \frac{K}{12} \times \frac{1}{V_{\text{DD}}} (V_{\text{DD}} - 2V_{\text{tn}})^3 \times \frac{\tau}{T}
\end{aligned} \qquad (18.12)$$

Thus, the static or the short-circuit power dissipation can be written as

$$P_{\text{static}} = V_{\text{DD}} \times I_{\text{mean}} = \frac{K}{12}(V_{\text{DD}} - 2V_{\text{tn}})^3 \times \frac{\tau}{T} = \frac{K}{12}(V_{\text{DD}} - 2V_{\text{tn}})^3 \times \tau \times f_{\text{clk}} \quad (18.13)$$

18.2.3 Leakage Power Dissipation

Leakage power dissipation in CMOS logic gates is due to the subthreshold current and the leakage current through pMOS and nMOS transistors.

We generally assume that pMOS and nMOS transistors are ON only when their gate-to-source voltage (V_{GS}) exceeds the threshold voltage (V_t). However, pMOS and nMOS transistors can have subthreshold current when they are operated in the subthreshold region ($V_{GS} < V_t$). When the input voltage is high, we assume that the nMOS transistor is ON and the pMOS transistor is OFF. However, the pMOS transistor works in the subthreshold region. Thus, a direct current path exists from the V_{DD} to the ground through the pMOS transistor operating in a subthreshold region and the nMOS transistor operating in a normal super-threshold region. The subthreshold current can be expressed as

$$I_{D,\text{subthreshold}} = Ke^{(V_{GS}-V_t)/nV_T}\left(1 - e^{-\frac{V_{DS}}{V_T}}\right) \qquad (18.14)$$

where K is a function of technology and $V_T = kT/e$ is the thermal voltage. The subthreshold slope (S_{th}) is the amount of voltage required to drop the subthreshold current by one decade. It is determined by taking the ratio of two points in the subthreshold region.

$$\frac{I_1}{I_2} = e^{\frac{(V_1 - V_2)}{nV_T}} \qquad (18.15)$$

This results in $S_{th} = nV_T \ln(10)$. At room temperature, the subthreshold slope lies between 60 and 90 mV/decade.

Another leakage component is the reverse diode leakage current. When the input voltage is high, the output of the CMOS inverter is low. Under this condition, the nMOS transistor is ON and the pMOS transistor is OFF. Though the pMOS transistor is OFF, a reverse saturation current flows through the drain–bulk junction of the pMOS transistor.

18.3 Power Reduction Techniques

A dynamic power dissipation depends on the power supply voltage (V_{DD}), load capacitance (C_L), clock frequency (f_{clk}), and switching activity (α) [see Eqn (18.5)]. Reducing all these parameters has direct impact on the dynamic power dissipation.

As the dynamic power dissipation has quadratic dependency on the power supply voltage, the reduction of power supply voltage significantly reduces the dynamic power dissipation in CMOS logic gates. But the reduction of power supply voltage also has a negative impact on the propagation delay in CMOS logic gates. Therefore, designers often have to trade-off between power dissipation and propagation delay to choose optimum power supply voltage.

The reduction of load capacitance can be achieved by using smaller gates in the output or having smaller number of fan-out gates. The small interconnect length at the output can also reduce the load capacitance. The reduction of transistor count and circuit nodes can achieve a smaller load capacitance. The reduction of transistor size is also an effective way of reducing load capacitance.

Reducing lower clock frequency and lower switching activity of the design also reduces the dynamic power dissipation in CMOS logic gates.

The static power or the short-circuit power dissipation can be reduced by using

- Fast rise/fall time of the input signal
- Low clock frequency
- Low power supply voltage

Using a small input capacitance can lead to fast rise/fall time and hence can reduce the static power dissipation. If the rise/fall time of the input pulse is slow enough, buffers can be used to make the rise/fall time faster before applying it to large gates.

The leakage power can be reduced by reducing power supply voltage.

18.4 Low Power Design Techniques

As the demand of portable devices is increasing, the need for low power devices is becoming very important. With the increase in the level of integration and complexity of systems, it has become extremely difficult to provide adequate cooling mechanism to the entire system. It not only increases the cost of the overall system, but also limits the functionality of the system. As the CMOS technology node shrinks down to 65 nm, the dynamic power dissipation is under control. However, the static or the leakage power dominates the dynamic power beyond 65 nm technology node. Therefore, minimizing the static or the leakage power dissipation has become a bottleneck in sub-nanometer technology nodes.

The overall low power design methodology having optimization at different abstraction levels are as follows:

1. System level—Partitioning, Power down
2. Algorithm level—Complexity, Concurrency, Regularity
3. Architecture level—Parallelism, Pipelining, Redundancy, Data encoding
4. Logic level—Logic styles, Energy recovery, Transistor sizing
5. Technology level—Threshold reduction, Multi-threshold devices

The dynamic or the switching power dissipation is due to charging and discharging of load capacitors driven by logic circuits. The reduction of power supply voltage is one of the most useful techniques for power optimization, because it can achieve considerable power saving due to the quadratic dependence of switching or the dynamic power dissipation on the power supply voltage (V_{DD}). However, lowering the supply voltage affects the speed of the circuit which is the major shortcoming of this approach. Therefore, both design and technological solutions must be adopted to compensate for the decrease in speed of the circuit introduced by reduced voltage. Some of the useful techniques used to reduce dynamic power are as follows:

- Designing logic circuits with transistors having multi-threshold voltage can be used to reduce leakage power at system level.
- Transistor resizing can be used to increase circuit performance and reduce power dissipation.
- The standby power can be reduced by using sleep transistors.

■ At the architecture level, use of parallelism and pipelining can reduce power significantly.

■ Power-down of selected logic blocks, adiabatic computing, clock disabling, and software redesign are some of the other techniques commonly used for low power design.

In the following subsections, we shall discuss some of the low power design techniques at the circuit level.

18.4.1 Adiabatic Logic Circuits

Adiabatic logic circuits utilize dynamic power source in which the energy drawn from the power source is restored. Adiabatic process is a thermodynamic process where there is no exchange of heat between the system and the environment and there is no energy loss in the form of heat. In logic circuits, the charge transfer between the various nodes of the circuit is considered as process. During the charge transfer process, the goal is to reduce the energy loss.

The basic adiabatic logic gate is an adiabatic buffer as shown in Fig. 18.6. It consists of two transmission gates, two load capacitances, and one dynamic power source. When the input (In) is high (V_{DD}), the left transmission gate (TG1) turns ON and the right transmission gate (TG2) turns OFF. The equivalent circuit of the adiabatic buffer is shown in Fig. 18.7(a) when the input (In) is high (V_{DD}).

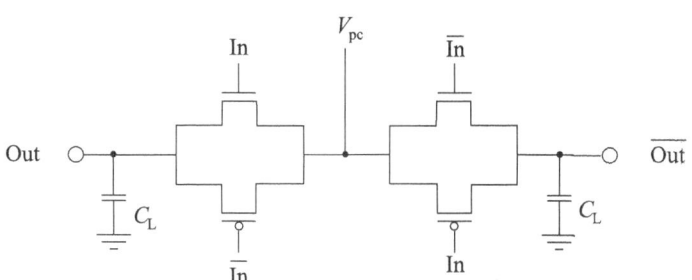

Fig. 18.6   Adiabatic buffer.

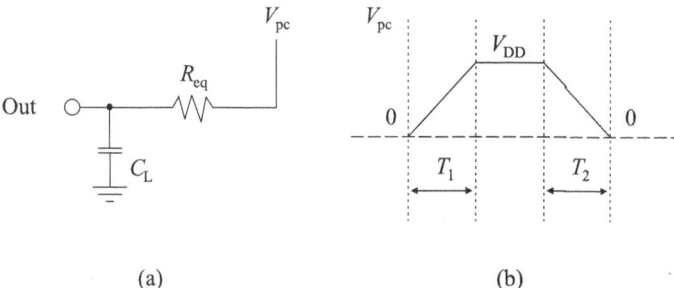

(a)   (b)

Fig. 18.7   (a) Equivalent circuit of Fig. 18.6. (b) Waveform of the dynamic power source.

Here, R_{eq} represents the equivalent resistance of the left transmission gate (TG1). Let us assume that the load capacitance (C_L) was initially discharged to 0 V. When the input (In) is high, the left transmission gate (TG1) turns ON, and the load capacitance will be charged according to the following equation:

$$V(t) = \frac{I \times t}{C_L} \tag{18.16}$$

where I is the constant charging current. The energy dissipated in the resistor R_{eq} for a period of time T is given by

$$E_{\text{dissipation}} = \int_0^T I^2 R_{eq} dt = I^2 R_{eq} T \tag{18.17}$$

Combining Eqs. (18.16) and (18.17), we can write

$$E_{\text{dissipation}} = \frac{R_{eq} C_L}{T} C_L V^2(t) \tag{18.18}$$

In conventional CMOS inverters, we have seen that the average power dissipation over a period of time T is CV_{DD}^2 / T. Therefore, from Eqn (18.18) we can conclude that the power dissipation using dynamic power supply will be less than in conventional CMOS logic circuits if the product $R_{eq} C_L$ is much less as compared to T. Hence, reducing the equivalent resistance of the transmission gate can reduce power dissipation. The increase in T also reduces power dissipation.

18.4.2 Logic Design for Low Power

A proper selection of logic styles is also an effective way for low power design. For example, choices between the static CMOS logic and the dynamic CMOS logic or between the conventional CMOS and the pass transistor logic are made during the design of a circuit. In static CMOS circuits, the short-circuit power dissipation is a significant percentage of the total power dissipation. However, in dynamic CMOS circuits this problem does not arise, as there is no short-circuit path from the supply voltage to the ground. In domino logic circuits, such a path exists, hence there is a small amount of short-circuit power dissipation. Similarly, one can use the pass transistor logic to explore reduced swing for lower power design (e.g., reduced bit-line swing in memory).

18.4.3 Reduction of Glitches

A dynamic power dissipation occurs due to the switching activity of logic circuits. Some of the undesirable switching activities are due to glitches in a logic circuit. Figure 18.8 illustrates a source for a glitch in a logic circuit. Let us assume that initially inputs A and B are at logic 0. Therefore, the output of the AND gate Y is at logic 0. Now, let us assume that inputs A and B switch to logic 1 simultaneously. The expected output of AND gate is logic 0. However, the input C takes some time to become logic 0 due to the propagation delay in the inverter. Hence, for a short duration the input of the AND gate is at logic 1 and the output of the AND gate

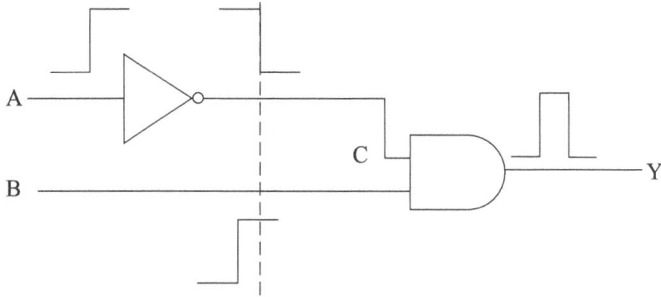

Fig. 18.8 A circuit showing glitch at the output.

is also at logic 1. When C becomes logic 0, the output Y becomes logic 0 again. So, there is unwanted switching of the output which results in a dynamic power dissipation.

For the low power design, these glitches must be minimized so that the unwanted switching activity can be minimized. The source of glitch in a logic circuit is the propagation delay in different paths of the logic circuit. Therefore, in order to minimize the dynamic power dissipation due to glitches, the timing of the circuit needs to be optimized in such a way that the number of glitches is minimized.

Different techniques have been developed to optimize the delay. Some of these techniques are as follows:

- Gate sizing—to optimize the dimensions of the transistors used to design the gate
- Gate triggering—to minimize the glitches by evaluating the output of a gate only when all of its inputs have stabilized
- Gate freezing—happens in the place-and-route phase where a suitable gate is selected based on delay and power optimization

18.4.4 Multiple Threshold MOS Transistors

A static or a leakage power dissipation occurs due to substrate currents and subthreshold leakage currents. For older technology nodes (>1 µm), the dynamic power dissipation was predominant. However for deep-submicron (DSM) technology nodes (<180 nm), the leakage power dissipation has become a dominant factor. The leakage power dissipation is a major concern in DSM and nanometer technology nodes. It critically impacts the battery lifetime. In order to minimize the leakage power dissipation, multiple threshold and variable threshold MOS transistors are used to design logic circuits. In the multiple threshold CMOS process, the technology provides two different types of MOS transistors having threshold voltage. Low threshold (low V_t) MOS transistors are used for the part of the circuit where a high-speed operation is required. This part circuit is fast but has a large power dissipation. High threshold (high V_t) MOS transistors are slow in speed but exhibit a low subthreshold leakage. So, they are used in the part of the chip where the operation can be slow. However, if the number of low V_t transistors becomes high as compared to high V_t transistors, the multiple threshold technique becomes ineffective.

18.4.5 Standby Mode: Sleep Transistors

It may happen that the entire or some part of the logic circuit is not functioning though it is ON. Under such cases, the circuit can be operated in two modes: one is called the normal active mode when the circuit is functioning, and the other is called the standby mode or the sleep mode when the circuit is ON but not functioning. So, when the circuit is in the standby mode the speed of the circuit is not a concern. Hence, high V_t transistors can be used in the standby mode. When the circuit is in the normal active mode, low V_t transistors can be used.

The transistors having high V_t are connected in series with the main logic circuit as shown in Fig. 18.9. These high V_t transistors are called sleep transistors. The control signal sleep is used to switch ON and OFF the sleep transistors. In the normal active mode, the signal sleep = 0 and the sleep transistors are ON. When the circuit is in the standby mode, the signal sleep = 1 and the sleep transistors are OFF. As the high V_t transistors appear in series with a low V_t circuit, the leakage current is determined by the high V_t transistors. Hence, the leakage power is very low. So, the net leakage power dissipation is reduced.

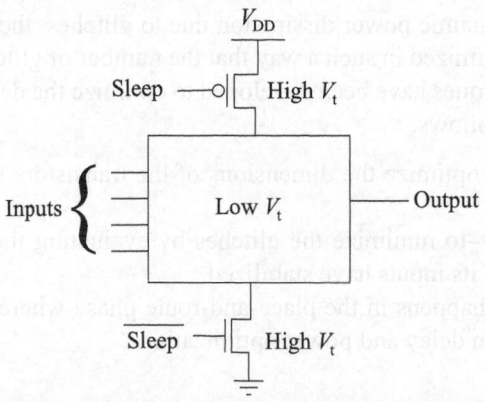

Fig.18.9   Circuit design with sleep transistors.

18.4.6 Variable Body Biasing

We know that the substrate bias has an effect on the threshold voltage of MOS transistors. Therefore, by suitably controlling the substrate bias, one can control the threshold voltage of MOS transistors dynamically. This way the shortcomings of multi-threshold technique can be solved.

In the standby mode, the substrate of nMOS transistors is negatively biased to increase their threshold voltage. Similarly, the substrate of pMOS transistors is positively biased to increase their threshold voltage in the standby mode. Therefore, the variable threshold circuits can solve the static or the leakage power problem. However, they require control circuits that can control the substrate voltage in the standby mode. A fast and accurate controlling of substrate bias with a control circuit is quite challenging, and it requires a cautiously designed closed-loop control system.

When the circuit is in the standby mode, the substrate of both the pMOS and the nMOS is biased by a third supply voltage to increase the threshold voltage of the MOS transistors as shown in Fig. 18.10. During the normal active mode of operation they are switched back to the V_{DD} and the ground to reduce the threshold voltage.

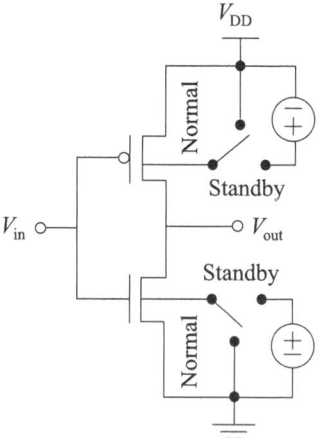

Fig.18.10   Variable substrate
biasing technique.

18.4.7 Dynamic Threshold MOS

In dynamic threshold CMOS (DTMOS) circuits, the threshold voltage of the MOS transistors is changed dynamically to suit the operating state of the circuit. A dynamic threshold CMOS circuit is designed by connecting the gate and substrate of MOS transistors together as shown in Fig. 18.11. We have learned already that a high V_t transistor in the standby mode gives low leakage current, whereas a low

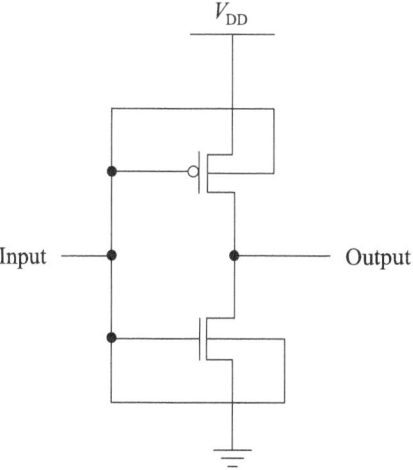

Fig. 18.11   DTMOS circuit.

V_t transistor provides higher current in the active mode of operation. The supply voltage of DTMOS is limited by the diode built-in potential in the bulk silicon technology. The PN diode between source and substrate should be reverse biased. Hence, this technique is only suitable for ultralow voltage (<0.6 V) circuits in the bulk CMOS technology.

SUMMARY

■ Power dissipation in CMOS circuits has three major components: dynamic power dissipation, static power dissipation, and leakage power dissipation.
■ A dynamic power dissipation depends on power supply voltage, switching activity, frequency of switching, and load capacitance.
■ A static power dissipation increases with the rise/fall time of the input pulse.
■ Adiabatic logic uses the dynamic power supply voltage.
■ Low power can be achieved at different design abstraction levels.

SELECT REFERENCES

Chandrakasan, A.P. and R. Brodersen R., *Low-Power CMOS Design*, IEEE Press, 1998.
Chandrakasan, A.P., S. Sheng, and R.W. Brodersen, *Low-power CMOS Digital Design, Solid-State Circuits, IEEE Journal* (Vol. 27, No. 4), 2002.
http://www.eeherald.com/section/design-guide/Low-Power-VLSI-Design.html
Kaushik, R. and Sharat C. Prasad, *Low Power CMOS VLSI Circuit Design*, John Wiley and Sons, 2000.

EXERCISES

Fill in the Blanks

1. Dynamic power does not depend on _____.
   (a) transistor dimensions   (b) load capacitance
   (c) power supply voltage   (d) switching activity
2. Short-circuit power dissipation does not depend on _____.
   (a) power supply   (b) load capacitance
   (c) rise/fall time   (d) transistor dimensions
3. Leakage power is due to _____.
   (a) subthreshold current   (b) leakage current
   (c) both (a) and (b)   (d) none of these
4. Standby mode _____ power dissipation.
   (a) increases   (b) decreases   (c) does not cause   (d) none of these
5. Battery operated devices must have _____ power dissipation.
   (a) more   (b) less   (c) zero   (d) none of these

Multiple Choice Questions

1. Low V_t transistors consume
   (a) more power   (b) less power   (c) no power   (d) none of these
2. High V_t transistors have
   (a) more speed   (b) less speed   (c) no speed   (d) all of these

3. Glitch in logic circuits
   (a) increases power dissipation
   (b) decreases power dissipation
   (c) does not have any effect on power dissipation
   (d) none of these
4. Adiabatic logic can achieve
   (a) low power and high speed
   (b) low power but low speed
   (c) high power but high speed
   (d) high power but low speed
5. Static power is more for
   (a) slow input rise/fall time
   (b) fast input rise/fall time
   (c) slow output rise/fall time
   (d) fast output rise/fall time

True or False

1. Sleep transistors are low V_t transistors.
2. DTMOS uses gate–drain connected MOS transistors.
3. Dynamic CMOS logic consumes more dynamic power.
4. Adiabatic logic requires dynamic power supply.
5. Dynamic power is independent of clock frequency.

Short Answer Type Questions

1. What are the major sources of power dissipation in CMOS circuits?
2. Derive the expression for switching power in CMOS circuits.
3. Discuss how the leakage power is dissipated in CMOS circuits.
4. Why static power dissipation occurs in CMOS circuits?
5. Briefly explain the power reduction techniques.

Long Answer Type Questions

1. Discuss the circuit level techniques to reduce power dissipation.
2. Show how a dynamic power depends on the switching activity of CMOS circuits.
3. Show how the standby mode and the normal active mode of operation can reduce overall power dissipation.
4. What is a sleep transistor? How is it used? What is a variable body biasing technique?
5. Discuss the methods of reduction of a short-circuit power dissipation.

Digital Design Using VHDL

A.1 Introduction to HDL

In this chapter, we discuss the logic design using hardware description language (HDL). HDL is a formal language that is used to design, synthesize, simulate, and model logic circuits. There are two popular HDL languages: VHDL and Verilog. VHDL is mostly used by academics, whereas Verilog is used by industry personnel. Both VHDL and Verilog are IEEE (Institute of Electrical and Electronic Engineers) standards. VHDL is discussed in this appendix and Verilog in Appendix B.

We have discussed various modelling styles of architecture with suitable examples. The syntax and semantics of VHDL language are discussed with examples. The main focus of this chapter has been the design of VLSI circuits using VHDL. For more details about the language, readers are recommended to refer to any textbook written exclusively on VHDL.

A.2 Introduction to VHDL

VHDL is an acronym of VHSIC hardware description language, where the term VHSIC stands for very high speed integrated circuit. It is a hardware description language that can be used to model a digital system at many levels of abstraction, ranging from algorithmic level to the gate level. The language has the following basic constructs:

- Sequential statements
- Concurrent statements
- Netlist statements
- Timing specifications
- Waveform generation

The hardware abstraction has two views:

- External
- Internal

Entity It represents the external view. It is the black box representation of the hardware to be designed with only port specifications. It specifies all the input and output ports of the design.

Architecture It represents the actual internal details of the hardware. It specifies the relation between the input and output ports. There are four different VHDL modelling styles of the architecture:

- Structural
- Behavioural
- Dataflow
- Mixed

In the *structural* modelling style, the hardware is represented with the intercon-nection of components where components are the building blocks of the hard-ware. In the *behavioural* modelling style, the outputs are specified in the form of behaviour of the hardware for each of the specific input combinations (in the form of a truth table). In the *dataflow* modelling style, the outputs are specified in the form of Boolean expressions. In the *mixed* modelling style, the architecture is specified with combinations of structural, behavioural, and/or dataflow model-ling styles.

A.3 VHDL Language

So far we have shown quite a number of designs using VHDL. Let us now explain the VHDL language syntax and semantics.

A.3.1 Basic Language Syntax

Comments: A line that starts with two consecutive '−' characters, is a comment line. For example:

— This is a comment line

The VHDL compiler skips comment lines while compiling the code. Comment lines can appear anywhere in the code.

VHDL is *case-insensitive*. You can use upper case, lower case, or a combina-tion of upper and lower case letters for writing VHDL programs. For example, 'ENTITY' and 'entity' represent the same identifier. Similarly, 'SIGNAL' and 'signal' represent the same identifier.

A.3.2 Data Objects

VHDL supports four different types of data objects:
- signal
- variable
- constant
- file

Syntax for Data Objects

A combination of alphanumeric characters along with the '_' underscore character is used for naming data objects. However, there are certain restrictions, as given below.
1. It must not be a VHDL keyword.
2. It must begin with an alphabet.
3. Underscore '_' cannot be the last character.
4. Two consecutive '_' underscores are not allowed.

For example
Allowed names: A, a1, abar2, a_bar, S1_2.
Not allowed names: _x, 1a, abar_, signal, x__y

Signal Data Object

Signal data objects represent the logic signals or the wires in a circuit. A signal is declared as

```
SIGNAL <signal name> : <:signal type> [:= <initial value>];
```

For example, to define a signal S1 with initial value '0' we write

```
SIGNAL S1 : BIT := '0';
```

The signal type can be any one of the following signal types:

- BIT
- BIT_VECTOR
- STD_LOGIC
- STD_LOGIC_VECTOR
- STD_ULOGIC
- SIGNED
- UNSIGNED
- INTEGER
- ENUMERATION
- BOOLEAN

Signals can be declared within (a) entity declaration section, (b) architecture declaration section, and (c) package declaration section.

Constant Data Objects

A constant data object is meant for fixed value objects. A constant data object is declared using the following syntax:

```
CONSTANT <const name> : <data type> := <value>;
```

For example, to define the constant for pi = 3.1414, we write

```
CONSTANT PI : REAL := 3.1414;
```

Constants are useful in writing codes for replacing a fixed numerical value. Suppose we design a memory of address of size 8-bit. Instead of hard coding 8 everywhere in the code, we can define a constant (address_size) and use it in writing the code. The advantage in using this approach is that if we want to change the address size, we can just change the value of the constant in one place instead of changing it throughout in the code. For example,

```
CONSTANT address_size : integer := 8;
CONSTANT PROP_DELAY : time := 10 ns;
```

Variable Data Objects

Variables are used to index the loops or to hold the values of computational results. They do not represent wires, unlike signal data objects. The syntax for variable declaration is

```
VARIABLE <variable name> : <data type> [= <value>];
```

For example, we can define a variable *j* for indexing a loop as follows:

```
variable j : integer
```

File Data Objects

The file data object is used for reading or writing files. The syntax for declaring file data objects is

```
FILE <file name> : <file type> [[open <mode>] is <string>];
```

The string contains the physical name of the file. The mode specifies the file operations such as (a) read, (b) write, or (c) append. For example,

```
FILE DATAIN : TEXT open READ_MODE is "C:\VHDL\infile.txt";
FILE DATAOUT : TEXT open WRITE_MODE is "C:\VHDL\outfile.txt";
```

Other Data Objects

There are some objects that are not declared explicitly. But they are declared within the *port* and *generic* sections within the *entity* declaration. For example,

```
entity myadder is
  generic SIZE : integer := 4;
  port ( a, b : in std_logic_vector(SIZE-1  downto 0);
     Cin : in std_logic;
     Sum : out std_logic_vector(SIZE-1 downto   0);
     Cout : out std_logic);
end myadder;
```

In the above example, ports a,b, cin, sum, and cout are signal data objects. Similarly, the generic SIZE is a constant data object.

There are two other data object types which are not declared explicitly: (a) the index variable used in FOR loop and (b) the index variable used in GENERATE statements. For example,

FOR **loop**   The variable i is not declared explicitly. But it is a data object of type integer which is implicitly declared.

```
for i in 1 to 10 loop
  fact := fact + i;
end loop;
```

GENERATE **statement**   The variable k is not declared explicitly. But it is a data object of type integer which is implicitly declared.

```
gk : for k in N-1 downto 0 generate
  a1 : fa port map(a(k), b(k), c(k),   sum(k), c(k+1));
end generate;
```

A.3.3 Data Object Values

The value of each signal is specified as 0 or 1 within single quote characters. For example, a signal *x* with value 1 is specified as

```
SIGNAL x : BIT := '1';
```

An array of signals is specified within double quote characters as

```
SIGNAL address : std_logic_vector := "1001";
```

Double quotes are also used to specify a binary number. For example, a variable j of value $(1010)_2 = (10)_{10}$ can be specified as follows:

```
variable j : integer := "1010";
```

A.3.4 Data Types

The data objects that are discussed in Section A.3.3 must always associate a data type. VHDL has several in-built data types. The new data types can also be defined. The syntax for defining a new data type is

```
TYPE <type name> is <type>;
```

For example, to define a new data type called HEXADECIMAL, we can write,

```
TYPE HEXADECIMAL is ('0', '1', '2', '3', '4', '5', '6', '7', '8',
'9', 'A', 'B', 'C', 'D', 'E', 'F');
```

The existing data types are discussed in the following subsections.

BIT and BIT_VECTOR Types

The objects of type BIT can have values '0' or '1'. Similarly, the objects of type BIT_VECTOR can have values array of '0's or '1's. For example,

```
SIGNAL x : BIT;
SIGNAL y : BIT_BECTOR(3 downto 0);

x <= '1';
y <= "1100";
```

STD_LOGIC and STD_LOGIC_VECTOR Types

The objects with type STD_LOGIC can have any one of the nine values shown in Table A.1. The STD_LOGIC_VECTOR data objects can have array of these values.

Table A.1   Different values of STD_LOGIC

0	Logic 0
1	Logic 1
Z	High impedance
—	Don't care
L	Weak 0
H	Weak 1
U	Uninitalized
X	Unknown
W	Weak unknown

The STD_LOGIC data type is defined in the *std_logic_1164* package in *IEEE* library. So, to use this data type, we must include the following two statements at the beginning of the VHDL program:

```
library IEEE;
use IEEE.std_logic_1164.all;
```

STD_ULOGIC Type

The data objects of type STD_ULOGIC take the same value as type STD_LOGIC. But the only difference between them is that STD_ULOGIC type does not allow multiple sources for the same signal, whereas STD_LOGIC type allows multiple sources for the same signal. If there are multiple sources for a signal, its correct value is evaluated based on a *resolution function* which is defined under the std_logic_1164 package. For example, if a signal z has two drivers called x and y, and if x produces a value Z and y produces a value 1, then resolution function assigns logic 1 to the signal z.

SIGNED and UNSIGNED Types

The SIGNED and UNSIGNED types are used to indicate the number representation schemes in VHDL program. A SIGNED type is used for signed numbers in the 2's complement form, whereas UNSIGNED type is used for unsigned numbers. These types can have same values as the STD_LOGIC_VECTOR data types.

Integer Type

A data object of type integer type can have only integer values within the range $-(2^{31}-1)$ to $+(2^{31}-1)$. The integer signal has 32 bits. The integer can be written using the underscore character '_' for better readability. For example, 1_50_421 is the same as 150421.

Boolean Type

The data object of type Boolean can have only two values TRUE and FALSE. The TRUE indicates '1' and FALSE indicates '0'. For example,

```
SIGNAL flag : BOOLEAN := FALSE;
```

Enumeration Type

This is a data type which the user can define. The syntax for defining enumeration type is

```
TYPE <type name> is (<name> {, <name});
```

For example, to define an enumeration type MONTH, we can write,

```
TYPE MONTH is (JAN, FEB, MAR, APR, MAY, JUN, JUL, AUG, SEP, OCT,
NOV, DEC);
```

The most common use of the enumeration data type is to specify the states of finite state machine (FSM). For example, if a FSM has five states called S0, S1, S2, S3, and S4, they can be defined as

```
TYPE states IS (S0, S1, S2, S3, S4);
SIGNAL x : states;
```

The signal x can have any one of the defined five states.

Floating Point Type

The data object of this type can have values within a specified range of real numbers. The range is $-1.0e+38$ to $+1.0e+38$. For example,

```
variable width : REAL := 1.11;
```

It provides six decimal digits of precision. Floating point numbers can be expressed in the exponential form as

```
3.0e+10, 1.6e-19
```

Floating point values must have the decimal point in the number. For example,

```
variable X : real := 2.0 ; correct
variable X : real := 2 ; incorrect
```

Physical Types

The data objects of physical type represent physical quantities, such as length, time, voltage, current, etc. There is a smallest unit which is called *base unit*. Any value is expressed as an integral multiple of the base unit. For example,

```
TYPE VOLTAGE is range 0 to 1e+12

Units
  nV;                  -- nano-volt (base unit)
  uV   = 1000 nV;      -- micro-volt
  mV   = 1000 uV;      -- milli-volt
  Volt = 1000 mV;      -- volt
  kV   = 1000 Volt;    -- kilo-volt
end units;
```

VHDL has a predefined physical type for physical quantity time as given by

```
TYPE TIME IS RANGE 0 to 3.6e+18

UNITS
  fs;                  --femtosecond
  ps  = 1000 fs;       --picosecond
  ns  = 1000 ps;       --nanosecond
  us  = 1000 ns;       --microsecond
  ms  = 1000 us;       --millisecond
  sec = 1000 ms;       --second
  min = 60 sec;        --minute
  hr  = 60 min;        --hour
END UNITS;
```

Array Type

Array type is defined to group a number of same type data objects into a single data object. The syntax for array type is

```
TYPE <type name> is array (<range>) of <type>;
```

For example, the BIT_VECTOR and STD_LOGIC_VECTOR defined in VHDL are given by

```
TYPE BIT_VECTOR is array (natural range <>) of BIT;
TYPE STD_LOGIC_VECTOR is array (natural range <>) of STD_LOGIC_
VECTOR;
```

The syntax natural range <> allows users to specify range. Users can define array types on their own. To define an array BYTE of size 8, we can write

```
TYPE BYTE is array (7 downto 0) of STD_LOGIC;
SIGNAL X : BYTE;
```

A two-dimensional array is defined as

```
TYPE Matrix is array (3 downto 0) of STD_LOGIC_VECTOR(7 downto 0);
SIGNAL Y : Matrix;
```

An alternate way of defining a two-dimensional array is

```
TYPE rom_data IS ARRAY(0 TO <size>, 0 TO <width>) OF std_logic;
```

Array elements are accessed as

$X(i)$ for one-dimensional array
$Y(i)(j)$ or $Y(i,j)$ for two-dimensional array

File Type

The data objects of this type represent the physical file types. The syntax for defining a file type is

```
TYPE <file type name> is file of <type name>;
```

For example, to define a file type BITS we can write,

```
Type BITS is file of BIT_VECTOR;
```

VHDL has a predefined file type TEXT in the **std.textio** package.

A.3.5 Operators in VHDL

VHDL supports the following operators:
- Boolean operator
- Arithmetic operator
- Shift operator
- Relational operator
- Miscellaneous operator

Boolean Operators

There are seven Boolean operators defined in VHDL. These are given by
- NOT
- AND
- OR
- NAND
- NOR
- XOR
- XNOR

The NAND and NOR operators are not associative; they must be used with parenthesis if used multiple times. For example,

```
X <= A nand B nand C;   -- is not allowed
X <= (A nand B) nand C; -- is allowed
```

Arithmetic Operators

The arithmetic operators defined in VHDL are shown in Table A.2.

Table A.2   Arithmetic operators

Sl. No.	Operator	Significance
1.	+	Addition
2.	−	Subtraction
3.	&	Concatenation
4.	*	Multiplication
5.	/	Division
6.	MOD	Modulus
7.	REM	Remainder

Shift Operators

The shift operators are given in Table A.3.

Table A.3   Shift operators

Sl. No.	Operator	Significance
1.	SLL	Shift left logical
2.	SRL	Shift right logical
3.	SLA	Shift left arithmetic
4.	SRA	Shift right arithmetic
5.	ROL	Rotate left
6.	ROR	Rotate right

Relational Operators

The relational operators are given in Table A.4.

Table A.4   Relational operators

Sl. No.	Operator	Significance
1.	=	Equality
2.	/=	Inequality
3.	<	Less than
4.	<=	Less than or equal
5.	>	Greater than
6.	>=	Greater than or equal

Miscellaneous Operators

VHDL supports other operators as given in Table A.5.

Table A.5   Miscellaneous operators

Sl. No.	Operator	Significance
1.	ABS	Absolute
2.	**	Exponentiation

Operator Precedence

VHDL operators' precedence is given in Table A.6.

Table A.6 Operator precedence

Precedence	Operator class	Operator
Highest	Miscellaneous	**, ABS, NOT
	Multiplication	*,/, MOD, REM
↓	Sign	+, −
	Addition	+, −, &
	Relational	=,/=, <, <=, >, >=
Lowest	Logical	AND, OR, NAND, NOR, XOR, XNOR

A.3.6 Hardware Modelling

In VHDL, each hardware module is modelled as an *entity*. The entity has two parts:
- Entity declaration
- Architecture body

Entity Declaration

The entity represents the external view of the hardware. It defines the external interfaces which are called *ports*. The ports are defined with the signal directions and their types. The syntax for entity declaration is given by

```
entity <entity name> is
  [generic <generic names and their types>;]
  [port <port names, directions, and types>;]
end [entity] [<ntity name>];
```

Let us consider a full-adder as shown in Fig. A.1. It has three input ports A, B, and C_{in}, and two output ports Sum and C_{out}.

The entity declaration for a full-adder is given by

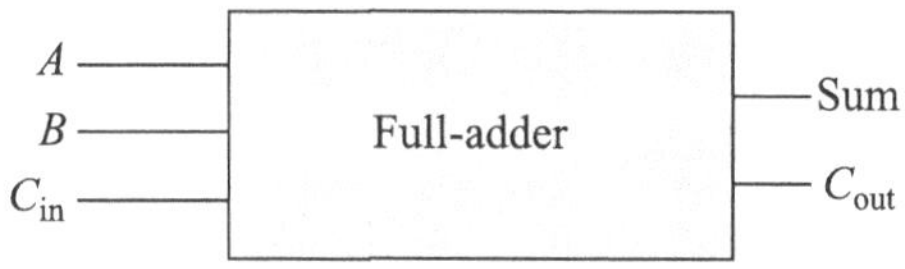

Fig. A.1   Block diagram of a full-adder

```
entity full-adder
  port( A, B, Cin : in STD_
  LOGIC;
    Sum, Cout : out STD_LOGIC);
end full_adder;
```

Port Direction

The port must have one of the following directions:
1. IN: a port with direction IN signifies that a signal is applied at this port.
2. OUT: a port with direction OUT signifies that a signal is taken out of this port.
3. INOUT: a port with INOUT direction signifies that signal is taken out and applied to this port.

4. BUFFER: a port with direction BUFFER signifies that the port is read and updated within the entity.

Architecture Body

The architecture of an entity represents the internal or functional view of the hardware. The syntax of defining an architecture body is given by

```
architecture <entity name> of <entity name> is
  [component declarations]
begin
  Concurrent statements;
  Sequential statements;
end [architecture] [<architecture name>];
```

For example, the architecture of a full_adder is given by

```
architecture dataflow of a full_adder is
begin
  Sum <= A xor B xor Cin;
  Cout <= (A and B) or (B and Cin) or (A and Cin);
end dataflow;
```

Signal Assignment Statements

The signal assignment statements are defined to assign values to the signals using the syntax:

```
<signal name> <= <expression> [after <delay value>];
```

For example, signal Sum of full-adder is written as

```
Sum <= A xor B xor Cin after 2 ns;
```

If any delay is not specified, a default delay is used.

Default Delay

When users do not specify any delay after the expression in the signal assignment statement, the simulator uses a default delay value. This default delay value is called *delta delay* (Δ), which is infinitesimally small time, but not equal to zero.

Figure A.2 illustrates the delay in signal assignment of signal Y for an input signal A.

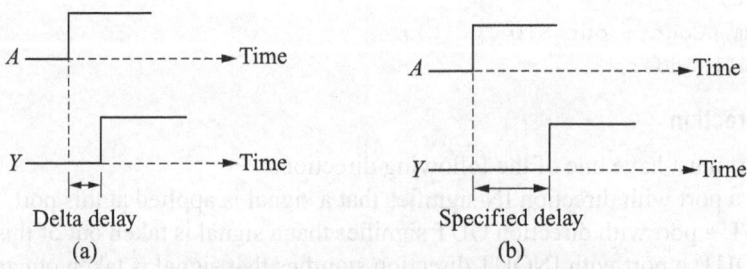

Fig. A.2 Delay in signal assignment: (a) delta delay; (b) specified delay

Variable Assignment Statement

The syntax for variable assignment statement is as follows:

```
<variable name> := <expression>;
```

For example,

```
count := count + 1;
```

Variables are similar to *static variable* in C language. They hold their values until the simulation is terminated.

Wait Statement

The wait statement is used to suspend a process from execution. There are three ways by which users can suspend a process. These are:
- wait on <sensitivity list>;
- wait until <Boolean expression>;
- wait for <time expression>;

For example,

```
WAIT ON X, Y;
WAIT UNTIL clock = '1' AND clock'EVENT;
WAIT FOR 2 ns;
```

A.3.7 Component Declaration

In the structural modelling style, first component must be declared and then they can be instantiated. The syntax for component declaration is given by

```
component <component name> [is]
  [port (list of ports);]
end component [<component name>];
```

For example, a two-input OR gate is declared as a component given by

```
component OR21 is
  port(a, b : in std_logic;
    c : out std_logic);
end component;
```

A.3.8 Component Instantiation

Once the components are declared, they can be used or instantiated in a design. The syntax for component instantiation is given by

```
<instance name> : <component name> [port map(<connectivity list>)];
```

For example, the two input OR gate is instantiated as follows:

```
U1 : OR21 port map(X, Y, Z);
```

A port can be left unconnected and it is modelled using the keyword `open` as

```
U2 : OR21 port map(X, open, Z);
```

In the above two examples, the connectivity of the signals to the ports of the component is according to the order in which they are specified. For example, in the instance U1, signal *X* is connected to port *a*, signal *Y* is connected to port *b*, and signal *Z* is connected to port *c*.

The connectivity information between the signals and the ports of the component can be mapped one-to-one. The syntax for the mapping is given by

```
<port name 1> => <signal name 1>, <port name 2> => <signal name 2>,
..., <port name n> => <signal name n>
```

For example, two-input OR gate can be instantiated as

```
U3 : OR21 port map (a => X, c => Z, b => Y);
```

Note that the ordering of the ports connecting to the signals is not important here.

A.3.9 Generic Declarations

The generics are declared within the entity of a design. These are used to define some parameters that are used to pass values to the design. These can be thought of as the arguments in a C program. The syntax is

```
entity OR_N is
  generic ( N : interger := 4);
  port( A : in BIT_VECTOR(1 to N);
    Y : out BIT);
end OR_N;
```

The generics can be used in component declaration and a generic map is used to pass the value of the parameter defined as generic. For example, a two-input OR gate is modelled using generic to define two delay values, TPHL (propagation delay for high-to-low transition), and TPLH (propagation delay for low-to-high transition) as

```
component OR21
  generic(TPHL, TPLH : TIME);
  port (A, B: in BIT; C : out BIT);
end component;
```

The two-input OR gate is instantiated with specifying the delay values as

```
U1 : OR21 generic map(4 ns, 3 ns) port map(X, Y, Z);
```

A.3.10 Statements in VHDL

Let us now describe different VHDL statements. There are two types of statements in VHDL: concurrent and sequential.

Concurrent Statements

Generally, for any programming language, the statements are executed one after another. But in VHDL, a new concept is introduced. The statements can be executed simultaneously. These types of statement are known as *concurrent statements*.

Concurrent Signal Assignment Statement

Concurrent signal assignment statements are specified in the architecture body of an entity. There can be many concurrent signal assignment statements in an architecture body. An example of concurrent signal assignment statements is given by

```
begin
  Sum <= A xor B xor Cin;
  Cout <= (A and B) or (B and Cin) or (A and Cin);
end dataflow;
```

As the concurrent signal assignment statements are executed simultaneously, their order is not important.

Conditional Signal Assignment Statement

These statements are used to assign a signal value based on some conditions. The syntax is given by

```
<signal> <= [<value> when <:condition 1> else]
            [<value> when <condition 2> else]
            . . .
            <value> [when <condition>];
```

For example, the *Y* output of 2:1 MUX can be written as

```
Y <= A when S = '0' else B;
```

Selected Signal Assignment Statement

It is used to select a signal value based on select expressions. The syntax is given by

```
With <expression> select
<signal> <= <expression 1> when <condition 1>,
            <expression 2> when <condition 2>,
            . . .
            <expression n> when <condition n>;
```

For example, 4:1 MUX is modelled as follows:

```
Signal S : integer;
With S select
  y <= i0 WHEN 0,
       i1 WHEN 1,
       i2 WHEN 2,
       i3 WHEN 3;
```

It is similar to the *case* statement.

Block Statement

The block statement allows users to partition the design into blocks. The syntax for the block statement is given by

```
<block label> : block [(<guard expression>)] [is]
  [<lock header>]
  [<block declaration>]
begin
  Concurrenpt statements;
end block [<block label>];
```

For example,

```
B1 : block (EN = '1')
begin
  Y <= guarded A;
end block B1;
```

When the guard expression (EN = '1') is True, signal *A* is assigned to signal *Y*.

Concurrent Assertion Statement

The assertion statement can be concurrent or sequential depending on its position. Its syntax is given by

```
assert <Boolean expression> [report <string>] [severity <expression>];
```

For example,

```
assert not (S = '1' and R = '1')
  report "Invalid inputs!"
  severity ERROR;
```

Sequential Statements

In addition to the concurrent statements, VHDL supports the sequential statements like any other programming language, C, C++, etc.

Process Statement

A process statement is used to write a set of sequential statements under it. The process statement appears in the architecture of an entity, and is a concurrent statement. The set of statements appearing under a process statement are sequential statements. Its general syntax is given by

```
[<process label>] : process [(<sensitivity list>)]
  [process item declarations >];
begin
  <sequential statements>;
end process [<process label>];
```

For example, a D flip-flop can be modelled as

```
process
begin
  if (clk = '1') then
    Q <= D;
  end if;
end process;
```

The above example can be rewritten as follows:

```
process(D, clk)
begin
  if (clk = '1') then
    Q <= D;
  end if;
end process;
```

When a process statement is specified with the signal names in parentheses, it specifies that the process block is evaluated when any signal appearing in the list changes. The list of signals specified within parentheses is called *sensitivity list*. When no sensitivity list is specified, the process must have at least one wait statement. Otherwise, the process will remain in infinite loop during the initialization phase of simulation.

If **Statement**

The syntax of if statement is given by

```
if < Boolean expression> then
   <sequential statements>;
elsif <Boolean expression> then
   <sequential statements>;
else
   <sequential statements>;
end if;
```

Example A.1   D flip-flop

```
entity DFF is
  port(CLK, RST, D : in std_logic;
    Q, QBAR : out std_logic);
end DFF:

architecture behv of DFF is
begin
  process(RST, CLK)
    variable state: std_logic := 'Z';
  begin
    if (RST = '0' ) then
      state <= 'Z';
    elsIf (clk = '1' and CLK'event ) then
      state <= D;
    end if;
    Q <= state;
    QBAR <= not state;
  end process;
end behv;
```

The input and output waveforms of the D flip-flop is shown in Fig. A.3.

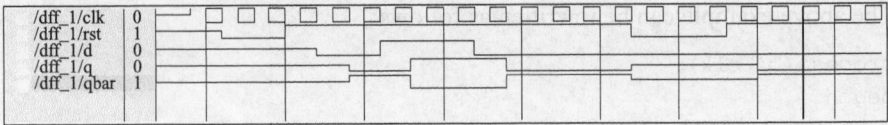

/dff_1/clk	0
/dff_1/rst	1
/dff_1/d	0
/dff_1/q	0
/dff_1/qbar	1

Fig. A.3   Simulated output of the D flip-flop

Case Statement

It must be within a process because it is a sequential statement. Its general syntax is given by Case <expression> is

```
when <choice> => <sequential statement>;
.
.
.
when others => <sequential statement>;
end case;
```

Example A.2   2:4 Decoder

```
library IEEE;
use IEEE.STD_LOGIC_1164.ALL;
use IEEE.STD_LOGIC_ARITH.ALL;
use IEEE.STD_LOGIC_UNSIGNED.ALL;

Entity decoder is

  port (S : in  std_logic_vector(1 downto 0);
    Q : out std_logic_vector(3 downto 0));
end decoder;

architecture behv of decoder is

begin
  process(S)
  begin
    case S is
      when "00" => Q <= (0 => '1', others => '0');
      when "01" => Q <= (1 => '1', others => '0');
      when "10" => Q <= (2 => '1', others => '0');
      when others => Q <= (3 => '1', others => '0');
    end case;
  end process;
end behv;
```

The input and output waveforms of the 2:4 decoder is shown in Fig. A.4.

| ⊞ | /decoder/s | 00 | 00 | 00 | 01 | 10 | 11 | 00 |
| ⊞ | /decoder/q | 0001 | 1000 | 0001 | 0010 | 0100 | 1000 | 0001 |

Fig. A.4   Simulated output of the 2:4 decoder

Generate Statement

VHDL supports two types of *generate* statements: (1) *if-generate* and (2) *for-generate*. The following example illustrates the `for-generate` statement:

Example A.3   Design a 16:1 multiplexer using 4:1 multiplexers.

Solution First, we have to design a 4:1 multiplexer. Let us design a 4:1 multiplexer using dataflow modelling style. The VHDL code for 4:1 multiplexer is given by

```
library ieee;
use ieee.std_logic_1164.all;

entity mux4to1 is

  port( I : in  std_logic_vector(0 to 3);
    Sel : in std_logic_vector(1 downto 0);
    Y : out std_logic);
end mux4to1;

architecture behv of mux4to1 is

begin
  Y <= ( (not Sel(1)) and (not Sel(0)) and I(0) ) or
       ( (not Sel(1)) and (Sel(0)) and I(1) ) or
       ( (Sel(1)) and (not Sel(0)) and I(2) ) or
       ( (Sel(1)) and (Sel(0)) and I(3) );
end behv;
```

The input and output waveforms of the 4:1 multiplexer is shown in Fig. A.5.

⊞	/mux4to1/i	0110	0110						1010
⊞	/mux4to1/sel	00	00	10	11	00	01	00	
	/mux4to1/y	0							

Fig. A.5   Simulated output of a 4:1 multiplexer

Now a 16:1 multiplexer is designed using 5 multiplexers as shown in Fig. A.6.

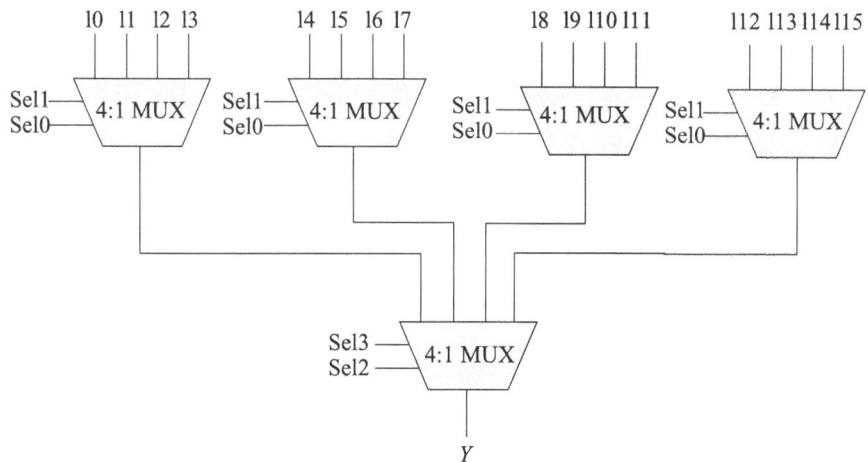

Fig. A.6   A 16:1 multiplexer constructed using four 4:1 multiplexers

The VHDL code for a 16:1 multiplexer using structural modelling style is given by

```
library ieee;
use ieee.std_logic_1164.all;
entity mux16to1 is
  port( I : in std_logic_vector(0 to 15);
    Sel : in std_logic_vector(3 downto 0);
    Y : out std_logic);
end mux16to1;

architecture structure of mux16to1 is
component mux4to1
  port( I : in std_logic_vector(0 to 3);
    Sel : in std_logic_vector(1 downto 0);
    Y : out std_logic);
end component;

  signal x : std_logic_vector(0 to 3);
begin
  gen1: for j in 0 to 3 generate
    IMUX : mux  4to1 port map( I((4*j) to (4*j+3)), Sel(1 downto
    0), x(j));
  end generate;
  IMUX5 : mux4to1 port map(x(0 to 3), Sel(3  downto 2), Y);
end structure;
```

The input and output waveforms of the 16:1 multiplexer is shown in Fig. A.7.

If-generate **Statement**

The syntax for if-generate statement is as follows:

```
<label> : if <expression> generate
  [declarations];
begin
  Concurrent statements;
end generate [<label>];
```

The following example illustrates the use of if-generate statement:

Example A.4   Design a 4-bit synchronous counter.

```
entity count4bit is

  port(clk, count  : in std_logic;
    Q : out std_logic_vector(3 downto 0));
end count4bit;

architecture struc of count4bit is

component
  port (D, CLK: in std_logic;
    Q : out std_logic);
end component;
```

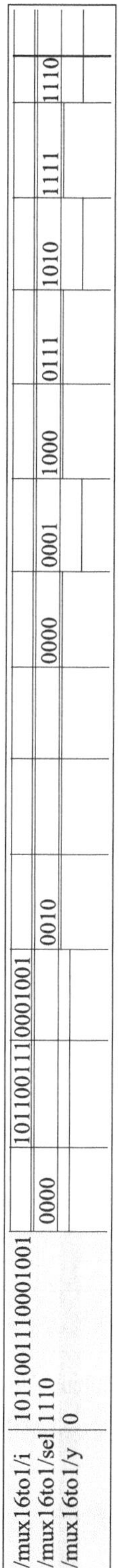

/mux16to1/i	1011001110001001	1011001110001001								
/mux16to1/sel	1110	0000	0010	0000	0001	1000	0111	1010	1111	1110
/mux16to1/y	0									

Fig. A.7 Simulated output of 16:1 multiplexer

```
begin
  g1: for j in 0 to 3 generate
    d0 : if j = 0 generate
      DFF : DFF port map(clk, count, Q(j));
    end generate d0;
      d1 : if j > 0 generate
      DFF : DFF port map(clk, Q(j-1), Q(j));
    end generate d1;
  end generate g1;
end struc;
```

Select Statements

The `select` statement in VHDL is used to select a signal value based on the selection criteria. The `select` statement is used along with the *when* statement. This is illustrated in the following example:

Example A.5 Design a 2:1 multiplexer.

Solution
```
library ieee;
use ieee.std_logic_1164.all;

entity mux2to1 is

  port(i0, i1, sel : in std_logic;
    Y : out std_logic);
end mux2to1;

architecture behv of mux2to1 is

begin
  with sel select
    Y <= i0 when '0',
      i1 when others;
end behv;
```

The input and output waveforms of the 2:1 multiplexer is shown in Fig. A.8.

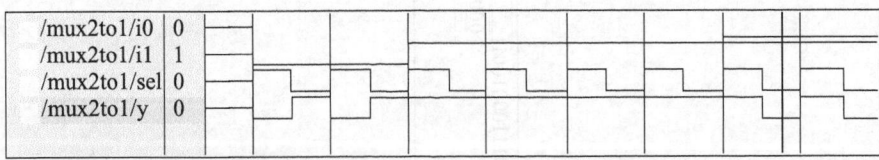

/mux2to1/i0	0
/mux2to1/i1	1
/mux2to1/sel	0
/mux2to1/y	0

Fig. A.8 Simulated output of the 2:1 multiplexer

The above example can also be implemented using only `when-else` statement as shown below.

```
architecture behv of mux2to1 is

begin
    Y <= i0 when '0' else i1;
end behv;
```

Loop Statement

The `loop` statement is used for repetitive operations. The syntax of `loop` statements is given by

```
[<loop label>:]
for <variable name> in <range> loop
  <statement>;
  .
  .
  .
end loop [<loop label>];
```

The following example illustrates the `loop` statement:

Example A.6   Design a circuit to count number of 1's in a stream of bits.

```
library ieee;
use ieee.std_logic_1164.all;

entity count1 is

  port( data : in std_logic_vector(0 to 3);
    Count : buffer integer range 0 to 3);
end count1;

architecture behv of count1 is

begin
  process(data)
  begin
    Count <= 0;
    for i in 0 to 3 loop
      if data(i) = '1' then
        Count  <= count + 1;
      end if;
    end loop;
  end process;
end behv;
```

Exit Statement

The `exit` statement is used inside a loop to break the loop. It is a sequential statement. The syntax of `exit` statement is given by

```
exit [<loop label>] [when <condition>];
```

For example,

```
for i in 0 to 10 loop
  exit when A(i) = B(i);
end loop;
```

Next Statement

The next statement is similar to the exit statement. The only difference is that an exit statement breaks the loop, whereas the next statement skips the current iteration and returns to the next iteration. Its syntax is given by

```
next [<loop label>] [when <condition>];
```

Assertion Statement

The assertion statements are used to check certain values and flag messages accordingly. It is very useful for checking constraints in the design. For example, to check for setup and hold times of flip-flops, the assertion statement is used. The syntax for assertion statements is given by

```
assert <Boolean expression> [report <string>] [severity <expression>];
```

The *string* is reported if the Boolean expression is False. There are four predefined severity level of assertion statement. These are given by
- NOTE
- WARNING
- ERROR
- FAILURE

Report Statement

The report statement is used to print a message. Its syntax is given by

```
report <string> [severity <expression>];
```

For example,

```
if A = '1' then
  report "Signal A is high";
end if;
```

A.3.11 Library

In VHDL, there is a concept of library. A library is a collection of packages and design entities. There is a pre-defined library called STD in VHDL. The STD library contains two packages given by
- STANDARD
- TEXTIO

There is another library called IEEE which contains a package STD_LOGIC_1164. This package defines the nine-value logic STD_LOGIC.

The working library is logically named as WORK. The current design is compiled under the library WORK. Users can create their own libraries and import into the WORK library. For example, MYLIB is a user created library, as shown in Fig. A.9.

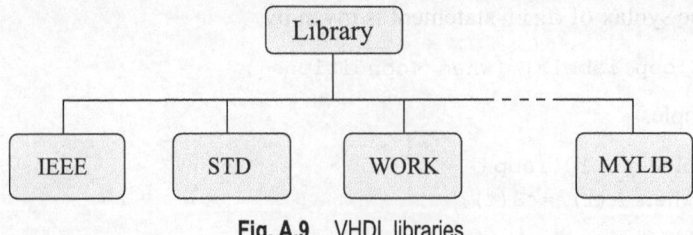

Fig. A.9 VHDL libraries

A.3.12 Package

A package contains functions, procedures, data types, declarations, etc. that are used in other design entities. A package has the following two parts:

- Package declaration
- Package body

A package declaration defines the interface to the package. It contains the declarations of the package components. The syntax for a package declaration is given by

```
package <package name> is
  package components;
  .
  .
  .
end [package] [<package name>];
```

A.3.13 Using Library and Package in VHDL

The library is included in a VHDL program with the following syntax:

```
library <list of library names>;
```

For example, to include libraries IEEE and STD in a VHDL program, we can write:

```
library IEEE, STD;
```

Including libraries enables a VHDL program to access predefined functions, procedures, object data types, etc., through the packages declared inside the libraries. The packages declared within the libraries are included using the following syntax:

```
use <library name>.<package name>.all;
use <library name>.<package name>.<component>;
use <library name>.<package name>;
```

If .all is specified at the end of the use statement, then it indicates that all components declared in the package are accessible. If .<component> is specified at the end of the use *statement*, it indicates only that component declared in the package is accessible. When no extension is specified, the components are accessed using the package name. The examples for these three cases explain the usage of the use *statement*.

```
library ieee;
use ieee.std_logic_1164.all;
entity OR21 is
  port(A, B : in std_logic;
    Y : out std_logic);
end OR21;

library ieee;
use ieee.std_logic_1164.std_logic;
entity OR21 is
  port(A, B : in std_logic;
    Y : out std_logic);
end OR21;
```

```
library ieee;
use ieee.std_logic_1164;
entity OR21 is
  port(A, B : in std_logic_1164.std_logic;
    Y : out std_logic_1164.std_logic);
end OR21;
```

SELECT REFERENCES

Bhasker, J. 2003, *A VHDL Primer*, Pearson Education.

Brown, S. and Z. Vranesic 2002, *Fundamentals of Digital Logic with VHDL Design*, Tata McGraw-Hill, New Delhi.

Ciletti, M.D. 2005, *Advanced Digital Design with the Verilog HDL*, Pearson Education, New Delhi.

Hayes, J.P. 1998, *Computer Architecture and Organization*, 3rd ed., McGraw-Hill International Editions.

Mano, M.M. 2001, *Computer System Architecture*, Prentice-Hall.

Perry, D.L. 2002, *VHDL: Programming by Example*, 4th ed., McGraw-Hill.

Smith, M.J.S. 2002, *Application Specific Integrated Circuits*, Pearson Education.

Weste, N.H.E., D. Harris, and A. Banerjee 2009, *CMOS VLSI Design: A Circuits and Systems Perspective*, 3rd ed., Pearson Education.

Digital Design Using Verilog

B.1 Introduction

In Appendix A, we have introduced the hardware description language (HDL). There are two widely used HDLs: VHDL and Verilog. As we have discussed VHDL in Appendix A, we shall discuss the digital design using Verilog HDL in the present chapter.

The Verilog model of a circuit is a description of the circuit functionality with its input and output ports. The functionality can be expressed either in structural or behavioural view. A structural view represents the circuit in terms of its blocks and their connections. The behavioural view normally represents the algorithm at register transfer level (RTL) or by a set of Boolean expressions.

B.2 Verilog Naming Conventions

Following are the naming conventions used for Verilog.
1. Verilog is a case-sensitive language.
2. The identifier in Verilog can be a combination of the following characters. [A-Z][a-z][0-9]_$. The identifier must not start with $ or a digit and must not exceed 1024 characters.
3. Each line must end with a semicolon (;) except the last line (*endmodule*).
4. Comments are specified in two ways: a pair of back slashes, // to represent in-line comments, and multiline comment is specified by symbol-pair /* followed by text followed by */, just like C++ language.

B.3 Some Other Verilog Naming Conventions

- An array of signals are represented by vectors in Verilog—A[3:0] represents A as 4-bit signal A(3)A(2)A(1)A(0).
- The interconnects or nets are represented by a keyword *wire* in Verilog. Default type of an identifier is *wire*.
- The edge-triggered flip-flops are treated as registers (*reg*) in Verilog. Registers are described by `always @(posedge clk)` statement.

B.4 Operators in Verilog

All the Verilog operators are listed in Table B.1.

Table B.1 Operators in Verilog

Sl. No.	Operator	Operation	Sl. No.	Operator	Operation
1.	~	Bit-wise complement	16.	!	Logical negation
2.	&	Bit-wise AND	17.	= =	Logical equal
3.	|	Bit-wise OR	18.	! =	Logical not equal
4.	~&	Bit-wise NAND	19.	&&	Logical AND
5.	~|	Bit-wise NOR	20.	||	Logical OR
6.	^	Bit-wise XOR	21.	<	Less than
7.	~ ^	Bit-wise XNOR	22.	<=	Less than or equal
8.	^ ~	Bit-wise XNOR	23.	>	Greater than
9.	+	Plus (addition)	24.	>=	Greater than or equal
10.	−	Minus (subtraction)	25.	? :	Conditional (if-then-else)
11.	<<	Logical shift left	26.	<<<	Arithmetic shift left
12.	>>	Logical shift right	27.	>>>	Arithmetic shift ight
13.	*	Multiply	28.	= = =	Case equal
14.	/	Divide	29.	! = =	Case not equal
15.	%	Modulus	30.	{ }	Concatenation

B.5 Verilog Data Types

There are several data types in Verilog. The most commonly used data types are as follows:

- *nets*—connects structural elements; nets are of four types: wire, tri, supply1, and supply 0.
- *registers*—used as abstract data storage elements
- *integer*—32-bit in 2's complement format
- *time-64*—bit in unsigned format
- *event*
- *real-64*—bit real number

The default initial value of wire is 'z'. A wire cannot store or hold a value. It must be driven by an assignment statement.

A *wire* or *reg* can be declared as vector with a specified range of bits. For example,

```
wire [7:0] DataBus; // declares DataBus as 8-bit bus
reg [15:0] A; // declares A as 16-bit register
```

Memory can be declared as array of registers as

```
reg [15:0] Ram[7:0]; /* declares 8 Ram locations with
                16-bit word size*/
```

B.6 Numbers in Verilog

Numbers can be of two types as follows:

- Integer constants
- Real constants

B.6.1 Integer Constants

Integer constants are written as

```
width'radix value
```

Radix indicates base of the number (d/D for Decimal, h/H for Hexadecimal, o/O for Octal, b/B for Binary).

B.6.2 Real Constants

Real constants are written in decimal or scientific notation as follows:

```
parameter LENGTH = 'h 10; // (10)H = (16)D
parameter WORD = 8'B 0010_1100 // (00101100)b = (44)d

real R1, R2, R3;
R1 = 2.718; // defines the value of e
R2 = 1.6e-19; // defines electronic charge
R3 = 0.314e+01 // defines the value of pi
```

B.6.3 Negative Numbers

Negative numbers are represented in 2's complement form as

```
parameter x1 = -12; // (-12)d = (FFFFFFF4)h
```

B.6.4 Strings

Strings are used to represent alphanumeric characters, such as,

```
parameter STR1 = "Hello"; // string constant
parameter STR2 = "Say \"Hello\""; // escape character is '\'
```

B.7 Four-value Logic

Verilog uses four-value logic for modelling constructs and truth tables. These are given in Table B.2.

Table B.2   Four-value logic in Verilog

Sl. No.	Value	Description
1.	0	Logic 0
2.	1	Logic 1
3.	x	Unknown (ambiguous)
4.	z	High impedance

B.8 Behavioural Modelling

The behavioural modelling is the style of modelling the design without going into actual hardware implementations. It merely describes the functionality of the design. Hence, this is the most suitable modelling style for prototyping a design quickly without knowing how it can be implemented.

B.8.1 Behavioural Modelling Using Boolean Expression

A Boolean expression can be used to express logic functionality in Verilog with a continuous assignment statement. For example, the two-input AND gate can be expressed as

```
module my_and (Y, A, B);
  input A, B;
  output Y;
```

```
 assign Y = A & B;
endmodule
```

B.8.2 Propagation Delay

Propagation delay can be modelled using continuous assignment statement as

```
module AOI_gate (Y, A, B, C);
  input A, B, C;
  output Y;

  wire #1 y1 = A & B;
  wire #2 y2 = B & C;
  wire #3 y3 = A & C;
  wire #2 Y = (y1 | y2 | y3);
endmodule
```

Example B.1 Write a Verilog program for Latch.

```
module Latch(Q, D, EN);
  input EN, D;
  output Q;

  assign Q = EN ? D : Q;
endmodule
```

B.9 Structural Modelling

Structural modelling starts with the schematic of the design which is the interconnection of its functional blocks. The schematic contains the logic blocks or gates, their interconnections, the input and output pins, and internal signal names. In Verilog, there are 26 predefined functional models which are called *primitives*. The primitives are the basic building blocks (e.g., AND, OR, XOR). The primitives are instantiated in a *module* which is a hardware abstraction of the design similar to the *entity* in VHDL (see Fig. B.1).

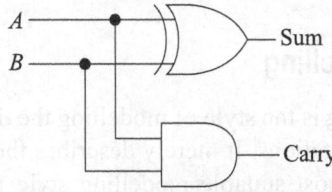

Fig. B.1 Schematic of a half-adder

Verilog code for the half-adder in structural modelling style can be written as

```
module Half_adder (Sum, Cout, A, B);
  input A, B;
  output Sum, Cout;
  xor (Sum, A, B);
  and (Cout, A, B);
endmodule
```

A *module* is an encapsulation of the functionality of the design and its interface to the external world. A module generally has three components:

- The functionality of the design.
- Input and output ports to interface the external world.
- The timing and other attributes of the design (such as physical area).

The design description starts with the keyword *module* and ends with the keyword *endmodule*. The name of the module can be specified by the user followed by the list of ports separated by comma, and enclosed by parentheses. The syntax is given below:

```
module <my_design> (port names);
<port declarations>
<functional descriptions>
.
.
.
endmodule
```

The ports of the module have to be associated with their modes. The *mode* of a port represents the direction of signal flow. These are

- Input signals applied to the module from outside.
- Output signals generated by the module to the external world.
- Inout bidirectional ports which signifies signal flow in either direction.

B.10 Delay Modelling in Verilog

The propagation delay through the logic gates can be modelled in Verilog. Propagation delay is the time delay between the input and output waveforms. By default, the time delay is assumed to be zero in Verilog. But the delay through a *primitive* (basic gate) or *wire* (net) can be defined. This is defined as follows:

```
module Half_adder (Sum, Cout, A, B);
  input A, B;
  output Sum, Cout;
  xor #1(Sum, A, B);
  and #1(Cout, A, B);
endmodule
```

The unit delay is specified by #1 before the instantiation of the primitives xor and and. The following statement at the beginning of Verilog program:

```
`timescale 1 ns/1 ps
```

signifies timescale directive which means time unit is *nanosecond* and the time resolution is *picosecond*.

B.10.1 Inertial Delay

It is the propagation delay of the *primitive* gates. For an input pulse to be propagated through, a *primitive* gate must have pulse width greater or equal to the inertial delay. If the input pulse width is less than the inertial delay of the gate, the pulse is not passed through the gate.

B.10.2 Transport Delay

It is the time delay through a *wire*. This delay model does not suppress any narrow input pulse. Every transition in the input pulse is propagated through the *wire* after a finite transport delay. Transport delay is declared along with the *wire* declaration. For example, a *wire* with *wire* name *my_wire* with transport delay of 3-time units can be specified as follows:

```
wire my_wire #3;
```

B.10.3 Min:Nom:Max Delay Modelling in Verilog

We have introduced the process, temperature, and voltage (*PTV*) corners in Chapter 3 while describing the library and cell characterization. The propagation delay of a logic gate depends on the *PTV* conditions. Typically, three sets of delay values are considered as explained in Table B.3.

Table B.3 Delay vs *PTV* conditions

Sl. No.	*PTV* condition	Process	Temperature	Voltage	Delay values
1.	Best case	Strong	Low	High	Minimum
2.	Nominal case	Nominal	Room	Nominal	Nominal
3.	Worst case	Weak	High	Low	Maximum

The propagation delay can also be different for rise and fall transitions. In Verilog, we can model the delay values for rise and fall transitions separately, and for three different *PTV* conditions. For example,

```
timescale 1ns/1ps
and #(1.0:1.5:2.0, 1.1:1.2:1.3) and2(c2, s1, Cin);
```

This example describes that the instance **and2** of the *primitive* **and** has delay values for rise transition as min delay = 1.0 ns, nom delay = 1.5 ns, and max delay = 2.0 ns. Similarly, the delay values for fall transition is given as min delay = 1.1 ns, nom delay = 1.2 ns, and max delay = 1.3 ns.

Example B.2   Design a full-adder with min:nom:max delay modelling.

```
module full_adder(A, B, Cin, Sum, Cout);
  input A;
  input B;
  input Cin;
  output Sum;
  output Cout;

  wire s1, c1, c2;

  xor #1.0 xor1(s1, A, B); //equal rise & fall delay
```

```
xor #(1.0:2.0:3.0) xor2(Sum, s1, Cin);
   //equal rise & fall delay

and #3.0 and1(c1, A, B); //equal rise & fall delay
and #(1.0:1.5:2.0, 1.1:1.2:1.3) and2(c2, s1, Cin);
   //different rise & fall delay

or or1(Cout, c1, c2);

endmodule
```

B.11 Truth Table Model with Verilog (User-defined Primitive)

Verilog supports *user-defined primitives* (UDP) in the form of a truth table. UDP is an alternative of *module* in Verilog. These are faster to simulate and requires less memory as compared to *modules*. UDPs are mostly used in ASIC cell libraries.

The syntax for defining a primitive is illustrated in the following example.

Example B.3   Design a UDP for a three-input XOR gate.

```
primitive XOR31_UDP(Y, A, B, C);
  output Y;
  input A, B, C;
  table
//A B C : Y
 0 0 0 : 0;
 0 0 1 : 1;
 0 1 0 : 1;
 0 1 1 : 0;
 1 0 0 : 1;
 1 0 1 : 0;
 1 1 0 : 0;
 1 1 1 : 1;
endtable
endprimitive
```

B.12 Assignment Statements

There are two types of assignment statements in Verilog:
■ Continuous
■ Procedural

A continuous assignment statement assigns a value to a *wire*. For example,

```
module my_inv1(a, y);
  input a;
  output y;
  assign y = ~ a; // continuous assignments statement
endmodule
```

The procedural assignment statement assigns a value to a *reg*. For example,

```
module my_inv2(a, y);
  input a;
  output y;
  always #1 assign y = ~ a; // procedural assignment
              // statement
endmodule
```

The procedural assignment statements must be under `always` or `initial` statements. These statements are executed sequentially.

B.13 Sequential Block

A sequential block is a set of statements within a `begin` and an `end` under `always` or `initial` statements. An `always` statements are executed repeatedly whereas, the `initial` statement is executed only once at the beginning of the simulation.

```
module clk_gen;
  reg clk, y;

  always
  begin: my_block
    @(negedge clk) #4 Y = 1;
    @(negedge clk) #4 Y = 0;
  end
  always #10 clk = ~ clk;
  initial y = 0;
  initial clk = 0;
endmodule
```

B.14 Wait Statement

In Verilog, there is `wait` statement which is used to suspend a *procedure* until a condition becomes true. For example,

```
module DFF(D, Q, CLK, Reset);
  input D, CLK, Reset;
  output Q;
  reg Q;
  wire D;
  always @(negedge CLK) if (Reset !== 1) Q = D;
  always
  begin
    wait (Reset == 1) Q = 0;
    wait (Reset !== 1);
  end
endmodule
```

The wait statement is level-sensitive—it only checks if the condition is true.

B.15 Procedures in Verilog

The procedure in Verilog consists of the following statements:
1. Always statement
2. Initial statement
3. A task
4. A function

A task is *procedure* which is called from another *procedure*. It has inputs and outputs but it does not return any value.

A function is a procedure (like a subroutine) which must have at least one input, and it returns a value. An example of a function is given below:

```
module my_module;
  reg [2:0] A, B;
  intial begin A = 1; B = 0;
  C = my_decode(A,B);

  function [2:0] my_decode;
    input [2:0] a, b;
    begin
      if (a <= b)
    my_decode = a;
      else
    my_decode = b;
    end
    endfunction
endmodule
```

B.16 Control Statements

Different control statements in Verilog are discussed in the following subsections.

B.16.1 Case Statement

The syntax for case statement is as follows:

```
module mux21(a, b, s, y);
  input a;
  input b;
  input s;
  output y;

  reg y;
  always begin
    case(s)
      0: y = a;
      1: y = b;
```

```
    default: y = 1'b0;
  endcase
    #1;
end

endmodule
```

The input and output waveforms are shown in Fig. B.2.

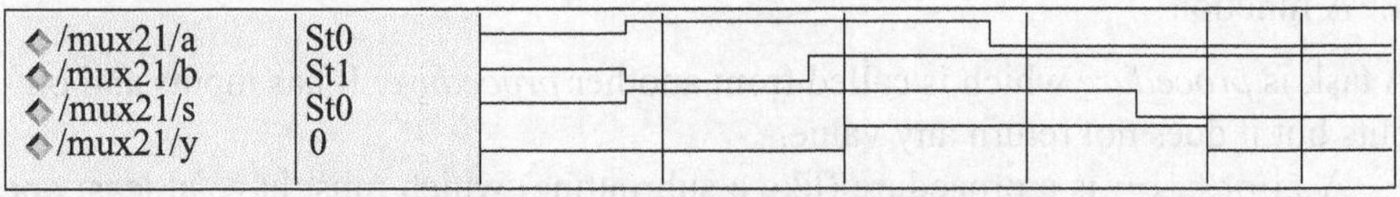

◇/mux21/a	St0										
◇/mux21/b	St1										
◇/mux21/s	St0										
◇/mux21/y	0										

Fig. B.2　Simulated output of 2:1 multiplexer

B.16.2 Loop Statement

A loop statement can be any of the following statements:

■ for
■ while
■ repeat
■ forever

Example of loop statements having for, while, repeat, and forever statements is given below:

```
module loop1();

  integer i;
  reg [31:0] A, B, C, D;
  initial A = 0;
  initial B = 64;
  initial C = 1023;
  initial D = 0;

  initial begin
    for(i=0; i <= 15; i=i+1) A[i] = 1;
    i = 30;
    while(i <= 31) begin B[i] = 1; i = i+1; end
    i = 0;
    repeat(3) begin C[i] = 0; i = i+1; end
    i = 5;
    forever begin : my_loop
      D[i] = 1;
        if (i == 15) #1  disable my_loop;
        i = i+1;

    end
  end
```

```
initial begin
  $display("A = %h", A);
    $display("B = %h", B);
  $display("C = %h", C);
    $display("D = %h", D);

  $finish;
end

endmodule
```

Output of the above program:
```
# A = 0000ffff
# B = c0000040
# C = 000003f8
# D = 0000ffe0
```

B.16.3 Disable Statement

A labelled sequential block can be disabled or stopped using a disable statement as explained in the previous example.

B.16.4 `If` Statement

An `if` statement is a conditional statement having two branches. For example,

```
if (select) Y = A; else Y = B;
```

B.17 Combinational Logic in Verilog

A combinational logic circuit can be designed using `always` statement. Following example describes a two-input AND gate using `always` statement:

```
module and21(a, b, y);
  input a;
  input b;
  output reg y;

  always @(*)
    y <= a & b;

endmodule
```

`always @(*)` evaluates the statements within the `always` block whenever any of the signals on the right-hand side of `<=` or `=` changes.

B.18 Sequential Logic in Verilog

Sequential logic circuits are described by `always` statement in Verilog.

B.18.1 Modelling Edge-sensitive Flip-flops

The keyword `always` is used to declare the edge triggered flip-flop as shown below.

```
module DFF(Q, QBAR, D, SER, RESET, CLK);
  input CLK, D, SET, RESET,
  output Q, QBAR;
  reg Q;

  assign QBAR = ~ Q;

  always @ (posedge CLK)
  begin
    if (RESET == 0) Q <= 0;
    else if(SET == 0) Q <= 1;
    else Q <= D;
  end
endmodule
```

The symbol <= is called concurrent assignment operator.

Example B.4   Write a Verilog program for mod-8 binary counter.

```
timescale 1ns/1ps
module counter;
  reg clock;
  integer count;

initial
  begin
    clock = 0;
    count = 0;
  end

always
    #10 clock = ~ clock;

always
  begin
    @ (negedge clock);
    if (count == 7)
      count = 0;
    else
      count = count + 1;
  end

endmodule
```

The input and output waveforms are shown in Fig. B.3.

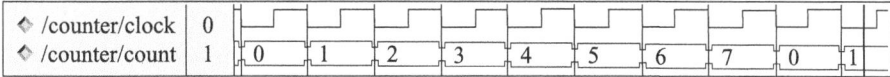

Fig. B.3 Simulated output of mod-8 counter

B.18.2 Blocking and Non-blocking Assignment Statement

There are two types of assignments within an `always` block:

- blocking assignment using the = symbol
- non-blocking assignment using the <= symbol

The blocking assignments are evaluated sequentially, whereas the non-blocking assignments are evaluated concurrently. Blocking assignments must be avoided within an `always` block while modelling sequential logic circuits.

B.19 Finite State Machines

A finite state machine (FSM) is a combination of sequential and combinational logic circuits which has a finite number of states and the machine makes transitions from one state to another depending on the present state and the input signals. There are two main types of FSMs:

- Moore machine
- Mealy machine

An FSM is called a Moore machine if the output depends only on the present state whereas the output of Mealy machine depends on the inputs as well as on the present state.

B.20 Test Benches in Verilog

A Verilog test bench is a Verilog program that is used to automate the testing of a module. A test bench program generally performs the following:

- Read a text file to get the test vectors
- Apply the test vector to the device under test (DUT)
- Check the results
- Report of any discrepancies

Example B.5 Write a Verilog test bench to test a 16-bit adder.

```
// adder program
module adder16bit (a, b, y);
  input [15:0] a;
  input [15:0] b;
  output [15:0] y;
  assign y = a + b;
endmodule
```

```
module my_testbench();
  reg [15:0] testvectors[100:0];
    reg clk;
    reg [10:0] N, err;
    reg [15:0] a, b, y;
    wire [15:0] y_t;

    adder16bit DUT(a,b,y_t);

    initial begin
      $readmemh("testvectors_adder16bit.txt", testvectors);
      N = 0; err = 0;
    end

    always begin
      clk = 0;
       #50;
      clk = 1;
       #50;
    end

  always @(posedge clk) begin
      a = testvectors[N*3];
       b = testvectors[N*3 + 1];
       y = testvectors[N*3 + 2];

    end

    always @(negedge clk) begin
      N = N + 1;
       if ( y_t !== y) begin
         $display("Inputs were %h, %h", a, b);
         $display("Expected %h but got %h", y, y_t);
         err = err + 1;
       end
    end

    always @(N) begin
      if ( N == 100 || testvectors[N*3] === 16'bx) begin
        $display("Completed %d tests with %d errors", N, err);
        $finish;
        end
    end

endmodule
Contents of testvectors_adder16bit.txt file:

0000
0000
```

```
0000
0001
0002
0003

000a
0001
000b

000a
0002
000f
```

The output

```
# Inputs were 000a, 0002
# Expected 000f but got 000c
# Completed 4 tests with 1 errors
```

In the last test case, we have deliberately made the output incorrect to catch if the error is flagged by the simulator.

B.21 Example Designs in Verilog

In the following subsections, we shall present few example designs using Verilog HDL.

B.21.1 Half-adder

The Verilog program for half-adder is given by

```
module half_adder(a, b, carry, sum);
  input a;
  input b;
  output carry;
  output sum;

  assign carry = a & b;
  assign sum = a ^ b;
endmodule
```

The input and output waveforms are shown in Fig. B.4.

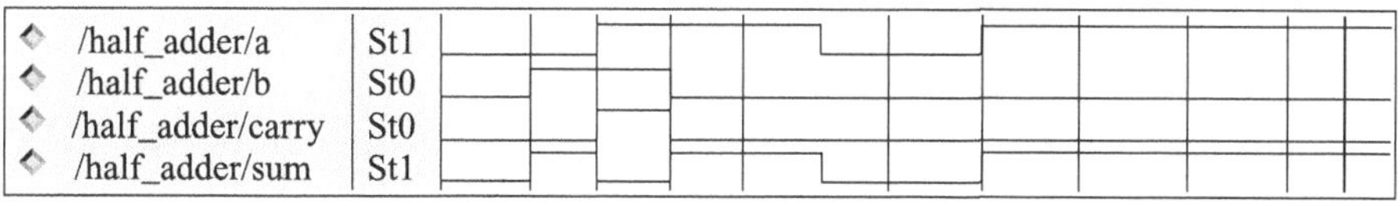

◇ /half_adder/a	St1	
◇ /half_adder/b	St0	
◇ /half_adder/carry	St0	
◇ /half_adder/sum	St1	

Fig. B.4　Input and output waveforms of half-adder

B.21.2 2:1 Multiplexer with 4-bit Bus Input

A 2:1 multiplexer is shown in Fig. B.5 with two-input data lines i0[3:0] and i1[3:0] and one select line sel. The output line is y[3:0].

Verilog code for 2:1 multiplexer is given by

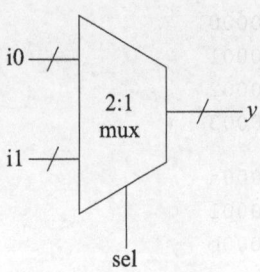

Fig. B.5 2:1 multiplexer with 4-bit bus input

```
module mux21(i0, i1, sel, y);
  input [3:0] i0;
  input [3:0] i1;
  input sel;
  output [3:0] y;

  assign y = sel ? i1 : i0;
endmodule
```

The input and output waveforms are shown in Fig. B.6.

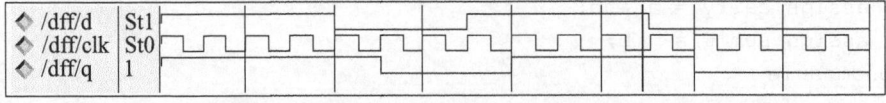

Fig. B.6 Input and output waveforms of 2:1 multiplexer

B.21.3 D Flip-flop

A D flip-flop designed in Verilog is given by

```
module dff(d, clk, q);
  input d;
  input clk;
  output reg q;

  always @(posedge clk)
      q <= d;
endmodule
```

The input and output waveforms are shown in Fig. B.7.

◇ /dff/d	St1
◇ /dff/clk	St0
◇ /dff/q	1

Fig. B.7 Input and output waveforms of D flip-flop

B.21.4 Eight-bit Register

An 8-bit register is designed in Verilog as

```
module reg8(d, clk, rst, q);
  input [7:0] d;
  input clk;
```

```
input rst;
output reg [7:0] q;

always @(posedge clk, posedge rst)
    if (rst) q <= 4'b0;
    else q <= d;
endmodule
```

The input and output waveforms are shown in Fig. B.8.

◇ /reg8/d	11110000	00011	100					11110000				
◇ /reg8/clk	St0											
◇ /reg8/rst	St0											
◇ /reg8/q	00000000	00011	100		00000000							11110000

Fig. B.8 Input and output waveforms of 8-bit register

B.21.5 4-bit Counter

A 4-bit counter modelled in Verilog with reset input is given by

```
module counter4bit(clk, rst, q);
  input clk;
  input rst;
  output reg [3:0] q;

  always @(negedge clk, posedge rst)
      if (rst)
          q <= 4'b0;
      else
          q <= q + 1;
endmodule
```

The input and output waveforms are shown in Fig. B.9.

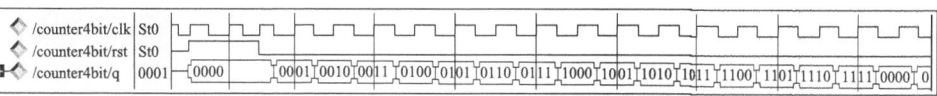

Fig. B.9 Input and output waveforms of 4-bit counter

B.21.6 Clock Generator

A clock generator and register with the generated clock is modelled in Verilog as

```
module ClkGen(clk, q);
  reg rst;
  output reg clk;
  output reg q;

    wire d;
    assign d = 1;
```

```
   always @(posedge clk, negedge rst)
      if (rst) q <= 0;
      else q <= d;

   initial begin rst = 1; #10 rst = 0; end
   initial
   clk = 0;
   always #5 clk = ~clk;
   initial
   begin
   #100; $finish;
   end

 always begin
      $display("T=%2g", $time, " RST=", rst, " D=", d, " CLK=",
clk, " Q=", q); #5;
   end
endmodule
```

The sample output of the clock generator program is as follows:

```
# T= 0 RST=1 D=x CLK=0 Q=x
# T= 5 RST=1 D=1 CLK=1 Q=x
# T=10 RST=0 D=1 CLK=0 Q=0
# T=15 RST=0 D=1 CLK=1 Q=1
# T=20 RST=0 D=1 CLK=0 Q=1
# T=25 RST=0 D=1 CLK=1 Q=1
# T=30 RST=0 D=1 CLK=0 Q=1
# T=35 RST=0 D=1 CLK=1 Q=1
# T=40 RST=0 D=1 CLK=0 Q=1
# T=45 RST=0 D=1 CLK=1 Q=1
# T=50 RST=0 D=1 CLK=0 Q=1
# T=55 RST=0 D=1 CLK=1 Q=1
# T=60 RST=0 D=1 CLK=0 Q=1
# T=65 RST=0 D=1 CLK=1 Q=1
# T=70 RST=0 D=1 CLK=0 Q=1
# T=75 RST=0 D=1 CLK=1 Q=1
# T=80 RST=0 D=1 CLK=0 Q=1
# T=85 RST=0 D=1 CLK=1 Q=1
# T=90 RST=0 D=1 CLK=0 Q=1
# T=95 RST=0 D=1 CLK=1 Q=1
```

B.21.7 Moore Machine

A Moore machine is a finite state machine in which the output depends only on the present state. It does not depend on the inputs. A 3-state Moore machine is shown in Fig. B.10. It can be modelled in Verilog as

```
module moore(clk, rst, y);
  input clk;
```

```
input rst;
output y;

reg [1:0] present_state, next_state;
  parameter S0 = 2'b00;
  parameter S1 = 2'b01;
  parameter S2 = 2'b10;

  always @ (posedge clk, negedge rst)
    if (rst) present_state <= S0;
      else present_state <= next_state;

  always @ (*)
    case (present_state)
      S0: next_state <= S1;
              S1: next_state <= S2;
              S2: next_state <= S0;
              default: next_state <= S0;
        endcase
  assign y = (present_state == S2);
endmodule
```

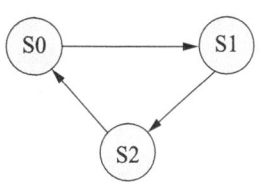

Fig. B.10 State diagram of a typical Moore machine

The input and output waveforms of the Moore machine is shown in Fig. B.11.

◇/moore/clk	St0											
◇/moore/rst	St0											
◇/moore/y	St1											
⊞◇/moore/present_state	10			00		01	10	00	01	10	00	01	10	00	01	10
⊞◇/moore/next_state	00			01		10	00	01	10	00	01	10	00	01	10	00

Fig. B.11 Input and output waveforms of a Moore machine

B.21.8 Mealy Machine

A Mealy machine is another finite state machine in which the output depends on the present state, as well as on the input. A three-state Mealy machine is described by a state diagram as shown in Fig. B.12. It has one input x and one output y.

The Verilog code for the Mealy machine shown in Fig. B.12 can be written as

```
module mealy(x, clk, rst, y);
  input x;
  input clk;
  input rst;
  output y;

  reg [1:0] present_state, next_
state;

  parameter S0 = 2'b00;
  parameter S1 = 2'b01;
  parameter S2 = 2'b10;
```

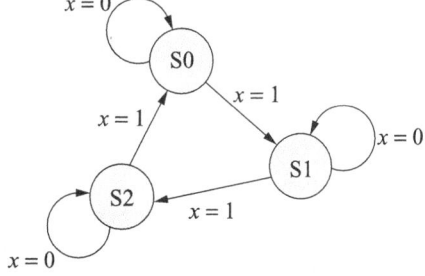

Fig. B.12 State diagram of a Mealy machine

```
    always @(posedge clk, negedge rst)
      if (rst) present_state <= S0;
        else present_state <= next_state;

    always @(*)
      case (present_state)
        S0:
            if (x == 1)
          next_state <= S1;
              else
                next_state <= S0;
        S1:
         if (x == 1)
           next_state <= S2;
              else
                next_state <= S1;
        S2:
         if (x == 1)
           next_state <= S0;
              else
                next_state <= S2;
        default: next_state <= S0;
      endcase
  assign y = (present_state == S2);
endmodule
```

The input and output waveforms are shown in Fig. B.13.

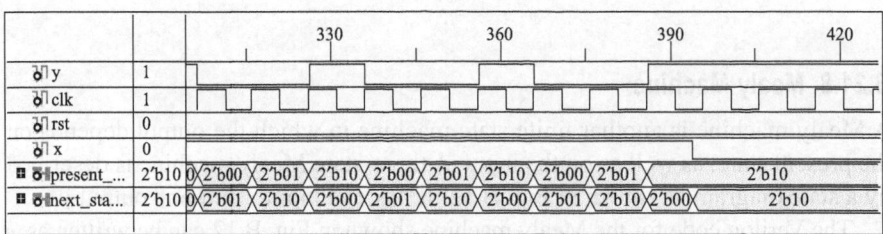

Fig. B.13   Input and output waveforms of the Mealy machine

SELECT REFERENCES

Ciletti, M.D. 2005, *Advanced Digital Design with the Verilog HDL*, Pearson Education, New Delhi.

Hayes, J.P. 1998, *Computer Architecture and Organization*, 3rd edn, McGraw-Hill International Editions.

Mano, M.M. 2001, *Computer System Architecture*, Prentice-Hall.

Smith, M.J.S. 2002, *Application Specific Integrated Circuits*, Pearson Education.

Weste, N.H.E., D. Harris, and A. Banerjee 2009, *CMOS VLSI Design: A Circuits and Systems Perspective*, 3rd ed., Pearson Education.

SPICE Tutorial

C.1 Introduction

Simulation program with integrated circuit emphasis (SPICE) is a software program to design and simulate circuits. It works in the following two modes:

1. Netlist mode in which the user writes a SPICE netlist to describe the circuit components along with their connectivity, excitations, analysis type, and outputs to be printed or plotted.
2. Schematic mode in which the user draws the circuit in the schematic editor, selects options, selects analysis type, simulates, and views the output waveforms.

C.2 Writing SPICE Netlist

This section describes how to write a SPICE netlist for a given circuit. Consider a resistive circuit as shown in Fig. C.1.

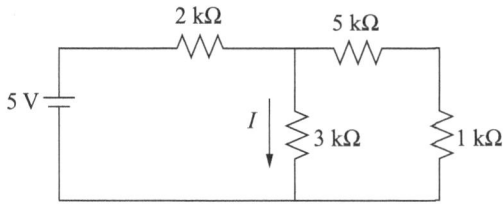

Fig. C.1   A resistive circuit with a DC source

Let us redraw the circuit by putting the node names and the element names as shown in Fig. C.2. We want to find out the current through the resistor 3 kΩ and the voltage across the resistor 1 kΩ.

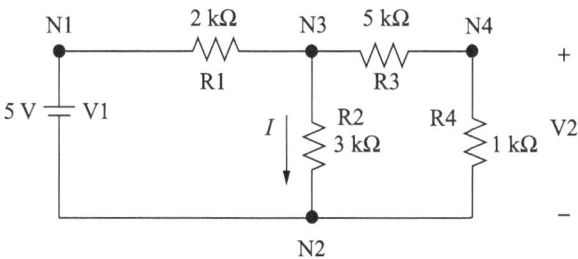

Fig. C.2   Resistive circuit shown in Fig. C.1 redrawn with node and element names

The SPICE netlist for the circuit is written as shown in Table C.1.

Table C.1 SPICE netlist for circuit shown in Fig. C.2

Line number	Statement
1	* This is a resistive circuit
2	
3	R1 N1 N3 2K
4	R2 N3 0 3K
5	R3 N3 N4 5K
6	R4 N4 0 1K
7	
8	V1 N1 0 5
9	
10	.op
11	
12	.end

The SPICE netlist has mainly three parts:
- The circuit and excitation description part
- The analysis part
- The output print or plot part

In the above example, the circuit description part is from line number 3–8. The first line is a netlist header statement. The simulator ignores the first line. This line should not contain any circuit description statement. The line number 10 describes the analysis type. The SPICE netlist must finish with an .end statement at the end of the file, as described by line number 12. SPICE supports comment lines. Any line that starts with a '*' specifies a comment line.

The node N2 represents the ground or reference node. In the SPICE netlist, a ground node is either written as 'GND' or '0'.

To test this circuit, copy paste the statements shown in Table C.1 into a text editor and save the file with any file name with a .sp or .cir extension.

After simulating the SPICE netlist, we get the following outputs:

```
DC ANALYSIS
v(N3)  =    2.5000e+000
v(N4)  =    4.1667e-001
v(N1)  =    5.0000e+000
i(V1)  =   -1.2500e-003

AC SMALL-SIGNAL MODELS
              R1           R2           R3           R4
R         2.00e+003    3.00e+003    5.00e+003    1.00e+003
VDROP     2.50e+000    2.50e+000    2.08e+000    4.17e-001
CURRENT   1.25e-003    8.33e-004    4.17e-004    4.17e-004
```

It prints the voltages for all nodes, current through all elements, and voltage drop across all resistors.

C.3 Syntax for Circuit Description

Any circuit element is described by the name of element, followed by the nodes between which it is connected, followed by the value of element. For example, a resistor R1 connected between nodes N1 and N2 of value 1 kΩ is described as:

R1 N1 N2 1K

Any line that starts with the letter 'R' specifies a resistor. The SPICE netlist is case-insensitive. You can use either upper case, or lower case, or mixed case alphabets for writing the SPICE netlist. Table C.2 lists the syntax for different circuit components.

Table C.2 Syntax for describing circuit elements in the SPICE netlist

Letter	Element	Nodes
R	Resistor	R<name> <N1> <N2> <value>
C	Capacitor	C<name> <N1> <N2> <value>
L	Inductor	L<name> <N1> <N2> <value>
K	Mutual inductor	K<name> <Inductor1> <Inductor2> <value of K>
V	Voltage source	V<name> <N+> <N-> <type> <value>
I	Current source	I<name> <N+> <N-> <type> <value>
M	MOSFET	M<name> <Drain> <Gate> <Source> <Bulk> <MOS Model> L=<value> W=<value>
D	Diode	D<name> <N+> <N-> <Diode Model>
Q	BJT	Q<name> <Collector> <Base> <Emitter> <BJT Model>
O	Lossy transmission line	O<name> <N1> <N2> <N3> <N4> <Model name>
X	Subcircuit	X<name> <N1> <N2> ... <Subckt name>
E	Voltage controlled voltage source	E<name> <N1> <N2> <NC1> <NC2> <Value>
G	Voltage controlled current source	G<name> <N1> <N2> <NC1> <NC2> <Value>
H	Current controlled voltage source	H<name> <N1> <N2> <Voltage source name> <Value>
F	Current controlled current source	F<name> <N1> <N2> <Voltage source name> <Value>
T	Lossless transmission line	T<name> <N1+> <N1-> <N2+> <N2-> Z0=<value> TD=<value>
S	Voltage controlled switch	S<name> <N1> <N2> <C1> <C2> <Switch Model>
W	Current controlled switch	W<name> <N1> <N2> <Voltage source name> <Switch Model>
J	JFET	J<name> <ND> <NG> <NS> <Model name> <AREA> <OFF> <IC=Value, Value> <TEMP=value>

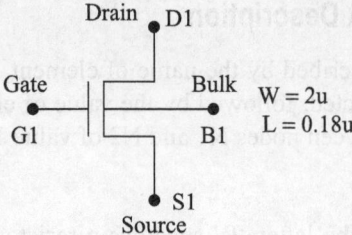

Fig. C.3 An nMOS transistor with four terminals

A MOSFET shown in Fig. C.3 is described as follows:

```
M1 D1 G1 S1 B1 nMOS W=2u L=0.18u AS=10u PS=14u AD=10u
+ PD=14u
```

nMOS signifies MOS model name
W=2u signifies width of nMOS as 2 micron
L=0.18u signifies length of nMOS as 2 micron
AS=10u signifies source area = $5 \times W \times \lambda$
PS=14u signifies source perimeter = $(2 \times W + 10) \times \lambda$
AD=10u signifies drain area = $5 \times W \times \lambda$
PD=14u signifies drain perimeter = $(2 \times W + 10) \times \lambda$

Note: A line is continued in next by putting '+' at the beginning of next line as shown in the above example.

Example C.1 Design a series *RLC* circuit with the following element values and obtain the pulse response: $R = 10\ \Omega$, $L = 10$ nH, $C = 5$ pF.

Solution The series *RLC* circuit is shown in Fig. C.4.

The SPICE netlist for *RLC* circuit shown in Fig. C.4 is given by

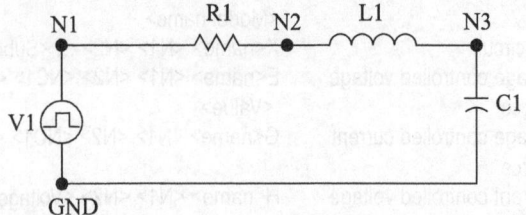

Fig. C.4 A series *RLC* circuit

```
* This is a resistive circuit
R1 N1 N2 10
L1 N2 N3 10n
C1 N3 GND 5p
V1 N1 GND PULSE 0 5 1n 1n 1n 5n 10n
.tran 10p 20n
.print tran v(N1) v(N3)
.end
```

The voltage source (V1) is a pulse-type voltage source with the following specifications:

   (i) Pulse low value = 0.0 V
   (ii) Pulse high value = 5.0 V
   (iii) Initial delay time of the pulse = 0.0 s
   (iv) Rise time of the pulse = 1ns
   (v) Fall time of the pulse = 1ns
   (vi) Pulse width = 5 ns
   (vii) Pulse period = 10 ns

The syntax for specifying a voltage source of pulse type is as follows:

```
V<name> <N+> <N-> PULSE <low value> <high value>
+ <initial delay> <rise time> <fall time>
+ <pulse width> <pulse period>
```

Any long statement in SPICE can be continued in the next line by putting the '+' character at the beginning, as described in the above statement.

The ".tran 10p 20n" statement specifies the analysis type. It signifies transient analysis with 10 ps; time step and simulation stop or end time is 20 ns. The syntax for transient analysis statement is as follows:

```
.tran <time step> <end time> [<start time>]
```

The start time is optional. If it is not specified, 0 s is used as default start time, as shown in Example C.1.

The input and output waveforms of the *RLC* circuit shown in Fig. C.4 are shown in Fig. C.5.

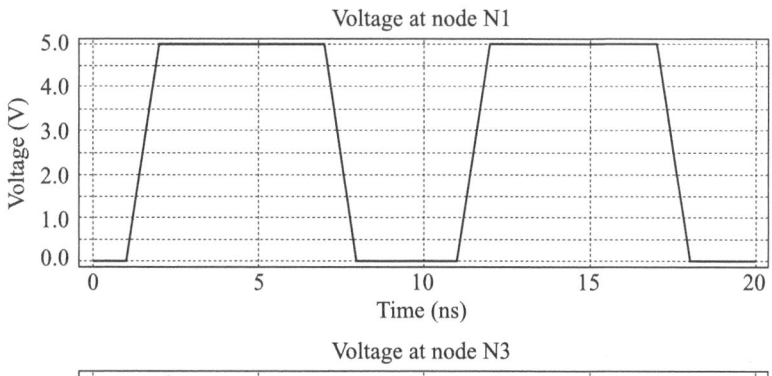

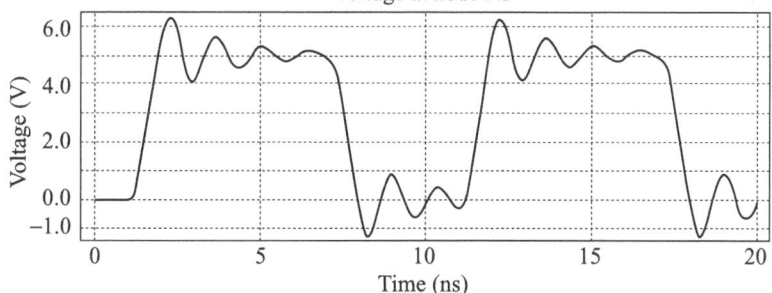

Fig. C.5   Response of *RLC* circuit

C.4 Analysis Types in SPICE

There are several analysis types supported by SPICE. The most commonly used analysis types are as follows:

- DC analysis
- AC analysis
- Transient analysis

C.4.1 DC Analysis

DC analysis is used to sweep the voltage value of DC source or a parameter value. It is specified by .dc statement. The syntax for a .dc statement is as follows:

```
.dc <sweep type> source/param <source/param name>
+ <start> <stop> <increment>
```

Sweep types are: Linear, Decade, and Octave.

Example

```
.dc lin source v1 1 5 2
.dc lin x 1 5 3
.dc lin param x 1 5 2
```

DC analysis is used to obtain V–I characteristics, transfer characteristics, and characteristics for different values of a circuit parameter.

C.4.2 AC Analysis

AC analysis is used to sweep the frequency of an AC source. It is specified by the .ac statement. The syntax for an .ac statement is as follows:

```
.ac <sweep type> <number of values> <start value>
+ <end value>
```

Sweep types are: Linear, Decade, and Octave.

Example

```
.ac dec 10 10 1000000
```

AC analysis is used to obtain the frequency response of a circuit.

C.4.3 Transient Analysis

Transient analysis is used to sweep the time. It is specified by the .tran statement. The syntax for the .tran statement is as follows:

```
.tran <time step> <end time> [<start time>]
```

Example

```
.tran 10p 20n start=1n
```

Transient analysis is used to find out the transient response of a circuit. If start time is not specified, a default 0 sec is used by the simulator.

C.5 Specifying Voltage Sources in SPICE

SPICE allows you to specify different types of voltage sources. Some of the commonly used voltage sources are discussed in the following sections.

C.5.1 DC Source

A DC source is specified with the following syntax:

```
V<name> <n+> <n-> <voltage value>
```

`<n+>` specifies the positive terminal of the source.
`<n->` specifies the negative terminal of the source. In general the negative terminal is ground.

Example A DC voltage source is written by

```
v1 n1 gnd 4
```

C.5.2 AC Source

An AC source is specified with the following syntax:

```
V<name> <n+> <n-> SIN (<voltage offset> <peak amplitude>
+ <frequency> <delay time> <damping factor>
+ <phase advance>)
```

Example The AC waveform, as shown in Fig. C.6 is written by

```
v2 n1 GND SIN (2 4 1e9 2n 5e8 30)
```

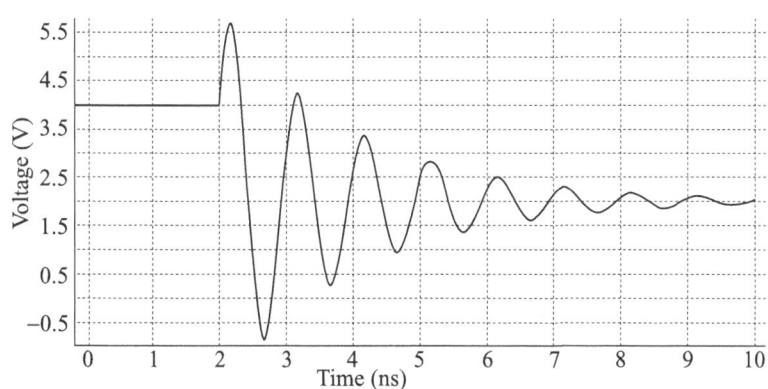

Fig. C.6   An example of voltage source of type sinusoidal

C.5.3 Voltage Source of Pulse Type

A voltage source of pulse type is specified as per the following syntax:

```
V<name> <n+> <n-> PULSE <low> <high> <TD> <TR> <TF>
+ <PW> <PER>
```

where

\<low\> specifies the pulse low value
\<high\> specifies the pulse high value
\<TD\> specifies the initial delay time
\<TR\> specifies the rise time of the pulse
\<TF\> specifies the fall time of the pulse
\<PW\> specifies the pulse width
\<PER\> specifies the pulse period

Example The pulse waveform as shown in Fig. C.7 is written by

```
V2 IN GND PULSE 0 5 1n 1n 1n 5n 10n
```

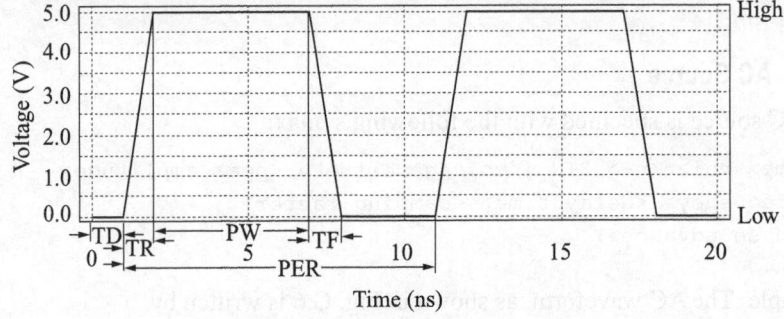

Fig. C.7 An example of pulse waveform

C.5.4 Piece-wise Linear Voltage Source

A piece-wise linear (PWL) waveform is specified by the following syntax:

```
V<name> <n+> <n-> PWL <t0> <v0> <t1> <v1> <t2> <v2>
+ <t3> <v3> <t4> <v4> ...
```

where

ti specifies the ith time instant.
vi specifies voltage value at the ith time instant.

Example A PWL waveform as shown in Fig. C.8 is written in SPICE by

```
v3 n3 GND PWL 0 0 1n 0 2n 2.5 3n 5 8n 5 9n 6 10n 5
+ 11n 2.5 12n 0 13n -0.5 14n 0
```

C.5.5 Voltage Source with Bit Pattern

A voltage source with a bit pattern is specified by the following syntax:

```
V<name> <N+> <N-> BIT {<bit pattern>} pw=<PW> lt=<LT>
+ ht=<HT> on=<high> off=<low> rt=<TR> ft=<TF> delay=<TD>
```

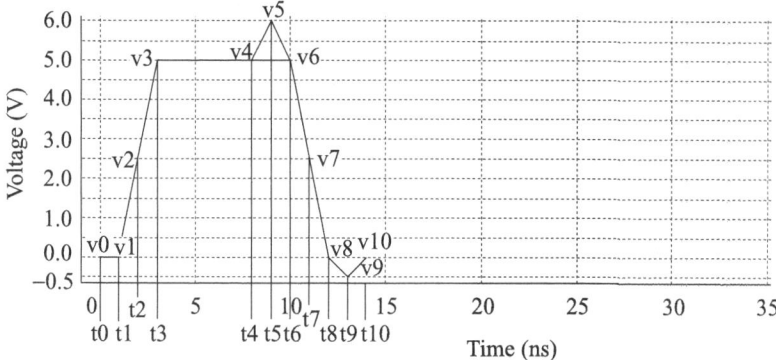

Fig. C.8 An example of a PWL waveform

Example A voltage source with bit pattern as shown in Fig. C.9 is written in SPICE as:

```
v5 n5 0 BIT ({11010110} pw=5n lt=4n ht=4n on=3 off=0 rt=1n ft=1n
delay=1n)
```

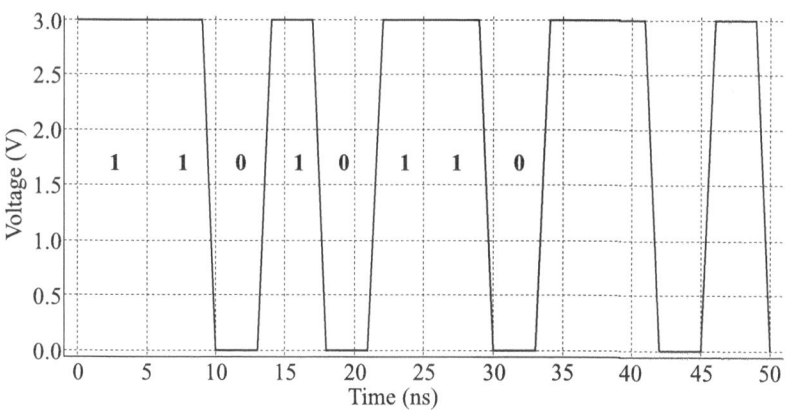

Fig. C.9 An example of a voltage source with a specified bit pattern

C.6 Specifying SPICE Model Parameters

SPICE model parameters are required for active elements, e.g., MOSFET, BJT, and JFET. For details about model parameters, refer to the BSIM SPICE model (Berkley website). We will concentrate mainly on the MOSFET model parameters. Any circuit containing either a pMOS, or nMOS, or both must specify the SPICE model parameters. These can be done by two ways:

- Using a .model statement
- Using a .include statement

The syntax of specifying the SPICE model parameters using the .model statement is as follows:

```
.model <Model Name> <Model Type> VTO=<value> KP=<value>
+ GAMMA=<value> LAMBDA=<value> PHI=<value> ...
```

Example The model for nMOS and pMOS can be specified as follows:

```
.model NCH NMOS VTO=0.7 KP=100u GAMMA=0.3 LAMBDA=0.02 PHI=0.7
.model PCH PMOS VTO=-0.7 KP=60u GAMMA=0.35 LAMBDA=0.04 PHI=0.8
```

Another method of specifying the SPICE model parameters is to write the model parameters in a text file, and include the file in the SPICE netlist. An example of such a file is shown in Table C.3.

Table C.3 A sample SPICE model (PTM website) file

```
* PTM 90nm NMOS

.model nMOS nMOS level = 54
+version = 4.0  binunit = 1  paramchk= 1  mobmod  = 0
+capmod  = 2  igcmod  = 1  igbmod  = 1  geomod  = 1
+diomod  = 1  rdsmod  = 0  rbodymod= 1  rgatemod= 1
+permod  = 1  acnqsmod= 0  trnqsmod= 0
+tnom    = 27  toxe= 2.05e-9  toxp = 1.4e-9  toxm    = 2.05e-9
+dtox = 0.65e-9  epsrox  = 3.9  wint  = 5e-009  lint  = 7.5e-009
+ll  = 0  wl  = 0  lln  = 1  wln  = 1
+lw  = 0  ww  = 0  lwn  = 1  wwn  = 1
+lwl  = 0  wwl  = 0  xpart  = 0  toxref  = 2.05e-9
+xl  = -40e-9
+vth0  = 0.397  k1  = 0.4  k2  = 0.01  k3  = 0
+k3b  = 0  w0  = 2.5e-006  dvt0  = 1  dvt1  = 2
+dvt2  = -0.032  dvt0w  = 0  dvt1w  = 0  dvt2w  = 0
+dsub  = 0.1  minv  = 0.05  voffl  = 0  dvtp0  = 1.2e-009
+dvtp1  = 0.1  lpe0  = 0  lpeb  = 0  xj  = 2.8e-008
+ngate  = 2e+020  ndep  = 1.94e+018  nsd  = 2e+020  phin  = 0
+cdsc  = 0.0002  cdscb  = 0  cdscd  = 0  cit  = 0
+voff  = -0.13  nfactor = 1.7  eta0  = 0.0074  etab  = 0
+vfb  = -0.55  u0  = 0.0547  ua  = 6e-010  ub  = 1.2e-018
+uc  = -3e-011  vsat  = 113760  a0  = 1.0  ags  = 1e-020
+a1  = 0  a2  = 1  b0  = -1e-020  b1  = 0
+keta  = 0.04  dwg  = 0  dwb = 0  pclm  = 0.06
+pdiblc1 = 0.001  pdiblc2 = 0.001  pdiblcb = -0.005  drout = 0.5
+pvag  = 1e-020  delta  = 0.01  pscbe1  = 8.14e+008  pscbe2 = 1e-007
+fprout  = 0.2  pdits  = 0.08  pditsd  = 0.23  pditsl  = 2.3e+006
+rsh  = 5  rdsw  = 180  rsw  = 90  rdw  = 90
+rdswmin = 0  rdwmin  = 0  rswmin  = 0  prwg  = 0
+prwb  = 6.8e-011  wr  = 1  alpha0  = 0.074  alpha1  = 0.005
+beta0 = 30  agidl  = 0.0002  bgidl  = 2.1e+009  cgidl  = 0.0002
+egidl  = 0.8
+aigbacc = 0.012  bigbacc = 0.0028  cigbacc = 0.002
+nigbacc = 1  aigbinv = 0.014  bigbinv = 0.004  cigbinv = 0.004
+eigbinv = 1.1  nigbinv = 3  aigc  = 0.012  bigc  = 0.0028
```

(Contd)

Table C.3 *(Contd)*

```
+cigc  = 0.002  aigsd = 0.012  bigsd   = 0.0028  cigsd   = 0.002
+nigc  = 1  poxedge = 1  pigcd  = 1  ntox = 1
+xrcrg1  = 12  xrcrg2  = 5
+cgso = 1.9e-010 cgdo = 1.9e-010 cgbo = 2.56e-011 cgdl = 2.653e-10
+cgsl  = 2.653e-10  ckappas = 0.03  ckappad = 0.03  acde  = 1
+moin  = 15  noff  = 0.9  voffcv  = 0.02
+kt1  = -0.11  kt1l  = 0  kt2  = 0.022  ute  = -1.5
+ua1  = 4.31e-009  ub1  = 7.61e-018  uc1  = -5.6e-011  prt  = 0
+at  = 33000  fnoimod = 1  tnoimod = 0
+jss  = 0.0001  jsws = 1e-011  jswgs   = 1e-010  njs = 1
+ijthsfwd= 0.01  ijthsrev= 0.001  bvs  = 10  xjbvs   = 1
+jsd  = 0.0001  jswd = 1e-011  jswgd   = 1e-010  njd = 1
+ijthdfwd= 0.01  ijthdrev= 0.001  bvd  = 10  xjbvd   = 1
+pbs  = 1  cjs  = 0.0005  mjs  = 0.5  pbsws   = 1
+cjsws   = 5e-010  mjsws   = 0.33  pbswgs  = 1  cjswgs  = 3e-010
+mjswgs  = 0.33  pbd  = 1  cjd  = 0.0005  mjd  = 0.5
+pbswd   = 1  cjswd   = 5e-010  mjswd   = 0.33  pbswgd  = 1
+cjswgd  = 5e-010  mjswgd  = 0.33  tpb  = 0.005  tcj  = 0.001
+tpbsw   = 0.005  tcjsw   = 0.001  tpbswg  = 0.005  tcjswg = 0.001
+xtis  = 3  xtid  = 3
+dmcg  = 0e-006  dmci  = 0e-006  dmdg  = 0e-006  dmcgt   = 0e-007
+dwj  = 0.0e-008  xgw  = 0e-007  xgl  = 0e-008
+rshg  = 0.4  gbmin   = 1e-010  rbpb  = 5  rbpd  = 15
+rbps  = 15  rbdb  = 15  rbsb  = 15  ngcon   = 1

* PTM 90nm PMOS

.model  pMOS  pMOS  level = 54

+version = 4.0  binunit = 1  paramchk= 1  mobmod  = 0
+capmod   = 2  igcmod   = 1  igbmod   = 1  geomod   = 1
+diomod   = 1  rdsmod   = 0  rbodymod= 1  rgatemod= 1
+permod   = 1  acnqsmod= 0  trnqsmod= 0

+tnom  = 27  toxe = 2.15e-009  toxp  = 1.4e-009  toxm  = 2.15e-009
+dtox  = 0.75e-9  epsrox  = 3.9  wint  = 5e-009  lint  = 7.5e-009
+ll  = 0  wl  = 0  lln  = 1  wln  = 1
+lw  = 0  ww  = 0  lwn  = 1  wwn  = 1
+lwl  = 0  wwl  = 0  xpart   = 0  toxref  = 2.15e-009
+xl  = -40e-9
+vth0  = -0.339  k1  = 0.4  k2  = -0.01  k3  = 0
+k3b  = 0  w0  = 2.5e-006  dvt0  = 1  dvt1  = 2
+dvt2  = -0.032  dvt0w   = 0  dvt1w   = 0  dvt2w   = 0
+dsub  = 0.1  minv  = 0.05  voffl   = 0  dvtp0   = 1e-009
+dvtp1   = 0.05  lpe0  = 0  lpeb  = 0  xj  = 2.8e-008
+ngate   = 2e+020  ndep  = 1.43e+018  nsd  = 2e+020  phin  = 0
+cdsc  = 0.000258  cdscb   = 0  cdscd   = 6.1e-008  cit  = 0
```

(Contd)

Table C.3 (*Contd*)

```
+voff  = -0.126  nfactor = 1.7  eta0  = 0.0074  etab  = 0
+vfb  = 0.55  u0 = 0.00711  ua = 2.0e-009  ub = 0.5e-018
+uc  = -3e-011  vsat  = 70000  a0  = 1.0  ags  = 1e-020
+a1  = 0  a2  = 1  b0  = 0  b1  = 0
+keta  = -0.047  dwg  = 0  dwb  = 0  pclm  = 0.12
+pdiblc1 = 0.001 pdiblc2 = 0.001 pdiblcb = 3.4e-008 drout = 0.56
+pvag = 1e-020 delta = 0.01 pscbe1 = 8.14e+008 pscbe2 = 9.58e-007
+fprout  = 0.2  pdits  = 0.08  pditsd  = 0.23  pditsl  = 2.3e+006
+rsh  = 5  rdsw  = 200  rsw  = 100  rdw  = 100
+rdswmin = 0  rdwmin  = 0  rswmin  = 0  prwg  = 3.22e-008
+prwb  = 6.8e-011  wr  = 1  alpha0  = 0.074  alpha1  = 0.005
+beta0  = 30  agidl  = 0.0002  bgidl  = 2.1e+009  cgidl = 0.0002
+egidl  = 0.8

+aigbacc = 0.012  bigbacc = 0.0028  cigbacc = 0.002
+nigbacc = 1  aigbinv = 0.014  bigbinv = 0.004  cigbinv = 0.004
+eigbinv = 1.1  nigbinv = 3  aigc  = 0.69  bigc  = 0.0012
+cigc  = 0.0008  aigsd  = 0.0087  bigsd  = 0.0012  cigsd  = 0.0008
+nigc  = 1  poxedge = 1  pigcd  = 1  ntox  = 1

+xrcrg1  = 12  xrcrg2  = 5
+cgso = 1.8e-010 cgdo = 1.8e-010 cgbo = 2.56e-011 cgdl = 2.653e-10
+cgsl  = 2.653e-10  ckappas = 0.03  ckappad = 0.03  acde  = 1
+moin  = 15  noff  = 0.9  voffcv  = 0.02

+kt1  = -0.11  kt1l  = 0  kt2  = 0.022  ute  = -1.5
+ua1  = 4.31e-009  ub1  = 7.61e-018  uc1  = -5.6e-011  prt  = 0
+at  = 33000

+fnoimod  = 1  tnoimod  = 0

+jss  = 0.0001  jsws  = 1e-011  jswgs  = 1e-010  njs  = 1
+ijthsfwd= 0.01  ijthsrev= 0.001  bvs  = 10  xjbvs  = 1
+jsd  = 0.0001  jswd  = 1e-011  jswgd  = 1e-010  njd  = 1
+ijthdfwd= 0.01  ijthdrev= 0.001  bvd  = 10  xjbvd  = 1
+pbs  = 1  cjs  = 0.0005  mjs  = 0.5  pbsws  = 1
+cjsws  = 5e-010  mjsws  = 0.33  pbswgs = 1  cjswgs = 3e-010
+mjswgs  = 0.33  pbd  = 1  cjd  = 0.0005  mjd  = 0.5
+pbswd  = 1  cjswd  = 5e-010  mjswd  = 0.33  pbswgd = 1
+cjswgd  = 5e-010  mjswgd = 0.33  tpb  = 0.005  tcj  = 0.001
+tpbsw  = 0.005  tcjsw = 0.001  tpbswg  = 0.005  tcjswg = 0.001
+xtis  = 3  xtid  = 3

+dmcg  = 0e-006 dmci  = 0e-006 dmdg  = 0e-006 dmcgt  = 0e-007
+dwj  = 0.0e-008  xgw  = 0e-007  xgl  = 0e-008

+rshg  = 0.4  gbmin  = 1e-010  rbpb  = 5  rbpd  = 15
+rbps  = 15  rbdb  = 15  rbsb = 15  ngcon  = 1
```

Suppose this file name is `spice_model.txt` and it is under `E:\SPICE_MODEL_DIR` folder/directory. It can be included in the SPICE netlist using the `.include` statement as given by

```
.include "E:\SPICE_MODEL_DIR\spice_model.txt"
```

C.7 Using Subcircuits in SPICE

Similar to the subroutines or functions of any other programming language, SPICE supports subcircuits which are very useful in writing the SPICE netlist in a hierarchical manner. For example, let us consider how to design a shift register circuit. The shift register circuit is a chain of D flip-flops. So, we can write a subcircuit for the D flip-flop and call the subcircuit as many times as we need in the SPICE netlist of shift register.

The syntax for writing a subcircuit is as follows:

```
.subckt <subckt name> <port names>
<circuit description statements>
.ends
```

For example, the CMOS inverter as shown in Fig. C.10 can be defined as a subcircuit as follows:

```
.subckt INV A Y
MP Y A VDD VDD PMOS w=5u l=0.18u
MN Y A 0   0   NMOS w=2u l=0.18u
.ends
```

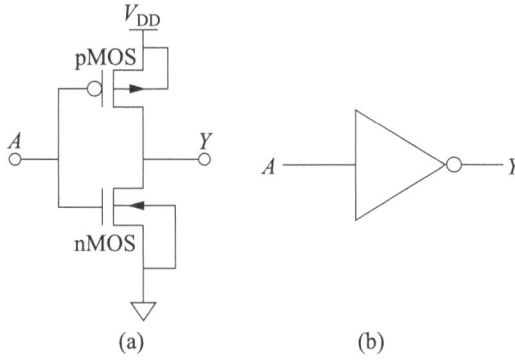

(a)   (b)

Fig. C.10   CMOS inverter: (a) schematic; (b) symbol

The subcircuit of the CMOS inverter defined now can be used to design a ring oscillator circuit as shown in Fig. C.11. The syntax of using the subcircuit is as follows:

```
X<name> <node names> <subcircuit name>
```

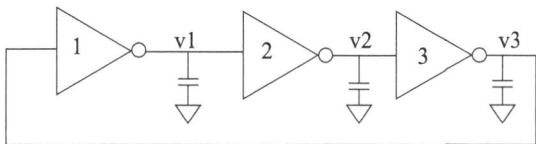

Fig. C.11   Ring oscillator

Example The subcircuit of the inverter can be called using *subcircuit call* as follows:

```
X1 In Out INV
```

The SPICE netlist for the ring oscillator circuit using subcircuit is already discussed in Section 6.19 (Chapter 6).

Example C.2 Design a 3-bit shift register as shown in Fig. C.12 using subcircuit feature of SPICE

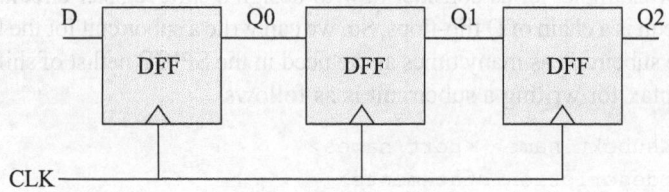

Fig. C.12 3-bit shift register

Solution A D flip-flop as shown in Fig. C.13 can be written as a subcircuit in SPICE as:

```
* SUBCKT of D flip-flop
.subckt DFF D CLK Q QBAR
Xinv1 CLK CLKB INV
XTG1 D CLK CLKB Q1 TG
Xinv2 Q1 QmB INV
Xinv3 QmB Qm INV
XTG3 Qm CLKB CLK Q2 TG
Xinv4 Q2 QBAR INV
Xinv5 QBAR Q INV
XTG2 Qm CLKB CLK Q1 TG
XTG4 Q CLK CLKB Q2 TG
.end
```

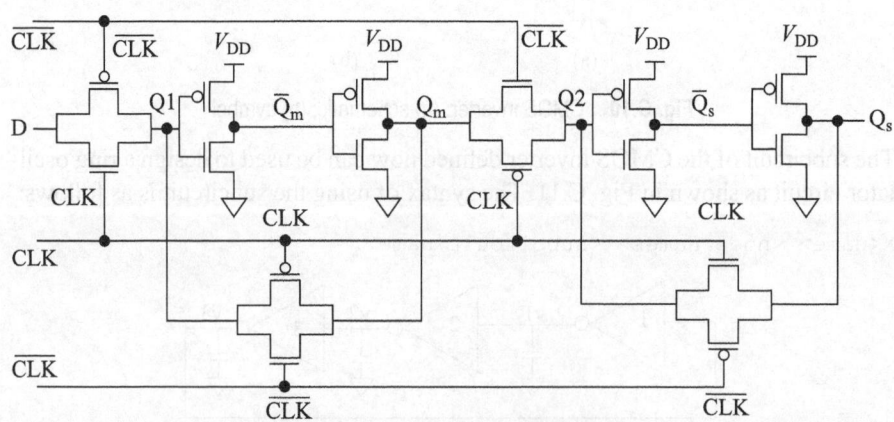

Fig. C.13 CMOS D flip-flop

In the above example of D flip-flop subcircuit, we have used subcircuits for the inverter and transmission gate. The subcircuit of the D flip-flop is instantiated or called 3 times to design a 3-bit shift register as shown below:

```
.include "E:\CMOS-CELLS\SPICE MODELS\tsmc018.md"
.global VDD

.subckt INV A Y
MP Y A VDD VDD PMOS w=5u l=0.18u
MN Y A 0   0   NMOS w=2u l=0.18u
.end
.subckt TG IN C CB OUT
MP OUT CB IN VDD PMOS w=5u l=0.18u
MN IN C OUT 0 NMOS w=2u l=0.18u
.end

.subckt DFF D CLK Q QBAR
Xinv1 CLK CLKB INV
XTG1 D CLK CLKB Q1 TG
Xinv2 Q1 QmB INV
Xinv3 QmB Qm INV
XTG3 Qm CLKB CLK Q2 TG
Xinv4 Q2 QBAR INV
Xinv5 QBAR Q INV
XTG2 Qm CLKB CLK Q1 TG
XTG4 Q CLK CLKB Q2 TG
.end

X1 D CK Q QBAR DFF
X2 Q CK Q1 QBAR1 DFF
X3 Q1 CK Q2 QBAR2 DFF

vd d 0 BIT ({1101011} pw=50n lt=8n ht=8n on=1.8 off=0
+ rt=1n ft=1n delay=3n)
VC CK 0 PULSE 0 1.8 2n 1n 1n 50n 100n
VDD VDD 0 1.8

.tran 10 1000n
.plot tran v(D) v(CK) v(Q) v(Q1) v(q2)
.end
```

The input and output waveforms of a 3-bit shift register is shown in Fig. C.14.

C.8 Other Useful Statements in SPICE

(i)   .global—defines the global nodes
(ii) .ic—sets the initial conditions

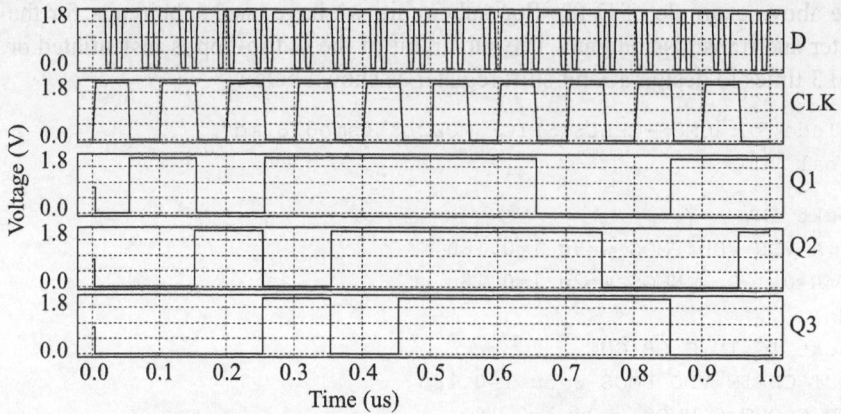

Fig. C.14 Input/output waveform of a 3-bit shift register

For example

```
.IC V<node1> = <value> V<node2> = <value>
```

(iii) .param — used to define parameter values

(iv) .temp — used to specify the analysis temperature

(iii) .measure — used to measure some output parameters, e.g., delay, power

(vi) .option — used to set analysis options

(vii) .op — used to output the analysis results

(viii) .alter — used to alter the circuit components

(ix) .modify — used to modify the circuit component values

(x) .scale — used to scale the dimensions of circuit elements, e.g., channel length and width of MOS devices

SELECT REFERENCES

Berkeley website, http://www-device.eecs.berkeley.edu/~bsim3/, last accessed on 30 Dec 2010.

PTM website, http://ptm.asu.edu/ Predictive Technology Model (PTM), last accessed on 30 Dec 2010.

Rashid, M.H. 1993, *SPICE for Circuits and Electronics Using PSPICE*, 2nd edn, Prentice-Hall.

Rashid, M.H. 2004, *Introduction to PSpice Using OrCAD for Circuits and Electronics*, 3rd edn, Pearson Prentice-Hall.

Answers to Objective-type Questions

Chapter 1

Fill in the blanks

1. b	2. a
3. c	4. d
5. a	

MCQs

1. d	2. d
3. a	4. d
5. d	

True or False

1. True	2. False
3. True	4. False
5. False	

Chapter 2

Fill in the blanks

1. a	2. c
3. c	4. b
5. d	

MCQs

1. d	2. b
3. a	4. b
5. a	6. c
7. b	8. a

True or False

1. True	2. True
3. True	4. False
5. True	

Chapter 3

Fill in the Blanks

1. Dynamic	2. less
3. large	4. switching
5. faster	

MCQ

1. a	2. b
3. a	4. a
5. d	

True or False

1. False	2. true
3. False	4. True
5. False	

Chapter 4

Fill in the blanks

1. a	2. a
3. c	4. a
5. b	

MCQs

1. d	2. d
3. a	4. a
5. e	6. b
7. a	8. a
9. b	

True or False

1. True	2. True
3. True	4. True
5. True	

Chapter 5

Fill in the blanks

1. d	2. a
3. b	4. b
5. c	

MCQs

1. a	2. d
3. c	4. b
5. a	

True or False

1. True	2. True
3. True	4. True
5. True	

Chapter 6

Fill in the blanks

1. b	2. a
3. a	4. b
5. a	

MCQs

1. a	2. a
3. d	4. a
5. c	

True or False

1. False	2. False
3. True	4. False
5. False	

Chapter 7

Fill in the Blanks

1. c	2. b
3. a	4. a
5. c	

MCQ

1. d	2. b
3. c	4. a
5. c	

True or False

1. False	2. False
3. False	4. false
5. False	

Chapter 8

Fill in the blanks

1. c	2. c
3. a	4. c
5. c	

MCQs

1. b	2. c
3. b	4. a
5. a	

True or False

1. True	2. False
3. True	4. True
5. True	

Chapter 9

Fill in the blanks

1. a	2. b
3. a	4. a
5. d	

MCQs

1. b	2. b
3. c	4. d
5. a	

True or False
1. True
2. True
3. True
4. True
5. True

Chapter 10

Fill in the blanks
1. a
2. a
3. a
4. d
5. b

MCQs
1. a
2. c
3. a
4. b
5. a

True or False
1. True
2. True
3. True
4. False
5. True

Chapter 11

Fill in the blanks
1. b
2. a
3. b
4. c
5. b

MCQs
1. d
2. a
3. a
4. c
5. c

True or False
1. True
2. True
3. True
4. False
5. True

Chapter 12

Fill in the blanks
1. a
2. a
3. d
4. a
5. a

MCQs
1. a
2. a or c
3. d
4. d
5. b

True or False
1. False
2. False

3. False
4. True
5. False
6. False

Chapter 13

Fill in the blanks
1. a
2. b
3. b
4. b
5. d

MCQs
1. a
2. d
3. b
4. a
5. c

True or False
1. False
2. False
3. True
4. False
5. True

Chapter 14

Fill in the blanks
1. b
2. c
3. a
4. a
5. a

MCQs
1. d
2. c
3. d
4. b
5. a

True or False
1. True
2. True
3. False
4. False
5. True

Chapter 15

Fill in the blanks
1. a
2. b
3. d
4. b
5. c

MCQs
1. b
2. d
3. c
4. a
5. a

True or False
1. True
2. False
3. False
4. True
5. False

Chapter 16

Fill in the blanks
1. a
2. a
3. a
4. a
5. b

MCQs
1. a
2. a
3. b
4. d
5. a

True or False
1. True
2. True
3. False
4. True
5. False

Chapter 17

Fill in the Blanks
1. A
2. b
3. b
4. b
5. B

MCQ
1. A
2. B
3. c
4. c
5. A

True or False
1. False
2. False
3. False
4. True
5. true

Chapter 18

Fill in the Blanks
1. a
2. d
3. c
4. c
5. b

MCQ
1. a
2. b
3. a
4. b
5. a

True or False
1. False
2. False
3. False
4. True
5. False

Index

About the Author

 **Debaprasad Das** is an Associate Professor and Head of the Department of Electronics and Telecommunication Engineering, Assam University, Silchar. Dr Das obtained his PhD from the School of VLSI Technology, Bengal Engineering and Science University (now Indian Institute of Engineering Science and Technology, IIEST—an institute of National importance), Shibpur, West Bengal. He has five years of working experience in ASIC Product Development Centre of Texas Instruments, Bangalore and about ten years of teaching experience at both undergraduate and postgraduate levels.

Dr Das has published several research articles in reputed international journals and conferences and also authored four books published by International Publishers.

Related Titles

DIGITAL ELECTRONICS
9780198061830

G.K. Kharate,
Matoshri College of Engineering and Research Centre, Nashik

This textbook is designed for an introductory course on digital electronics or digital system and design.

Key Features

- Establishes a balance between the theoretical concepts of digital electronics and its applications
- Includes separate chapters on the number system, synchronous and asynchronous circuits, and K-maps
- Includes a large number of solved examples and numerous illustrations supporting the text for better understanding of complex topics

POWER ELECTRONICS: DEVICES, CIRCUITS, AND INDUSTRIAL APPLICATIONS
9780195670929

V.R. Moorthi,
Formerly at Delhi College of Engineering

This textbook provides comprehensive coverage of various power electronic devices with emphasis on the thyristor.

Key Features

- Provides detailed and easy-to-understand derivations
- Includes detailed coverage of micro-computer control of industrial equipment
- Contains a separate chapter on industrial applications

MICROPROCESSORS AND MICROCONTROLLERS
9780198066477

N. Senthil Kumar,
Mepco Schlenk Engineering College, Sivakasi;
M. Saravanan,
Thiagarajar College of Engineering, Madurai;
S. Jeevananthan,
Pondicherry Engineering College, Puducherry

This textbook is designed as a comprehensive textbook for undergraduate engineering students to build a strong foundation in the basic principles, functioning, and applications of microprocessors and microcontrollers.

Key Features

- Includes case studies on traffic light control, washing machine control, and elevator control to enable students appreciate the applications of processors
- Includes discussions on advanced processors, such as 80186, 80286, 80386, 80486, Pentium, PowerPC, and PIC 16F877
- Contains a section on the advent of high-level language programming in 8051 processors with examples illustrated using the most popular language C

OTHER RELATED TITLES

9780198084570 **D.K. Bhattacharya and Rajnish Sharma:** *Solid State Electronic Devices, 2e*

9780199457052 **David A. Bell:** *Pulse, Switching and Digital Circuits, 5th edition*

9780198079064 **N. Senthil Kumar, M. Saravanan, S. Jeevananthan, S.K. Shah:** *Microprocessors and Interfacing*

9780198063575 **Satish Shah:** *8051 Microcontrollers: MCS 51 Family and Its Variants*